Magnesium Technology 2013

TMS2013

142nd Annual Meeting & Exhibition

Magnesium Technology 2013

Proceedings of a symposium sponsored by the
Magnesium Committee of the Light Metals Division of
The Minerals, Metals & Materials Society (TMS)

Held during the
TMS 2013 Annual Meeting & Exhibition
San Antonio, Texas, USA
March 3-7, 2013

Edited by

Norbert Hort
Suveen N. Mathaudhu
Neale R. Neelameggham
Martyn Alderman

Editors
Norbert Hort
Suveen N. Mathaudhu
Neale R. Neelameggham
Martyn Alderman

ISBN 978-3-319-48579-9 ISBN 978-3-319-48150-0 (eBook)
DOI 10.1007/978-3-319-48150-0

Chemistry and Materials Science: Professional

Published by Springer International Publishers, Switzerland, 2016
Reprint of the original edition published by John Wiley & Sons, Inc., 2013, 978-1-11860-552-3

Printed on acid-free paper

This Springer imprint is published by Springer Nature
The registered company is Springer International Publishing AG
The registered company address is: Gewerbestrasse 11, 6330 Cham, Switzerland

TABLE OF CONTENTS
Magnesium Technology 2013

Magnesium Technology 2013

Plenary

Creep

Cast Materials

Primary Production and Shaping

Texture and Twinning

Wrought Materials I

Corrosion I

Nanostructured Materials

Corrosion II

Joining and Friction Processing

Wrought Materials II

Phase Formation

Solidification

Nanocomposites and Metal Matrix Composites

Wrought Materials III

Microstructure

Twin Roll Casting

PREFACE

Magnesium was produced first in 1808, more than 200 years ago, by Sir Humphry Davy. It took a few decades to develop and establish stable production processes, and in the middle of the 19th century magnesium was available in larger quantities and ready to be used. In fact it met a new technology at these times: photography. Magnesium was utilized in flashlights and this endured until the end of the 20th century. Of course, flashlights are not a structural application. But with time, in the first years of the 20th century again a step in the technical development took place: the Wright brothers started to fly. It was found out early that wooden airplanes would not do their job at all. Light metal alloys had to be used, and magnesium alloys were part of aeronautics from the beginning. In fact an engine for aircrafts made out of a magnesium alloy was shown 1909 in Frankfurt, Germany, at the first "Internationale Luftschiffahrt Ausstellung," the first international aerospace exhibition. While magnesium was important for aerospace 100 years ago, today it is still looking for acceptance in aerospace industries.

In difference to the slight acceptance in aerospace, magnesium alloys are fairly well accepted in automotive industries as well as for consumer, computer, and communication applications. Due to its excellent specific properties magnesium alloys can compete especially in transportation with aluminium and even with steel leading to tremendous weight savings. This directly leads either to an increase in payload or a fuel reduction (saves money for the end user), or reduces green house gas emissions. Today automotive industries are showing lots of examples for the use of magnesium alloys in various applications, but an average standard car only contains a few kilograms of magnesium alloys. To enhance the use of magnesium even in this field it is still necessary to develop alloys and to optimize processes.

The development of processes for primary production, alloy development and the processing routes started in the early 20th century and continued over the years. Several authors compiled the state of the art. Among them are J. L. Haughton and W. E. Prytherch (1938), A. Beck (1939, reprinted 2001), L. M. Pidgeon, et al. (1946), W. H. Gross (1949), G. V. Raynor (1959), C. S. Roberts (1960), E. F. Emley (1966), R. S. Busk (1986), and in recent years M. A. Avedesian and H. Baker (1999), C. Kammer (2000), K. U. Kainer (2003), and H. E. Friedrich and B. L. Mordike (2004). In all of these books the state of the art is shown and the progress in magnesium alloy and technology development is described. But these books only present a view into the past.

The first Magnesium Technology symposium at the TMS meetings took place at the TMS Annual Meeting in Nashville, Tennessee, 2000. Since this time the symposium is showing the lively research of the active magnesium community in the world and it presents also a look into the future. This year is the 14th edition of this meeting, and it will be held in San Antonio, Texas, from March 3-7, 2013. The symposium is arranged into an opening plenary session and a number of technical sessions which address the full scope of Magnesium research and development. These sessions include primary production, casting and solidification, deformation mechanisms and texture, alloy and microstructural design, corrosion, high temperature behaviour, processing-microstructure-property relationships, and advanced processing and joining

Last, the organizing of these proceedings would not have been possible without the dedication and assistance of TMS staff. In particular, we would like to recognize Ms. Patricia Warren for her tireless patience and hard work.

Over the past 13 years, the Magnesium Technology proceedings, through its high quality technical content, has become the primary proceedings for the magnesium industry and technical community. We hope that this year's volume will serve you just as well.

Norbert Hort (Helmholtz-Zentrum Geesthacht, Germany)
Suveen N. Mathaudhu (U.S. Army Research Office, USA)
Neale R. Neelameggham (IND LLC, USA)
Martyn Alderman (Magnesium Elektron, UK)

ABOUT THE LEAD EDITOR

Norbert Hort
MAGNESIUM TECHNOLOGY 2013 LEAD EDITOR

Norbert Hort is the head of the Magnesium Processing Department at the Magnesium Innovation Centre (MagIC) within the Helmholtz-Zentrum Geesthacht Zentrum für Material- und Küstenforschung, Geesthacht, Germany (formerly the GKSS Research Centre). Concurrently he is lecturer at the Leuphana University, Lüneburg, Germany. He studied Materials Sciences at the Clausthal University of Technology (CUT), Germany, where he has been involved in magnesium research since the early 1990s. His diploma thesis (1994) dealt with gas-atomized, hot-extruded magnesium alloys. During 1994-95 he worked as a researcher at the Institute of Materials Sciences (CUT) and he joined the HZG in 2000. He got his Ph.D. degree in Materials Sciences in 2002 from the Clausthal University of Technology. In the MagIC he is responsible for the development of new creep resistant magnesium alloys, grain refinement and the castability of magnesium alloys (viscosity and density of melts, fluidity, and mould filling). A major focus of his research is the development of new magnesium alloys for biodegradable implants in a close collaboration with biologists and clinicians. He is co-author of more than 150 peer reviewed journal papers and contributions to conference proceedings. In recent years he was involved in the organizing committees of the conference series "Magnesium Alloys and their Applications" and of the conference "Light Metal Technologies 2011." - Since 2009 he also has been the chairman of the technical committee "Magnesium" of the German Society of Materials (DGM).

ABOUT THE EDITORS

Suveen Nigel Mathaudhu serves as the Program Manager for Synthesis and Processing of Materials with the U.S. Army Research Office (ARO), Materials Science Division. He received his B.S.E. in Mechanical Engineering from Walla Walla University in 1998, and Ph.D. in Mechanical Engineering from Texas A&M University in 2006. In his current position, he manages programs which focus on the use of innovative approaches for processing high performance structural materials reliably and at lower costs. He also concurrently serves as an Adjunct Assistant Pofessor in the Department of Materials Science and Engineering at North Carolina State University, where his research interests include ultrafine-grained and nanostructured materials by severe plastic deformation, consolidation of metastable particulate materials and processing-microstructure-property relationships of refractory metals and lightweight metals, and thermally stable nanocrystalline materials. He has co-authored over 60 technical publications in these areas.

Neale R. Neelameggham is 'Guru' at IND LLC, involved in technology marketing and consulting in the field of light metals and associated chemicals, [boron, magnesium, titanium, lithium and alkali metals], rare earth elements, battery and energy technologies, etc. He has over 38 years of expertise in magnesium production technology from the Great Salt Lake brine in Utah. He was involved in Process Development of its startup company, NL Magnesium, through the presently known US Magnesium LLC, and was its Technical Development Scientist from where he retired. Dr. Neelameggham's expertise includes all aspects of the magnesium process, from solar ponds through the cast house including solvent extraction, spray drying, molten salt chlorination, electrolytic cell and furnace designs, lithium ion battery chemicals and by-product chemical processing. In addition, he has an in-depth and detailed knowledge of alloy development as well as all competing technologies of magnesium production, both electrolytic and thermal processes worldwide. Dr. Neelameggham holds 13 patents and a pending patent on boron production, and has several technical papers to his credit.

As a member of TMS, AIChE, and a former member of American Ceramics Society, he is well-versed in energy engineering, bio-fuels, rare-earth minerals and metal processing, and related processes. Dr. Neelameggham has served in the Magnesium Committee of LMD since its inception in 2000, chaired it in 2005, and has been a co-organizer of the Magnesium symposium since 2004. In 2007 he was made a permanent co-organizer for the Magnesium symposium. He has been a member of the Reactive Metals Committee and Recycling Committee, and a Programming Committee representative of LMD. In 2008, LMD and EPD created the Energy Committee following the symposium on CO_2 Reduction Metallurgy symposium initiated by him. Dr. Neelameggham was selected as the inaugural Chair for the Energy Committee with a two-year term. He received the LMD Distinguished Service Award in 2010 and has been a co-editor of the Energy Technology symposium proceedings volume. Dr. Neelameggham holds a doctorate in extractive metallurgy from the University of Utah.

Martyn Alderman is Divisional Director of Technology for the Magnesium Elektron Group Worldwide. He obtained a Masters Degree in Material Science from the University of Cambridge (United Kingdom) in 1979 and for the next 20 years he worked in aluminium rolling mills producing and developing Al-Cu-Zr, Al-Mn, Al Mg-Mn, Al-Zn-Mg, and Al-Li-Cu alloys for Superplastic forming. He also has significant operational experience direct chill casting and in the extrusion of high strength aluminium alloys.

In 2003 he joined Magnesium Elektron during the acquisition of their large Madison, Illinois rolling operation in the United States. For the next few years he was travelling between the United States and Europe encouraging the wider use of magnesium sheet in transport applications, in particular by use of superplastic forming, and later in managing the post acquisition integration of the Revere North American Graphic Arts business. He chaired the 2010 International Magnesium Association Conference in Hong Kong, and for the past two years had been involved in editing and producing a 270 page *Handbook on Designing with Magnesium Alloys.*

As a member of TMS, SAE, and AMS, he is a regular attendee at worldwide magnesium conferences keeping abreast of global technology development with the aim of maintaining the Magnesium Elektron's position as a world leader in magnesium alloys and their end use applications.

MAGNESIUM TECHNOLOGY 2013
Session Chairs

Plenary	**Norbert Hort** Helmholtz-Zentrum Geesthacht
Primary Production and Casting	**Neale Neelameggham** Ind LLC
	Adam Powell Metal Oxygen Separation Technologies
Mechanical Properties	**Alan Luo** General Motors Global Research & Development
	Menahem Bamberger Technion
Corrosion	**Liming Peng** Shanghai Jiaotong University
	Carlos Caceres The University of Queensland
Texture and Wrought Materials	**Martyn Alderman** Magnesium Elektron
	Alok Singh National Institute for Materials Science
Wrought Materials II	**Julian Rosalie** National Institute for Materials Science
	Jennifer Hay Agilent Technologies
Grain Refinement, Twinning, and Composites	**Karl Kainer** Helmholtz-Zentrum Geesthacht
	Suveen Mathaudhu U.S. Army Research Office
Phase Formation	**Alok Singh** National Institute for Materials Science
	Alan Luo General Motors Global Research & Development
Poster Session	**Eric Nyberg** Pacific Northwest National Laboratory

Magnesium Technology 2013

Plenary

Magnesium Technology 2013
Edited by: Norbert Hort, Suveen N. Mathaudhu, Neale R. Neelameggham, and Martyn Alderman
TMS (The Minerals, Metals & Materials Society), 2013

A BRIEF HISTORY OF THE DEVELOPMENT OF GRAIN REFINEMENT TECHNOLOGY FOR CAST MAGNESIUM ALLOYS

D. H. StJohn[1,2], P. Cao[4], M. Qian[1,2,3], M. A. Easton[1,5]

[1]CAST Cooperative Research Centre
[2]Centre for Advanced Materials Processing and Manufacturing (AMPAM)
[3]ARC Centre of Excellence for Design in Light Metals
The University of Queensland, School of Mechanical and Mining Engineering, St Lucia, Queensland, Australia
[4]University of Auckland, Auckland, New Zealand
[5]Monash University, Materials Engineering, Clayton, Victoria, Australia, 3069

Keywords: Magnesium, Grain refinement, Solidification

Abstract

Grain size control during casting is important for ensuring the best and most consistent mechanical and corrosion performance is achieved in magnesium alloy components. Also, it is important for the fabrication of wrought Mg products with desired surface quality. Industrially, this means employing grain refinement methods. The important role of grain refinement was realized in the 1930s. However, developing a good understanding of the solidification mechanisms for improved refinement technologies involved in successful refinement has been the subject of ongoing research ever since. This paper provides a brief history of developments both scientifically and industrially, and summarizes current issues related to the development of improved grain refiners.

Introduction

Grain refinement is essentially a nucleation process. The origins of our understanding about nucleation are well summarized by Kelton and Greer in their book 'Nucleation in Condensed Matter'[1]. As understanding of the deformation behavior of alloys improved, relationships between grain size and properties were developed such as the Hall-Petch relationship developed in the 1950s which predicts the effect of grain size on the yield strength of alloys, including Mg-Al alloys [2]. Thus the casting process itself became a focus for improving the properties of components and this was primarily achieved by stimulating the nucleation process. There are parallels between the developments of grain refining technologies for Al [3, 4] and Mg [5, 6] alloys. Both developments were first reported in the 1930s. The Al-5Ti-1B master alloy developed in the early 1970s works well for most Al alloys [3, 4]. However, for Mg alloys there is no equivalent that provides reliable, repeatable as-cast grain sizes in a cost effective manner. Since then much research effort has been focused on understanding the mechanisms of grain refinement.

Mg alloys are cast by a large variety of casting processes including ingot, sand, gravity die, low pressure die and high pressure die (HPDC) casting methods. Most Mg components produced by HPDC are Mg-Al alloys. However, complex castings for applications such as gearbox housings use precision sand casting as well as HPDC. Alloy ingot and billet are commonly produced by direct chill casting and gravity die casting. Grain refinement technology is not needed for HPDC but is important for other casting processes and can help improve creep resistance for certain Mg alloys used at elevated temperatures.

Given the limited length of a conference paper we will cover highlights rather than attempt to provide a detailed history. However, we hope that there is sufficient information provided so that the reader can gain an appreciation of the pathways followed by founders and researchers in an attempt to develop effective and reliable grain refining technologies. This paper firstly presents a summary of the range of grain refining technologies and associated mechanisms developed to explain refinement. A model that attempts to predict grain size outcomes is then described. Finally, current issues that remain unresolved are presented.

Grain Refinement Technologies

Tables 1 and 2 summarize the key advantages and disadvantages of the developed grain refinement technologies with reference to relevant patents while Table 2 summarizes the grain refinement mechanisms proposed to explain what controls grain refinement. Table 1 is subdivided into Mg-Al alloys, Mg-Zn and RE alloys that do not contain Al, and methods applicable to all alloys.

The development of refinement technologies has more or less followed assumptions derived from observation of casting outcomes. The following puts these developments in the context of four elements, Fe, Mn, C and Zr, that underpin the assumptions used to develop particular refinement methods.

The idea of ***grain refinement by Fe*** stems from early work on superheating [7]. ***Superheating*** is an important early grain refining method developed for Mg alloys containing Al, and this process was widely used prior to and during World War II in the commercial production of Mg-Al alloys [8]. The process involves heating a Mg alloy to a temperature about 180-300 °C above its liquidus, holding it for a required period, and then cooling quickly to the pouring temperature [7]. It was believed that superheating was caused by Fe picked up from the crucible. The idea that Fe particles acted as nuclei for Mg grains resulted in the invention of an Fe inoculation process called the ***Elfinal process*** or $FeCl_3$ process [9]. The Elfinal process consists of plunging 0.4-1.0% of anhydrous $FeCl_3$ powder at temperatures between 740 °C and 780 °C into a molten Mg alloy [10]. The first application of the Elfinal process was on Mg-Al-Zn alloys [10] and later Al-free Mg-Zn alloys. Jessup and Petch [11, 12] of Magnesium Elektron Limited (MEL) reported that grain refinement of Mg-Zn-RE-Mn alloys was achievable using a Zn-7%Fe pre-alloy as an Fe additive replacing $FeCl_3$. Nelson pointed out that the Elfinal process didn't work for Mg-Al alloys without Mn [8].

Table 1. A summary of the approaches taken to grain refine Mg alloys with the advantages and disadvantages of each approach.

Process	Advantages	Disadvantages
Mg – Al alloys		
Solute additions	Solute additions can lead to significant grain size reduction even without the addition of a nucleant.	There is a limit to the grain refinement that can be achieved and it is not effective in casting alloys which already have high solute contents. Alloy composition changes and mechanical properties may not be desirable.
Superheating	Significant grain refinement can be achieved.	Difficult to control. High temperatures (~850°C) must be used that can cause melt handling problems. Not clear what the nucleant particles are.
Elfinal	Produces a degree of refinement.	Adds Fe which decreases corrosion resistance. Chlorine is produced which has health implications.
C inoculation	Significant grain refinement can be achieved.	The only approach used commercially at this stage. However, the most effective way of adding C is under debate.
Native grain refinement	The native particles present in high purity melts are more potent than those present in commercial alloys.	No direct evidence of particles and their composition. Pure alloys are more expensive. Limited to Mg-Al based alloys.
SiC	Works reasonably well.	Grain refinement is not as effective as desired. Also, SiC is unstable in the melt and appears to be another C-inoculation technique.
Fe, Mn, Ti additions	Reported that some grain refinement achieved. However, poisoning occurs when Fe and Mn are both present.	Results vary suggesting that the reason for the grain refinement cannot be controlled. It is not understood completely why they work.
Borides	These are known nucleants in Al-based alloys and are available as master alloys. Reports of good grain refinement.	Borides have been tried in CAST and other work and have been found to show inconsistent results. Settling of particles is an issue for commercial application.
AlN, ZnO	Crystallographic edge-to-edge matching (*e2em*) predicts AlN and ZnO are possible nucleants.	ZnO is unstable in melts and a previous attempt at using AlN particles did not perform as well as expected.
High pressure die casting	High cooling rate driven grain refinement without any change of alloy chemistry.	Not suited for the production of basic shapes (billet and slab); having dendritic grains in most cases.
Mg – Zn, RE alloys that do not contain Al, Mn, Si, Fe		
Mg-Zr master alloys	Very effective, producing grain sizes down to < 50 microns.	Expensive, high degree of waste Zr in sludge that forms from settling of Zr particles. Reaction with Fe-based crucibles can also consume Zr. Large agglomerates affect mechanical integrity of castings.
Al additions	Small Al additions appear to grain refine some Mg-RE alloys through the formation of pro-peritectic Al_2RE particles that act as nuclei.	More information is required on which alloys this approach can be applied to and the subsequent effect on properties.
Applicable to all Mg alloys		
Ultrasonic Treatment	Depending on the alloy composition Ultrasonic Treatment can significantly reduce the grain size.	Ultrasonic transducers do not function at elevated temperatures and ultrasonic attenuation is a concern. May not be as effective for large castings.
Electromagnetic stirring	An effective and clean process that can produce fine grain sizes.	Electromagnetic equipment adds cost. The process appears to be more suited to casting basic shapes (billet and slab) rather than components.
Melt Conditioning	Melt conditioning is able to increase the number of nucleation events, significantly reducing the grain size.	Melt conditioning equipment adds cost to casting process. More data is needed on the performance of castings produced this way.

In 1945 Tiner [13] studied the superheating of Mg alloys and the influence of Fe and Mn on the response of superheating. He found that Fe favours grain refinement by superheating. Partridge [14], in 1948, observed obvious grain refinement of Mg-Al alloys using a new but slightly rusty mild steel crucible. He concluded that the loosely adherent rusty Fe had produced the observed grain-refining effect. Emley [5], however, suggested that the observed grain refinement was due to the uptake of carbon from the crucible, as the surface layer of a new mild steel crucible contains more carbon than does the surface layer of a repeatedly used crucible. Regarding the Elfinal process, Emley suggested that Fe containing intermetallic particles or Al carbide (Al_4C_3) particles were possibly the nucleants [5]. Emley apparently preferred the Al_4C_3 hypothesis, as he believed that the hydrolysis of $FeCl_3$ gave rise to copious HCl fumes, which could attack steel crucibles to liberate carbon from the surface layers [5].

Tamura et al. [15] found that the presence of Fe increases the grain size of high purity AZ91E alloys. They presented evidence showing that the ternary Al-C-O particles, which are thought to be effective nucleants, would transform to ineffective quaternary Al-C-O-Fe compounds in the presence of Fe [15]. They thus concluded that Fe was an inhibiting element for the grain refinement of Mg-Al alloys, as it poisoned the potency of the Al-C-O nucleants by transforming them to Al-C-O-Fe [15]. Gao et al. [16] then found that the grain size of scrap AZ91 decreased with decreasing Fe content from 0.024% to 0.0009%, and suggested that Fe acts as an inhibiting element for grain refinement. This apparently contradicts the basis of the Elfinal process. On the other hand, it is interesting to note that although the Elfinal process has been known since the early 1940s [8-10, 17, 18], no micrographs or data can be found in the literature (including the original patent on this process [10]) as to the effectiveness of this process. It is likely that some experiments were conducted but for some reason the results were not published.

The influence of Fe on the grain refinement of high purity Mg-Al alloys was investigated using various methods of Fe addition [19, 20]. The addition of Fe in the form of anhydrous $FeCl_3$ produced obvious grain refinement in high-purity Mg-3%Al and Mg-9%Al alloys. Grain refinement could also be achieved through the uptake of Fe from the crucible surfaces; however, the addition of Fe in the form of ALTAB^{TM} Fe75 compact did not cause grain refinement. The use of anhydrous $FeCl_3$ also leads to grain refinement in high-purity Mg-Al alloys when these alloys are melted in carbon-free aluminium titanite crucibles, suggesting that the Elfinal process has little to do with the Al_4C_3 hypothesis proposed by Emley. In contrast, Fe- and Al-rich intermetallic particles were observed in many Mg grains when the Elfinal process was used, suggesting that these particles are possibly nucleants for Mg grains.

The above contradictions illustrate the problem in understanding the mechanisms operating during solidification. Although the use of Fe results in good grain refinement via the Elfinal process [10-12], it has been found that Mg-Al alloys with ultra-low levels of Fe have a naturally fine grain size compared to similar alloys that contain normal impurity levels of Fe. Taking another approach Nelson [8] observed that AZ92 and AZ63 alloys containing <0.001% Fe had a finer grain size than the same alloys containing 0.02% Fe. Tamura et al. [15, 21, 22] confirmed Nelson's observations. The grain refinement by control of impurity levels is referred to as ***native grain refinement***. They also showed particles containing Al, C and O in the central regions of certain Mg grains in high purity Mg-Al alloys and attributed native grain refinement to the presence of Al_4C_3 or Al-C-O particles.

Native grain refinement was observed exclusively in high purity Mg-Al alloys. The grain size of Mg-9%Al alloys was found to increase with increasing proportion of commercial purity up to 100% Mg [20]. Little Fe and Mn is present in the high purity Mg-Al alloys and as there was no uptake of any other impurities from the crucible used in this study, it was assumed that the nucleant particles involved in native grain refinement of Mg-Al alloys are Al_4C_3 or Al-C-O particles as concluded by Tamura et al [15, 22]. In commercial purity Mg-Al based alloys the presence of Fe or Mn degrades the potency of Al_4C_3 or Al-C-O by forming quaternary Al-C-O-Fe or Al-C-O-Mn compounds.

Investigations conducted since the 1930s have been unable to provide consensus on the ***influence of Mn on the grain refinement*** of Mg-Al alloys. It was believed that superheating is very closely related to the presence of impurity elements such as Fe, Mn and C, in the alloy melts. Tiner [13] found that high-purity Mg-Al alloys with Mn content less than 0.02% showed no appreciable superheating effect, while Mg-Al alloys containing 0.19% or more Mn demonstrated some grain refinement during superheating. Tiner [13] further claimed that binary Mg-Al alloys that can be refined by superheating need to contain either Fe or Mn in excess of their solubilities at the liquidus temperature of the alloy. Nelson [8] subsequently investigated the grain refinement of an AZ92 (Al: 9% and Zn: 2%) alloy with 0.4% or more Mn, and found that alloys with such a high Mn content could not be grain-refined by superheating unless the excessive Mn was settled out by holding at low temperatures (650 – 700°C) and then removed from the melt by sludging out. Nelson's [8] observations contradict Tiner's [13], who obtained full grain refinement in Mg-9%Al alloy containing 0.98% Mn.

Tamura et al. [15, 22] re-examined the influence of minor impurity elements on the grain refinement of high purity Mg-9%Al alloys (Fe < 0.001%; Mn < 0.001%). They used an Al-10%Mn master alloy to add Mn. Contrary to Tiner's observations, Tamura et al. [15] found that the average grain size of a chill bar sample increased progressively from about 40 μm to 160 μm with increasing Mn content from less than 0.001% to 0.47%, where Fe was kept as low as 0.002%. Al-C-O particles were observed in the Mg grains before Mn additions. However, it was found that, in the presence of Mn, these ternary particles transformed to quaternary Al-C-O-Mn. Tamura et al. [15] therefore concluded that Mn was a grain refinement inhibitor, as it poisoned the potency of the Al-C-O nucleants by transforming them into Al-C-O-Mn. Particles that contained Al, Mn, Fe, O, and occasionally Si, were also observed at the centers of Mg grains. To add to the complexity of the effect of Mn, they later published an article about the formation of Mn-containing particles in a commercial AZ91E alloy using a single-roll rapid solidification method [23]. Interestingly, they found that cross-shaped Al-Mn(-Fe) particles were frequently observed when the AZ91E melt was rapidly cooled from superheating temperatures (e.g. 900 or 950°C) to 700°C. They then claimed that the cross-shaped Mn-containing particles produced effective nucleation sites for Mg grains. These conclusions differ from those they made before, where they claimed that the nucleant particles are Al-C-O particles and that Mn is a poisoning element for the grain refinement of Mg-Al alloys. Byun et al. [24] observed the existence of $Al_8(Mn,Fe)_5$ particles within the α-Mg grains in an AZ91 slurry sample containing 0.23% and more Mn, suggesting that $Al_8(Mn,Fe)_5$ particles act as the heterogeneous nucleation sites. On the basis of Mn solubility calculations, Byun et al. [24] suggested that high Mn concentrations favour grain refinement because supersaturated Mn is likely to precipitate to form $Al_8(Mn,Fe)_5$ particles where primary Mg grains nucleate.

A recent study [25] showed that the addition of 0.1% to 1.0% Mn introduced in the form of an Al-60%Mn master alloy splatter, resulted in obvious grain refinement of high purity Mg-3%Al, Mg-6%Al and Mg-9%Al, as well as commercial AZ31 alloys. The majority of the Mn-containing particles observed in the commercial AZ31 base alloy were of the Al_8Mn_5 type, whereas after addition of extra Mn the predominant group of intermetallic particles changes to the ε-AlMn type. The metastable hexagonal ε-AlMn phase is likely to be a more effective nucleant for Mg grains than the stable rhombohedric Al_8Mn_5. It was proposed that the increase in grain size after a long time of holding at a given melt temperature is due to the transformation of the metastable ε-AlMn to the stable Al_8Mn_5. This hypothesis for grain refinement by Mn can also explain the mechanism of grain refinement by superheating.

To overcome the shortcomings of the superheating process (e.g. high operating temperature and applicability to small melts only), post-war research of grain refinement of Mg alloys shifted to developing an alternative method applicable to large melts and at lower operating temperatures. These efforts led to the discovery of the ***carbon refining process*** [26, 27].

In principle, any materials containing carbon can be used as inoculating additives, and do have a grain-refining effect when properly inoculated. These additives can be categorized into three groups: carbonaceous gases, solid carbon additives and organic chlorides. The use of carbon monoxide (CO) or carbon dioxide (CO_2) was first reported in a US patent [28] to give satisfactory grain refinement when bubbled through the melt. Subsequently a number of carbonaceous gases were reported to have a grain-refining ability to Mg alloys, such as acetylene, methane, and natural gas [8, 27, 29-31]. Solid carbon additives include carbon powder, lampblack, magnesite [28], calcium carbide [32, 33], calcium cyanamide, paraffin wax, silicon carbide [34, 35].

From an operational aspect, the carbonaceous gases and organic chlorides can be more easily introduced into Mg melts than the solid carbon additives. In addition, bubbling a melt also helps to degas the melt. The addition of carbon in the form of C_2Cl_6 or Cl_2-CCl_4 blend was widely used before the 1970s because of the combined action of grain refining and degassing. The major problem of using carbonaceous gases, particularly chlorine-based gases, is the emission of chlorinated hydrocarbons (CHC). Considerable efforts have been made to produce a commercially reliable carbon-based grain refiner for Mg-Al alloys. For example, Renger and Simon [36] reported an effective grain refiner named Nucleant 5000, which appeared to work satisfactorily on AZ91 alloys. Waxfluorspar-carbon grain refiners were developed early in England [37] and in Canada [38]. Liu et al. [39] reported an Al_4C_3-SiC-Al alloy suitable for grain refinement of Mg-Al alloys such as AZ31 and AZ63. Motegi et al. [40, 41] disclosed a carbon grain refiner that contains either pure carbon powder or a mixture of carbon and Nb_2O_5 or V_2O_5 powder. However, to date carbon-based grain refiners prove unreliable and produce inconsistent grain refinement in Mg-Al alloys.

There have been many other particles tried in Mg-Al alloys including borides [42, 43], ZnO [44] and AlN [45]. This is understandable because the presence of foreign particles will always tend to facilitate the heterogeneous nucleation of Mg grains as long as they are not completely unwettable. Whilst many of these particles have shown promise there is usually an issue to do with their use, be it instability in the melt, rapid settling or that the grain refinement efficiency is not sufficient enough.

Table 2 groups the mechanisms of grain refinement in Mg-Al alloys into four categories, solute additions, impurity poisoning, C additions and other nucleant particles. A recent review [6] by the authors has shown that broadly all of the grain refining techniques that have been proposed so far can be grouped into one of these categories. This begins to simplify the issues related to grain refinement of Mg-Al alloys, but more work is required to confirm these hypotheses.

The only commercially successful technology is ***Zr grain refinement by Mg-Zr master alloys*** while carbon inoculation is sometimes used for sand and gravity die casting of Rare Earth (RE) alloys for aerospace and automotive applications [5]. Zr is almost the perfect grain refiner except that it is not able to grain refine Al containing alloys due to the affinity of Al for Zr reacting to form Al_3Zr. Some research has been undertaken to understand why Zr is such a remarkably effective grain refiner, although further work is required to optimize these alloys. It has been shown that the performance of Mg-Zr master alloys is related to the Zr particle number density of particles less than 5 microns in size as the larger particles settle out during alloying [46, 47]. Recently, it has also been shown that small Al additions are able to effectively grain refine some Mg-RE containing alloys [48].

There is increasing interest in processes that may assist the nucleation of grains such as Ultrasonic Treatment (UT) and Melt Conditioning. The results to date have shown that they are effective in producing considerable refinement. However, little industrial application has occurred to date and thus the focus of future research is on their practical implementation in casting operations [49].

More details on research into the grain refinement of Mg alloys undertaken to date can be found in two reviews by the authors [50].

Theoretical Developments

As highlighted in the previous section and Tables 1 and 2, the proposed theories of grain refinement of Mg alloys have essentially been explanations of the observed, and often contradictory, refinement effects with little definitive evidence of particle characteristics such as type and size distribution. The underlying assumptions such as the role of Fe and the presence of C, were used to develop new refining methods. There has been much effort focused on identifying potential nucleant particles [51] or speculation on the type of native or impurity nucleant particles [20]. It is known that grain size is affected by the Al content in Mg-Al alloys but alloy composition was not taken into account when analyzing grain refinement outcomes until about 2005 [50].

In 2001 a theory was proposed that brought together alloy constitution and particle potency showing that grain size is related to the growth restriction factor Q [52]. The recent Interdependence Model, equation 1, is a rigorous predictive equation that highlights the factors affecting the as-cast grain size and it incorporates elements of the Free Growth Model [53] developed in the 1990s for the calculation of ΔT_n, the undercooling required for nucleation on a particle of a particular size [54].

$$d_{gs} = \frac{D.z.\Delta T_{n-min}}{v.Q} + \frac{4.6.D}{v}\left(\frac{C_l^* - C_o}{C_l^*.(1-k)}\right) + x_{Sd} \qquad (1)$$

where D is the diffusion coefficient, v the rate of growth of the solid-liquid (S-L) interface, C_l* the composition of the liquid at the S-L interface at x_{cs}, $\Delta T_{n\text{-}min}$ the undercooling required for nucleation on the largest available particle, and $z\Delta T_{n\text{-}min}$ is the incremental amount of undercooling required to activate the next

Table 2. A summary of the theories proposed to explain the mechanisms of grain refinement of Mg alloys with the weaknesses of each theory.

Process	Related approaches	Future Considerations
Mg – Al alloys		
Solute additions	It is well established that growth restriction by alloying elements contributes to grain size reduction. However, it should be noted that alloys that show a high growth restriction effect may not always be commercially attractive or useful.	Lack of physical data to ensure reliable prediction of grain size using the developed predictive equations. It is also clear that whilst this is a significant effect it is not the most important one, particularly in casting alloys.
Impurity poisoning	It appears that the superheating effect, the Elfinal process and Native grain refinement are all related to the effect of impurities precipitating on the surface of particles.	Whilst there is circumstantial evidence for this occurring, it is difficult to obtain direct evidence. Also the wide variety of effects means that it is difficult to explain all observations using this approach.
C inoculation and SiC	Carbon-based inoculants are the most effective nucleants known and the ones that are used commercially. There have been many different approaches proposed. A recent breakthrough has been the understanding that carbon reacts with Al to form Al_4C_3 particles and Al_2MgC_2 which can both act as nuclei. Al_2CO may be a nucleant in some situations.	An effective commercial approach has not as yet been able to be developed and the effectiveness of grain refinement is significant rather than very good.
Other nucleant particles	No other attractive nucleants have been identified yet, despite many trials.	A potent nucleant such as Zr for Al-free Mg alloys may exist or may not. However, given the commercial importance of Mg-Al based alloys, it is worth further effort to identify a potent grain refiner for these alloys.
Mg – Zn, RE alloys that do not contain Al, Mn, Si, Fe		
Mg-Zr master alloys	Both soluble and insoluble Zr contribute to grain refinement.	Prediction capability to take into account simultaneous increases in solute Zr and Zr particle number density needs further refinement.
Al additions	Small Al additions appear to grain refine some Mg-RE alloys through the formation of pro-peritectic Al_2RE particles that act as nuclei.	More information is required on which alloys this approach can be applied to and the subsequent effect on properties.
Applicable to all Mg alloys		
Mechanical approaches	Includes approaches such as ultrasonic treatment, electromagnetic stirring processes and melt conditioning.	For most of these approaches there is no agreed theory. Depending on the technique, issues such as cavitation and fragmentation of impurity particles and α-Mg grains are proposed as generators of many nucleants.

nucleation event as the temperature gradient moves towards the thermal center of the casting. The constant *4.6* is a cut-off factor for the solute profile in front of the S-L interface where $(C_l(x')-C_o)/(C_l^*-C_o) = 1\%$ [54].

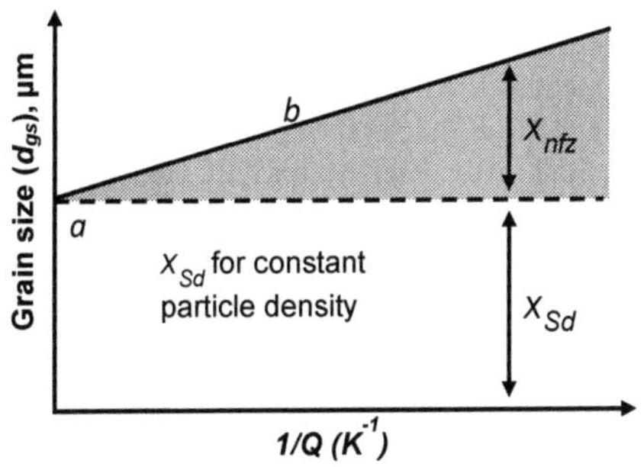

Figure 1: Representation of a grain size vs $1/Q$ plot showing the contribution of the nucleation-free zone x_{nfz} and the distance to the next most potent particle x_{Sd} to the final as-cast grain size.

The first two terms in equation 1 are determined from the alloy characteristics and the casting conditions defining the formation of the nucleation-free zone (x_{nfz}) as insufficient CS is developed within this zone to cause a nucleation event to occur [54]. Figure 1 illustrates the change in x_{nfz} with Q highlighting that for low solute alloys x_{nfz} significantly restricts refinement by preventing nucleation on suitable particles. The third term x_{Sd} is defined by the distance to the next most potent particle, assumed to be constant in Figure 1. The value of x_{Sd} can be decreased by increasing the particle number density, for example, by adding more Zr master alloy.

Plots of the measured grain size vs $1/Q$ present a straight line that provides a useful tool for the analysis of the mechanisms of grain refinement and has been successfully applied to Al, Ti and Mg alloys [54, 55] . The application of the grain size vs $1/Q$ plots has revealed the mechanisms of refinement. For example, superheating of commercial Mg-Al alloys and native refinement of high purity alloys are related [20, 56]. It now appears that superheating removes the effect of impurities allowing the more potent native nucleants whether they are carbides or oxide particles, to be available for nucleation thus producing a finer grain size. Also, a comparison of high and commercial purity Mg-Al alloys has highlighted that a major factor controlling grain size is the nucleation-free zone which reflects the effect of alloy chemistry (Q) in producing large grain sizes for AZ31 and finer sizes for AZ91 [54]. It has also shown the improvements in nucleation efficiency on the addition of SiC to Mg-Al alloys [57].

These plots have been applied to analyzing the performance of Mg-Zr master alloys showing that Zr particle number density is the key factor differentiating different types of master alloy [46, 47]. Also, the linear form of d_{gs} vs $1/Q$ has been shown to apply to a wide range of casting conditions and the application of Ultrasonic Treatment highlighting the important role of Q [6].

Issues Remaining

Current understanding can be summarized as: (1) high purity Mg-Al alloys exhibit native grain refinement with Al carbide or oxide particles being the nucleant particles and addition of either Fe or Mn individually leads to further refinement; (2) commercial Mg-Al alloys exhibit poor refining performance due to an interaction between Al, Fe and Mn forming an intermetallic layer of lower nucleation potency on the native nucleants; (3) Zr remains the best refiner for Mg -Zn and -RE alloys without Al. However, there is no ideal grain refining technology for Mg alloys. Even the superior Mg-Zr master alloy produces sludge wasting considerable Zr that does not contribute to refinement. In almost every case there are issues with commercial implementation and considerable gaps in our knowledge of the mechanisms of grain refinement.

For Mg-Al alloys poisoning by impurity and minor elements appears to be a major issue preventing the development of a reliable grain refiner and it is probable that a different approach to developing a grain refining system is needed. For all Mg alloys the formation of a nucleation-free zone reduces grain refining performance particularly in lean alloys. Strategies for tackling these issues are discussed in a recent publication [6].

The recent investigations of mechanical methods such as ultrasonic treatment, electromagnetic stirring and melt conditioning show promise. However, proving that these methods will be practical and cost effective for industrial application is an important next step.

Research on the grain refinement of Mg alloys has been very active but that has been largely driven by researchers in universities and institutes over the last two decades. Closer collaboration between academia and industry will help address these issues and more importantly, increase the impact on industrial practice.

Acknowledgements

Support was provided by the CAST CRC, ARC Discovery Grants DP120101672 (DStJ, MQ) and DP2010000071 (ME), ARC CoE for Design in Light Metals (MQ) and the Centre for Advanced Materials Processing and Manufacturing (AMPAM) at UQ (DStJ, MQ).

References

[1] K.F. Kelton, A.L. Greer, Nucleation in Condensed Matter, 1st ed., Pergamon, 2010.
[2] G. Nussbaum, et al., Scripta Metall., 23 (1989) 1079.
[3] M.Easton, D.H.StJohn, Metall. Mater. Trans., 30A (1999) 1613.
[4] D.G. McCartney, Int. Mater. Rev., 34 (1989) 247.
[5] E.F. Emley, Principles of Magnesium Technology, Pergamon Press, Oxford, 1966.
[6] D.H. StJohn, et al., Metall. Mater. Trans. A, (2012) in press.
[7] I. Farbenindus,British Patent GB359,425 (1931).
[8] C.E. Nelson, AFS Trans., 56 (1948) 1.
[9] R.T. Wood, The Foundryman, 46 (1953) 98.
[10] I. Farbenindus,Belgian Patents 444,757 (1942).
[11] A.C. Jessup, J.H.T. Petch,British Patent 757,326 (1956).
[12] A.C. Jessup, et al.,US Patent 2,829,973 (1958).
[13] N. Tiner, AIME Tech Pub, 12 (1945) 1.
[14] G.B. Partridge, Production of Magnesium Castings, Kennedy Press, Manchester, 1948.
[15] Y. Tamura, et al., Mater. Sci. Forum, 350-351 (2000) 199.
[16] H. Gao, et al., Mater. Sci. Eng. A, 368 (2004) 311.
[17] F.A. Fox, Metal Treatment, 14 (1947) 169.
[18] Y.A. Tsenter, Tsvetnye Metally, 19 (1946) 55.
[19] P. Cao, et al., Scripta Mater., 51 (2004) 125.
[20] P. Cao, et al., Scripta Mater., 53 (2005) 841.
[21] T. Haitani, et al., J. Jpn. Inst. Light Met., 51 (2001) 403.
[22] Y. Tamura, et al., Mater. Trans., 43 (2002) 2784.
[23] Y. Tamura, et al., Mater. Trans., 44 (2003) 552.
[24] J.Y. Byun, et al., in: K.U. Kainer (Ed.) Magnesium alloys and their applications, Wiley-VCH, 2003, pp. 713.
[25] P. Cao, et al., Scripta Mater., 54 (2006) 1853.
[26] R. Hultgren, D.W. Mitchell, Trans. AIME, 161 (1945) 323.
[27] V.C.F. Holm, A.I. Krynitsky, The Foundry, 75 (1947) 81.
[28] C.E. Nelson, G.E. Holdeman,US Patent 2,380,863 (1945).
[29] J.A. Davis, et al., Trans. AFS, 53 (1945) 352.
[30] Q. Jin, et al., Scripta Mater., 49 (2003) 1129.
[31] C.H. Mahoney, et al., Trans. AIME, 161 (1945) pp.328.
[32] V.B. Kurfman, Trans. A. I. M. E., 221 (1961) 540.
[33] K. Makino, et al., Japan Keikinzoku, 42 (1992) 810.
[34] H. Hu, Scripta Mater., 39 (1998) 1015.
[35] A. Luo, Can. Metall. Q., 35 (1996) 375.
[36] K. Renger, R. Simon, in: A. Dahle (Ed.) 1st International Light Metals Technology Conference, Brisbane, 2003, p. 231.
[37] F.A. Fox,British Patent, GB608941 (1948).
[38] D.O. Karlsen, et al., in: Proc. Int. Sym. Light Metals Process. Appl., Quebec, Canada, 1993, pp. 397.
[39] Y. Liu, et al., Mater. Lett., 58 (2004) 1282.
[40] T. Motegi, et al.,US Patent 6,616,729 (2003).
[41] T. Motegi, et al.,EP 1,282,780 (2003).
[42] G. Klosch, et al., in: A. Luo, et al. (Eds.) Magnesium Technology 2006, TMS, San Antonio, 2006, p. 69.
[43] Y. Wang, et al., Scripta Mater., 54 (2006) 269.
[44] H.M. Fu, et al., J. Alloys Compd., 456 (2008) 390.
[45] H.M. Fu, et al., J. Alloys Compd., 478 (2009) 809.
[46] M. Sun, et al., Adv. Eng. Mater., (2012) submitted.
[47] M. Qian, et al., Scripta Mater., 50 (2004) 1115.
[48] D. Qiu, et al., Acta Mater., 57 (2009) 3052.
[49] EXOMET, in, http://www.materials.imdea.org/Research/Projects/EXOMET/tabid/4236/Default.aspx.
[50] D.H. StJohn, et al., Metall. Mater. Trans. A, 36 (2005) 1669.
[51] M.X. Zhang, P.M. Kelly, Acta Mater., 53 (2005) 1085.
[52] M.A. Easton, D.H. StJohn, Acta Mater., 49 (2001) 1867.
[53] A.L. Greer, et al., Acta Mater., 48 (2000) 2823.
[54] D.H. StJohn, et al., Acta Mater., 59 (2011) 4907.
[55] D.H. StJohn, et al., Mater. Sci. Forum, 690 (2011) 206.
[56] P. Cao, et al., Scripta Mater., 56 (2007) 633.
[57] M.A. Easton, et al., Scripta Mater., 55 (2006) 379.

Magnesium Technology 2013

Creep

Magnesium Technology 2013
Edited by: Norbert Hort, Suveen N. Mathaudhu, Neale R. Neelameggham, and Martyn Alderman
TMS (The Minerals, Metals & Materials Society), 2013

COMPRESSIVE CREEP PROPERTIES OF WROUGHT HIGH TEMPERATURE MAGNESIUM ALLOYS IN AXIAL AND TRANSVERSE ORIENTATION – A NEUTRON DIFFRACTION STUDY

D. Sediako[1] , L. Bichler[2], M. van Hanegem[2] and S. Shook[3]
[1]Canadian Neutron Beam Centre, Chalk River, Canada
[2]University of British Columbia – Okanagan, School of Engineering, Kelowna, Canada
[3] TH Magnesium Inc., Farmington Hills, MI, USA

Keywords: Magnesium alloys, high-temperature creep, neutron diffraction

Abstract

Magnesium alloys developed for high temperature applications may become a material of choice in structural components exposed to elevated temperatures, for example, around automotive front end structures. Performance of wrought materials, however, often depends on the direction of the load with respect to the extrusion direction.

This study focused on investigating compressive creep response of three extruded magnesium alloys: AE42, EZ33 and ZE10 in the longitudinal and transverse directions. In-situ neutron diffraction experiments on evolution of primary and secondary creep strain at 175°C, 50MPa load for the duration of 24 hours was studied for several crystallographic planes.

The results suggest that the AE42 alloy exhibited significant strain fluctuations on the basal (0002) and $(10\bar{1}2)$ planes, possibly related to plastic deformation processes of grain boundary sliding or twinning. In contrast, EZ33 and ZE10 alloys had a significantly lower and more uniform strain deformation in both elastic and plastic regime. Microstructure analysis revealed that the AE42 alloy consisted of large $Al_{11}RE_3$ particles with $Mg_{17}Al_{12}$ phase forming around Al-RE particles and in intergranular regions, while the aluminum-free alloys contained complex intermetallics on the grain boundaries, as well as Zn-RE nanoscale dispersoids throughout the matrix.

Introduction

The primary challenge with utilizing more Mg alloys in vehicles stems mainly from their relatively high cost (~50% higher than that of aluminum alloys) and their poor creep resistance above 120 °C. The loss of strength at elevated temperatures has been attributed to grain boundary movement and plastic deformation, which ultimately lead to inter-granular failure [1]. Enhancing the creep resistance of magnesium alloys has therefore been the subject of intense international research efforts.

Magnesium with its HCP crystal structure is known to have only a limited number of slip systems. In general, slip most readily takes place on the close-packed basal plane (0002) followed by a secondary slip on the $(10\bar{1}1)$ planes [1]. The addition of rare earth elements (REs) to aluminum containing magnesium alloys was seen to significantly increase creep resistance due to the formation of thermally stable intermetallic compounds along the grain boundaries [2]. The low solubility of REs in Mg results in the formation of $(Mg,Al)_xRE_y$ complex intermetallics, which decreases the Al saturation in the matrix and decrease the possibility of undesirable β-phase formation [3].

The addition of REs to Mg-Zn alloys has been also seen to increase creep resistance due to precipitation hardening [4]. Mg_xRE_y and Mg_xZn_y intermetallic compounds readily form in EZ magnesium alloys due to the low solid solubility of REs in Mg. Additional RE_xZn_y intermetallic compounds on the grain boundaries were seen to remain stable at temperatures as high as 420°C [5]. With the increasing concentration of rare earths, presence of fine precipitates dispersed throughout the matrix was reported. The role of these precipitates, however, on the creep performance required further study [6].

In this research, the longitudinal and transverse compressive creep strain of AE42, EZ33 and ZE10 wrought alloys was measured using an *in-situ* neutron diffraction technique. The creep performance was subsequently related to the alloy microstructure and the observed creep strain anisotropy is discussed.

Experimental Setup

Alloy samples were produced by Timminco Corporation Canada via proprietary controlled-cooling permanent mold casting followed by hot extrusion. Quality of all cast and extruded samples was inspected with respect to inclusions, cracking and porosity. The composition of the alloys conformed to the ASTM composition limits for each alloy.

The extruded alloys were machined to cylindrical specimens according to the ASTM E139-70 standard for compressive creep testing of Mg alloys. The longitudinal axis of the cylindrical test specimen was aligned either with the extrusion axis (EA) of the billet, i.e., along the longitudinal (L) direction parallel to EA or along the transverse (T) direction perpendicular to EA. Compressive creep strain and microstructure were studied in these two directions.

Neutron Diffraction

Compressive creep strain *in-situ* measurements were carried out at the National Research Council-Canadian Neutron Beam Center (NRC-CNBC) in Chalk River, ON. Material samples were maintained at 175°C with infrared heaters, and then loaded in compression at 50MPa over a 24 hour period. Elastic creep microstrain was measured using the peak-shift method for the $(10\bar{1}0)$, (0002), $(10\bar{1}1)$, $(10\bar{1}2)$ and the $(2\bar{1}\bar{1}0)$ planes. In addition to measuring crystallographic strain, a traditional extensometer was used to measure the total (plastic and elastic) compressive creep strain during the experiment. A detailed description of the neutron diffraction experiments, set-up and methods of calculating compressive strain has been published by the authors earlier [7].

Optical and Electron Microscopy

Alloy samples were prepared for microstructure analysis using standard polishing procedures for Mg alloys. AE42 alloy was etched with acetic glycol to reveal the grain size and general microstructure. EZ33 and ZE10 alloys were etched using dilute nitric acid to reveal second phases. Optical microscope (Zeiss AxioObserver) and scanning electron microscope (Tescan Mira-3) with XEDS capability were used to study the alloy microstructure and chemical composition.

Results and Discussion

Neutron Diffraction Creep Microstrain Measurement

The results of compressive creep measurements using an extensometer are provided in Figure 1. The values show the combined plastic and elastic creep strain during the test. In both orientations (L and T), the AE42 alloy experienced the greatest compressive creep, while EZ33 and ZE10 alloys showed an order of magnitude smaller strain. Further, the results suggest that the AE42 alloy continued to creep with time, while the EZ33 and ZE10 alloys reached a stable creep strain after a short primary creep stage, and remained at this creep level for the remainder of the test. Consequently, the results suggest that aluminum-free alloys were significantly more stable than current industry high-temperature benchmark AE42 alloy.

The duration of the creep test ranged from 12 – 24 hours for the samples tested in the L direction, and from 23 - 100 hours for the samples tested in T direction. By extrapolating the data points of the secondary creep stage (linear portion with constant slope on the creep curve in Figure 1) the creep strain at 100 hrs was estimated for all the samples. In the case of the AE42 alloy, the creep strain in the transverse direction was ~38% greater than in the longitudinal direction, with a total alloy average strain of 11.8% after 100 hrs. In contrast, at 100 hrs, both EZ33 and ZE10 alloys outperformed the AE42 alloy in creep resistance, with an average creep strain of only 0.07% and 0.17%, respectively, as seen in Figure 2a.

Figure 2b summarizes the relative creep resistance of each alloy based on the ratio of creep strain in the L- and T-directions. The results indicate that the AE42 and ZE10 alloys exhibited comparable strain isotropy in both directions, while the EZ33 alloy deformed readily in the transverse direction (0.12%), but exhibited good creep resistance in the longitudinal direction (0.03%). Thus, the response of the EZ33 alloy was highly anisotropic. This result will be related to the alloy's microstructure, as will discussed in subsequent sections.

The results of *in-situ* elastic creep strain measurement via neutron diffraction are presented in Figure 3 (Longitudinal direction) and Figure 4 (Transverse direction). The data presented in Figures 3 and 4 indicates that all the samples generally come to the steady state creep (secondary creep) within the first two hours of loading. Following this, the elastic microstrains obtained in the neutron diffraction experiments remain relatively constant. This validates the strain-extrapolation approach adopted in this study to estimate the 100-hr strains presented in Figure 2.

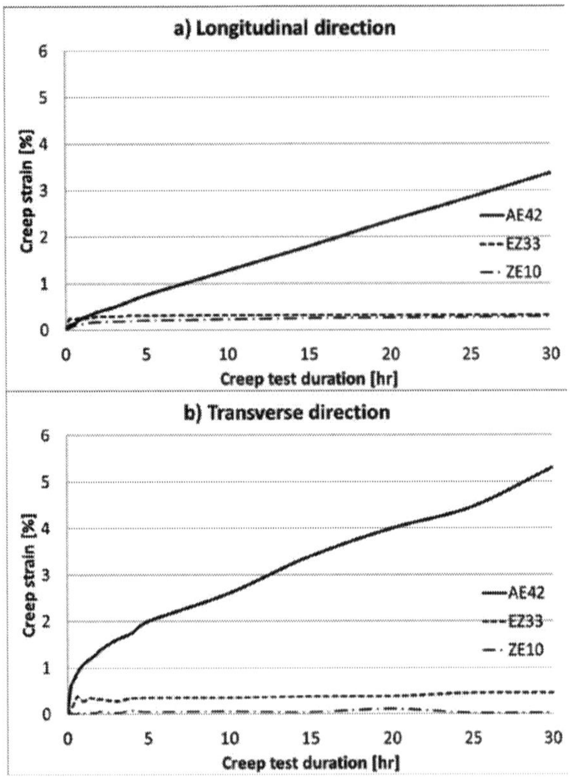

Figure 1: Compressive total (plastic + elastic) creep strain in: a) Longitudinal direction, and b) Transverse direction

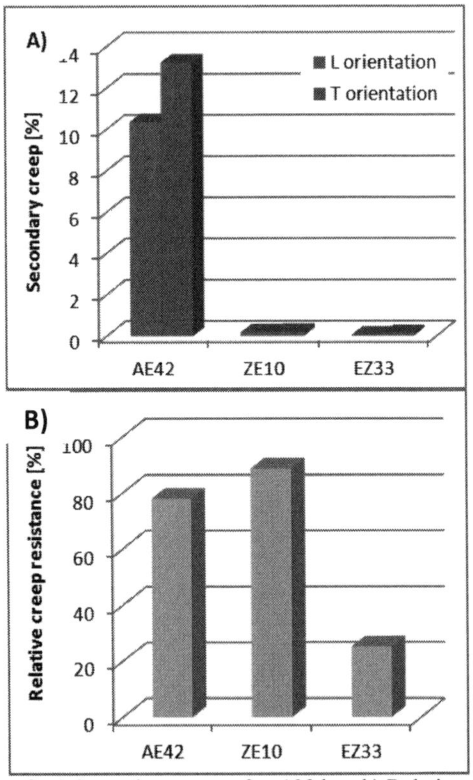

Figure 2: a) Secondary creep after 100 hrs; b) Relative creep anisotropy ($\varepsilon_{trans}/\ \varepsilon_{long}$)

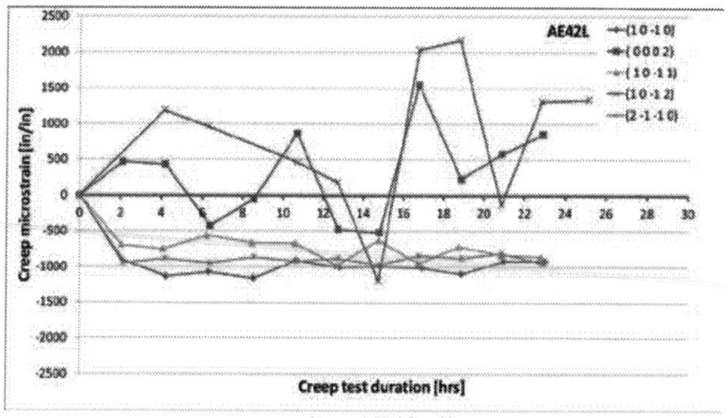

a) AE42 alloy

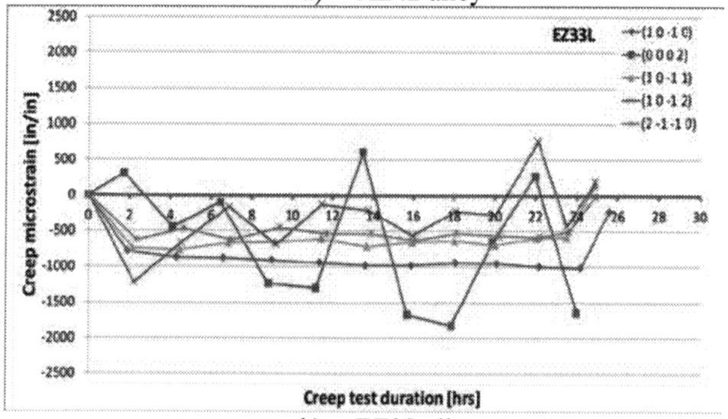

b) EZ33 alloy

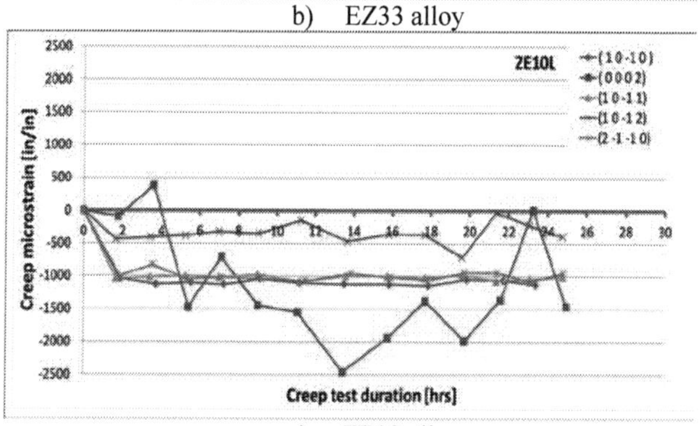

c) ZE10 alloy

Figure 3: Creep microstrain in the longitudinal direction.

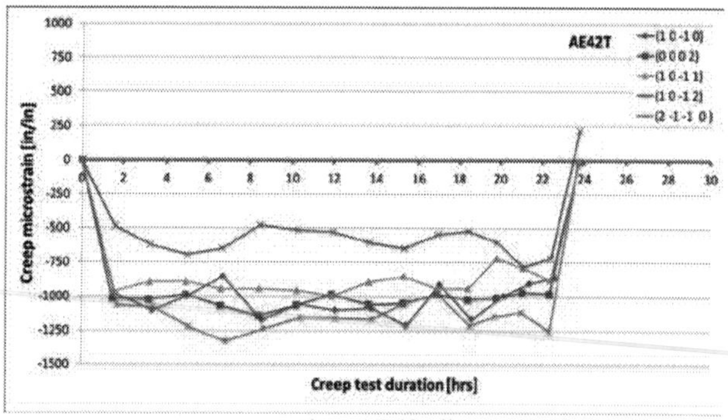

a) AE42 alloy

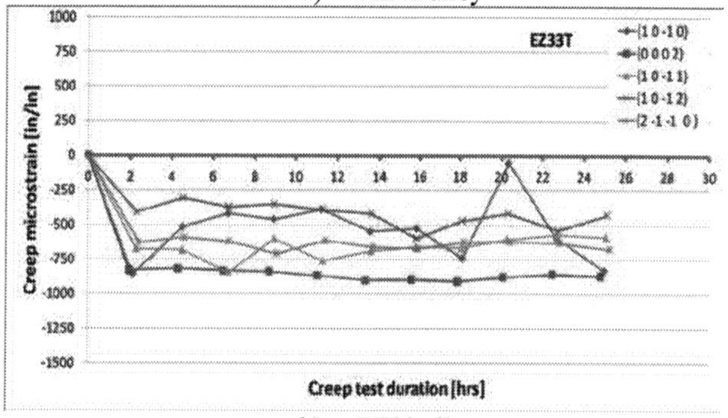

b) EZ33 alloy

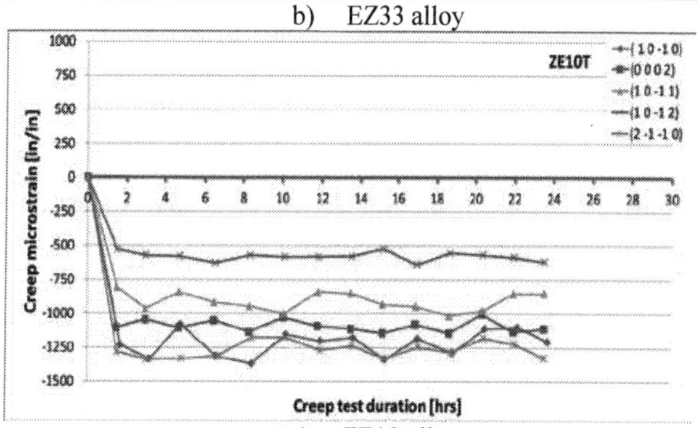

c) ZE10 alloy

Figure 4: Creep microstrain in the transverse direction

Longitudinal Direction

It was observed that the creep strain on the close packed basal plane (0002) in all alloys exhibited the greatest fluctuations during the creep test. This result was related to the strong basal texture frequently observed in the extruded samples, with a very high tendency of the (0002) poles to align perpendicular to the EA. The texture may get further modified as a result of creep-induced deformation [8]. Although the load on the cylindrical specimens remained compressive at all times during the test, the (0002) basal planes often exhibited tensile – compressive - tensile strain fluctuations, particularly in the case of the AE42 alloy. These fluctuations were likely an indicator of plastic deformation processes via grain reorientation (due to grain boundary sliding) and/or due to localized twinning. The authors have reported on twinning of the AE42 alloy during compressive creep earlier [9].

The presence of twinning was also suggested by observing the microstrain profile for the $(10\bar{1}2)$ plane, which is a preferred twin plane in Mg alloys, and was seen to exhibit significant strain reversal oscillations. The strain on the remaining crystallographic planes was seen to remain relatively uniform. In the case of the aluminum-free alloys EZ33 and ZE10, the creep microstrain remained uniform, with the exception of (0002) basal plane fluctuations. These fluctuations, however, remained predominantly in the compressive state, unlike in the AE42 alloy where tension evolved. It is of interest to note, that twinning was not observed in the EZ33 or ZE10 alloys, which is in agreement with the uniform strain on the $(10\bar{1}1)$ and $(10\bar{1}2)$ planes.

Transverse Direction

The elastic strain response of the studied crystallographic planes varied depending on the sample loading direction, as seen in Figure 4, a-c. The compression-tension (0002) strain fluctuations observed in the L-samples were not readily observed in the T-samples. It is suggested that as grain boundary sliding likely occurred parallel to the *c*-axis of the grains (due to the strong extrusion-type basal texture), the basal plane strain was less affected and remained relatively constant for the T-samples.

The magnitude of the (0002) microstrain in the AE42 alloy was ~1000 [in/in], which was of the same order of magnitude as the strain on the remaining crystal planes. The only exception was the microstrain on the $(10\bar{1}2)$ plane, which was less stable and reached only -700 [in/in]. Such relative decrease in elastic strain values frequently indicates reduced yield strength of the material, resulting in enhanced plastic deformation of the material [10]. The lower values of microstrain on $(10\bar{1}2)$ planes may therefore indicate higher levels of plastic deformation on these planes. It is likely that this plastic deformation, however, was not necessarily accommodated by twinning, since the level of microstrain remained relatively constant. Similar observation was made for the other alloys as well.

The range and magnitude of the microstrain recorded for the L-oriented samples are comparable to the values recorded for the T-oriented samples (see Table 1). The close proximity of the results for the AE42 and ZE10 suggests relative deformation isotropy for the two alloys in both orientations. In contrast, significant reduction of microstrain for the T-oriented sample of EZ33 alloy as compared to the values for L-oriented sample indicate anisotropy in plastic deformation of this alloy in the two orthogonal orientations, also observed in the analysis of the relative creep resistance of the alloys (Figure 2). This observation may have far-reaching practical implications, indicating limitation in the application of this alloy in the structural components subjected to elevated temperatures and directional load.

Table 1: Range of microstrain for $(10\bar{1}0)$, $(10\bar{1}1)$ and $(2\bar{1}\bar{1}0)$ crystallographic planes.

Alloy	Microstrain range for the L-oriented samples	Microstrain range for the T-oriented samples
AE42	-600 -1100	-800 -1250
EZ33	-500 -1000	-500 -700
ZE10	-1000 -1100	-800 -1350

Alloy Microstructure

Optical micrographs of the post-creep alloys are provided in Figure 5 and 6. The evident extrusion texture observed in the AE42 and EZ33 alloys reveals that second phases strongly segregated between the elongated grain boundaries in the longitudinal direction. In contrast, ZE10 alloy had intermetallics relatively uniformly dispersed throughout the microstructure in both directions.

The AE42 alloy contained massive Al-RE intermetallics (Black particles in Figure 5a and 6a), which exhibited severe fracturing as a result of creep testing, as seen in Figure 7. These phases consisted of Mg–(32 wt%Al, 27wt%Ce, 21 wt%La, 6wt% Nd) and traces of Pr, as determined from SEM-XEDS analysis. It has been established that La:Nd ratio below 0.7 promotes formation of Al_2RE phase, whereas higher La:Nd ratios promote formation of $Al_{11}RE_3$ phase [3]. Thus, these large acicular particles were likely the metastable $Al_{11}RE_3$ phase, which possibly began softening and degrading as a result of elevated temperature exposure. In addition, the metastable $Mg_{17}Al_{12}$ (β)-phase was also observed at the grain boundaries and in the vicinity of the $Al_{11}RE_3$ particles (medium gray in Figure 5a), possibly suggesting that exposure to the test conditions was sufficient to activate Al atom mobility. Consequently, the eutectic β-phase formed with its relatively low melting temperature and contributed to grain boundary sliding and reduced the effectiveness of grain boundary pining by the Al-RE complex intermetallics during creep. The grain interior of the AE42 alloy contained extensive lens-shape twins, as illustrated in Figure 8. Twinning was not observed in any of the aluminum-free alloys.

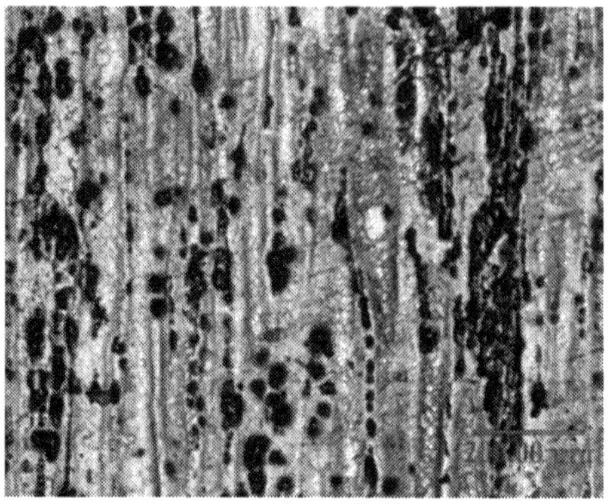

a) AE42

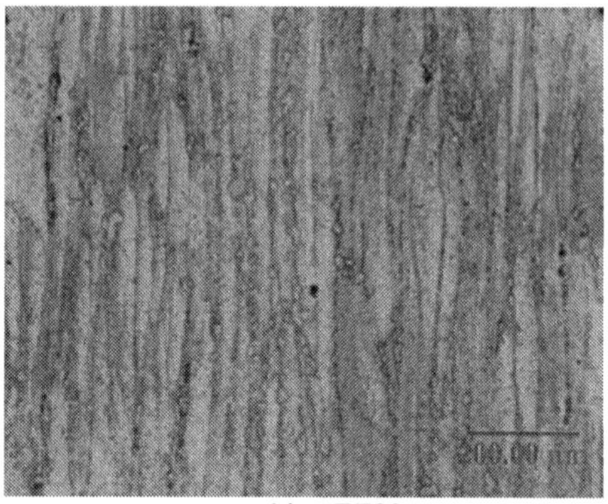

b) EZ33

c) ZE10

Figure 5: Alloy microstructure in longitudinal direction

Similar to AE42, the EZ33 alloy had intermetallics agglomerated in the intergranular regions as well. However, the intermetallics were significantly smaller in size and more uniformly distributed, as seen in Figure 6b and 9. The EZ33 alloy contained three dominant types of intermetallic compounds, as seen in Figure 9: "A" – large gray particles consisting of Mg-(13wt%Zn, 11wt%La, 14wt%Ce, 2wt% Nd), "B" – fine white particles of Mg-(13wt%Zn, 6wt%La, 13wt%Ce, 3wt%Nd,7wt% Zr), and "C" – nanoscale dispersoids of Mg-(3wt%Zn, 0.9wt%Ce, 0.3wt%Nd).

As seen in Figure 9, the nanoscale particles were homogeneously dispersed throughout the microstructure in both directions. The precipitates "A" and "B" appeared to be present along most of visible grain boundaries in the longitudinal alloy samples.

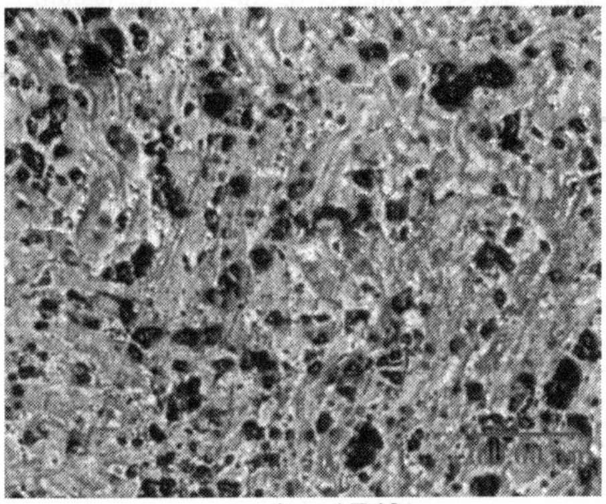

a) AE42

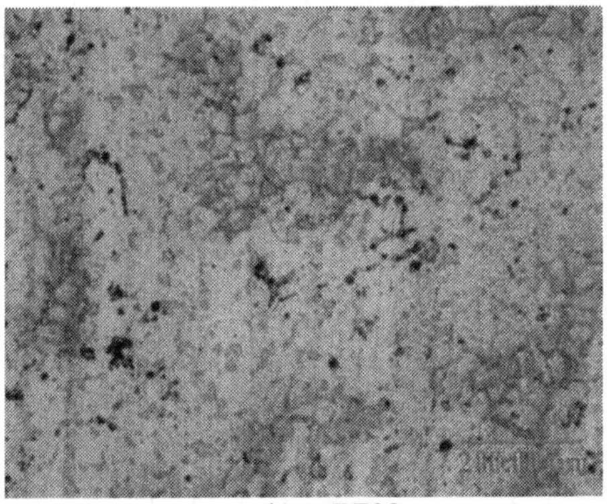

b) EZ33

c) ZE10

Figure 6: Alloy microstructure in transverse direction

In contrast, these particles were not homogeneously distributed in the transverse direction, and only fractions of grain boundaries contained these intermetallics. This observation possibly supports the high strain anisotropy recorded in the EZ33 alloy with the extensometer (Figure 2b). In the longitudinal direction, grain boundary sliding was inhibited by the pinning effect of the precipitates, as also confirmed by the ND results in Figure 4 (elastic strain relaxation due to plastic deformation was more prominent in the transverse sample). In contrast, deformation in the transverse direction was relatively easier due to available grain boundaries where grain boundary pinning was not successfully achieved.

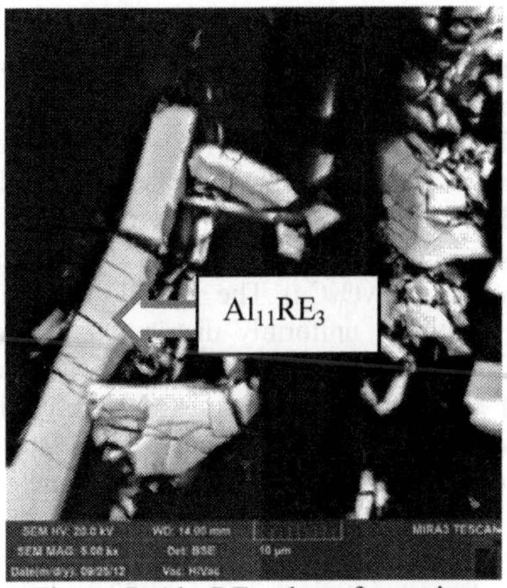

Figure 7: $Al_{11}RE_3$ phase fracturing

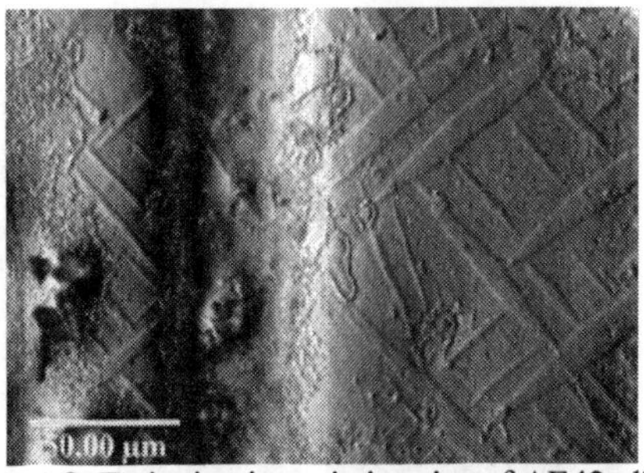

Figure 8: Twinning in grain interior of AE42 alloy

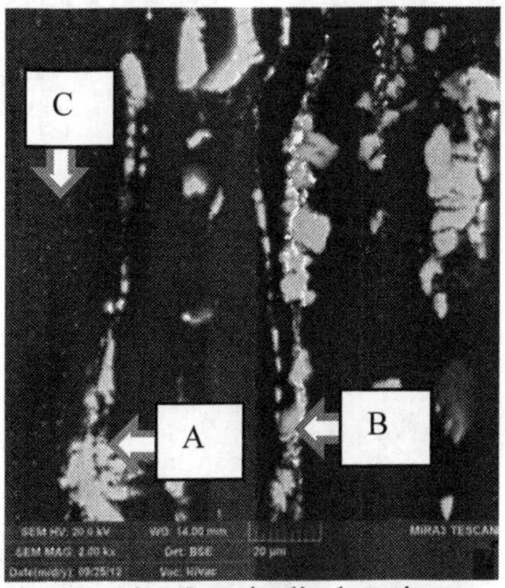

a) Longitudinal section

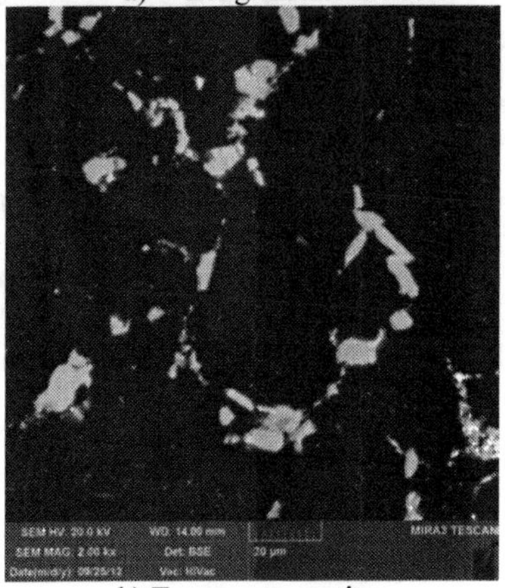

b) Transverse section

Figure 9: Microstructure of EZ33 alloy

The ZE10 alloy's microstructure appeared to be very homogeneous in both directions (L and T) as seen in Figure 5c and 6c. Complex intermetallic aggregates as seen in Figure 10 were identified via SEM/XEDS: "A" - dark gray particles with Mg-(8wt%Zn, 8wt%Zr, 7wt%La, 11wt%Ce, 2wt%Nd), "B" – fine white particles with Mg-(3wt%Zn, 36wt%Zr, 3wt%La, 4wt%Ce and 1wt%Nd) and "C" – nanoscale dispersiods with Mg-(3wt%Zn, 1wt%La, 2wt%Ce). The second phases appeared to be smaller and relatively uniformly distributed in both (L and T) directions. Consequently, the relatively uniform creep response (Figure 2b) in both directions was likely the result of homogeneous microstructure and dispersion strengthening by precipitates on the order of 0.5 – 2 μm in diameter. Presence of these very fine dispersoids (as in the EZ33 alloy) would significantly inhibit plastic flow within the grains or any twining. Observation of grain interiors in the EZ33 and ZE10 alloys did not reveal presence of twins.

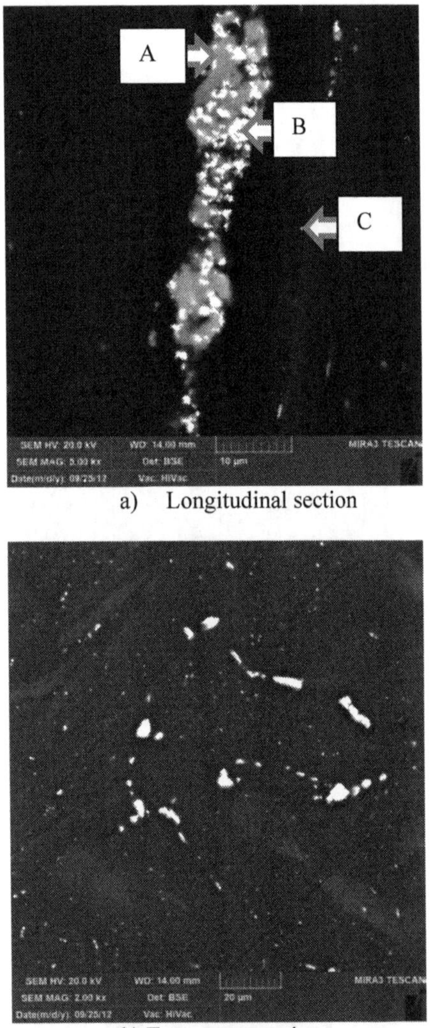

a) Longitudinal section

b) Transverse section

Figure 10: Microstructure of ZE10 alloy

Conclusions

The findings of this research demonstrate that neutron diffraction successfully enabled quantitative characterization of anisotropic creep strain in selected high-temperature magnesium alloys. The results suggest that:

1) The total compressive creep strain for the AE42 alloy was an order of magnitude greater than for EZ33 and ZE10 alloys tested at 175 °C and 50 MPa.
2) The EZ33 alloy had a typical extrusion-type microstructure. Also, the microstrain response was highly anisotropic in the T- direction.
3) The ZE10 alloy, although extruded, had a relatively homogeneous microstructure, with fine second phases. Further, this alloy had uniformly dispersed Mg-Zn-RE nanoscale particles, which likely prevented twinning and plastic deformation during creep.

Acknowledgement

The authors acknowledge the financial support from the National Science and Engineering Research Council (NSERC) Discovery grant. We also thank the staff at NRC-Canadian Neutron Beam Center for their valuable assistance in performing neutron diffraction experiments.

References

1. Haughton, J.L. and Prytherch, W.E., *Magnesium and its alloys* (London, Dept. of Scientific and Industrial Research, 1937).
2. K. M. Asl, A. Tari, and F. Khomamizadeh, "The effect of different content of Al, RE and Si element on the microstructure, mechanical and creep properties of Mg-Al alloys", *Materials Science and Engineering,* (A 523) (2009), 1-6.
3. B. R. Powell *et al.* "Microstructure and creep behavior in AE42 Magnesium die-casting alloy", *J. of Metals*, 54(8) (2002), 34-38.
4. J. Yan et al, "Creep deformation mechanism of magnesium-based alloys", *J. of Material Science*, 43 (2008) 6952-6959.
5. T. Ryspaev et. al. "Microstructure of superplastic QE22 and EZ33 magnesium alloys", *Materials Letters*, 62 (2008) 4041-4043
6. T. Zhou, T., H. Xia, and Z.H. Chen "Effect of Ce on microstructures and mechanical properties of rapidly solidified Mg-Zn alloy" *Mat. Sci. and Technology*, 27(7) (2011) 1198-1205.
7. D. Sediako et al. "Application of neutron diffraction in characterization of texture evolution during high-temperature creep in magnesium alloys" Can. Nuclear Soc. Meeting 2010
8. M. Celikin et. al., " Compressive creep behavior of cast Mg under stresses above the yield strength and the resultant texture evolution" *Can. Metallurgical Quart*., 48(4) (2009) 419-432.
9. M. Fletcher et. al. "Evolution of strain and microstructure during creep of wrought AE42 and ZE10 magnesium alloys", *Trans. Of Indian Institute of Metals*, DOI: 10.1007/s12666-012-0200-3 (In press)
10. D. Sediako and S. Shook "Application of ND in in-situ studies of stress evolution in high temperature creep testing of magnesium alloys" *Magnesium Technology 2009*, TMS (2009) 255-259.

Magnesium Technology 2013
Edited by: Norbert Hort, Suveen N. Mathaudhu, Neale R. Neelameggham, and Martyn Alderman
TMS (The Minerals, Metals & Materials Society), 2013

CREEP BEHAVIOUR OF Mg BINARY SOLID SOLUTIONS

Saeideh Abaspour, C.H. Cáceres*
ARC-Centre of Excellence for Design in Light Metals
Materials Engineering, School of Engineering
The University of Queensland, QLD 4072, Australia

Keywords: Solid solution strengthening, short range order, Mg alloys, creep

Abstract

Specimens of cast Mg-0.8 at.%Gd, Mg-2.2 at.%Zn and Mg-2.5at.%Al alloys were tested in compression at an initial strain rate of ~$1.5x10^{-5}s^{-1}$ at room temperature and 180°C. The Mg-Zn alloy, and more so the Mg-Gd alloy, exhibited a largely athermal behaviour, in contrast with the Mg-Al alloy which softened considerably at the higher temperature. The athermal behaviour of the Zn- and Gd-containing alloys can be accounted for by their strong tendency to developing short range order.

Introduction

A plot of the strength of a solid solution versus temperature normally takes the shape of Figure 1. At low temperatures the strength decreases rapidly, whereas at intermediate temperatures it becomes virtually insensitive to temperature. The behaviour can be rationalised by dividing the strength into two components [1-3]:

$$\tau = \tau_i + \tau^*$$

where τ_i is the athermal component of the stress arising from long range stress fields and which cannot be overcome through thermal activation, and τ^* is the thermally activated component connected with short range obstacles that can be overcome by thermal energy. At temperatures above two-thirds of the melting point the flow stress decreases rapidly again, approaching the critical resolved shear stress of the pure metal, a behaviour attributed to an increase in the mobility of the solute atoms. The extent of the athermal regime determines the resistance to dislocation creep of any solid solution [4]. Ideally, the athermal regime should extend well past the service temperature of the alloy.

The strength of a solid solution is determined by a number of solute-dislocation interactions, among them [1-3,5,6]:

- Elastic interactions involving the shear modulus and atomic size misfit (random solid solution effects). These are considered short range interactions, i.e., amenable of thermal activation.
- Short range order (SRO). This is a typical long range, i.e., athermal, hardening mechanism [7].

Random solid solution effects dominate the strength of Mg-Al [8]; SRO, in turn, has been confirmed by x-ray scattering Mg-In, Mg-Gd, Mg-Er and Mg-Sn [9-11], whereas theoretical and experimental evidence suggests that it is present in Mg-Zn alloys as well [8,12]. On a parallel presentation in this symposium [13] thermodynamical arguments are used to rank Mg-based binary alloys considering their relative tendency to develop SRO. Taking Mg-Al, Mg-Zn and Mg-Gd as model alloys, it is predicted that Gd in solution should develop the strongest SRO, hence it should produce the most creep resistant alloy, followed by Zn and Al, in that order. The object of this paper is to present experimental evidence in support of those predictions.

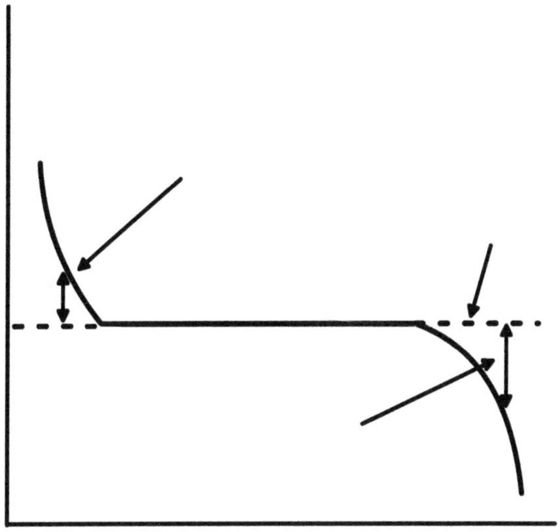

Fig. 1: The temperature dependence of the flow strength (schematic).

2. Experimental Details

Castings with nominal compositions 1.5 at%Zn, 2.5 at%Al and 1 at% Gd were produced. The amount of solute for each alloy was set to approximately match the solubility at 200°C in order to limit precipitation hardening effects during the testing at the higher temperature. The actual compositions are listed in Table 1.

Commercially pure Mg, Zn, Al and a Mg-Gd master alloy (2:3 Gd:Mg) were melted in steel crucibles coated with boron nitride in an electric furnace under a SF6+CO2 protective atmosphere.

* Corresponding author. E-mail: c.caceres@uq.edu.au ; Fax:+ 61-7-33467015.

The liquids were stirred mechanically for 20 minutes to ensure the dissolution of the solute and subsequently poured at 735-750°C under an argon atmosphere into steel moulds of size 80 x 80 x 10 mm3, preheated to 300°C. The cast plates were solution heat treated under Ar atmosphere as per the times and temperatures given in Table 1, and quenched into water. Compression tests were carried out on cylindrical specimens, 18 mm in height and 9 mm in diameter, on a screw driven machine at 25°C and 180°C in a temperature controlled chamber. The crosshead speed was 0.016 mm/min for all the tests (initial strain rate = 1.5*10-5s-1).

Table 1. Chemical composition of the alloys studied determined by inductively coupled plasma atomic emission spectroscopy (ICP-AES), and the respective grain sizes and solution heat treatment schedule.

Alloy	Solute (at.%)	Grain size (μm)	Time (h)	Temperature (°C)
Mg-1.5%Zn	2.2	123	4	470
Mg-1%Gd	0.8	119	4	535
Mg-2.5%Al	-	129	10	413

3. Results

Figures 2-4 show the deformation flow curves of Mg-2.5Al, Mg-2.2Zn and Mg-0.8Gd, respectively, at room and high temperature. Serrated flow was observed in Mg-Zn at room temperature (Fig. 3) and at 180°C in Mg-Gd (Fig. 4). Figure 5 compares the alloys' behaviour at room temperature and Figure 6 at 180°C.

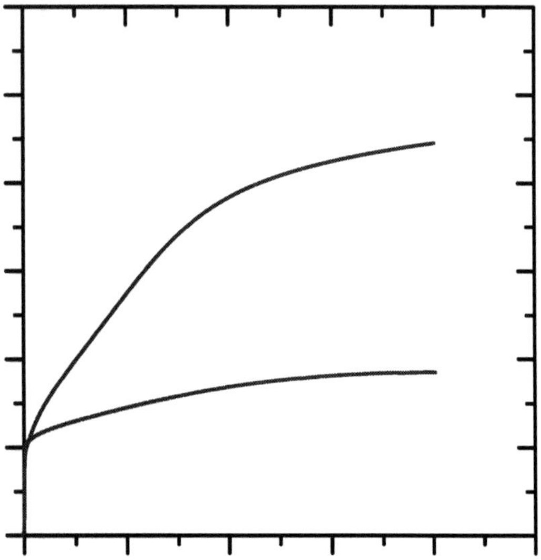

Figure 2. Compressive flow curves of Mg-2.5Al.

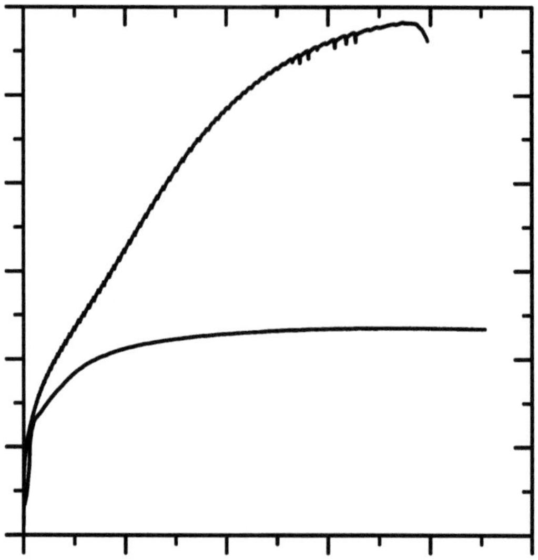

Figure 3. Compressive flow curves of Mg-2.2Zn.

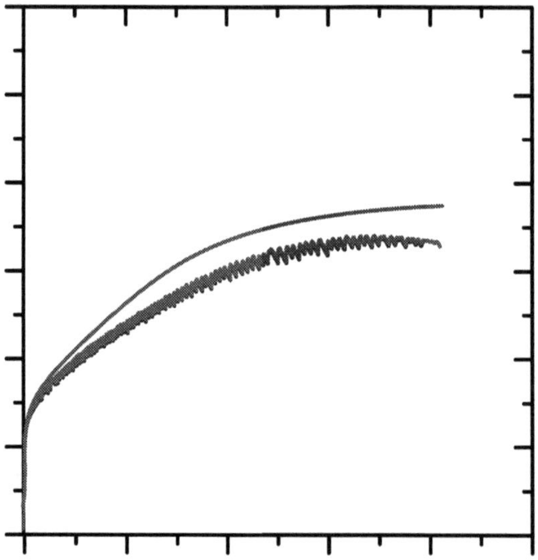

Figure 4. Compressive flow curves of Mg-0.8Gd.

At room temperature all three alloys exhibited an extended linear hardening regime, which reached saturation at about 10 % strain. The saturation stress was largest in the Mg-Zn alloy. At 180°C the loss of strength was very limited in the Mg-Gd alloy, whereas it was extensive in the Mg-Al alloy. The Mg-Zn alloy exhibited a short linear hardening regime, saturating at a strength level which was matched by the Mg-Al alloy at large strains. Figure 7 compares the flow stress at 0.05 strain, showing that the behaviour of the Mg-Gd is nearly athermal, followed by Mg-Zn and Mg-Al, in that order, and line with the expectations from the theoretical analysis of the companion publication [13].

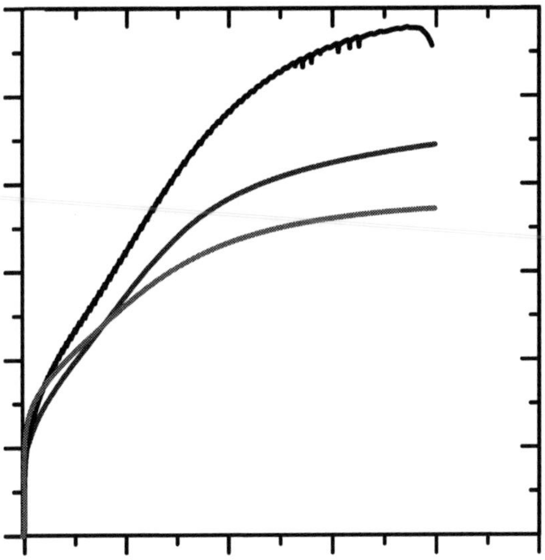

Figure 5. The flow behavior of the alloys studied, at 25°C. The crosses identify the strength values plotted in Figure 5.

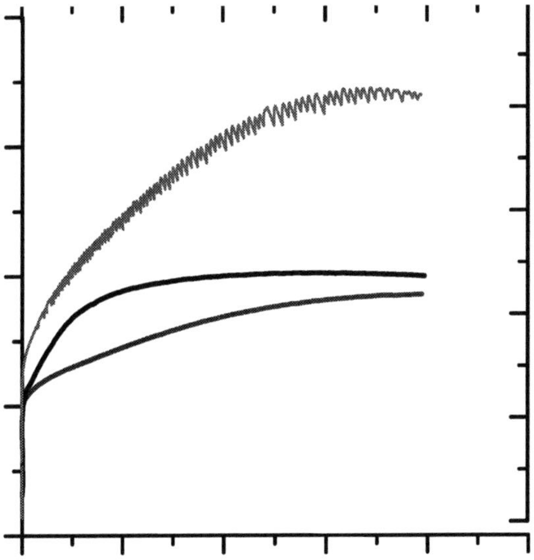

Figure 6. The flow behavior of the alloys studied, at 180°C. The crosses identify the strength values plotted in Figure 7.

4. Discussion

The remarkably athermal behaviour of the Mg-Gd alloy, indicated both by the extended linear hardening regime [1,14] at both temperatures in Figures 2-4 and the small decrease in strength in Figure 7, constitutes strong evidence of the presence of SRO which remains stable at the higher temperature. The Mg-Zn alloy partly reproduces the behaviour of the Mg-Gd, with a shorter linear hardening stage and a larger drop in strength at the higher temperature. This behaviour is consistent with a lower intensity of SRO, which makes it less thermally stable in comparison with the Mg-Gd. The Mg-Al alloy, as expected from its virtual absence of SRO, although it still exhibits a linear hardening at the higher temperature, the hardening rate is much lower and it suffers a substantial drop in strength despite its larger alloy content.

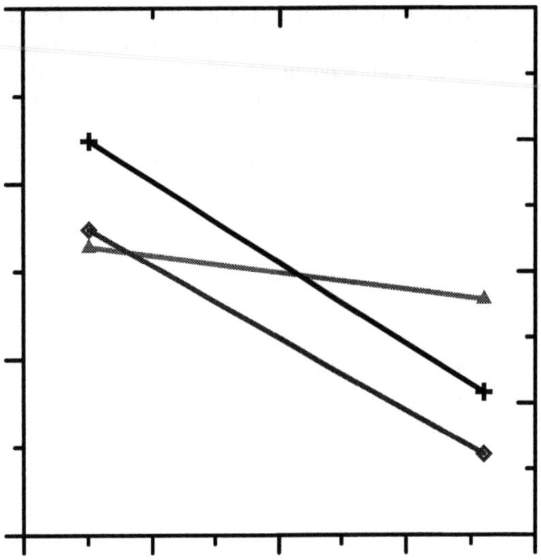

Figure 7. The strength of the alloys studied, at 25 and 180°C, at a strain of 5% (the crosses in Figures 3 and 4).

4. Conclusions

Mg-0.8at.%Gd, Mg-2.2at.%Zn and Mg-2.5at.%Al solid solutions were tested at RT and 180°C.

The solute contents were kept deliberately low to prevent precipitation hardening at the higher testing temperature.

The Mg-Gd lost very little strength at the higher temperature. The loss of strength was larger for the Mg-Zn and substantial for the Mg-Al.

The extent of the athermal behaviour is consistent with the relative strength of short range order in each of the alloys, largest in the Mg-Gd and lowest in the Mg-Al.

References

[1] Kuhlmann-Wilsdorf D., Wilsdorf, H. G. F., Wert, J. A., "LEDS theory of workhardening stages and "planar" versus "distributed" glide," *Scripta Metallurgica et Materialia*, 31 (1994), 729-734.

[2] Hull D., Bacon, D. J., *Introduction to dislocations*. 3rd ed.; (Pergamon Press: Oxford, 1984).

[3] Seeger A. "The mechanism of glide and work hardening in face-centered cubic and hexagonal close-packed metals," In: Fisher, J.C., Johnston W.G., Thomson R., Vreeland T., eds.; "*The mechanism of glide and work hardening in face-centered cubic and hexagonal close-packed metals,*" (London,Chapman and Hall: 1957) 243-329.

[4] Nabarro F. R. N., de Villiers, H. L., *The physics of creep.* (Taylor and Francis London, 1995).
[5] Akhtar A., Teghtsoonian, E., "Substitutional solution hardening of magnesium single crystals," *Phil Mag*, 25 (1972), 897-916.
[6] Chino Y., Kado, M., Ueda, T., Mabuchi, M., "Solid solution strengthneing for Mg-3.0 mass pct. (2.71 at.pct) Al and Mg-0.06 mass pct (0.036 at. pct) Ca alloys," *Metallurgical and Materials Transactions A*, 42 (2011), 1965-1973.
[7] Akhtar A., Teghtsoonian, E., "Solid solution strengthening of magnesium single crystals - I. Alloying behaviour in basal slip," *Acta Metallurgica*, 17 (1969), 1339-1349.
[8] Cáceres C. H., Blake, A, "The strength of concentrated Mg-Zn solid solutions," *physica status solidi*, 194 (a) (2002), 147-158.

[9] Gencheva D. S., Katsnel'son, A. A., Rokhlin, L. L., Silonov, V. M., Khavadzha, F. A., "Investigation of short-range order in magnesium alloys with erbium and gadolinium," *Fiz metal metalloved*, 51 (1981), 788-793.
[10] Safronova L.A., Katsnel'son, A.A., Sveshnikov, S.V., L'Vov, Yu.M., "Short-range order in Mg-In solid solutions," *Fiz metal metalloved*, 43 (1977), 76-80.
[11] Henes S., Gerold, V., *Zeitschrift fur Metallkunde*, 53 (1962), 703-708.
[12] Jha N., Mishra, A. K., "Thermodynamic and surface properties of liquid Mg-Zn alloys," *Journal of Alloys and Compounds*, 329 (2001), 224-229.
[13] Caceres C. H. , Abaspour, Saeideh. In: TMS 2013, 2013.
[14] Kocks U. F., Mecking, H., "Physics and phenomenology of strain hardening: the FCC case," *Progress in Materials Science*, 48 (2003), 171-273.

Magnesium Technology 2013
Edited by: Norbert Hort, Suveen N. Mathaudhu, Neale R. Neelameggham, and Martyn Alderman
TMS (The Minerals, Metals & Materials Society), 2013

INFLUENCE OF YTTRIUM ON CREEP BEHAVIOR IN NANO-CRYSTALLINE MAGNESIUM USING MOLECULAR DYNAMICS SIMULATION

M. A. Bhatia[1], K.N. Solanki[1]
[1]School of Energy, Matter and Transport; Arizona State University; Tempe, AZ 85287-6106, USA

Keywords: Magnesium, Yttrium, Creep, Creep rate, Grain boundary

Abstract

Due to their light weight, magnesium (Mg) and its alloys have great potential for reducing vehicular mass and energy consumption. However, the use of Mg alloys is currently restricted to low-temperature automotive components application. This work strives to gain a better understanding of the effect of yttrium (Y) (up to 3at%) on creep behavior of columnar nanocrystalline Mg with a grain size of 5nm and 10nm. Using molecular dynamics (MD) simulations, nanocrystalline Mg with various local concentrations of Y was subjected to a constant-stress loading, ranging from 0-500MPa, at different initial temperatures, ranging from 473-723K. Our simulations reveal that the secondary stage creep rate ($\dot{\varepsilon}$) decreases by 71% with the addition of only 1at% Y at 500MPa and 623K. With the addition of alloying elements such as Y, the creep rate in the secondary region decreases and the creep deformation mechanism is changed from the void nucleation, growth, and coalescence to GB rotation/sliding. The analyses of the diffusion coefficient and energy barrier reveal a stronger contribution to the overall deformation by the grain boundary diffusion at the low-temperature (423K) and by the lattice diffusion at the higher-temperature (723K).

Introduction

Increasing global demand for energy-efficient yet safer material requires development of new lightweight alloys. Among structural materials, Magnesium (Mg) has the highest strength-to-weight ratio. Moreover, Mg-alloys have excellent recyclability and are ideal for transportation and biological material applications, with enormous potential for energy savings [1-4]. However, Mg has certain limitations when compared with other metallic systems and polymers [5]. Since Mg has a hexagonal close packed (hcp) structure, it has a limited number of active slip systems at room temperature [6-7], which leads to failure in polycrystalline Mg-alloys at ambient conditions before dislocation glide can occur [8-9]. In an effort to overcome this material performance shortcoming, recent studies have considered the effects of adding to Mg certain alloying elements, such as Al-Zn, Al-Si, Al-Ca, Al-Sr, Sn-Ca, Zr-Y, and minute traces of other rare-earth (RE) elements. The objective of these studies has been to qualitatively identify the principles governing solute strengthening and ductility while improving manufacturability as well as lowering production costs of these alloys (e.g., [10-11]). These studies have shown that the alloying elements can cause Mg-alloys to either harden or soften by interacting with dislocations/twins/interface and/or forming favorable phases [10]. Rare earth elements in particular have been found to generate the most favorable effects on the mechanical and physical properties of Mg-alloys, including improved creep and corrosion resistances [10, 16]. However addition of RE is cost prohibitive [12], so other alternatives are needed. In order to develop additional, cost-effective alloys, a better understanding of the deformation mechanisms governing Mg-alloys' performance is necessary.

The high-temperature deformation mechanisms, such as creep in metallic systems, have been extensively studied over the years, and for a detailed review, please refer to [14]. In fact, it is a well-established fact that with increasing temperature, creep reduces the load-bearing capacity of Mg-alloys. During creep, the strain is accumulated either by diffusional flow through the interfaces (Nabarro-Herring creep [22], [23]), diffusional flow along interfaces (Coble creep [27]), interface sliding (Raj and Ashby [24]), or dislocation motion (climb/glide [25]). In general, these creep mechanisms/models have been vigorously discussed as well (e.g., see [14]), mostly due to the inability of certain models to accurately predict creep behavior across materials with varying microstructures. In a recent experimental study, it has been shown that microstructural attributes such as grain size and solutes/phases segregated to the grain boundary (GB) play an important role in creep behavior of Mg-alloys [26]. It was found that with increasing grain size, the creep rate decreased with an attendant loss in room temperature yield strength, as predicted by the Hall-Petch equation [25]. On the other hand, grain-size refinement can be controlled by the addition of micro-alloys, such as Zr [26]. Thus, with a balance between goals for optimal grain size and room-temperature strength, creep deformation behavior potentially could be enhanced by elucidating the deformation mechanisms necessary to engineer GBs with solutes such as yttrium (Y).

In this paper, extensive MD simulation results are presented to elucidate the effect of grain size (5nm-15nm) and GB with various Y concentrations (1-3%at) on diffusional creep and deformation mechanism of nanocrystalline Mg.

Simulation Methodology

A parallel molecular dynamics code (LAMMPS) [18] that incorporates spatial decomposition was used to deform the nanocrystalline Mg samples of 5nm-15nm grain dimension as shown in Fig 1. The Voronoi tessellation method [21] was used to construct $[11\bar{2}0]$ columnar samples containing 5 grains of identical hexagonal shape within periodic simulation cells. After generating the desired Mg nanocrystalline model, the GB of the model was locally doped with different (1~3%at) percentages of Y. The total number of atoms in a simulation cell was in the range of 250000-1 million. The samples were first relaxed at the desired temperature using an NVT (conserving number of atoms, volume and temperature) ensemble for 15 ns, followed by an independent relaxation in three directions using an NPT (conserving number of atoms, pressure and temperature to mimic bulk behavior) ensemble for another 10 ns. Atomistic simulations were carried out using an MD time step of 1 fs. A periodic boundary condition was adopted in all directions. A constant stress, ranging from 0-500MPa, was applied along y-direction at different temperatures ranging from 423k – 723k (150 °C – 450 °C), whereas deformation in the other two directions was carried out by maintaining a zero pressure according to the Parrinello-Rahman method [19].

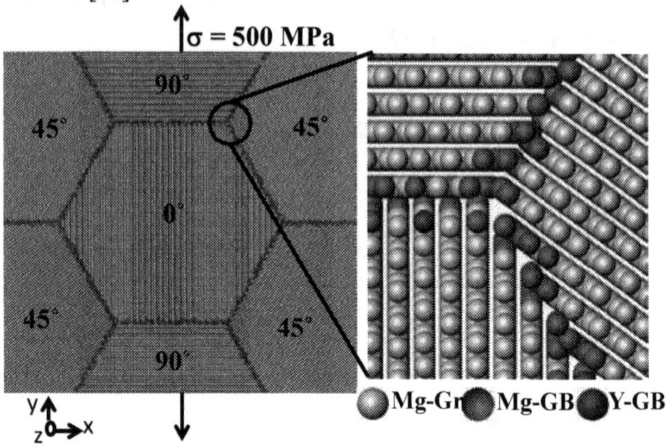

Fig 1 a) An equilibrated, undeformed simulation model at 473K with different misorientation angles with respect to the loading (y) direction. Viewing direction is along the $[11\bar{2}0]$ columnar z-axis. b) Triple point showing Mg lattice atoms (yellow), Mg GB atoms (Red), and doped Y atoms (blue).

The molecular dynamics simulations in this work employed the embedded atom method (EAM) potential of Sheng [20]. The interatomic potential parameters were derived from a fitting method using isothermal compressibility data up to 2000 K (1727 °C). Moreover, the potential was developed by also fitting the vacancy migration energy as well as with different Mg-Y crystal structures. It is important to note here that a recent study [21] that investigated the effect of precipitate in aluminum alloys proposed that thermally-activated vacancy movement plays a central role in grain growth mechanism.

Results and Discussions

Diffusional analysis with varying grain sizes at 0 MPa

To study the diffusional mechanisms of nanocrystalline Mg, equilibrated structure with different grain sizes were annealed for 500 ps in the temperature range of 473 K – 723 K under zero applied load condition. These simulations were used to calculate the mean square displacement (MSD) values as

$$MSD = \sum_{i=1}^{N} (r_i(t) - r_i(0))^2$$

at different constant temperatures, where N is the number of atoms, $r_i(t)$ is the position of an atom at any instant t, and $r_i(0)$ is the initial position of an atom at t=0. From the MSD values, we computed the diffusion coefficient with the help of Einstein's equation as follows:

$$MSD = 6Dt,$$

where D is the diffusion coefficient, and t is the time. Figure 2 shows Arrhenius plots for self-diffusion of Mg atoms in nanocrystalline Mg with different grain sizes that are given by following:

$$D = D_o e^{-Q/KT},$$

where D_o is the maximum diffusion coefficient at infinite temperature, Q is the activation energy, K is the Boltzmann constant, and T is the temperature.

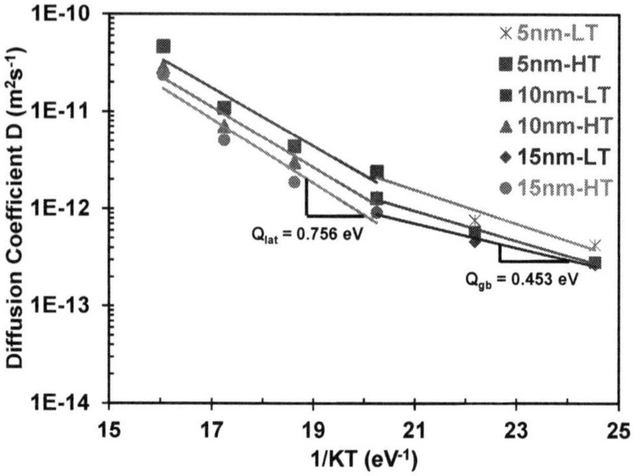

Fig 2 Arrhenius plots for self-diffusion of Mg atoms in nanocrystalline Mg with different grain sizes showing two distinct slopes corresponding to the lattice (at high temperature) and GB diffusions (at low temperature). The transition temperature from the GB-dominant to lattice-dominant diffusion mechanism was found to be at 573K. LT – Low Temperature HT- High Temperature

As the grain size increases, the activation energy for GB diffusion and for lattice diffusion increases. Table 1 lists the respective activation energies for nanocrystalline Mg with different grain sizes. It is to be noted that the fraction of GB atoms increases as the grain size decreases. Hence, with smaller grain sizes, the contribution from the GB diffusion increases and dominates the creep deformation mechanism as predicted by the activation energy ratio $Q_R = Q_{gb}/Q_{lattice}$, where Q_{gb} is the GB diffusion activation energy, and Q_{gb} is the lattice diffusion activation energy. For the nanocrystalline Mg with 5nm grain size, Q_R is equal to 0.558 and hence the activation energy of GB diffusion (0.384 eV) is much lower than activation energy of lattice diffusion (0.688 eV). As the grain size increases, the activation energy also increases, but the rate of increase in activation energy for the lattice diffusion is slower than for the GB diffusion. Moreover, the fraction of GB atoms' contribution to total diffusion also decreases. Hence, at 15nm grain size, Q_R is equal to 0.599. Further work is required to find out the grain size at which there is a transition from the GB diffusion to the lattice diffusion mechanism.

Table 1. Activation energies of the GB diffusion and the lattice diffusion in Mg nanocrystalline with different grain sizes. $Q_R = Q_{gb}/Q_{lattice}$ is the parameter used to identify dominant diffusional mechanisms. A smaller ratio indicates less activation energy for GB diffusion, and GBD dominates.

Grain Size	Fraction of GB atoms	Lattice diffusion	GB diffusion	Ratio
	f_{gb}	Q_{lat} (eV)	Q_{gb} (eV)	Q_R
5nm	0.03528	0.688	0.384	0.558
10nm	0.02465	0.714	0.413	0.578
15nm	0.02013	0.756	0.453	0.599

Diffusional analysis with varying concentrations of Yttrium at 0 MPa

To monitor the diffusion paths of the GB atoms, the equilibrated 5nm grain nanocrystalline structure with different initial concentration of Y was annealed for 250 ps under zero applied stress in the temperature range 423–723 K. These equilibrated structures were used to calculate mean square displacement (MSD) at given temperatures of interest. Figure 3 shows the MSD evolutions for the Mg and Y atoms for different temperatures. These results suggest a linear increase in MSD values for both Mg and Y, with Mg being more mobile than Y.

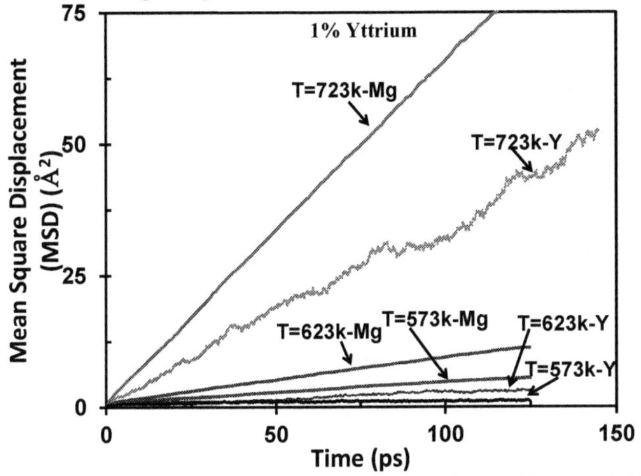

Fig 3 The mean square displacement evolutions of Mg and Y atoms for a 5nm grain size Mg-1%Y nanocrystalline at different temperatures in the absence of external stress loading. These results suggest a linear increase in MSD values for both Mg and Y, with Mg being more mobile than Y.

Figure 4 shows the Arrhenius plots for self-diffusivity of Mg and Y in 5nm-grain size Mg-1%Y nanocrystalline sample. In the case of Mg-1%Y, for a low temperature range, the self-diffusivity activation energy of Y, 0.094 eV, is lower than that of Mg, 0.279 eV. In contrast, for a higher temperature range, the self-diffusivity activation energy of Y, 0.93 eV, is higher than that of Mg, 0.659 eV. However, in the case of Mg-3%Y, for high temperature, the self-diffusivity activation energy of Mg and Y decreases when compared with Mg-1%Y. Table 2 lists the respective self-diffusivity activation energies of Mg and Y in Mg nanocrystalline with different Y concentrations.

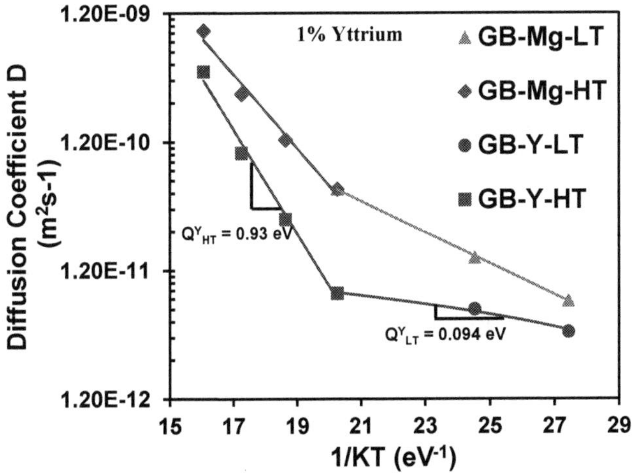

Fig 4 Arrhenius plots for self-diffusive GB Mg and Y atoms in a 5nm-grain size Mg-1%Y nanocrystalline sample. At lower temperature, the activation energy of Y, 0.094 eV, is lower than that of Mg, 0.279 eV. In contrast, at higher temperature, the activation energy of Y, 0.93 eV, is higher than that of Mg, 0.659 eV.

Table 2 The respective self-diffusivity activation energies of Mg and Y in Mg nanocrystalline with different Y concentrations.

	Activation Energy Q (eV)		
	0% at. Y	**1% at. Y**	**3% at. Y**
Mg-GB LT	0.180	0.279	0.340
Mg-GB HT	0.593	0.659	0.588
Y-GB LT		0.094	0.180
Y-GB HT		0.930	0.894

Diffusional analysis with varying grain sizes at 500 MPa

To study the effect of applied loads, a constant applied stress of 500 MPa was applied in a y-direction, and lateral dimensions were held at 0 MPa. The magnitude of the applied stress was low enough to ensure that no dislocations were nucleated from the GBs. Figure 5 shows the strain as a function of time at 623 K with different grain sizes. Creep behavior in Figure 5 is divided into three regions, which is observed in many materials. In Region I (primary creep), there is an instantaneous response, followed by a decreasing rate of strain with increase in time. In Region II (secondary creep), the strain increases linearly with time. In Region III (tertiary creep), the strain increases rapidly with time. Most of the damage or fracture is experimentally observed in Region III due to voids and crack nucleations. We also observe, as shown in Figure 5, that the secondary creep in the 5nm-grain sample is short-lived in comparison to the 10nm- and 15nm-grain nanocrystalline samples. Based on these results, we used the 10nm-grain nanocrystalline sample to further explore the effect of Y on the steady state creep regime. Our main objective here was to determine whether alloying elements such as Y can delay tertiary creep by prolonging the secondary creep regime.

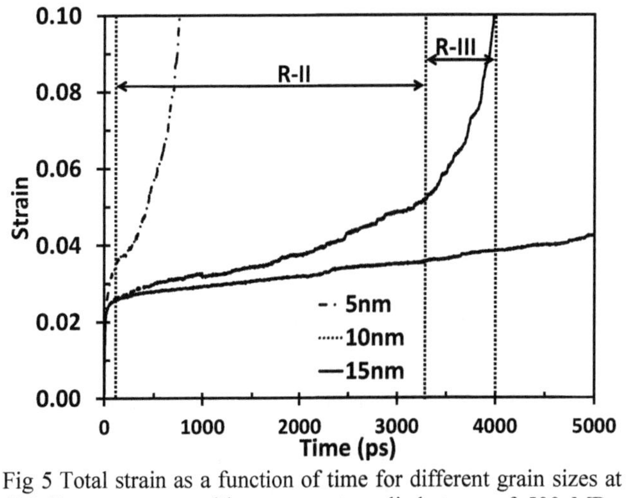

Fig 5 Total strain as a function of time for different grain sizes at 623 K temperature with a constant applied stress of 500 MPa. After initial elastic region, the secondary creep regime (R-II) for

5nm-grain exists for about 200 ps. In contrast, for the 10nm-grain, Region II exists for a longer period (3800 ps) of time. Figure shows R-II and R-III for 10nm grain.

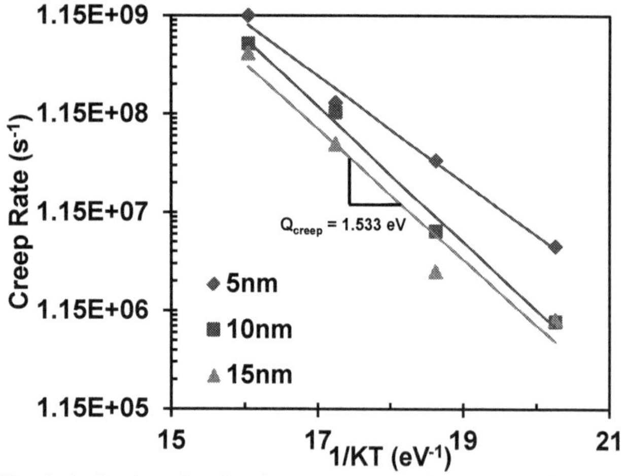

Fig 6 Arrhenius plot for the creep rates of nanocrystalline Mg with different grain sizes. The creep activation energy for 5nm- and 15nm-grain sizes was found to be 1.047 eV and 1.533 eV, respectively.

An empirical equation which describes the strain-time relation is given by

$$\dot{\varepsilon} = \frac{C\sigma^m}{d^b} e^{-Q/KT}$$

where C, m, and d are material constants, $\dot{\varepsilon}$ is the secondary creep rate, σ is the applied stress, d is the mean grain size, Q is the creep activation energy. Figure 6 shows an Arrhenius plot for the secondary creep rates of nanocrystalline Mg with different grain sizes for a temperature range 573–723 K. The increase in the creep rate with increasing temperature indicates that the deformation is thermally activated, indicative of a diffusion-driven creep mechanism. The creep activation for 5nm- and 15nm-grain sizes was found to be 1.047 eV (101 kJ/mol) and 1.533 eV (148 kJ/mol), respectively, which are in good agreement with the experimental range of 106–220 kJ/mol reported in literature [10] for different applied stresses and temperature ranges.

Diffusional analysis with varying concentrations of Yttrium at 500 MPa

Figure 7 shows strain as a function of time for a 10nm-grain nanocrystalline sample with different concentrations of Y at 673K and with an applied constant stress of 500MPa. Initially, for the pure Mg nanocrystalline sample, we observed that the material deformed elastically until t = 20 ps. Thereafter, the strain increases linearly at a constant creep rate, as shown in Figure 7. In the tertiary creep region (R-III), there is sudden increase in the strain rate that results in nucleation of cracks and voids, leading into an intergranular fracture as shown in Figure 8a. With the addition of alloying elements such as Y, the creep rate in the secondary region (R-II) decreases, indicative of GB stabilizations, as shown in Figures 8b and 8c. Moreover, with the addition of Y, the creep deformation mechanism is changed from void growth and coalescence to grain shrinkage and rotation, as shown in Figure 8. Further work is required to study grain rotation and stretch at different concentrations of Y to enhance our understanding of Y's effect on creep behavior as it is shown in Figure 8. Note that the creep rate for the tertiary creep region with 1% Y is lower than that for 2% and 3%at Y, as seen in Figures 7 and 9. Finally, Figure 9 shows an Arrhenius plot for the creep rates of nanocrystalline Mg with different concentrations of Y for a range of temperatures 573–723 K. We observe a slight change in the overall creep rate, with creep activation energy increasing from 1.128-1.154 eV for 0-3% at Y as shown in Figure 9.

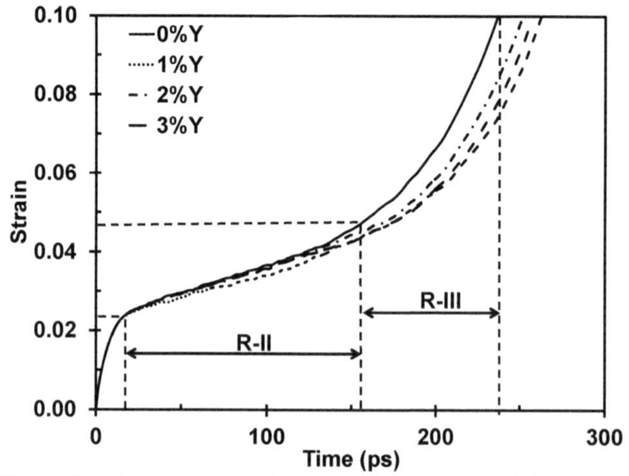

Fig 7 Total strain as a function of time for a 10nm-grain nanocrystalline sample with different concentrations of Y at 673 K and an applied constant stress of 500 MPa. After the initial elastic region, the creep rate decreases to a steady state that persists throughout the secondary creep region (R-II), where small voids are formed in the pure Mg polycrystalline model. In the tertiary creep region (R-III), the creep rate increases, resulting in an intergranular fracture in the pure Mg nanocrystalline sample. Figure shows R-II and R-III for 0% Yttrium sample.

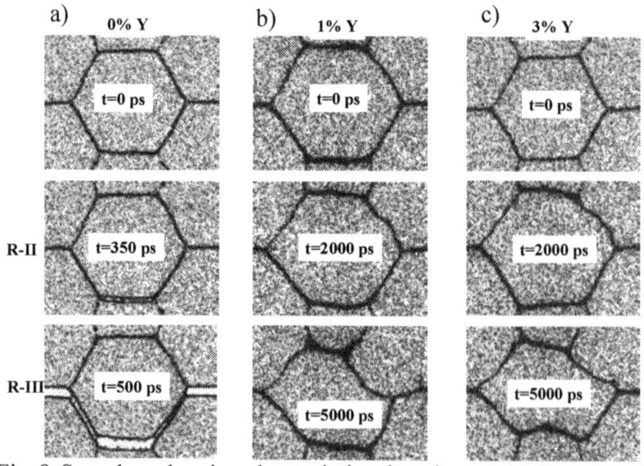

Fig 8 Snapshot showing the variation in microstructure under a constant stress of 500 MPa and at 623 K temperature. a) Micro cracks are formed in the pure Mg nanocrystalline model in the R-II region (secondary creep) at GB, which coalesces in the R-III region (tertiary creep), resulting in an inter-granular fracture due to thermal stresses. For b) and c), no nano-cracks were formed at the GB in the R-II region, and the deformation proceeds with GB migration where the central grain shrinks in tertiary creep region (R-III). Atoms are filtered with potential energy higher than -1.35 eV.

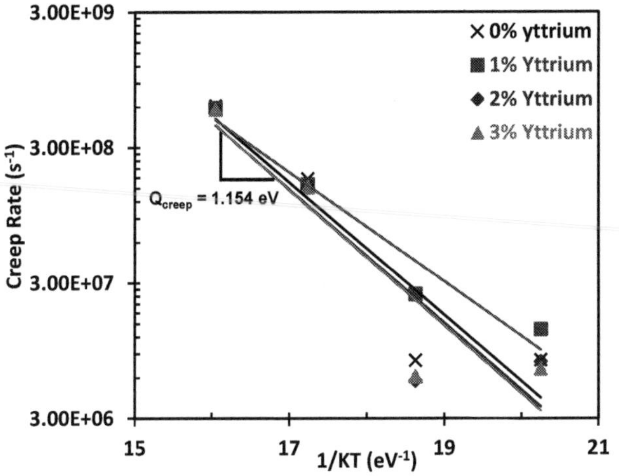

Fig 9 Arrhenius plot for the creep rates of nanocrystalline Mg with different local concentrations of Y. The creep activation energy increases from 1.128-1.154 eV for 0-3% Y.

Conclusion

In this paper, we present the high-temperature deformation behavior and mechanisms of nanocrystalline Mg with and without Y for a temperature range of 573–723 K under constant stress loadings. Our simulations show that with the addition of alloying elements such as Y, the creep rate in the secondary region decreases, indicative of GB stabilizations, as shown in Figures 8b and 8c. Moreover, with the addition of Y, the creep deformation mechanism is changed from void growth and coalescence to grain shrinkage and rotation as shown in Figure 8. The following conclusions can be drawn from this work:

1. For Mg nanocrystalline in the absence of external load, we observed the transition temperature from the GB-dominant to lattice-dominant diffusion mechanism at 573K. Similarly, in the absence of external load for the case of Mg-1%Y, at a lower temperature range, the self-diffusivity activation energy of Y, 0.094 eV, is lower than that of Mg, 0.279 eV. In contrast, for a higher temperature range, the self-diffusivity activation energy of Y, 0.93 eV, is higher than that of Mg, 0.659 eV. However, in the case of Mg-3%Y, for high temperature, the self-diffusivity activation energy of Mg and Y decreases when compare with Mg-1%Y.
2. In the case of constant applied stress of 500MPa on the Mg nanocrystalline sample with different grain sizes, we observe, as shown in Figure 5, that the secondary creep in the 5nm-grain is short lived as compared to 10nm- and 15nm-grain nanocrystalline samples.
3. The creep activation for 5nm- and 15nm-grain sizes in pure Mg nanocrystalline samples was found to be 1.047 eV (101 kJ/mol) and 1.533 eV (148 kJ/mol), respectively, which are in good agreement with the experimental range of 106–220 kJ/mol reported in literature [10] for different applied stresses and temperature ranges.
4. With the addition of alloying elements such as Y, the creep rate in the secondary region decreases and the creep deformation mechanism is changed from the void nucleation, growth, and coalescence to GB rotation/sliding.
5. Finally, further work is required to study grain rotation and stretch at different concentrations of Y to enhance our understanding of Y's effect on creep behavior as seen in Figure 8.

Acknowledgement

The authors are grateful for the financial support for this work from the School for the Engineering of Matter, Transport, and Energy (SEMTE) at Arizona State University. We also appreciate the Fulton High Performance Computing at Arizona State University for enabling us to conduct our simulations.

References

1. S. N. Mathaudhu and E. A. Nyberg, "Magnesium alloys in US military applications: Past, Current and Future Solutions", Magnesium Technology 2010: Proceedings of a Symposium Sponsored by the Magnesium Committee of the Light Metals Division of TMS, 2010.
2. A. A. Luo, "Recent magnesium alloy development for automotive powertrain application", Materials Science Forum, vol. 419-422, pp. 57-66, 2003.
3. K. N. Solanki, A. Moitra, and M. Bhatia, "Effect of substituted aluminum in magnesium tension twin", Magnesium Technology 2010, pp. 325-329, 2010.
4. J. A. Yasi, L. G. Hector Jr., and D. R. Trinkle, "First principles data for solid-solution strengthening of magnesium: From geometry and chemistry to properties", Acta Materialia, vol. 58, no. 17, pp. 5704-5713, Oct. 2010.
5. T. M. Pollock, "Weight loss with magnesium alloys", Science, vol. 328, no. 5981, pp. 986-987, May 2010.
6. J. W. Christian and S. Mahajan, "Deformation twinning", Progress in Materials Science, vol. 39, no. 1–2, pp. 1-157, 1995.
7. A. H. Cottrell and B. A. Bilby, "A mechanism for the growth of deformation twins in crystals", Philosophical Magazine Series 7, vol. 42, no. 329, pp. 573-581, 1951.
8. D. L. McDowell and G. B. Olson, "Concurrent design of hierarchical materials and structures", Scientific Modeling and Simulations, vol. 68, S. Yip and T. D. Rubia, Eds. Dordrecht: Springer Netherlands, 2008, pp. 207-240.
9. W. A. Curtin, D. L. Olmsted, and L. G. Hector, "A predictive mechanism for dynamic strain ageing in aluminum-magnesium alloys", Nature Materials, vol. 5, no. 11, p. 875, Oct. 2006.
10. N. D. Saddock, "Microstructure and creep behavior of Mg-Al alloys containing alkaline and rare earth additions", PhD Dissertation-2008.
11. Y. C. Lee, A. K. Dahle, and D. H. StJohn, "The role of solute in grain refinement of magnesium", Metallurgical and Materials Transactions A, vol. 31, no. 11, pp. 2895-2906, Nov. 2000.
12. S. Agnew, "NSF workshop: Magnesium alloys science and technology-Fundamental research issues", Arlington, Virginia, 2011.
13. V. Vitek and M. Igarashi, "Core structure of 1/3<1120> screw dislocations on basal and prismatic planes in hcp metals: An atomistic study", Philosophical Magazine A, vol. 63, no. 5, pp. 1059-1075, May 1991.
14. H. V. Swygenhoven, "Grain boundaries and dislocations", Science, vol. 296, no. 5565, pp. 66-67, Apr. 2002.

15. M. Tschopp, K. N. Solanki, and M. Horstemeyer, "A generalized framework for quantifying material structure-property relationships and uncertainties: Applicability to ICME", Models, Databases, and Simulation Tools Needed for the Realization of Integrated Computational Materials Engineering: Proceedings of the Symposium Held at Materials Science & Technology 2010, October 18-20, 2010 Houston, Texas, USA, 2011.
16. N. Kashefi, R. Mahmudi, "The microstructure and impression creep behavior of cast AZ80 magnesium alloy with yttrium additions", Materials and Design, vol 39, pp. 200-210, 2012.
17. Tapan G. Desai, Paul C. Millett and Dieter Wolf, "Molecular dynamics study of diffusional creep in nanocrystalline UO2", Acta. Materialia, vol 56, pp. 4489-4497, September 2008.
18. S. J. Plimpton, "Fast parallel algorithm for short-range molecular dynamics", J Comp Phys. vol. 117, pp. 1-19, 1995.
19. M. Parrinello and A. Rahman, "Polymorphic transitions in single crystals: A new molecular dynamics method", J. Appl. Phys. vol. 52, 1981.
20. https://sites.google.com/a/gmu.edu/eam-potential-database/Home/MgY. Manuscript under review.
21. P. S. De, J. Q. Su and R. S. Mishra, "A stress-strain model for a two-phase ultrafine-grained aluminum alloy", Scripta Materialia (2011).
22. F. R. Nabarro, "Deformation of crystals by the motion of single ions", in Report of a conference on strength of solids, held at the HH Wills Physical Laboratory, University of Bristol, on 7-9 July 1947, 1948, p. 75.
23. C. Herring, "diffusional viscosity of a polycrystalline solid", Journal of Applied Physics, vol. 21, no. 5, pp. 437-445, May 1950.
24. R. Raj and M. F. Ashby, "On grain boundary sliding and diffusional creep", Metallurgical Transactions, vol. 2, no. 4, pp. 1113-1127, Apr. 1971.
25. G. E. Dieter, Mechanical metallurgy. McGraw-Hill, 1986.
26. S. M. He, X. Q. Zeng, L. M. Peng, X. Gao, J. F. Nie, and W. J. Ding, "Microstructure and strengthening mechanism of high strength Mg-10Gd-2Y-0.5Zr alloy", Journal of Alloys and Compounds, vol. 427, no. 1–2, pp. 316-323, Jan. 2007.
27. R. L. Coble, "A model for boundary diffusion controlled creep in polycrystalline materials", Journal of Applied Physics, vol. 34, no. 6, pp. 1679-1682, Jun. 1963.

Magnesium Technology 2013
Edited by: Norbert Hort, Suveen N. Mathaudhu, Neale R. Neelameggham, and Martyn Alderman
TMS (The Minerals, Metals & Materials Society), 2013

AGING BEHAVIOR AND MICROSTRUCTURAL EVOLUTION IN MG-3ND-0.2ZN-0.5ZR ALLOY

Amirreza Sanaty Zadeh[1], Xiangyu Xia[1], Alan A. Luo[2], Joseph E. Jakes[3], Donald S. Stone[1]

[1]Materials Science and Engineering Department; University of Wisconsin, Madison, WI, 53706
[2] Chemical and Materials Systems Laboratory; General Motors Research and Development Center; Warren, MI, 48090
[3]Performance Enhanced Biopolymers, Forest Products Laboratory, Madison, WI 53726

Keywords: Mg-Zn-Nd; Precipitation hardening; Transmission Electron Microscopy

Abstract

The aging behavior and microstructure evolution of Mg-0.2Zn-3Nd-0.5Zr alloy were studied in this paper. Transmission Electron Microscopy investigations showed four sets of precipitates formed during aging. These precipitates are β" which forms on prismatic planes during early stages of precipitation, γ' which forms on basal plane, and β' which forms on prismatic and pyramidal planes. In addition, some unknown precipitates in the form of dark spots were detected in 8 hours aged sample. The results from nanoindentation showed an activation volume in the range 80-100 b^3 that is consistent with prismatic glide as the rate controlling mechanism in the system. Furthermore, a small amount of dislocation climb along with dynamic recovery occur at room temperature upon deformation in this system.

Introduction

There has been significant attention paid to magnesium and its alloys due to their high specific strength. The application of magnesium products are in aerospace and automotive industries for reducing energy consumption. However, poor ductility of magnesium due to its structure (HCP) limits its applications. Modification of texture is reported to improve the ductility of magnesium alloys greatly [1-3]. This is done by alloying magnesium with rare-earth elements such as Ce, Nd, and Gd [4, 5]. It has also been reported that the addition of rare earth elements has a significant effect on creep resistance of Mg alloys both at room and elevated temperatures. Among the rare earth elements, neodymium (Nd) has relatively high solid solubility at the eutectic point and is more suitable for texture randomization and age hardening. Also, Nd plays an important role in precipitation strengthening of magnesium alloys. The strengthening can be enhanced by controlled addition of third element such as zinc. It has been reported that small addition of Zn to Mg–Nd binary system causes an increase in the precipitation hardening effect and the mechanical properties. It should be mentioned that the amount of zinc seems to change the sequence and nature of precipitation. It has been reported that higher amounts of Zn (>1 wt.%) reduces the precipitation hardening effect by restricting the precipitation of Mg-Nd phase. Therefore, combining the effect of solid solution strengthening from zinc as well as precipitation strengthening form Nd would result in improved mechanical properties. These effects have been reported in previous works [6-8].

The sequence of precipitation in binary Mg-Nd alloys is somehow different from ternary Mg-Zn-Nd alloys. In Mg-Nd alloys, the sequence is reported to be: supersaturated solution –G.P. zones- β"- β'- β, which was modified later with addition of another intermediate phase β_1 (fcc) forming during β' precipitation [9]. In Mg-2.8wt%Nd-1.3wt%Zn alloy, the sequence is reported to be supersaturated solution- low-temperature reaction- γ" - γ' - γ [7, 8]. The nature of low-temperature reaction has not been fully elucidated but it is believed to be –G.P. zones. The results showed that the ratio of Nd/Zn affects the sequence and kinetics of precipitation, which further affect the mechanical properties.

Due to limited research on the precipitation and the nature of precipitates in Mg-Zn-Nd system, there is not a comprehensive understanding of precipitation in this system. In addition, although there are reports on the characterization and mechanical properties of this alloy [10-19], there is not a detailed study on the deformation mechanisms involved at room temperature in this system. Therefore, in this research, it is aimed to further investigate the precipitation of metastable phases and deformation mechanisms at room temperature in this alloy. For this purpose, Transmission Electron microscopy and nanoindentation including Broadband Nanoindentation Creep (BNC) [20, 21] were used to characterize the microstructure and generate mechanical properties data.

Experimental

The as-cast ingot of Mg-3Nd-0.2Zn-0.5Zr alloy was prepared by mixing and casting of pure magnesium, zinc, zirconium, and Mg-20Nd master alloy ingots in an electric furnace and under protection of SF_6/CO_2. Cubic samples of 1×1×0.5 cm^3 were cut from the ingot, polished, and cleaned in alcohol using ultrasonic cleaning. Samples then were wrapped in Ta foil and encapsulated in glass ampules under He atmosphere. Heat treatment of the samples was done using a horizontal tube furnace. Samples were first solution-treated at 540 °C for 4 hours followed by quenching in water at 20 °C. After solid solution treatment, the samples were placed in fridge for 24 hours. The aging of the samples was done using the same sample preparation method but at 200 °C for different aging times. For hardness measurements, Scanning Electron Microscope (SEM), and nanoindentation tests, samples were prepared using Allied Multiprep machine and diamond lapping films (15 μm, 3 μm, 1 μm, and 0.1 μm). Hardness tests were performed using a Wilson Tukon 1102/1202 micro-hardness tester (Wilson Hardness, Norwood, MA). SEM and EDS analyses were done using a LEO 1530 equipped with an EDS detector. Nanoindentation tests were conducted using a Hysitron TI 900 TriboIndenter (Hysitron, Minneapolis, MN). Transmission Electron Microscopy (TEM) studies were performed on a high resolution Philips CM200. The wedge TEM samples were prepared on an Allied Multiprep machine using diamond lapping films with different grid sizes from 15 μm down to 0.1 μm. The wedge samples were further thinned using Fischione 1010 ion milling instrument.

Results and Discussion

Microstructure

Fig. 1 shows the results of age hardening experiments of NZ30K alloy. Also the hardness values are compared with the results in previous publications [7, 11, 13]. As can bee seen, the peak hardness in this work occurs at around 8 hours aging time. This agrees well with Ma et al [7], although Xingwei et al [11] and Ding et al [13] showed peak hardness at about 12 and 16 hours, respectively. It is noted that the maximum hardness in this work is about 10 HV higher than the values in the previous reports [7, 11, 13]. The reason can be attributed to the fact that the samples were pre-aged at 273 K for 24 hours in this study, which causes some pre-existing clusters in the solid solution samples to grow into critical nucleation sizes for precipitation. Another reason could be the different quenching conditions used. In previous reports, samples were quenched in hot water (~70 °C), while in this work the samples were quenched in cold water (20 °C), which results in faster quenching rate and higher vacancy concentration in the sample. These vacancies play significant role in precipitation, especially in early stages of precipitation and in the formation of G.P. zones [6, 8].

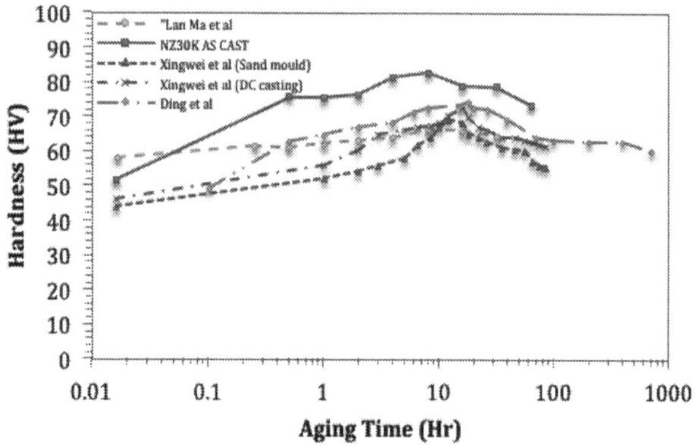

Fig. 1. Aging curve for NZ30K (Mg-3Nd-0.2Zn-0.5Zr).

Fig. 2 shows the XRD patterns obtained from selected aged samples. As can be seen, in the as cast sample, the microstructure is mainly composed of α Mg, $Mg_{12}Nd$ and $Mg_{41}Nd_5$. These phases are concentrated on grain boundaries, as shown in the elemental maps of the as-cast sample in Fig. 3.

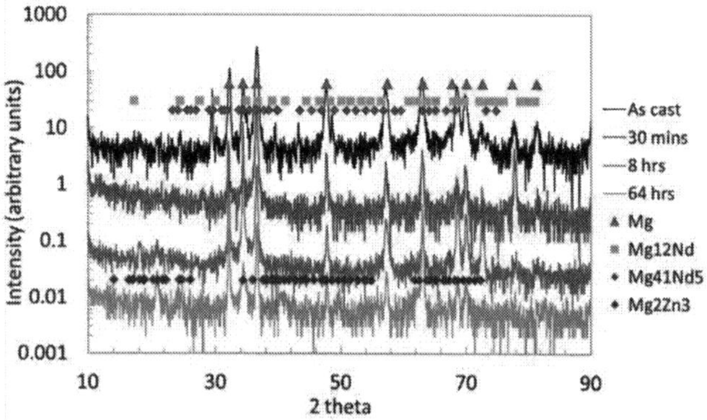

Fig. 2. XRD patterns of NZ30K samples aged at 200 200 °C for different length of time.

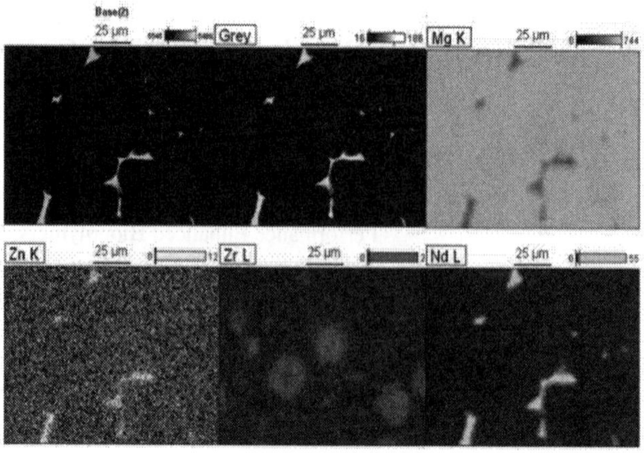

Fig. 3. EDS map of NZ30K as cast sample

Fig. 4 shows the elemental maps of the as-cast sample after solid solution treatment. The maps show no evidence of eutectic phase on grain boundaries indicating good dissolution of the eutectic phase during solution treatment. In addition, in some grains, a concentration of some small and large precipitates was observed. These precipitates, as confirmed by EDS, are zirconium concentrated phases, which form during the solidification of the ingot. Due to the fact that the formation of these precipitates occur at around 700 °C, the solution treatment does not affect these precipitates. No evidence of the presence of Zr could be detected on grain boundaries.

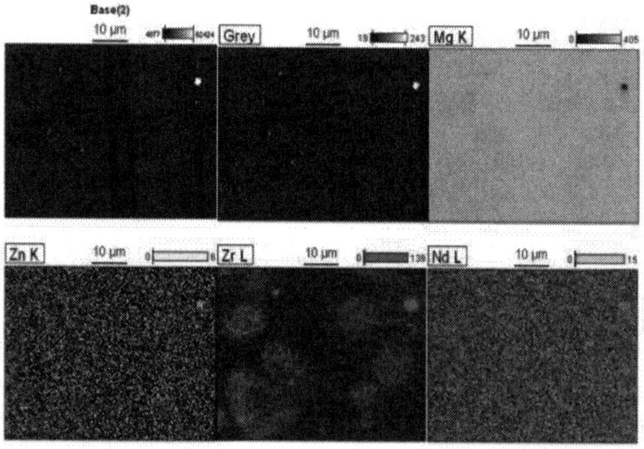

Fig. 4. EDS maps of NZ30K sample solid solution treated at 540°C

Also, the XRD patterns for aged samples at 200 °C for different times show that the peaks related to $Mg_{12}Nd$ and $Mg_{41}Nd_5$ phases are somewhat weak and absent in some cases. It should be pointed out that for the 8 hours aged sample some Mg_2Zn_3 peaks can be detected, which become stronger in 64 hours aged sample. The TEM micrograph taken from a grain boundary (see Fig. 5) confirms the appearance of these precipitates on grain boundaries after 8 hours aging. Precipitation of these phases has been observed in previous researches [10]. The existence of other intermediate phases could not be detected from the XRD patterns; therefore, TEM was employed to study these intermediate phases in the selected samples.

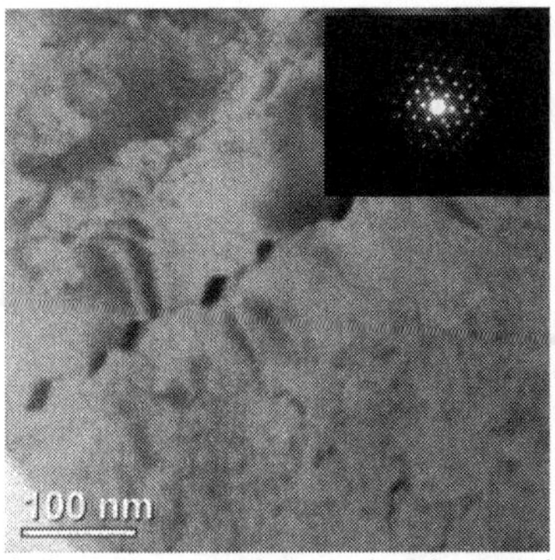

Fig. 5. Grain boundary precipitates in 8 hrs aged sample of NZ30K

An investigation on 30 minutes aged sample showed no apparent precipitation throughout the sample, instead, some coherent metastable and intermediate phases were observed precipitated randomly in some grains. Fig. 6 shows BF TEM micrographs of these precipitates in the 30 minutes aged sample. The beam is parallel to [11-20] direction. As can be seen, there are 3 sets of precipitates visible in these micrographs. In Fig. 6 a, there are two sets of precipitates on prismatic and basal planes. In Fig. 6 b, in addition to the first two sets, there are some unknown dark spots that do not have any clear orientation with the matrix.

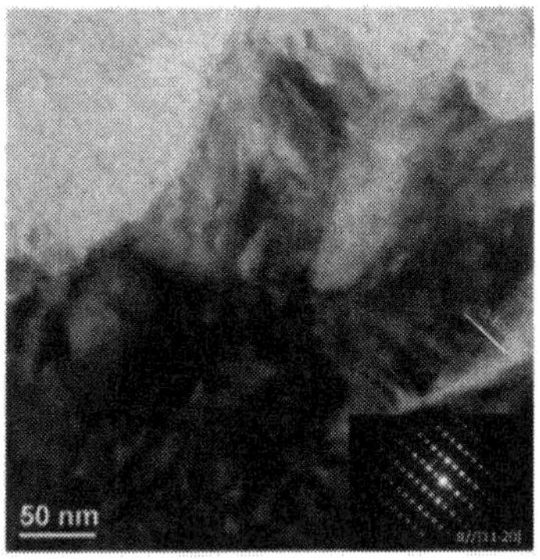

(a)

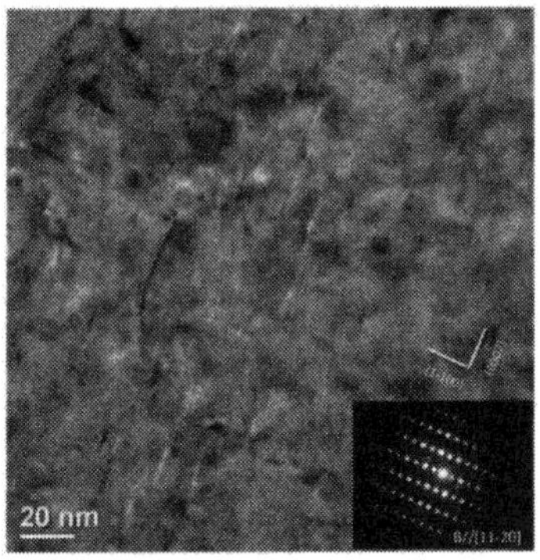

(b)

Fig. 6. - TEM micrographs showing coherent β″ precipitates on {1-100} prism planes, coherent γ′ precipitates on basal planes, and unknown dark spots. The beam is parallel to [11-20].

In the corresponding diffraction pattern, there are some extra spots which are at ½ 0 0 position. These spots were used for dark field imaging in order to study their origin. As can be seen in Fig. 7, these spots belong to sets of precipitates, which lie on {01-10} prism planes.

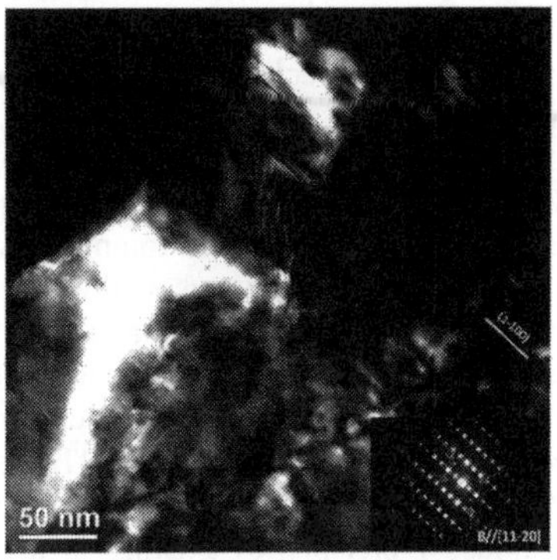

Fig. 7. DF image of the precipitates taken from extra spots in diffraction pattern. The beam is parallel to [11-20] direction.

Fig. 8 shows a high-resolution image of these precipitates. The precipitates are 5- 20 nm long and about 2-3 atomic layers thick with the aspect ratio of 20. According to Ref. 6, 7 and 21, these precipitates are likely either coherent G.P. zones or coherent metastable β″ which form in early stages of precipitation and on {01-10} or {11-20} prims planes. The small size of these precipitates does not allow us to perform a SAED from individual precipitates. The β″ precipitates have D019 structure with $a_{\beta''}=2a_{Mg}$ and $c_{\beta''}=c_{Mg}$. Since the structure of G.P. zones is reported to be disordered D019 structure in this system [22], it is still not clear to the authors whether these precipitates are G.P. zones or coherent β″. However, since in previous researches on G.P. zones, streaks in diffraction patterns are related to the existence of G. P. zones, these precipitates are designated as β″ due to the fact that no streaks could be detected in the corresponding diffraction pattern. With regard to the fact that aging was performed at a short time, this could indicate that the formation of G.P. zones in this system at 200 °C is either not feasible or too fast and cannot be detected. Further studies are underway, using DSC and resistivity measurements, to investigate this matter in Mg-Zn-Nd system.

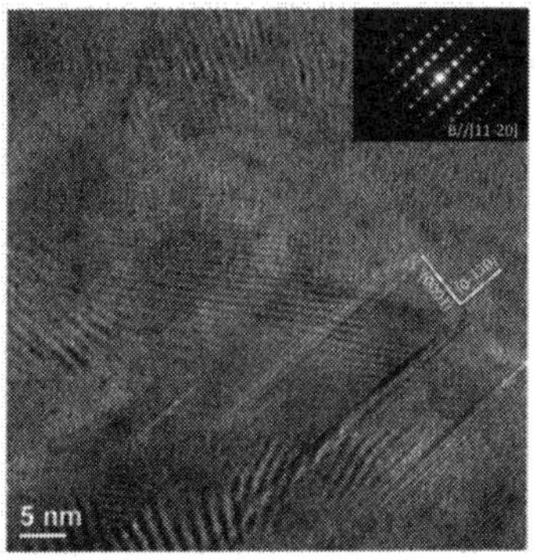

Fig. 8. HRTEM image of β″ precipitates. The beam is parallel to [11-20] direction.

Because the second set of precipitates are on basal planes and there are no additional reflections in the SAED pattern, acquired from the α-Mg matrix containing both sets of precipitates, these precipitates are designated to be coherent γ′. According to Nie et al [23], the structure of γ' is disordered hexagonal with $a_{\gamma'} = 0.321$ nm and $c_{\gamma'} = 0.781$ nm. These precipitates are Mg-Zn rich precipitates, which further increase the hardness of the alloy by pinning dislocations on basal planes. The combination of these two sets of precipitates, which act as barriers to dislocation motion on basal and prismatic planes, results in further increase in the hardness of the alloy. The ratio of Nd/Zn is the factor controlling the formation of precipitates and therefore changes the precipitation nature.

Fig. 9 shows BF images from the 8 hours aged sample. As can be seen the density of precipitates in this sample is higher and the precipitates are uniformly distributed within the grains. The precipitates are on {01-10} prism planes and are 50-80 nm long and 5-7 nm thick. These precipitates are designated as β′ with base centered orthorhombic structure and lattice parameters of $a_{\beta'} = 0.64$ nm, $b_{\beta'} = 2.22$ nm, $c_{\beta'} = 0.64$nm. HRTEM studies of this region, Fig. 10, reveal that beside prism plane precipitates, there are two other kinds of precipitates existing in the image taken. The first one is the presence of unknown precipitates in the form of dark spots, which do not have specific orientation with the matrix and are uniformly distributed in the grains. The second set of precipitates is the ones lying on {0-111} pyramidal planes. This orientation with the matrix has not been observed in this system before. However, this orientation has been reported for Mg-Sn-Zn system [24]. The volume fraction of these precipitates was decreased by addition of Ce into the system [24]. The nature of the precipitates is not still clear. However, since the morphology and dimensions of these precipitates are similar to β′ and these precipitates were just observed in the 8 hours aged sample, these precipitates are designated as β′. Also, the diffraction pattern taken from the region containing these precipitates did not show any additional reflection. The existence of the new set of precipitates on pyramidal planes can further pin the dislocation movement, which results in an increase in hardness, as seen earlier in Fig. 1 for the 8 hours aged sample.

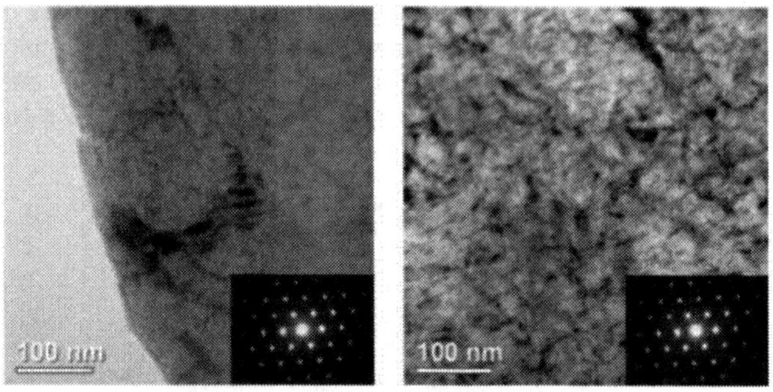

Fig. 9. BF images of NZ30K aged for 8 hours at 200 °C. The beam is parallel to [1-21-3] direction.

Mechanical Properties

In order to study the mechanical properties of the alloy, nanoindentation and Broadband Nanoindentation Creep (BNC) [20, 21] were performed on 4 different samples of solid solution treated, under-aged (30 minutes aging time), peak-aged (8 hours aging time), and over-aged (64 hours ageing time) samples.

The results are shown in Fig. 11 in the form of hardness as a function of indentation load. As can be seen, the data shows an indentation size effect (a decrease in the hardness with an increase in the load). The results of micro-hardness are also added to the graph for comparison. The hardness values are not completely consistent with the values obtained from microhardness for the 8 hours aged sample. This can be attributed to the fact that the nanoindentation results are more sensitive to thermo-mechanical treatment and surface preparation than the microhardness data, causing the observed discrepancy. For both methods, the over-aged sample has the lowest hardness value.

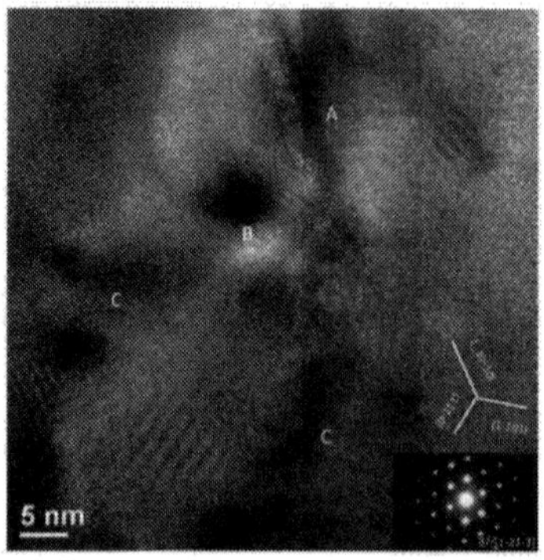

Fig. 10. HRTEM image of the 8hours aged sample. The beam is parallel to [1-21-3]. There are 3 different type of precipitates visible in this image which are marked as: a) prism b) unknown precipitates in the form of a dark spot, and C) pyramidal precipitates.

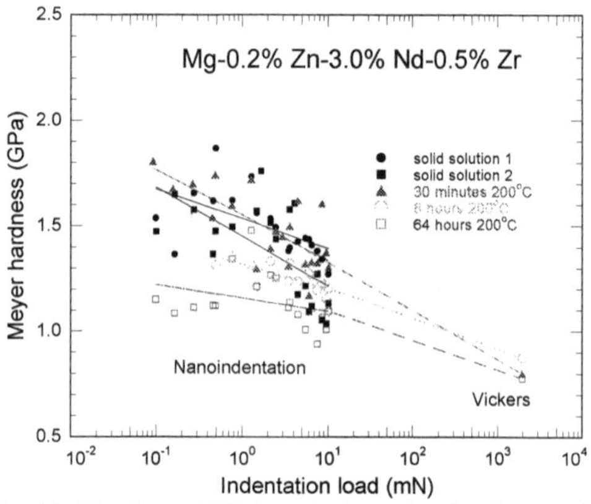

Fig. 11. Hardness (GPa) vs. indentation load (mN) for Mg-0.2Zn-3Nd-0.5Zr alloy. The microhardness values added to the graph for comparison.

The BNC experiments were designed to study the mechanisms involved in the deformation of different samples. The procedure was such that the indenter was loaded into the specimen as rapidly as possible and then was held for as long as possible (until significant drift). The data was corrected for electronic delay during rapid loading. A schematic of load-depth curve (with and without instrumental corrections is shown in Fig. 12 [20].

Fig. 13 shows BNC results in the form of hardness (GPa) vs. strain rate (1/s). The strain rate is extended to 5 orders of magnitude. For all the samples, the hardness tends to increase as the strain rate increases. Also further inspection of the curves reveals that for all the curves, there is a slight downward curvature. This phenomenon is probably caused by dynamic recovery during deformation [25, 26] and a small amount of dislocation climb at room temperature [27]. In addition activation volumes calculated from the BNC data range from 80-100 b^3, which is consistent with prismatic glide as the rate-controlling mechanism in this system.

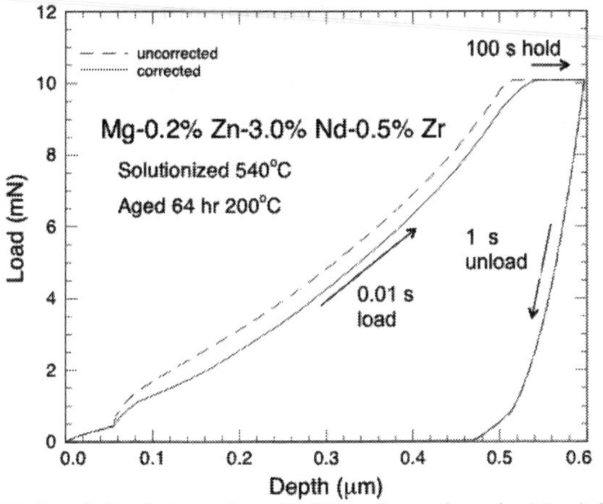

Fig. 12. Load-depth trace from BNC performed on the Mg-0.2Zn-3Nd-0.5Zr

It should be pointed out that the shift in the position of the curves with respect to each other is due to the fact that precipitation hardening mechanism introduces an athermal component (i.e. constant, independent of the strain rate) that contributes to the flow stress of the aged alloy and results in shifting the curves.

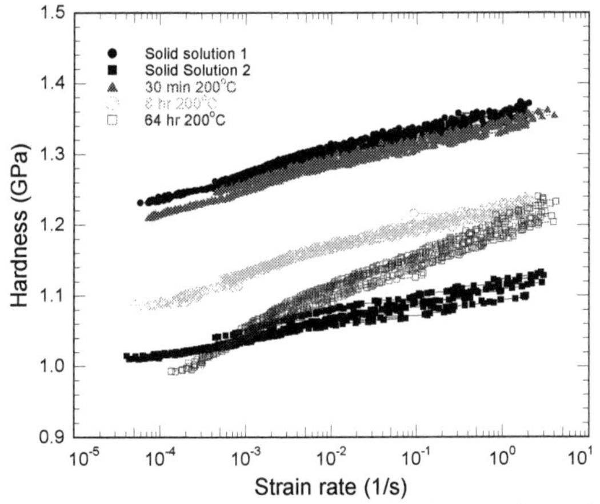

Fig. 13. Hardness vs. Strain Rate curve for the tested samples at different aging times.

Conclusion

1. Four sets of precipitates formed during aging of Mg-0.2Zn-3Nd-0.5Zr alloy at 200 °C which are β" , β', γ' and an dark spots.

2. β" precipitates form on prismatic planes during early stages of precipitation, along with γ' precipitates which form on basal planes. β' precipitates form at higher aging times and precipitate on both prismatic and pyramidal planes. In addition, some unknown precipitates in the form of dark spots were detected at higher aging times. These precipitates had no specific orientation with the matrix.

3. BNC experiments showed an activation volume in the range 80-100 b^3 that is consistent with prismatic glide as the rate controlling mechanism in the system. Furthermore, a small amount of dislocation climb along with dynamic recovery occur at room temperature upon deformation in this system.

Acknowledgement

This work was fully supported by National Science Foundation, DMR GOALI Program, Grant No. 1005762.

References

1. N. Stanford, M. Barnett, "Effect of composition on the texture and deformation behavior of wrought Mg alloys' Scripta mater. 58 (2008) 179-182.

2. R. Gehrmann, M. M. Frommert, G. Gottstein, "Texture effects on plastic deformation of magnesium" J. Mater. Sci. Eng. A 395 (2005) 338-349.

3. A. Sanaty-Zadeh, P. K. Rohatgi, "Comparison between current models for the strength of particulate-reinforced metal matrix nanocomposites with emphasis on consideration of Hall-Petch effect" J. Mater. Sci. Eng. A 531 (2012) 112-118.

4. K. Hantzsche, J. Bohlen, J. Wendt, K. Kainer, S. Yi, D. Letzig, "Effect of rare earth additions on microstructure and texture development of magnesium alloy sheets" Scripta Mater. 63 (2010) 725-730.

5. R. K. Mishra, A. K. Gupta, P. R. Rao, A. K. Sachdev, A. Kumar, A. Luo, "Influence of cerium on the texture and and ductility of magnesium extrusions" Scripta Mater. 59 (2008) 562-565.

6. T. J. Pike, B. Noble, "The formation and structure of precipitates in a dilute magnesium-neodymium alloy" J. the Less-Common Met. 30 (1973) 63-74.

7. L. Ma, R. K. Mishra, M. P. Balogh, L. Peng, A. A. Luo, A. K. Sachdev, W. Ding, "Effect of Zn on the microstructure evolution of extruded Mg-3Nd(-Zn)-Zr (wt.%) alloys" J. Mater. Sci. Eng. A 543 (2012) 12-21.

8. P. A. Nuttall, T. J. Pike, B. Noble, "Metallography of Dilute Mg-Nd-Zn alloys" Metallography 13 (1980) 3-20.

9. K.Y. Zheng, J. Dong, X.Q. Zeng, W.J. Ding, "Precipitation and its effect on the mechanical properties of a cast M-Gd-Nd-Zr alloy" J. Mater. Sci. Eng. A 489 (2008) 44-54.

10. Zheng Xingwei, Dong Jie, Liu Wencai, Ding Wenjiang, "Microstructure and mechanical properties of NZ30K alloy by semi-continuous direct chill and sand mould casting processes" Res. Dev. 8 (2011) 41-46.

11. Xingwei Zheng, Alan A. Luo, Jie Dong, Anil K. Sachdev, Wenjiang Ding, " Plastic flow behavior of a high-strength magnesium alloy NZ30K" Mat. Sci. Eng. A 532 (2012) 616– 622.

12. Jun Dai, Jian Huang, Min Li, Zhuguo Li, Jie Dong, Yixiong Wu, "Effect of heat treatments on laser welded Mg-rare earth alloy NZ30K" Mat. Sci. Eng. A 529 (2011) 401– 405.

13. W.J. Ding, P.H. Fu, L.M. Peng, H.Y. Jiang, X. Q. Zeng, "Study on the microstructure and mechanical property of high strength Mg-Nd-Zn-Zr alloy" Mat. Sci. Forum 546-549 (2007) 433-436.

14. Xingwei Zheng, Jie Dong, Yazhen Xiang, Jianwei Chang, Fenghua Wang, Li Jin, Yingxin Wang, Wenjiang Ding, "Formability, mechanical and corrosive properties of Mg-Nd-Zn-Zr magnesium alloy seamless tubes" Mater. Des. 31 (2010) 1417–1422.

15. Wu Wen-Xiang, Jin Li, Dong Jie, Ding Wen-Jiang, "Prediction of flow stress of Mg-Nd-Zn-Zr alloy during hot compression" Trans. Nonferrous. Met. Soc. China 22 (2012) 1169−1175.

16. Ding Wenjiang, Li Daquan, Wang Qudong, Li Qiang, "Microstructure and mechanical properties of hot-rolled Mg-Zn-Nd-Zr alloys" Mater. Sci. Eng. A 483–484 (2008) 228–230.

17. Qiang Li, Qudong Wang, Yingxin Wang, Xiaoqin Zeng, Wenjiang Ding, "Effect of Nd and Y addition on microstructure and mechanical properties of as-cast Mg-Zn-Zr alloy" J. Alloys Comp. 427 (2007) 115–123.

18. Xiao Zhou, Haitao Zhou, Zhendong Zhang, Ruirui Liu, Libin Liu, "Tensile properties of Hot Extruded Mg-Zn-Nd-Y-Zr alloy at elevated temperatures" Adv. Mat. Res. 415-417 (2012) 1157-1163.

19. Yang Lin, Gao Xiao-dan, Lin Li, Zou Peng, Chen Li-jia, Liu Zheng, "Microstructural evolution of rolled Mg-5Zn-3Nd(-Zr) alloy" Trans. Nonferrous Met. Soc. China 20 (2010) s498-s502.

20. J. E. Jakes, R. S. Lakes, D. S. Stone Broadband nanoindentation of glassy polymers: Part II. Viscoplasticity J. Mater. Res. 27 (2012) 475-484.

21. J. B. Puthoff, J. E. Jakes, H. Cao, D. S. Stone, Investigation of thermally activated deformation in amorphous PMMA and Zr-Cu-Al bulk metallic glasses with broadband nanoindentation creep J. Mater. Res. 24 (2009) 1279-1290.

22. R. Wilson, C. J. Bettles, B. C. Muddle, J. F. Nie, "Precipitation hardening in Mg-3wt%Nd(-Zn) casting alloys" Mater. Sci. Forum 419-422 (2003) 267-272.

23. J. F. Nie, K. Oh-ishi, X. Gao, K. Hono, "Solute segregation and precipitation in a creep-resistant Mg-Gd-Zn alloy" Acta Mater. 56 (2008) 6061-6076.

24. C. Weili, P. S S, T. Weineng, Y. B S, K. B H, "Influence of rare earth on the microstructure and age hardening response of indirect-extruded Mg-5Sn-4Zn alloy, J. Rare Earths 28 (2010 785-789.

25. A. Couret and D. Caillard, "an in-situ study of prismatic glide in magnesium-I. The rate controlling mechanism" Acta Metall. 33 (1985) 1447-1454.

26. P. Lukâĉ and Z Trojanová, "Study of thermally activated dislocation motion in AJ51 and AE42 magnesium alloys" J. Phys.: Conference Series 240 (2010) 012-019

27. D. S. Stone, "Scaling laws in dislocation creep" Acta Met. et Mat. 39 (1991) 599-608.

Magnesium Technology 2013

Cast Materials

Magnesium Technology 2013
Edited by: Norbert Hort, Suveen N. Mathaudhu, Neale R. Neelameggham, and Martyn Alderman
TMS (The Minerals, Metals & Materials Society), 2013

FE Modelling of Tensile and Impact Behaviours of Squeeze Cast Magnesium Alloy AM60

Sante DiCecco, William Altenhof and Henry Hu

Mechanical, Automotive & Materials Engineering
University of Windsor, 401 Sunset Ave.
Windsor, Ontario, Canada N9B 3P4

Keywords: Squeeze casting, Magnesium, AM60, FE analysis, Tensile properties, Impact behaviour

Abstract

In response to the need for reduced global emissions, the transportation industry has been steadily increasing the magnesium content in vehicles. This trend has resulted in experimental documentation of numerous alloy and casting combinations, while comparatively little work has been done regarding the development of numerical material models for vehicle crashworthiness simulations. In this study, material mechanical behaviour was implemented into an existing material model within the nonlinear FEA code LS-DYNA to emulate the mechanical behaviour of squeeze cast magnesium alloy AM60 with a relatively thick section of 10 mm thickness. Model validation was achieved by comparing the numerical and experimental results of a tensile test and Charpy impact event. Validation found an average absolute error of 5.44% between numerical and experimental tensile test data, whereas a relatively large discrepancy was found during Charpy evaluation. This discrepancy has been attributed to the presence of microstructure inhomogeneity in the squeeze cast magnesium alloy AM60.

Introduction

In an effort to maintain a competitive, fuel efficient, product line, automotive manufacturers have been increasingly incorporating magnesium into the design of various vehicle subsystems. This has led to the rapid development of numerous magnesium alloy and casting combinations [1]. However, despite these advances, the use of magnesium alloys in a load-bearing capacity has been primarily restricted to thin-walled casting applications due to the high porosity and coarse microstructure traditionally associated with the casting of relatively thick components [2]. Furthermore, while a great deal of effort has been focused on the experimental development of new alloys, comparatively little work has been done regarding the development of corresponding numerical material models for vehicle crashworthiness simulations; thus further inhibiting the use of the alloys in a load bearing capacity. In an effort to alleviate these issues, past studies have been performed regarding the experimental characterization and numerical analysis of magnesium alloy AM50.

In the work of Zhou, M et al. [2], the mechanical performance of a relatively thick squeeze casting of magnesium alloy AM50 was investigated. In their work, tensile testing of squeeze cast and high pressure die cast (HPDC) specimens was performed. Results from the study found that the squeeze casting process demonstrated considerable improvements in porosity, density and tensile performance over conventional HPDC, when casting section thicknesses were above 10 mm.

In the work of Altenhof, W. et al. [3], FE (Finite Element) modelling of HPDC AM50 was investigated within the nonlinear FEA code LS-DYNA. In their work, tensile testing and Charpy impact testing were simulated based on the experimental performance of the alloy, as obtained from a tensile test, using the piece linear plasticity model, MAT_24. Results from the study found that the tensile simulation demonstrated excellent agreement with experimental results, while the Charpy simulation correlated sufficiently well with experimental data. Discrepancies in the results of the Charpy simulation were primarily attributed to increased porosity within the Charpy specimens relative to the tensile specimens.

To further the use of relatively thick magnesium alloy castings in a load bearing capacity, the objective of this study was to simulate popular magnesium alloy AM60 in a 10 mm thick squeeze cast configuration, under tensile (static) and Charpy impact (dynamic) loading conditions, and to evaluate the results relative to experimental observations. This paper presents the results of the aforementioned testing, as well as a thorough documentation of the experimental and numerical procedures followed during the course of the study. It is anticipated that the results will provide the foundation for the development of future high fidelity numerical models of squeeze cast magnesium alloy AM60.

Experimental Procedure

Alloy Preparation and Analysis

To obtain tensile and Charpy impact specimens, cylindrical coupons of a conventional AM60 alloy (Table 1) were squeeze cast with a diameter of 95 mm and a section thickness of 10 mm, and sectioned, as necessary. Parameters employed during the squeeze casting process included a molten alloy temperature of 690°C, a die temperature of 275°C and an applied pressure of 30 MPa. Subsize rectangular tensile specimens were prepared in accordance with ASTM standard B557M [4]; however, thickness of the tensile specimens was inherently larger than the maximum thickness recommended by ASTM, due to the nature of the study. Full size Charpy specimens were prepared in the unnotched condition, in accordance with ASTM standard E23 [5].

Table 1: Composition of Magnesium alloy AM60

Alloy Symbol	Al (wt.%)	Mn (wt.%)	Zn (wt.%)	Ni (wt.%)
AM60A	5.89	0.335	0.0068	0.0004

In addition to tensile and impact specimens, a sample cylindrical coupon was also sectioned for density measurement and porosity evaluation. Density was evaluated following the Archimedes principal of ASTM standard D3800 [6], while porosity was calculated (Equation 1) based on the theoretical density of AM60 relative to that which was experimentally determined. In

Equation 1, D_t is the theoretical density of AM60, 1.8 g/cm^3, and D_e is the experimentally determined alloy density.

$$\%Porosity = [(D_t - D_e)/D_t] \times 100\% \quad [1]$$

Tensile Testing

Tensile testing was performed at ambient temperature using an Instron 8562 universal testing machine equipped with a computer data acquisition system. Acquired load-displacement data was normalized relative to the nominal dimensions of the tensile specimens to obtain the experimental stress-strain data of the alloy. This data was subsequently used to obtain 0.2% offset yield strength (YS), ultimate tensile strength (UTS) and elongation to failure (EF); values which were later used in the development of the alloy's numerical material model, as described under FE Modelling of a Tensile Simulation.

Instrumented Charpy Impact Testing

A Riehle Impact Testing Machine was instrumented and calibrated following ASTM standards E2298 [7] and E23 [5] for the purpose of Charpy impact testing. Instrumentation of the machine was done by placing Omega KFG series 350 ohm strain gauges on each side of the striker. Subsequently, the striker was loaded under static conditions, and the load as a function of voltage output of the gauge was acquired using a PCB load cell rated for 90 kN and a NI 9237 data acquisition module. This data was then used during impact testing to obtain load as a function of time, from the striker strain gauge voltage output. Additionally, a 300 mm range laser displacement transducer was used during testing to obtain striker displacement as a function of time. The resulting data acquired was cross-plotted to obtain striker load as a function of displacement during each impact event. The use of load-displacement data was required to provide a more robust validation metric against numerical data. Given that temperature dependency of the alloy was not under investigation, testing was performed at room temperature.

FE Modelling

Tensile Simulation

LS-PPREPOST was used to develop the FE model of the tensile specimen. The model consisted of a total of 2100 fully integrated plane stress shell elements (Figure 1), which corresponded to a mesh density of approximately 4.7 mm^2 within the gauge region of the specimen. Model dimensions were selected to coincide with experimental tensile specimen geometry, to maintain consistency between experimental and numerical testing. Similarly, boundary conditions were employed to match experimental testing. Thus, during simulation of the tensile test, the nodes of one end of the numerical model were restricted from motion, while the other end was displaced by 5.4 mm, ensuring that the gauge region of the model reached the experimentally determined failure strain.

Material model type 24 of the nonlinear FE software LS-DYNA was used to emulate the mechanical-material response of the 10 mm thick squeeze cast AM60 alloy. This material model is a piecewise linear plasticity model that is widely used within the automotive industry [3]. It was selected for this simulation due to its ease of implementation. This material model utilizes a von Mises yield criterion with common input parameters of the model including material density, elastic modulus, Poisson's ratio and yield stress. These were input into the model based on experimental testing (Table 2). Additionally, to simulate the plasticity of the alloy, a series of data points representing the true stress as a function of effective plastic strain was required for input. A total of 20 pairs of data points were inserted into the model, based on experimental testing, to ensure a sufficient amount of detail exists within the plasticity curve to capture any significant nonlinearities.

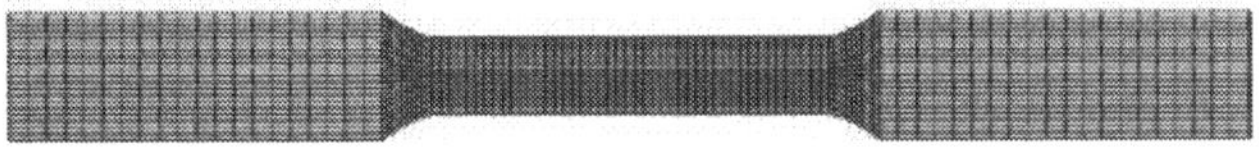

Figure 1: Numerical model of AM60 tensile specimen.

Charpy Impact Simulation

A FE Charpy impact model was developed for the simulation of the Charpy impact event. The model consisted of a Charpy impact test specimen, a striker and the supporting anvils (Figure 2). The Charpy specimen was developed in LS-PREPOST, and had a total of 236,425 under integrated solid elements. Hourglass control was implemented through the use of a Flanagan-Belytschko stiffness form, with an hourglass coefficient of 0.02. Dimensions of the specimen were selected to correspond with experimental specimen dimensions. The striker and supporting anvils were previously developed within the research group [3], using TrueGrid. These parts were discretized appropriately to capture the curvature of the striker and the anvil corners. Additionally, they were modelled as rigid to simplify analysis and reduce simulation time. Nodes of the anvils were fully constrained from motion, while the nodes of the striker were constrained to the direction of the motion of the striker; no other translations or rotations were permitted.

The material model implemented within the Charpy impact simulation corresponds to a modified version of the material model used during the tensile test simulation. All material parameters from the previous model were unchanged (e.g. density, elastic modulus); however, in addition to the initial piecewise linear plasticity curve, a second curve was included to account for strain rate effects. This decision was made based on information available in the literature suggesting that the mechanical performance of AM60 is strain dependent for the high strain rates expected to be seen during a Charpy impact test (>1000 s^{-1}) [8-9]. The additional piecewise curve corresponds to a modified version of the initial curve where effective plastic strain and true stress have been scaled to reflect the results presented in reference [9]. With these two curves, the material model is capable of addressing strain rates between 0.01 s^{-1} and 1000 s^{-1}. If the strain rate of an element falls outside of this range, the model defaults to the most appropriate stress-strain curve. If the strain rate of an element falls within this range, interpolation between curves is performed.

In addition to strain rate effects, simulation of the Charpy impact event required inclusion of an element failure criterion. This was facilitated by invoking the *MAT_ADD_ERROSION command within LS-DYNA; a command which allows for the specification of various pre-designated failure criteria. In this study, the failure criterion selected was that the maximum principal stress within an element must exceed the true stress ultimate tensile strength of the material for the high strain rate condition (233.6 MPa). Once this criterion was met, the element was deleted from the simulation.

Figure 2: Numerical model of AM60 Charpy impact specimen, striker and supporting anvils.

Validation Procedure

In an effort to validate the tensile and Charpy impact numerical models discussed throughout this study, a rigorous error analysis was completed using Equation 2. Equation 2 calculated the average absolute error between two functions, over a specified range interval. In the case of the tensile model, average absolute error was calculated between the experimental and numerical force curves as a function of strain. The range in which this analysis was performed covered the entire spectrum of experimental tensile strains. Similarly, in the case of the Charpy impact model, average absolute error was calculated between the experimental and numerical force curves as a function of striker displacement. The range in which this analysis was performed encompasses the entire range of the striker displacement during experimental testing. In addition to the error analysis performed using Equation 2, a validation metric [10] (Equation 3) was employed over the same range of tensile strains and Charpy impact displacements. This metric was intended to serve as an additional method of verifying model accuracy. Using this approach, a perfect overlay of two functions would yield a validation metric value of '1'. In both equations, d represents displacement when evaluating Charpy data and strain when evaluating tensile data, while F represents force for either Charpy or tensile validation.

$$Error = \frac{1}{d_2 - d_1} \cdot \int_{dt_1}^{d_2} \left| \frac{F_{Exp}(d) - F_{Num}(d)}{F_{Exp}(d)} \right| dd \quad [2]$$

$$V = 1 - \frac{1}{d_2 - d_1} \cdot \int_{d_1}^{d_2} \tanh\left(\left| \frac{F_{Exp}(d) - F_{Num}(d)}{F_{Exp}(d)} \right| \right) dd \quad [3]$$

Results and Discussion

Experimental Observations

Porosity: Following procedures of the Archimedes principal, the experimental density of the 10 mm thick squeeze cast alloy was found to be 1.787 g/cm^3. Subsequently, this value was used in conjunction with Equation 1 to determine that the porosity of the squeeze cast 10 mm thick alloy was 0.727%. This value compares favourably with porosity measurements that have been reported regarding die cast AM60 using similar casting parameters [2], and should be attributable to the fact that the applied pressure during squeeze casting suppresses gas nucleation, while enabling melt penetration within areas of microshrinkage.

Tensile Data: Figure 3 shows a representative stress-strain curve of the AM60 alloy following uniaxial tension testing. As can be seen, the AM60 alloy first deforms elastically, then, after yielding has occurred, plastic deformation of the alloy takes place. Also, it should be noted that since there is no single obvious yielding point, the 0.2% offset strain method was required to determine the elastic modulus of the alloy. True stress and strain values of the YS, UTS and EF are as follows: 58.3 MPa, 191.2 MPa and 5.89%. A summary of these properties and other relevant material data may be found in Table 2.

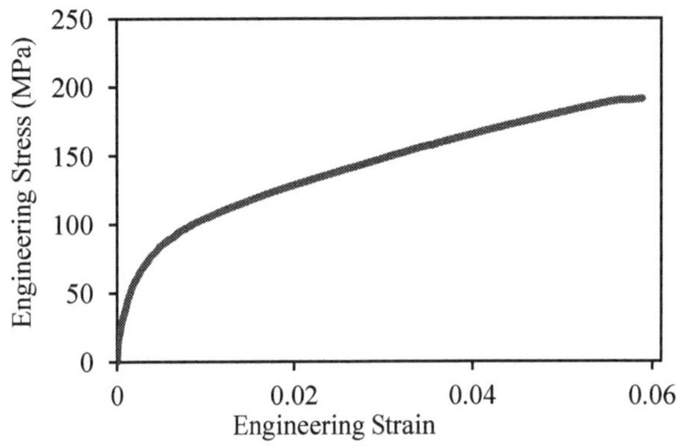

Figure 3: Engineering stress versus engineering strain of AM60 alloy.

Table 2: Summary of relevant AM60 properties. YS, UTS and EF are presented in true stress and strain, respectively.

Density (g/cm^3)	Poisson's Ratio	Elastic Modulus (MPa)	Yield Strength (MPa)	Ultimate Tensile Strength (MPa)	Elongation to Failure (%)
1.787	0.35	34.45	58.3	191.2	5.89

Charpy Impact Testing: Figure 5 illustrates a (smoothed) representative plot of the load-displacement performance of the AM60 alloy, as encountered during testing. An average value of the impact energy absorbed prior to specimen failure was calculated based on the area under the load-deflection curves obtained from instrumented Charpy impact testing, as well as some readings obtained from the Riehle dial gauge, prior to instrumentation. This value was found to be 9.66 J ± 1.94 J.

Numerical Analysis and Validation

Tensile Simulation: Presented in Figure 4 are the force-strain curves obtained from experimental and numerical uniaxial tensile testing of the 10 mm thick squeeze cast AM60 alloy. Overall, it can be seen that the two curves demonstrate good agreement. This is further reinforced by the average absolute error between the two curves, which is 0.5437, or 5.437%. Similarly, the validation metric presented in Equation 3 yielded a value of 0.946.

One source of error within the material model originates from the calculation of the elastic modulus of the AM60 alloy. Due to the curvature of the stress-strain data in the vicinity of the yield point, the best fit elastic modulus had a coefficient of determination (R^2 value) of only 0.93. This low value indicates that the elastic modulus used did not perfectly represent the elastic region of the AM60 alloy. This issue may be remedied in the future by taking the average of multiple stress-strain plots, to obtain a more definable yield point.

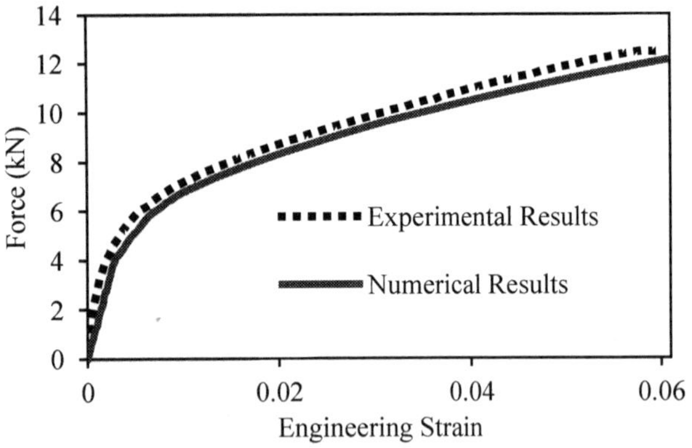

Figure 4: Experimental and numerical load-strain curves of AM60 alloy, obtained from uniaxial tension testing/simulation.

Charpy Impact Simulation: Experimental and numerical load-deflection curves of the performance of the AM60 alloy are found in Figure 5. As can be seen, the numerical model significantly over-predicts the experimental Charpy impact testing data. This is quantified by the average absolute error, which was determined to be 156.01%, while the validation metric was calculated to be 0.1. There are several possible explanations for this discrepancy, two of which are discussed below.

(1) The inclusion of rate effects was done in a basic way, based on data available in the literature for an AM60 alloy that was not necessarily processed in the same manner as that which was used in this study. Therefore, it is not fully known if this assumption was fully correct. To verify this issue, high strain rate (> 1000 s^{-1}) tensile testing is required.

(2) While the porosity of a sample casting was found to be low, this may not be indicative of the overall porosity of the cast coupon. This is especially true if any form of microstructural inhomogeneity exists within the specimen. To verify this issue, SEM fractography analysis was previously performed on a Charpy impact specimen. Results from this analysis found that there was a certain amount of localized porosity present within the specimen. To quantify this degree of porosity, porosity analysis, as previously described, was performed on four fractured Charpy impact specimens. The results from these measurements showed that the average porosity of the tested impact specimens was 1.669%, with a standard deviation of 0.03%. This level of porosity was significantly greater than the 0.727% porosity associated with the tensile data used in establishing the numerical models of this study, and could have a detrimental effect on the impact performance of the tested alloy. The overestimation of the impact behaviour of the squeeze cast AM60 alloy by the numerical model should be attributed to the increased level of porosity within the Charpy specimens, relative to the tensile specimens.

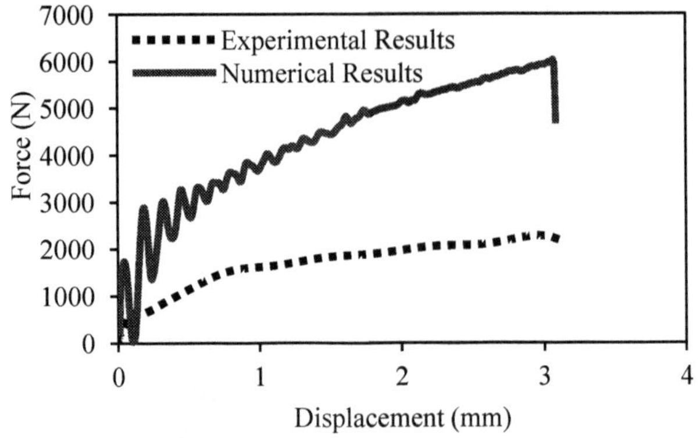

Figure 5: Experimental and numerical Charpy impact event load-deflection curves.

While the cause of the discrepancy is not yet fully understood, the model provides a good first step towards the simulation of thick squeeze cast AM60 alloy castings under dynamic loading conditions. In an effort to resolve this issue in the future, a more thorough analysis of specimen inhomogeneity should be performed, as well as an increase in Charpy impact sample testing size, to fully understand the range of impact energies that might be produced from different levels of specimen porosity.

Conclusions

Numerical simulation of a 10 mm thick squeeze cast AM60 magnesium alloy was performed under tensile (static) and Charpy impact (dynamic) testing conditions. Results from the tensile simulation were validated against experimental uniaxial tensile test data, with a corresponding average absolute error of 5.437%. Conversely, the numerical simulation of the Charpy impact event was found to significantly over-predict experimental observations; thus preventing validation of the model for dynamic loading conditions. Despite this, the study, as a whole, provides a sound methodology regarding the simulation of the AM60 alloy under different loading conditions.

Acknowledgements

The authors would like to take this opportunity to thank the Natural Sciences and Engineering Research Council (NSERC) of Canada for supporting this work. One of the co-authors, Sante DiCecco, also wishes to acknowledge NSERC and University of Windsor for financial support in the form of an NSERC CGS M award and a University of Windsor Tuition Scholarship, respectively.

References

[1] H. Hu, A. Yu, N. Li and J.E. Allison, "Potential Magnesium Alloys for High Temperature Die Cast Automotive Applications: A Review," *Materials and Manufacturing Processes*, 18 (5) (2003), 687-717.

[2] M. Zhou, H. Hu, N. Li and J. Lo, "Microstructure and Tensile Properties of Squeeze Cast Magnesium Alloy AM50," *Journal of Materials Engineering and Performance*, 14 (4) (2005), 539-545.

[3] W. Altenhof, A. Raczy, M. Laframboise, J. Loscher and A. Alpas, "Numerical Simulation of AM50A Magnesium Alloy Under Large Deformation," *International Journal of Impact Engineering*, 30 (2004), 117-142.

[4] ASTM Standard B577M, 2007e1, "Standard test methods for tension testing wrought and cast aluminum- and magnesium-alloy products (metric)," *ASTM International*, 2007, DOI: 10.1520/B0557M-07E01.

[5] ASTM Standard E23, 2007a, "Standard Test Methods for Notched Bar Impact Testing of Metallic Materials", *ASTM International,* 2002, DOI: 10.1520/E0023-07AE01.

[6] ASTM Standard D3800, 1999 (2010), "Standard Test Method for Density of High-Modulus Fibers", *ASTM International*, 2002, DOI: 10.1520/D3800-99.

[7] ASTM Standard E2298, 2009, "Standard Test Method for Instrumented Impact Testing of Metallic Materials", *ASTM International*, 2002, DOI: 10.1520/E2298-09.

[8] W.Q. Song, P. Beggs and M. Easton, "Compression Strain-Rate Sensitivity of Magnesium-Aluminum Die Casting Alloys," *Materials and Design*, 30 (2009), 642-648.

[9] S. Simunovic, D. Erdman and J. M. Starbuck, "High Strain Rate Characterization of Magnesium Alloys," *Oak Ridge National Laboratory*, (2012). http://thyme.ornl.gov/Mg_new/

[10] W. L. Oberkampf and T. G. Trucano, "Verification and validation in computational fluid dynamics," *Progress In Aerospace Sciences*, 38 (2002), 200-272.

Magnesium Technology 2013
Edited by: Norbert Hort, Suveen N. Mathaudhu, Neale R. Neelameggham, and Martyn Alderman
TMS (The Minerals, Metals & Materials Society), 2013

HIGH TEMPERATURE DEFORMATION OF MAGNESIUM ALLOY TX32-0.4Al-0.8Si

C. Dharmendra[1], K.P. Rao[1], N. Hort[2], K.U. Kainer[2]

[1]Department of Mechanical and Biomedical Engineering
City University of Hong Kong, 83 Tat Chee Avenue, Kowloon, Hong Kong SAR
[2]Helmoltz-Zentrum Geesthacht, Max-Planck Str. 1, Geesthacht 21502, Germany

Keywords: Mg-Sn-Ca-Al-Si alloy, Hot deformation, Processing map, Kinetic analysis

Abstract

The effect of aluminum and silicon as micro-alloying elements in TX32 magnesium alloy on its hot deformation behavior has been studied by conducting uniaxial compression tests at various combinations of temperatures and strain rates in the ranges 300-500 °C and 0.0003-10 s^{-1}. A processing map has been developed and the effect of processing conditions on deformation behavior has been analyzed. Two processing windows that enable good hot working of the alloy are identified at (1) 390-500 °C/0.0003-0.005 s^{-1} (Domain 1), (2) 430-500 °C/0.3-10 s^{-1} (Domain 2). The kinetic analysis is obeyed in these two domains and the relevant apparent activation energy values are found to be 215 and 170 kJ/mole respectively. These are higher than that for self-diffusion in magnesium suggesting that intermetallic particles present in the matrix generate back stress.

Introduction

Magnesium alloys have become very attractive in applications such as automotive, aerospace and railway industries due to their light weight and high specific strength. In spite of having better mechanical properties than cast alloys, wrought Mg alloys still have limited applications due to their poor workability at lower temperatures [1,2] and their strong anisotropic character [3]. These limitations are due to their hexagonal close-packed (hcp) crystal structure and therefore insufficient number of slip systems [1,2]. At room temperature, basal slip is the principal slip system. Due to the temperature dependency of critical resolved shear stress of other slip systems, non-basal slip systems such as prismatic and pyramidal are activated at higher temperatures only [4,5]. Magnesium develops pronounced crystallographic textures after deformation operations which is not desirable for further processing [6]. In practice, formability can be improved by working at higher temperatures, alloying additions and finer grain sizes [7]. All of these can be rationalized in terms of non-basal slip activation [8].

Mg-Sn-Ca (TX series) alloys are new magnesium wrought alloys that are designed to offer good corrosion and creep resistance [9]. Among them, Mg-3Sn-2Ca (TX32) alloy is identified as the most promising one as it exhibited good creep resistance [9,10]. The addition of aluminum and silicon at micro-alloying levels are considered to further enhance the properties of this alloy. Aluminum improves room temperature mechanical properties of Mg by solid solution strengthening [11] due to large atomic size difference (16%) while silicon forms intermetallic particles which can enhance creep strength.

The aim of this work is to evaluate the hot workability of TX32-0.4Al-0.8Si alloy in as-cast condition by applying kinetic analysis and developing a processing map using flow stress data obtained in compression tests. The standard kinetic rate equation relating the strain rate ($\dot{\varepsilon}$) and temperature (T) to the steady state flow stress (σ) is given by [12]:

$$\dot{\varepsilon} = A\sigma^n \exp\left[-\frac{Q}{RT}\right] \quad (1)$$

where A = constant, n = stress exponent, Q = activation energy, and R = gas constant. The rate-controlling mechanisms are identified on the basis of the activation parameters n and Q. The technique of processing maps is based on the dynamic materials model [13,14]. The efficiency of power dissipation occurring through microstructural changes during deformation is given by:

$$\eta = 2m/(m+1) \quad (2)$$

where m is the strain rate sensitivity of flow stress. Further, the extremum principles of irreversible thermodynamics as applied to continuum mechanics of large plastic flow [15] have been explored to define a criterion for the onset of flow instability given by:

$$\xi(\dot{\varepsilon}) = \frac{\partial \ln[m/(m+1]}{\partial \ln \dot{\varepsilon}} + m \leq 0 \quad (3)$$

These equations can be employed to find out the effect of temperature and strain rate on deformation processing and microstructural evolution.

Experimental Details

Mg – 3 wt.% Sn – 2 wt.% Ca – 0.4 wt.% Al-0.8 wt.% Si (TX32 – 0.4Al-0.8Si) alloy was prepared using 99.99% pure Mg, 99.96% pure Sn, 98.5% pure Ca, 99.9% Al and 99.9% Si. The alloy was molten at about 720 °C under a protective cover of Ar+3% SF_6 mixed gas atmosphere, and was poured into pre-heated permanent molds to obtain cylindrical billets of 100 mm diameter and 350 mm length. For compression testing, cylindrical specimens of 10 mm diameter and 15 mm height were machined from the as-cast billet. A 1 mm diameter hole was drilled at mid height of the specimen for inserting a thermocouple to measure the specimen temperature as well as the adiabatic temperature rise during the compression test.

The compression tests in the temperature range 25-250 °C were conducted at a low strain rate of 0.0001 s^{-1} using a computer-controlled servo-hydraulic testing machine to evaluate strength of the alloy. The data for developing processing maps were obtained from uniaxial compression tests conducted at constant true strain rates in the range 0.0003 – 10 s^{-1} and temperature range 300 – 500 °C. Details of the test set-up and procedure are described in an earlier publication [16]. Constant true strain rate during a test was achieved using an exponential decay of actuator speed in the servo hydraulic machine. Graphite powder mixed with grease was used as the lubricant in all the experiments. The specimens were deformed up to a true strain of about 1.0 and then quenched in water. The load - stroke data were converted into true stress - true strain curves using standard equations. The flow stress values were corrected for the adiabatic temperature rise. The deformed specimens were sectioned in the center parallel to the compression axis and the cut surface was mounted, polished and etched for metallographic examination.

Results and Discussion

The initial microstructure of as-cast TX32-0.4Al-0.8Si alloy is shown in Figure 1(a). The average grain diameter was about 300 μm. The intermetallic particles are identified as CaMgSi and Ca_2Sn phases based on X-ray diffraction (XRD) and electron probe micro analysis (EPMA), and are marked in the scanning electron micrograph as shown in Figure 1(b).

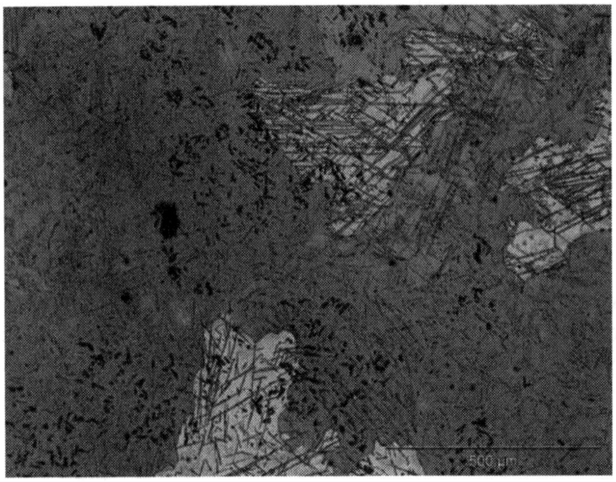

(a)

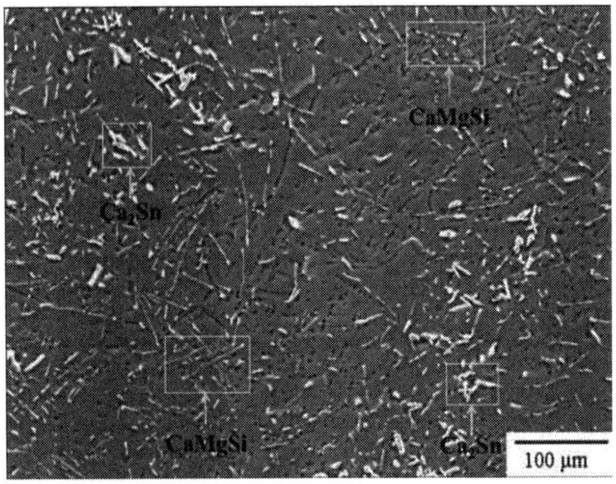

(b)

Figure 1. (a) Optical and (b) Scanning electron micrographs of TX32-0.4Al-0.8Si alloy, the latter showing CaMgSi and Ca_2Sn phase particles.

The variation of the ultimate compressive strength with temperature in the range 25 – 250 °C at a strain rate of $0.0001 s^{-1}$ obtained from compression tests of TX32-0.4Al-0.8Si alloy is shown in Figure 2. A strain rate slower than that used for generating processing maps is selected for obtaining the strength of the alloy in this temperature range. Figure 2 shows that the strength drops continuously with increase in temperature. The values for the alloy Mg-3Sn-2Ca-0.4Al (with-out Si addition) are also plotted in this figure for the purpose of comparison and to evaluate the effect of Si. The reduction in strength with Si addition may be attributed to the presence of CaMgSi intermetallic particles. The Si addition often promotes the precipitation of Si-containing particles which form during solidification and are quite detrimental to the mechanical properties of Mg alloys [17, 18]. Asl et al. [11] have indicated that addition of Si to some aluminum containing magnesium alloys improves creep resistance while reducing the room temperature mechanical properties. The role of Si on mechanical and creep properties of such alloy systems is not fully understood. The present results, however, clearly show that 0.8% Si addition is not useful for improving the strength of the TX32 base alloy containing 0.4%Al.

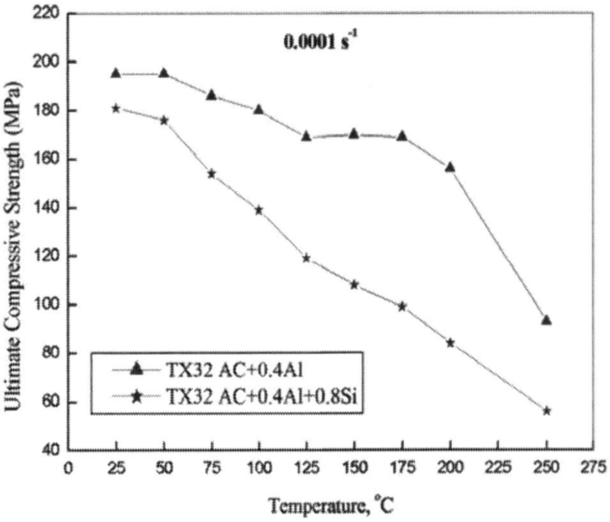

Figure 2. Variation of ultimate compressive strength (UCS) values with temperature for TX32-0.4Al and TX32-0.4Al-0.8Si alloys.

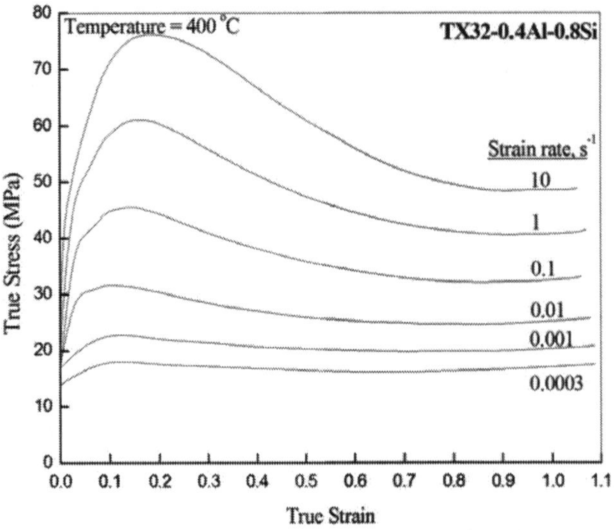

Figure 3. True stress - true strain curves obtained at 400 °C on TX32-0.4Al-0.8Si alloy in compression at different strain rates.

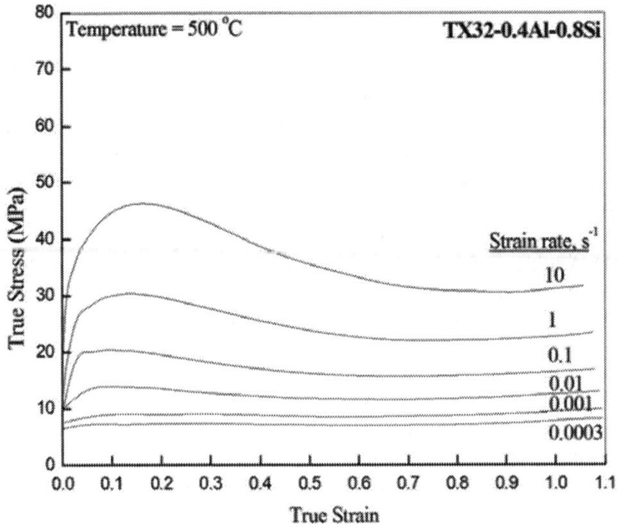

Figure 4. True stress - true strain curves obtained at 500 °C on TX32-0.4Al-0.8Si alloy in compression at different strain rates.

The true stress-true strain curves obtained at the test temperatures of 400 °C and 500 °C are shown in Figure 3 and Figure 4, respectively. At 400 °C and strain rates above 0.1 s^{-1}, the flow stress rapidly increased with increasing strain and reached a peak value before flow softening occurred. At lower strain rates, the flow curves exhibited near steady state deformation. At 500 °C and strain rates lower than 0.01 s^{-1}, the flow curves are of steady-state type from the beginning of deformation. Flow stress decreased significantly with increase in deformation temperature from 400 °C to 500 °C.

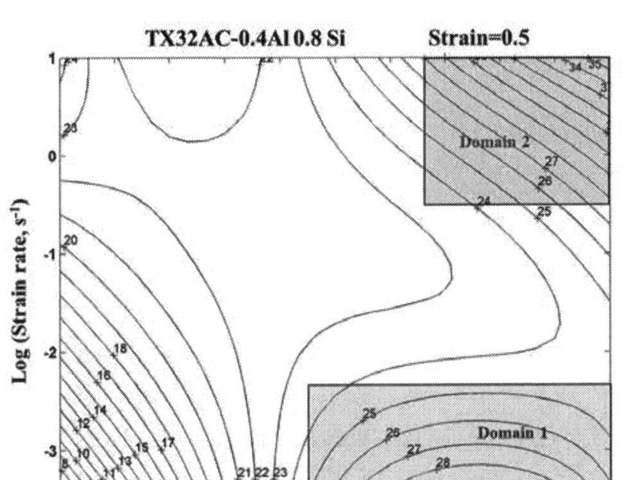

Figure 5. Processing map for TX32-0.4Al-0.8Si alloy. The numbers against the contours represent efficiency of power dissipation in percent.

The processing map obtained at a strain of 0.5 (near steady state flow) is shown in Figure 5. The map exhibits two domains in the temperature and strain rate ranges: (1) 390-500 °C/0.0003-0.005 s^{-1} with a peak efficiency of 30% at 460 °C/0.0003 s^{-1} (Domain 1), and (2) 430-500 °C/0.3-10 s^{-1} with a peak efficiency of 35% at 500 °C/10 s^{-1} (Domain 2). The resultant microstructure of the specimen compressed at 500 °C/10 s^{-1} (Domain 2 peak efficiency condition) is shown in Figure 6 indicate that the alloy undergoes dynamic recrystallization (DRX) during hot working replacing the as-cast microstructure by wrought microstructure.

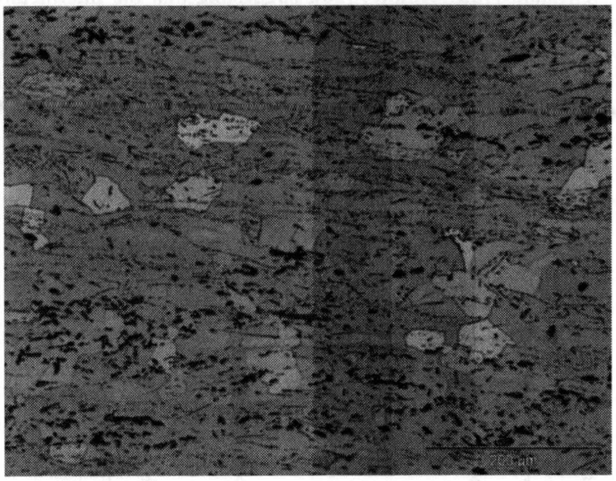

Figure 6. Microstructure of Mg-3Sn-2Ca-0.4Al-0.8Si alloy deformed at 500 °C/10 s^{-1} exhibiting recrystallized grain structure.

It is generally accepted that as-cast alloy exhibit near-random texture. At the start of plastic deformation at lower temperatures (300-350 °C), basal and prismatic slip systems are more likely to operate. When the continued slip on these slip systems causes a rotation of the slip planes towards the normal to the compression axis, favorable slip systems get exhausted and further deformation becomes difficult. This may be the reason for the absence of workability domains at lower temperatures. However, at the deformation temperatures above 400 °C, pyramidal slip systems get activated and contribute to deformation. Also, dynamic restoration processes, like DRX may also occur simultaneously, which help to soften the material. The recovery mechanisms leading to DRX are strain rate dependent; lattice self-diffusion at lower strain rates and grain boundary diffusion at higher strain rates. Thus, while pyramidal slip occurring during DRX cause steady-state flow, two different DRX domains appear in the processing map; one at lower strain rates (0.0003-0.005 s^{-1}) and another at higher strain rates (0.3-10 s^{-1}) at temperature above 430 °C. The stacking fault energy on the second order pyramidal slip systems is reported to be about 173 mJ/m^2 [19] and cross-slip can be the recovery mechanism for nucleating DRX.

Using the standard kinetic rate equation (Eq. 1) to represent the steady-state flow under hot working conditions [20], the activation parameters (stress exponent (n) and apparent activation energy (Q)) are evaluated. A plot of normalized flow stress with strain rate, i.e., ln (σ/μ) vs. ln ($\dot{\varepsilon}$), where μ is shear modulus of Mg is shown in Figure 7. As a single straight line fit over the entire strain rate range is not apparent, the data are separated into two sets of conditions; one for the Domain 1 and another for Domain 2. In the lower strain rate domain (Domain 1), the value of the stress exponent is about 5.88 while it is about 5.68 in the higher strain rate domain (Domain 2).

Arrhenius plot, which relates the normalized flow stress ln (σ/μ) with inverse of temperature, is shown in Figure 8. The data at strain rates of 0.0003 s^{-1} and 0.001 s^{-1} (between 400-500 °C), corresponding to Domain 1 in the processing map, exhibited a linear fit giving an apparent activation energy of 215 kJ/mole. On the other hand, the data at strain rates of 10 and 1 s^{-1}, i.e., in Domain 2 of the processing map, gave a linear fit only at higher temperatures and yielded apparent activation energy of 175 kJ/mole. The apparent activation energy values obtained for both the domains are much higher than that for self-diffusion in pure magnesium (135 kJ/mole) [20], suggesting that the intermetallic particles in the matrix cause significant back stress. Domain 2 which is occurring at higher strain rates is the recommended window for processing of this alloy as these strain rates can be easily obtained in the commercial deformation processes like rolling and extrusion.

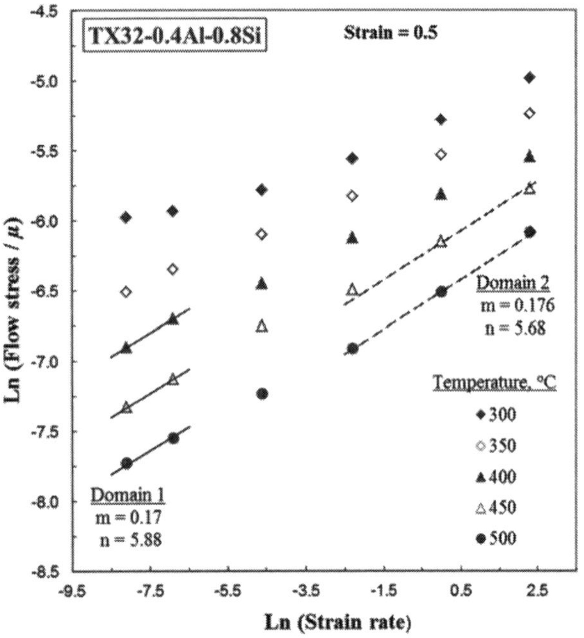

Figure 7. Variation of normalized flow stress values (at a strain of 0.5) with strain rate at different test temperatures for TX32-0.4Al-0.8Si alloy.

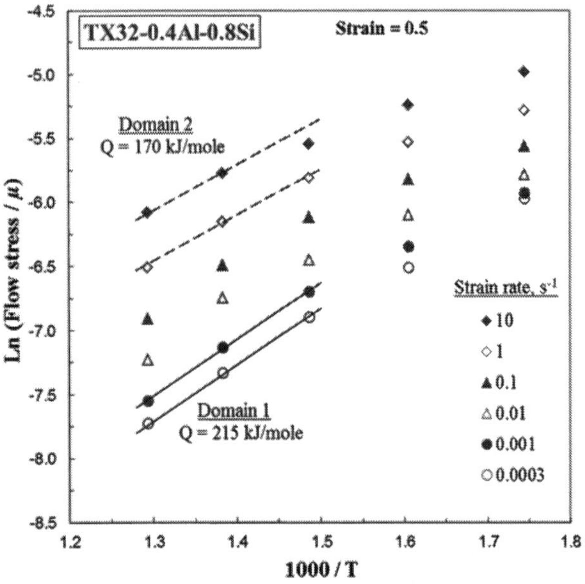

Figure 8. Arrhenius plot showing the variation of normalized flow stress values with inverse of temperature at different strain rates.

Summary and Conclusions

The compressive strength of TX32-0.4Al-0.8Si alloy in as-cast condition in the temperature range 25-250 °C and its hot deformation behavior in the temperature range 300-500 °C and strain rate range 0.0003-10 s^{-1} have been studied using hot compression tests. The flow stress data was used for the kinetic analysis and to develop the processing map for the alloy. The following conclusions are drawn from this study.

(1) Addition of 0.8 wt% Si reduced the compressive strength at all the temperatures.
(2) The stress-strain curves exhibited flow softening after the appearance of a peak stress at high strain rates while near steady-state flow has been observed at lower strain rates.
(3) The processing map exhibited two hot workability domains: (i) 390-500 °C/0.0003-0.005 s^{-1} and (ii) 430-500 °C/0.3-10 s^{-1}, the latter one being suitable for industrial processing.
(4) The apparent activation energy values in the two domains are higher than that for self-diffusion, suggesting the generation of back stress by the intermetallic particles in the matrix.
(5) Dynamic recrystallization is the softening mechanism during hot working of the alloy and is controlled by cross-slip at higher strain rates.

Acknowledgement

This work was supported by a grant (Project #115108) from the Research Grants Council of the Hong Kong Special Administrative Region, China.

References

1. I.J. Polmear, *Metallurgy of the Light Metals*, 3rd edn, Edward Arnold, London, 1995.

2. J.F. Stohr, J.P. Poirier, *Philos. Mag.*, 25 (1972), 1313-1329.

3. F. Kaiser, J. Bohlen, D. Letzig, A. Styczynski, Ch. Hartig, K.U. Kainer, *Mater. Sci. Forum*, 419-422 (2003), 315-320.

4. B.C. Wonsiewicz and W.A. backofen, *Trans. Metall. Soc. AMIE*, 239 (1967), 1422-1431.

5. R.E. Reed-Hill and W.D. Robertson, *Acta Metall.*, 5 (1957), 728-737.

6. S. Suwas, G. Gottstein, R. Kumar, *Mater. Sci. Eng. A*, 471 (2007), 1-14.

7. J.A. Chapman and D.V. Wilson, *J. Inst. Met.*, 91 (1962), 39-40.

8. J. Koike, T. Kobayashi, T. Mukai, H. Watanabe, M. Suzuki, K. Maruyama and K. Higashi, *Acta Mater.*, 51 (7) (2003), 2055-2065.

9. T. Abu Leil, K.P. Rao, N. Hort, Y. Huang, C. Blawert, H. Dieringa and K.U. Kainer, in: *Magnesium Technology 2007*, eds. R.S. Beals, A.A. Luo, N.R. Neelameggham, M.O. Pekguleryuz, TMS (2007), pp.257-262.

10. N. Hort, K.P. Rao, T. Abu Leil, H. Dieringa, Y.V.R.K. Prasad, K.U. Kainer, in: *Magnesium Technology 2008*, eds. M. Pekguleryuz, E. Nyberg, R.S. Beals and N. Neeramegghan, TMS (2008), pp.401-406.

11. K.M. Asl, A. Tari, F. Khomamizadeh, *Mater. Sci. Eng. A*, 523 (2009), 1-6.

12. J.J. Jonas, C.M. Sellars, W.J. McG. Tegart, *Metall. Rev.*, 14 (1969), 1-24.

13. Y.V.R.K. Prasad, T. Seshacharyulu, *Inter. Mater. Rev*, 43 (1998), 243-258.

14. Y.V.R.K. Prasad, S. Sasidhara, *Hot Working Guide: A Compendium of Processing Maps,* Materials Park, Ohio: ASM International; 1997.

15. H. Ziegler, in: I.N. Sneddon, R. Hill (Eds.) *Progress in Solid Mechanics*, Vol. 4, Wiley, New York, (1965), pp. 91-193.

16. Y.V.R.K. Prasad, K.P. Rao. *Mater. Sci. Eng. A*, 374 (2004), 335-341.

17. D.H. kang. S.S. Park, N.J. Kim, *Mater. Sci. Eng. A*, 413-414 (2005), 555-560.

18. M.S. Dargusch, A.L. Bowles, K. Pettersen, P. Bakke, G.L. Dunlop, *Metall. Mater. Trans.*, 35A (2004), 1905-1909.

19. J.R. Morris, J. Schraff, K.M. Ho, D.E. Turner, Y.Y. Ye, M.H. Yoo, *Phil. Mag.*, 76 (1997), 1065-1077.

20. H.J. Frost, M.F. Ashby, *Deformation-mechanism Maps*, (Oxford: Pergamon Press) (1982), p.44.

Magnesium Technology 2013

Primary Production and Shaping

Magnesium Technology 2013
Edited by: Norbert Hort, Suveen N. Mathaudhu, Neale R. Neelameggham, and Martyn Alderman
TMS (The Minerals, Metals & Materials Society), 2013

Doing Projects in a Foreign Language – Communications Management, Issues and Strategies

Deling Xian
Hatch Ltd., 5 Place Ville Marie, Bureau 200, Montreal, Quebec, Canada H3B 2G2

Keywords: China, QSLIC, Communications, Culture

Abstract

The Qinghai Salt Lake Industry Co. Ltd. (QSLIC) smelter project involves a Chinese state-owned client, a Canadian engineering company, their American technology partner, several Chinese design institutes and international vendors. Proper technical communication plays an important role during the development of a project in a foreign language and across cultures. Using the QSLIC project as an example, this paper presents the role of Communications Manager and personal qualifications required, as well as technical communications management, issues and strategies, lessons learnt while doing smelter projects in China and Chinese business culture and etiquette.

Introduction

QSLIC

QSLIC, a Chinese state-owned enterprise, whose business is founded on access to extensive chloride-based mineral resources present in the Qarhan salt lake, located approximately 40 kilometers northeast of Golmud in the province of Qinghai. Already China's largest potash producer, QSLIC recognized the economic potential of using the magnesium chloride brine by-product from the potash process as a raw material for the production of commercial magnesium metal and alloys. Importantly, QSLIC's raw material is a chloride, so that chlorine gas is a by-product from the magnesium production process. This availability of chlorine inspired QSLIC to develop a synergistic integrated flow sheet for an industrial complex, the heart of which would be an electrolytic magnesium smelter.
The proposed industrial complex includes eight process plants as well as a utilities facility and produces saleable products of magnesium and magnesium alloys, polyvinyl chloride and polypropylene, calcium chloride and soda ash, the most economically important of these being the magnesium metal products and polyvinyl chloride (PVC). This industrial complex is known as the QSLIC Phase 3 expansion and it is envisioned that a further expansion (Phase 4) of the entire complex will follow the successful start-up of Phase 3.

Hatch

Hatch is providing engineering, equipment custom design and project implementation for QSLIC. Hatch started to work for QSLIC by performing basic engineering for the proposed 100,000-tpa primary magnesium plant. Afterwards, QSLIC engaged Hatch to continue with detailed process technology design and support.
Furthermore, QSLIC has separately awarded Hatch with larger contracts for the calcium carbide (CaC_2) smelter project providing furnace technology and engineering design associated with adjacent process plants that are part of this same integrated facility.

American technology partner

During the course of the design, Hatch partnered with Carbide Industries LLC (C.I.) of Louisville, Kentucky, USA. Carbide Industries' operation provides the foundation for the scale-up and its experience with a proven operation is essential to the success of the CaC_2 furnace technology.

Design Institutes (D.I.)

Generally speaking, there is one Chinese D.I. in charge of design coordination for each plant of the whole QSLIC industrial complex. In Mg smelter and CaC_2 smelter projects, there are also several Chinese D.I. performing engineering in utilities and other areas.

Vendors

The vendors involved in QSLIC project are worldwide.

The present document describes the role of the communications manager employed by the engineering firm.

Role of Communications Manager

Within Hatch, the Communications Manager reports directly to the Program Director. Broadly, this role is responsible for all technical translation, technical interpreting and communications services for the suite of projects being executed for QSLIC by Hatch. This includes dealing with the client, local design institutes and vendors, which is shown in Figure 1.

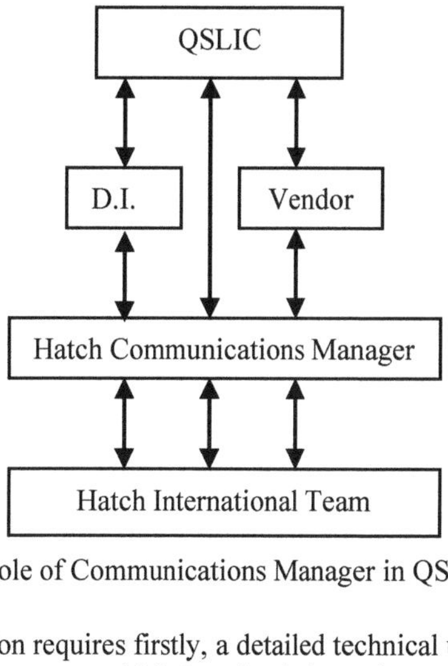

Figure 1 Role of Communications Manager in QSLIC Project

This position requires firstly, a detailed technical understanding of the subject matter which is gained through careful study of the project technical documents (drawings, technical reports, specifications, data sheets, etc.) along with in depth briefings with the design team. The role often requires off-line explanation to external parties about the detail of the technical subject matter and either direct response or coordination of a response to queries that will arise.

Functional management capability is also necessary to develop comprehensive communications procedures and execute them through a careful coordination between Hatch, client, local design institute and vendor organizations.

Finally, interpersonal skills are required to assist, and at times guide, the various project managers and their leadership teams in development of productive and positive relationships with the various external Chinese parties.

Communications Management

The communications manager develops procedures and forms the team, including acting as hiring manager to staff technical translators and external contract translation companies. Additionally, the communications manager establishes training for all project team members in the use of the Communications team on the project. Directing the work of some 4-5 technical translators and 2-3 external translation companies, the communications managers detailed responsibilities include:

- Receive verbal or written communications to/from International Project Team for translation and forwarding to intended recipient(s).
- Manage and coordinate all translation services. Maintain translation log of items to be translated as per assigned priorities.
- Manage and coordinate all interpreting services for meetings and teleconferences via on-phone or live attendance.
- Manage and coordinate all interpreting services for site visits, vendor shop inspections and other such in-country activities.
- Personally provide interpreting services at key client face-to-face meetings in varying locations (Golmud, Xining, Tianjin, Shanghai, Beijing).
- Develop and maintain procedures outlining workflows and quality management steps for:
    - o Translation of messages to/from the International Project Team received verbally or by email
    - o translation of documents
    - o translation of data from the design tools
    - o provision of interpreting services.
- With the Project Systems Manager, specify requirements and coordinate administration of Hatch Execution Technologies used for Communications Services.
- With the Project Systems Manager, identify and implement 3rd party software required for the execution of Communication Services.
- With Hatch Corporate Learning Center, sponsor the development and delivery of Chinese cultural training for international team members.
- Assemble and train a team of in-house translation specialists and interpreters to facilitate requirements in these areas and act as coaches for international team members in matters of Chinese culture and customs.
- Contract external translation services based on evaluation of competencies and negotiation of rates.
- Review and approve external translation services invoices for final approval by the Project Manager.
- Establish budgets and manage man-hours and costs for all Communications services (either provided by internal resources or external service companies).
- Refer matters of a serious nature (especially related to difficult Client concerns and questions) to the Project Manager for agreement on handling.
- Effectively Report on the progress for Communication Services activities. Work with the Project Manager to resolve translation issues that impact on the successful delivery of the project

Communications Issues & Strategies

There are multiple communication issues related to both communications in strictest sense and significant differences in business culture. Table 1 describes these issues and strategies to mitigate their impact.

Table 1 Communications Issues and Strategies in QSLIC Project

Issues	Strategies
Complicated China business culture	• Communications team present Chinese cultural training material to international team members. • Interpreters attend business meetings and provide cultural context to comments and behaviors.
Difficulty in accurate translation of technical documents	• Obtain help from in-house Chinese-speaking engineers in different disciplines worldwide. • Build a Project Dictionary for team use. • Check the translation for every document. • Get Chinese-speaking engineers to review technical document translation.
Large translation workload (reports,	• Develop the Translation Log to keep track of all translation request. • Hire competent and responsible

memos, minutes, correspondence etc.)	translators. • Contract external translation services to help for peak workloads. • Schedules need to consider translation duration.
Various and frequent Requests from client, D.I. and vendors	• Develop Request List and follow up. • Set notification to remind responsible person.
Battery limit with Chinese D.I.	• Responsibility matrix need to be detailed explained at the beginning of the project to avoid possible confusions, and further modified and detailed as project goes on.
Understanding Chinese design standard	• Get Chinese Standard translated and circulated to the project team. • Obtain Chinese Standard Supplement document for further explanation. • Communicate during weekly teleconference and face-to-face meeting with D.I. to obtain local interpretation of codes, norms, rules and regulations. • Meet with China Government Authority to evaluate the design.
Different understanding of engineering level of details	• Understand China engineering standard and definition. • Consult local experienced engineers. • Explain Hatch's traditional project execution Strategy to Client.
Complicated procurement bidding procedure	• Get experienced local procurement support. • Understand Client's procurement procedures.
Different project procedures between Hatch and China Client	• Translate Hatch procedure and explain to the Client. • Understand Client's project approval and decision making procedure.
Confidentiality issue	• Sign Confidentiality Agreement with Client, D.I., review committee, vendors and explain the importance of IP. • Use invited bidding instead of public bidding.

Chinese Business Culture and Etiquette

China is infamously known as a country of etiquette and ceremonies. The unique character of the Chinese is built on a strong sense of pride in their ancient history and culture. Understanding the basic Chinese cultural, ethical and business values is paramount to any organization wanting to conduct business in today's rapidly progressing China [1] .

Guanxi

Thousands of years ago Confucius talked about "reciprocity", Yi, the root of Guanxi. In literal terms, this central concept in Chinese culture means "relationships" or "connections". Guanxi is a network of elaborate relationships promoting trust and cooperation. For centuries, Guanxi was the main way of accomplishing everyday tasks [1]. Establishing a sincere, supportive relationship based on mutual respect is a fundamental aspect of Chinese culture. In the world of business, possessing the right guanxi is crucial to ensure minimum difficulties and frustrations that are often encountered. Chinese business relationships inevitably becomes a social relationship after a while.

A key role of the communications manager is to guide top managers to ensure that they build the correct relationship with their Chinese counterparts.

Mianzi

An important issue that should be considered throughout business interactions with the Chinese is the concept of 'mianzi' or 'face'. Face is a mark of personal pride and forms the basis of an individual's reputation and social status. In Chinese business culture "saving face", "losing face", "giving face" and "repairing face" are vital for successful business. Causing someone to lose face through public humiliation or inappropriate allocation of respect to individuals within the organization can seriously damage business discussions and relationship. On the other hand, praising someone in moderation before their colleagues is a form of 'giving face' and can earn respect, loyalty and aid negotiations [1].

The communications manager should be present during key meetings to provide insights and guide the international team in the appropriate behavior.

Seniority

Seniority is very important to the Chinese especially if you are dealing with a State owned or government body [2]. The importance of seniority goes through all Chinese culture. Instead of addressing the other party as Mr. Wang, it is always appropriate to address the other party by his designation or title, i.e. Wang Director or Wang Engineer. Senior members generally lead the negotiations and will direct the discussion. The appropriate respect must be given according to rank and seniority. For example, when giving out business cards or brochures, make sure the most senior person is started with before moving down the line. If gifts are being bought, buy better gifts for the senior managers instead of buying similar gifts across the board. Similarly, sitting positions in a meeting room or a dining table is accorded to rank, importance and seniority.

The communications manager should provide advice before a first meeting with Chinese business contacts to avoid making the wrong move.

Chinese Meeting Etiquette [3]

Arrive on time. Arriving late is considered rude. If you do arrive late, apologizing for your tardiness is a must.

After handshakes and the exchange of business cards, guests take their seats. Business cards are always exchanged on first meeting a new contact. Cards are held in both hands when exchanging and then introduce in detail. It is recommended to have your card printed in Chinese on the reverse and always offer it Chinese-side up. Treat the card with great respect as the card is the man.

The seating is typically arranged by rank and status. If the meeting is held around a large conference table, all the Chinese delegation may sit on one side of the table and foreigners on the other. This is especially true for formal meetings and negotiations.

Meetings usually begin with small talk to help both sides feel more comfortable. After a few moments of small talk, there is a short welcoming speech from the host (normally from the most senior person) followed by discussion of the meeting's topic.

During any conversation, the Chinese will often nod their heads or make affirmative utterances. These are signals that they are listening to what is being said and understand what is being said. These are not agreements to what is being said.

Do not interrupt during the meeting. Chinese meetings are highly structured and interjecting beyond a quick remark is considered rude. Wait to speak when it's your turn.

Respect should be shown to age, seniority, the history and traditions of China, political sensitivities, the company, the region etc. Stand up when a senior person enters the room, offer the seat of honor and be attentive.

Chinese negotiations are process-oriented and time-consuming. They want to determine if relationships can develop to a stage where both parties are comfortable doing business with the other. Relationship between people means more than the business itself. It is rare that the Chinese will deal with people they don't know or trust. Establishing a proper introduction with your counterparts is vital before entering into business.

When hosting guests from China for a business meeting, set out water, coffee, tea, fruit or other refreshments in the room. Make it seem like you have put effort into making them feel welcome and comfortable.

As a courtesy, when the meeting is over, walk your guests to their vehicles, taxi, or at least to the main door of the office building.

Chinese Dining Etiquette [4]

The seating arrangements at a banquet are very complicated and linked to perceptions of rank and status. If you are invited, normally you will be shown where to sit. However if you are the host it is probably best to get some local advice on the best seating plan if you want to avoid insulting anybody.

The meal is usually coming to an end when the fruit is served and the hot towels given out.

Alcohol will in invariably be consumed in quite large quantities during a formal banquet — mainly either beer or local clear alcohols which can be very strong. Drinking alcohol is very important to develop Guanxi. However, you may use a medical reason to decline.

Eat well to demonstrate that you are enjoying the food and appreciate the hospitality of the host.

Make toasts to your guests, by standing up, turning to your guest, and saying a few sentences that will honor him/her (ex. I am honored to be working with you, and look forward to many more successful projects together). When making a toast, hold lower your cup against the more senior guests' to show respect. Toast more to the most important guests.

Dress Codes

Appearance is important within Chinese business circles. Successful people are expected to look successful. Wealth is admired, so wear good quality clothes, watches etc.

In China, many men wear suits and ties and women tend to wear skirts and blouses of a modest cut. It is advisable to have business attire with you when visiting.

Be aware of the variety of the Chinese climate, which varies from sub-tropical to freezing. Make sure you check the weather conditions before you travel and check the climatic zone of the city you are visiting.

Gift Giving

Gift giving is an everyday part of Chinese business culture. Giving and receiving gifts helps to cement Guanxi.

Gifts are rarely opened in front of the giver.

Gifts could be foreign cigarettes, cognac, fine whisky, quality wine etc. Do not give clocks, cut flowers, or white objects; these objects carry association with death. Red and gold are the best wrapping paper.

Key Learnings

Here are some Chinese Project Paradigms from our Program Director:

1. Relationships make things work
2. Focus on outcomes not specific procedures or forecasts
3. Control behaviors not numbers
4. Expect role and scope changes as project matures
5. Patience and perseverance in all communications
6. Maintain a long term vision (build long-term goals and objectives in your proposal)

Tips for Working with Chinese Clients

Here are some communication tips for working with Chinese Clients:

- You have no status until you have earned it
- Hard work and persistence is highly valued
- Never lose your temper, always be in control
- Never refute outright what another says (Note: as they are considered impolite. Instead of saying "no", answer "maybe" or "I'll think about it." It's important that do not make people "lose face" in public)
- It is difficult for the Chinese to say "no" directly. Anything other than a direct "yes" could mean 'no". Be circumspect and reflect on seeming agreements reached. Has an agreement actually been reached?
- Expect to get back exactly what you give
- Respect status, traditions and cultural sensitivities.
- Group/team always more important than the individual
- When working with new clients and new cultures, no work should be started without formal approval from the client. And be sure to know what is "formal approval" for the client.
- Learning a little Chinese goes a long way. While it's not necessary to master Chinese, learning to say a few phrases will help break the ice.
- Do not assume comprehension. It is often useful to go over the same point several times from different angles in order to aid comprehension.

Tips for Working With Interpreters and Translators

The use of translation and interpretation are common services used by project in a foreign language. They are essential in order to adequately communicate and relay information with our client, other parties within China, and between teams in Canada.

- When shifting topics of conversation, announce what topic you are now discussing to avoid confusion.
- To the greatest extent, provide background data or information to the interpreter prior to your meeting. Discuss the goals and proposed outcomes from the meeting. This will raise efficiency and save time.
- When speaking through an interpreter, pause after one point is completed, thus allowing the interpreter to convey your meaning as clearly, and orderly as possible. Do not try to convey too much information at once.
- Try to provide short but full sentences when addressing complicated technical terms, thus enabling the translators to follow.
- At meetings and teleconferences, do not to speak to the interpreter at the same time when the client is speaking to him or her. Allow a pause to ensure that they are ready to proceed.
- Leave enough time for translators (normally 1 - 2 hours for one A4 page) when planning your documentation preparation and issue schedule. Allow enough allow time to get documents translated.

Conclusions

Communications play a very important role in international project development. In a broader sense, the communications manager need to deliver accurate information to each party involved in an efficient way. The communications manager requires strong technical background knowledge and general local business culture and etiquette and interpersonal skills.

References

[1] Doing business in China, www.communicaid.com

[2] Business Culture and Etiquette in China, http://www.kwintessential.co.uk/

[3] Chinese meetings, http://www.worldbusinessculture.com/

[4] Successful Entertaining in China, http://www.worldbusinessculture.com/

Magnesium Technology 2013
Edited by: Norbert Hort, Suveen N. Mathaudhu, Neale R. Neelameggham, and Martyn Alderman
TMS (The Minerals, Metals & Materials Society), 2013

IMPACT OF SITE ELEVATION ON MG SMELTER DESIGN

Phillip W. Baker B.E., M.Eng.Sc.
Hatch Ltd. 61 Petrie Tce., Brisbane, QLD, Australia

Keywords: magnesium, smelting, engineering design

Abstract

Site elevation has many surprising and significant impacts on the engineering design of metallurgical plant of all types. Electrolytic magnesium smelters maybe built at high elevation for a variety of reasons including availability of raw material, energy or electric power. Because of the unit processes they typically involve, Mg smelters can be extensively impacted by site elevation. In this paper, generic examples of the design changes required to adapt a smelter originally designed for sea level to operate at 2700 m are presented. While the examples are drawn from a magnesium plant design case, these changes are generically applicable to all industrial plants utilizing similar unit processes irrespective of product.

Introduction

The earth's atmosphere is made up of a mixture of gases nominally 78% nitrogen, 21% oxygen, 0.9% Ar and 0.04% CO_2 with the remainder comprised of fractional percentages of some 12 gas species. The atmosphere of also contains a variable amount of water vapor ranging from 1-4 % depending on altitude, location and weather conditions.

Although there is no definite position that delineates the boundary of the earth's atmosphere it is commonly regarded as being 100 km above the sea level. However, the atmosphere is highly stratified with more than 80% its total mass within the lowest stratum called the Troposphere. The troposphere extends some 10 km above sea level and its properties and behavior are the only ones considered relevant to terrestrial engineering design procedures.

Within the troposphere the decrease in pressure with increasing elevation is approximately linear, a phenomenon that is emulated by temperature.

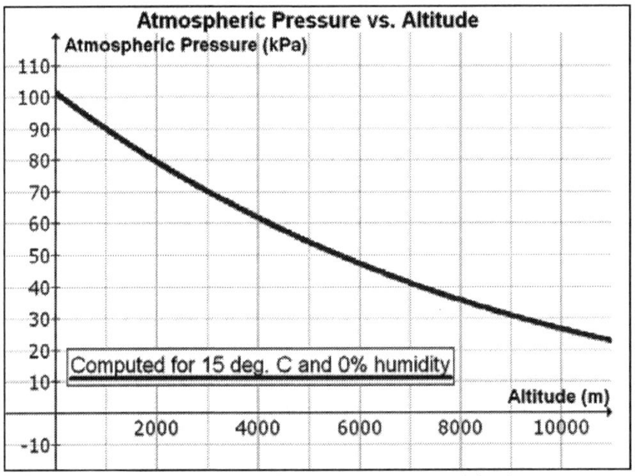

Figure 1. Atmospheric pressure Vs Altitude

So for conventional engineering design purposes the atmosphere may be considered as a column air 10 km high in which the pressure, temperature and density of the gas vary almost linearly with elevation.

Within the range of site elevation changes encountered in engineering projects, there can be significant impact on the design of engineering processes that involve any of the following;

- Phase changes
- Heat and/or mass transfer in gas phase
- System pressure drop
- System volumetric flows
- High voltage power

Magnesium Smelter Flowsheet

Unlike aluminum smelters the Flowsheet of a magnesium smelter is often unique to the feedstock and site conditions. Furthermore, magnesium smelter flowsheets are not all based on the same core technology and can represent unique technical solutions. Nevertheless the unit processes employed are often generic even though their configuration and detail design may be unique. We will consider the flowsheet shown below.

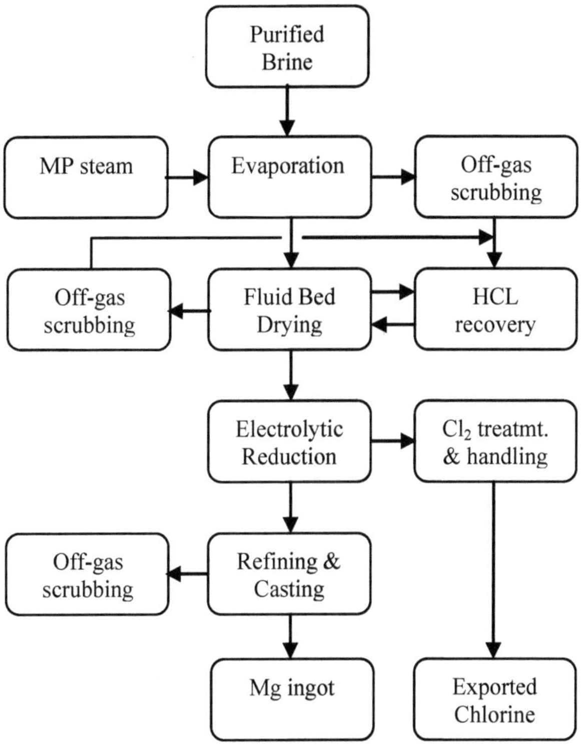

Figure 2. Process Block Diagram

The flowsheet described by the block flow diagram above involves the continuous evaporation and concentration of $MgCl_2$ brine followed by cooling and transformation into a solid hydrate intermediate. This intermediate product is then dehydrated in two stages of fluidized beds with the final stage being performed in an environment of HCL gas. The HCL lost to scrubbers and bled from the system must be recovered purified and returned to the process. All off-gases must be cleaned of contaminants before release to the environment.

The anhydrous magnesium chloride solid is used as feedstock for the electrolytic reduction process which produces liquid magnesium and chlorine gas. Melt refining and casting processes also require off-gas handling and treatment for environmental reasons.

The integrated chemical process includes a wide range of smaller chemical process equipment including;

- Pumps
- Fans, blowers and compressors
- Heat exchangers

Brine Evaporation Process

Industrial Evaporators are designed in a variety of forms and configurations but all depend on the boiling behavior of the process media.

The objective of evaporation unit process in this flowsheet is to a produce hot liquid at a critical concentration (refer to Cc in Figure 3) which when cooled will transform into the desired hydrate phase. It is essential for plant operation that after the required concentration has been reached no premature precipitation of solid will occur in the pipe work leading to the cooling process. This requires an acceptable range of temperature between boiling and solid transformation in which the concentrate can be maintained as a stable liquid.

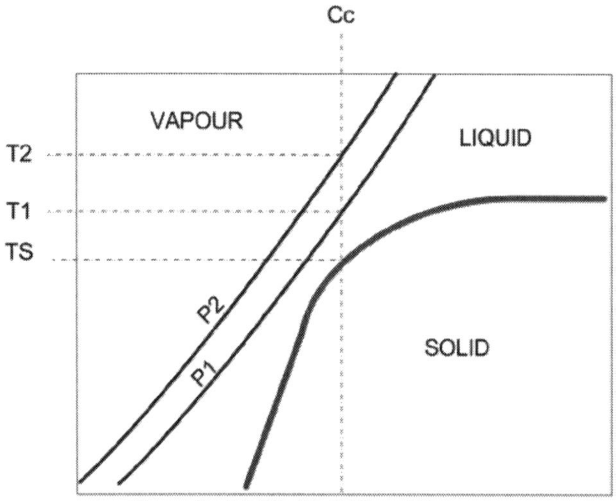

Figure 3. $MgCl_2$-H_2O Phase Diagram

P2 and P1 shown in Figure 3 above represent the Liq-Vap transition line at sea level and site elevations respectively. At sea level this range is wider, T2-TS Vs T1-TS, because the liq-sol transition temperature is unaffected by pressure but the boiling temperature is reduced at the increased elevation.

The question for the process engineer is can the temperature be reliably controlled in the narrower temperature range or should the process be operated at elevated gauge pressure simulating sea level conditions. Lower temperature operation offers additional energy savings but risks operational robustness unless reliable control is possible.

Fluidized Bed Drying Process

The objective of the fluidized bed drying stages is to remove the water of hydration to produce solid anhydrous MgCl2 free of oxide and hydroxychloride. To accomplish this, a substantial quantity of water must be transferred to the fluidizing gas and substantial energy must be transferred in order to raise the temperature of the feedstock to the reaction temperature. Both heat and mass transfer operations are dependent on the density of the fluidizing gases, air and HCL, and are therefore directly impacted by the site elevation.

Maintaining equivalent gas mass flow rates at an elevation of 2700 m means volumetric flow rates are increased by a factor of 1.40. Without increasing vessel diameter the increased gas velocities will change the fluidizing behavior of the bed and unacceptably increase elutriation losses which then increase the contamination of the scrubber liquor.

In this case the design engineer must choose between increasing bed diameter and adding additional units to meet the throughput and capacity requirements. In either option, sensitive process parameters inside the beds such as temperature profile, residence time and fluidization mode must be adhered to or problems such as feedstock accretion, melting and oxidation can occur.

Compressors, Blowers and Fans

The process flowsheet under discussion is supported by many machines that provide the motive force to move gases at the correct flow rates and pressures. High elevation projects require a reevaluation of all of these machines.

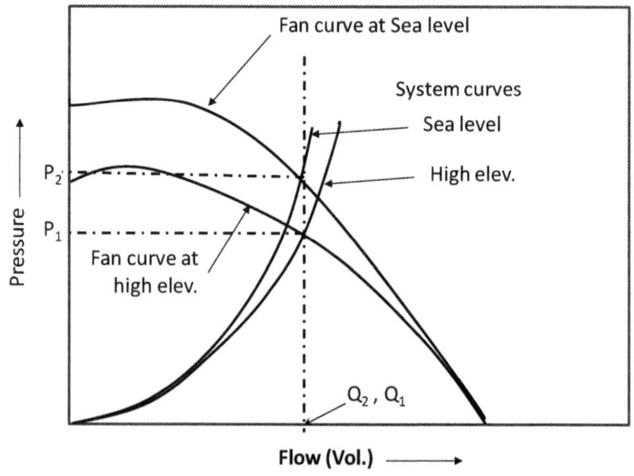

Figure 4. Change of Gas Density

Figure 4 shows that if only the gas density is changed then a fan operating at sea level will deliver the same volumetric flow rate in the same system as at high elevation but at lower pressure. That the flow rate is unchanged is to be expected as fans/blowers are volumetric machines but less intuitive is that head generated is proportional to fluid density.

Consider the case of the fluid bed dryer: the pressure drop across a fluidized bed at minimum fluidization velocity is equal to the weight of the bed and is therefore independent of air density. Therefore to fluidize the same bed at high elevation requires the same pressure drop but from Figure 4 we can see that $P_1 < P_2$. However, if we must also maintain the velocity as constant then the fan cannot simply operate at higher speed but instead should use a larger diameter impeller to produce more pressure at the same speed but with narrower vanes displacing the same volume of gas with each rotation as the originally specified fan. Other solutions can be formulated using dampers but none are energy efficient.

An additional concern is moving the operating point of the fan towards its region of instability where the flow rate is sensitive to small head change can lead to "hunting" (varying flow rate).

In situations where mass flow of gas must be maintained such as in heat transfer applications, higher volumetric flows must be produced. This cannot always be accomplished by increasing the number of fans. Consider the case of heat extraction from the process buildings where a higher volumetric flow will require a higher pressure drop through the building because pressure drop varies linearly with gas density but according to the square of the velocity as follows;

$$\Delta p = k.\rho.v^2$$

For constant mass flow at different gas densities:

$$\rho_1 . v_1 = \rho_2 . v_2$$

$$v_1 = \left(\frac{\rho_2 . v_2}{\rho_1}\right)$$

$$\frac{\Delta p_1}{\Delta p_2} = \left(\frac{k.\rho_1.(v_1)^2}{k.\rho_2.(v_2)^2}\right)$$

Substituting for v_1;

$$\frac{\Delta p_1}{\Delta p_2} = \left(\frac{\rho_2}{\rho_1}\right) = \left(\frac{v_1}{v_2}\right)$$

The pressure drop required to maintain the same mass flow through the same building is increased in proportion to the increase in volumetric flow rate. Hence cooling fans must support both a higher flow rate and pressure to maintain cooling effectiveness at higher elevations. In the case of sea level versus 2700 m, a 40% increase in both volumetric flow rate and pressure drop can be expected where no changes are made to the flow path.

Cooling by Natural Convection

Natural convection is important in heat loss even where forced convection is superimposed the local impact of buoyancy is substantial. In smelters heat loss to the air by natural convection is important to much of the process equipment including fluidized bed, reduction cells, high amperage bus bar and refining furnaces. The impact on the bus bar has been studied in more detail using CFD modeling under these conditions and is presented in another Hatch authored paper at TMS 2013.

Air behaves as an ideal gas; its density varies linearly with absolute pressure and inversely with absolute temperature. Thermal buoyancy forces drive velocities and are a consequence of the different masses of unit volumes of gas. Lower density air at high elevations means reduced driving force, less velocity and so reducing heat transfer. Furthermore the mass flow rate of cooling air past the hot surface which is already decreased due to lower velocity is further reduced by its lower density resulting higher film temperatures as the air rises past the hot surface.

More precisely, the Prandtl number (Pr) is relatively unaffected by the change in barometric pressure, since specific heat, dynamic viscosity and thermal conductivity are nearly independent from pressure. Air density, however, is not, and the drop in barometric pressure translates into a considerable impact on buoyancy. Since natural convection depends on Grashof number (Gr), the impact of density on heat transfer coefficients is significant. Refer to the calculated heat transfer coefficients in the table below for sea level and smelter site elevation.

		72.4 kPa	101.3 kPa
Prandtl, *Pr*, [-]		0.7160	0.7162
Grashof, *Gr*, [-]		1.239E+10	2.424E+10
Raleigh, *Ra*, [-]		8.871E+09	1.736E+10
Nusselt *f* [-]	Single vertical surface	207.0	258.9
	Horizontal, facing down	82.86	98.01
	Horizontal, facing up	217.4	271.9
h_c [W/m².K]	Single vertical surface	4.516	5.649
	Horizontal, facing down	1.808	2.138
	Horizontal, facing up	4.741	5.932

Depending on surface orientation, h_c is calculated to be 15% to 20% lower at the proposed smelter site altitude than at sea level. This conclusion is supported by other work in Avionics design. [1]

For critical facilities within the smelter Hatch has carried out CFD modeling of ventilation designs where forced convection design supports natural convection to ensure adequate cooling performance will be achieved.

Pumps

Pumps in process plants are designed to move a wide range of liquids across a range of temperatures. They must be specified and installed in such a way that Cavitation is avoided. Cavitation is the result of the fluid being exposed to pressures lower than its vaporization pressure and if prevalent can rapidly damage pump impellers and housings and other system components.

High site elevation reduces ambient pressure moves all fluids closer to their vaporization point for given process temperatures. The design engineer must consider the location of the pump and its design to avoid

In this application many circumstances existed where hot and often corrosive liquids must be pumped. In all such cases proper determination of net positive suction head (NPSH) was required to ensure that Cavitation corrosion does not occur and allow the leakage of hazardous and corrosive fluids.

Vacuum Transfer

To extract the liquid metal from the electrolytic cells a vacuum line is connected to a mobile ladle equipped with a siphon tube. The tube is lowered into the cell and the vacuum applied to lift the metal and electrolyte into the ladle. The difference in pressure between the ladle under vacuum and the ambient pressure provides the required driving force to accomplish this transfer.

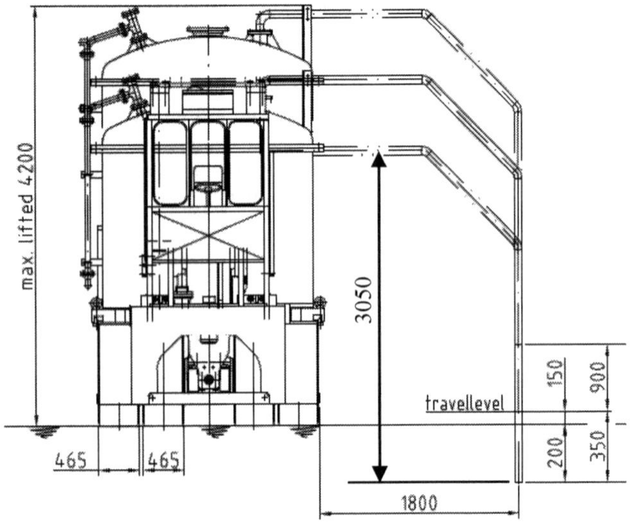

Figure 5. Tapping Geometry

Commercially available tapping trucks have a maximum static head height of more than 3 m requiring a pressure differential of;

- 46.0 kPa for liquid Mg ($\varrho_{Mg} = 1.60\ g/m^3$)
- 53.5 kPa for electrolyte ($\varrho_{elec.} = 1.82\ g/m^3$)

Because salt is always drawn through the siphon tube first, the higher value must be used for design. Dynamic head loss required to cause flow is not included in the above and adds a further Δp to provide for rapid lifting of the salt/metal to prevent freezing within the siphon pipe.

Simple vacuum systems based on single stage air ejectors that are commonly used to generate vacuum for cell tapping cannot reliably produce absolute pressures lower than 20 kPa. The available differential pressure at a site elevation of 2700 m with a single stage ejector will be 52.4 kPa (72.4 – 20), insufficient overcome the static head required to lift salt from the cell into the vacuum ladle and borderline for liquid magnesium even without accounting for dynamic head loss requirements. Consequently design engineers must select higher performance vacuum systems able to generate higher vacuums unless radical changes are made to the design of the tapping trucks.

HCL Recovery

As shown on Figure 2, the HCL returned from the drying and scrubbing processes is recovered and regenerated as dry HCl gas before being returned to the process. The recovery of HCl from aqueous solution involves distillation and breaking of the azeotrope to recover HCl from solutions less than 20% HCl. Technologies available to do this include vacuum and pressure swing technology as well as extractive rectification by the addition of a reboiler circuit of a strongly hydrophilic salt such as $CaCl_2$.

The fundamental processes of HCl recovery include distillation and reboiling are both impacted by the change in elevation from sea level to 2700 m. Additionally, the change to the volumetric flows of gaseous and vapor phases must be accounted for in the design and specification of the process equipment.

Scrubbing and Off-Gas Systems

The process flow sheet includes single and multi-stage scrubber systems designed to handle the off-gas products emanating for the various process areas. The off-gas ductwork systems, fans/blowers, baghouses and the scrubber units themselves are impacted in similar ways to other process facilities that involve significant gas flows and must be designed specially to suit the lower atmospheric pressure at the high site elevation.

High Voltage Facilities

High voltage power is delivered to the Air Insulated Switchgear (AIS) at 330 kV. The dielectric rating of air varies with the density of the air such that lower air density means a reduced electrical insulation effect as follows;

$$BIL_{site} = BIL_{sea\ level} / f$$

Where:

BIL standard lightning impulse withstand voltage
f altitude correction factor

At 2700 m elevation $f = 0.8$ so that;

$$BIL_{site} = \frac{1175}{0.8} = 1468\ kV$$

This increased design potential impacts the spacing of the HV equipment such as the transformer rectifiers must be more widely spaced at high elevations that at sea level. This increase the footprint of the switchgear facilities overall and impacts the layout and design of DC bus bar from the rectifiers to the potroom.

Conclusion

The location of smelters and indeed industrial facilities in general, at high elevations has far reaching consequences for the design of those facilities. These different atmospheric conditions impact all processes involving gas flows used in mass and heat transfer, phase changes, liquid pumping, off gas handling, vacuum transfer and high voltage facilities design. These effects require detailed consideration by the design engineers and the suppliers of equipment destined for high elevation process plants.

Reference

1. Doron Bar-Shalom, *Altitude Effects on Heat Transfer Processes in Aircraft Electronic Equipment Cooling,* (M.Sc. thesis, MIT, Dept of Aeronautics, Feb. 1989)

Magnesium Technology 2013
Edited by: Norbert Hort, Suveen N. Mathaudhu, Neale R. Neelameggham, and Martyn Alderman
TMS (The Minerals, Metals & Materials Society), 2013

Purification of Highly Contaminated Magnesium Melt

Byoung-Gi Moon[1], Bong-Sun You[1], and Ki-Ho Koh[2]
[1]Korea Institute of Materials Science, 797 Changwondaero, Changwon, Gyeongnam, 642-831, Korea
[2]Inje University, 607 Obang-dong, Gimhae, Gyeongnam, 621-749, Korea

Keywords: Magnesium, Purification, Contamination, Polymer, Recycle

Abstract

The steering wheel core is chosen as the first target for the development of a purification technology for highly contaminated magnesium melt, because it contains abundant foreign matter such as polyurethane, copper electrodes, and steel inserts, which have high potential to form non-metallic inclusions and to deteriorate the corrosion resistance of recycled alloys. Various melt treatment technologies have been investigated for refining AM50 magnesium alloy contaminated with polyurethane. The NMI content in magnesium alloy scrap contaminated with polyurethane was effectively reduced by a sequential refining process consisting of filtration, fluxing, and gas bubbling treatments. The filtration step reduced most large inclusions such as carbon residues from the decomposed polyurethane. The subsequent fluxing and gas bubbling treatments effectively removed the small inclusions such as carbonates and oxides.

Introduction

The application of magnesium in the automotive industry has been expanded by the growing necessity of automotive weight restrictions and reduction [1-4]. For example, individual automotive components such as steering wheels, gear boxes, instrument panels, seat components, wheel rims, rear flaps, reflectors, door frames, oil pans, and crank cases have been made of magnesium alloys. The European directive on end-of-life vehicles demands that by 2015 95% of the total car weight must be reused and recovered [5]. The rates specified for reuse and recycling are 85%, excluding energetic recovery. This means that by 2015 85% of the parts from a car of average weight must be "used for the same purpose for which they were conceived" or that they must be "reprocessed in a production process of the waste materials for the original purpose or for other purposes" [5-6]. The Korean Government also launched a directive on end-of-life vehicles according to which 95% of the vehicle's weight will have to be reused and recycled by the beginning of 2015, keeping pace with the EU's legislation. Therefore, with the increased use of magnesium for automotive components, the development of more effective magnesium recycling technology for all types of scrap becomes more important. Thus far, magnesium recycling for magnesium alloy production is done only for clean scrap, which is also referred as Class-1 scrap, consisting of sprues and runners from magnesium die casters. Most end-of-life scrap has been used for desulphurization of steel and the alloying of aluminum [2,7]. There is significant motivation to investigate recycling technology of end-of-life magnesium scrap.

Among the automotive components using magnesium alloys in the Korean automotive industry, the steering wheel has the longest history of magnesium alloy use. The steering wheel contains abundant foreign matter such as a polyurethane cover, copper electrodes, and steel inserts, which have high potential to form non-metallic inclusions (NMIs) and to deteriorate the corrosion resistance of recycled alloys. In this research, a purification technology for highly contaminated magnesium melt was developed. Various melt treatment technologies for refining AM50 magnesium alloy scrap contaminated with polyurethane were investigated.

Experimental procedures

By analyzing steering wheel assemblies collected from various ELVs, the mass percentages of foreign matters in AM50 magnesium alloy frame were determined as 28.7%, 10.8%, and 0.5% for the polyurethane (PU) cover, steel hub, and copper electrode, respectively. The mass percentage of the PU cover in the U-shaped groove (Fig. 1) relative to the whole magnesium alloy frame reached 7.5%. Because of the existence of adhesive bonding between the magnesium alloy frame and the PU cover as well as the structural joining by a U-shaped groove, it is difficult to separate the PU cover from the magnesium frame by the crusher. To simulate the refining process with similar conditions to the actual situation, the mixing ratios of PU to the magnesium alloy were determined as 1, 4, 7, and 10%.

In order to estimate the amount of thermally reduced PU and to eliminate ignition risk, the mass reduction of PU by heating was measured using a thermal gravity analyzer (TA Instrument SDT2960). 2kg of a mixture of AM50 magnesium alloy scraps and a predetermined weight of PU scraps was melted in a crucible electric resistance furnace under the protection of a mixed gas atmosphere of SF_6 (50 cc/min) and CO_2 (500 cc/min). Before melting the metal, slow heating between 300°C and 550°C was carried out until fume discharging was completed. The furnace was then heated to 730°C followed by various melt treatments as shown in Table 1: (1) filtering with a stainless steel mesh (D_{wire}=0.45 mm, hole=2mm x 2 mm); (2) flux treatment with a

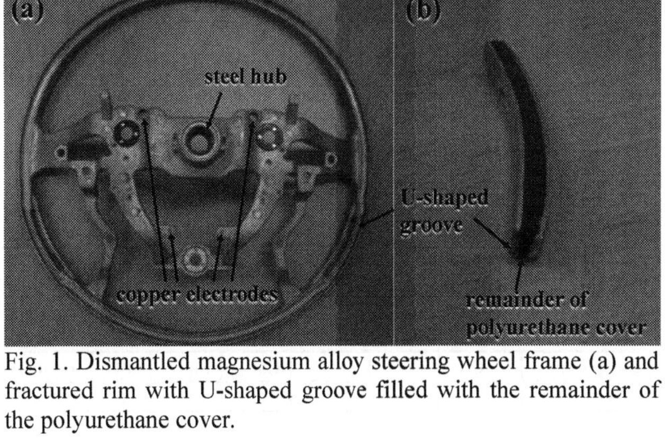

Fig. 1. Dismantled magnesium alloy steering wheel frame (a) and fractured rim with U-shaped groove filled with the remainder of the polyurethane cover.

commercial flux RJ-5 (amount of flux=1% of metal weight, rotational speed of impeller=120 rpm, stirring time=30 min, holding time=30 min) as shown in Table 2; and (c) gas bubbling with 99.9% argon (rotational speed of impeller=120 rpm, gas flow rate=0.5 and 1.0 liter/min, stirring time=30 min, holding time=30 min).

After each melt treatment, a small quantity of melt was poured with a ladle into a coupon-shaped steel mold of 50 mm diameter and 15 mm height. The shapes and compositions of NMIs were analyzed using scanning electron microscopy and energy dispersive X-ray spectroscopy (JSM-6610 LV), and the area fraction of NMIs was evaluated using image analyzing software (i-Solution DT) on optical micrographs sequentially obtained 150 times with 100 magnification from each polished specimen. In order to verify the NMI content, analyses by an optical emission spectroscopic (OES) method and by a particle analyzer using Oxford INCAFeature software in conjunction with scanning electron microscopy (JSM-6480 LV) were also performed.

Results and discussion

Because the PU is a thermoset polymer, the PU covering the magnesium alloy frame cannot be removed by melting. The PU mixed with magnesium scraps consequently causes serious melt contamination during the refining process. Fig. 2 shows the mass changes of the PU collected from the steering wheel scrap by heating at a rate of 10°C/min and isothermally holding at 300°C and 400°C, respectively. The decomposition of PU began and finished at 267°C and 365°C, respectively, and the weight of remaining material decreased to 20% of the initial mass at 500°C. However, slow heating of the mixed scrap is necessary, as fast heating can cause ignition of the magnesium. By isothermal holding at 300°C, the weight of the PU remainder decreased to about 30% of the initial mass after holding for 50 min, and no

Table 1. Processing conditions used for remelting and refining AM50 magnesium alloy scrap containing polyurethane.

Sample code	PU addition (%)	Filtration	Flux addition (%)	Gas flow rate (ℓ/min)
P1-R	1	×	0	0
P1-F	1	○	0	0
P1-FX	1	○	1	0
P1-FX-B0.5	1	○	1	0.5
P1-FX-B1.0	1	○	1	1.0
P4-R	4	×	0	0
P4-F	4	○	0	0
P7-R	7	×	0	0
P7-F	7	○	0	0
P7-FX	7	○	1	0
P7-FX-B0.5	7	○	1	0.5
P7-FX-B1.0	7	○	1	1.0
P10-R	10	×	0	0
P10-F	10	○	0	0

Table 2. The chemical composition of the RJ-5 flux.

Component	$MgCl_2$	KCl	$BaCl_2$	CaF_2	MgO	H_2O	Insoluble
Amount (%)	30±3	26±3	22±3	22±3	<1.5	0.5	<1.5

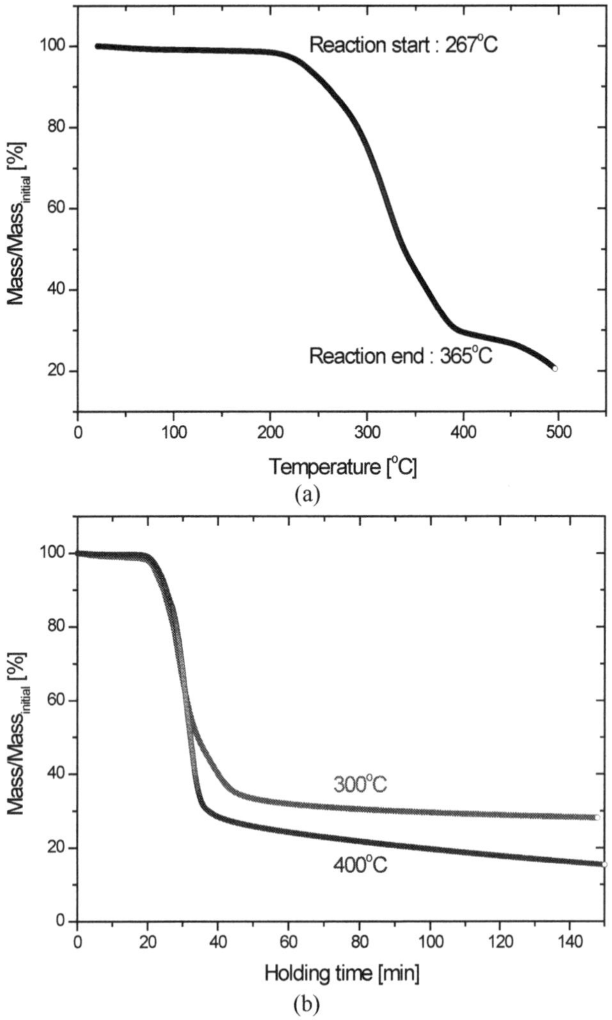

Fig. 2. The mass change of polyurethane by heating at a rate of 10°C/min (a) and isothermal holding at 300 and 400°C (b).

Fig. 3. The remaining tar and char on the surface of magnesium alloy scraps during decomposition of polyurethane scraps by heating at 400°C in air.

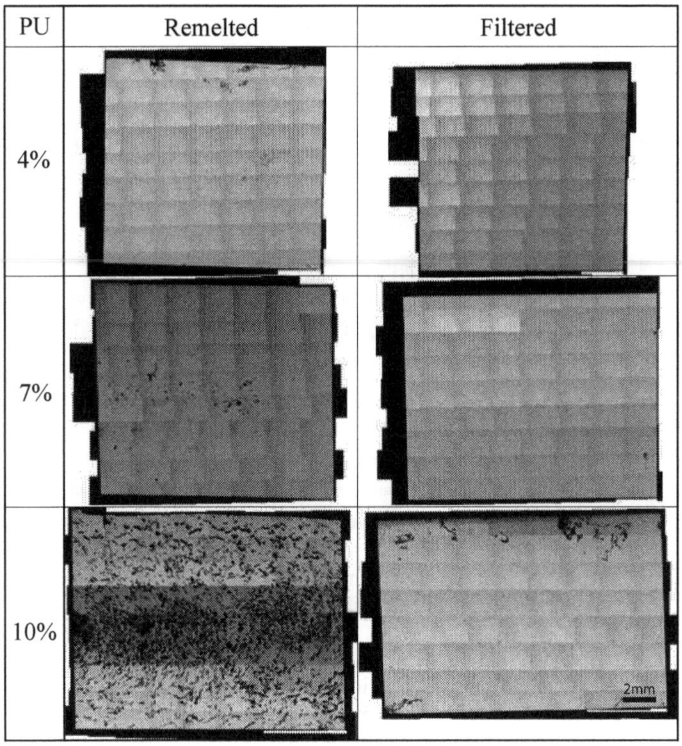

Fig. 4. Optical microscopic images of remelted and filtered ingots made of AM50 magnesium alloy scraps initially containing 4, 7, and 10% of polyurethane scraps.

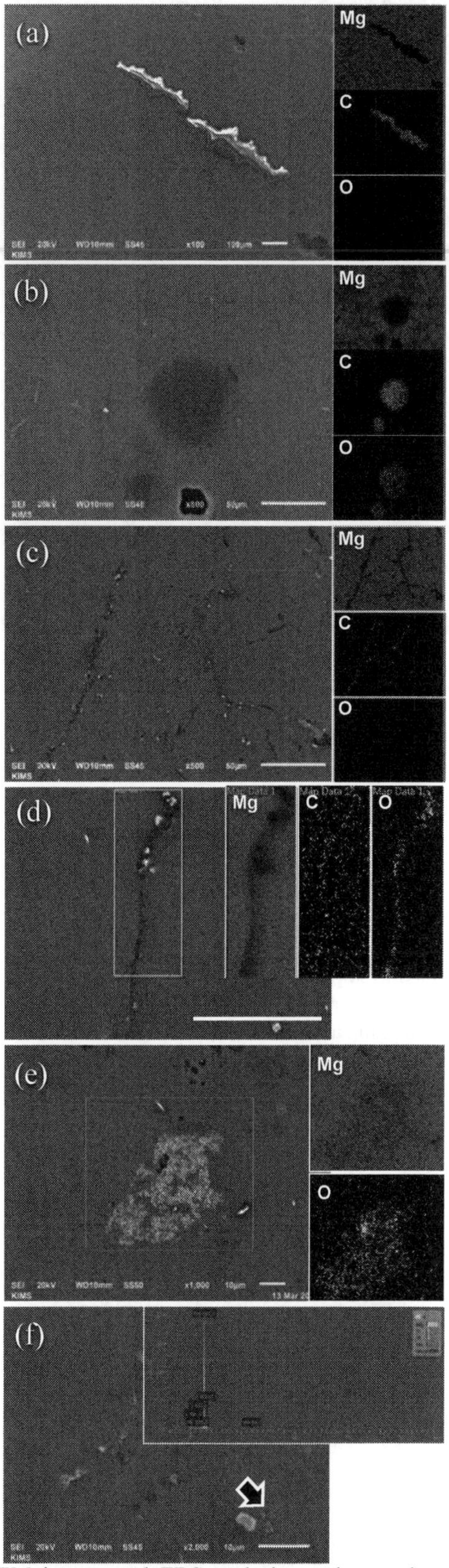

Fig. 5. SEM images and EDS analysis results on the second phases in remelted ingot of AM50 magnesium alloy scrap initially containing 7% of polyurethane scrap representing carbon residues (a), carbonate inclusions (b), fluoride films (c), oxide films (d), oxide clusters (e), and oxide particles (f).

further decrease was achieved. On the other hand, the weight decrease of the PU remainder to 20% of the initial mass after holding for 60 min was achieved by isothermal holding at 400°C. Therefore, preheating at 400°C for more than 60 min is suitable for decomposing the PU before melting the magnesium alloy. Fig. 3 shows the mixed scrap during heating at 400°C for decomposing the PU. Tar and char, which are residues of PU, cover the surface of the magnesium alloy scrap. They discharge a huge amount of fume which may contain harmful gasses such as dioxin, NO_x, HCN, and NH_3 [8]. Most of the remaining material was mixed with the magnesium melt during the remelting process, and caused serious contamination of solidified ingots.

Fig. 4 shows the microstructure of remelted and filtered ingots made of AM50 magnesium alloy scraps initially containing 4, 7, and 10% of PU scraps. The filtered ingots show large differences from the remelted ingots. Most large inclusions were removed from the ingot initially containing 4% of PU scrap by filtration, whereas the ingots initially containing more than 7% of PU scrap had a considerable amount of large inclusions. In order to classify the NMIs, SEM and EDS analyses were carried out. Fig. 5 shows SEM images and EDS results for the NMIs in remelted ingot of AM50 magnesium alloy scrap containing 7% of PU scrap (P7). The large inclusions as shown in Fig. 5 (a) were mainly composed of carbon, which originates from the remaining char of PU. Some carbon inclusions were found as carbonate, as shown in Fig. 5 (b). Also, typical inclusions found in magnesium alloy such as fluoride and oxide films, oxide clusters, and small oxide particles were detected as shown in Fig. 5 (c), (d), (e) and (f), respectively. From these results, it was concluded that the remaining tar and char after decomposition of PU were not eliminated from the magnesium melt by heating up to 730°C, and they were present in

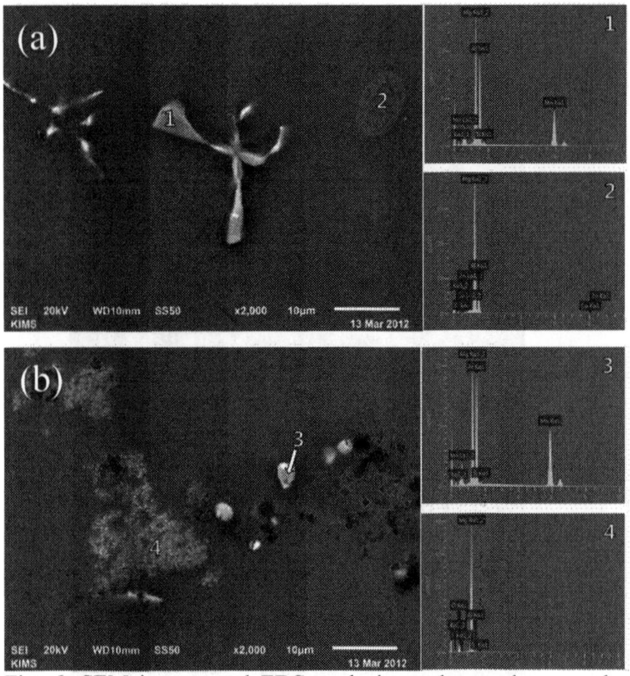

Fig. 6. SEM images and EDS analysis results on the second phases in filtered ingot of AM50 magnesium alloy scrap initially containing 7% of polyurethane scrap: (a) Al-Mn intermetallics and (b) oxide inclusions in the forms of cluster and particle.

the forms of carbon and carbonate inclusions in the solidified ingot. Fig. 6 shows SEM images and EDS results for the NMIs of filtered ingots of the same scrap (P7-F). By means of filtration of the contaminated melt, the amount of large inclusions was considerably reduced. Most of the inclusions after filtering were oxides in the forms of lumps and particles. Although they could be expected to improve the filtration effect on smaller inclusions, filters with smaller openings were not used in this study in order to avoid plugging and choking by metal [9].

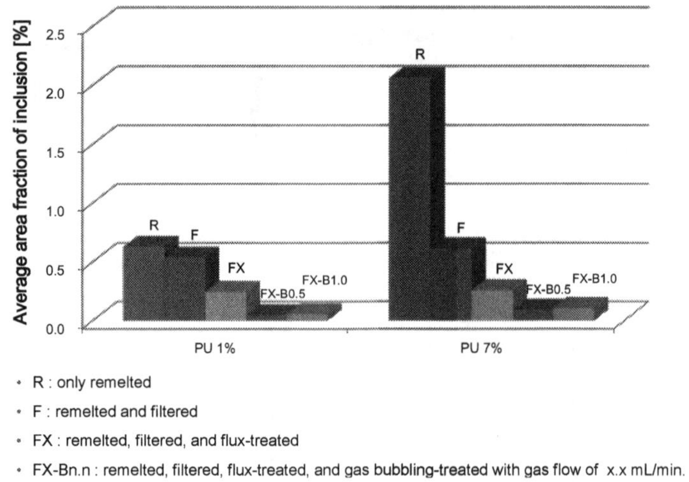

Fig. 8. Variation in the average area fraction of non-metallic inclusions in remelted and refined ingots of AM50 magnesium alloy scraps initially containing 1 and 7% of polyurethane scrap according to the melt treatment.

Fig. 8 shows the variation in the average area fractions of NMIs in remelted and refined AM50 ingots initially containing 1 and 7% PU according to the refining process. In the case of 1% PU specimens, the filtration was not effective in terms of decreasing the amount of NMIs, whereas the followed fluxing and gas bubbling treatments showed better refining effects. In the case of 7% PU specimens, the filtration was more effective than in the former case, and the effects of subsequent treatments were similar to observations made in the former case. In order to verify the type of inclusions that were mainly reduced by each refining treatment, other NMI analysis methods were introduced. Fig. 9 (a) shows the variation in the weight percentages of NMIs according to the refining process in refined ingots initially containing 7% PU, which were measured using an OES method. Fig. 9 (b) shows the

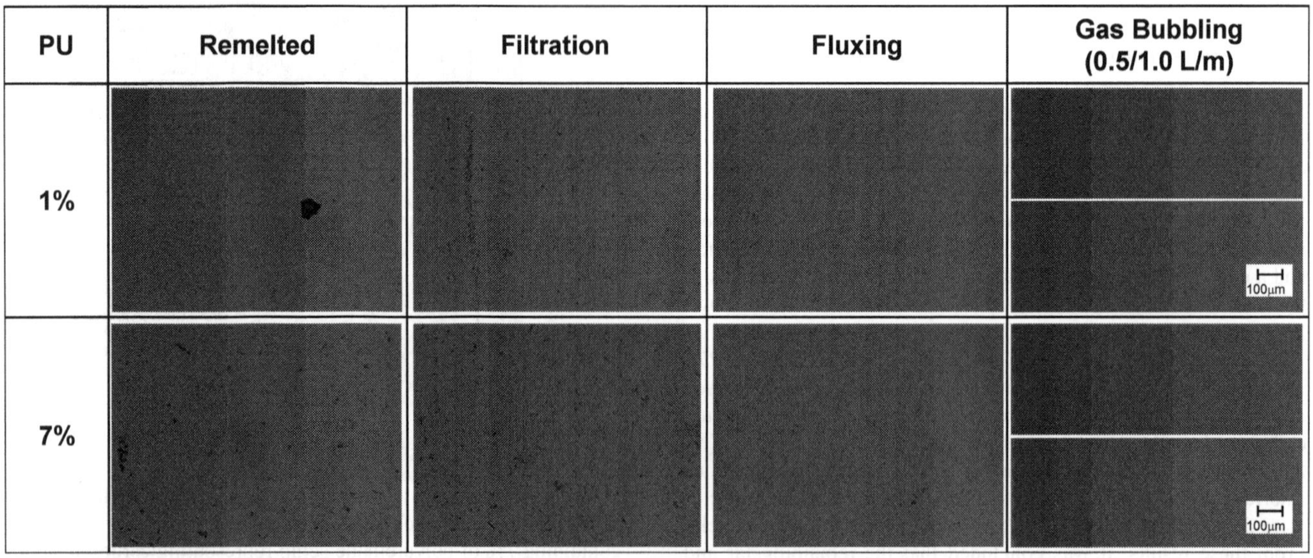

Fig. 7. Variation in microstructures of remelted and refined ingots of AM50 magnesium alloy scraps initially containing 1 and 7% of polyurethane scraps according to the melt treatment.

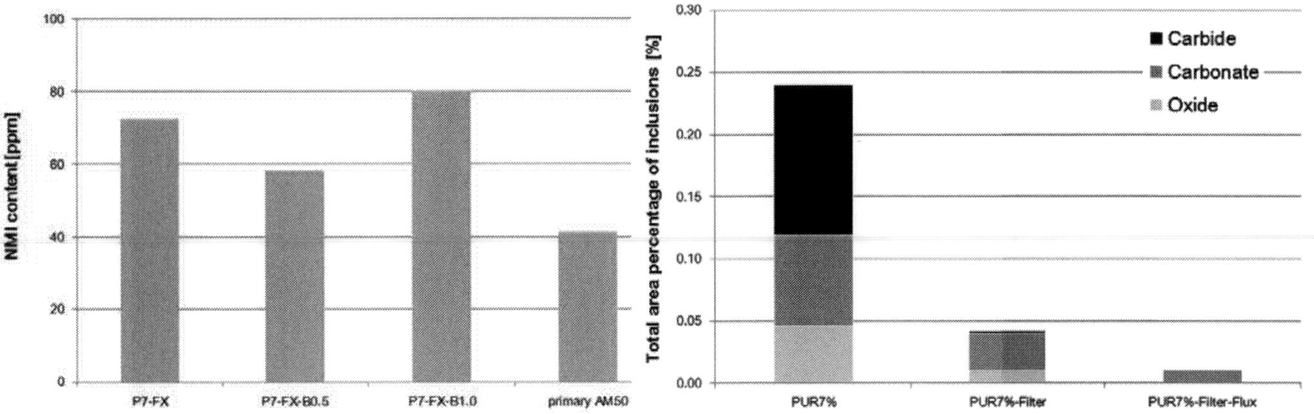

(a) (b)

Fig. 9. Variations in NMI contents (a) and area percentages of inclusions (b) by types in the remelted and refined ingots of AM50 magnesium alloy scraps initially containing 7% of polyurethane scrap according to the melt treatment.

variation in the average area fractions of NMIs according to the refining process in the same ingots, which were measured using an automated particle analyzer on the EDS mapping images. From the OES analysis results, it was concluded that the NMI contents did not show a significant change according to the refining treatment. Also, the NMI content of refined ingots was slightly higher than that of the remelted AM50 primary ingot. Because the OES equipment used in this study can distinguish oxide-type inclusions only, the carbon, carbide, or carbonate content were not included in these results. On the other hand, the results from a particle analyzer include the classification of NMIs according to the chemical composition regardless of the type. However, carbon residues could not be distinguished from carbides due to the inherent limit of the EDS analysis and particles smaller than 3 μm could not be identified due to the resolution limit of the 100 times magnified EDS mapping image. Therefore, it can be concluded that most of the carbon or carbide inclusions were reduced by filtration, whereas the carbonate and oxide inclusions were gradually reduced by the sequential refining processes.

Conclusions

The NMI content in magnesium alloy scrap contaminated with polyurethane was effectively reduced by a sequential refining process consisting of filtration, fluxing and gas bubbling treatments. The filtration could reduce most large inclusions such as carbon residues originating from the decomposed polyurethane. The following fluxing and gas bubbling treatments effectively removed the small inclusions such as carbonates and oxides. However, a separation process before the remelting and refining process should be employed to avoid discharging a huge amount of fume which may contains harmful gasses such as dioxin during the remelting process of polymer-contaminated scrap.

Acknowledgements

This study was supported by the R&D Center for Valuable Recycling (Global-Top Environmental Technology Development Program) funded by the Korean Ministry of Environment. (Project No.: 11-B19-MD)

References

1. G. Hanko, H. Antrekowitsch, and P. Ebner, "Recycling Automotive Magnesium Scrap", JOM, 54, 2 (2002) 51-54.
2. D. Fechner, N. Hor, C. Blawert, W.Dietzel, and K.U. Kainer, "Magnesium Recycling: State-of-the-Art Developments, Part I", Light Metal Age, 67, 4 (2009) 42-46.
3. H. Antrekowitsch, G. Hanko, P. Ebner, "Recycling of Different Types of Magnesium Scrap", Magnesium Technology 2002, 43-48.
4. B.G. Moon, K.H. Kim, and H.M. Lee, "Effect of Strontium Addition on the Microstructures and Mechanical Properties of Mg-5Al-3Ca Alloys", Mater. Trans. 52, 6 (2011) 1181-1185.
5. C. Wögerer, G. Klammer, and M. Kettner, "Separation and Recycling Technologie of Mixt Al and Mg Scrap", Light Metals 2007, TMS (2007) 1211-1219.
6. Directive 2000/53/EC of the European Parliament and of the Council of 18 September 2000 on end-of-life vehicles, Official Journal L 269, 21/10/2000, 0034-0043.
7. A.J. Gesing and A. Dubreuil, "Recycling of Post-Consumer Mg Scrap", Proceedings of 65th Annual World Magnesium Conference, Warsaw, Poland, IMA, 2008.
8. K.S. Kim, S.B. Kim, E.S. Chang, S.J. Lee, S.D. Kim, and H.C. Ahn, "Pyrolysis Characteristics of Waste Polyurethane", Abstract Book of 2005 Korean Society of Environmental Engineering (2005) 1326-1327.
9. S.E. Housh and V. Petrovich, "Magnesium Refining: A Fluxless Alternative", SAE 920071 (1992).

Magnesium Technology 2013
Edited by: Norbert Hort, Suveen N. Mathaudhu, Neale R. Neelameggham, and Martyn Alderman
TMS (The Minerals, Metals & Materials Society), 2013

Selective Laser Melting of Magnesium and Magnesium Alloys

Matthias Gieseke[1], Christian Noelke[1], Stefan Kaierle[1], Volker Wesling[1], Heinz Haferkamp[1]
[1]Laser Zentrum Hannover e.V., Hollerithallee 8, Hannover, D-30419, Germany

Keywords: Selective Laser Melting, Magnesium, Biodegradable Implants

Abstract

Selective Laser Melting (SLM) offers the possibility to create three dimensional parts by having full freedom of design. Therefore prototypes can be produced faster and conventionally manufactured parts can be shaped individually, including an optimized design regarding potential loads and parts weight. The manufacturing of biocompatible metals like 316L and TiAl6V4 is already industrially established. Because of the corrosive and mechanical properties of magnesium and the advantages of the SLM process, using magnesium is of great interest for manufacturing individual biodegradable implants. Recent investigations on SLM of magnesium have not led to successful operation so far. Due to the low vaporizing temperature, manufacturing non-porous and three dimensional parts from magnesium was not possible yet. Following a new strategy, using an industrial SLM system with an overpressure building chamber, investigations on SLM of magnesium are now carried out in order to overcome these difficulties and produce fully dense three dimensional parts.

Introduction

Additive manufacturing is an automated and repeating process where parts are built from layers. This enables the creation of three-dimensional parts with full freedom of design. The Selective Laser Melting (SLM) process is an additive manufacturing technique, which produces fully dense parts from metallic powder material. The 3-D part is first “sliced” into layers, which are manufactured in a two step process. Metal powder is deposited in the first step, and melted by laser radiation in the second step. Afterwards, the build-up platform is moved, and the process is repeated until the part is finished (see figure 1) [1].

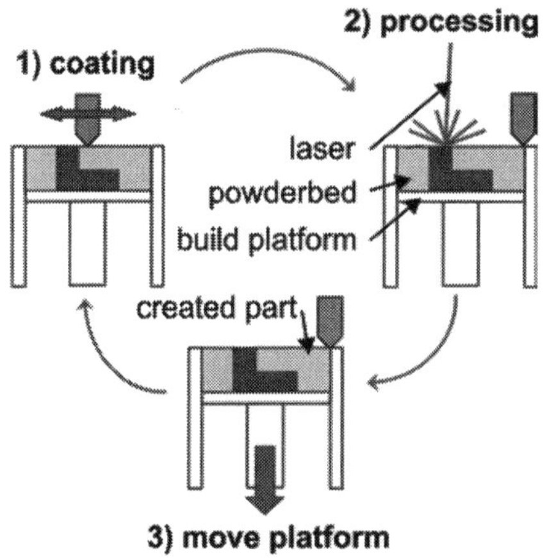

Figure 1. Sketch of an SLM process (according to [2])

The technology was originally used for manufacturing prototypes, and is also known as ‘rapid prototyping’. Nowadays, further improvements on the process have been made so that prototypes, individual parts and small series can be produced faster [3], including an optimized design regarding potential loads and the part weight [4].

Since the Selective Laser Melting process is already established for implant materials like 316L stainless steel and TiAl6V4 [5], manufacturing individual implants for surgery is of high interest [3]. One of the future challenges is the manufacturing of biodegradable, biocompatible metal implants. These implants, e.g. for osteosynthesis applications or stents, must not be removed through additional surgery, after the healing process has taken place. Magnesium alloys are the preferred materials for producing such implants, because of their mechanical and corrosive properties.

Selective Laser Melting of magnesium alloys therefore offers the possibility to create complex shaped individual biodegradable implants for osteosynthesis. The great challenges are to establish magnesium alloys in the SLM process, and to control the degradation rate and the resulting mechanical integrity in the human body.

To accomplish this, a research project funded by the Deutsche Forschungsgemeinschaft (German Research Foundation, DFG) was launched in 2012, to manufacture biodegradable implants for cranial bone replacement using the SLM process. The main component of the future implant is a biodegradable scaffold made of magnesium alloys. A surrounding biodegradable polymeric coating made of poly(3-hydroxybutyrate) (P(3HB)) controls the degradation behaviour. Additionally, a titanium component can be added for long term stability (see figure 2). Before implanting the device, it is pre-vitalized with human cells, to support the growth of bone (on the implant). During the healing process, the magnesium scaffold degrades, and the growing bone closes the gap.

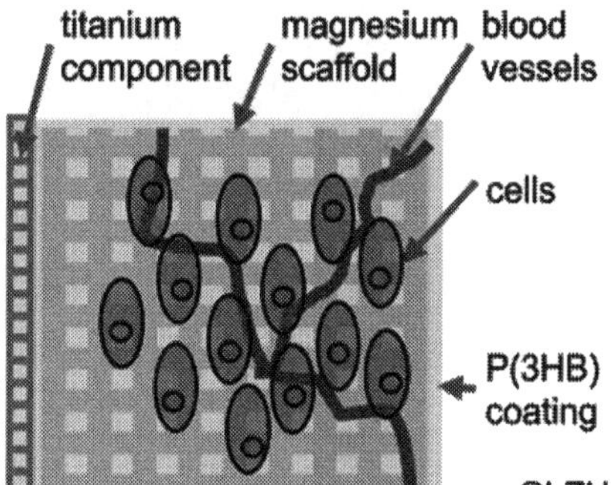

Figure 2. Pre-vitalized hybrid implant

State of the art

Several investigations on Selective Laser Melting of magnesium and magnesium alloys have been carried out during the last few years. In 2010, Ng et al. investigated Selective Laser Melting of magnesium to produce scaffolds for bone substitution. Using a miniature SLM system constructed at the Hong Kong Polytechnic

University, single tracks in a shielding gas atmosphere were built up at ambient pressure. For first trials, coarse particles with the size of 75 µm - 150 µm and spherical particles with the size of 5 µm - 45 µm were used, which only lead to satisfactory results. The produced structures showed sintered powder around the track [6]. In 2011 they investigated the SLM process with magnesium under different processing conditions, using the same setup as in 2010. Magnesium was processed using pulse mode and continuous wave mode. Continuous wave irradiation led to a disrupted surface with regular beads, whereas pulsed mode irradiation led to a smooth surface and flat surface morphology with pores within the tracks (see figure 3) [7]. Later they also discovered that an increase of the laser energy density resulted in a decrease of cooling rate, and therefore grain coarsening in the melted zone [8].

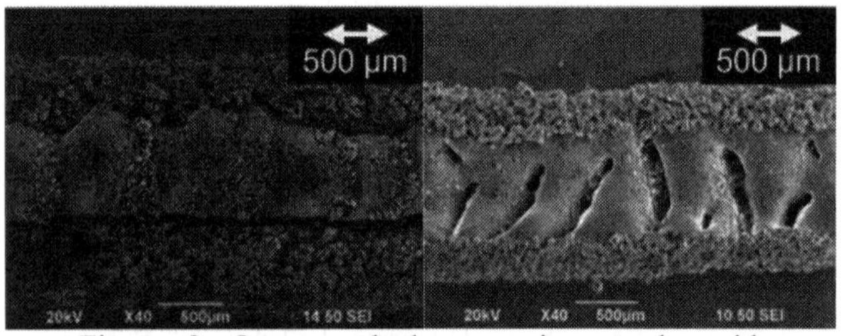

Figure 3. Laser melted magnesium tracks with continuous wave radiation (left) and pulsed mode laser radiation (right) (according to [7])

In 2011, Zhang et al. investigated Selective Laser Melting of a magnesium powder mixture containing 9 wt% of aluminum. Using an MCP Realizer 250 II^{SLM} (MCP HEK Tooling GmbH, Germany) machine with a process pressure of 7 Pa, structures with 82 % density were produced. During the experiments they observed balling phenomena (see figure 4) and powder evaporation, due to the low boiling point of magnesium using high energy densities [10].

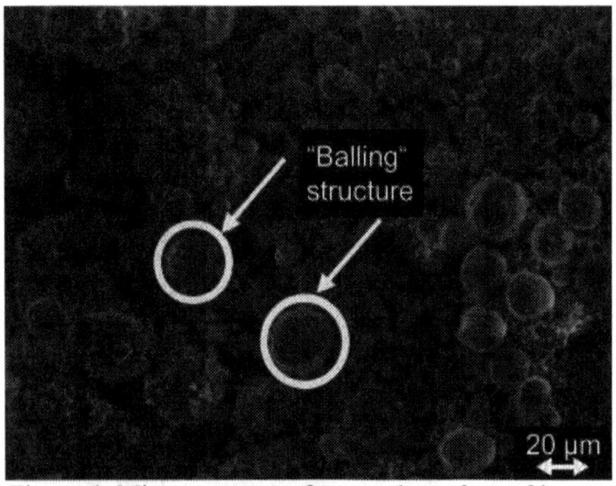

Figure 4. Microstructure of a sample surface of laser melted magnesium aluminum powder material (according to [10])

Materials and methods

One major problem is the low boiling point of magnesium of 1,093 °C, compared to the melting point of 650 °C and the low evaporation heat of 5.272 kJ/kg at ambient pressure [9]. Zhang et al. produced porous structures with a maximum density of 82 % by working at 7 Pa process pressure. Ng et al. had better results by working at an ambient pressure atmosphere. To overcome this problem, the boiling point has to be increased by working in an overpressure atmosphere. Another problem is the low dynamic viscosity of molten magnesium, which is 1.5 Pa·s, and therefore lower than molten titanium (2.2 Pa·s) and iron (6.93 Pa·s) [11] alloys which are well established in the SLM process [5]. Since the viscosity of the melt depends on the temperature [11], the viscosity decreases, due to high energy input during laser treatment. This can lead to problems regarding the track formation.

Preliminary tests and results

In 2009, the first unpublished investigations on SLM of magnesium powder material were carried out with an SLM test setup at the LZH. The challenge of the low boiling point and evaporation heat was known at that time, so the SLM setup was equipped with a vacuum and overpressure process chamber and a pressure control valve (see figure 5). Rectangular, single layer magnesium samples with a size of 4.5 mm x 4.5 mm were manufactured in a 2 mm thick powder bed using a 50 W fiber laser. Pure spherical magnesium powder in grain size < 75 µm with 80 % < 45 µm provided by Tropag Oscar H. Ritter Nachf. GmbH Hamburg, Germany was used. In several tests, the effect of laser power, scanning speed, shielding gas and process atmosphere were investigated.

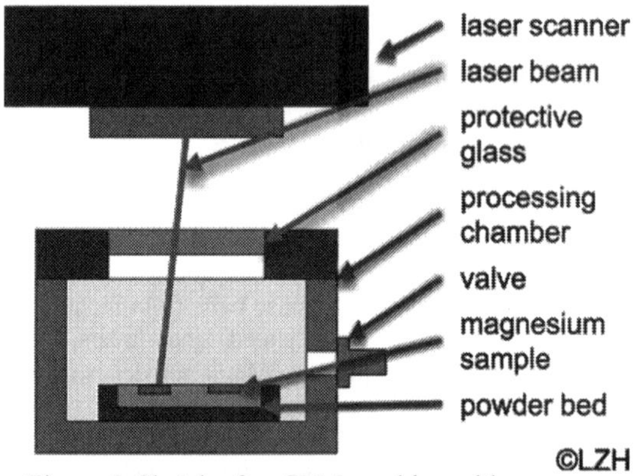

Figure 5. Sketch of an SLM machine with overpressure process chamber

Using the maximum feasible overpressure of 0.13 MPa (absolute pressure), stable structures of magnesium were manufactured (see figure 6). Processing under an argon atmosphere led to precipitation of process emissions on the protective glass, which could easily be removed. This effect could not be observed using a helium atmosphere.

Figure 6. Laser melted magnesium sample

The preliminary results showed that it is possible to use magnesium powders in the SLM process. Nevertheless, a smooth surface with definite straight edges could not be achieved, due to the evaporation of the powder material. Therefore, the boiling point had to be increased by increasing the process pressure, and the process parameters had to be improved.

Concept for the SLM machine for magnesium processing
Consequently, a Selective Laser Melting Machine with an overpressure process chamber based on an SLM 125^{HL} was developed by SLM Solution GmbH, Lübeck, Germany in 2012, to produce magnesium parts using the SLM process. The build-up chamber is designed for an absolute pressure of 0.3 MPa, which increases the boiling point of magnesium of 127 °C compared to the investigations of Ng et al. 2010 and 2011, to a boiling temperature of 1,220 °C (see figure 7).

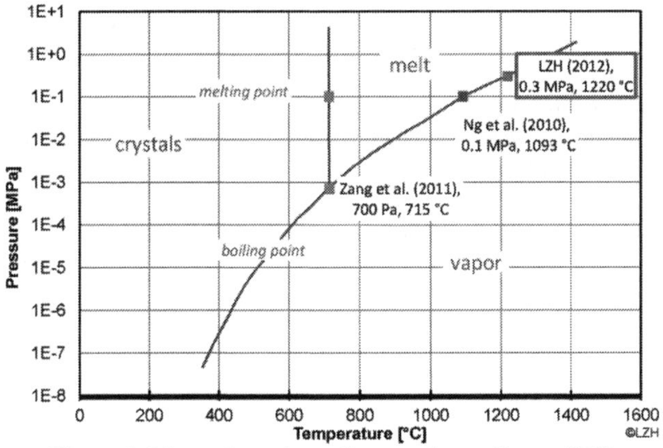

Figure 7. Magnesium phase diagram (according to [12])

For security reasons, the build-up platform is reduced to 50 mm x 50 mm and a maximum part height of 50 mm, in order to limit the maximum amount of magnesium powder to 500 g. To enable a Selective Laser Micro-Melting (SLµM) process which creates structures < 100 µm, the focus spot size is minimized to 70 µm.

Focus of future investigations
First, the investigations will focus on the effects of key process parameters like laser power, scan speed and scan strategy as well as the process atmosphere and process pressure. A first parameter set for manufacturing scaffolds needs to be identified.

Second, the evolving microstructure has to be investigated. As Ng et al. showed in 2011, the SLM process can be used to adjust the grain size in the material according to the process parameters [8]. Since the grain size has an important effect on the mechanical and chemical properties, it can be adjusted according to the requirements [9].

Magnesium alloy selection
Initially, the behavior of magnesium in the SLM process will be investigated using spherical magnesium powder ATOULTRA 325 with a grain size of 10 µm - 45 µm provided by SFM SA, Matigny, Switzerland (see figure 8). Since pure magnesium leads to an increase of the pH-value during *in* vivo and *in* vitro degradation, it is not suitable as an implant material [13].

Since no biodegradable magnesium implant is available on the market [2], and since no magnesium alloy has approval as a biodegradable implant material, several magnesium alloys will be investigated for the intended purpose.

Figure 8. Magnesium powder ATOULTRA 325 (SFM SA)

As a first magnesium alloy for the required scaffolds, a magnesium alloy containing 0.8 % calcium will be used, since the amount of alloying elements is small. This MgCa0.8 alloy showed good biocompatibility for an osteosynthesis implant in *in vivo* test, but insufficient initial mechanical strength [14]. Since low mechanical stability is required, it is suitable as an implant material for the hybrid implants.

Later, further magnesium alloys will be established in the SLM process. Since the final hybrid implant will be an osteosynthesis implant closely meshed with blood vessels, it cannot be compared with the osteosynthesis implants investigated in the various publications. Several alloys showing good results regarding biocompatibility, degradation and mechanical stability in vascular, abdominal and osteosynthesis applications have to be taken into account.

The WZ21 magnesium alloy containing biocompatible alloying and rare earth elements showed good biocompatibility, and slow and homogenous degradation in the abdominal cavity of miniature pigs [15], and stable mechanical integrity in the femora of rats [16].

The WE43 alloy containing yttrium, rhenium and zirconium is the best alloy for biodegradable stent manufacturing. Results from using this alloy as an osteosynthesis implant material are varied. Krause et al. observed unpredictable and inhomogeneous degradation behavior in 2010 [14], whereas Castellani et al. observed high implant stability and good osseo-integration in 2010 [17].

Last, the metallic glass $Mg_{60}Zn_{35}Ca_5$ containing 60 % magnesium, 35 % zinc and 5 % calcium will be investigated. It showed less hydrogen evolution than other alloys, and a good biocompatibility in the abdominal wall of domestic pigs [18].

Conclusion

Biodegradable implants can have a significant impact on the future of osteosynthesis applications. Magnesium alloys, which have been thoroughly investigated by several research groups in recent years, are the preferred material for manufacturing such implants, because of the load bearing properties, compared to biodegradable polymers. Selective Laser Melting of magnesium alloys can be used to create individual complex shaped biodegradable implants.

Recent investigations on Selective Laser Melting of magnesium and magnesium alloys did not lead to the desired success, due to the low boiling point of the materials. Porous structures and single tracks with adhering powder particles were manufactured, and it was not possible to create three-dimensional parts. In a new approach, an industrial SLM machine setup equipped with an overpressure working chamber up to 0.3 MPa was developed to increase the boiling point up to 1,220 °C. Previous investigations at the LZH showed that using SLM in an overpressure atmosphere can result in improved sample properties.

In order to establish the SLM process for magnesium powder materials, the parameters have to be extensively investigated, to prevent evaporation and a further decrease of the viscosity due to high energy input. As recent investigations have shown, the energy input also changes the induced cooling rate, and therefore the evolving grain sizes in the SLM part. This offers the possibility to adjust the mechanical and chemical properties, which depend on the grain size.

Taking all these issues into account, the SLM process in an overpressure atmosphere is a challenging, but promising way to successfully generated scaffolds made of different magnesium alloys for the production of new hybrid implants for cranial bone replacement.

Acknowledgement

The authors would like to thank the Deutsche Forschungsgemeinschaft (German Research Foundation, DFG) for their funding and support of the project listed under support code HA 1213/77-1.

References

1. Andreas Gebhardt, *Understanding Additive Manufacturing* (Munich: Carl Hanser Verlag GmbH & Co. KG, 2012), 31-63

2. M. Gieseke et al., "Additive Manufacturing of Drug Delivery Systems," *46th annual conference of the German Society for Biomedical Engineering (BMT 2012)*, 2012, 425-428.

3. Ian Gibson, David W. Rosen, and Brent Stucker, *Additive Manufacturing Technologies* (New York Heidelberg Dotrecht London: Springer, 2010), 1-12, 385-399

4. Claus Emmelmann, and André Goeke "Laser freeform fabrication of aircraft components" *Proceedings of the 2nd Internation Workshop on Aircraft System Technologies* (Aachen: Shaker, 2009), 243-248

5. SLM Solutions GmbH, "Discover the variety" (company homepage SLM Solutions GmbH, Lübeck, Germany, http://www.slm-solutions.com/cms/upload/pdf/120923_SLM_Materialien.pdf, 09/26/2012)

6. C.C. Ng et al., "Layer manufacturing of magnesium and its alloy structures for future applications," *Virtual and Physical Prototyping*, 5 (2010), 13-19

7. C.C. Ng et al.," Fabrication of magnesium using selective laser melting technique," *Rapid Prototyping Journal*, 17 (2011), 479-490

8. C.C. Ng. et al., "Microstructure and mechanical properties of selective laser melted magnesium," *Applied Surface Science*, 257 (2011), 7447-7454

9. Catrin Kammer, *Magnesium Taschenbuch* (Düsseldorf: Aluminium-Verlag Marketing & Kommunikation GmbH, 2000), 642-649

10. B. Zhang et al., "Effects of processing parameters on properties of selective laser melting Mg–9%Al powder mixture," *Materials & Design*, 34 (2011) , 753 – 758

11. Herbert Pfeifer, *Taschenbuch industrielle Wärmetechnik: Grundlagen, Berechnungen* (Essen: Verfahren Vulkan-Verlag, 2007) 139-140

12. Stephan Hasse, *Gießerei Lexikon* (Berlin: Fachverlag Schiele & Schön GmbH, 2007, 19) 207

13. K. Lips et al., "Elektrochemische Korrosionsuntersuchungen an Magnesiumlegierungen AZ91: Beschreibung kritischer Parameter und deren Einfluss auf die Angriffsmechanismen auf NRC-Proben," *Materials and Corrosion*, 55 (2004), 5-17

14. Krause et al., "Degradation behavior and mechanical properties of magnesium implants in rabbit tibiae," *Journal of Material Science*, 45 (2010), 624-632

15. A.C. Hänzi et al., "On the in vitro and in vivo degradation performance and biological response of new biodegradable Mg-Y-Zn alloys," *Acta Biomaterialia*, 6 (2010), 1824-1833

16. T. Kraus et al., "Magnesium alloys for temporary implants in osteosynthesis: In vivo studies of their degradation and interaction with bone," *Acta Biomaterialia*, 8 (2010), 1230-1238

17. C. Castellani et al. "Bone-implant interface strength and osseointegration: Biodegradable magnesium alloy versus standard titanium control," *Acta Biomaterialia*, 7 (2010), 432-440

18. B. Zberg et al., "MgZnCa glasses without clinically oberservable hydrogen evolution for biodegradable implants," *Nature Materials*, 8 (2010), 887-891

Magnesium Technology 2013
Edited by: Norbert Hort, Suveen N. Mathaudhu, Neale R. Neelameggham, and Martyn Alderman
TMS (The Minerals, Metals & Materials Society), 2013

Evolution of the Magnetherm Magnesium Reduction Process

James C. Sever[1], P.E., Marlyn Ballain[2]

[1] Alpha / Omega Engineering Inc.
17321 N. Colton St. Colbert, WA 99005
severjc@msn.com
[2]Ballain Consulting Inc.
661 S. Williwaw Dr, Palmer, AK 99645

Keywords: Magnetherm, Silicothermic, Magnesium

Abstract

The magnetherm process was developed and implemented as an improvement of the solid state silicothermic magnesium reduction process. This process was used commercially for over 35 years. The reduction plant operation and equipment are described as they existed at the time of the industry wide curtailment.

Introduction

The chemical reaction for all silicothermic methods [1] for producing magnesium from dolomite is the same:

$$2CaOMgO + Fe\,Si\,75\% \rightarrow Ca2SiO4 + 2Mg\uparrow + FeSix$$

The major differences are in the method of heating the mass to the reaction temperature, >1200 deg. C., the pressure in the reactor, and the condenser design. Under these conditions of temperature and pressure, magnesium is release in a gaseous form and condensed in either a liquid or solid phase.

The Magnetherm process was commercially operated from 1964 until 2005. Over the years of operation significant improvements were made to the process. The following is a brief description of that evolution.

Equipment Description

Raw Material Feed System

Northwest Alloys Inc. Shut-down of commercial operation was the fall of 2001.[2] At that time, the entire furnace assembly required a footprint of approx. 90' H x 50' W x 75'D. The furnace and feed tanks extended over 47 feet in height. The other plus 40 feet were required for the movement of the radio controlled cranes and the loaded feed containers that filled the furnace feed containers. Typically there were three raw material feed tanks.

Each feed tank rested on a load cell system in order to monitor the dynamic change in the mass of the tank as the process runs through its cycle of operation. Below each tank was an isolation valve and a feeder. The ferrosilicon tank had a screw feeder. All other feed tanks used vibrating feeders. Pipes extend from each feeder discharge to a common collection vessel which had a single pipe leading to the furnace.

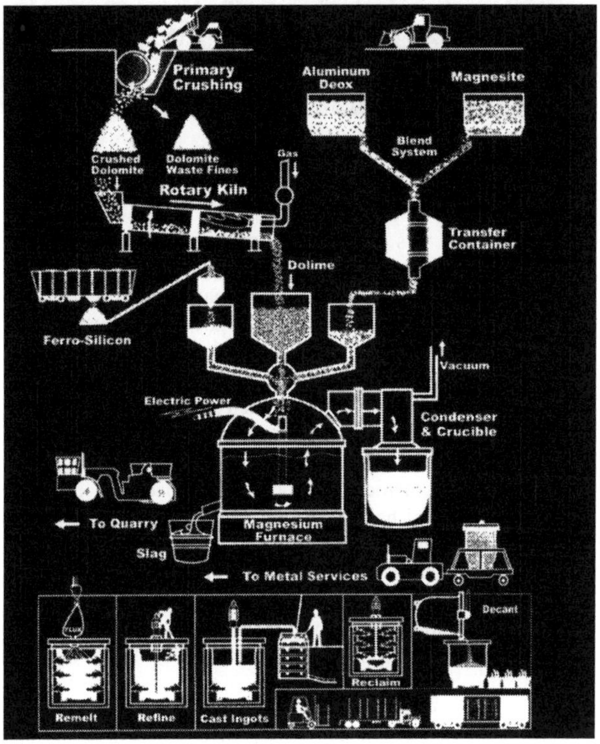

Figure 1. Schematic of Magnetherm Operations

Furnace

The furnace, a large cylindrical stainless steel vessel was lined with insulating refractory followed by carbon blocks for the inner lining. Carbon paste was rammed between the refractory bricks and carbon blocks tying the refractory components together. The top of the furnace was a refractory lined hemisphere. A water cooled copper electrode entered at the top of the furnace and was suspended above the carbon hearth. The side of the furnace nearest the raw material system had a flanged opening that allowed the feed system to be attached to the furnace. Directly opposite this opening was an exhaust port or tuyere from which the magnesium vapor exited to the condenser assembly.

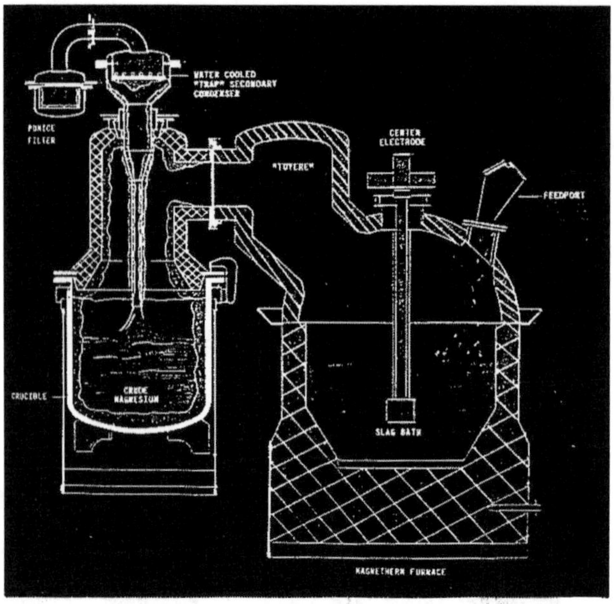

Figure 2. Cross Section of Magnetherm Furnace and Condenser

The furnace was heated by passing an electrical current through the molten slag. A single phase AC transformer provided electrical power to the furnace. The top electrode was connected to one output of the power transformer. Six water cooled electrodes (sole electrodes) were radially located in the prebaked carbon blocks constituting the floor of the furnace. They were connected to the other output of the power transformer. Great care was taken in establishing the length and position of the buss connecting the electrodes to achieve the correct impedance ensuring proper electrical balance. Power input was controlled by changing the winding output (tap) of the transformer. The transformer was a 15 MVA, rated at 90,000 amperes and 165 volts AC. A multiple of 3 operating furnaces were needed to maintain the phase balance on the utility company power supply transformer.

To remove slag from the furnace, two tap ports were located at the base of the furnace shell. A silicon carbide sleeve penetrated from the taping port flange into the body of the furnace providing an exit path for the molten slag and residual ferrosilicon.

The entire exterior of the furnace was water cooled by a combination of sprays, over flow weirs and water cooled flanges to maintain a freeze zone of slag within the furnace. Water cooling was critical in maintaining the freeze zone of slag and to protect the outer shell from penetration by molten ferrosilicon. Periodic sandblasting of the operating furnace shell was required to remove scale buildup and ensure proper heat transfer.

Condenser

Magnesium vapor produced during furnace operation was condensed in a water cooled crucible (3 inch thick wall). A transition unit, called the elbow, joined the crucible to the furnace. It consisted of a double walled steel cylinder bolted to the top of the crucible. Mating flanges joined the elbow to the furnace. A support assembly on the side of the cylinder permitted the unit to be suspended on a wheeled cart that allowed the entire condenser assembly to be moved into or away from the furnace. A water tank, hung on the flange of the elbow-crucible assembly, provided the water cooling for the steel crucible.

A steel pipe extending from the top of the elbow to a point just below the crucible flange allowed for non-condensable vapors to flow through the elbow assembly to the vacuum system (see figure 2). The funnel shaped top of this unit (trap) extended above the top of the elbow to function as a secondary magnesium condenser. The cone was topped with a domed lid (trap lid) that had an internal baffle to deflect residual magnesium vapor to the cool surfaces of this vessel. Vapors leaving the trap passed through the pumice filter, which protected the vacuum system from magnesium metal and pyrophoric metals such as sodium and potassium condensing in the vacuum lines. A large diameter flexible braided stainless steel pipe extended from the side of the filter canister attaching the condenser assembly to the vacuum system.

Vacuum System

. The entire furnace system from the raw material container lid to the exhaust on the vacuum pump was operated to maintain 70 torr measured at the furnace. Two particulate filters (90 micron and 10 micron) were located in series prior to the vacuum pump. Each furnace had a dedicated positive displacement vacuum pump that maintained pressure throughout the operating cycle. For the most part rubber "o" rings made the seal between mating flanges. If the system was exposed to heat, the flanges were water cooled.

Argon was fed to the raw material tank to maintain a pressure gradient from the top of the feed system to the feed port, to the vacuum system. This gas also assisted in directing the magnesium to the condenser on start up.

Center Electrode

A single electrode provided power to the furnace. This consisted of a flange for sealing to the furnace, a copper pipe and a graphite head attached to the base by cast bronze. To prevent melting the copper, an inner pipe carried cooling water down the center discharging cool water to the electrode head, flowing up the sides of the copper shaft, and out at a side discharge at the top of the electrode. Refractory blocks placed around the exterior of the electrode provided added protection.

Operation

The process was a batch operation consisting of a two part cycle. The first half consisted of feeding out the mass of dolime with a stoichiometric amount of other feed materials in the fixed containers. The process interruption termed minor consisted of: bringing the furnace and dolime fixed container back to atmospheric pressure, refilling the dolime container, taping slag and drawing the system back down to operating pressure for restart. The 2nd half operation consisted of feeding out the recharged dolime and its associated feed materials and stopping operations to allow the furnace assembly to be reconditioned. This major downtime included bringing the furnace pressure to atmospheric, the tapping of the slag, removal of the full condenser and elbow assembly, and refurbishing the furnace with additional feed materials, clean parts, and a fully pressure checked system ready for a new operation cycle.

At the beginning of the furnace cycle, the furnace was at its minimum slag level. A new condenser was in place, vacuum prefilter tubes were cleaned, the raw material tanks were full and all ports were cleaned, closed, tightened.

The procedure for verifying that a furnace was ready for operation consisted of: holding the furnace at caloric power input and opening the valve to the vacuum system. The furnace assembly was evacuated to the target pressure for leak evaluation. The entire system was then isolated and held for a period of time. The pressure inside the system was observed and if it remained within a target level, the furnace system was deemed to be free of air leaks and it was ok to proceed with the cycle. If a leak was indicated, the system was once again, drawn down to target pressure and held. At this time the isolation valve located between the raw material feed system and the furnace was closed. The pressure in each zone was then observed and the one exhibiting a leak was identified. The system was then pressurized to slightly above atmospheric and the operators physically inspected the area for the source of the leak. Once found the problem was resolved and the furnace was again pressure tested. After passing the pressure test, the system was subjected to another drawdown evaluation.

Once the system was confirmed as leak free, the power was increased to the furnace, argon was fed to the dolime tank and the feeders activated to move raw material from the storage tanks into the furnace. The power and the rate that material was fed to the furnace were steadily increased until a target level was reached or the condenser performance showed that its maximum potential had been reached. Raw material input continued at this steady state until all the dolime was emptied from the dolime tank and the furnace was full of slag. During this steady state portion of the cycle, furnace and condenser conditions were monitored for changes in pressure and/or temperature, indicating problems. The feed system was monitored to ensure the material draw down rate was consistent. Deviations from the expected rate of weight loss could indicate blockages in the feed system. These incidents would necessitate opening the feed system to clear the blockage.

With the first half of the feed cycle completed, the power fed to the furnace was decreased to calorific level or shut off. The furnace was brought up to atmospheric pressure by isolating the system from the vacuum system and switching to a low pressure-high flow argon source. Once the furnace reached atmospheric pressure, the door at the base of the furnace and the feed port at the top of the furnace were opened. The clay plug sealing the furnace was drilled followed by use of an oxygen lance to burn a path into the molten furnace slag. Once slag flow was established, the slag level in the furnace was monitored through the feed port. Once the desired level was reached, a clay plug was forced into the tap hole stopping slag flow and a final slag level was measured. The flanges on the open ports were clean and sealed, readying the furnace for continuing operation. The dolime feed tank was refilled with hot dolime from the kiln while the slag tap was in progress.

Once the furnace was closed and all work completed, the furnace was once again tested for pressure integrity. If all was well, the second half of furnace operations began. Exposure of the condenser to air, while tapping slag and refilling the dolime tank after the first half of the cycle resulted in the degradation of condenser efficiency. This reduced the rate at which the furnace could be fed during the 2nd half of the cycle. As the condenser filed the raw material feed rate had to be decreased to accommodate the reduced efficiency.

If all went well, the dolime tank would be empty and the furnace would be full of slag for the second time in the cycle. Power was turned off or reduced to low level, the vacuum system was isolated from the furnace and argon was used to raise the furnace pressure. Many refurbishing operations occurred simultaneously: the feed port was opened; the tapping door opened; the raw material tanks opened and the condenser backed away from the furnace. The condenser assembly was removed to be serviced. The condenser crucible full of magnesium was covered and transported to the ingot casting department (Metal Services) for further processing. The raw material tanks were refilled. The slag was tapped. The feed port and the tuyere openings were cleaned of buildup. Once all ports were cleaned and opened to the desired size, the slag level in the furnace lowered to the desired level and the raw material tanks full, a cleaned condenser assembly placed on the furnace and all ports closed, the cycle was ready to be repeated.

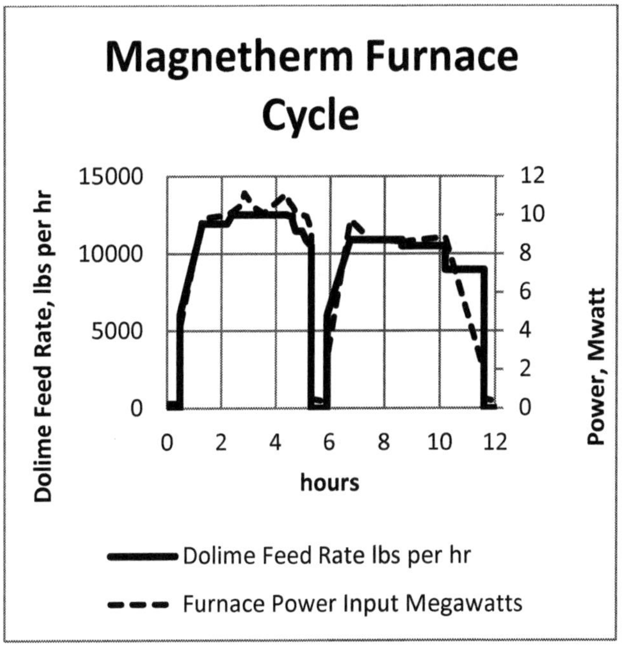

Figure 3. A typical magnetherm cycle of power and dolime feed to the furnace over time.

Process Evolution

Early Days

Work began in France in 1948 to improve the silicothermic reduction process to produce magnesium.[3,4,5] The aim was to utilize the raw materials directly without having to produce a briquette and to be able to operate with less manpower and better thermal efficiency. This led to the development of a resistance furnace where current passing through the slag provided the source of heat. By adding alumina to the raw material mix a lower melting point slag was obtained.

The first successful furnace began operation in 1955 and produced 200 kg of magnesium per day. The unit connecting the furnace to the collection crucible was the condenser which was preheated to 675 deg. C by means of a internal electric heating element. This allowed liquid magnesium to be collected in the crucible which was cooled with water sprays. Evacuation took place off the back of the crucible.

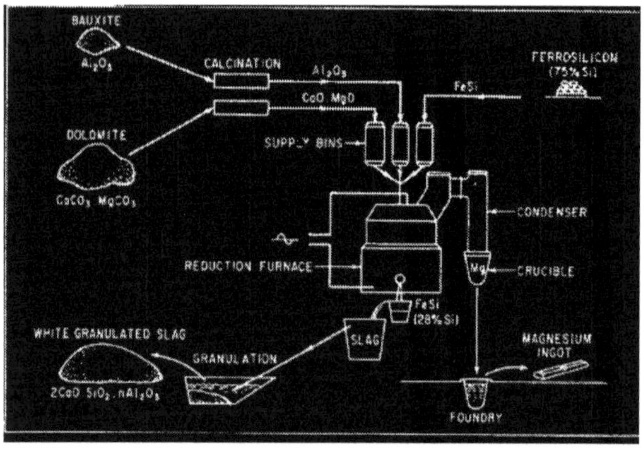

Figure 4. Original process schematic

An attempt was made to change from a single phase furnace power source to a three phase power source in 1956. This was unsuccessful in that it was difficult to balance the phases.

From 1957 through 1964 three separate furnaces were constructed. Each time the power supply was increased in size resulting in higher productivity. During this time frame the ability to use hot dolime directly from the kiln was incorporated into the design.

Prior to 1962 feed rates were volumetrically controlled. In 1962 material feed rates were monitored by overall tank weight changes. for improved raw material control. At the same time bauxite was substituted for the alumina.

Throughout the development period, furnace lining was undergoing frequent change. Owing to the aggressive nature of the slag and the reaction of magnesium with carbon, it was determined that a frozen layer of slag provided a self-healing furnace lining minimizing migration of residual ferrosilicon to the shell and production of hot spots. Addition of cooling to the shell improved this issue and extended the life of the furnace.

Commercial Operation

Work on the magnetherm process moved from Beaudean to Marignac France. In 1964 the first commercial scale furnace was brought on line. It had a power supply rated at 2.3 Megawatts and produced 3,400 kg of magnesium per 24 hour cycle.[6] The next generation furnace was installed in 1970. The power input was increased to 4.5 megawatts and a daily production of 7,500 kg per furnace day was achieved. About 1970, Alcoa acquired a license from Pechiney to produce magnesium via the magnetherm process at Addy, Washington.

Northwest Alloys Operation

In 1976, magnethermic furnace operations began. By 1977, nine furnaces were in operation.[7,8] These furnaces had the potential of supplying 8 megawatts of power. The operating pressure was 35 torr. A pellet formed from alumina and dolime was attempted. The idea was to feed hot pellets of alumina-dolime from the kiln to the furnace along with a separate hot dolime feed. This was altered to using hot alumina-lime calcined pellets feed via one feed tank and all the dolime being provided as a separate feed material through another feed tank. The targeted operating cycle time was intended to be 21 hours but owing to operating difficulties the cycle was 24 hours and the production rate was only 7,500 Kg per furnace per day.

The formative years of the 1980's realized the following advantages:

- Corrosion resistance and crack resistance of a stainless steel furnace shell;
- Inflatable gaskets on the slag tapping doors and the furnace tuyere;
- Development of a jacking car for the elbow assembly to allow quick alignment and connection of the elbow assembly to the furnace;
- Development of the motorized slag plugging machine which allowed for more positive assurance that a slag tap would be stopped and a good clay plug achieved each time a slag tap was made;
- Improved welding techniques reduced the difficulties with air leaks in assembled components and especially with center electrodes head and shaft repairs;
- Redesign of the elbow cleaning room to improved safety cleaning efficiency and fume capture;
- Washing and bake-out of elbow parts improved operational reliability of these components;
- Design and implementation of a track mounted jack hammer to provide better cleaning of elbow components;
- Furnace operating pressure changed from 35 to 70 torr;
- Improved computer control algorithms;
- Work was also performed evaluating aluminum as a reductant. [9, 10]

Other feed materials were also improved:

- Ferrosilicon lump was altered to granulated ferrosilicon, Gransil;
- Dead burned magnesite was used to enrich the MgO content of the feed;
- Hard-burned dolime was evaluated as a replacement for regular dolime feed.

By 1989 operations had improved and the cycle time was reduced to 18 hours. Raw material feed quality was improved to decrease the amount of dusty material fed to the furnace and resulted in improved furnace condenser operations. However, to take advantage of these improvements higher feed rates were required which required more power. The increase power requirements caused center electrode life to drop from 28 days to as little as 10 days in some instances. Pressure surges within the furnace caused by feed material piling on the slag and variations in feed material feed rates caused magnesium vapor to migrate into the feed port area and plug the feedport. These necessitated periodic shut downs mid cycle for cleaning, further contaminating of the condenser surfaces. By increasing the liquid portion of the two phase slag, an active slag surface resulted thus reducing the frequency of this problem

Status Prior to Curtailment

Furnace cycle time was reduced from 18 hours to 12 hours with the potential of a 10 hour cycle as a consequence of incremental improvements in the operation and equipment.

Furnace power increased via installation of higher voltage and higher ampere transformers. In actuality, only a little over 12 megawatts were available for use due to: high slag resistivity at the start; low slag resistivity at the end of operations; and condenser limitations. The high performance transformers did have the advantage of allowing high voltage usage at the start of a run to establish good operations and the use of high ampere near the end of each half cycle to maintain good material feed rates at the end of each half cycle when the resistivity would drop significantly due to high slag levels in the furnace. If the raw material feed rate were increased to the point that the secondary condenser on the top of the elbow was condensing a great deal of liquid magnesium, the drain for the secondary condenser, which also acted as the outlet for non-condensable gas could become momentarily vapor locked. If this occurred, the feed system would be blocked with magnesium vapor being forced up the feed port area. Work was underway on a revised condenser design to alleviate the problem.

The premature electrode failure occurred due to three mechanisms of failure:

1. Damage to the electrode flange due to arcing resulting in air leak at the flange perimeter gasket.
2. Erosion of the shaft and head of the copper electrode leading to water leakage into the furnace.
3. Electrode head failures due to excessive power.

These problems were partially addressed with: improved construction materials and design around the electrode flange; establishing low chloride standard for feed materials; iron cladding on the electrode shaft; increased water cooling for the electrode; use of larger diameter electrode heads to allow original current density levels to be reestablished. Electrode life continued to be a significant problem and a significant safety concern up to curtailment.

A major improvement was achieved by addressing the quality of the raw materials used as feed to the reduction furnaces. Initially a soft-burn dolime, alumina-lime pellet, and lump ferrosilicon were used as feed materials. These were replaced with hardburned dolime, dead-burned magnesite, aluminum shot (deox), and Gransil. The purchased materials (Gransil, deox, and dead-burned magnesite) were greatly improved through tightening of material specification and inspection throughout the delivery chain. Thus ensuring no contamination occurred through storage and handling and that material specifications were met. The materials were closely monitored up to the point of actual usage by the furnace.

Kiln performance was improved by changing the procedure used to feed the material from the stockpile. Blending the outside of a stockpile (naturally coarse by gravity segregation) with the interior of another stockpile (naturally fine by gravity segregation) the variation in particle size distribution was reduced and the porosity of the preheater stabilized.

The dolime kiln refractory was revised to allow operation at higher temperatures (1450 C – 1470 C) to produce hardburned dolime. Use of this material greatly reduced magnesium furnace pressure spikes and feedport blockages resulting in an immediate improvement in furnace uptime and overall metal recovery. Dolime samples taken from the kiln were monitored for product quality using nitrogen BET evaluation. Acceptable hardburned dolime was set at equal to or less than 1 square meter per gram with a desired value of 0.7 square meters/gram. Unacceptable dolime was scrapped.

Increasing the amount of magnesium oxide fed per hour improved not only the productivity but also appeared to increase the uniformity of furnace operations. The amount of aluminum shot fed per hour was increased to the point that a separate alumina addition was no longer needed. It was now possible to charge dead burned magnesite to the furnace along with hardburned dolime. The percentage of magnesite to dolime was started at 6% by mass but was steadily increased to 12%. Higher levels were under test at the time of curtailment.

The increased density of magnesium vapor emanating from the furnace appeared to improve the pressure gradient throughout the system. To maximize production, efforts were made to improve the condenser capacity. The crucible surface area was increased and elbow components were thoroughly cleaned after each major. Work was begun on a new secondary condenser elbow design providing a higher surface area for the secondary condenser which would allow more liquid magnesium to return to the crucible without causing backpressure problems in furnace operations.

A significant step in attaining process stability was achieved through revision of the process control algorithm. Initially a predictive approach was used to set the raw material feed ratios for each cycle. It was found that small changes based on feedback, the performance of the furnace over previous cycles, increased furnace stability and productivity.

The equipment used to clean the furnace tuyere opening at each major was significantly changed. In place of manual cleaning of the tuyere with jackhammers, a rotary cleaner with a carbide tipped head devised from a mining/tunnel boring machine was used. This quickly removed any residue decreasing the tuyere opening, while improving working conditions for the furnace attendants and reducing furnace operator exertion and exposure. It also reduced the thermal stresses on the furnace and provided a uniform standard size exhaust opening.

The feed port was completely redesigned. The simple inlet was converted to a double wall port that mechanically opened up and allowed removal of the inner liner. At the end of each cycle a clean feed port liner was inserted into the furnace thus maintaining the maximum opening for feed material on the next cycle and improving operator safety conditions through less exposure to blowbacks from the furnace.

Vacuum leak detection was also improved. A tool was found that allowed the furnace system to be monitored while in operation. It was a very sensitive directional microphone and amplifier. By pointing it at a portion of the furnace, a telltale hiss could be heard if there was a vacuum leak. This could then be verified by pressure testing at the end of a cycle. Adoption of this tool increased furnace up time and decreased loss of product due to re-oxidation in the furnace.

In an effort to increase production through more uniform furnace and condenser operation two programs were initiated: 1) liquid phase operation capable of vacuum or atmospheric operation and 2) plasma furnace operation at atmospheric pressure. Both were terminated before successful results were achieved.

Pechiney operated a commercial liquid phase condenser for a number of months under vacuum conditions. [11]

Mintek operated a test furnace and liquid phase splash condenser at or near atmospheric conditions. The reduction furnace used a DC plasma power supply with a movable electrode. By operating at a higher temperature, they were able to successfully demonstrate the production of magnesium at atmospheric conditions using a DC plasma furnace coupled with a splash condenser. The reported results were published Oct. 2006 by M. Abdellatif of Mintek.[12]

Conclusion

A number of improvements were made from 1964 to curtailment in 2001. While these changes improved operations and productivity, market forces dictated closure in 2001.

This paper mentions only a few of the things accomplished by the people at NWA, Alcoa, and our technological associates in France. These improvements allowed this plant to establish itself and to meet the rigorous environment requirements placed upon this organization. Significant changes were made from the initial concept furnace to the 12 Megawatt production furnace used at the turn of the century. These changes resulted in substantial productivity and efficiency gains, significant process improvements, major safety improvements, and environmental gains. Many more milestone improvements could have been attained if operations had continued.

References

1. E. F. Emley, Principles of Magnesium Technology (London, England: Pergamon Press 1966) 47-50
2. Alcoa, "Alcoa Shuts Down Northwest Alloys Magnesium Smelter", *News Release*, Pittsburgh, June 22, 2001
3. C. Faure and J. Marchal, "Magnesium by the Magnetherm Process", *Journal of Metals,* 9 (1964) 721-723.
4. P. Lugnagne, "The Magnetherm Process for the Production of Magnesium", *Erztmetall,* Vol31 (7/8), 1970, pp 105 – 112.
5. F. Trocme, "The Development of the "Magnetherm" Process", (Paper Presented AIME 100th Annual Meeting, New York, New York, Feb 26 to March 4 1971).
6. F, Trocme, "The Magnatherm Process, Margnac, France", *Advances in Extractive Metallyrgy and Refining*, M.J. Jones Ed. (London: Inst. Min. Met., 1972), 517-520.
7. P. Lagagne, "Information on the Magnesium Plant of Northwest Alloys", (Paper Presented at the International Club of Magnesium, Rome, Italy, October 3 – 5, 1973.)
8. J.R. Chapman, "Operations and Processes of Northwest Alloys Plant at Addy, Washington", (Paper Presented at the AIME 1976 Pacific Northwest Metals and Minerals Conference, Coeur d'Alene, Idaho, April 8 – 10, 1976.)
9. R.A.Christini & M.D. Ballain: " Magnetherm Atmospheric Pressure Operation: Aluminum Reactivity in a Silicate Slag"; Light Metals TMS, Reactive Metals, February 1991.
10. R.A.Christini & M.D.Ballain: " Aluminothermic Magnetherm Development of Aluminum Skim and Aluminum Shot as a Reductant" - Proceedings of Reinhardt Schumann International Symposium 1986, pgs. 965-986.
11. P. Faure, "Method for Condensation of Metal Vapors to the Liquid State", US Patent 5,540,754, July 30, 1996.
12. M. Abdellatif, "Pilot Plant demonstration of the Mintek Thermal Magnesium Process", (Paper Presented at Conference of Metallurgists, Montreal, Quebec, Canada, 1 - 4 October, 2006.)

Magnesium Technology 2013
Edited by: Norbert Hort, Suveen N. Mathaudhu, Neale R. Neelameggham, and Martyn Alderman
TMS (The Minerals, Metals & Materials Society), 2013

Research on new type materials preparation for magnesium production by Silicothermic process

WEN Ming, ZHANG Ting-an, DOU Zhi-he, REN Xiao-dong, ZHANG Rui, ZHOU Lian

(School of Materials and Metallurgy of Northeastern University, Key Laboratory of Ecological Utilization of Multi-metal Inter grown Ores of Education Ministry, Shenyang, China 110819)

Key words: Dolomite; Bentonite; Composite binder; Silicothermic process

Abstract

During the magnesium preparation process of the silicothermic process, dolomite was made into pellets after calcination process, about 5% fine materials would be produced which can't be used and lead to high energy consumption and high-cost. Based on in-situ reduction of dolomite-based desulfurization theory, this paper proposed a new technology that dolomite was made into the pellets first and then was calcined to prepared the pellets with higher reactivity and stability. The results indicate that: In contrast with dolomite pellets prepared by single binder, dolomite pellets prepared by composite binder are dominant in functions without fine materials produced, the size of pellets is uniform, the falling strength can reach 2.5 times per 0.5 m, and the compressive strength gets 80N as well.

Introduction

Magnesium metal and its alloys have many properties such as high specific strength, high thermal and electric conductance, high impact resistance, good electromagnetic shielding, ease of machining and recovering. So that it has become the third widely applied metallic engineering material[1~6]. The production methods of metal magnesium can be divided into two processes: molten salt electrolysis and silicothermic process. In contrast, silicothermic process has become the main production method because the rich and high-grade ore of dolomite and magnesite. [6~12]

During the magnesium preparation process of the silicothermic process, about 5% fine materials would be produced – in the dolomite calcination step which can not be used and leads to high energy consumption and high-cost. This paper proposed a new technology of preparing magnesium metal that dolomite was made into the pellets along with the reducing agent first and then was calcined to prepared the pellets. Dolomite pellets was mainly composed of dolomite, reducing agent and binder. Reducing agent was 75Si-Fe, a certain amount of binder was added into the dolomite to improve the strength and the sintering performance of the pellets as well. The pellets have higher activity and stability to improve the resource utilization and to reduce the energy consumption of calcining. Pelletizing experiments with bentonite, organic binder and organic/inorganic composite binder were conducted. According to the mixing effects of the dolomite in the binder and the quality of the pellets, the reasonable choice of the binders was put forward.

Experiment

Material

The chemical composition and main phase of dolomite are shown in table Ⅰ and Figure1.

The chemical component of 75Si-Fe of dolomite are shown in table Ⅱ.

The experimental equipment needed can be divided into two main parts: the material preparation equipment and the pelletizing equipment.

The material preparation equipment: Crusher, Rod Mill, Vibration Rod, Abrasive Tool, Mortar, Sieve,

Electronic Balance, Platform Scale of 10 kg, Mixing Bowl, Mixing Machine,101# Electric Blast Drying Oven.

Table Ⅰ Chemical composition of dolomite

Compound	MgO	CaO	Al_2O_3	SiO_2	Na	K	Fe
Content / %	21.73	31.05	0.16	0.44	0.02	0.005	0.07

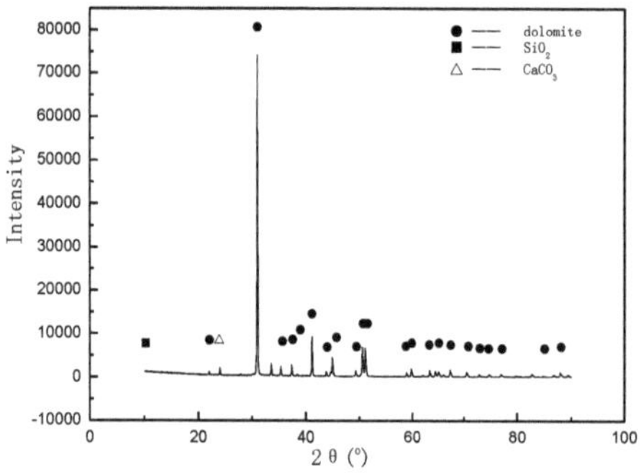

Figure 1 The main phases of dolomite

Table II Chemical component of 75Si-Fe				
Element	Si	Al	S	C
Content / %	75.6	1.24	0.091	0.015

The pelletizing equipment: Disc Pelletizer, Atomization Sprinkler, Scraper.

The pelletizing equipment: Disc Pelletizer, Atomization Sprinkler, Scraper.

The main experiment process: the raw ore including dolomite, reducing agent and fluorite catalyst are crushed and ground, and the finely powder are mixed with binder according to a certain ratio first; then the mixture is made into pellets by disk balling machine and the pellets are calcined in sintering furnace; the influence of binder type on the quality of pellets are investigated.

Results and discussions

Pelletizing capabilities of dolomite

Not all materials are capable of pelletizing, the superior sphericity is needed[13]. In the dolomite pellets, the ingredient of dolomite is more than 80% both in mass fraction and volume fraction, so the sphericity of dolomite was mainly considered.

The sphericity of finely ground materials is the performance of water dripping into pellets and its mechanical action density in the natural state. It can be considered as the formation and growth rate in the pelletizing process. So that the sphericity is a composite indicator, it can be determined by an index (k).

$$k = \frac{W_a}{W_b - W_a} \tag{1}$$

In the equation:

W_a——The largest molecule wet capacity of the fine material, %;

W_b——Capillary wet capacity of the fine material, %。

The sphericity of fine material can be divided into the following series:

k<0.20 non-sphericity

k=0.20~0.35 low-sphericity

k=0.35~0.60 medium-sphericity

k>0.80 high-sphericity

The largest wet capacity of the fine material (%) can be calculated by the following equation:

$$W_a = \frac{(Q_1 - Q_2) \times 100}{Q_2} \tag{2}$$

In the equation:

Q_1——pressurized sample quality, g;

Q_2——dried sample quality, g。

Capillary wet capacity of the fine material:

$$W_a = \frac{Q_3 \times 100}{Q_4} \tag{3}$$

In the equation:

W_a——Capillary wet capacity, %;

Q_3——sample quality water absorption, g;

Q_4——sample quality dried, g。

The table III indicates that dolomite is with the sphericity of k=0.90 itself which can be regarded as good quality in pelletizing. So that dolomite pelletizing is feasible.

Table III Performance test results into a ball							
Q_1 g	Q_2 g	W_a %	Quality of ball after water absorption	Q_4 g	Q_3 g	W_b%	*k*
5.712	5.508	3.70	7.573	7.036	0.537	7.63	0.94
5.601	5.402	3.68	8.074	7.518	0.556	7.40	0.99
5.472	5.275	3.73	6.452	5.981	0.471	7.87	0.90

The particle size distribution of pellets made by different binders

The composite binder is mixed by organic and inorganic binders by certain proportion. In the pelletizing process, investigation on 0.5kg materials with the pelletizing 1h, sieved by a group of circular screens with the aperture 4, 8, 10, 15, 20, 25mm and the particle size analysis was done.

The figure 2 indicates that there is just a little difference in particle sizes between the pellets prepared by bentonite and prepared by pure water. And there is no uniform size in pellets prepared by bentonite. And the pellets prepared by organic binder had a larger particle size, which can not meet the requirement of sintering. The pellets prepared by composite binder had a uniform particle size which is mainly between 10-15mm with excellent effect.

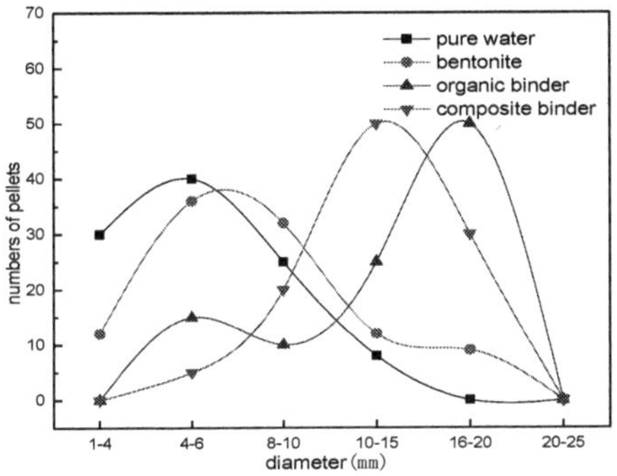

Figure 2 Particle sizes distribution of pellets prepared by different binder

Water content of pellets prepared by different kinds of binders

Dried fine materials cannot be rolled into pellets without a binder, so water must be added in the pelletizing process. Water is quite a significant factor in the pelletizing process. Figure 3 indicates that the addition of binder can improve the water content of the pellets. When the water content of pellets reaches the limitation of the bound water content, the capillaries in the pellets would be thinner, so that better plasticity and better quality of the pellets can be reached.

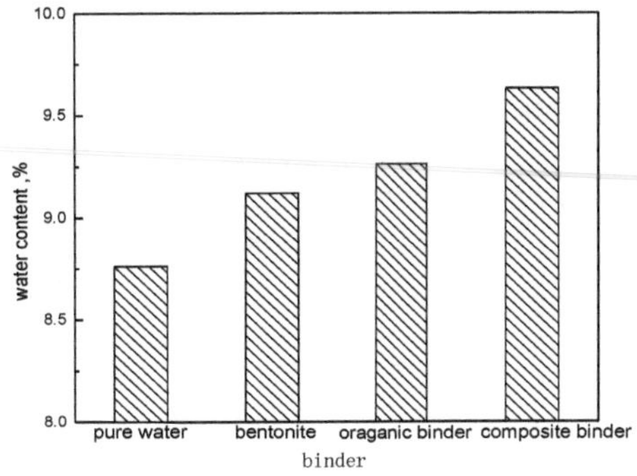

Figure 3 Water content of pellets prepared by different binder

Dropping strength of pellets prepared by different kinds of binders

The figure 4 indicates that there is just a little difference in dropping strength between wet-pellets and dried-pellets which are prepared using the same binder. But there is quite a big difference in dropping strength among pellets prepared by different binders Both the wet-pellets and dried-pellets prepared by composite binders have better dropping strength which can get 2.5times per 0.5m or above.

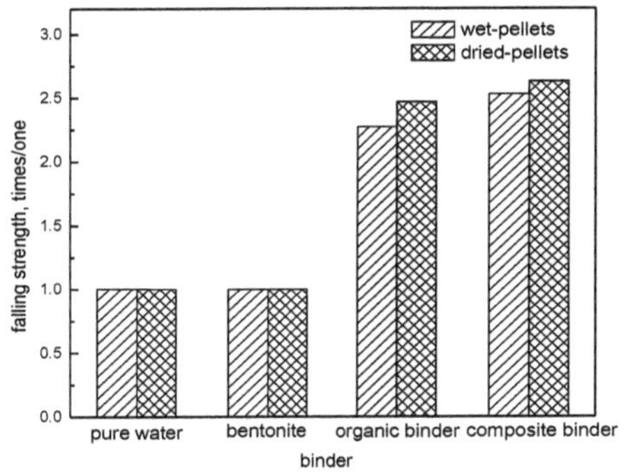

Figure 4 Falling strength of pellets prepared by different binder

Compressive strength of pellets prepared by different binder

Figure 5 indicates that the binders have a quite effect on compressive strength (whether wet-pellets or dried-pellets). The pellets prepared by pure water have the lowest compressive strength; the pellets prepared by composite binders have the highest compressive strength in all. The dried-pellets prepared by composite binders can reach the compressive strength of 80N per one or more, it's mainly because the water-soluble substance in the composite binder, during the drying-process, the water-soluble substance is forced to the surfaces of the green pellets via the effect of capillary force, relatively compact shells are formed quickly after drying which can improve the compressive strength obviously.

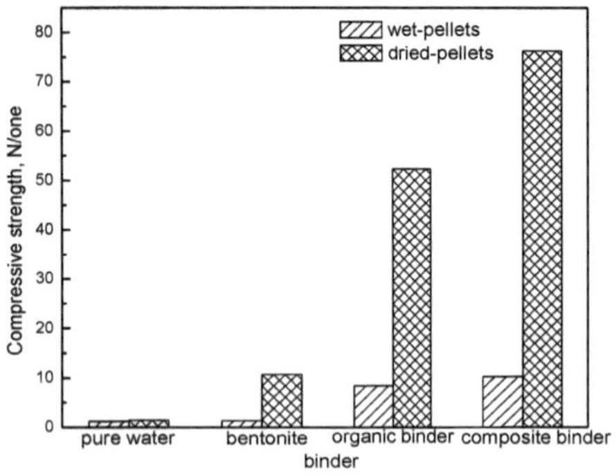

Figure 5 Compressive strength of pellets prepared by different binder

Magnesium loss in the pellets in different sintering temperatures

Sintered the pellets prepared by composite binder in sintering furnace, and investigated the magnesium loss ratio. Shown in figure 6 which indicates the ratio can reach 0.2%, and it is far less than the 5% caused by sintering pure dolomite. And due to the addition of binders in the pellets, no fine materials were formed in the sintering-process, the source utilization was improved and the environment was protected as well.

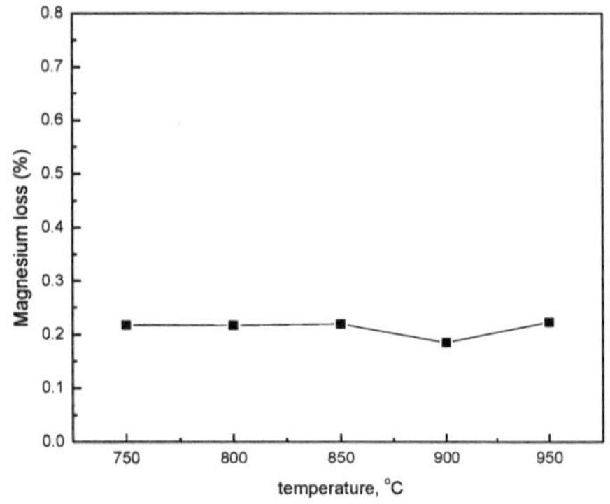

Figure 6 Magnesium loss in different sintering temperatures

The figure 7 indicates that there was no $CaCO_3$ or $MgCO_3$ in the sintered pellets; CaO and MgO were formed in the sintering-process via XRD detection (during the process of acid-dissolution, there was no bubble formed, also proved there was no $CaCO_3$ or $MgCO_3$ from other side). Moreover, the detection result indicates that silicon was not oxidized in the process which is to the benefit of subsequent magnesium reduction.

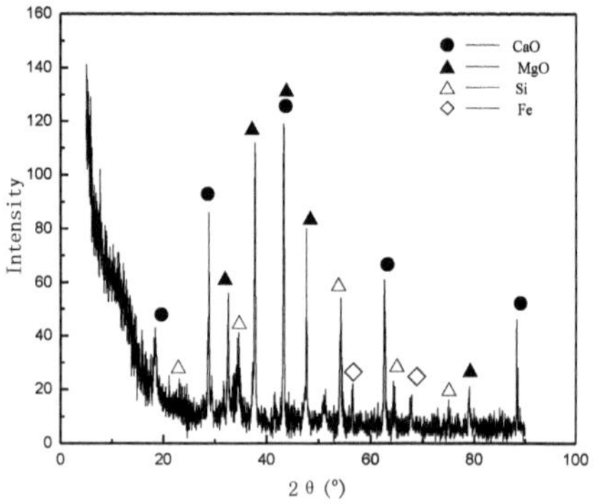

Figure 7 The phase of sintered dolomite pellets

Conclusion

1. Dolomite is with the sphericity of k=0.90 itself which can be regarded as good quality in pelletizing. So that dolomite pelletizing experiment is feasible.

2. The pellets prepared using a composite binder had uniform particle size; bentonite alone is not suitable for dolomite pelletizing; there is no fine material formed in the pellets prepared using a composite binder; the dropping strength of pellets prepared by composite binder can reach 2.5 times per 0.5m and the compressive strength can reach 80N per one.

3. The magnesium loss ratio can reach 0.2% during the sintering-process and no silicon is oxidized.

4. Composite binder which has little impurity, fine and uniform particle, high surface area, good expansibility, cohesiveness and plasticity can improve the quality of the dolomite pellets.

ACKNOWLEDGEMENT

This research was supported by National Natural Science Foundation(NSFC) (51274064,51074044, 51002025).

References

1. BAO Rong Hua, ZHOU Jin Sheng. World magnesium metal supply and demand situation and coping strategies [J]. China Mining Magazine, 2009, 18(12), 7-9.

2. Mustafa Kemal Kulekei. Magnesium and its alloy applications in automotive industry[J]. Int. J. Adv. Manuf. Technol., 2008, 39: 851-865.

3. F.H. Froes, D. Eliezer and E. Aghion. The science, Technology, and Applications of Magnesium[J]. Light Metals, 1998, (9): 30-34.

4. XU Ri Yao. Magnesium metal production technology [M]. Central South University Press, 2003, 1-11.

5. XU Ri Yao. Silicothermic process magnesium metal production technology [M]. Central South University Press, 2003, 4-15.

6. XU Ri Yao. The extraction metallurgy manual of non-ferrous metals (magnesium) [M]. Beijing: Metallurgical Industry Press. 1992, 4-6.

7. Mustafa Kemal Kulekei. Magnesium and its alloy applications in automotive industry[J]. Int. J. Adv. Manuf. Technol., 2008, 39: 851-865.

8. F.H. Froes, D. Eliezer and E. Aghion. The science, Technology, and Applications of Magnesium[J]. Light Metals, 1998, (9): 30-34.

9. B.Humes. Vacuum Engineering as Related to the Dolomite Ferro-Silicon Process, Reduction and Refining of Non-Ferrous Metals[J]. Trans A.I.M.E, 1944, 159: 353.

10. J.R. Wynnyckyj, L.M. Pidgeon. Equilibria in the Silicothermic Reduction of Calcined Dolomite[J]. Metallurgical Transactions, 1971, 2(4): 979-985.

11. J.M. Toguri, L.M. Pidgeon. High-temperature studies of metallurgical process-The thermal reduction of calcined dolomite with silicon[J]. Canadian J. Chem., 1962, 40: 1769-1776.

12. D. Minić, D.Manasijević, Jelena Đokić, et al. Silicothermic Reduction Process in Magnesium Production-Thermal analysis and characterization of the residue[J]. Journal of Thermal Analysis and Calorimetry, 2008, 93(2): 411-415.

13. ZHANG Yi Min. Theory and technology of pellets [M]. Beijing: Metallurgical Industry Press. 1997，76-130.

Magnesium Technology 2013

Texture and Twinning

Magnesium Technology 2013
Edited by: Norbert Hort, Suveen N. Mathaudhu, Neale R. Neelameggham, and Martyn Alderman
TMS (The Minerals, Metals & Materials Society), 2013

Effect of grain size and basal texture on tensile properties and fracture characteristics of extruded AZ31 alloy

Hsiang-Ching Chen, Truan-Sheng Lui, and Li-Hui Chen

Department of Materials Science and Engineering, National Cheng Kung University, Tainan, TAIWAN 701

With different tilting angles corresponding to the basal planes of the extruded Mg-3%Al-1%Zn alloy (0°, 30°, 45°, 60°, 90°), uniaxial tensile tests of the extruded alloy were conducted at room temperature to discuss the effects of basal texture and grain size on its tensile properties and fracture characteristics. Compared to the coarse-grained sample, the fine-grained sample showed obvious increase in the yield stress since the tilting angles were 0° and 90°. However, the total elongation of the fine-grained sample was higher than that of the coarse-grained specimen since the tilting angles ranged from 30° to 60°. According to the observation of fracture surface, a dimple feature was observed in the fine-grained samples. In the contrast, a plate-like feature was recognized on the fracture surface of the sample with coarse grains.

Keywords: AZ31 Mg alloy, tilting angle, grain size, tensile mechanical properties

1. Introduction

Magnesium alloys are regarded as poor formability due to their limited available slip systems at room temperature. Because the magnesium is hexagonal close packed (HCP) crystal structure, the c/a ratio is about 1.624. The previous experimental study [1], the tensile ductility of extruded Mg alloy at room temperature is rather poor due to its (0002) basal plane parallel to the extrusion direction, which is not beneficial to inhibit the occurrence of deformation twins. In general, the formability of AZ31 Mg alloy, twinning is an important in keeping strain compatibility, especially when strained at room temperature. The two twinning systems {10-12}<-1011> and {10-11}<10-1-2> are anticipated in Mg alloy, the former is named as tension twin that is associated with tensile loading along the c-axis, and the latter is named as compression twin that corresponds to compressive loading along the c-axis[2-4].

In this study, the main aim is to discuss the deformation behavior of extruded AZ31 of different grain size and different c-axis tilting angles with tensile direction. Meanwhile, the influence of fracture surface and the microstructure of specimens deformed on the tensile ductility is investigated as well.

2. Experimental procedures

AZ31 Mg alloy with nominal chemical composition of Mg-3.26 mass% Al-0.6 mass% Zn- 0.6 mass% Mn is selected in this investigation as shown in Table1. An as-received billet with above-mentioned composition is extruded to 6-mm thickness and 350-mm width, this as-fabricated plate is machined to the dimension. The extruded plate has grain size of coarse condition due to choose the different temperature during extruded process, also has fine grain size in the extruded plate. In this study, the grain size and basal texture are recognized that on tensile properties and fracture characteristics of extruded AZ31 alloy. The coarse grain of extruded AZ31 is named as EX-1 and the fine grain of extruded AZ31 as

EX-2 in the following content.

Table 1 Chemical composition of AZ31 alloy.

	Al	Zn	Mn	Mg
mass %	3.26	0.58	0.61	Bal.

Microstructure samples for optical microscopy are polished and then etch with a solution of 4.2 g picric acid, 70 mL ethanol, 5 mL acetic acid and 5 mL water. The information of texture and precipitation of the particles is identified by X-ray diffraction spectra.

The tensile specimens are machined to tensile specimens shown in Fig. 1. The tensile tests with different tilting angles (0°, 30°, 45°, 60°, 90°) corresponding to the c-axis of the extruded AZ31. In this study, tensile tests of the specimens are performed at an initial strain rate of $1.67\times10^{-3}s^{-1}$ at room temperature. All tensile data are obtained from more than 3 samples.

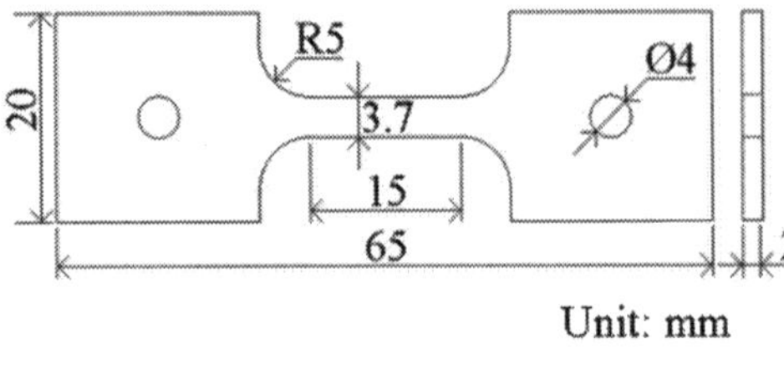

Fig.1 Schematic illustration of the dimensions of tensile specimens.

3. Results and discussion

The microstructure of EX-1 is shown in Fig.2, and that tend to coarse grain size. The average grain size of EX-1 is 184 μm. The average grain size of EX-2 is 13μm (see in Fig. 3).

X-ray diffraction patterns of the normal direction (ND) and transverse direction (TD) are demonstrated in Fig. 4(a) and 4(b) for EX-1 and EX-2 respectively. The result of the TD section in EX-1 and EX-2 are exhibit high (0002) intensity while the ND section indicates higher (10-10) and (10-11) intensity. The extruded possesses a texture in which (0002) basal plane is tended to align parallel to the extrusion direction (ED) and c-axis tend to align parallel to TD.

It is obvious that the grain size of the EX-2(13μm) is great smaller than EX3 as in Fig.2. The yield stress is reciprocal of the square root of the grain size from the Hall-Petch relationship σy = σy0 + kyd-1/2, where σy0 is the resistance of the lattice to dislocation motion or regarded as the yield stress of the single crystal, and ky is a strengthening coefficient relevant to the difficulty to slip across the grain. The empirical relationship is constructed in Fig.5(a) to deduce the grain size dependence of yield stress. It can be seen that the tensile stress of EX-1 with tensile direction parallel with TD (tilting angle 0°, EX-1- 0°) is remarkable lower than of EX-2, and the values of EX-1 and EX-2 become similar at the tilting angle 30°, 45°and 60°.

It is suggested by Armstrong [5] that σ_{y0} is effected by the critical resolved shear stress of the most easily initialized slip system and k_y is related to the critical resolved shear stress of the most difficulty initialized slip system or the twinning system for the metal with the Hexagonal Close Packed (HCP) crystal structure. The expression can be corrected as the following: $\sigma_y = M\tau_c + Mk_sd^{-1/2}$, where M is Taylor orientation factor, τ_c is critical resolved shear stress of the single crystal, k_s is a constant of the shear stress. The Hall-Petch relationship for the grain size and the yield stress of Mg alloy needs to be modified due to the anisotropy of the materials.

The microstructures of the tensile specimens deformed exhibit a lot of twin with tilting angles 0° shown as Fig.5(b). On the other hand, it is found that the deformation twins in the specimens of other tilting angle only can be observed in a few coarse grains

(Fig.6 and Fig.7). But EX-1-0° of yield stress is lowest (Fig.5(a)). This suggests the possibility that coarse grains structure of extruded AZ31 may in fact have low yield strength of the EX-1-0° by some as yet unknown mechanism.

The results of the tensile test with tilting angles 0° are shown that fine grain is no great favor of the elongation on tensile test as Fig.5(b). The better elongation can be acquired on tensile test while (0002) basal plane non-parallel to tensile direction, therefore the basal slip system and/or twinning system will contribute to the increase of tensile flow elongation [6,7]. Furthermore, extruded AZ31 of fine grain characterize anisotropic condition on elongation, but no remarkable changes to the coarse grain. The fracture surface of EX-1 and EX-2 characteristic are respectively shown as Fig.8 and Fig.9. The results of the tilting angle 0° exhibit the form of plate-like and the microvoids. The other tilting angles (30°, 45°, 60°) of EX-1 obviously observe that plate-like characteristic in Fig.8 suggest that fracture either along twin boundaries or shear bands in this coarse grain of AZ31 alloys. The deformation microstructure of EX-2 have exhibit few plate-like structure to the brittle fracture with tilting angles 30° 、45°and 60°, and the microstructure various parts shows ductile fractures on dimple structures in Fig.9. This fine grain of AZ31 alloys suggest that fracture can increase elongation with tilting angles 30° 、45°and 60°.

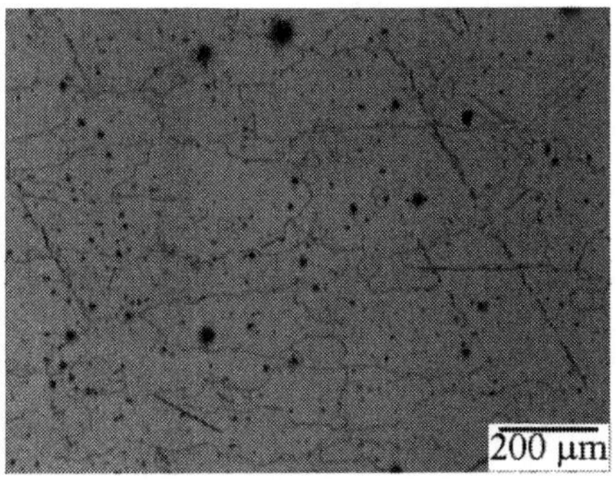

Fig.2 The microstructure of EX-1, the average grain size of EX-1 is 184 μm.

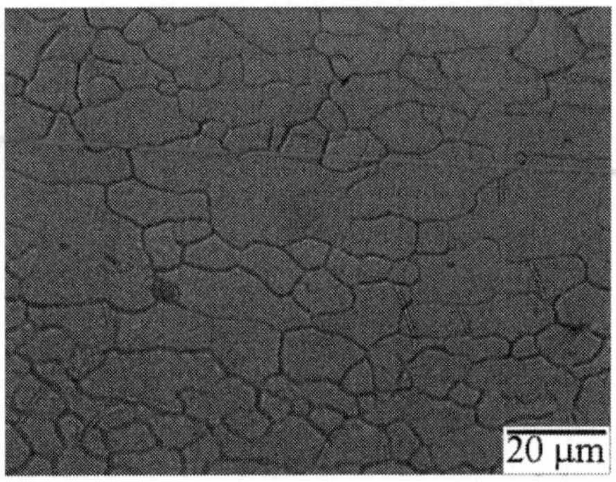

Fig.3 The microstructure of EX-2, the average grain size of EX-2 is 13μm.

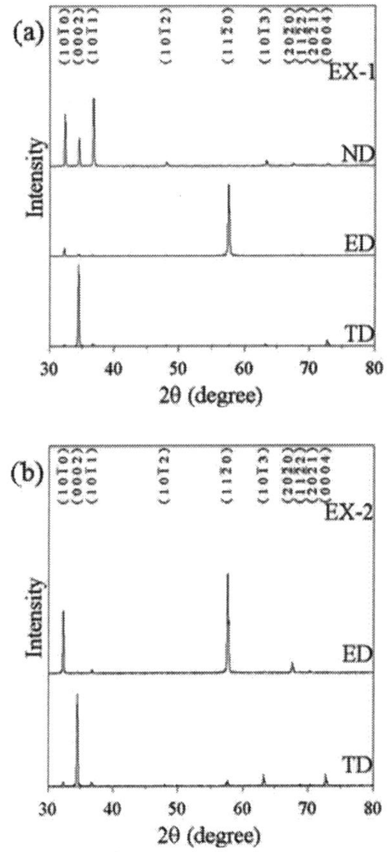

Fig.4 X-ray diffraction patterns : (a) EX-1, (b)EX-2.

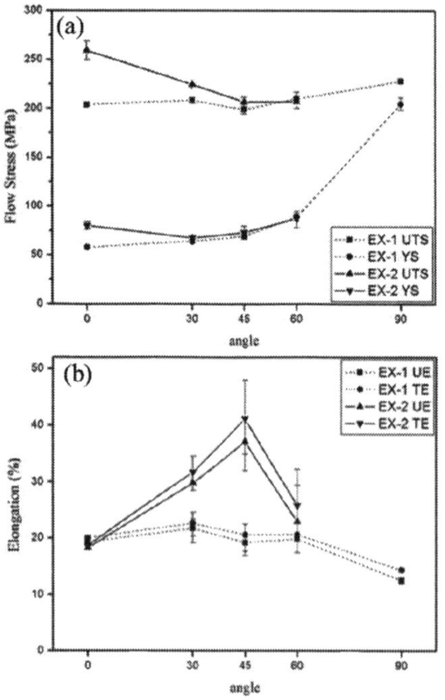

Fig.5 The tensile properties of EX-1 and EX-2: (a)flow stress, (b)elongation.

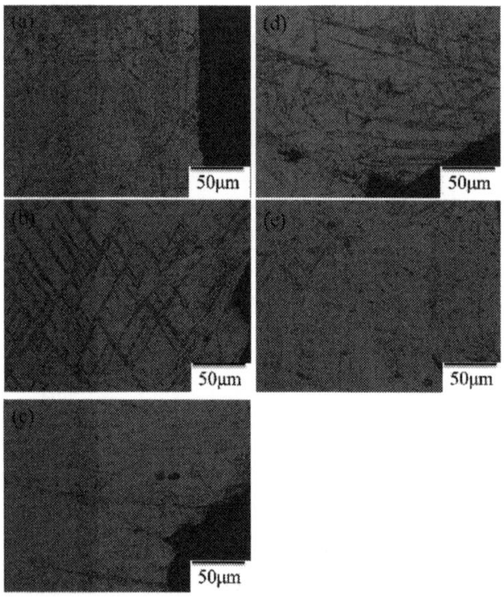

Fig.6 Deformation microstructures of EX-1 with different tilting angle:(a) 0°, (b) 30°, (c) 45°,(d) 60°, (e) 90°.

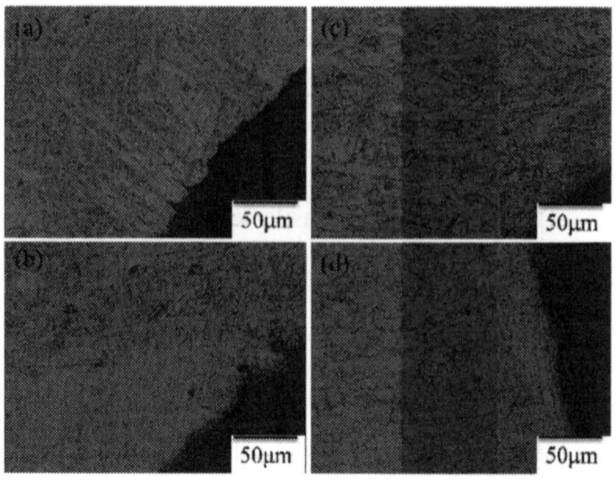

Fig.7 Deformation microstructures of EX-2 with different tilting angle:(a) 0°, (b) 30°, (c) 45°,(d) 60°.

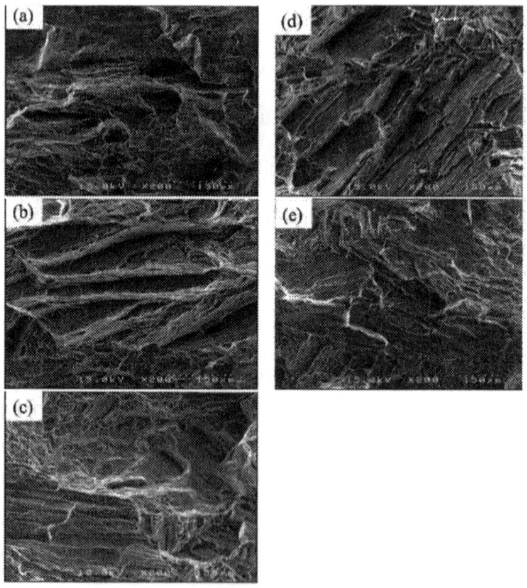

Fig.8 The fracture surface of EX-1 with different tilting angle :(a) 0°, (b) 30°, (c) 45°,(d) 60°, (e) 90°.

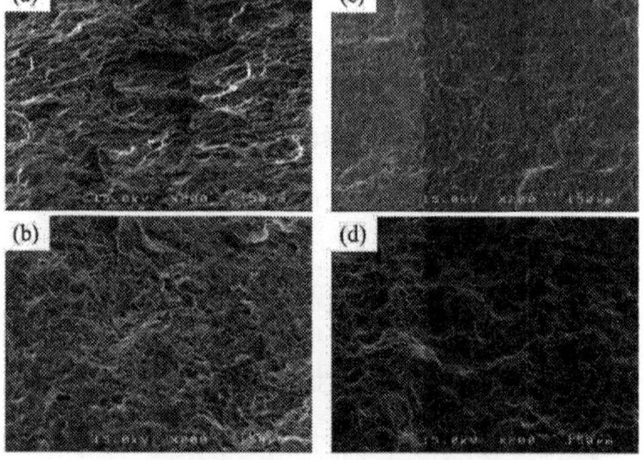

Fig.9 The fracture surface of EX-2 with different tilting angle :(a) 0°, (b) 30°, (c) 45°,(d) 60°.

4. Conclusion

The effects of the grain size and the tilting angle on the microstructure and tensile deformation behavior of the AZ31 magnesium alloy can be summarized as follows:

- The tensile stress of EX-1 with tensile direction parallel with TD (tilting angle 0°, EX-1- 0°) is remarkable lower than of EX-2, and the values of EX-1 and EX-2 become similar with the tilting angle 30°, 45°and 60°.
- The fine grain is characterized anisotropic condition for elongation of extruded AZ31, but no remarkable changes to the elongation of coarse grain with tilting angles 30°、45°and 60°.
- The fracture surface of EX-2 have exhibit few plate-like structure to the brittle fracture with tilting angles 30°、45°and 60°, and the microstructure various parts shows ductile fracture on dimple structures.
- This coarse grains structure of extruded AZ31 may in fact have low yield strength of the EX-1-0° by some as yet unknown mechanism.

Acknowledgements

This paper has been supported by the Chinese National Science Council (contract: No. NSC 101-2221-E-006 -116 -), for which the authors are grateful.

References

[1] H. W. Lee, T. S. Lui and L. H. Chen,"Studies on the improvement of tensile ductility of hot-extrusion AZ31 alloy by subsequent friction stir process",J. *Alloy. Compd.*, 475 (2009) 139-144.

[2] J. Koike, Y. Sato and D. Ando, "Origin of the Anomalous {10-12} Twinning during Tensile Deformation of Mg Alloy Sheet" *Mater. Trans.*, 49 (2008) 2792-2800.

[3] Z. Keshavarz and M. R. Barnett, "EBSD analysis of deformation modes in Mg–3Al–1Zn", *Scr. Mater.* 55 (2006) 915-918.

[4] M. H. Yoo"Slip, Twinning, and Fracture in Hexagonal Close-Packed Metals", *Metall. Trans. A*, 12 (1981) 409-418.

[5] R. W. Armstrong, "Theory of the Tensile Ductile-Brittle Behavior of Polycrystalline H.C.P. Materials, with Application to Beryllium", *Acta Metall.*, 16 (1968) 347-355.

[6] H. Li, E. Hsu, J. Szpunar, R. Verma and J. T. Carter, "Determination of Active Slip/Twinning Modes in AZ31 Mg Alloy near Room Temperature", *J. Mater. Eng. Perform.*, 16 (2007), 321-326.

[7] S. Kleiner and P. J. Uggowitzer, "Mechanical Anisotropy of Extruded Mg-6%Al-1%Zn Alloy", *Mater. Sci. Eng. A*, 379 (2004), 258-263.

Magnesium Technology 2013
Edited by: Norbert Hort, Suveen N. Mathaudhu, Neale R. Neelameggham, and Martyn Alderman
TMS (The Minerals, Metals & Materials Society), 2013

Interaction between a $Mg_{17}Al_{12}$ precipitate and $\{10\bar{1}2\}<10\bar{1}\bar{2}>$ twin boundary in magnesium alloys

B. Li [1], S.N. Mathaudhu [2]
[1] Center for Advanced Vehicular Systems, Mississippi State University, 200 Research Blvd. Starkville, MS 39759, USA
[2] Weapons and Materials Research Directorate, U.S. Army Research Laboratory, Aberdeen Proving Ground, MD 21005, USA

Keywords: Magnesium alloy; twinning; dislocation; precipitate

Abstract

Interactions between $Mg_{17}Al_{12}$ precipitates and $\{10\bar{1}2\}<10\bar{1}\bar{1}>$ twin boundaries (TBs) in magnesium were studied by molecular dynamics simulations. The results obtained agree well with experimental observations in which precipitates can be entirely engulfed by $\{10\bar{1}2\}<10\bar{1}\bar{1}>$ twins without being sheared. Structural analysis of the TBs in the atomic scale shows that the TBs are extremely incoherent during twin growth and highly deviate from the $\{10\bar{1}2\}$ twinning plane as previously observed in a number of experiments. The simulation studies indicate that $\{10\bar{1}2\}<10\bar{1}\bar{1}>$ twinning was accomplished solely by atomic shuffling that converts the parent lattice to the twin lattice without involving twinning dislocations, resulting in zero shear strain at the TBs.

Introduction

Precipitate strengthening is an important approach in alloy design. For metals with cubic structures, precipitate-dislocation interactions play a crucial role in strengthening. But for metals with hexagonal close-packed (hcp) structures such as Mg, precipitates strengthen Mg alloys in two aspects: interactions with dislocation slip and interactions with twin boundaries (TBs). It has long been recognized that precipitate strengthening in Mg alloys is much less effective than in Al alloys with a face center cubic (fcc) crystal structure. This was attributed to the facts that (1) in simple ageing of Mg alloys, the spacing between the equilibrium precipitate plates is not small enough to cause dislocations to shear the precipitates. The lack of shearing of the precipitates reduces the strengthening effect [1]; (2) the precipitate plates are parallel to the basal plane, resulting in reduced effect in blocking the basal slip that is considered the easiest slip system in Mg alloys. Accordingly, there has been effort to modify the orientations of the precipitates such that the plates are inclined with the basal plane [2]. It was anticipated that precipitate plates non-parallel to the basal plane strengthen Mg alloys more effectively.

Interactions between precipitates and twin boundaries in Mg alloys have received little attention. Twinning is generally mediated by twinning dislocations [3] at TBs. Hence, we would expect considerable interactions between precipitates and the twinning dislocations. But the most recent experimental observations [4,5], showed that the TBs of $\{10\bar{1}2\}<10\bar{1}\bar{1}>$ twinning interact with β-intermetallics ($Mg_{17}Al_{12}$) in an unexpected manner: the precipitates are engulfed by $\{10\bar{1}2\}<10\bar{1}\bar{1}>$ twins without being sheared. The motivation of this work is to investigate the interactions between a $Mg_{17}Al_{12}$ precipitate and $\{10\bar{1}2\}<10\bar{1}\bar{1}>$ TBs, using molecular dynamics simulations.

Simulation method

The interatomic potentials for Mg-Al binary alloys used in this work were developed by Liu et al. [6]. The potentials are embedded atom method (EAM) type [7,8].

Figure 1 shows the cross-section view of the initial configuration in which a Mg single crystal with a β-$Mg_{17}Al_{12}$ precipitate was constructed. For simplicity of analysis of the simulation data, we neglected the effect of alloying element (Al) on the motion of a TB and just used pure Mg rather than a Mg-Al solid solution to simulate the interaction between a TB and a $Mg_{17}Al_{12}$ precipitate. We only show a thin slice (~1.5 nm thick) of the system in the middle of through thickness direction. The basal planes of the matrix are perpendicular to the figure plane, and are colored in yellow and blue to show the *ABABAB…* hcp stacking from left to right. This color scheme in 2-D view was kept unchanged throughout the simulation such that the evolution in microstructure during deformation can be recognized with clarity.

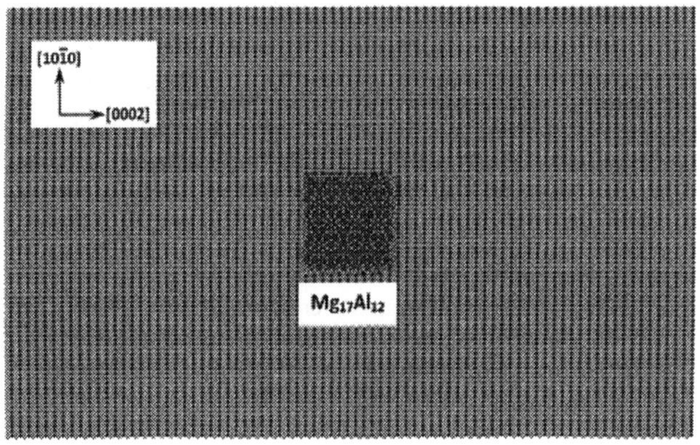

Figure 1. Cross-section view of the initial configuration of a Mg single crystal with a β-$Mg_{17}Al_{12}$ precipitate. Only a thin slice (~ 1.5 nm thick) in the middle of through thickness is shown. The basal planes of the matrix are perpendicular to the figure plane, and are colored in yellow and blue to show the *ABABAB…* stacking from left to right. The matrix and the precipitate satisfy the Burgers orientation relationship: $(0001)_M \parallel (011)_P$, and $[2\bar{1}\bar{1}0]_M \parallel [1\bar{1}1]_P$. The system contains over one million atoms. A tensile strain was applied on the left and the right end of the box along the [0002] direction.

The system contains over one million atoms, with dimensions $40\times20\times20$ nm^3. The crystallographic orientation of the Mg matrix is indicated in Figure 1. The $Mg_{17}Al_{12}$ precipitate has dimensions 5×5×5 nm^3, with a bcc structure and located in the center of the Mg matrix. The precipitate is colored differently from the matrix such that it can be readily identified during deformation. The matrix and the precipitate satisfy the predominant Burgers orientation relationship: $(0001)_M \parallel (011)_P$, and $[2\bar{1}\bar{1}0]_M \parallel [1\bar{1}1]_P$. Although other orientation relationships were also reported [9-11], only the Burgers orientation relationship was considered in this work.

In our simulations, no pre-existing twin boundaries (TBs) were introduced into the system before the simulations started. The simulation scheme is such that twin nucleation and growth is initiated by an external strain. To obtain an extension twin, a tensile strain was applied by moving the atoms on the right-side end at a constant displacement rate of a/10,000 per time step (3.0 fs) along the [0002] direction, where a = 0.32094 nm is the lattice constant of Mg, corresponding to a strain rate about 8.0×10^8 s^{-1}. In this scenario, the $\{10\bar{1}2\}<10\bar{1}\bar{1}>$ extension twinning is favored. The resolved shear stresses on the basal plane and the prismatic plane are zero, so only pyramidal slip can be activated. The simulation temperature was kept constant at 10 K. Free surfaces were applied to all three dimensions.

In addition to 2-D plots, 3-D plots were made to reveal the migration of the TB and the interaction between the TB and the precipitate. These plots are necessary to obtain unambiguous interpretation of the simulation results.

Results and discussions

During straining along the [0002] direction, multiple pyramidal dislocations were produced. The 2-D and 3-D views of the pyramidal dislocations and the resultant stacking faults were shown in Figure 2a. These dislocations are on $\{10\bar{1}1\}$. Li and Ma [12] showed that $\{10\bar{1}1\}<10\bar{1}\bar{2}>$ pyramidal slip is energetically more favorable than $\{11\bar{2}2\}<11\bar{2}\bar{3}>$ for hcp metals because its Burgers vector is smaller and the interplanar spacing is larger [13-15].

As the tensile strain increases, deformation twinning starts to take place. A $\{10\bar{1}2\}$ twin nucleated near the pyramidal dislocations. The twinned region is enclosed by the stacking faults and an incoherent TB delineated by the dashed pink line in Figure 2a can be observed. Morris et al. [16] show that pyramidal dislocations can be energetically preferred for twin nucleation because the nucleation energy barrier is reduced near the dislocation core. By incoherent, we mean that the TB does not match the $\{10\bar{1}2\}$ twinning plane. Notably it can be observed that, in the 2-D view, the TB appears hugely deviating from the $\{10\bar{1}2\}$ twinning plane. Such a deviation can be appreciated from the fact that part of the TB is nearly parallel to the basal plane of the matrix, whereas the theoretical twinning plane $\{10\bar{1}2\}$ should be at 43.1° with the basal plane. To better reveal the orientation relationship between the twin and the matrix, a circled area in Figure 2a that comprises both the matrix and the twin was magnified and shown in Figure 2b. In the twinned region (bounded by the pink dashed line), the stacking of the basal planes was marked with capital letters *ABAB...* from bottom to top, whereas the stacking of the basal planes of the matrix was also marked with *ABAB...* but from left to right. Clearly, the basal plane of the matrix was reoriented by nearly 90°, typical of $\{10\bar{1}2\}<10\bar{1}\bar{1}>$ extension twinning. An important observation is that, after twinning, the originally flat basal planes, i.e., the blue and yellow columns of atoms, convert to the corrugated prism planes of the twin, but the atoms remain nearly in the original vertical direction. In other words, the basal plane of the matrix did not experience a shear which would otherwise deflect the direction of the columns of atoms. To reveal this interesting phenomenon, we drew a vertical red line that crosses the atoms of a basal plane in the matrix and extends into the twin. In the twinned region, the yellow atoms reside on a double-layered prism plane which is geometrically not a flat plane, but on average, the double-layered prism plane nearly parallels to the matrix basal plane. Such a pattern unequivocally demonstrates that no shear strain was produced by the $\{10\bar{1}2\}<10\bar{1}\bar{1}>$ twinning. 3-D view of the motion of the TB was plotted in Figure 2c. The TB is extremely incoherent in 3-D and does not coincide with a strictly defined crystallographic plane. In Figure 2c, part of the TB (indicated by the block arrow) is passing over the precipitate. As the strain increases, the TB expands, and the precipitate is being engulfed while the matrix material in the front is being twinned. Meanwhile, more stacking faults were generated in the matrix.

As the tensile strain further increases, the twinned region continues to expand and reaches the surfaces, as shown in Figure 3, which is the final stage when the precipitate was engulfed by the twin. In the 2-D view, at the top and the bottom interfaces between the precipitate and the matrix, a thin layer of retaining matrix material can be seen, although the precipitate is entirely embedded inside the twin. In 3-D, the TB has swept over the precipitate without causing shear deformation to it. Some of the defects generated during deformation in the matrix prior to twinning were "cleaned" up by twinning.

Our simulation results confirm that a $Mg_{17}Al_{12}$ precipitate can indeed be engulfed by $\{10\bar{1}2\}$ deformation twins without being sheared while the TB is passing the precipitate. Our simulations also reveal that the TBs are extremely incoherent and can hugely deviate from the $\{10\bar{1}2\}$ twinning plane. In other words, the twinning plane of this predominant twinning mode for hcp metals is far from strictly defined. This phenomenon vastly differs from other twinning modes in hcp, fcc and bcc metals in which a twin boundary has to match the twinning plane at least in the atomic scale, although microscopically small deviations no more than a few degrees are allowed due to the presence of twinning dislocation loops at the TB and strain accommodation. It is required that a TB match the twinning plane because a twinning plane has to be an "invariant plane" during straining [3] and the glide of twinning dislocations is strictly confined in the twinning plane. If $\{10\bar{1}2\}<10\bar{1}\bar{2}>$ twinning were mediated by twinning dislocations, as suggested previously [17-18], we would expect a strong interaction between the precipitate and the twinning dislocations. The twinning dislocations would either shear the precipitate if $Mg_{17}Al_{12}$ is deformable, or be impeded at the precipitate/matrix interface if $Mg_{17}Al_{12}$ is non-deformable. Either of these two cases would result in strengthening to the material, especially if the intermetallic precipitate is non-deformable. But these interactions did not occur in experiments [4-5] and in our atomistic simulations.

Partridge and Roberts [19] first reported extremely incoherent TBs in Zn and Mg in their interesting experiments. Large deviations between TBs and the $\{10\bar{1}2\}$ twinning plane were also observed by Zhang et al. [20] in their transmission electron microscopy (TEM) analysis on deformed pure Co and

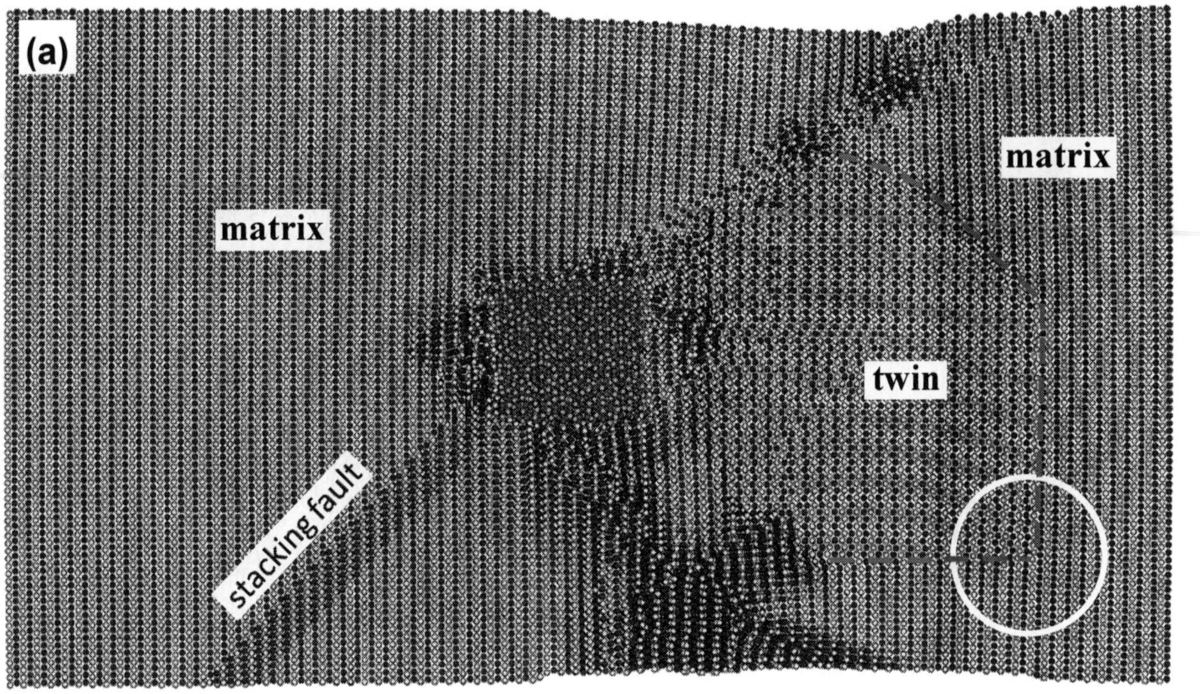
(a)
matrix
matrix
twin
stacking fault

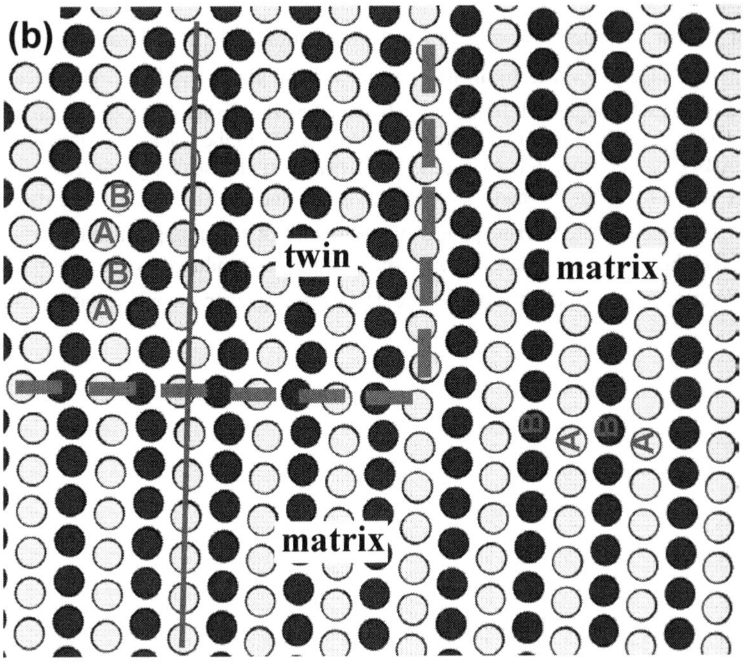
(b)
B
A
B
A
twin
matrix
B
A
B
A
matrix

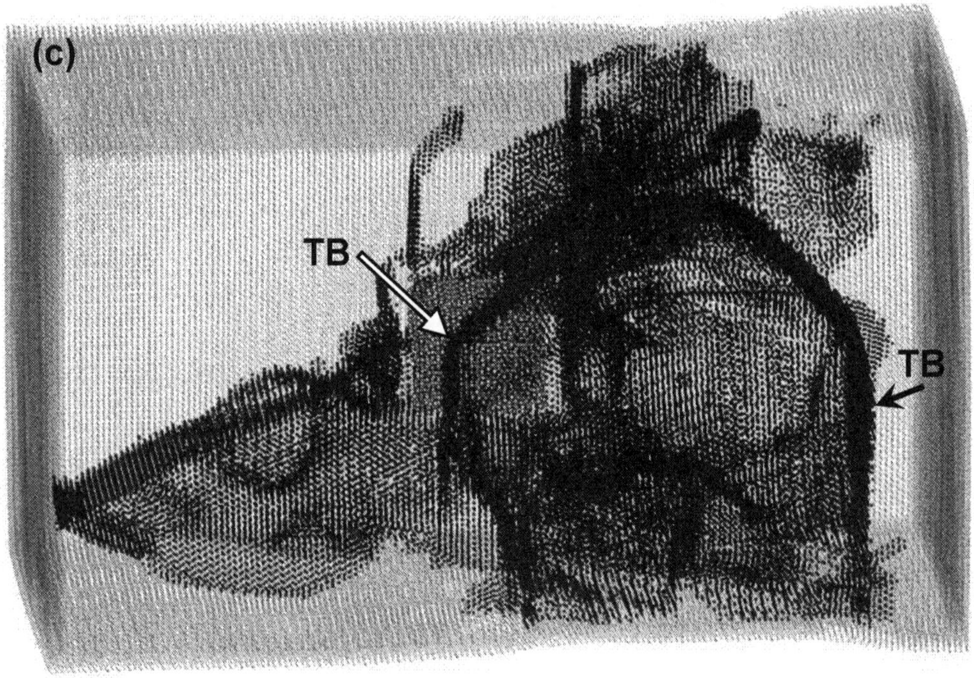

Figure 2. (a) A $\{10\bar{1}2\}$ twin nucleated near the pyramidal dislocations. The twin boundary (TB) is delineated by the dashed pink line. The TB hugely deviates from the $\{10\bar{1}2\}$ twinning plane. (b) The circled area in (a) that comprises both the matrix and the twin was magnified. In the twin (bounded by the dashed line), the stacking of the basal planes is marked with capital letters ABAB… from bottom to top, distinctive from the stacking of the basal planes of the matrix (also marked with ABAB… but from left to right). The change in the stacking sequence indicates that twinning reoriented the matrix lattice by about 90° and the $\{10\bar{1}2\}$ twinning indeed took place. After twinning, the initially flat basal planes convert to the corrugated prism planes of the twin, but the atoms remain nearly in the same vertical direction. A vertical straight line demonstrates that zero shear strain was produced by the $\{10\bar{1}2\}$ twinning. (c) 3-D view of the TB. The TB is extremely incoherent. Part of the TB is passing over the precipitate, as indicated by the block arrow.

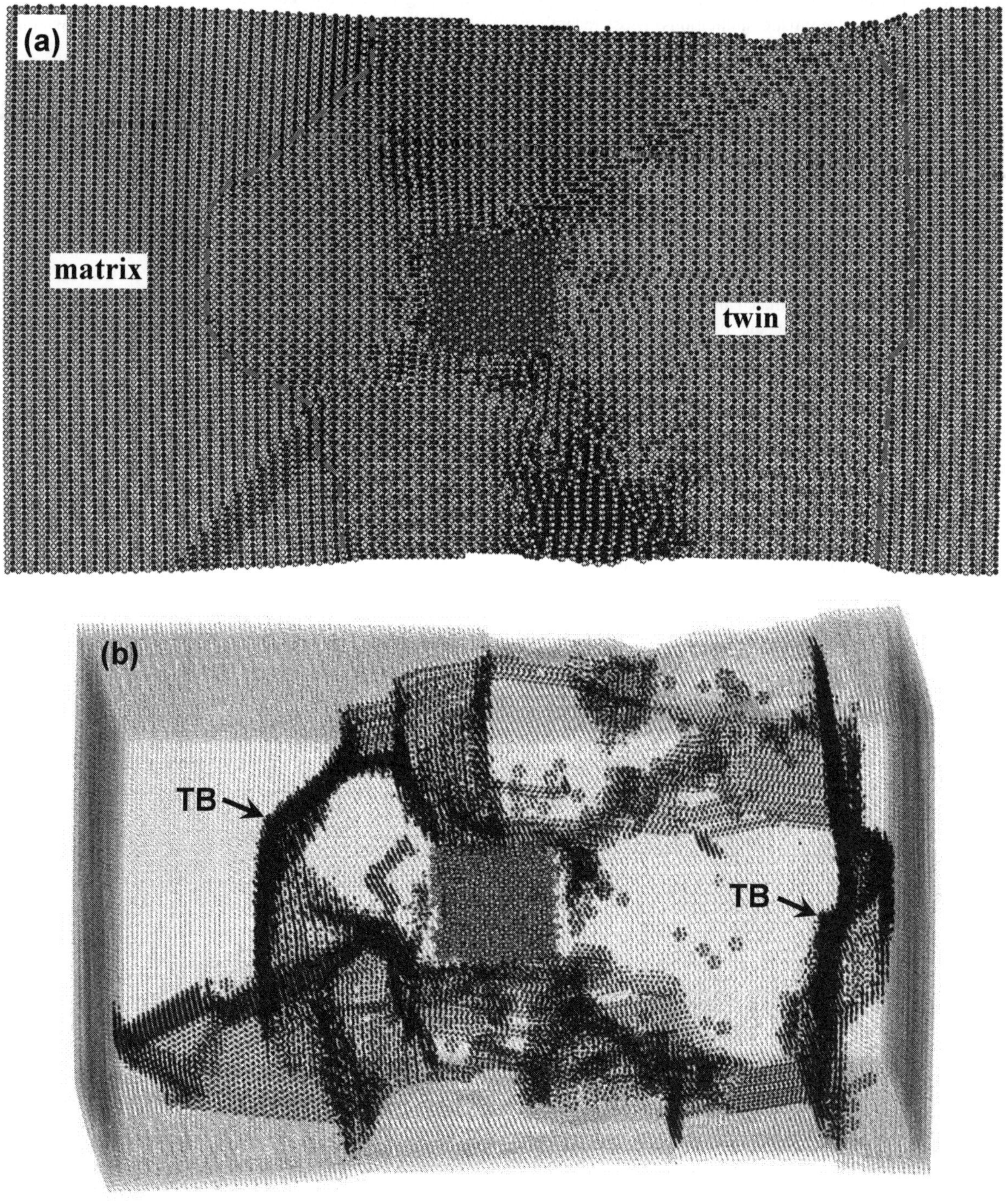

Figure 3. (a) The precipitate is entirely engulfed by the twin without being sheared. The TBs are extremely incoherent. (b) 3-D view. The twinned region expands, digesting the defects generated in the matrix during deformation.

Mg samples. The TBs were observed to deviate from the $\{10\bar{1}2\}$ twinning plane with a magnitude greater than 45°, similar to the deviations observed in our atomistic simulations. Most recently, Liu et al. [21] reported *in-situ* TEM observations of tensile tests on single crystal pure Mg and found that in both twinning and detwinning no shear strain was produced on the specimen.

The experimental observations and our simulations strongly suggest that $\{10\bar{1}2\}<10\bar{1}\bar{1}>$ twinning in hcp metals is not mediated by twinning dislocations. In fact, if $\{10\bar{1}2\}<10\bar{1}\bar{1}>$ twinning were controlled by twinning dislocations, the extensively observed reversible twinning in Mg [22-24] should not happen because dislocation-dominated plastic deformation in crystalline materials is irreversible.

Bilby and Crocker [25] calculated the Burgers vectors and the required atomic shuffling for the twinning modes in hcp lattices. Christian and Mahajan [3] described in detail shear and shuffling of individual twinning modes in various lattices. Shuffling is necessary for most of the twinning modes in hcp metals because a homogeneous shear cannot carry the parent lattice to the twin lattice. Local adjustments are needed to accomplish twinning with a correct lattice structure and orientation relationship.

For the predominant twinning mode in hcp metals, i.e., $\{10\bar{1}1\}<10\bar{1}\bar{2}>$, the magnitude of the theoretical Burgers vector of an elementary twinning dislocation [26] equals 0.024 nm, which is less than one tenth of the Burgers vector of the basal dislocations. Thompson and Millard [26] first suggested that this twinning dislocation comprises two twinning planes in the same time, i.e., a two-layer zonal dislocation. Even so, the overall Burgers vector would still equal 0.05 nm, much smaller than any known Burgers vectors of matrix dislocations. Such an unusually tiny Burgers vector of the twinning dislocation strongly suggests that the controlling mechanism of this particular twinning mode vastly differs from other modes. In fact, this small Burgers vector implicitly conveys important structural information: the twin lattice almost exists in the parent lattice already, before an external load is applied or any twinning shear is initiated.

Li and Ma [27] demonstrated how atomic shuffling can accomplish the $\{10\bar{1}2\}<10\bar{1}\bar{1}>$ twinning, without involving well-defined twinning dislocations. In their model, the lattice conversion between the twin and the parent is achieved by atomic shuffling that converts the parent basal planes to the twin prism planes, and the parent prism planes to the twin basal planes. This direct lattice conversion requires atomic shuffling to create correct hcp lattice and twin orientation relationship. No twinning shear is involved at all. Because of the shuffling dominated twinning, the twin boundaries can pass precipitates, but leaving the precipitates unsheared. Also, because no twinning dislocations are involved, the actual twin boundaries do not have to match the $\{10\bar{1}2\}$ twinning plane and can be extremely incoherent, as seen in our simulations.

Conclusions

We simulated interactions between a $Mg_{17}Al_{12}$ intermetallic precipitate and a $\{10\bar{1}2\}<10\bar{1}\bar{1}>$ deformation twin in Mg, using molecular dynamics. The results reveal that the precipitate can be entirely engulfed by the twin, consistent with experimental observations. The TB sweeps over the precipitate without shearing it. The TBs in the atomistic simulations were observed extremely incoherent and hugely deviate from the $\{10\bar{1}2\}$ twinning plane, also consistent with experimental observations. These observations suggest that $\{10\bar{1}2\}<10\bar{1}\bar{1}>$ twinning is not mediated by twinning dislocations.

A mechanism that is able to account for the zero shear strain and other properties of $\{10\bar{1}2\}<10\bar{1}\bar{1}>$ twinning was discussed. In this mechanism, atomic shuffling dominates the twinning and no twinning dislocations are needed at TB. Zero shear strain is produced by atomic shuffling.

Acknowledgements

This work was supported at Center for Advanced Vehicular Systems (CAVS), Mississippi State University, by W911NF-12-1-0023. B. Li gratefully acknowledge the support from CAVS, Mississippi State University.

References

[1] J.B. Clark, Acta Met 16 (1968) 141.
[2] J.F. Nie, Scripta Mater 48 (2003) 1009.
[3] J.W. Christian, S. Mahajan, Prog. Mater Sci 39 (1995) 1.
[4] N. Stanford, M.R. Barnett, Mater Sci Eng A 516 (2009) 226.
[5] J.D. Robson, N. Stanford, M.R. Barnett, Acta Mater 59 (2011) 1945.
[6] X.Y. Liu, J.B. Adams, F. Erocolessi, J.A. Moriarty. Modell. Simul Mater Sci Eng 4 (1996) 293.
[7] M.S. Daw, M.I. Baskes, Phys Rev B 29 (1984) 6443.
[8] M.S. Daw, M.I. Baskes, Phys Rev Lett 50 (1983) 1285.
[9] D. Duly. Acta Metall Mater 41 (1993) 1559.
[10] D. Duly, M.C. Cheynet, Y. Brechet, Acta Metall Mater 42 (1994) 3843.
[11] S. Celotto, Acta Mater 48 (2000) 1775.
[12] B. Li, E. Ma, Phil Mag 89 (2009) 1223.
[13] S.R. Agnew, J.A. Horton, M.H. Yoo. Metall Trans A 33A (2002) 851.
[14] T. Obara, H. Yoshinaga and S. Morozumi, Acta Metall. 21 (1973) 845.
[15] J.F. Stohr and J.P. Poirier, Phil Mag 25 (1972) 1313.
[16] J.R. Morris, K.M. Ho, K.Y. Chen, G. Rengarajan, M.H. Yoo, Modell. Simul. Mater. Sci. Eng. 8 (2000) 25.
[17] A. Serra, D.J. Bacon, Mater Sci Eng A 400-401 (2005) 496.
[18] J. Wang, J.P. Hirth, C.N. Tome, Acta Mater 57 (2009) 5521.
[19] P.G. Partridge and E. Roberts, Acta Met 12 (1964) 1205.
[20] X.Y. Zhang, B. Li, X.L. Wu, Y.T. Zhu, Q. Ma, Q. Liu, P.T. Wang, M.F. Horstemeyer, Scripta Mater 67 (2012) 862-865.
[21] B.Y. Liu, Z.W. Shan, X.Y. Zhang, J. Sun, E. Ma, "In-situ quantitative tension and compression study on twinning and detwinning in submicron-sized Mg crystals inside a transmission electron microscope", TMS 2012, Orlando, USA.
[22] L. Wu, A. Jain, D. W. Brown, G.M. Stoica, S.R. Agnew, B. Clausen, D.E. Fielden, P.K. Liaw, Acta Mater 56 (2008) 688.
[23] A. Jain, S.R. Agnew, Mater Sci Eng A 462 (2007) 29.
[24] C.H. Caceres, T. Sumitomo, M. Veidt, Acta Mater 51 (2003) 6211.
[25] B.A. Bilby, A.G. Crocker, Proc Roy Soc. A, 288 (1965) 240.
[26] N. Thompson, D.J. Millard, Phil Mag 43 (1952) 422.
[27] B. Li and E. Ma, Phys Rev Lett 103 (2009) 035503.

Magnesium Technology 2013
Edited by: Norbert Hort, Suveen N. Mathaudhu, Neale R. Neelameggham, and Martyn Alderman
TMS (The Minerals, Metals & Materials Society), 2013

MICROSTRUCTURE AND TEXTURE EVOLUTION IN A MAGNESIUM ALLOY DURING EXTRUSION AT VARIOUS EXTRUSION SPEEDS

Q. Ma[1], S.J. Horstemeyer[1], B. Li[1], Z. McClelland[1], P.T. Wang[1], M.F. Horstemeyer[1,2]

[1]. Center for Advanced Vehicular Systems, Mississippi State University, Starkville, MS 39759, USA
[2]. Department of Mechanical Engineering, Mississippi State University, Starkville, MS 39762, USA

Keywords: Magnesium; Texture; Twinning; Recrystallization; EBSD.

Abstract

An AM30 magnesium alloy was extruded by using a lab-scale flat die at ~450 °C and various ram speeds: 5 mm/min, 10 mm/min, 20 mm/min, 30 mm/min, and 50 mm/min, respectively. Microstructure and texture in the representative locations inside the die and the extrudate of the AM30 at different ram speeds were examined by electron backscatter diffraction (EBSD). Significant dynamic recrystallization (DRX) occurred inside the die, whereas static recrystallization (SRX) took over in the extrudate outside the die. Profuse $\{10\bar{1}2\}\langle10\bar{1}\bar{1}\rangle$ extension twinning activated during extrusion at low ram speed of 10 mm/min, but twinning was hardly observed at high ram speed of 50 mm/min. DRX and SRX led to different microstructure evolution at different extrusion speeds. Possible mechanisms that govern the DRX and the SRX were analyzed.

Introduction

Magnesium (Mg) alloys are potential lightweight materials for automotive and aerospace industry. But the low ductility of Mg and its alloys at ambient temperature hinders their applications. Wrought Mg alloys are usually subjected to deformation at high temperature, e.g. extrusion at temperatures typical of 350-450 °C [1-3]. Extrusion is an important thermomechanical processing for Mg alloys in industry. Though extensive studies have been performed on simple compression, tension, rolling and extrusion of Mg alloys at elevated temperatures, there are limited studies focused on how microstructure and texture evolve during extrusion inside the die [3]. In extrusion, deformation and dynamic recrystallization (DRX) inside the die, and the subsequent static recrystallization (SRX) outside the die dominate the final microstructure, texture and property of the product. Since the stress state in extrusion differs from uniaxial compression, tension, and rolling, the deformation, DRX and SRX mechanisms during extrusion should exhibit different characteristics. Therefore, understanding the mechanisms of deformation and recrystallization of Mg alloy during extrusion is of significance for process optimization.

It is well known that twinning, basal slip, prismatic slip, <c+a> slip and DRX are strongly affected by temperature and strain rate [4-5]. Conceivably, DRX and SRX will be influenced by extrusion temperature and extrusion speed as well. However, at typical extrusion temperature of around 450 °C, how the extrusion ram speed change influences the microstructure and texture of the extrudate of Mg alloy is still not fully understood.

The aim of this work is to investigate the effect of extrusion speed on the microstructure and texture evolution of a commercial Mg alloy during a lab-scale flat-die extrusion at 450 °C.

Experimental

A commercial extruded Mg alloy billet AM30 (Al 2.54%, Mn 0.40%, balance Mg, wt%) with a diameter of 178 mm from Timminco Metals Corporation was selected as the experimental material. The cylindrical lab-scale extrusion billet was cut from the as-extruded AM30 billet with the axis parallel to the extrusion direction (ED). The lab-scale billet had a diameter of 31.8 mm and a height of 25.4 mm. An in-house lab-scale flat die was used in this study. The extrudate has a diameter of 6.35 mm. The extrusion ratio is 25.1. The flat die was fixed in a furnace chamber that held the sample stage of the INSTRON 8850 machine. A high temperature graphite paste GPL-420 was used as lubricant during extrusion. The lab-scale extrusion billet was heated up to 450 °C inside the die and soaked for 30 minutes. Then the billet was indirectly extruded at different ram speeds as: 5 mm/min, 10 mm/min, 20 mm/min, 30 mm/min, and 50 mm/min, respectively. After extrusion, the AM30 extrusion butt inside the flat die and the extrudate outside the die were cut and cold mounted to study the microstructure. The cold mounted samples were polished using series SiC papers down to a grit number of 4000. Then the AM30 samples were polished using alumina (down to 0.05 μm) suspended in ethylene glycol. The polished samples were etched using a solution of HNO_3 and acetic acid [5]. The etched samples were then examined by electron backscatter diffraction (EBSD) scan in the different representative positions inside the die or in the extrudate outside the die. Some samples and the extrusion butts were also examined by optical microscope (OM) to characterize the microstructure change of the AM30. Figure 1 presents the lab-scale flat die and the extrusion fixture in this study.

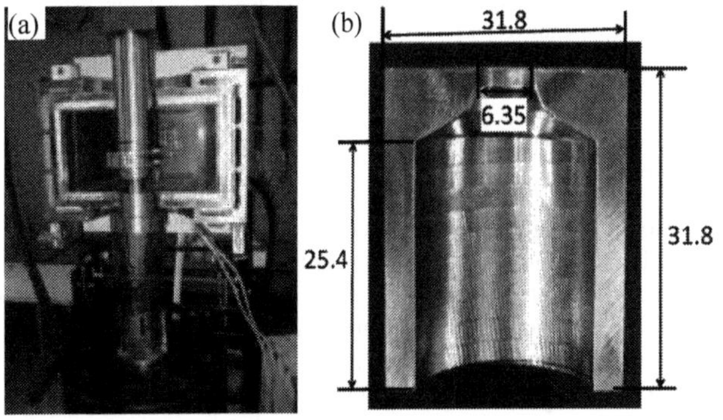

Figure 1: Lab-scale extrusion fixture inside the furnace chamber on the INSTRON 8850 machine (a), and a magnified picture of the flat die section (b).

Results

After extrusion, the extrusion butt of the AM30 inside the die was cut along ED with the cut section comprising the die axis. Figure 2 presents typical optical microscope (OM) pictures at different representative positions termed as Top, Middle, and Exit positions inside the die of AM30. The microstructure in the extrudate of AM30 exit far outside the die is defined as the Extrudate position.

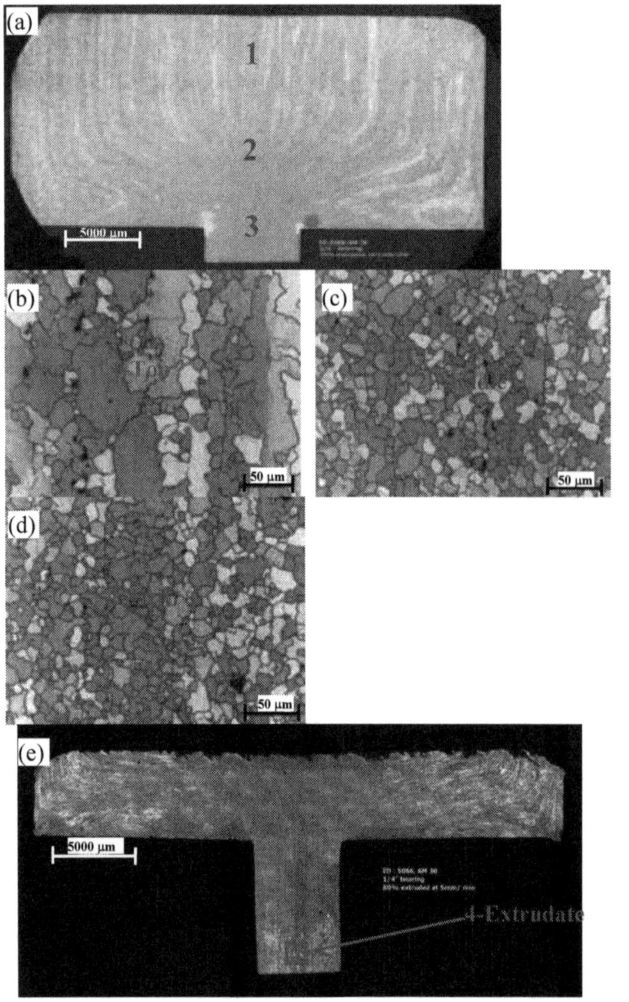

Figure 2: Optical microscope microstructure of the AM30 during extrusion at different representitive positions inside the die and the extrudate extruded at 450 °C and ram speed of 5 mm/min. (a) representitive positions inside the flat die.1-Top, 2-Middle, 3-Exit; (b) Top microstructure; (c) Middle microstructure; (d) Exit microstructure; and (e) the Extrudate position and the extrsuion butt. Extrusion flow lines and different microstructures can be seen in (a) and (b)-(e), respectively.

As shown in Figure 2, microstructure dynamically changes during extrusion. The big grain zone appears in the extrudate due to grain growth after extrusion by the meta-dynamic recrystallization or SRX. Microstructure and texture scanned by EBSD in different representative positions of the AM30 extruded at ram speed of 10, 20, 30, and 50 mm/min are presented in Figures 3-6.

At ram speed of 10 mm/min, both microstructure and texture significantly change during extrusion (Figure 3). The texture in the Middle and the Exit position differ, but texture in the Exit and the Extrudate position are similar where $\langle 10\bar{1}0\rangle$ and $\langle 2\bar{1}\bar{1}0\rangle$ parallel to ED oriented grains (designated as $\langle 10\bar{1}0\rangle$ grains and $\langle 2\bar{1}\bar{1}0\rangle$ grains, respectively hereafter) appear (Figures 3e,3f).

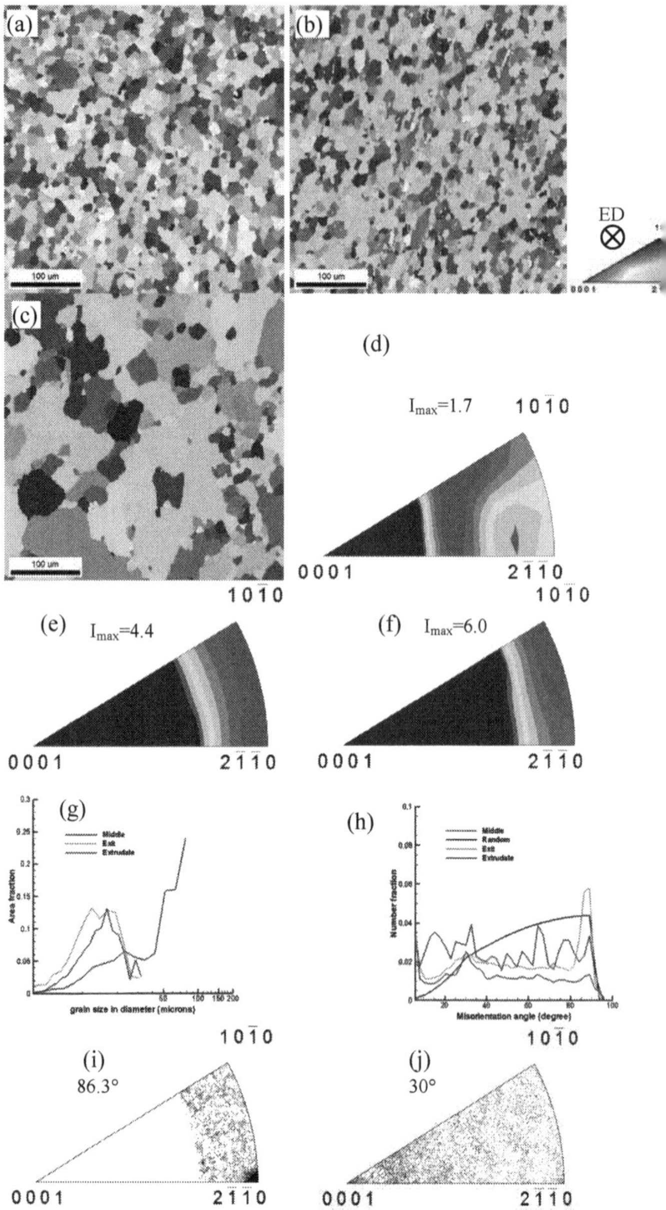

Figure 3: Microstructure and texture in different representative positions of the AM30 extruded at 450 °C and a ram speed of 10 mm/min. EBSD inverse pole figure (IPF) maps: (a) Middle; (b) Exit; and (c) Extrudate. The inverse pole figure of the extrusion direction (ED) in representative positions: (d) Middle; (e) Exit; and (f) Extrudate. (g) Grain size distribution; and (h) misorientation distribution and rotation axis/angle pair distribution at 86.3±2.5° (i) and 30±2.5° (j) in the Exit position. All these results are based on EBSD results in Figures 3(a)-(c). The color triangle represents the extrusion direction (ED).

Similar to 5 mm/min, grain dramatically grows in the Extrudate position as well (Figure 3g). The misorientation distribution and the rotation axis/angle pairs show that profuse twinning occurs at the Exit position. But twinning weakens in the Extrudate outside the die (Figure 3h). Twinning can be verified by the 86.3°<$1\bar{2}10$> orientation relationship which is the characteristic of $\{10\bar{1}2\}\langle 10\bar{1}\bar{1}\rangle$ twin boundary. Note that a misorientation peak appears at around 30°, and a visible preferred rotation axis clustering at <0001> in the Exit position of the AM30 (Figures 3h,3j).

After extrusion, there is an evident transition area where some grains quickly grow at the expense of other grains. Figure 4 shows a typical transition area that includes both big grains and small grains at the ram speed 20 mm/min. In this case, two main texture components of $\langle 10\bar{1}0\rangle$ and $\langle 2\bar{1}\bar{1}0\rangle$ can be seen (Figures 4b,4c). Furthermore, $\langle 2\bar{1}\bar{1}0\rangle$ grain seems to grow preferentially over other orientations, e.g. the $\langle 10\bar{1}0\rangle$ grain. Grain size distribution shows a bimodal (Figure 4d). Misorientation distribution shows a high density of low angle grain boundaries (<6°), and a 30° and a 90° peak (Figure 4e). The misorientation and rotation axis/angle pair distribution show that $\{10\bar{1}2\}$ twins still exist in the Extrudate of the AM30 extruded at 450 °C and a ram speed of 20 mm/min (Figures 4e,4f). The misorienation 30° peak and the rotation axis/angle <u v w>δ° pair imply that <0001> 30° orientation relations may play roles on texture evolution (Figures 4e,4g).

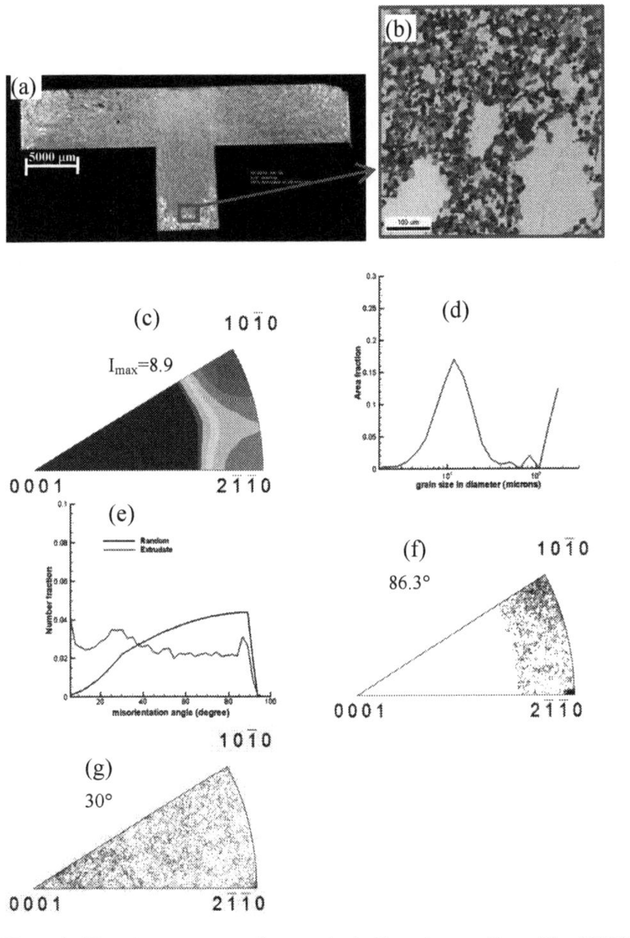

Figure 4: The microstructure and texture in the Extrudate position of the AM30 extruded at 450 °C and ram speed of 20 mm/min. (a) the microstructure of the Extrudate; (b) EBSD inverse pole figure map of the transition area from small grains to big grains; (c) inverse pole figure of ED; (d) grain size distribution; and (e) misorientation distribution and rotation axis/angle pair distribution at 86.3±2.5° (f) and 30±2.5° (g).

As ram speed increases to 30 mm/min, the Middle position texture appears different from the Exit position texture (Figures 5a-d). But grains grow not as much (Figure 5e). Misorienation distribution in the Middle looks similar to that in the Exit position. A high peak of low angle boundaries exists in the Middle and the Exit position. Strikingly, few twins are seen at the ram speed 30 mm/min according to misorientation and rotation axis/angle pair distribution (Figures 5f,5g). Similar to the case of ram speed of 10 and 20 mm/min, <0001>30° orientation relationship seems to play roles in microstructure and texture evolution during extrusion (Figures 5f,5h).

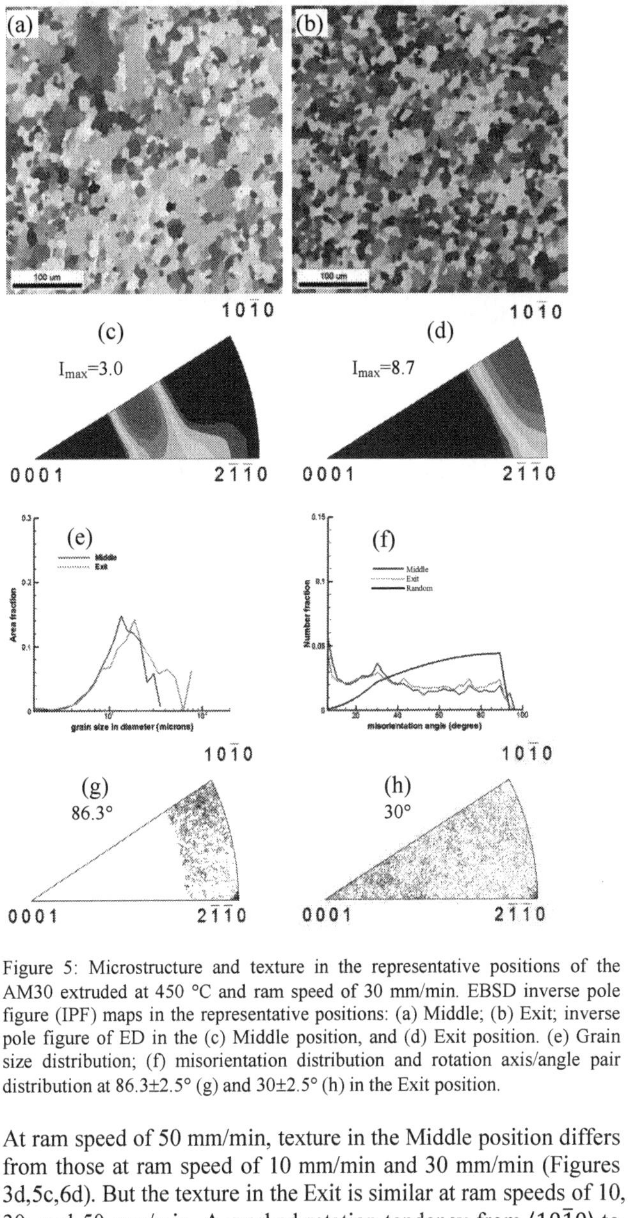

Figure 5: Microstructure and texture in the representative positions of the AM30 extruded at 450 °C and ram speed of 30 mm/min. EBSD inverse pole figure (IPF) maps in the representative positions: (a) Middle; (b) Exit; inverse pole figure of ED in the (c) Middle position, and (d) Exit position. (e) Grain size distribution; (f) misorientation distribution and rotation axis/angle pair distribution at 86.3±2.5° (g) and 30±2.5° (h) in the Exit position.

At ram speed of 50 mm/min, texture in the Middle position differs from those at ram speed of 10 mm/min and 30 mm/min (Figures 3d,5c,6d). But the texture in the Exit is similar at ram speeds of 10, 30, and 50 mm/min. A marked rotation tendency from $\langle 10\bar{1}0\rangle$ to $\langle 2\bar{1}\bar{1}0\rangle$ can be seen during extrusion at ram speed of 50 mm/min (Figures 6d-f). In the Extrudate position of the AM30, abnormal grain growth can be observed where $\langle 2\bar{1}\bar{1}0\rangle$ grains rapidly grow at the expense of other oriented grains (Figures 6c,6g). Misorientation and rotation axis/angle pair show that almost no $\{10\bar{1}2\}$ twinning activates during extrusion at the ram speed 50 mm/min. The <0001>30° orientation relation again be verified based on the misorienatation and rotation axis/angle distribution (Figures 6h,6j).

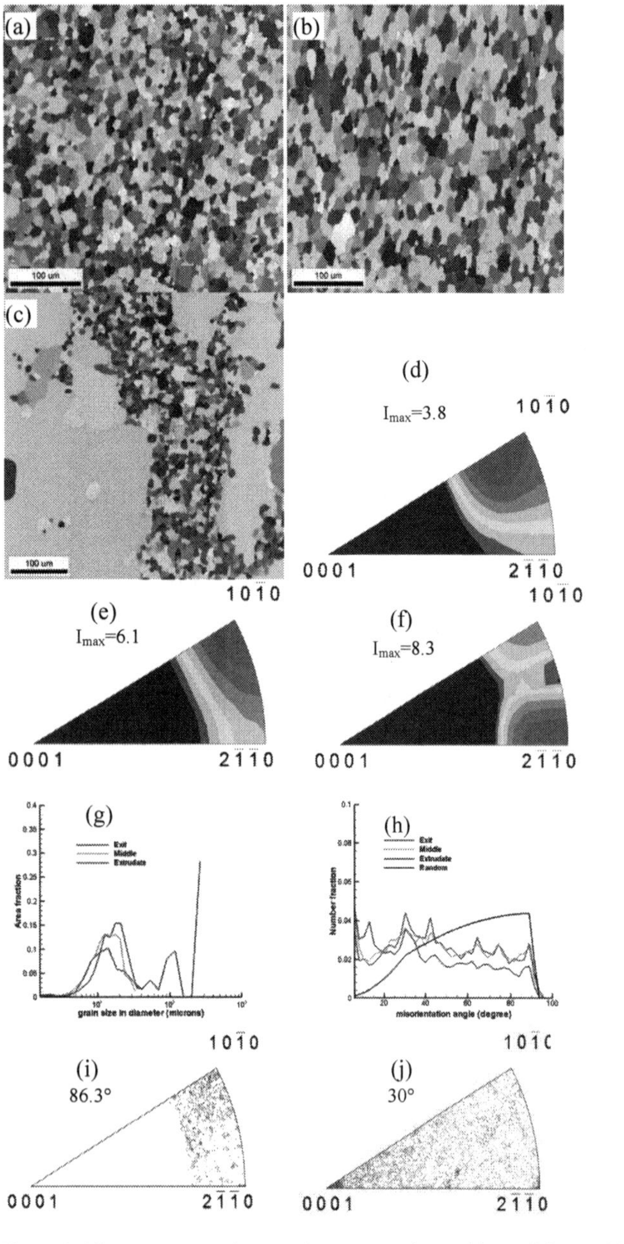

Figure 6: Microstructure and texture in representative positions of the AM30 extruded at 450 °C and ram speed of 50 mm/min. EBSD inverse pole figure (IPF) maps in the representative positions: (a) Middle; (b) Exit; (c) Extudate. The ED inverse pole figure at different positions: (d) Middle, (e) Exit; and (f) Extrudate. (g) Grain size distribution, and (h) misorientation distribution and rotation axis/angle pair distribution at 86.3±2.5° (i) and 30±2.5° (j) in the Extrudate position.

Discussion

During extrusion at high temperature of 450 °C, not only common deformation modes such as basal slip $\{0001\}<11\bar{2}0>$ and extension twinning $\{10\bar{1}2\}\langle10\bar{1}\bar{1}\rangle$ will be activated, but also non-basal slips, including prismatic $\{10\bar{1}0\}<11\bar{2}0>$ slip and $<c+a>$ pyramidal slips may also be activated . Meanwhile, DRX occurs and causes softening.

Extension $\{10\bar{1}2\}\langle10\bar{1}\bar{1}\rangle$ twinning is predominant in Mg alloys; contraction twinning $\{10\bar{1}1\}\langle10\bar{1}\bar{2}\rangle$ and double twinning $\{10\bar{1}1\}-\{10\bar{1}2\}$ are also often observed in Mg alloy near fracture [6,7]. Extension $\{10\bar{1}2\}\langle10\bar{1}\bar{1}\rangle$ twins were observed in Mg alloys at 450 °C during extrusion and uniaxial compression [4,5]. In this study, $\{10\bar{1}2\}$ twinning is still activated at the ram speed of 10 mm/min (Figures 3h,3i). However, the $\{10\bar{1}2\}$ twinning subsides when the ram speed increases to 30 mm/min (Figures 5f,5g), and it is almost suppressed at the high ram speed 50 mm/min (Figures 6h,6i). Obviously, the $\{10\bar{1}2\}$ twinning behavior at 450 °C during extrusion differs from room temperature twinning. It is widely accepted that twinning is favored at low temperature and high strain rate [8,9]. However this former thought related to twinning may be not always the case. The critical resolved shear stress (CRSS) of slip is sensitive to temperature and strain rate [8]. This explains why texture is different in the same position of the Middle at different ram speeds (Figures 3d,5c,6d). However, twinning is considered athermal and the CRSS of twinning should be insensitive to temperature and strain rate [10]. At low temperatures, twinning easily initiates and accommodates plastic strain due to very limited number of easy slip systems [9]. However at high temperature, non-basal slips become more active because their CRSSs dramatically decrease [11].Thus slips and twinning compete to accommodate plastic strain at high temperature deformation at low strain rate. On the other hand, dislocation rearrangement and elimination, subgrain formation and growth induce strong recovery at high temperature in Mg [12]. Thus the matrix dislocation density may vary at different strain rates. The defect type and density within matrix may strongly affect the twin nucleation and growth. For example, in some bcc metals, twinning is totally suppressed by prestraining at high temperature [9]. Profuse twinning activated in AZ91 Mg alloy compressed at 300 °C at low strain rate of 0.05 s^{-1}, 0.1 s^{-1}, and 0.5 s^{-1}, but almost no twinning was found at high strain rate of 1 s^{-1} [13]. Thus twinning is hindered at high ram speed extrusion, e.g. 50 mm/min (high strain rate), but twinning is still activated at low speed extrusion (Figures 3h,3i). However, if strain rate is high enough that most slips are hard to activate, twinning could dominate deformation again. This interesting issue needs further study to clarify the underlying mechanism.

During extrusion, the original grains in the Middle position begin to deform and recrystallization occurs through grain boundary "bulging" and/or subgrain rotation inside matrix grain [12]. Thus the Middle position grain is finer than the original grain (Figures 2-6). Some grains coarsen in the Exit position due to grain growth via DRX. The driving force of DRX is the dislocation density difference across neighboring grains, and the grains grow via grain boundary migration. That means that the stable or "hard" oriented grains consume the unstable or "soft" oriented grains because the former has less dislocation density than the latter. In other words, the DRX texture of extruded AM30 exhibits a similar deformation texture without DRX. In extrusion, grains inside the die experience two main stresses: a compression stress along the radial direction (RaD) of the die and a tension stress along the extrusion direction (ED). The compression stress will force the *c*-axis of the grains parallel to compression axis due to the dominant basal slip, while the tension stress will force the $\langle2\bar{1}\bar{1}0\rangle$ or $\langle10\bar{1}0\rangle$ parallel to tensile axis because two prismatic slip systems have the same Schmid factor [14]. Thus during extrusion, only grains with the <0001> parallel to RaD and $\langle2\bar{1}\bar{1}0\rangle$ or $\langle10\bar{1}0\rangle$ parallel to ED are the stable grains.

Accordingly, grains in these orientations will survive and develop during extrusion. While other oriented grains will be consumed sooner or later by the stable grains. This explains why $\langle 2\bar{1}\bar{1}0\rangle$ grains or $\langle 10\bar{1}0\rangle$ grains mainly exist in the Exit position at all ram speeds extrusions (Figures 3-6). Due to activity of the prismatic slip, the rotation axis is <0001> and the rotation angle is 30° between $\langle 2\bar{1}\bar{1}0\rangle$ and $\langle 10\bar{1}0\rangle$. This is the reason the rotation axis/angle of <0001>30° exists at all ram speeds of the extrusions of AM30 at 450 °C.

When the AM30 exits the die, the $\langle 2\bar{1}\bar{1}0\rangle$ grains and $\langle 10\bar{1}0\rangle$ grains are the stable grains that have lower dislocation density due to the prismatic slip. Thus, the $\langle 10\bar{1}0\rangle$ and $\langle 2\bar{1}\bar{1}0\rangle$ grains grow at the expense of other oriented grains (Figure 3c). However, at high speed extrusions, more dislocations are reserved within the deformed grains due to anterior uncompleted DRX. Thus SRX well develops at rapid ram speed extrusion and abnormal grain growth is seen (Figures 4b,6c). Based on Figures 4b and 6c, some $\langle 2\bar{1}\bar{1}0\rangle$ grains abnormally grow by consuming other grains. It seems that at high ram speed extrusion, $\langle 2\bar{1}\bar{1}0\rangle$ grains are more stable than <10-10> grains. The transition from $\langle 10\bar{1}0\rangle$ to $\langle 2\bar{1}\bar{1}0\rangle$ was also reported in an extruded AZ31 Mg alloy during annealing at high temperature [15]. The reason was ascribed to the lower internal misorientation (less defects) in the $\langle 10\bar{1}0\rangle$ grains than other small grains. If this is the case, some $\langle 2\bar{1}\bar{1}0\rangle$ grains happen to form a domain that quickly involves a big grain via SRX. The big $\langle 2\bar{1}\bar{1}0\rangle$ grain rapidly grows and only large $\langle 2\bar{1}\bar{1}0\rangle$ grains dominate the final microstructure and texture in the extrudate (Figures 4b,6c).

Conclusions

A commercial AM30 Mg alloy was extruded at 450 °C and various ram speeds of 5, 10, 20, 30, and 50 mm/min in a lab-scale flat die. Microstructure and texture evolution during extrusion were examined by EBSD in different representative positions. Some conclusions can be drawn based on experimental results and analysis.

1) Deformation and DRX simultaneously occurred during extrusion inside the flat die. SRX started in the AM30 after it exited the die. The deformation, dynamic and static recrystallization dominate the final structure and texture of the extrudate.

2) $\{10\bar{1}2\}\langle 10\bar{1}\bar{1}\rangle$ extension twinning initiated at 450 °C during extrusion. Profuse twinning activated at low ram speed of 10 mm/min, but a few twins initiated at high ram speed of 50 mm/min. The reason could be that denser defects in matrix due to high speed extrusion hindered twinning nucleation and propagation.

3) Stress state of extrusion resulted in the two stable oriented grains: $\langle 2\bar{1}\bar{1}0\rangle \parallel$ ED and $\langle 10\bar{1}0\rangle \parallel$ ED grains. These two oriented grains grew at the expense of other oriented grains via grain boundary migration during DRX.

4) At high speed extrusion, the reserved defects within deformed grains served as driving force of SRX. The $\langle 2\bar{1}\bar{1}0\rangle$ grains had priority of abnormal growth at expense of other neighboring fine grains due to $\langle 2\bar{1}\bar{1}0\rangle$ grains being the most stable orientation during extrusion.

Acknowledgements

The authors are grateful to the financial support from the Department of Energy, Contract No. DE-FC-26-06NT42755, and the Center for Advanced Vehicular Systems (CAVS) at Mississippi State University.

References

1. A. Sadeghi and M. Pekguleryuz, "Recrystallization and texture evolution of Mg-3%-1%Zn-(0.4-0.8)%Sr alloys during extrusion", *Materials Science and Engineering A*, 528(2011) 1678-1685.
2. A.A. Luo, R.K. Mishra and A.K. Sachdev, "High-ductility magnesium-zinc-cerium extrusion alloys", *Scripta Materialia*, 64(2011) 410-413.
3. J.D. Robson, et al., "Effect of extrusion conditions on microstructure, texture, and yield asymmetry in Mg-6Y-7Gd-0.5 wt%Zr alloy", *Materials Science and Engineering A*, 528(2011) 7247-7256.
4. Q. Ma, et al., "Twinning-induced dynamic recrystallization in a magnesium alloy extruded at 450 °C", *Scripta Materialia*, 65(2011) 823-826.
5. Q. Ma, et al., "Strain rate dependence of twinning at 450 °C and its effect on microstructure of an extruded magnesium alloy", *Materials Science and Engineering A*, doi:10.1016/j.msea.2012.08.104, in press.
6. Q. Ma, et al., "Twinning effects in a rod-textured AM30 magnesium alloy", *International Journal of Plasticity*, 29(2012) 60-76.
7. Q. Ma, et al., "Twinning and double twinning upon compression of prismatic textures in an AM30 magnesium alloy", *Scripta Materialia*, 64(2011) 813-816.
8. M.A. Meyers, O. Vöhringer and V.A. Lubarda, "The onset of twinning in metals: A constitutive description", *Acta Materialia*, 49(2001) 4025-4039.
9. J.W. Christian and S. Mahajan, "Deformation twinning", *Progress in Materials Science*, 39(1995) 1-157.
10. A. Jain and S.R. Agnew, "Modeling the temperature dependent effect of twinning on the behavior of magnesium alloy AZ31B sheet", *Materials Science and Engineering*, A 462 (2007), 29-36.
11. A. Chapuis and J.H. Driver, "Temperature dependency of slip and twinning in plane strain compressed magnesium single crystals", *Acta Materialia*, 59 (2011), 1986-1994.
12. S.E. Ion, F.J. Humphreys and S.H. White, "Dynamic recrystallization and the development of microstructure during the high temperature deformation of magnesium", *Acta Metallurgica*, 30 (1982), 1909-1919.
13. L. Liu and H. Ding, "Study of the plastic flow behaviors of AZ91 magnesium alloy during thermomechanical process", *Journal of Alloys and Compounds*, 484(2009) 949-956.
14. T. Mayama, et al., "Crystal plasticity analysis of texture development in magnesium alloy during extrusion", *International Journal of Plasticity*, 27(2011) 1916-1935.
15. S. Yi, H.G. Brokmeier and D. Letzig, "Microstructural evolution during the annealing of an extruded AZ31 magnesium alloy", *Journal of Alloys and Compounds*, 506(2010) 364-371.

Magnesium Technology 2013
Edited by: Norbert Hort, Suveen N. Mathaudhu, Neale R. Neelameggham, and Martyn Alderman
TMS (The Minerals, Metals & Materials Society), 2013

THE TEXTURE AND MICROSTRUCTURE EVOLUTION OF Mg-Zn-Ce ALLOYS

M. Sanjari[1], A. Farzadfar[1], T. Sakai[2], H. Utsunomiya[2], E. Essadiqi[3], In-Ho Jung[1], S.Yue[1]
[1]Department of Materials Engineering, McGill University, Montreal, Canada H3A 2B2
[2] Division of Materials and Manufacturing Science, Graduate School of Engineering, Osaka University
[3] Université Internationale de Rabat (UIR) Technopolis, Rabat Morocco

Keywords: Rare-earth elements, Rollability, Texture weakening

Abstract

The texture evolution in four Mg-Zn-Ce alloys was compared to that of Mg-3Al-1Zn (AZ31) alloy following rolling and subsequent isothermal annealing. All the as-cast and homogenized alloys were rolled through two stages that can be characterised as rough rolling and finish rolling, respectively. To investigate the effect of finish rolling temperature, one finish rolling pass with 65% reduction in thickness was performed at 300 °C and 450 °C. Of the studied compositions, the Mg-1Zn-1Ce, which had the highest Ce/Zn ratio, showed the weakest as-rolled texture and homogenous shear banding/twinning. Changing the Zn content changed particle size and, in alloys subject to texture weakening, the static recrystallization mechanism altered. On annealing, the maximum intensity of basal pole figures decreases as recrystallization progresses. The Mg-1Zn-1Ce (with the highest Ce/Zn), texture weakening is maintained even after full recrystallization, when grain coarsening occurs. However, in the Mg-4Zn-1Ce and AZ31 alloys, texture strengthening occurs when grain coarsening occurs, and the double split basal peak is replaced by a single peak. In these two alloys, grain coarsening is also accompanied by a bimodal grain size whereas in the Mg-1Zn-1Ce alloy, the grain coarsening leads to a uniform grain size. It is concluded that the differences between the Ce bearing alloys is related to Zn; increasing Zn decreases the solubility of Ce, which can influence the texture changes.

Introduction

The application of magnesium and magnesium alloys in the automotive industry can reduce vehicle weight, and, consequently, fuel consumption can be reduced. Sheet Mg is one way to increase the usage of Mg [1]. One of the major obstacles of its use is that a preferred crystallographic orientation (texture) develops in wrought alloys [2-4], which substantially limits the subsequent formability, especially in flat-rolled sheets [2, 5-7]. It has recently been found that the addition of rare earth (RE) elements, such as yttrium, cerium or neodymium, significantly weakens the rolling or extrusion texture of Mg alloys [4-9]. The texture weakening mechanism has been associated with different mechanism such as particle stimulated mechanism (PSN) [10, 11], deformation or shear bands containing twins [5], or retardation of dynamic recrystallization (DRX) [12]. Although the crystallographic texture evolution of Mg alloys has already been studied as a function of alloying elements and different thermomechanical processing parameters, there is still a lack of work regarding whether the RE-textures are mainly associated with changes in the deformation texture or also with changes in the recrystallization texture.

In our previous studies on Mg–Y alloy and Mg-Zn-Ce the possibility of attaining texture weakening by DRX suppression and SRX promotion was studied. It was hypothesised, in the previous work of the author, that by suppressing the DRX, the deformation cannot be accommodated in the soft DRX region; instead deformation is accommodated by compression and double twinning in basal parent grains, leading to the formation of deformation bands [24]. Despite these observations, the orientation relationship between activated deformation mechanisms and recrystallized grains remains unclear.

In this work, the nucleation and growth of recrystallized grains is studied in both as deformed and annealing state in Mg-Zn-Ce systems and compared with AZ31 as a reference alloy. In addition the effect of precipitates is investigated in the Mg-Zn-Ce alloys systems by changing the Ce and Zn contents in the alloy.

Experimental procedure

The following four alloys in the Mg-Zn-Ce system were cast into ingots of 70 mm x 110 mm x 500 mm and machined into plates 70 mm x 110 mm x 6 mm thickness:

1) Mg-1Zn-0.5Ce
2) Mg-1Zn-1Ce
3) Mg-2Zn-1Ce
4) Mg-4Zn-1Ce

The as-cast material was homogenized at 450 °C for 24 hr. To compare the results with the commercial AZ31 alloy, an AZ31 alloy with the following chemical composition (%wt): 3% Al, 0.9% Zn, 0.67% Mn and Mg (balance) was cast as a plate with 6 mm thickness and homogenized at 350 °C for 4 hr.

The rolling experiment was conducted through two steps: rough rolling and finish rolling. The purpose of rough rolling was to decrease the grain size from about 1mm to about 130 μm in preparation for high speed rolling. Rough rolling was performed at 15 m/min using a low-speed, two-high, mill with ϕ 320 mm rolls. The plates were rolled with three consecutive passes of 12-15% reduction per pass at 450 °C and the thickness reduced from 6mm to 3 mm. The samples were reheated for 10 min between each passes to keep the rolling temperature constant. After cooling in air to room temperature the rough rolled samples were annealed at 450°C for 15 min to fully recrystallize the sheet. For the finish rolling, low speed rolling (LSR) and high speed rolling (HSR) was performed in a single pass operation with reduction in thickness of about 60% at two temperatures, 300 °C and 450 °C, at rolling speeds of 15 and 1000 m/min respectively. LSR was performed in the roughing mill. HSR was performed in a two-high laboratory HSR mill with ϕ530mm rolls. Prior to finish rolling, the specimens were reheated for 10 min at the corresponding rolling temperature. Details of the experiments and the high speed rolling machines can be found in previous work [18].

The samples were cut in RD-ND planes, mounted and ground to 1200 grit and then polished with alcohol based 3 and 1 micron diamond suspensions for microstructural examination by optical microscopy. The samples were etched with an acetic-picral solution (10 mL acetic acid + 4.2 g picric acid + 10 mL water + 70 mL ethanol (99.5 pct)). The macro texture was evaluated in a Siemens D-500 X-ray diffractometer equipped with a texture goniometer using Co radiation in TD-RD plan. The data were analyzed to calculate orientation distribution functions and to recalculate the pole figures.

Orientation imaging microscopy (OIM) was used to study grain orientation via EBSD in a Hitachi S-3000 FE-SEM at 20 kV, 70° tilt angle, and a step size between 0.3 to 0.5 µm, depending on deformation condition. For EBSD analysis, the polished samples samples were electropolished at 20 V using 10% nital solutions cooled to -10 to -30°C.

Results and discussions

Finish rolled and annealed microstructure

Figure 4 shows the microstructure of finish rolled AZ31B and Mg-Zn-Ce alloys rolled at 300°C and annealed at 450 °C for 15 min.

In terms of microstructure evolution for the AZ31B, as can be seen from Fig. 1 the as rolled structure appears to have undergone little recrystallization, revealing a 'pancaked' structure with shear bands. In this case it seems that the shear bands act as crack propagators, since only a few shear bands forms in AZ31 and strain is strongly localized in them. Macroscopic bands or shear zones have been observed in rolled magnesium, inclined to the rolling plane along surfaces that support high shear stresses, and act frequently as sites for the origin of recrystallization during subsequent annealing. It is generally found that as the temperature of deformation increases, i.e. at 450 °C, the deformation becomes more homogeneous due to an increase in the number of operating slip systems.

As can be seen in Fig. 4, in the Mg-Zn-Ce alloys, the general features of microstructure are almost the same as AZ31; the main difference is that shear bands are somewhat more difficult to observe, although in Mg-2Zn-1Ce and Mg-4Zn-1Ce the shear bands can be distinguished. However, by increasing the alloying level, the shear bands become much more difficult to discern. It seems that shear bands formed in RE-containing alloys are less intense than those formed in AZ31. The preceding studies show that the severity of shear banding is reduced when adding RE solute.

With respect to annealing, it can be seen from Fig. 1 that, in the case of AZ31, considerable recrystallization takes place, as indicated by the equiaxed structure. However addition of rare earth alloying elements has obviously changed the recrystallization behaviour and, as a general conclusion, it appears that the Mg-Zn-Ce alloys retard recrystallization. The Mg-Zn-Ce alloys (except Mg-4Zn-1Ce) start to recrystallized after 30 s (as indicated in the bottom graph of Fig. 1); however recrystallized grains can be seen in the microstructure of AZ31 and Mg-4Zn-1Ce after 30s annealing. By increasing the Zn to 4 wt% (Mg-4Zn-1Ce) the recrystallized grain size increases in this alloy compared to the alloy with lower alloying levels.

After 15 min annealing the minimum average grain size is obtained for Mg-1Zn-1Ce and larger grain sizes are obtained for AZ31 and Mg-4Zn-1Ce with the minimum Ce/Zn ratio.

Deformation Texture of rough and finish rolled sheets

The initial texture of the finish rolled samples is depicted in Fig. 2. In the case of AZ31 at 300 °C the samples shows the typical basal texture of rolled Mg alloys, with the majority of c-axes of the grains aligned with the ND. However, by increasing the temperature to 450 °C the PF slightly elongates along the RD and the maximum intensity slightly decreases. All the Mg-Zn-Ce alloys show that the basal poles are split in RD direction even at 300 °C. The minimum basal texture intensity is exhibited by Mg-1Zn-1Ce at both 300 °C and 450 °C, which has the highest Ce/Zn ratio.

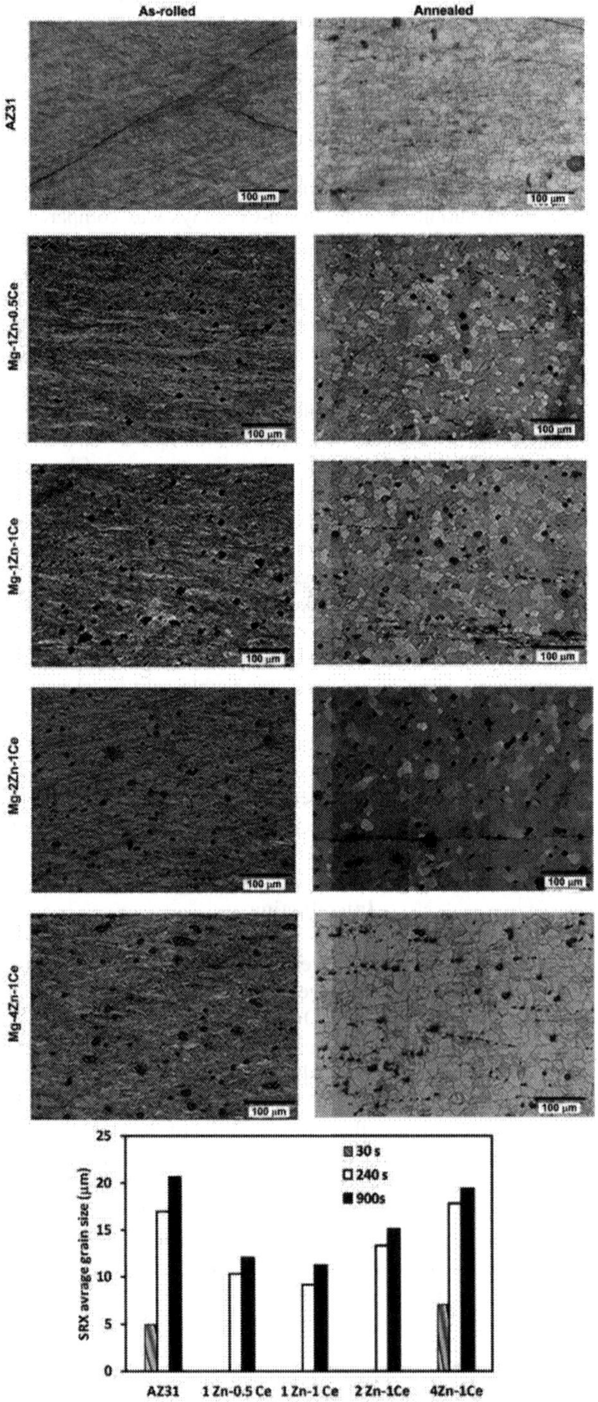

Fig. 1 The microstructure of the finish rolled samples of the Mg-Zn-Ce alloys rolled at 300°C in the as rolled condition (left side) and annealed for 15 min at 450°C (right side). The effect of

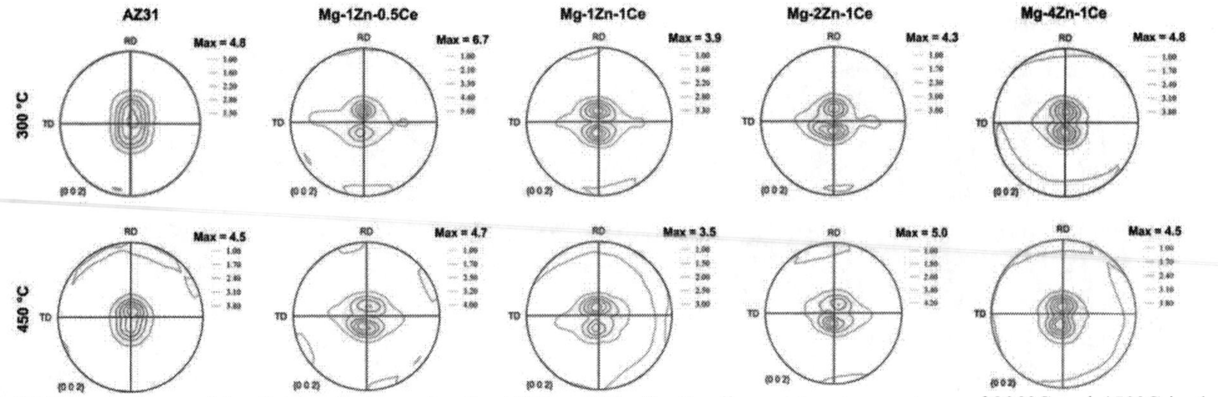

Fig. 2 {0002} pole figure of the finish rolled samples for AZ31 and Mg-Zn-Ce alloys at two temperatures of 300°C and 450°C in the as-rolled condition.

Microstructure and texture evolution during annealing

To study the effect of the Ce/Zn ratio on the texture and microstructure evolution during annealing, Mg-1Zn-1Ce and Mg-4Zn-1Ce were selected as the highest and lowest Ce/Zn ratio respectively and compared with AZ31. The texture evolution during annealing at 300 °C for different annealing times is shown in Fig 3. As can be seen, after 60 min annealing, all the textures are weaker than the as-rolled textures regardless of alloy. However, the three alloys show two different types of texture evolution during annealing. AZ31 and Mg-4Zn-1Ce are similar and they both exhibit a sudden decrease in texture intensity after 30 s anneal, followed by a plateau trough and then an increase at 6 min for AZ31 and 10 min for Mg-4Zn-1Ce. Both alloys reach a maximum after about 30 min, which plateaus with longer times. AZ31 exhibits much higher intensities outside the trough, but the trough intensities are similar. Mg-1Zn-1Ce, on the other hand, exhibits a steady decrease after 60 s annealing; after 60 min, the texture intensity is lower than the trough values of the other alloys. After about 20 min annealing the texture intensity is does not change significantly.

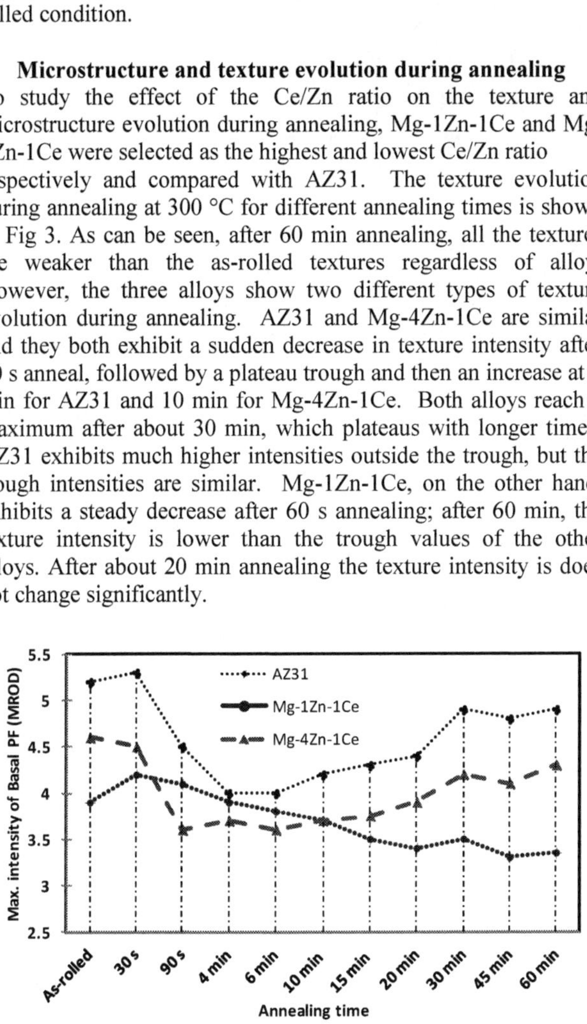

Fig 3. The intensity (in multiples random of distribution or MROD) in the basal pole figures plotted as a function annealing time at 300 °C for the AZ31, Mg-1Zn-1Ce and Mg-4Zn-1Ce alloys finish rolled at 300 °C.

The microstructure evolution with annealing time is shown in Fig. 4. Following annealing up to 90 s, both AZ31 and Mg-4Zn-1Ce are fully recrystallized, whereas Mg-1Zn-1Ce is not. For both AZ31 and Mg-4Zn-1Ce, full recrystallization coincides approximately with the beginning of the sharp decrease in hardness, as indicated by the arrows on the graph. After 10 min annealing, both AZ31 and Mg-4Zn-1Ce exhibit a bimodal microstructure consisting of small and large grains, but generally having a size less than 10 μm. The hardness value after this time follows a plateau trough. The slightly further decrease in hardness and further grain growth as well as evidence of abnormal grain growth can be observed in the microstructure (arrowed) after increasing the annealing time to 60 min.

On the other hand for Mg-1Zn-1Ce, there is no sign of recrystallization in the microstructure up to 4 min annealing time. After 60 min annealing, the recrystallized grain size is smaller in comparison with the other two alloys and there is no sign of abnormal grain growth in the microstructure.

Texture evolution of AZ31 and Mg-4Zn-1Ce during annealing

In the case of AZ31 deformed at 300 °C with reduction ≈60% in a single pass, there were few signs of DRX after rolling. The deformed microstructure was completely replaced by SRXed grains after 90s annealing at 300 °C, (Fig. 4). At this stage, as was seen in Fig. 3, the maximum intensity also decreases from 5.3 to 4.5 multiples random of distribution (MROD). Therefore; it seems that decrease in texture intensity after 90 s annealing time in the AZ31 is corresponding to recovery or the beginning of SRX.

As was seen in Fig. 4, the new SRX grains started forming at bands. Sandlobes et al. [15] reported that such shear bands contain a high volume fraction of compression and secondary twins.

The orientation changes associated with twin recrystallization are not understood well yet. In the previous work of the authors [17], the misorientation of new recrystallized grains on secondary twin (S-twins) was investigated with respect to the primary twin hosts in partially recrystallized twins. It was concluded that the high frequency of ≈ 77° misorientation illustrated that the orientations of the recrystallized grains are closer to secondary (tension twin 86°), rather than of the primary, compression twins (C-twins). Therefore; the orientation of recrystallized grains has much more wider distribution than that of initial basal parent. For this reason, it seems that rapid recrystallization on twins contributes to the texture weakening during first 90 s of annealing for both AZ31 and Mg-4Zn-1Ce.

After the initial sudden drop for both alloys the maximum basal intensity reach to a plateau trough. By further annealing the SRX grain size increase in normal way up to about 10 min for both alloys. After this time some grains start to grow in abnormally, as can be seen in the microstructure of AZ31 annealed for 10 min in Fig.4.

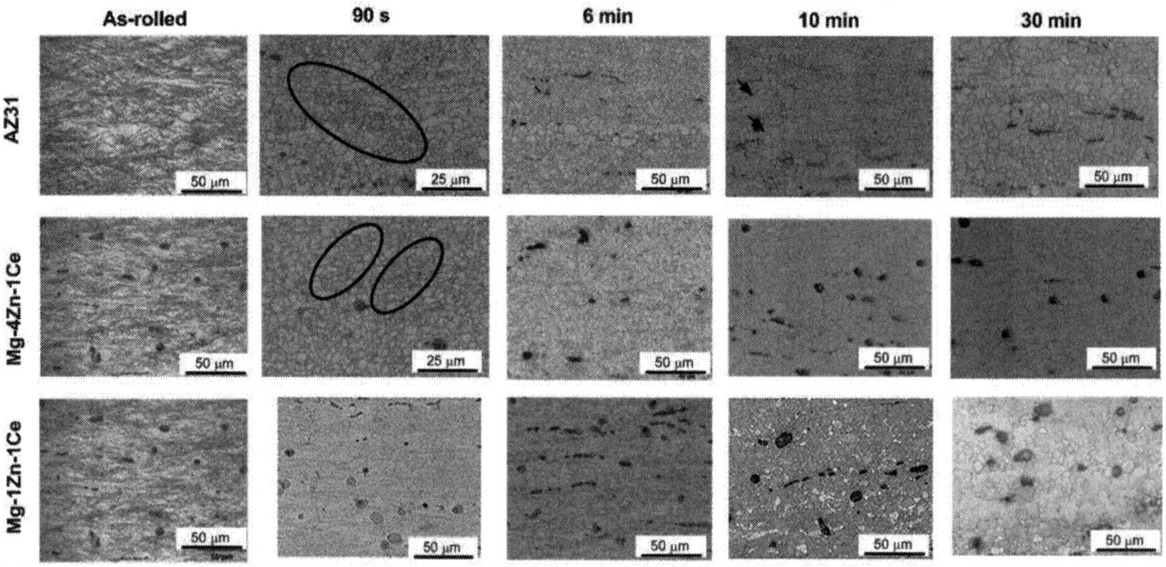

Fig. 4 Optical micrographs for AZ31, Mg-4Zn-1Ce and Mg-1Zn-1Ce rolled at 300 °C followed by annealing at 300 °C for different times

Close inspection of the EBSD data in the annealed samples of AZ31 at different times revealed that the size distribution of recrystallized grains exhibits a principal peak representing the average size, and for samples annealed more than 10 min, secondary peaks can be detected (Fig 5).

As the PF for the grains larger than the average shows in Fig.5, the orientation of these recrystallized grains shows more basal orientation. Therefore; it seems by increasing the annealing time the area fraction of these basal grains increases and the basal intensity increased.

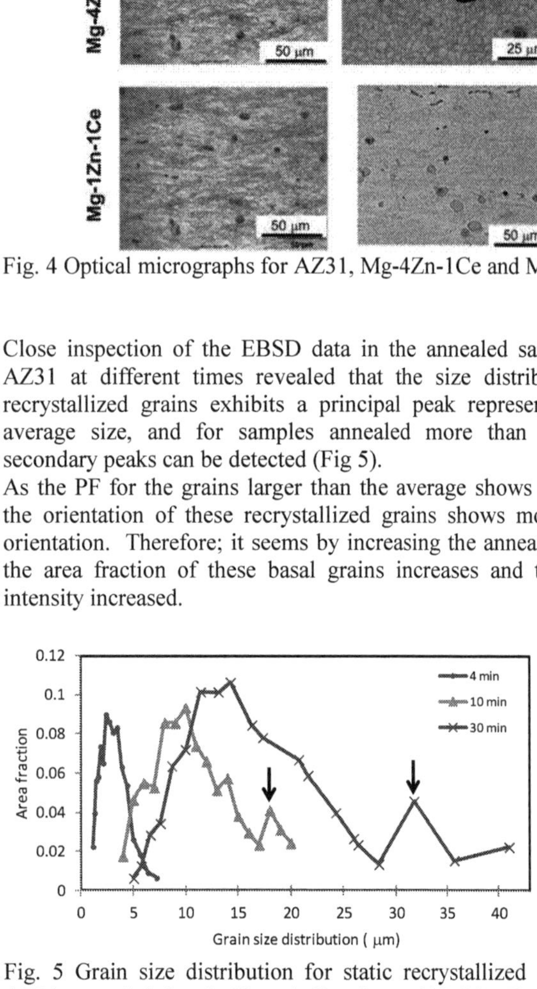

Fig. 5 Grain size distribution for static recrystallized grains in AZ31 annealed for 4, 10 and 60 min at 300 °C. The samples rolled at 300 °C.

Texture evolution of Mg-1Zn-1Ce during annealing

As was seen in Fig. 3 the basal texture is continuously dropping during annealing. By adding Ce to the alloys, the DRX and SRX mechanisms can be influenced by changing both nucleation and growth. As was observed in the as-deformed microstructure in Fig. 4 and also reported by other researchers [4, 8, 13], by adding rare earth elements more C-twins and S-twins form in the material. By increasing the annealing time the volume fraction of SRX increases, which corresponds to weakening of the basal texture and an increase in the inclination of the basal pole toward the rolling direction (RD) from ~12° to ~25°. Close inspection of Fig. 4 shows that, although there appears to be a continuous drop of texture intensity, in fact the intensity effectively reaches a plateau at 20 min, which corresponds to the end of SRX in this alloy (Fig. 4). Therefore it seems that the different SRX rates of these three alloys is the main reason for the different texture weakening behaviour as will be explained in more detail in following section.

Possible texture weakening mechanism for Mg-1Zn-1Ce

To study the SRX mechanism in this alloy a rough rolled sample, which was not dynamically recrystallized and had large grains, was annealed at different times. The microstructure and texture evolution were tracked for the same area. As can be seen in Fig. 6a, the rough rolled microstructure of Mg-1Zn-1Ce contains many twins. These twins are zones of deformation localization and can pass through several grains and become deformation bands, as also mentioned in the recent study of Sandlobes et al. [23]. After 4 min annealing at 300°C, new non basal grains recrystallized in bands and individual twins. Figure 6b depicts the growth of one SRX grain (arrow) into a basal parent region, and the initial twin nucleation site of this grain is shown in Fig. 6a. As is seen in Fig. 6b and c, this grain coarsened and some parts expanded into a basal parent grain of the basal matrix. The growth of the highlighted grains continues with further annealing (Fig. 6 d). Static recrystallization also occurs at basal parent grains, giving rise to grains exhibiting mostly basal orientations, as is illustrated by the grains numbered 1 and 2 in Figs. 6 b, c and d. As shown in Fig. 6 during the progress of SRX, the intensity of the basal texture continues to drop.

Stanford et al [12] suggested that the dislocation/solute interactions are an important aspect in the rare earth effect on the texture. They reported that Ce and La are the most effective texture modifiers because of their large atomic radii. It was suggested that there is a strong interaction of RE solutes with

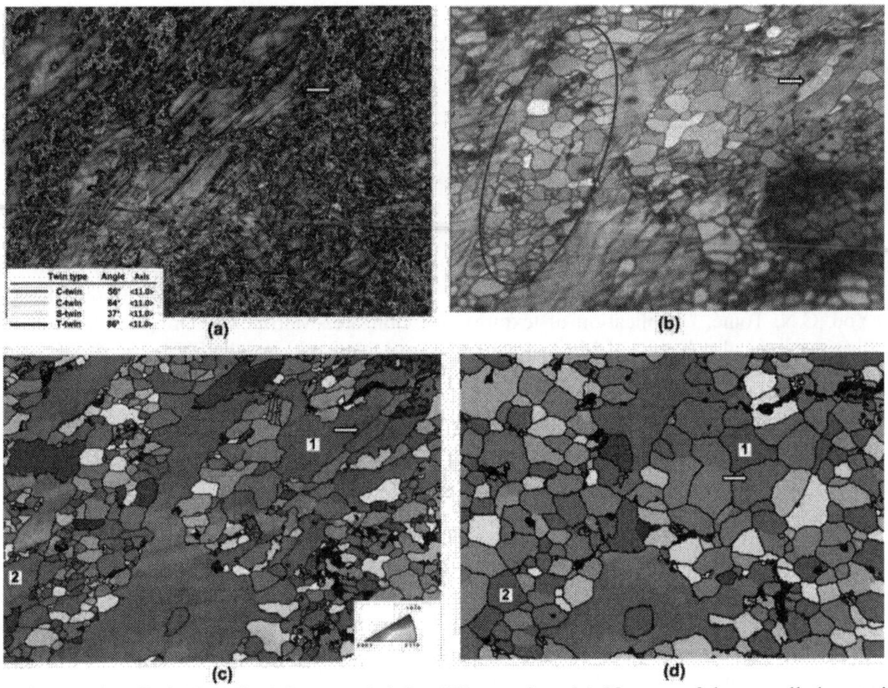

Fig. 6 EBSD results for the rough rolled Mg-1Zn-1Ce annealed for different time (a) IQ map of the as-rolled material with the boundaries corresponding to different twins (Color figure online) (b) IQ map for the sample anneal for 4 min and (c) IPF map for the sample anneal for 4 and (d)10 min. Twin nucleation site (a) and the growth of this grain during annealing is arrowed (b, c and d).

dislocations and boundaries in magnesium-based alloys [4].

To study the effect of Zn content on the solubility of Ce in magnesium, thermodynamic calculations were carried out using the FactSageTM thermodynamic software (Fig. 7). Based on these results, by increasing the Zn in the Mg-Zn-Ce alloy systems, the solubility of Ce in Mg decreases. This was supported by the observed increase in Ce-rich precipitates in Mg-4Zn-1Ce. Therefore, it can be concluded that the grain boundary mobility increased in the alloys with lower Ce/Zn ratio. Therefore, it seems that higher solubility of Ce in Mg-1Zn-1Ce magnesium may be a key reason for the weaker basal texture in both deformed and recrystallized conditions. It was hypothesised, in the previous work of the author, that by suppressing the DRX, the deformation cannot be accommodated in the soft DRX region; instead deformation is accommodated by compression and double twinning in basal parent grains, leading to the formation of deformation bands [24].

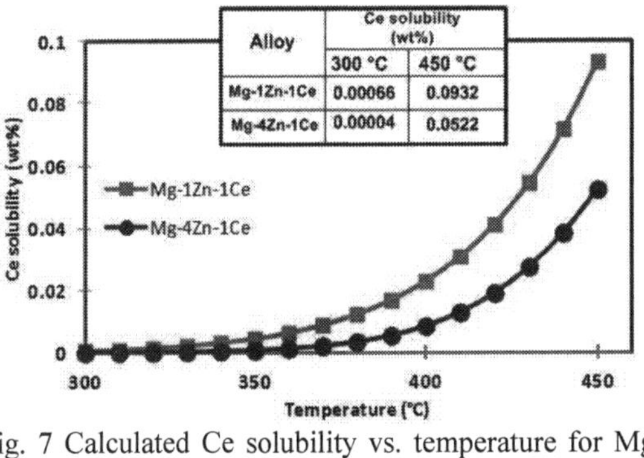

Fig. 7 Calculated Ce solubility vs. temperature for Mg-1Zn-1Ce and Mg-4Zn-1Ce alloys using FactSage^{TM} thermodynamic software.

Conclusions

Microstructure and texture development in four Mg-Zn-Ce alloys and Mg-3Al-1Zn were studied following rolling and subsequent isothermal annealing. The main results are summarized as follows.

-The addition of Ce weakens the basal texture in both the as-hot rolled and annealed conditions. This appears to be related to an increase in deformation twins during hot rolling, which is due to Ce in solid solution.

Increasing Zn decreases Ce in solid solution. Hence the alloy with the higher Ce/Zn ratio exhibited weaker basal textures.

The maximum intensity of basal pole figures decreases as static recrystallization progresses for all the alloys. This probably corresponds to the formation of SRX grains with wider orientation spread at twins and deformation bands.

The basal texture intensifies after the end of SRX because of abnormal grain coarsening of some grains with orientation close to basal.

Acknowledgments

The authors would like to thank Mr Miyamoto, Mr Hattori and Mr Muraoka from his kind support for rolling experiments at Osaka University. Thanks to Pierre Vermette at McGill University and Amjad Javaid, Howard Webster and David Saleh in the Natural Resources Canada's CANMET materials technology laboratory for casting the alloys. This study was supported by the NSERC Magnesium Strategic Research Network (MagNET) and MEDA Fellowship from Faculty of Engineering of McGill University.

References

[1] H.B. M. Avedesian, *ASM specialty handbook. Magnesium and magnesium alloys.* (ed., Materials Park, Ohio: ASM International, 2000).

[2] M. Shamsi, M. Sanjari, A.Z. Hanzaki, "Study of fractional softening of AZ31 magnesium alloy under multistage hot deformation," *Materials Science and Technology*, 25 (2009) 1039-1045.

[3] T. Al-Samman, G. Gottstein, "Dynamic recrystallization during high temperature deformation of magnesium," *Materials Science and Engineering A*, 490 (2008) 411-420.

[4] S.R. Agnew, M.H. Yoo, C.N. Tomé, "Application of texture simulation to understanding mechanical behavior of Mg and solid solution alloys containing Li or Y," *Acta Materialia*, 49 (2001) 4277-4289.

[5] L.W.F. Mackenzie, M. Pekguleryuz, "The influences of alloying additions and processing parameters on the rolling microstructures and textures of magnesium alloys," *Materials Science and Engineering: A*, 480 (2008) 189-197.

[6] A. Styczynski, C. Hartig, J. Bohlen, D. Letzig, "Cold rolling textures in AZ31 wrought magnesium alloy," *Scripta Materialia*, 50 (2004) 943-947.

[7] K. Hantzsche, J. Bohlen, J. Wendt, K.U. Kainer, S.B. Yi, D. Letzig, "Effect of rare earth additions on microstructure and texture development of magnesium alloy sheets," *Scripta Materialia*, 63 (2010) 725-730.

[8] J. Bohlen, M.R. Nürnberg, J.W. Senn, D. Letzig, S.R. Agnew, "The texture and anisotropy of magnesium-zinc-rare earth alloy sheets," *Acta Materialia*, 55 (2007) 2101-2112.

[9] R. Cottam, J. Robson, G. Lorimer, B. Davis, "Dynamic recrystallization of Mg and Mg-Y alloys: Crystallographic texture development," *Materials Science and Engineering: A*, 485 (2008) 375-382.

[10] L.W.F. Mackenzie, M.O. Pekguleryuz, "The recrystallization and texture of magnesium-zinc-cerium alloys," *Scripta Materialia*, 59 (2008) 665-668.

[11] N. Stanford, "Micro-alloying Mg with Y, Ce, Gd and La for texture modification--A comparative study," *Materials Science and Engineering: A*, 527 (2010) 2669-2677.

[12] H. Yan, S.W. Xu, R.S. Chen, S. Kamado, T. Honma, E.H. Han, "Twins, shear bands and recrystallization of a Mg-2.0%Zn-0.8%Gd alloy during rolling," *Scripta Materialia*, 64 (2011) 141-144.

[13] E.A. Ball, P.B. Prangnell, "Tensile-compressive yield asymmetries in high strength wrought magnesium alloys," *Scripta Metallurgica et Materiala*, 31 (1994) 111-116.

[14] L.W.F. Mackenzie, G.W. Lorimer, F.J. Humphreys, T. Wilks, "Recrystallization behaviour of two magnesium alloys," *Journal*, 467-470 (2004) 477-482.

[15] N. Stanford, M.R. Barnett, "The origin of "rare earth" texture development in extruded Mg-based alloys and its effect on tensile ductility," *Materials Science and Engineering: A*, 496 (2008) 399-408.

[16] T. Al-Samman, X. Li, "Sheet texture modification in magnesium-based alloys by selective rare earth alloying," *Materials Science and Engineering A*, 528 (2011) 3809-3822.

[17] H. Kon, T. Sakai, H. Utsunomiya, S. Minamiguchi, "Deformation and texture evolution during high-speed rolling of AZ31 magnesium sheets," *Materials Transactions*, 48 (2007) 2023-2027.

[18] H. Utsunomiya, T. Sakai, S. Minamiguchi, H. Koh, "High-speed heavy rolling of magnesium alloy sheets," *Journal*, 2006 (2006) 201-204.

[19] G. Hamada, T. Sakai, H. Utsunomiya, "Effect of rolling speed on deformability and microstructure in rolling of AZ31B magnesium alloy," *Journal*, 89-91 (2010) 227-231.

[20] T. Sakai, Y. Watanabe, H. Utsunomiya, "Microstructure and texture of AZ80 magnesium alloy sheet rolled by high speed warm rolling," *Journal*, 618 619 (2009) 483-486.

[21] R.K. Mishra, A.K. Gupta, P.R. Rao, A.K. Sachdev, A.M. Kumar, A.A. Luo, "Influence of cerium on the texture and ductility of magnesium extrusions," *Scripta Materialia*, 59 (2008) 562-565.

[22] S. Sandlöbes, S. Zaefferer, I. Schestakow, S. Yi, R. Gonzalez-Martinez, "On the role of non-basal deformation mechanisms for the ductility of Mg and Mg-Y alloys," *Acta Materialia*, 59 (2011) 429-439.

[23] Y.B. Chun, J. Geng, N. Stanford, C.H.J. Davies, J.F. Nie, M.R. Barnett, "Processing and properties of Mg–6Gd–1Zn–0.6Zr: Part 1 – Recrystallisation and texture development," *Materials Science and Engineering: A*, 528 (2011) 3653-3658.

[24] H. Li, E. Hsu, J. Szpunar, H. Utsunomiya, T. Sakai, "Deformation mechanism and texture and microstructure evolution during high-speed rolling of AZ31B Mg sheets," *Journal of Materials Science*, 43 (2008) 7148-7156.

[25] S.L. Couling, F. Pashak, L. Sturkey, *Trans. ASM*, 51 (1959) 94-107.

[26] B.H. Lee, W. Bang, S. Ahn, C.S. Lee, "Effects of temperature and strain rate on the high-temperature workability of strip-cast Mg-3Al-1Zn alloy," *Metallurgical and Materials Transactions A: Physical Metallurgy and Materials Science*, 39 A (2008) 1426-1434.

[27] S. Yi, I. Schestakow, S. Zaefferer, "Twinning-related microstructural evolution during hot rolling and subsequent annealing of pure magnesium," *Materials Science and Engineering: A*, 516 (2009) 58-64.

Magnesium Technology 2013
Edited by: Norbert Hort, Suveen N. Mathaudhu, Neale R. Neelameggham, and Martyn Alderman
TMS (The Minerals, Metals & Materials Society), 2013

Twin boundary migration creating zero shear strain: In-situ TEM observations and atomistic simulations

B.Y. Liu [1], B. Li [2], Z.W. Shan [1]
[1] Center for Advancing Materials Performance from the Nanoscale (CAMP-Nano), State Key Laboratory for Mechanical Behavior of Materials, Xi'an Jiaotong University, Xi'an, 710049, China
[2] Center for Advanced Vehicular Systems, Mississippi State University, Starkville, MS 39759, USA

Keywords: Twinning; Twin boundary; TEM; Atomistic simulation

Abstract

Atomistic simulations were conducted to study the migration of $\{10\bar{1}2\}<10\bar{1}\bar{1}>$ twin boundary. A bi-crystal that satisfies the twin orientation relationship was constructed and a tensile strain was applied in parallel to the <*c*> axis of one crystal. Under the tensile strain, the twin boundary starts to migrate but the migration (twin growth) does not produce any observable shear strain on the bi-crystal. In-situ transmission electron microscopy (TEM) observations of a single crystal Mg under tension and compression confirm that during twinning and detwinning, no shear strain is produced. The specimen uniformly elongates and narrows during twinning, and widens across the width during detwinning.

Introduction

Twinning plays a vital role in metals with hexagonal close-packed (hcp) crystal structures. It accommodates plastic strain along the <*c*>-[0001] direction during deformation [1-3]. Compared with twinning in metals with cubic crystal structures, where twin nucleation and twin growth are mediated by well-defined partial dislocation at the twin/matrix interface, i.e., twin boundaries (TBs) [4-5], twinning in hcp metals is much more complicated in the following aspects:

(1) In cubic metals, twinning can be accomplished by a homogeneous shear mediated by slip of Shockley partials. However, in hcp metals, for most of the twinning modes, a homogeneous shear is unable to complete the twinning process and atomic shuffling is required to achieve correct twin orientation relationship [6]. Shuffling is usually considered having negligible effect on twinning but recent studies show that this is not the case [7]. In fact, the relative magnitude of elementary twinning dislocation and atomic shuffling determines the configuration of twinning dislocations, which in turn determines the migration of TBs.

(2) Multiple twinning modes can take place. There are four major twinning modes with $\{10\bar{1}2\}\langle10\bar{1}\bar{1}\rangle$ mode being the predominant in all hcp metals. The properties of individual twinning modes vastly differ from one mode to another. For example, the $\{10\bar{1}2\}\langle10\bar{1}\bar{1}\rangle$ mode presents reversibility or pseudoelasticity [8-10], non-Schmid effect [11], penetration of twin variants through grain boundaries [12], zero shear strain [13], and extremely incoherent TBs [14]. These properties cannot be accounted for by any model of twinning dislocations [15-16]. Recently, Li and Ma [7] proposed that this particular twining mode is dominated by atomic shuffling and no twinning dislocations are involved. This theory properly resolves the puzzles regarding the mechanism of the predominant twinning mode in hcp metals. As to $\{10\bar{1}1\}\langle10\bar{1}\bar{2}\rangle$ twinning, shear dominates and minor shuffles are involved, and zonal twinning dislocations can be well defined at TBs [17].

In this work, we performed in-situ transmission electron microscopy and atomistic simulations on twinning and detwinning in single crystal Mg. The results obtained show that indeed, $\{10\bar{1}2\}\langle10\bar{1}\bar{1}\rangle$ twinning produces zero shear strain during reversible twinning.

Methods

Tension specimens were fabricated from a single crystal Mg by a FEI Helios 600 Dual Beam Focused Ion Beam (FIB) with rectangular cross sectional area and an aspect ratio 4:1. The dimensions of the specimens are shown in Figure 1a, with a thickness 125 nm and a width 420 nm. The c-axis of the specimen is exactly in the axial direction. A uniaxial tension was first carried out along the c-axis. After twinning, one end of the specimen was removed by FIB, and the load was reversed to compression. In-situ mechanical experiments were performed with a Hysitron PicoIndenter inside a JEOL 2100 TEM operating at 200 keV. All tension and compression tests are run in displacement control mode.

Atomistic simulations were conducted by applying a tensile strain to a bi-crystal in which the two individual crystals satisfy $\{10\bar{1}2\}\langle10\bar{1}\bar{1}\rangle$ twin orientation relationship. The loading axis was parallel to the *c*-axis of one crystal. After twinning, the tensile load was reversed to compressive.

Results and discussions

A tensile load was first applied along the c-axis, i.e. [0002] direction. $\{10\bar{1}2\}\langle10\bar{1}\bar{1}\rangle$ twinning was favored and activated during tension. Figure 1a and b shows the sample geometry before and after tension. The entire gauge length was twinned, as confirmed by the selected area diffraction (SAD) analysis. The *c*-axis was reoriented by ~90 degrees, typical of $\{10\bar{1}2\}\langle10\bar{1}\bar{1}\rangle$ twinning. Since the electron beam is parallel to the $\{10\bar{1}2\}$ twinning plane and the $<1\bar{2}10>$ zone axis, any shear deformation in the single crystal can be readily captured. Surprisingly, during and after twinning, the specimen uniformly elongated along the loading direction and narrowed across the width, leaving no observable traces of shear offset on the sample surface.

A high resolution transmission electron microscopy (HRTEM) image was acquired and the structure of the twin boundary is shown in Figure 2. In the atomic scale, the twin boundary is extremely incoherent and far off the $\{10\bar{1}2\}$ twinning plane. Such a huge deviation between the twinning plane and the twin boundary violates the requirement that a twin boundary should match the twinning plane if the twinning is controlled by twinning dislocations.

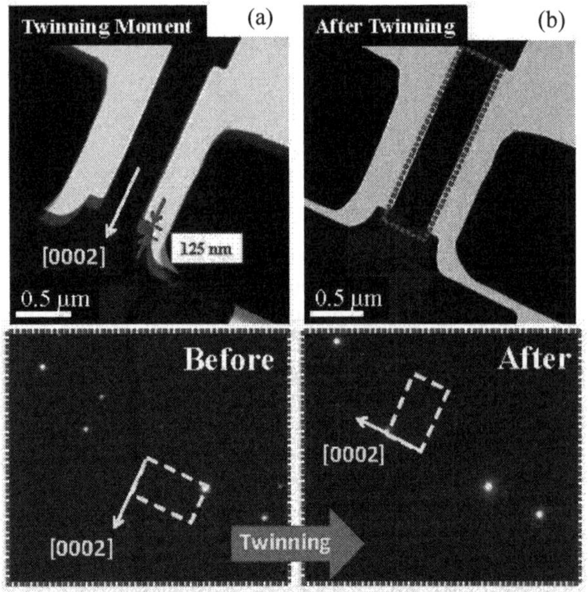

Figure 1. (a) Specimen geometry before tensile loading. (b) After loading along the c-axis, twinning occurred but the specimen uniformly elongated without observable shear offset. The diffraction patterns show that after twinning, the orientation of the twin is ~90 ° with the parent.

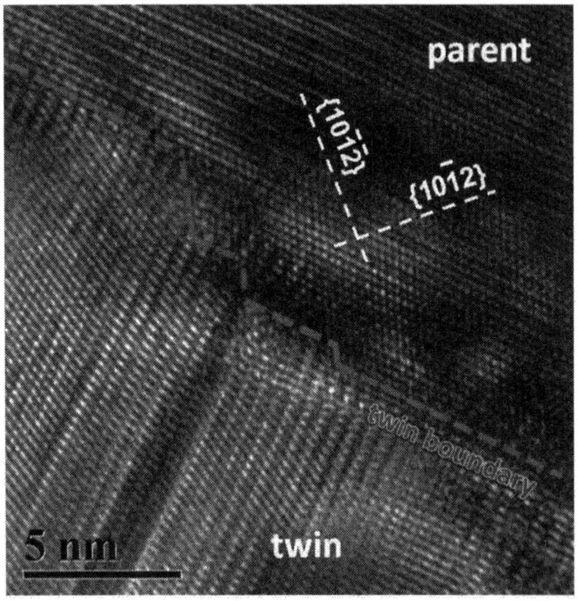

Figure 2. High resolution TEM shows the lattice fringes of the twin and the parent. The twin boundary is extremely incoherent and hugely deviates from the two possible twinning planes $\{10\bar{1}2\}$ and $\{10\bar{1}\bar{2}\}$.

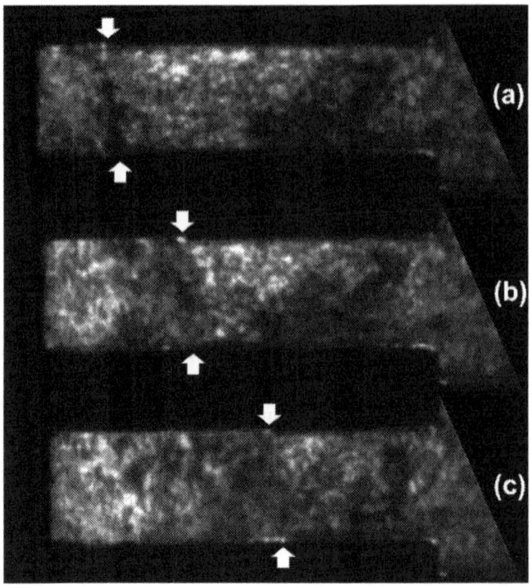

Figure 3. After twinning, the tensile load was reversed to compression and detwinning occurred. The specimen uniformly widens across the width, leaving no shear offset on the specimen as well.

It has been extensively observed that $\{10\bar{1}2\}\langle 10\bar{1}\bar{1}\rangle$ twinning is reversible. If an external load is removed or reversed, deformation twins shrink or completely disappear [8-10]. Indeed, detwinning was observed in our *in-situ* TEM observations. Figure 3 shows in time sequence the geometry evolution when the tensile load was reversed to compressive. The specimen uniformly widens toward both sides. SAD analysis confirms that the *c*-axis of the twinned region is reversed to the original orientation before twinning.

To understand the mechanism of the astonishing *in-situ* TEM observations, we also conducted atomistic simulations. The results of twinning under tension and detwinning under compression are shown in Figure 4 and 5. In Figure 4a-c, during tension, the incoherent twin boundary migrates toward left. The twinned region uniformly narrows. After the twin boundary moved close to the left end, the tensile load was reversed to compression. The migration of the twin boundary is shown in Figure 5a-c. The twin boundary reverses its moving direction while detwinning occurs. A perfect match between our experiments and simulations can be seen. During twinning and detwinning, no measurable shear strain was produced while the sample surfaces remained atomically flat after the twin boundary traverses, but the sample uniformly narrows during twinning and widens during detwinning.

Theoretically, the twinning shear *s* for $\{10\bar{1}2\}\langle 10\bar{1}\bar{1}\rangle$ equals $(\gamma^2 - 3)/(\sqrt{3} \cdot \gamma)$. For Mg, $s = -0.129$. Hence, a considerable shear strain should be generated during twinning. But in our *in-situ* TEM observations and atomistic simulations, no shear strain was produced. On the other hand, a reversible twinning should invoke a coherent, elastic interface free of dislocations because dislocation slip is irreversible once the dislocations reach free surfaces. Our results strongly suggest that this particular twinning mode involves no twinning dislocations and results in zero shear strain.

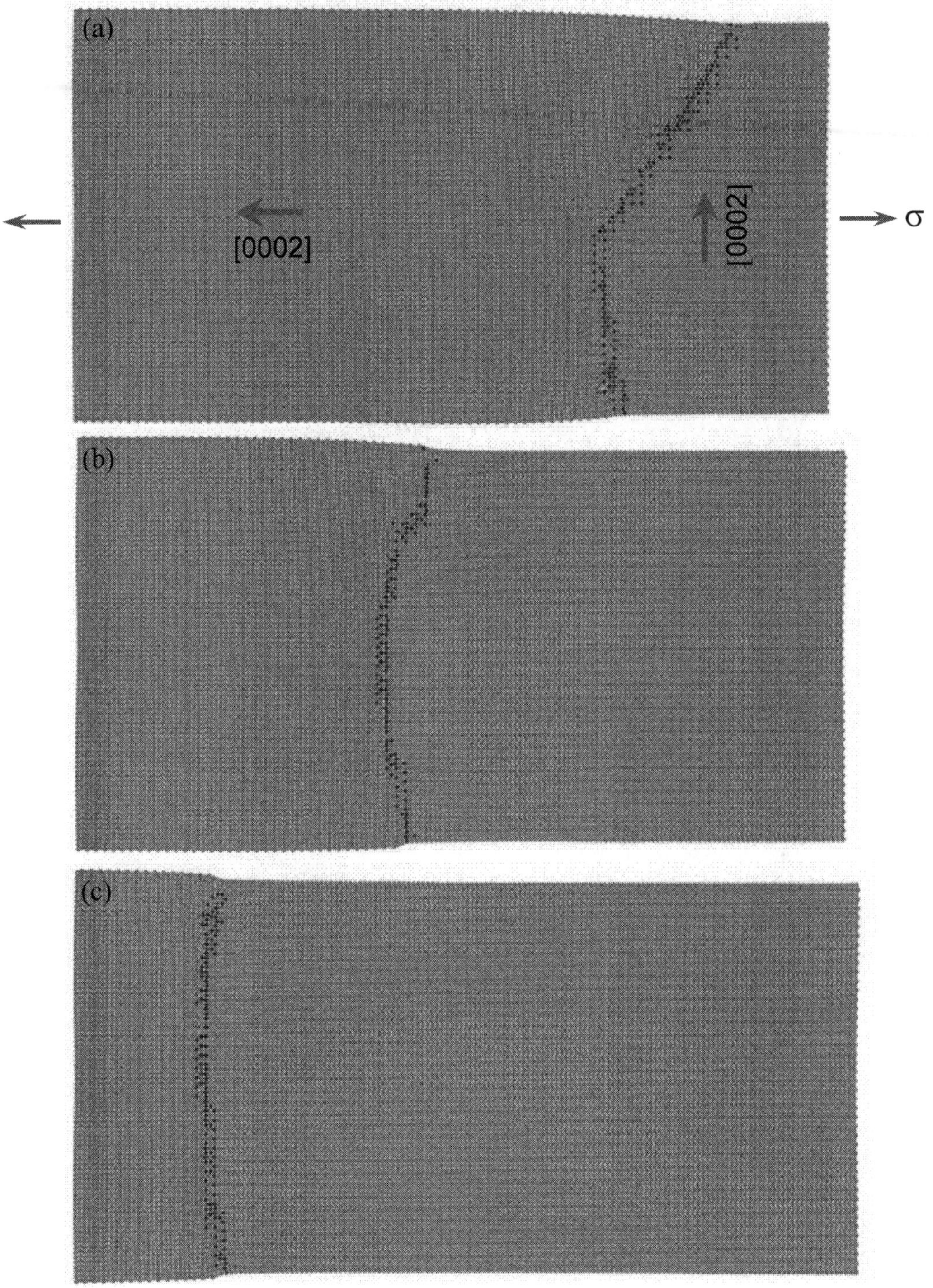

Figure 4. Molecular dynamics simulations of twin boundary migration under a tensile load. The twin boundary migrates toward left and the twin boundary hugely deviates from the twinning plane. The twinned region uniformly narrows without being sheared.

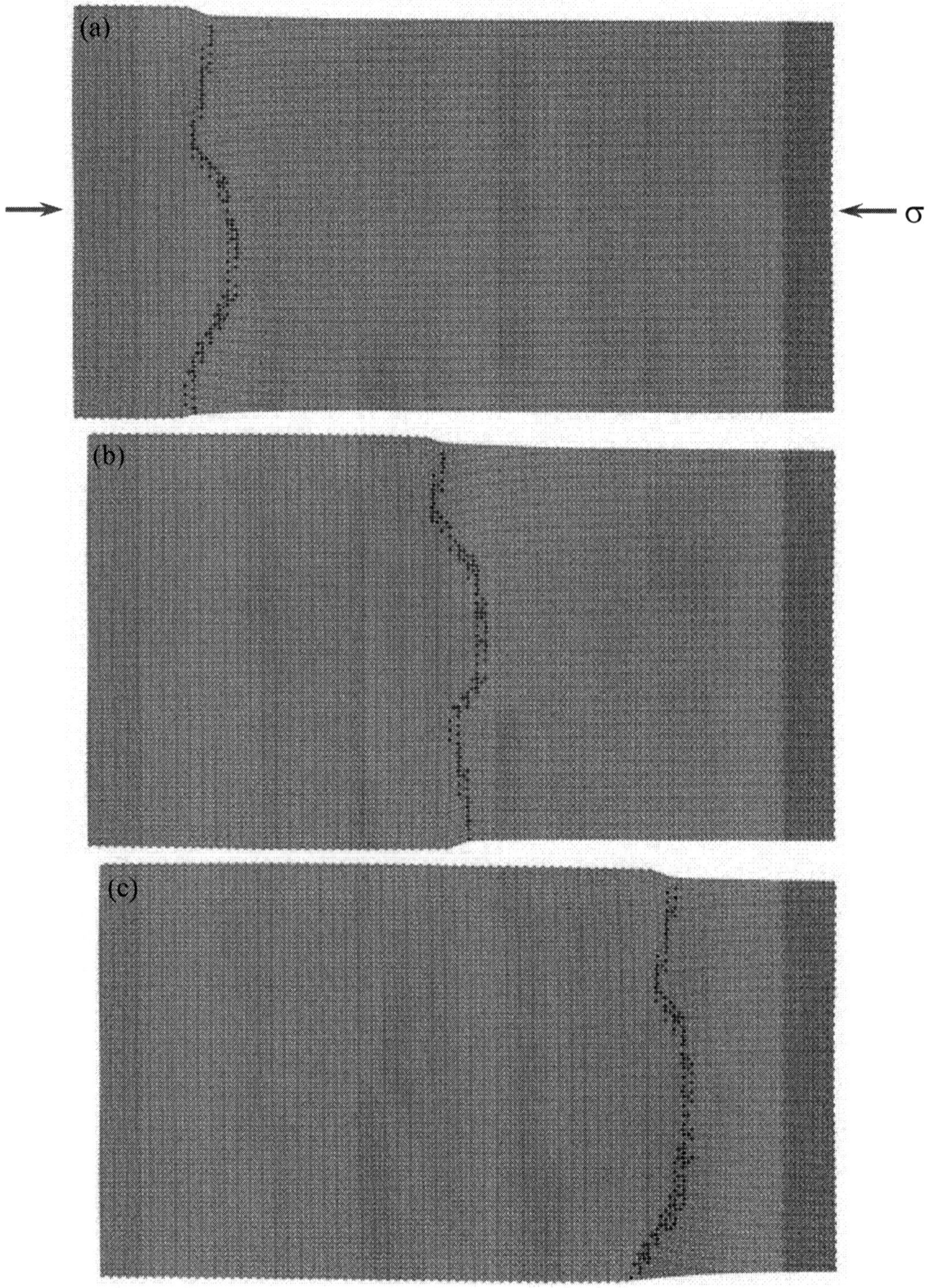

Figure 5. After the twin boundary approaches the left end, the tensile load is reversed to compression. The twin boundary reverses its moving direction and detwinning occurs. During detwinning, the specimen widens uniformly on both sides, different from a shear deformation.

For the predominant twinning mode in hcp materials, i.e., $\{10\bar{1}2\}\langle 10\bar{1}\bar{1}\rangle$, the theoretical Burgers vector of the twinning dislocation equals $(a(3-\gamma^2))/(2\sqrt{3+\gamma^2})$, where a is the lattice parameter and γ the c/a ratio [3]. For magnesium (Mg), the Burgers vector equals 0.024 nm, which is exceptionally small. Such a small Burgers vector indicates that atomic shuffling should dominate $\{10\bar{1}2\}\langle 10\bar{1}\bar{1}\rangle$ twinning [7].

The tiny, theoretical Burgers vector of the twinning dislocation (0.024 nm for Mg) indicates that, structurally, the twin lattice almost exists in the parent lattice before a homogeneous shear is activated. Li and Ma [7] demonstrated how a twin lattice can be directly reconstructed from a parent hcp lattice without involving any twinning dislocation. In their model, only atomic shuffling is required to accomplish $\{10\bar{1}2\}\langle 10\bar{1}\bar{1}\rangle$ twinning. Through concerted shuffles, the correct hcp stacking and *c*/*a* ratio are achieved. Immediately, the reversible twinning is well explained: because no twinning dislocations are involved, the twinning process is reversible, and the twin boundary does not have to coincide with the twinning plane, resulting in the huge deviations observed in Figure 2 and in the atomistic simulations.

Conclusions

We performed in-situ TEM observations on uniaxial tension and compression of a Mg single crystal. During extension twinning, the specimen uniformly elongated without being sheared; whereas during compression, detwinning occurred and the specimen uniformly widened across the width without observable shear offset. These observations were supported by our atomistic simulations in which both twinning and detwinning were accomplished by atomic shuffling.

Acknowledgements

The authors thank Dr. Xiyan Zhang at the Chongqing University, China for providing the Mg single crystal that was used in this work. B. Li is grateful for the support from the Center for Advanced Vehicular Systems (CAVS) at Mississippi State University.

References

[1] J.P. Hirth, J. Lothe: Theory of Dislocations, 2nd edition, John Wiley &Sons, New York, NY, 1982.
[2] M.M. Avedesian, H. Baker, Magnesium and magnesium alloys, ASM specialty handbook, 1999.
[3] J.W. Christian, S. Mahajan, Prog Mater Sci 39 (1995) 1.
[4] S. Mahajan, G.Y. Chin, Acta Mater 21 (1973) 1353.
[5] B. Li, B.Y. Cao, K.T. Ramesh, E. Ma. Acta Mater 57 (2009) 4500.
[6] B.A. Bilby, A.G. Crocker, Proc Roy Soc. A, 288 (1965) 240.
[7] B. Li, E. Ma, Phys Rev Lett 103 (2009) 035503.
[8] C.H. Caceres, T. Sumitomo, M. Veidt, Acta Mater 51 (2003) 6211.
[9] L. Wu, A. Jain, D. W. Brown, g. M. Stoica, S. R. Agnew, B. Clausen, D.E. Fielden, P.K. Liaw, Acta Mater 56 (2008) 688.
[10] A. Jain, S. R. Agnew, Mater Sci Eng A 462 (2007) 29.
[11] M.R. Barnett, Z. Keshavarz, A.G. Beer, X. Ma, Acta Mater 56 (2008) 5.
[12] B. Li, E. Ma, K.T. Ramesh, Metall Mater Trans A 39A (2008) 2607.
[13] B.Y. Liu, Z.W. Shan, X.Y. Zhang, J. Sun, E. Ma, " In-situ quantitative tension and compression study on twinning and detwinning in submicron-sized Mg crystals inside a transmission electron microscope", TMS 2012, Orlando, USA.
[14] P.G. Partridge, E. Roberts, Acta Met 12 (1964) 1105.
[15] A. Serra, D.J. Bacon, Mater Sci Eng A 400-401 (2005) 496.
[16] J. Wang, J.P. Hirth, C.N. Tome, Acta Mater 57 (2009) 5521.
[17] B. Li, E. Ma, Acta Mater 57 (2009) 1734.

Magnesium Technology 2013
Edited by: Norbert Hort, Suveen N. Mathaudhu, Neale R. Neelameggham, and Martyn Alderman
TMS (The Minerals, Metals & Materials Society), 2013

Microstructure Characterization of Weakly Textured and Fine Grained AZ61 Sheet

T.D. Berman[1], W. Donlon[1], C.K. Hung[1], P. Milligan[1], R. Decker[2], T.M. Pollock[3], J.W. Jones[1]

[1]Materials Science and Engineering, University of Michigan, 2300 Hayward St; Ann Arbor, MI, 48109, USA
[2]nanoMAG, LLC., 13753 Otterson Court; Livonia, MI, 48150, USA
[3]Materials Department, University of California, Santa Barbara; Santa Barbara, CA, 93106-5050, USA

Keywords: AZ61L, Microstructures, Texture, Mechanical Properties, Thixomolding, Thermomechanical Processing

Abstract

Formability in magnesium alloy sheet is strongly limited by a strong basal texture in the as-rolled material, which is difficulty to remove by thermal processing. We introduce a new process to the control of texture by combining Thixomolding and Thermomechanical Processing (TTMP). Plates of AZ61L with a divorced β-$Mg_{17}Al_{12}$ eutectic are produced by Thixomolding, resulting in a non-textured, fine grained (2.8 μm) precursor. Sheet produced from the plate by single pass warm-rolling exhibits a weaker texture, and more isotropic tensile deformation than generally observed in AZ-series alloy sheet. Recrystallization annealing produces a further reduction in texture and average grain size (2.3 μm) and results in nearly isotropic room temperature deformation, a yield strength of $\sim$ 220 MPa, and an elongation of $\sim$ 23%. Particle stimulated nucleation of new grains by the β-phase during both dynamic and static recrystallization, is critical for achieving the low levels of texture. The influence of β-phase distribution in microstructure development is discussed.

Introduction

Magnesium alloy sheet has many potential applications in reducing the weight of automotive, aerospace, and consumer electronics. However, the limited number of slip systems active at room temperature necessitate warm forming, which can make material costs prohibitive to widespread use. Several mechanisms have been employed to improve both the strength and formability of magnesium alloy sheet. Grain refinement is often employed to achieve a combination of high strength and high ductility [1]. Another important mechanism to improve ductility is texture control. Pronounced basal texture strongly reduces the formability of magnesium due to the inactivity of $\langle c+a \rangle$ slip [2]. It has been shown in weakly textured ECAPed material that the elongation to failure was over twice as high as that of the conventional AZ31 [3]. Stress-strain curves are orientation dependent in a textured material, and more isotropic behavior is produced from weaker textured material [4].

In order to produce fine-grained, weakly textured material a Thixomolded AZ61L plate was used as a precursor. As the dynamically recrystallized grain size depends on initial grain size in AZ series alloys [5], it is logical to start with the fine initial grain size from Thixomolding to produce sheet. Thixomolding also results in a distribution of relatively fine β-$Mg_{17}Al_{12}$ particles. The presence of these particles suggests opportunities to control the recrystallization texture and microstructure by adjusting their size, distribution, and volume fraction [6]. Second phase particles can also suppress grain growth during hot working.

Table 1: Composition of received AZ61L in wt %

Al	Zn	Mn	Si	Fe	Mg
6.5	0.46	0.14	0.01	0.003	bal.

Experimental

Preheated Thixomolded plates, with dimensions of 200 mm x 200 mm x 3 mm were preheated for 5 minutes at 315 °C and then warm-rolled in a single pass with a roll temperature near 200 °C. After rolling, the sheets were allowed to cool to room temperature in ambient air. The final sheet thickness was 1.8 mm, a thickness reduction of approximately 40%. Table 1 gives the composition of the AZ61L plates. Annealing treatments were done in a Techne SBL-2D fluidised bath. Annealed samples were water quenched.

Dogbone tensile specimens with a gauge length of 25.45 mm and cross section of 6.35 mm x 1.8 mm were machined from the sheets with the tensile axis (TA) at 0 °, 45 °, or 90 ° from the rolling direction (RD). Tensile tests were conducted at room temperature; samples were preloaded to 20 MPa and deformed at a displacement rate of 0.60 mm/min. At least three specimens were used for each condition. An extensonometer was used to measure tensile elongation. Samples used for r-value measurements were strained to 10%.

All microstructures examine the plane of the sheet. SEM samples were prepared by polishing to 1 μm diamond paste followed by a chemical etch. The as-Thixomolded samples were etched for 10 s at room temperature in a 1:9 solution of o-phosphoric acid in ethanol. Wrought samples were etched for 5 s at room temperature in an solution of 10 ml water, 10 ml acetic acid, 4.3 grams of picric acid, and 70 ml of ethanol. SEM examination was conducted with a Philips XL30 FEG.

XRD pole figure analysis was conducted on the sheet cross section in order to be able to capture texture information that would fall at angles greater than 75 ° from the normal direction in the planar orientation. Specimen preparation involved polishing mounted cross sections to 1200 grit SiC paper with ethanol. A Rigaku rotating anode XRD system was used to make XRD pole figure measurements. XRD volume fractions were determined by using the whole pattern fitting routine in the Jade (2010) software package.

Results

Evolution of Mechanical Properties

Table 2 illustrates the evolution of mechanical properties as we go from the as-Thixomolded plate, to the as-rolled sheet, to sheet subjected to post-deformation annealing at 285 °C for 10 minutes. The 3 mm thick molded plate has a mean ductility of 16% and a mean yield strength of 145 MPa.

The rolling process leads to a material twice as strong as the Thixomolded precursor. However, there is tensile anisotropy in the plane of the sheet; the ultimate tensile strength is ~25 MPa stronger along the RD than along the transverse direction (TD) (Figure 1). The average Lankford coefficient ($\bar{r}$) is 1.4 for the as-rolled sheet, with the planar anisotropy of r-value (Δr) being 0.3.

After annealing at 285 °C for 10 minutes the yield strength of the sheet drops nearly 120 MPa to 225 MPa along the RD, however, the tensile deformation becomes much more isotropic in the plane of the sheet (Figure 1). In addition, the elongation along the RD increases to 24.7%. The improved isotropy is also evident in the Lankford values, with $\bar{r}$ being 1.09, and Δr being 0.12 in the recrystallized sheet.

Microstructures

Thixomolded Precursor The rapid solidification rate during Thixomolding results in a fine grain size and weak crystallographic texture. The microstructure is composed of fine α-Mg grains surrounded by a divorced β eutectic (Figure 2). The average grain size of the as-Thixomolded material, as measured near the surface of the sheet, is 2.8 μm. In addition to controlling the grain size, the rapid cooling rate leads to a supersaturation of aluminum; the average volume fraction of the β-phase is only 3% as determined by XRD and area fraction calculation. The β-particles are heterogeneously distributed throughout the microstructure, primarily along grain boundaries. The β-particles have a mean area of 31 nm with over 95% of the particles being less than 1 μm in area.

As-Rolled Sheet The as-rolled microstructure largely consists of heavily deformed grains. The β-phase network is elongated along the RD (Figure 3). The extended time at temperature during the rolling process leads to further precipitation of the β-phase. Due to

Anneal	Angle	YS (MPa)	UTS (MPa)	Max Strain (%)	r	$\bar{r}$	Δr
as-Molded	0	145.7 ± 6.4	272.8 ± 21.4	16.0 ± 6.5			
	0	344 ± 10	371.4 ± 4.1	14.6 ± 4.3	1.48		
as-Rolled	45	271 ± 14	343.8 ± 5.1	16.7 ± 5.1	1.59		
	90	260 ± 12	344.8 ± 1.1	15.0 ± 2.8	1.10		
	Mean	292	353	15.4		1.44	0.30
	0	225 ± 4	315.5 ± 0.3	24.7 ± 4.5	0.96		
285 °C	45	212 ± 4	310.0 ± 2.5	26.1 ± 3.0	1.14		
	90	219 ± 10	321.6 ± 8.9	19.0 ± 3.0	1.09		
	Mean	219	316	23.3		1.09	0.12

Table 2: Tensile properties of the Thixomolded plate and TTMP AZ61L sheet. The annealing treatment at 285 °C was 10 minutes in duration. The angle is measured between the rolling and tensile directions. Samples strained to 10% were used for determination of the Lankford value.

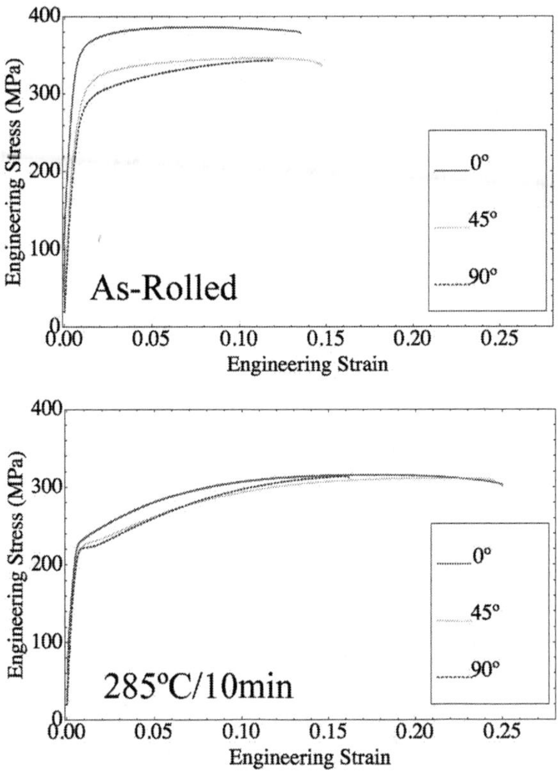

Figure 1: Engineering stress vs. engineering strain for as-rolled and annealed sheet.

a strong texture, it is difficult to measure the volume fraction of the β-phase via X-ray diffraction, however based on area fraction measurements of SEM images, the as-Rolled material has a β-phase volume fraction of 4.6%. β-particles in the as-rolled sheet can be classified into 2 groups, (1) large particles that existed in the as-Thixomolded microstructure and were deformed and coarsened during rolling and (2) fine spheroidal and rod-like particles that precipitated during rolling or during post-deformation cooling (Figure 4).

Annealed Sheet Post-deformation annealing at 285 °C for 10 minutes leads to a fully recrystallized sheet the a average grain size of 2.3 μm (Figure 5). The grain size distribution is nearly identical to the original as-Thixomolded distribution (Figure 6). The β-phase network remains elongated in the RDs and the volume fraction of β is nearly unchanged from the as-rolled condition at 4.8% as measured by XRD.

Deformed Sheet Examination of one of the as-rolled tensile samples, which failed at 13% strain in the RD, reveals damage accumulation in the β-phase in the form of cracks and voids (Figure 7). The cracks run perpendicular to the TA and RD.

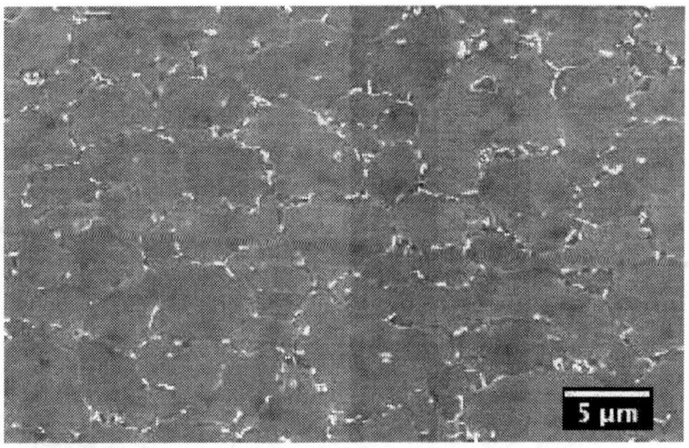

Figure 2: Microstructure of the as-Thixomolded plate. β-$Mg_{17}Al_{12}$, the brighter phase, forms a discontinuous network along the grain boundaries.

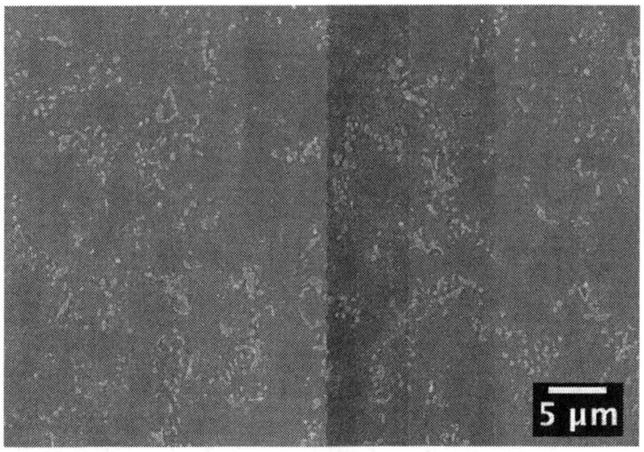

Figure 3: Microstructure of the as-rolled condition. The volume fraction of the β-$Mg_{17}Al_{12}$, has increased and the β-phase network is elongated in the RD (vertical).

Texture Evolution
The as-Thixomolded sheet has a near random intensity in both the 0002 and $10\bar{1}0$ pole figures (Figure 8). XRD pole figures reveal that the as-rolled sheet has a near basal texture in the plane of the sheet, with double averages located $\sim$ 20 ° from the basal direction. The intensity of the basal texture is greatly diminished after annealing at 285 °C for 10 minutes. In addition, the peaks tilt further away from the basal pole, to near 45 °. The texture is elongated along the TD after rolling, as indicated by the $10\bar{1}0$ pole figures.

Discussion

The texture of the rolled sheet is distinctive among AZ alloys. Figure 8 illustrates the weak (max = 3.1 times random), double-peaked basal texture. 10 times

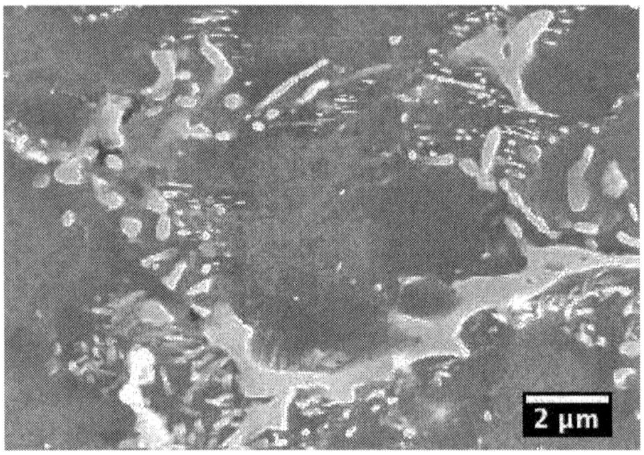

Figure 4: Microstructure of fine β-particles precipitated during the rolling process. Spherical and rod-like particles are evident.

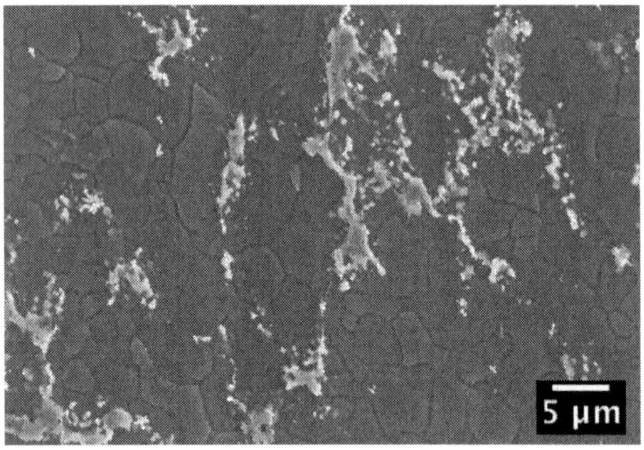

Figure 5: SEM microstructure of sample annealed for 10 minutes at 285 °C. The β-phase network remains elongated in the rolling (vertical) direction

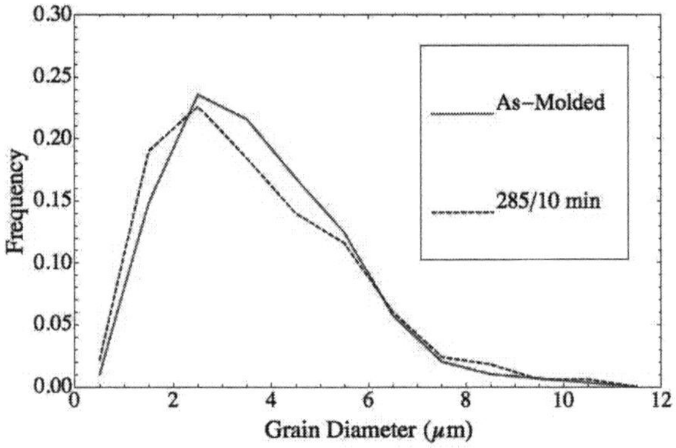

Figure 6: Grain size distribution of the as-Molded and recrystallized (285C/10min) materials.

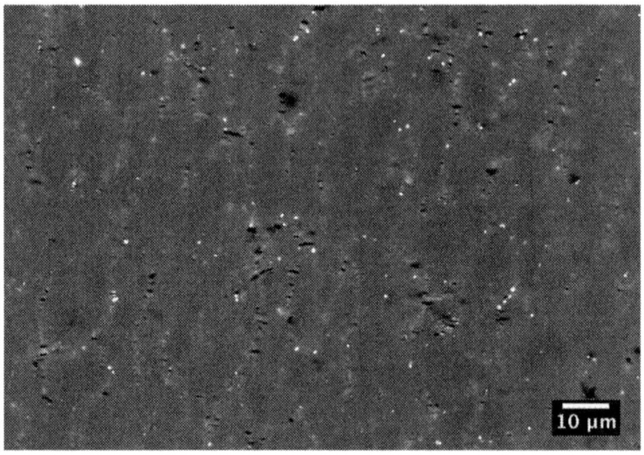

Figure 7: BSE image of damage accumulation in the as-rolled material when the TA is along the RD (vertical). It is clear that the damage is initiating in the β-phase.

random is typical for a basal pole figure intensity in the AZ series [7–11]. The peaks are tilted $\sim 20°$ from the basal direction. A splitting, or elongation of the basal peak in the RD is commonly observed [7, 9, 12–15], this split likely results from increased activity of the $\langle \mathbf{c}+\mathbf{a} \rangle$ slip mode [16]. Li el al have shown that AZ61 and and AZ91 exhibit weaker texture after rolling due to particle stimulated nucleation [17]. The notable feature of the texture of TTMP AZ61L is that there is a preferential orientation of the $\langle 10\bar{1}0 \rangle$ directions, other reports on AZ series alloys that provide the $10\bar{1}0$ pole figure show a random orientation of **a** [7–9, 15]. Though not observed in AZ series alloys, textures similar to that of TTMP AZ61L, with a broader spread of the basal poles along the TD than in the RD, have been reported in RE containing wrought Mg sheets [18]. The mechanisms behind the development of the RE textures are still not understood [19].

Another interesting feature of the texture of TTMP AZ61L is that it weakens during static annealing. Texture generally does not change during static annealing [4, 6]. Masoumi et al have shown texture reduction during the annealing of twin-roll cast AZ31, which they attributed to particle stimulate nucleation arising from a higher volume fraction of secondary phases [20]. This mechanism is likely active in the TTMP AZ61L sheet as well. Another postulate is that continuous recrystallization, which involves rotation of grains and subgrains, may become the dominate recrystallization mechanism in the TTMP AZ61L as pinning of grain boundaries and dislocations by β-particles hinders dynamic recrystallization. Similar observations have been made during dynamic recrystallization of RE alloys [21, 22]. Recrystallization occurring via lattice

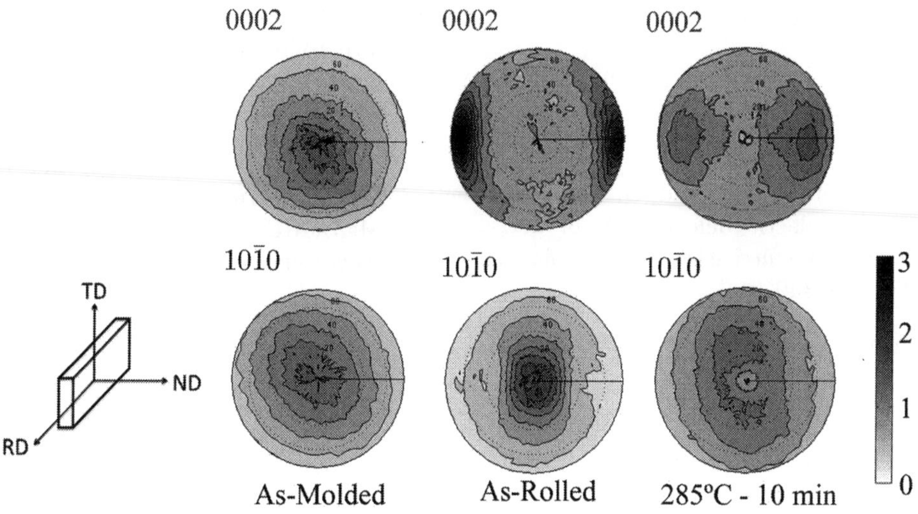

Figure 8: Measured pole figures for the material after molding, warm-rolling, and annealing at 285 °C for 10 min.

rotations at grain boundaries can lead to a change in recrystallization texture [6].

The texture of the TTMP AZ61L leads to encouraging mechanical behavior. The average Lankford r-value ($\bar{r}$) is low for Mg-alloy sheet at 1.4. Average room temperature r-values in the range of 1.7 - 4.2 have been reported in AZ31 [7, 9, 10, 12, 13, 23]. Huang et al did measure low r-values (1.1 and 1.2) in AZ31 produced by multiple roll processing, with finishing passes at 525 °C and 555 °C [12]. Interestingly, in all of these studies, it was found the the r-value was higher in the TD than in the RD, whereas in the as-rolled TTMP AZ61L sheet the r-value was found to be higher in the RD than the TD. Similar observations have published on sheets of RE containing Mg-alloys and are a result of the unique textures exhibited by these materials [18]. Another mechanical property of the TTMP AZ61L that results from the RE-like texture is that the UTS of the material is higher in the RD than in the TD due to flow stresses being lower along the TD than in the RD.

A unique feature of the TTMP AZ61L is the β-phase network in the final sheet product. AZ-series alloys are usually solution treated, removing the brittle β-phase. Though the secondary phase provides for particle stimulated nucleation and grain size stability during annealing, it is clearly illustrated in Figure 7 that damage does initiate at either the α/β phase boundary or inside the β-particles. In both the as-rolled and recrystallized sheets we observed that the elongation is significantly lower along the TD. It has yet to be determined if this is due to the elongation of the β-phase network along the RD, which could allow for easier propagation of cracks perpendicular to the TA when the TD is parallel to the TA. Future work will be addressing the orientation dependence of damage accumulation in the TTMP AZ61 material at fixed strain increments.

Summary

Warm-rolling of an AZ61L Thixomolded plate produces a Mg-alloy sheet with high strength and reasonable formability. The TTMP sheet exhibits a weaker texture, and more isotropic tensile deformation than generally observed in AZ-series alloy sheet. Annealing at 285 °C for 10 minutes leads to a recrystallized sheet with a average grain size of 2.3 μm and an even weaker basal texture. These characteristics result in nearly isotropic room temperature tensile deformation, a yield strength of $\sim$ 220 MPa and a elongation of $\sim$ 23%. The β-phase plays an important roll in both dynamic and static recrystallization, keeping the as-deformed texture weak and allowing for texture reduction during annealing.

Acknowledgements

This material is based upon work supported by the National Science Foundation under Grant No. 0847198. Thanks to Dr. Anthony Rollett for suggestions regarding texture analysis.

References

1. J. Chapman and D. Wilson, "The Room-Temperature Ductility of Fine-Grain Magnesium," *Journal of the Institute of Metals*, 91 (1962), 39–40.
2. R. Gehrmann, M. Frommert, and G. Gottstein, "Texture effects on plastic deformation of magnesium," *Materials Science and Engineering A*, 395 (1-2) (2005), 338–349.
3. T. Mukai, "Ductility enhancement in AZ31 magnesium alloy by controlling its grain structure," *Scripta Materialia*, 45 (1) (2001), 89–94.
4. S. Agnew et al., "Enhanced ductility in strongly textured magnesium produced by equal channel angular processing," *Scripta Materialia*, 50 (3) (2004), 377–381.
5. H. Watanabe et al., "Grain Size Control of Commercial Wrought Mg-Al-Zn Alloys Utilizing Dynamic Recrystallization," *Materials Transactions*, 42 (7) (2001), 1200–1205.
6. F. Humphreys and M. Hatherly, *Recrystallization and Related Annealing Phenomena* (Elsevier, Netherlands, 2004), 2 ed.
7. S. Agnew and O. Duygulu, "Plastic anisotropy and the role of non-basal slip in magnesium alloy AZ31B," *International Journal of Plasticity*, 21 (6) (2005), 1161–1193.
8. A. Jain et al., "Grain size effects on the tensile properties and deformation mechanisms of a magnesium alloy, AZ31B, sheet," *Materials Science and Engineering: A*, 486 (1-2) (2008), 545–555.
9. S. Yi et al., "Mechanical anisotropy and deep drawing behaviour of AZ31 and ZE10 magnesium alloy sheets," *Acta Materialia*, 58 (2) (2010), 592–605.
10. Y. Chino, K. Sassa, and M. Mabuchi, "Enhanced stretch formability of Mn-free AZ31 Mg alloy rolled by cross-roll rolling," *Journal of Materials Science*, 44 (7) (2009), 1821–1827.
11. M. Masoumi, F. Zarandi, and M. Pekguleryuz, "Microstructure and texture studies on twin-roll cast AZ31 (Mg3wt.%Al1wt.%Zn) alloy and the effect of thermomechanical processing," *Materials Science and Engineering: A*, 528 (3) (2011), 1268–1279.
12. X. Huang et al., "Improvement of stretch formability of Mg3Al1Zn alloy sheet by high temperature rolling at finishing pass," *Journal of Alloys and Compounds*, 509 (28) (2011), 7579–7584.
13. H. Zhang et al., "Influence of initial texture on formability of AZ31B magnesium alloy sheets at different temperatures," *Journal of Materials Processing Technology*, 211 (10) (2011), 1575–1580.
14. L. Jin et al., "Effects of hot rolling processing on microstructures and mechanical properties of Mg3%Al1%Zn alloy sheet," *Materials Science and Engineering: A*, 527 (7-8) (2010), 1970–1974.
15. J. Koike, "Enhanced deformation mechanisms by anisotropic plasticity in polycrystalline Mg alloys at room temperature," *Metallurgical and Materials Transactions A*, 36 (7) (2005), 1689–1696.
16. S. Agnew, M. Yoo, and C. Tome, "Application of texture simulation to understanding mechanical behavior of Mg and solid solution alloys containing Li or Y," *Acta Materialia*, 49 (20) (2001), 4277–4289.
17. X. Li et al., "Influence of second-phase precipitates on the texture evolution of Mg Al Zn alloys during hot deformation," *Scripta Materialia*, 66 (2012), 159–162.
18. J. Bohlen et al., "The texture and anisotropy of magnesium zinc rare earth alloy sheets," *Acta Materialia*, 55 (2007), 2101–2112.
19. S. Agnew and J. Nie, "Preface to the viewpoint set on: The current state of magnesium alloy science and technology," *Scripta Materialia*, 63 (7) (2010), 671–673.
20. M. Masoumi, F. Zarandi, and M. Pekguleryuz, "Alleviation of basal texture in twin-roll cast Mg3Al1Zn alloy," *Scripta Materialia*, 62 (11) (2010), 823–826.
21. J. P. Hadorn et al., "Role of Solute in the Texture Modification During Hot Deformation of Mg-Rare Earth Alloys," *Metallurgical and Materials Transactions A*, 43 (April) (2012), 1347–1362.
22. K. Hantzsche et al., "Effect of rare earth additions on microstructure and texture development of magnesium alloy sheets," *Scripta Materialia*, 63 (7) (2010), 725–730.
23. M. Barnett et al., "Deformation mechanisms in Mg alloys and the challenge of extending room-temperature plasticity," *Jom*, 61 (8) (2009), 19–24.

Magnesium Technology 2013

Wrought Materials I

Magnesium Technology 2013
Edited by: Norbert Hort, Suveen N. Mathaudhu, Neale R. Neelameggham, and Martyn Alderman
TMS (The Minerals, Metals & Materials Society), 2013

FLOW BEHAVIOR AND HOT WORKABILITY OF PRE-EXTRUDED AZ80 MAGNESIUM ALLOY

Lei Gao[1], Alan A. Luo[2], Shiyi Wang[3], Xiaoqin Zeng[3]
[1]General Motors China Science Lab, 56 Jinwan Road, Pudong, Shanghai 201206, China
[2]General Motors Global Research and Development Center, Warren, MI, USA
[3]National Engineering Research Center of Light Alloy Net Forming, Shanghai Jiao Tong University, Shanghai 200240, China

Keywords: Magnesium alloy, Hot deformation, Processing map, Dynamic recrystallization, Flow instability

Abstract

The hot deformation behavior of pre-extruded AZ80 magnesium alloy has been studied using the processing map technique. Compression tests using Gleeble-3800 thermal simulator were performed in the temperature range of 250-450°C and the strain rate range of 0.001-10 s^{-1}. The flow stress data were used to develop processing maps at true strains of -0.1 to -0.8 according to the well-known dynamic material model and instability criterion. A single dynamic recrystallization (DRX) domain occurs in the range of 420-450°C and 0.1-1.0 s^{-1}, which are the optimum forming conditions for the hot working of this alloy. There are two flow instability regimes occurring at 250-450°C and 0.004-10 s^{-1} and 433-450°C and 0.002-0.014 s^{-1}. The former occurs at low temperatures and/or high strain rates and is associated with adiabatic shear bands or cracks, flow localization, and deformation twinning, while the latter at higher temperatures and lower strain rates is due to abnormal grain growth and wedge cracking.

Introduction

While high-pressure die casting is the dominant process for current magnesium automotive applications, wrought magnesium and their manufacturing processes are receiving increasing attention from academia and industries due to their higher strength and ductility than the cast counterparts [1]. However, wrought magnesium alloys have much less workability than steel or aluminum, especially at low temperatures (<300°C), due to the hexagonal close-packed crystal structure of magnesium [1, 2]. AZ80 (Mg-8.5Al-0.5Zn-0.2Mn, wt.%) is a typical magnesium alloy used in forging and extrusion with high strength, good plasticity and toughness, and good corrosion resistance [3]. The aim of the present investigation is to study the hot workability of AZ80 alloy using the processing map technique and metallography to explore its potential for automotive applications. The optimum temperature-strain rate windows for manufacturing and the flow instability regimes are identified.

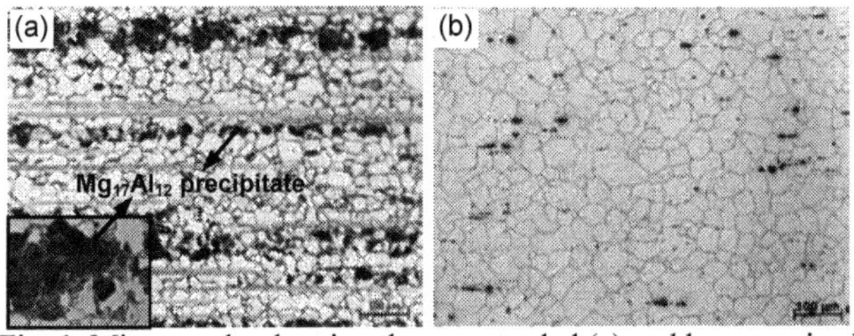

Fig. 1. Micrographs showing the as-extruded (a) and homogenized (b) microstructures of AZ80 alloy. Extrusion direction is horizontal.

The basis for the processing map of a material is the dynamic materials model (DMM) [4], where the workpiece undergoing hot deformation is considered to be a dissipater of power. The strain rate sensitivity (m) of flow stress is the factor that partitions power between deformation heat and microstructural changes. The efficiency of power dissipation, a dimensionless parameter η, occurring through microstructural changes is derived by comparing the non-liner power dissipation occurring instantaneously in the work-piece with that of a linear dissipater for which the m value is unity, and is given by [5]:

$$\eta = \frac{2m}{m+1} \qquad (1)$$

The power dissipation map is obtained by plotting iso-efficiency contours of the parameter η in a temperature-strain rate frame at a constant strain. A continuum instability criterion [6, 7] based on the extreme principles of irreversible thermodynamics as applied to large plastic flow [8, 9] is used to identify the regimes of flow instabilities. According to the criterion, flow instability will occur during hot deformation if:

$$\xi(\dot{\varepsilon}) = \frac{\partial \ln[m/(m+1)]}{\partial ln\dot{\varepsilon}} + m < 0 \qquad (2)$$

The variation of instability parameter $\xi(\dot{\varepsilon})$ with deformation temperature and strain rate constitutes an instability map which delineates regimes of instability where ξ is negative.

A superimposition of the instability map on the power dissipation map constitutes a processing map. Processing maps help in identifying temperature-strain rate windows for hot working where the intrinsic workability of the materials is a maximum (e.g. DRX, DRV, and superplasticity) and also in avoiding the regimes of flow instabilities (e.g. adiabatic shear bands or flow localization) [10, 11]. The processing map technique has been used earlier to study the hot deformation mechanisms in pure Mg [12] and Mg alloys [9, 13-16].

Experimental procedure

Pre-extruded commercial AZ80 billets 63 mm in diameter obtained from Magnesium Elektron North America were used as the initial material in this study. The chemical composition of the alloy billets is shown in Table 1. The billets were homogenized at 400°C for 24 h before machining the specimens for workability testing.

Table 1. Chemical composition of AZ80 alloy (in wt.%).

Alloy	Al	Mn	Zn	Fe	Ni	Cu	Zr	Y	Mg
AZ80	7.8	0.14	0.26	<0.01	<0.01	<0.01	<0.01	<0.01	Balance

Cylindrical specimens 10 mm diameter and 15 mm tall were machined from the billets with their longitudinal axis parallel with the pre-extrusion direction. Uniaxial compression testing for hot workability analysis was carried out using a Gleeble-3800 machine at a constant strain rate ranging from 0.001 to 10 s^{-1} and at an initial temperature between 250 and 450°C. The specimens were heated at a rate of 2°C/s and soaked for 3 min at the preset temperature in low vacuum to allow the temperature to equalize throughout the specimen. Graphite flakes were used as the lubricant in all the experiments. The specimens were deformed up to a true strain of -1.0 (height reduction ~63%) and rapidly quenched into water. The true stress-strain (σ_T-ε_T) curves were derived from the nominal stress-strain (σ_N-ε_N) curves obtained in compression according to the following formulas: $\sigma_T=\sigma_N(1+\varepsilon_N)$, $\varepsilon_T=\ln(1+\varepsilon_N)$. The processing maps were developed according to Eqs. (1) and (2). Selected samples were deformed to different final true strains of -0.2 and -0.5.

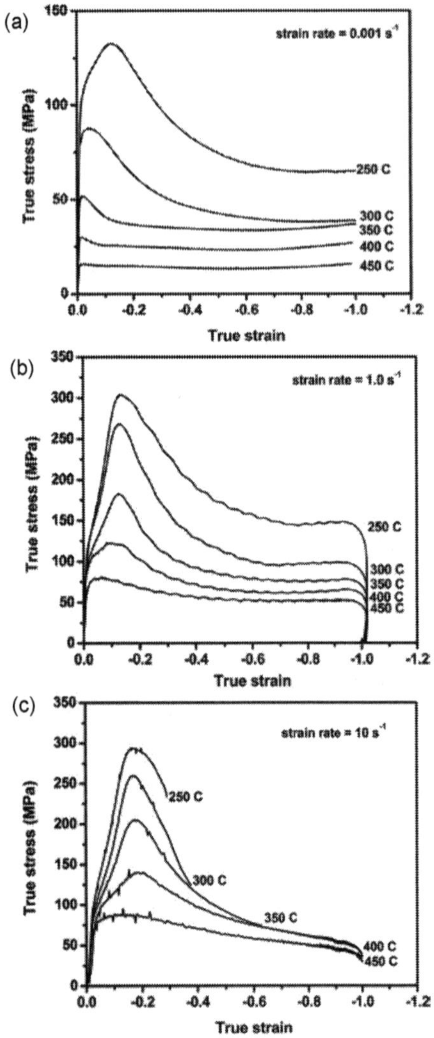

Fig. 2. Typical true stress-true strain curves of AZ80 magnesium alloy at different strain rates and temperatures. (a) 0.001 s^{-1}, 250-450°C; (b) 1.0 s^{-1}, 250-450°C; (c) 10 s^{-1}, 250-450°C.

For metallographic examination, the initial and deformed specimens were sectioned parallel to the compression axis and the cut surfaces were prepared for microstructure observation. Microstructures were observed by optical microscopy (OM) and scanning electron microscopy (SEM). Samples were etched with a solution of 150 ml H_2O, 1 g oxalic acid, 1 ml HNO_3, and 1 ml acetic acid for 15-30 s. The microscopy was constrained to the central part of the section parallel to the compressive direction. The average grain size (d) was determined using the mean linear intercept method, where $d = 1.74L$; L is the linear intercept size.

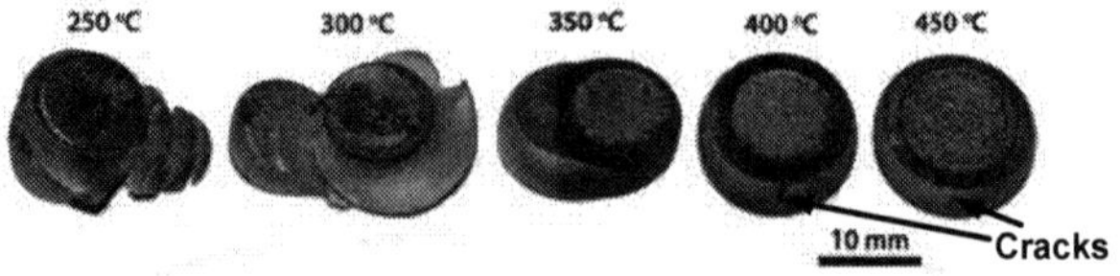

Fig. 3. Macro pictures of the AZ80 samples compressed to ε= -1.0 at 10 s^{-1} and 250-450°C showing shear fracture ($T \leq$ 350 °C) and cracks (400-450°C).

Results and discussion

Initial microstructure

The as-received and homogenized microstructures are shown in Fig. 1. The as-extruded microstructure (Fig. 1(a)) consists of recrystallized fine grains and lamellar β-$Mg_{17}Al_{12}$ precipitates (black particles) along deformation bands. It is evident that the precipitates are dissolved into the magnesium matrix after homogenization (Fig. 1(b)), which is in accordance with the Mg-Al phase diagram [17]. The homogenized grain size for AZ80 is about 59 μm (Fig. 1(b)).

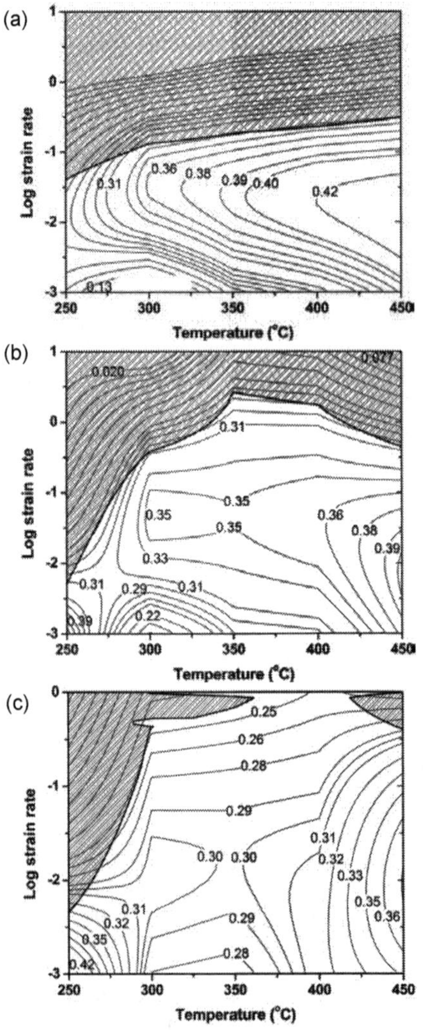

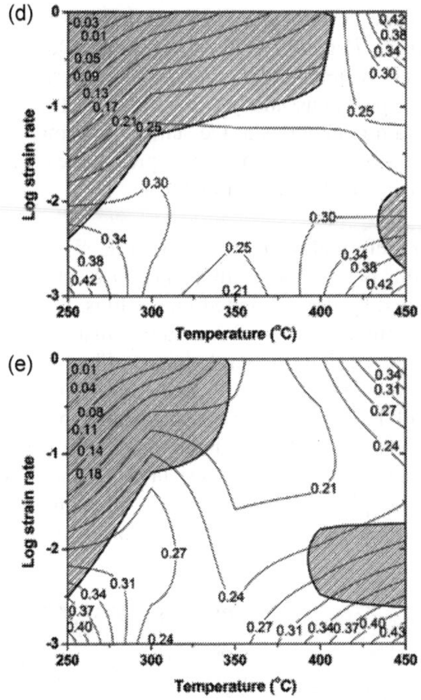

Fig. 4. Processing maps of AZ80 magnesium alloy at true strains of: (a) $\varepsilon = -0.1$, (b) $\varepsilon = -0.2$, (c) $\varepsilon = -0.3$, (d) $\varepsilon = -0.5$, and (e) $\varepsilon = -0.8$. The efficiency of power dissipation (η value) is marked against each contour. The shaded regions correspond to flow instability ($\xi < 0$). Note the difference in the strain rate range (vertical axis) between (a, b) and (c-e).

Flow stress behavior

Fig. 2 shows the corresponding true stress and true strain curves of AZ80 specimens deformed at 0.001, 1.0, and 10 s^{-1} and at different temperatures. The flow curves (Figs. 2(a) and (b)) at lower strain rates ($\dot{\varepsilon} \leq 1.0$ s^{-1}) show three stages of strain hardening with a single peak, strain softening, and a steady state as strain increases. However, at a higher strain rate of 10 s^{-1} (Fig. 2(c)), the flow behavior is distinctly different and exhibits severe work hardening followed by continuous flow softening. At T $\leq$ 350°C, fracture occurs before or (shortly) following the peak stresses of the curves. All specimens deformed at the strain rate of 10 s^{-1} exhibit shear fracture or cracks, as shown in the macro pictures of the compressed samples in Fig. 3. This type of cracking may be triggered by the onset of intense flow instabilities [18], which is subsequently supported by the processing maps of the AZ80 magnesium alloy.

Processing maps and microstructural interpretation

Fig. 4 shows processing maps of AZ80 alloy obtained at true strains of -0.1, -0.2, -0.3, -0.5, and -0.8. The numbers against the contours represent the efficiency of power dissipation (η value) and the shaded regions correspond to flow instability ($\xi < 0$). Due to premature failure of the material (Fig. 3), the stress-strain curves at a strain rate of 10 s^{-1} are not well defined to delineate the processing map at strains of $\varepsilon >$ -0.2. Note the difference in the strain rate range (vertical axis) between Figs. 4(a)-(b) and 4(c)-(e).

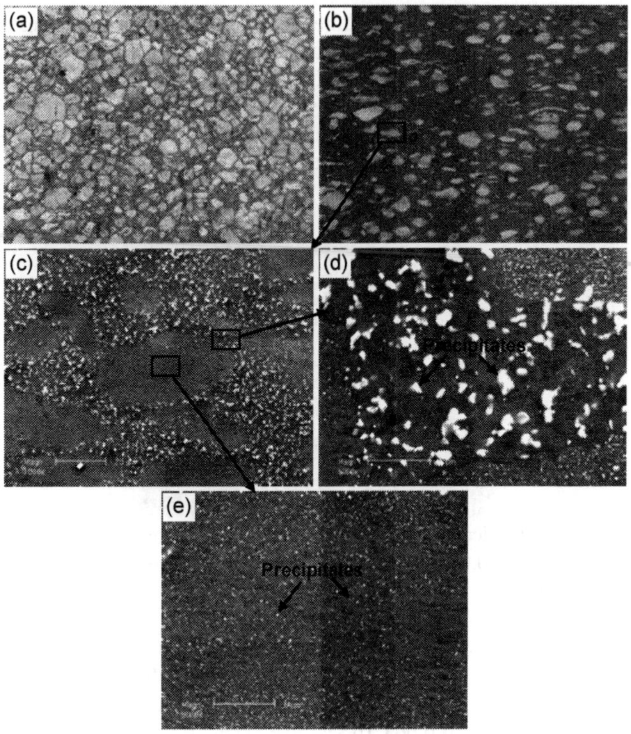

Fig. 5. Optical (a-b) and SEM (c-e) images of AZ80 specimens deformed to $\varepsilon = -0.5$ (a) and $\varepsilon = -1.0$ (b-e) at 250°C and 0.001 s^{-1} showing strain induced precipitates (light particles).

At a strain of -0.1 (Fig. 4(a)), the map exhibits a single domain in the temperature range of 300-450°C and in the strain rate range of 0.003-0.03 s^{-1} with a peak efficiency of 44% occurring at 450°C and 0.01 s^{-1} (*Domain I*). As the strain is increased to -0.2 (Fig. 4(b)), this domain is retained, but with a slight decrease in peak efficiency of power dissipation. In addition, a new domain appears at 250-270°C and 0.001-0.003 s^{-1} with a peak efficiency of 40% occurring at 250°C and 0.001 s^{-1} (*Domain II*). With further deformation to a strain of -0.3 (Fig. 4(c)), *Domain I* is noted to shift to high temperatures T≥ 400 °C, and finally disappears at a strain of -0.5 (Fig. 4(d)). Two new domains appear: one in the range of 420-450°C and 0.1-1.0 s^{-1} with a peak efficiency of about 47% at 450°C and 1.0 s^{-1} (*Domain III*); and the other with a peak efficiency of about 49% at 450°C and 0.001 s^{-1} (*Domain IV*).

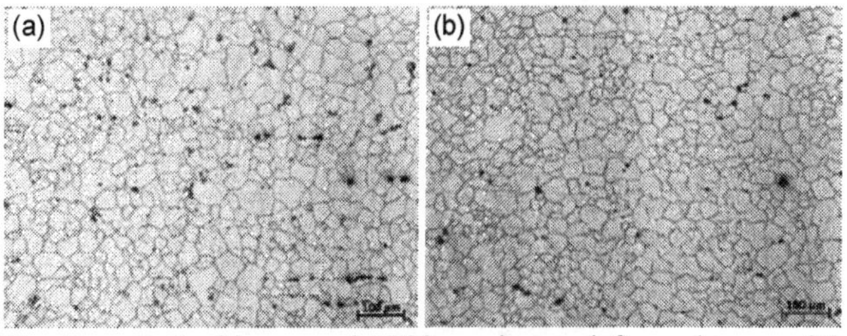

Fig. 6. Optical images of AZ80 specimens deformed to $\varepsilon = -0.2$ (a) and $\varepsilon = -1.0$ (b) at 450°C/1.0 s^{-1} showing fine DRXed grains.

Figs. 5-7 shows the microstructures obtained for specimens deformed at 250°C/0.001 s^{-1}, 450°C/1.0 s^{-1}, and 450°C/0.001 s^{-1}, corresponding to *Domains II-IV*, respectively. Homogenized DRXed grains of ~39 μm are observed in *Domain III* (Fig. 6). However, abnormal grain growth and wedge cracking at grain boundaries occur in *Domain IV* (Fig. 7) due to lower rates of deformation at high temperatures [19]. Grain-boundary sliding is a possibility at lower strain rates and causes wedge cracking which could, in turn, also dissipate energy with high efficiency [4]. No evident DRX is observed in the case of *Domain II*, as shown in Fig. 5. In contrast, a large amount of $Mg_{17}Al_{12}$ particles discontinuously precipitate near the initial grain boundaries, which may also dissipate energy with high efficiency.

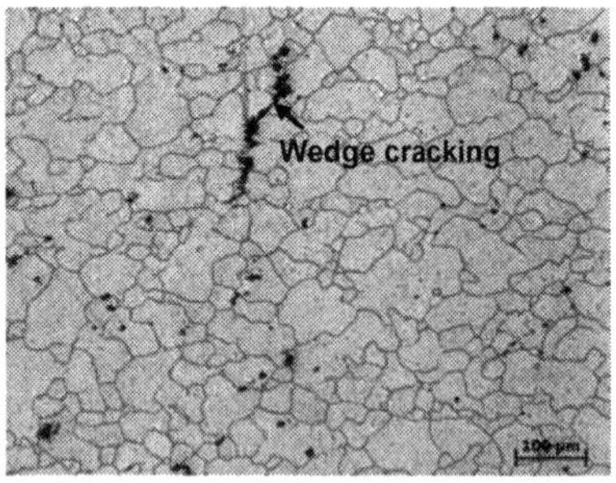

Fig. 7. Optical image of AZ80 specimens deformed to ε = -1.0 at 450°C and 0.001 s^{-1} showing abnormal grain growth and wedge cracking.

Referring to Figs. 4(a)-(c), AZ80 alloy exhibits a single regime of flow instability at the start of the plastic deformation:

(1) *Regime I* occurs in the temperature range of 250-450°C and strain rate range of 0.004-10 s^{-1}.

With increasing strain to -0.5 (Fig. 4(d)), *Regime I* is partially replaced by *Domain III*, and a new regime of flow instability appears:

(2) *Regime II* occurs in the temperature range of 433-450°C and strain rate range of 0.002-0.014 s^{-1}.

With further deformation to strains up to -0.8 (Fig. 4(e)), *Regime I* is reduced to temperatures lower than 350°C, while *Regime II* extends to the temperature range of 393-450°C and strain rate range of 0.002-0.018 s^{-1}.

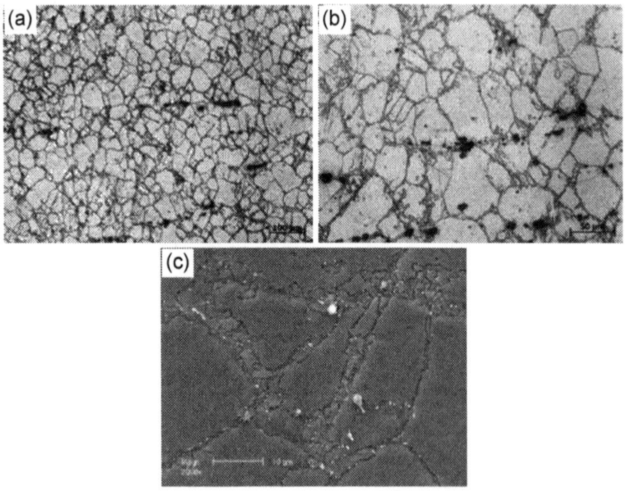

Fig. 8. Optical (a-b) and SEM (c) images of AZ80 specimens deformed to ε = -0.2 at 300°C and 1.0 s^{-1} showing deformation twinning and DRX at wavy grain boundaries.

Regime I corresponds to the specimens that have undergone shear fracture or cracking (Fig. 3) at the highest strain rate (10 s^{-1}) because of the possible occurrence of adiabatic shear bands [19]. Figs. 8 and 9 show bands of flow localization at 300°C/1.0 s^{-1} which should be described as unstable deformation regions also corresponding to *Regime I* of flow instability. Due to the limited slip systems at this temperature and a coarse microstructure in the homogenized specimens, many twins occur in the original grains with wavy grain boundaries at the start of deformation, as shown in Figs. 8(a)-(b). DRX and fine precipitates are observed preferentially inside the twins (Fig. 8(c)).

With further deformation to ε = -1.0 (Fig. 9), the compressed sample exhibits highly uneven microstructure, where *Zones A* and *B* are delineated by the red dotted line in Figs. 9(a) and (b). *Zone A* exhibits a similar "necklace" structure to that of Fig. 8, but with a larger fraction of deformation twinning (Figs. 9(c) and (e)). As for *Zone B*, the matrix is crushed by local shear stresses into isolated small blocks (Fig. 9(d)) surrounded by newly formed DRXed grains (Fig. 9(f)). A large number of fine $Mg_{17}Al_{12}$ precipitates are seen to be distributed at the DRXed grain boundaries and play an important role in pinning the DRXed grains from growing. The average grain size is measured to be ~2 μm.

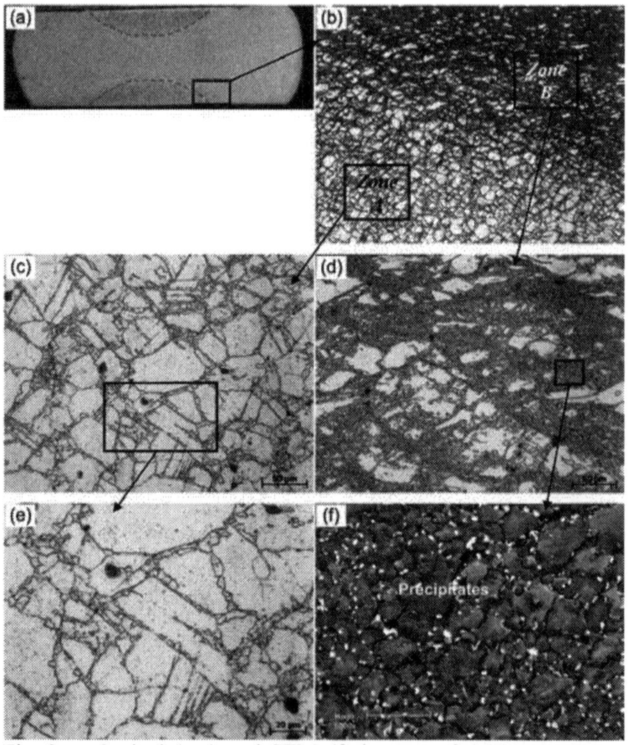

Fig. 9. Optical (a-e) and SEM (f) images of AZ80 specimens deformed to ε = -1.0 at 300°C and 1.0 s^{-1} showing "necklace" structure and deformation twinning in *Zone A* and shear bands and precipitates in *Zone B*. The red dotted lines in (a) and (b) delineate the interface between *Zones A* and *B*.

Fig. 10 shows the microstructure of the specimen deformed to ε = -1.0 at 450°C and 0.01 s^{-1}, corresponding to *Regime II* of flow instability. Abnormal grain growth (Fig. 10b) and wedge cracking (Fig. 10c) at grain boundaries are detected in the matrix, which are not desirable. Similar microstructures are also observed in *Domain IV* (Fig. 7), which may suggest that *Regime II* extends into *Domain IV* at strain ε> -0.8. These temperatures and strain rates should be avoided in processing the AZ80 alloy.

Conclusions

The processing maps of AZ80 alloy exhibit a single DRX domain in the range of 420-450°C and 0.1-1.0 s^{-1} with a peak efficiency of about 47% at 450°C and 1.0 s^{-1}, which are the optimum forming conditions for the hot working of this alloy. There are two flow instability regimes occurring at 250-450°C and 0.004-10 s^{-1} and 433-450°C and 0.002-0.014 s^{-1}. The former occurs at low temperatures and/or high strain rates and is associated with adiabatic shear bands or cracks, flow localization, and deformation twinning, while the latter at higher temperatures and lower strain rates is due to abnormal grain growth and wedge cracking.

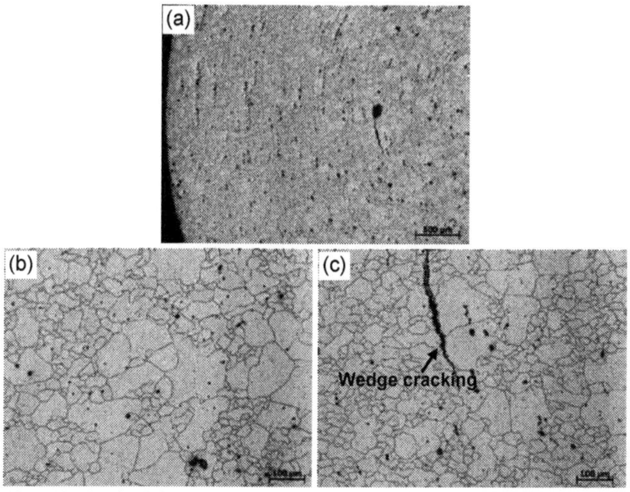

Fig. 10. Optical images (a-c) of AZ80 specimen deformed to $\varepsilon = -1.0$ at 450°C and 0.01 s^{-1} showing abnormal grain growth (b) and wedge cracking (c).

Acknowledgments

The authors would like to thank Magnesium Elektron North America for providing the materials. Prof. Rongshi Chen, Jilin Li, and Fa Huang of Institute of Metal Research, Chinese Academy of Sciences, are gratefully acknowledged for carrying out the Gleeble tests. Drs. Jeff Wang and Anil K. Sachdev of GM Global R&D are gratefully acknowledged for technical discussions and constructive review of this report.

References

[1] A.A. Luo, "Wrought magnesium alloys and manufacturing processes for automotive applications", *SAE Technical Paper*, 2005, no. 2005-01-0734.

[2] Y. Lou, L. Li, J. Zhou, L. Na, "Deformation behavior of Mg–8Al magnesium alloy compressed at medium and high temperatures", *Mater Charact*, 62 (2011), 346-353.

[3] Elektron Wrought Alloys, Magnesium Elektron, http://www.magnesium-elektron.com/.

[4] Y. Prasad, H. Gegel, S. Doraivelu, J. Malas, J. Morgan, K. Lark, D. Barker, "Modeling of dynamic material behavior in hot deformation: Forging of Ti-6242", *Metallurgical and Materials Transactions A*, 15 (1984), 1883-1892.

[5] K.P. Rao, Y.V.R.K. Prasad, K. Suresh, N. Hort, K.U. Kainer, "Hot deformation behavior of Mg-2Sn-2Ca alloy in as-cast condition and after homogenization", *Materials Science and Engineering: A*, 552 (2012), 444-450.

[6] F. Garofalo, "An empirical relation defining the stress dependence of minimum creep rate in metals", *Trans Metall Soc AIME*, 227 (1963), 351-356.

[7] Y.V.R.K. Prasad, "Recent advances in the science of mechanical processing", *Indian J. Technol*, 28 (1990), 435-451.

[8] Y.V.R.K. Prasad, T. Seshacharyulu, "Modelling of hot deformation for microstructural control", *International Materials Reviews*, 43 (1998), 243-258.

[9] J. Luo, M.Q. Li, B. Wu, "The correlation between flow behavior and microstructural evolution of 7050 aluminum alloy", *Materials Science and Engineering: A*, 530 (2011), 559-564.

[10] K.P. Rao, Y.V.R.K. Prasad, N. Hort, K.U. Kainer, "Hot workability characteristics of cast and homogenized Mg-3Sn-1Ca alloy", *J Mater Process Tech*, 201 (2008), 359-363.

[11] H.T. Zhou, Q.B. Li, Z.K. Zhao, Z.C. Liu, S.F. Wen, Q.D. Wang, "Hot workability characteristics of magnesium alloy AZ80-A study using processing map", *Materials Science and Engineering: A*, 527 (2010), 2022-2026.

[12] O. Sivakesavam, I.S. Rao, Y.V.R.K. Prasad, "Processing map for hot working of as cast magnesium", *Mater Sci Tech*, 9 (1993), 805-810.

[13] G.Z. Quan, T.W. Ku, W.J. Song, B.S. Kang, "The workability evaluation of wrought AZ80 magnesium alloy in hot compression", *Materials Design*, 32 (2011), 2462-2468.

[14] X.W. Zheng, A.A. Luo, J. Dong, A.K. Sachdev, W.J. Ding, "Plastic flow behavior of a high-strength magnesium alloy NZ30K", *Materials Science and Engineering: A*, 532 (2012), 616-622.

[15] N. Srinivasan, Y.V.R.K. Prasad, P. Rama Rao, "Hot deformation behaviour of Mg-3Al alloy-A study using processing map", *Materials Science and Engineering: A*, 476 (2008), 146-156.

[16] L. Li, X. Zhang, "Hot compression deformation behavior and processing parameters of a cast Mg-Gd-Y-Zr alloy", *Materials Science and Engineering: A*, 528 (2011), 1396-1401.

[17] T.B. Massalski, P.R. Subramanian, H. Okamoto, L. Kacprzak, Binary alloy phase diagrams, 2nd ed., *ASM International*, Materials Park, OH, 1990.

[18] T. Seshacharyulu, S.C. Medeiros, W.G. Frazier, Y.V.R.K. Prasad, "Hot working of commercial Ti-6Al-4V with an equiaxed α-β microstructure: materials modeling considerations", *Materials Science and Engineering: A*, 284 (2000), 184-194.

[19] S. Anbuselvan, S. Ramanathan, "Hot deformation and processing maps of extruded ZE41A magnesium alloy", *Materials Design*, 31 (2010), 2319-2323.

Magnesium Technology 2013
Edited by: Norbert Hort, Suveen N. Mathaudhu, Neale R. Neelameggham, and Martyn Alderman
TMS (The Minerals, Metals & Materials Society), 2013

Inverse strain rate sensitivity of bendability of an AZ31 sheet in three-point bending

B. Li, S.J. Horstemeyer, A.L. Oppedal, P.T. Wang, M.F. Horstemeyer
Center for Advanced Vehicular Systems, Mississippi State University, 200 Research Blvd., Starkville, MS 39759

Keywords: Strain rate sensitivity; magnesium; bending; deformation

Abstract

Three-point bending tests were performed on as-rolled and annealed (at 150 °C) AZ31 sheet specimens at various displacement rates (1.0, 5.0 and 50.0 mm/min) at room temperature. The as-rolled specimens present a negative sensitivity, i.e., the bending angle decreases as the strain rate increases; however, the annealed specimens show a positive sensitivity, i.e., the bending angle increases as the strain rate increases. Such an inverse strain rate sensitivity of sheet bending may significantly impact the sheet forming of Mg alloys.

Introduction

Magnesium (Mg) alloys are of interest as potential replacement materials for heavier aluminum and steel alloys in transportation industries due to their low densities (1740 kg/m3 for pure Mg) and high strength-to-weight ratios. The inherently low ductility and fracture toughness of wrought Mg alloys at room temperature (RT), consequences of their hexagonal-close packed (HCP) crystal structures, have largely limited applications to cast materials. The number of easy slip systems is insufficient to satisfy the von Mises criterion [1]. This requires five independent slip systems to accommodate the compatibility strain at grain boundaries, despite the multiplicity of slip systems on basal, prismatic, and pyramidal planes, with the basal slip being the easiest. Irrespective of c/a ratio, HCP metals twin profusely during plastic deformation. The main twinning systems in HCP metals are either on the first order or the second order pyramidal planes [2,3], allowing twinning to accommodate the strain in the direction perpendicular to the basal plane. Alternatively, type <*a*> dislocations (on basal and prismatic planes) are unable to do the same. Hence, twinning plays an important role in plastic deformation of HCP metals. Twin-slip interactions give rise to unique mechanical behavior in Mg alloys that vastly differs from other alloys with high symmetry crystal structures in which dislocation slip dominates deformation [4-6].

Due to the relatively poor ductility of wrought Mg alloys at room temperature, conventional forming processes (e.g., stamping) require elevated temperatures at which dislocation slip and dynamic recrystallization are dominant [7-13]. Concerns about hot forming of Mg alloys (T~ 450 °C) stem from high energy consumption and manufacturing costs. One approach to better understanding RT deformation mechanisms in Mg is through the study of tests involving simple strain paths. Among these is bending, which occurs in numerous sheet forming processes [14-15].

We recently uncovered highly localized twin bands that developed during quasi-static, three-point bending of a wrought Mg AZ31 alloy at RT using in-situ electron backscatter diffraction (EBSD) [16]. Inside the twin bands, a high density of deformation twins was present; however, outside of the twin bands, twins were completely absent. Although the microstructure and mechanical properties of wrought Mg AZ31 have been extensively studied [17-25] over the past few decades, understanding of what occurs in the microstructure during bending and how the microstructure evolves require further investigation. Previous efforts have, in fact, focused on mechanical properties of AZ31 Mg alloys in uniaxial tension or compression [18-25] at various temperatures and strain rates. But no effort has been made to directly study strain rate dependence in sheet bending in which the stress state is different from that in uniaxial deformations. The purpose of this study is to investigate strain rate dependence (3.0×10^{-4} s^{-1} - 1.5×10^{-2} s^{-1}) of an AZ31 Mg alloy during three-point bending at room temperature.

Experimental Methods

The starting material was a hot rolled Mg AZ31 tooling plate with a composition of Mg-3%Al-1%Zn (wt. %) and 6.0 mm thickness. The plate was heated to 400 °C by an infrared heating source and rolled at an average of 23% reduction per pass with a total of six passes to reach the final 1.0 mm [26]. Specimens with a dimension $60\times10\times1$ mm^3 were cut from the rolled sheet for three-point bending tests. Some of the specimens were annealed at 150 °C for an hour in a vacuum furnace filled with Argon gas, and then cooled to ambient temperature inside the furnace to prevent oxidation. Annealing at other temperatures (100, 200, and 300 °C) was also conducted, but the results will be presented elsewhere. The purpose of annealing is to study how initial microstructure affects the bending behavior.

The three-point bending tests were conducted on an Instron 5882. A fixture for three-point bending was designed and machined with a span of 25 mm between the two supporting points. The bending tests were performed with displacement control. Three displacement rates were used in our tests: 1.0 mm/min, 5.0 mm/min, and 50.0 mm/min, corresponding to maximum strain rates about 3.0×10^{-4} s^{-1}, 1.5×10^{-3} s^{-1} and 1.5×10^{-2} s^{-1}, respectively. In this study, all bends were parallel to the transverse direction (TD). Our results reveal a surprising phenomenon in annealed specimens.

The microstructures of the as-rolled and the annealed specimens were examined by optical microscopy. Metallographic observations were made on the through-thickness cross section following standard procedures. The polished specimens were etched for 3 seconds with a solution of 70 ml ethanol, 10 ml distilled water, 10 ml acetic acid, and 4 g picric acid.

The textures of the as-rolled and the annealed AZ31 sheets were examined by X-ray diffraction (XRD) using a Regaku SmartLab X-ray diffractometer with a Cu-Kα radiation source, 40 kV operating voltage, and a 30 mA current. Six pole figures, i.e., $\{10\bar{1}0\}$, $\{10\bar{1}1\}$, $\{0002\}$, $\{10\bar{1}2\}$, $\{11\bar{2}0\}$ and $\{10\bar{1}3\}$ were measured.

The orientation distribution functions (ODFs) were calculated using the measured pole figures after corrections for background and defocusing effects. The recalculated pole figures based on the ODFs were plotted by MTEX [27]. The texture was measured such that the XRD data were collected from the same specimen. First, a piece of the as-rolled material was cut and the surface was slightly polished. Then XRD was performed on the polished surface. After the XRD measurement, the same specimen was annealed at 150 °C, and then the texture of the annealed specimen was measured on the same polished surface.

The texture of the as-rolled and the annealed specimens is shown in Figure 2. Both pole figures are typical of rolled Mg alloys with a strong basal pole (the basal plane is approximately parallel to the rolling direction). An extra intensity spot appears in the $\{10\bar{1}0\}$ pole figure after annealing. The maximum intensity remains almost identical. No significant changes in the texture occurred after annealing, although the defect structure may have changed in terms of dislocation density and configurations due to recovery.

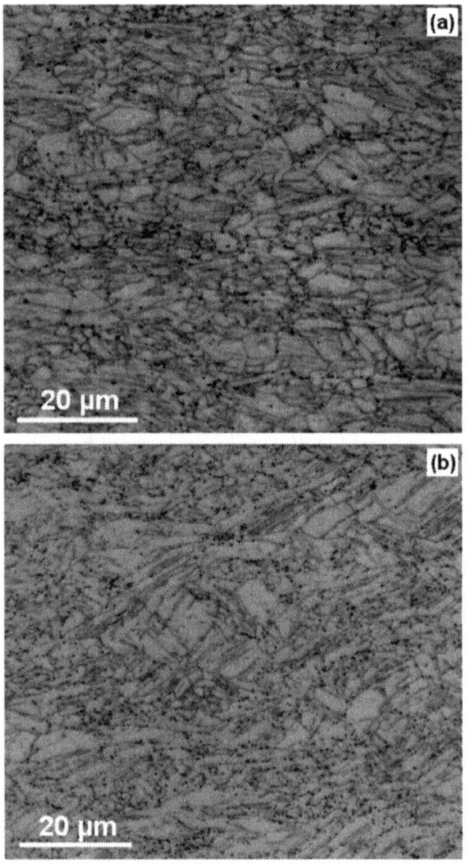

Figure 1. Optical metallography of the microstructures of the AZ31 Mg alloy sheet. (a) As-rolled Mg AZ31 alloy sheet. High density twins can be observed. (b) Annealed Mg AZ31 alloy sheet at 150 °C for an hour. No significant change in grain size occurred after annealing.

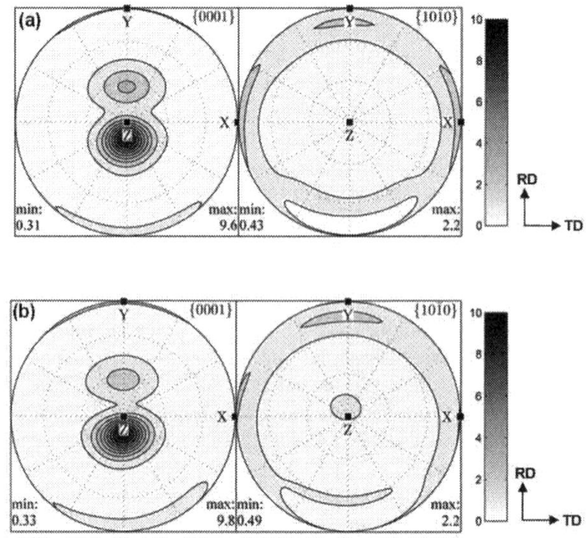

Figure 2. X-ray diffraction analysis of the texture in the Mg AZ31 sheet materials. (a) Pole figure of the as-rolled Mg AZ31. A strong basal texture can be observed, typical of rolled sheets of Mg alloys. A split in the rod texture is present. (b) Pole figure of the Mg AZ31 annealed at 150 °C for an hour before bending tests. No significant changes in the texture are evident.

Results and Discussions

Figure 1 shows the microstructures of both the as-rolled and the annealed Mg AZ31 sheet specimens. The microstructures were examined on the through-thickness cross-sections. In Figure 1a, the initial microstructure is comprised of a high density of $\{10\bar{1}2\}<10\bar{1}\bar{1}>$ twins, which were generated during hot rolling at a relatively high strain rate and large strains [23]. After annealing at 150 °C for an hour, no measurable change in grain size occurred (Figure 1b).

The load-displacement profiles from the three-point bending tests are shown in Figure 3 (the loads are negative because the bending was performed in compression mode of the machine). For the as-rolled materials, all specimens failed (fractured) at all displacement rates, i.e., 1.0, 5.0, and 50.00 mm/min. As expected, cracks initiated in the tension zone of each specimen. In general, for the as-rolled Mg AZ31 sheet material, the displacement to failure (the plummet in the load-displacement profiles) decreases as the displacement rate increases, as seen in the Figure 3a. In sharp contrast, the load-displacement behavior of the annealed specimens revealed much different behavior. At displacement rates of 1.0 and 5.0 mm/min, the annealed specimens failed, even though at 5.0 mm/min, the displacement to failure was somewhat larger than that at 1.0 mm/min, corresponding to a larger bending angle. Surprisingly, the annealed specimen did not fail as the displacement increased to 50.0 mm/min, and three-point bending continued until reaching the limit of the bending fixture, creating an unexpectedly large bending angle (Figure 3b). In this scenario, the gradual decrease in the load as the displacement increases is due to the sliding of the sheet specimen over the two supporting points as the bending angle increased to a certain value. Sliding also caused the serrations in the load-displacement curves.

To better reveal this interesting bending behavior, we recorded the three-point bending process, and selected snapshots are shown in Figure 4. At 1.0 and 5.0 mm/min displacement rates, both specimens failed (see 4d and 4e) although the bending angle at 5.0 mm/min was slightly larger than that at 1.0 mm/min. At 5.0 mm/min, the specimen did not fail and a large bending angle was achieved (Figure 4c). Visual examination revealed that cracks were initiated in the tension zone at displacement rates 1.0 and 5.0 mm/min (Figure 4d, 4e); however, no cracks were observed in the tension zone of specimens deformed at 50.0 mm/min (Figure 4f).

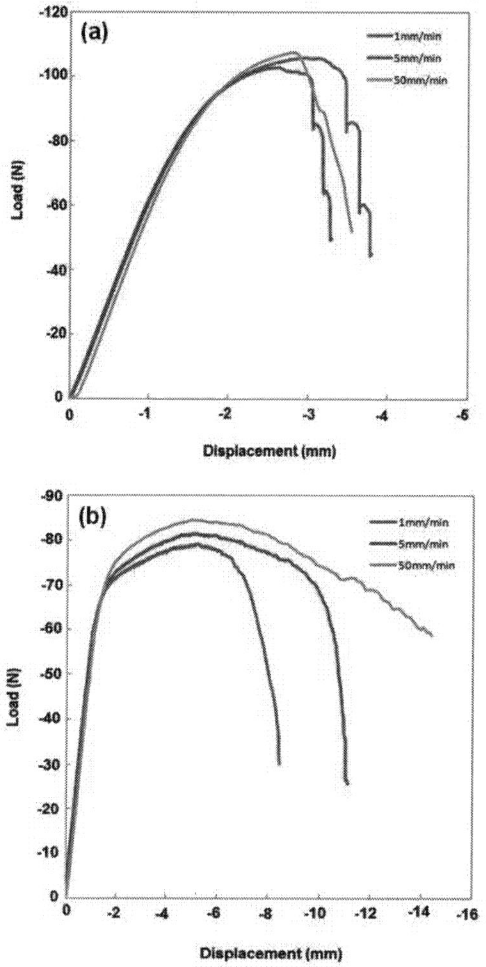

Figure 3. Load-displacement curves from the room temperature bending tests. All the bends are parallel to the transverse direction (TD). (a) As-rolled specimens. All the specimens failed at all strain rates, but the slower strain rates generate larger bending angles. (b) Specimens annealed at 150 °C for an hour. At lower strain rates, 1.0 mm/min and 5.0 mm/min, both specimens failed, but the 5.0 mm/min specimen failed with a larger displacement. At 50.00 mm/min, no specimen failed. An inverse strain rate sensitivity can be observed.

We repeated our bending tests with another set of specimens which were cut in the same direction and annealed under the same condition, and we found the observed behavior is repeatable.

Bends parallel to the rolling direction (RD) were also carried out, and no specimen failed at all strain rates. Whether similar inverse strain rate sensitivity exists in this direction requires further investigations.

Compared with the as-rolled specimens (Figure 3a), the annealed specimens show improved bending angle because of the change in the defect structures due to the heat treatments. The dislocation density decreases in the parent grains due to recovery, contributing to the decrease in strength (Figure 3). At the same displacement rate, the displacement to failure of an annealed specimen is larger than that of the as-rolled specimen. However, the bending behavior of the annealed specimens displays what we refer to as inverse strain rate sensitivity in bending: as the displacement rate increases, the bending angle increases. We have not found any indication of this phenomenon being previously reported in the literature on Mg alloys. This inverse strain rate sensitivity of the annealed sheet during bending observed in our experiments is counter-intuitive. One potential explanation is that the ductility of the annealed specimens is improved as the strain rate increases. But, available data in previous reports show that, in uniaxial tension, the ductility of Mg alloys presents a negative strain rate sensitivity [16-18]. The strain to failure in uniaxial compression behaves in a similar manner. Recently, Khan et al. [16] conducted extensive testing on a Mg AZ31 alloy in uniaxial tension, compression, and shear loading over a range of temperatures and strain rates. They showed that the material presents positive strain rate sensitivity in terms of flow stress at all strain rates at room and elevated temperatures. At a given temperature, their data suggests that the ductility decreases as strain rate increases. Maksoud et al. [17] studied deformation of a Mg AZ31 alloy under tension at different strain rates and temperatures. They showed that the ductility of the material increases as the strain rate decreases. Yin et al. [18] observed similar behavior in uniaxial tensile tests at different temperatures and strain rates in a hot-rolled Mg AZ31 Mg alloy. Hence, it would be inappropriate to conclude that the inverse strain rate sensitivity of the bending behavior of the annealed Mg AZ31 sheet is due to improved ductility.

It is worth noting that the stress state in three-point bending is quite different from uniaxial tension or compression. In three-point bending, a gradient of strain through the specimen thickness is created and the strain reverses signs at the neutral axis. Hence, it can be anticipated that the deformation modes, i.e., slip and twinning in the tension and the compression zones of the as-rolled and the annealed specimens are significantly different. In the as-rolled specimens, the microstructure is nearly saturated with deformation twins, as shown in Figure 1a. Although the stress state in the compression zone favors the $\{10\bar{1}2\} < 10\bar{1}\bar{1} >$ twinning in bending [25], deformation by twinning should be insignificant. Slip by pyramidal dislocations is activated because the grains are oriented favorably for pyramidal slip (the shear factors for basal and prismatic slip are close to zero). In the tension zone, detwinning of the pre-existing twins first occurs, and dislocation slip dominates plastic deformation. Hence, we would expect a negative strain rate sensitivity of bending because hardening by dislocation interactions in the tension zone has a positive strain rate sensitivity [16].

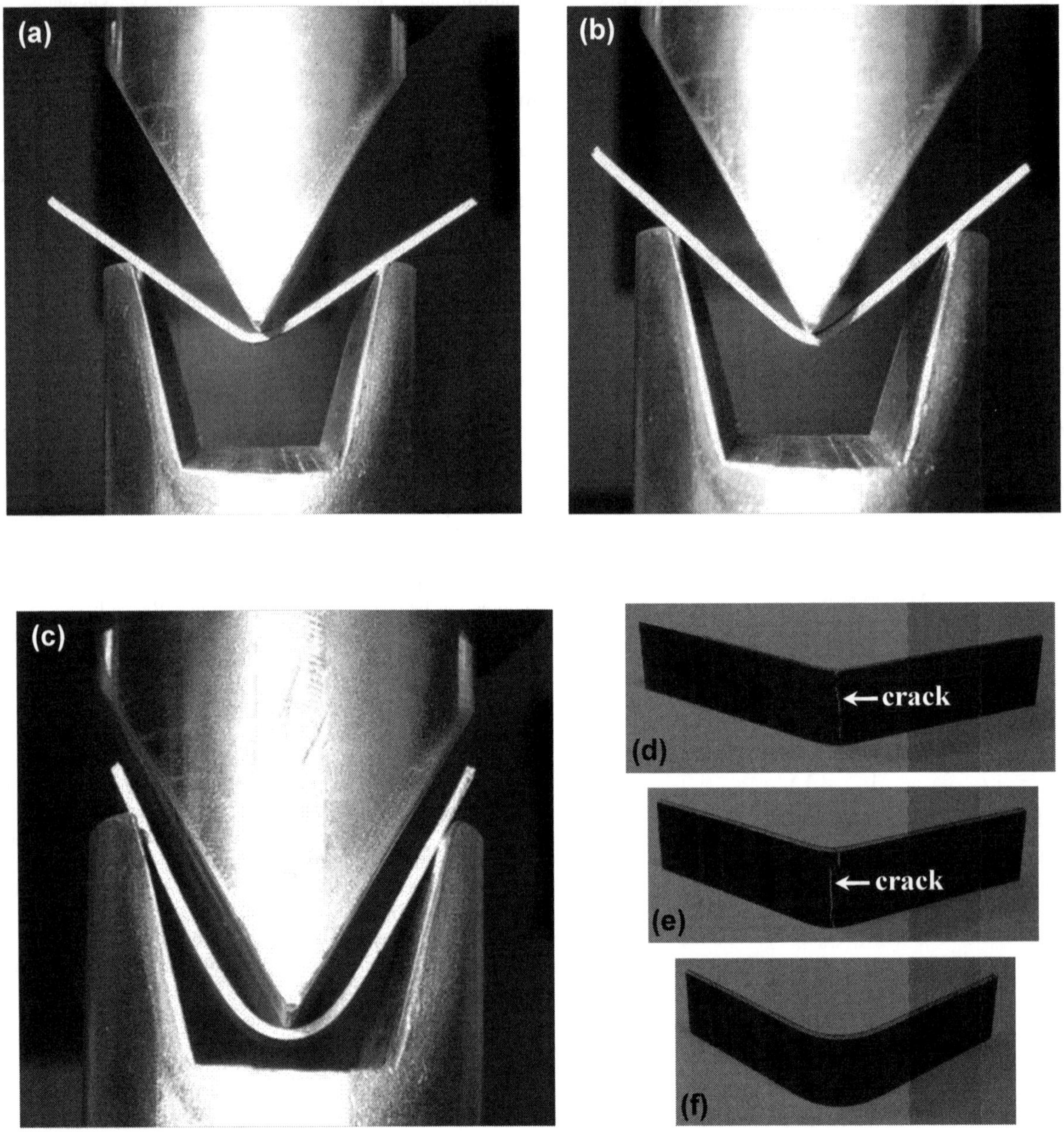

Figure 4. Room temperature bending tests at various displacement rates of the Mg AZ31 sheets annealed at 150 °C for an hour. (a) 1.0 mm/min, the specimen failed with cracks initiated in the tension zone; (b) 5.0 mm/min, the specimen failed but with a larger bending angle than that in the 1.0 mm/min test; (c) 50.0 mm/min, the specimen did not fail and was bent to a large angle; (d-f) Visual examinations: (d) 1.0 mm/min, cracks can be seen in the tension zone; (e) 5.0 mm/min, cracks can be seen in the tension zone; (f) 50.0 mm/min, no cracks were initiated.

In the case of the annealed specimens, recovery occurs and hence the deformation mechanisms differ from the as-rolled specimens. Dislocation density and configurations changed as a result of annealing. How these microstructural changes affect the activation of the deformation modes needs further investigations. Although slip and twinning are both strain rate dependent, to our surprise, the bending of the annealed specimens presents a positive strain rate sensitivity. Hence, there must be an unknown mechanism behind the anomalous bending behavior observed reported here. Determining what mechanism accounts for the inverse strain rate sensitivity of the bending of the annealed AZ31 sheet deserves a thorough investigation.

Summary

We performed three-point tests on sheet specimens of a Mg AZ31 alloy at various strain rates. The as-rolled specimens presented a negative strain rate sensitivity in which the bending angle decreased as the strain rate increased. However, in the annealed specimens (annealing temperature 150 °C), a positive strain rate sensitivity was observed, i.e., the bending angle increased as the strain rate increased. To the best of our knowledge, such abnormal bending behavior has not been previously reported for Mg alloys.

Acknowledgements

B. Li, S.J. Horstemeyer, A.L. Oppedal, Y. Hammi, P.T. Wang, M.F. Horstemeyer are grateful for the support from the Center for Advanced Vehicular Systems (CAVS) at Mississippi State University.

References

[1] G.I. Taylor, J Inst Metals 62 (1938): 307-338.
[2] B.A. Bilby, A.G. Crocker, Proc Roy Soc. A, 288 (1965) 240-255.
[3] J.W. Christian, S. Mahajan, Prog Mater Sci 39 (1995) 1-157.
[4] Q. Ma, B. Li, E.B. Marin, S.J. Horstemeyer, Scripta Mater 65 (2011) 823-826.
[5] J. Bohlen, M. R. Nurnberg, J. W. Senn, D. Letzig, S. R. Agnew, Acta Mater 55(2007) 2101-2112.
[6] B. Li, S.P. Joshi, O. Almagri, Q. Ma, K.T. Ramesh, T. Mukai, Acta Mater 60 (2012) 1818-1826.
[7] E. Taleff, L.G. Hector, Jr., R. Verma, P.E. Krajewski, J.-K. Chang, J. Materials Engineering and Performance 19 (2010) 488-494.
[8] L.G. Hector, Jr., P.E. Krajewski, E.M. Taleff, J.T. Carter, Key Engineering Materials 433 (2010) 197-210.
[9] R. Verma, L.G. Hector, Jr., P. Krajewski, E. Taleff, JOM 61 (2009) 29-37.
[10] T. Al-Samman, G. Gottstein, Mater. Sci. Eng. A 490 (2008) 411-420.
[11] M.M. Myshlyaev, H.J. McQueen, A. Mwembela, E. Konopleva, Mater. Sci. Eng. A 337 (2002) 121-133.
[12] A.G. Beer, M.R. Barnett, Metall. Mater. Trans A. 38A (2007) 1856-1867.
[13] Q. Ma, B. Li, A.L. Oppedal, W. Whittington, S.J. Horstemeyer, E.B. Marin, P.T. Wang, M.F. Horstemeyer, Mater Sci Eng A, 2012, in press.
[14] S.L. Couling, J.F. Pashak, L. Sturkey, Trans of the ASM 51 (1959) 94-107.
[15] L. Zhang, G.S. Huang, H. Zhang, B. Song, Journal of Materials Processing Technology, 211 (2011) 644–649.
[16] J.C. Baird, B. Li, S. Yazdan Parast, S.J. Horstemeyer, L.G. Hector Jr., P.T. Wang, M.F. Horstemeyer, Scripta Mater 67 (2012) 471-474.
[17] M.R. Barnett, Z. Keshavarz, M.D. Nave, Metall Mater Trans A 36A (2005) 1697-1704.
[18] M. Knezevic, A. Levinson, R. Harris, R. K. Mishra, R. D. Doherty, and S. R. Kalidindi, Acta Materialia, 58 (2010) 6230-6242.
[19] A. S. Khan, A. Pandey, T. Gnäupel-Herold, and R. K. Mishra, International Journal of Plasticity, 27 (2011) 688-706.
[20] I.A. Maksoud, H. Ahmed, and J. Rödela, Materials Science and Engineering A, 504 (2009) 40–48.
[21] D.L. Yin, K.F. Zhang, G.F. Wang, and W.B. Han, Materials Science and Engineering A, 392 (2005) 320–325.
[22] J. Koike, T. Kobayashi, T. Mukai, H. Watanabe, M. Suzuki, K. Maruyama, and K. Higashi, Acta Materialia, 51 (2003) 2055–2065.
[23] M. R. Barnett, Materials Science and Engineering A, 464 (2007) 1-7.
[24] J. Koike, T. Kobayashi, T. Mukai, H. Watanabe, M. Suzuki, K. Maruyama, and K. Higashi, Acta Materialia, 51 (2003) 2055–2065.
[25] S. R. Agnew and O. Duygulu, International Journal of Plasticity, 21 (2005) 1161–1193
[26] J. A. Horton, C. A. Blue, T. Muth, A. L. Bowles, and S. R. Agnew, Magnesium Technology, Edited by Neale R. Neelameggham, Howard I. Kaplan, and Bob R. Powell, TMS 2005.
[27] R. Hielscher, H. Schaeben, J Appl Crystallography 41 (2008) 1024.
[28] M.H. Yoo, Met Trans A 12A (1981) 409-418.

Magnesium Technology 2013
Edited by: Norbert Hort, Suveen N. Mathaudhu, Neale R. Neelameggham, and Martyn Alderman
TMS (The Minerals, Metals & Materials Society), 2013

ANNEALING OF COLD AND WARM ROLLED AZ31B MAGNESIUM ALLOY SHEETS

Litzy L. Choquechambi Catorceno[1], Denise Adorno Lopes[1]
[1]PMT-EPUSP Departamento de Engenharia Metalúrgica e de Materiais da Escola Politécnica da Universidade de São Paulo, USP, Brazil);
2463 Av Prof Mello Moraes; São Paulo, SP, BRAZIL

Keywords: Magnesium, AZ31B, precipitation, $Mg_{17}Al_{12}$

Abstract

Annealing experiments have been performed on warm and cold rolled AZ31B magnesium alloy sheets at 200°C in several time intervals. The effect of the amount of deformation, the annealing time, and the presence of intermetallic phases were evaluated to obtain a microstructure constituted of fine recrystallized grains. Microstructural evolution was followed using optical microscopy, scanning electron microscopy combined with energy dispersive spectroscopy (SEM-EDS), X-ray diffraction and Vickers hardness measurements. After annealing for 60 min, it was found that static recrystallization mostly takes place on cold-rolled sheets with high thickness reduction, refining the grain size below 10 μm. Twins that were observed in warm rolled specimens nearly disappear and microstructure became suitable for further rolling process. Coarse phases of $Mg_{17}Al_{12}$ were found inside grains and fine precipitation of Al_4Mn and AlMn were detected at grain boundaries, which presented strong pinning effects to prevent the grain growth.

Introduction

Magnesium alloys have great potential as lightweight structural materials in automotive and electronics applications [1] because of their superior performance such as high strength, high strength-to-weight ratio, good thermal conductivity and shock-absorbing characteristics [2–3]. However, magnesium alloys show poor plasticity due to their limited slip´s systems (hcp structure) at room temperature [1–3]. Then in cold rolling process, plastic deformation takes place by basal slips and twinning. Meanwhile, at high rolling temperature, above 200 °C, non-basal plane slips, dynamic recrystallization and preferred growth of grain are activated [4]. In this sense, during warm rolling process, thermodynamically unstable microstructures containing defects, dislocations and sub-grains are produced, depending greatly of rolling temperature. On the other hands, the precipitate formation (second phase) and nodes of dislocation network also affect the final microstructure improving or getting worse. A lot volume of big precipitates are commonly associated with crack's formation during rolling process, but fine precipitates can offer pinning effects on grain boundary motilities and can promote dislocation slip on non-basal planes via cross-slip, moreover, they can generate a grain´s refinement and texture´s modifications.

It is known, that rolling process of Mg alloys at temperatures lower than 200°C can be improved by reduction of grain size because it increases ductility [5–6]. Consequently, the microstructure and mechanical responses of rolling process is potentially improved by annealing treatments in order to homogenized microstructure, minimizing the big size of second phase precipitates (manganese-rich particles), refining grain size, reducing segregation, increasing solid solubility, enhancing precipitate nucleation within the matrix, and generating a distribution of fine precipitates [7-8].

In this study, the effects of previous deformation before annealed treatment, the annealing time and the intermetallic phase´s formation during annealed of AZ31 rolled sheets are investigated in order to obtain a microstructure constituted of fine recrystallized grains, which will be enable to posterior rolling with high reduction.

Experimental details

Material

The material used in this research was the same sheet used in the study of effects of rolling temperature on the AZ31B magnesium alloy [10]. The nominal composition was approximately 3.34–3.63 wt% Al, 0.45–0.53 wt% Zn, 0.27–0.29 wt% Zn and balance Mg. At room temperature, two small strips (1 cm×5cm×2mm) were rolled in single pass and without any lubricant, one with 10% of thickness reduction and the other one with 15%. Larger rolling reduction would cause extensive side crack, then was chosen 250 °C as rolling temperature in order to reach 15% and 25% of thickness reduction. All samples were rolled, maintaining 3.4 m/min as a rolling speed. After rolling, small pieces of rolled samples were annealed at 200°C for different time intervals, beginning from 30 minutes and increasing to 240 minutes. Every annealed sample was cooled rapidly in cool water to avoid ($Mg_{17}Al_{12}$)-phase precipitation. Annealed responses were study measuring Vickers hardness with a 300g load and studying the microstructure evolution using optical and scanning electron microscopy (SEM). Metallographic preparations included polishing and etching surfaces with standard etchant (acetic picral). The lattice structure was determined from X-ray diffractogram (XRD) and precipitates were analyzed by scanning electron microscopy (SEM) equipped with an energy dispersive X-ray (EDX) spectrometer.

Result and discussions

Rolling processes

In previous studies, warm-rolled magnesium alloys have shown a moderate hardening due to precipitate's formations, which interfered on properties and microstructure of rolled samples depending on their sizes and distributions [10]. As a follow-up to that study, it was evaluated the effect of the intermetallic phases presence, amount of previous deformation, before annealing and annealing time. To begging, it is interesting to note that AZ31B samples showed similar hardening after cold and warm rolling deformation (figure 1). A thickness reduction of 10% increased the value hardness in 48%, this hardening was attributed to extensive formation of twinning and dense dislocation pile-ups within grains, especially, on bigger grains where twins were more prominent, resulting in reorientations and activations of basal planes slips. A thickness reduction over 15% could cause the fractures on AZ31B magnesium strips; hence 15% represents the upper limit of maximum deformation under compression at room temperature. In high rolling temperature (250 °C), samples with 10% and 25% of thickness reduction showed similar hardness values due to the interaction and competitions of different deformation mechanisms like recovery, dynamic recrystallization and formation of intermetallic phases. In figure 2, it was evidenced from microstructures of cold and warm rolled samples that plastic deformation strain generates large amounts of twinning, shear bands, microscopy voids and cracks inside precipitates, resulting in inhomogeneous deformations. The voids were formed preferentially at triple points and at those grain boundaries along which there was a high shear stress. In some research [9], these void formations were associated with grain boundary sliding and new grains formed by dynamic recrystallization during warm deformation.

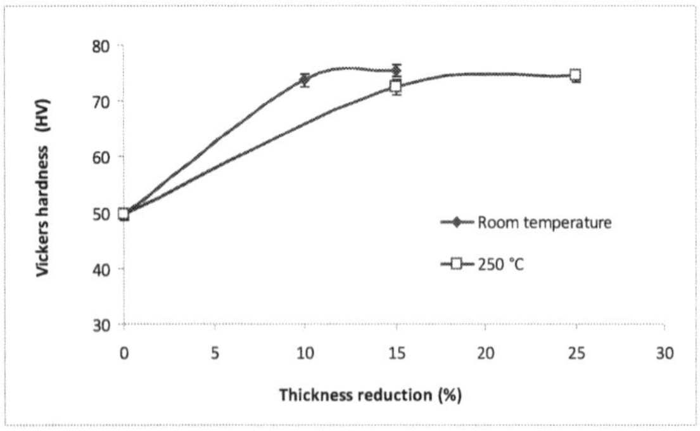

Figure 1 Microhardness values of rolled samples at room temperature and 250 °C.

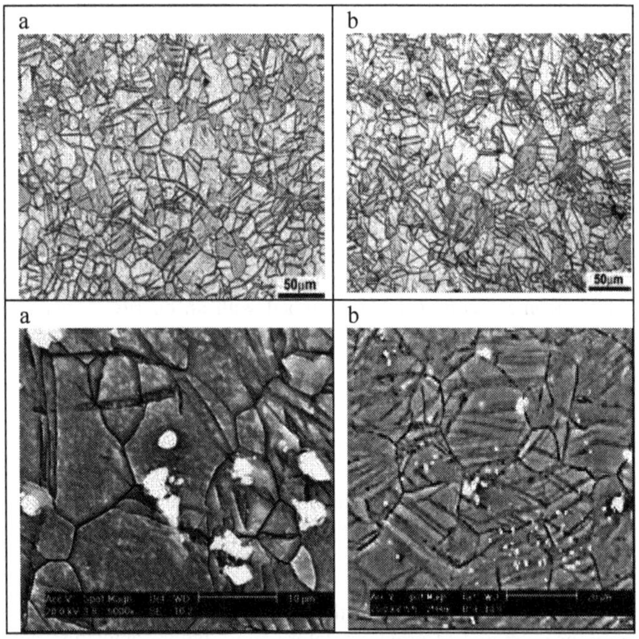

Figure 2 Micrographs of (a) cold and (b) warm rolled samples.

In the micrograph of cold rolled samples (figure 2a), it possible to see that grain size within the shear zones is somewhat smaller than in the regions without shear bands. Fine grains were formed by the higher strain rate in the shear zones, producing bimodal distributions of grain sizes with smaller grains about 5 μm and coarser ones about 25 μm. In warm rolling samples were also formed shear bands zones with fine recrystallized grains follow by precipitate formation that was not strongly affected by thickness reduction.

Annealing Treatment

The annealing times were showed remarkable effect on the twinning, the grain size's average and the precipitate's distribution. A period of 30 minutes of annealing treatment was not enough to disappear totally the twinning formed in rolling process (Figure 3-a1). Some big grains still showed traces of twinning; however, for a period of 60 minutes of annealing treatment, twins were growing or shrinking, leaving completely untwined grains. The stored energy has been lowered by static recovery and recrystallisation. The grain size within the shear bands was somewhat smaller than of the rest of grains due to the higher strain rate in the shear zones that produce finer structure in that regions.

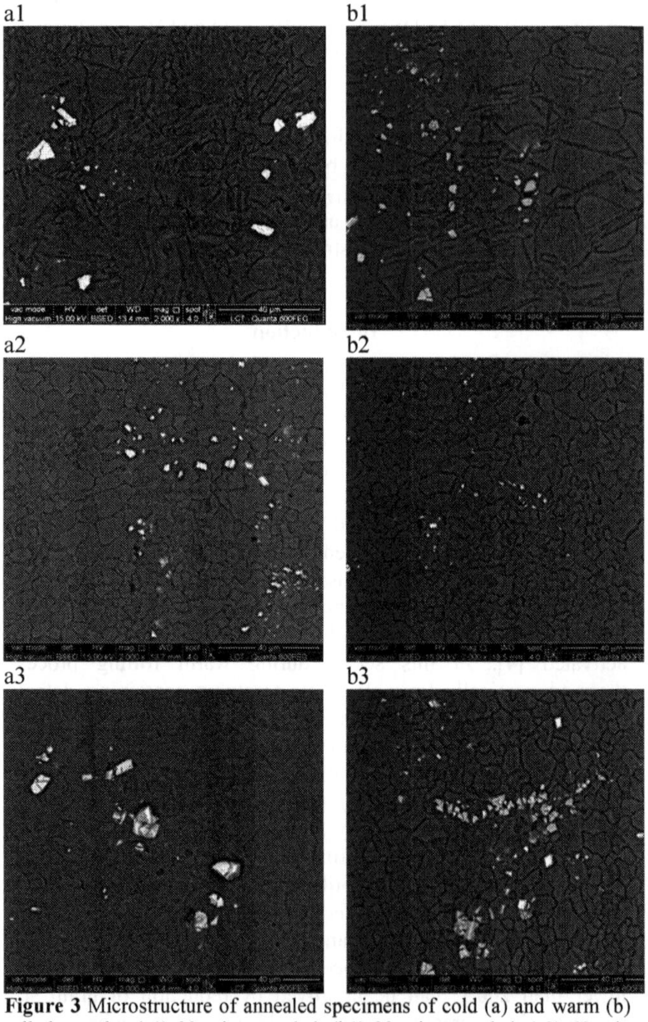

Figure 3 Microstructure of annealed specimens of cold (a) and warm (b) rolled samples (a1) 30 min annealed, (b1) 30 min annealed; (a2) 60 min annealed, (b2) 60 min annealed; (a3) 240 min annealed, (b3) 240 min annealed.

In micrographs of annealed samples (Figure 3), it was evidenced that 30 minutes of annealing treatments of cold rolled samples formed big-size precipitations which were distributed heterogeneously. However, the warm rolling applied prior to annealing increased the precipitate nucleation and formed fine precipitates. The thickness reductions (10% and 15%) affected slightly on hardness values of annealed samples, they showed different grain size and precipitate's distributions. In the Figure 4, it is evidenced that high thickness reductions significantly affects the hardening response due to precipitates formed during annealing treatment, especially after 2 hours of annealing. Since warm rolled samples contained small precipitates of $Mg_{17}Al_{12}$ and Al_4Mn, they were better redistributed after 60 minutes of annealing. In the Figure 3-c2, it possible to see a fine recrystallized grains structure with fine intermetallic particles. It is means that previous warm rolling significantly improved the precipitates redistribution after 1 hour of annealing treatments. Nevertheless, in longer annealing periods the size precipitates could increase a lot.

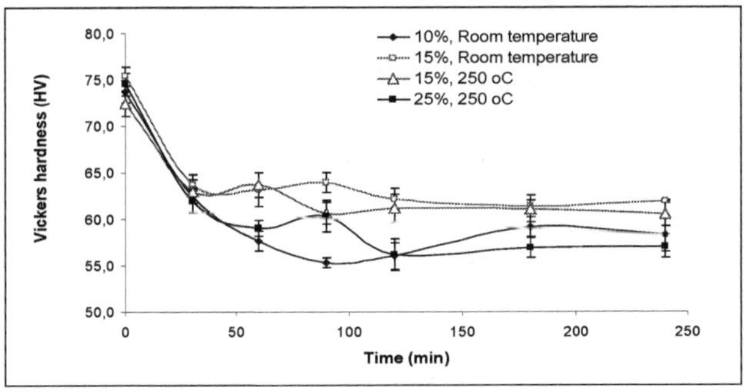

Figure 4 Microhardness values of annealed specimens of cold and warm rolled samples (at room temperature and 250 °C)

The hardening results of rolled and annealed samples suggest that intermetallic phases were partially solved and redistributed within matrix grains during warm rolling and annealing treatments, affecting on mobility of grain boundaries, recrystallization and precipitates redistribution.

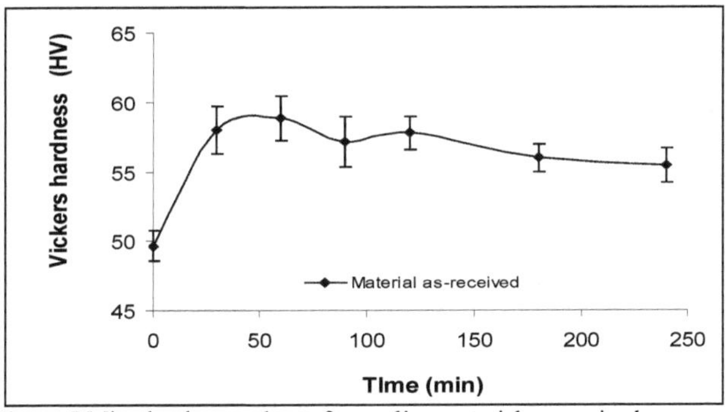

Figure 5 Microhardness values of annealing material as-received.

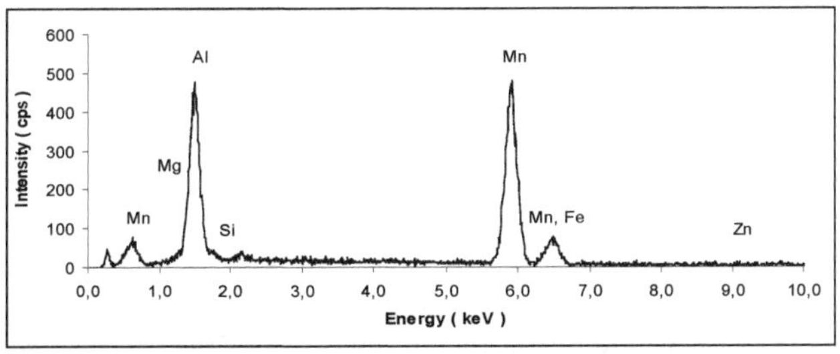

Figure 6 EDX precipitates spectrum at surface of warm rolled AZ31B-magnesium alloy.

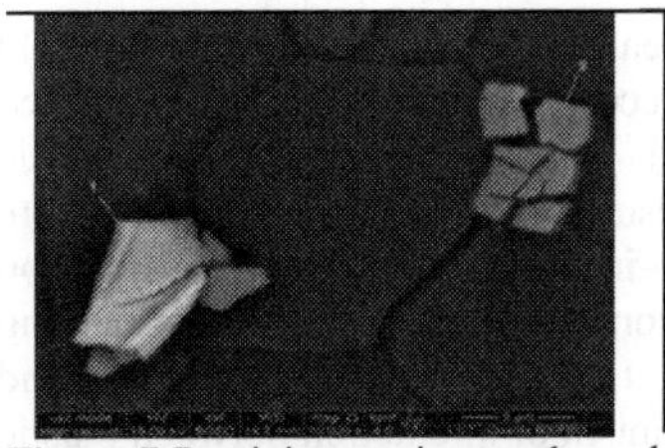

Elem	Wt%	At%
MgK	2.36	3.88
AlK	31.39	46.47
SiK	2.16	3.08
MnK	61.67	44.84
FeK	2.42	1.73
Total	100.00	100.00

Figure 7 Precipitates micrographs and EDX spectrum result at surface of warm rolled AZ31B-magnesium alloy.

In the figure 7, it is shown precipitates on the grain boundaries and their SEM-EDX analysis showed that precipitates in AZ31B rolled samples contained roughly about 4 at.%Mg, 46 at.%Al and about 45 at.%Mn, which changed in composition in about 60at % Mg and 30 at% Al and about 9 at% Mn after annealed treatments. And, some fine particles may be formed by almost pure Al and Mn elements due to their limited solubility in magnesium (Figure 8).

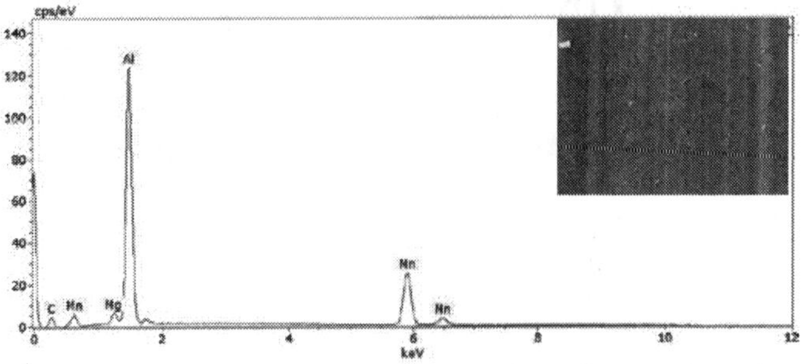

Figure 8 EDX precipitates spectrum at surface of annealed AZ31B-magnesium alloy.

The X-ray diffraction patterns of cold and warm rolled samples are shown in the figure 9. The intensity's differences of peaks were resulted by interaction of plastic deformation, dynamic recrystallization and transformation of precipitates into magnesium matrix.

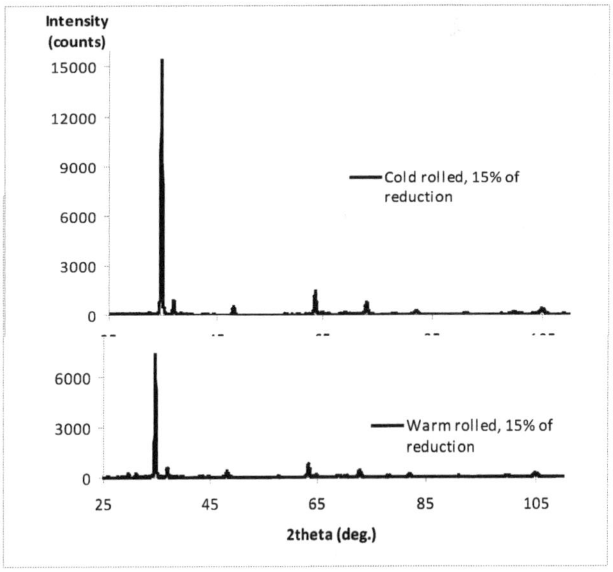

Figure 9 XRD diffraction patterns of AZ31B Mg alloy for rolled samples.

Comparing X-ray diffraction peaks of annealed samples with cold rolled sample and material as received (Figure 10), was evidenced that intensity of as-received material decreased by plastic deformation in cold rolling, resulting on broad peaks diffraction, meanwhile, after annealed, intensity of peak's diffraction increased due to volume fraction of precipitates grows, changing the lattice constants (Table 1). During rolling process and annealed treatments, the mechanism of interaction between plastic deformation and the formation of the eutectic precipitates ($Mg_{17}Al_{12}$, Al_4Mn and AlMn) for the AZ31 Mg alloys are still not totally clear. Meanwhile, it is possible to verify the presence of Mg_{17} Al_{12} and Al_4Mn precipitates with reference patters which coincided well with the diffractogram's peaks of rolled samples.

a.) b.)

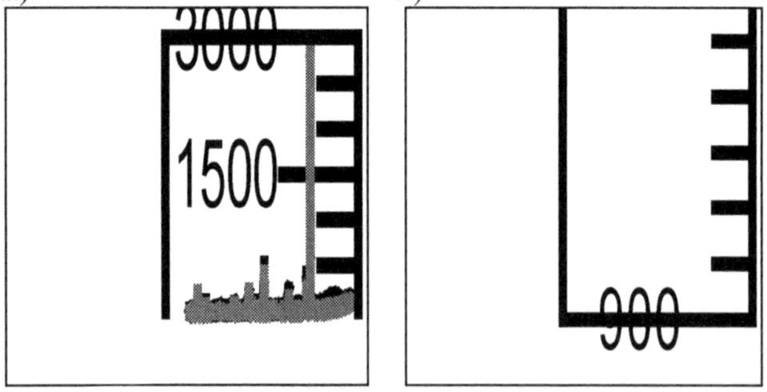

Figure 10 XRD diffraction patterns of (a) cold rolled sample and data peak's of $Mg_{17}Al_{12}$; (b) AZ31B Mg alloy for rolled and annealed samples

One way to analyze the phase's diffusion is by calculating the lattice parameter "a" and "c". Noting the parameter values in the Table 1, interparticle spacing on the basal plane 'a' was greater than material as received and 'c' was smaller. And increasing the rolling temperature, this behavior is inverted, suggesting diffusion phenomena in the magnesium alloy. As on annealed samples, continuous precipitation was found to occur preferentially in the regions of the grains that were richer in manganese or aluminum, causing nonuniform precipitations. The regions of magnesium matrix that are higher in manganese or aluminum concentrations provided a higher driving force for precipitation.

Table 1 Values of lattice constants "a" and "c" determined by Rietveld method.

a) Samples cold and warm rolled of AZ31B magnesium alloy

Temperature (°C)	Thickness reduction (%)	a (nm)	c (nm)
	As received	3.198431	5.195763
25	10	3.202474	5.194697
25	15	3.204011	5.191890
250	15	3.204313	5.192003
250	25	3.201473	5.194611

b) Annealed samples at 200 °C for 4 hours

Temperature (°C)	Thickness reduction (%)	a (nm)	c (nm)
	As received	3.198930	5.195658
25	10	3.205553	5.193650
25	15	3.205378	5.193723
250	15	3.199044	5.196323
250	25	3.199793	5.196505

Conclusions

During cold rolling big precipitates were gradually broken by the large deformation and partly dissolved during warm rolling and annealing treatments, resulting in a redistribution of precipitates.
The recovery, recrytallization, grain growth and precipitation of second phases interacted in annealing treatments at 200°C, and the annealing microstructure change to homogeneous fine grains from mixed deformed big grains. If AZ31B Magnesium Alloys contain big precipitates with heterogeneous distribution, the grain sizes grow fast and the Vickers hardness decrease, especially to cold rolled samples with low thickness reduction. Since static recovery and recrystalization are develops on annealing process, the grain refinement could be achieved by 1 hours of annealing treatment of warm rolling.
Annealing treatments can improve the ductility by grains' refinements, by fine precipitate formation with homogeneous distribution.
The grain size and second-phase precipitates ($Mg_{17}Al_{12}$, Al_4Mn and AlMn) affect on mechanical properties, particularly hardness and their effects depends on level, size and precipitate's distribution, a high level of fine $Mg_{17}Al_{12}$ precipitates impeded extension twinning and enhanced dynamic recrystallization, resulting in a fine grain microstructure.

References

[1] Polmear I J. "Magnesium alloys and applications", *Materials Science and Technology,* (1994), 1−14.

[2] Avedesian M M, Baker H. "*ASM speciality handbook — Magnesium and magnesium alloys. ASM Internationall,* (1999), 274.

[3] Humphreys F.J., Hatherly M., "Recrystallization and Related Annealing Phenomena", *Pergamon, Oxford,* (1995)

[4] Matsumoto R, Osakada K. "Development of warm forging method for magnesium alloy", *Materials Transactions, 2004,* 45(9):2838−2844.

[5] Peng-Xi Li, Bo-Ping Zhang, Jing-Yi Chen, Yi-Fan Tu, "Microstructure and Mechanical Properties of AZ31B Magnesium Alloy Sheets Prepared by Extruding and Rolling," *Advanced Materials Research Vols. 11-12,* (2006), 653-656.

[6] J. Koike, T. Kobayashi, T. Mukai, H. Watanabe, M. Suzuki, K. Maruyama, K. Higashi, "The activity of non-basal slip systems and dynamic recovery at room temperature in fine-grained AZ31B magnesium alloys," *Acta Materialia 51,* (2003), 2055–2065.

[7] Chino Y, Sassa K, Kamiya A, Mabuchi M. "Enhanced formability at elevated temperature of a cross-rolled magnesium alloy sheet". *Materials Science and Engineering A 441,* (2006), 349−356.

[8] Chino Y., Sassa K., Kamiya A., Mabuchi M. Stretch, "Formability at elevated temperature of a cross-rolled AZ31 Mg alloy sheet with different rolling routes", *Materials Science and Engineering A,* (2008) 195−200.

[9] Ion S. E., Humphreys F. J., White S. H., "Dynamic recrystallisation and the development of microstructure during the high temperature deformation of magnesium", *acta metall.* Vol. 30. (1982), 1909 -1919.

[10] Catorceno L., Padilha A. F., "Effect of Rolling Temperature on the AZ31B Magnesium Alloy Microstructure", Magnesium Technology - TMS (The Minerals, Metals & Materials Society) (2012), 283-287.

Magnesium Technology 2013
Edited by: Norbert Hort, Suveen N. Mathaudhu, Neale R. Neelameggham, and Martyn Alderman
TMS (The Minerals, Metals & Materials Society), 2013

GAS-PRESSURE BULGE FORMING OF MG AZ31 SHEET AT 450°C

Alexander J. Carpenter[1], Jon T. Carter[2], Louis G. Hector, Jr.[2], and Eric M. Taleff[3]
[1]Southwest Research Institute; Engineering Dynamics; 6220 Culebra Rd.; San Antonio, TX 78238-5166, USA
[2]General Motors Corp.; Research and Development; MC 480-106-224; 30500 Mound Rd.; Warren, MI 48090-9055, USA
[3]The University of Texas at Austin; Department of Mechanical Engineering; 1 University Station C2200; Austin, TX 78712-0292, USA

Keywords: Magnesium, AZ31, high temperature, bulge forming, grain growth

Abstract

Magnesium (Mg) sheet materials, such as wrought AZ31, possess low densities and high strength- and stiffness-to-weight ratios. These properties suggest that the use of Mg sheet is viable for reducing vehicle weight, an important goal of the automotive industry. Magnesium exhibits poor ductility at room temperature, but high-temperature forming processes may be used to manufacture complex vehicle closure panels. Tensile tests are the most common method of characterizing the plastic deformation of sheet materials. However, gas-pressure bulge tests may be more representative of the stress states that occur during the manufacture of sheet metal components. This study investigates the plastic deformation of AZ31 sheet during both biaxial and plane-strain gas-pressure bulge forming at 450°C. The heights and thicknesses of formed specimens are measured and compared. The deformation behaviors of the AZ31 sheet are related to observations of grain growth and cavitation that occur during forming.

Introduction

The automotive industry is interested in magnesium alloys, such as Mg AZ31 rolled sheet, for potential application in vehicle closure panels [1-6]. The low density of magnesium makes it an attractive material for reducing vehicle mass to increase both fuel efficiency and overall vehicle performance [7-10]. However, wrought Mg alloys exhibit poor ductility at room temperature [11], which is primarily a result of a large disparity between the ease of basal and prismatic slip, coupled with the prevalence of twinning [12-13]. Elevated-temperature forming processes, such as superplastic forming (above 500°C) [14-17] and quick plastic forming (~450°C) [18-20], are required to form currently-available wrought Mg sheet materials into complex automotive components. Gas-pressure bulge tests are useful for characterizing the deformation of sheet materials under these conditions, as the stress states from these tests are characteristic of those that occur during the manufacture of automotive components [21].

At 450°C, AZ31 deforms primarily by grain-boundary-sliding (GBS) creep at slow strain rates and five-power dislocation-climb (DC) creep at fast strain rates [22-24]. For a constant temperature, the strain rate from GBS creep, $\dot{\varepsilon}$, is related to the applied stress, σ, by

$$\dot{\varepsilon} = \frac{A_{GBS}}{d^p}\sigma^{n_{GBS}} , \quad (1)$$

where A_{GBS}, n_{GBS}, and p are material constants, and d is the average grain size of the material [25-26]. As d increases, $\dot{\varepsilon}$ decreases for a constant σ. Thus, the deformation behavior of AZ31 sheet may be strongly affected by both static and dynamic grain growth. Static grain growth during preheating of the AZ31 sheet prior to deformation is thought to slow deformation during GBS creep [23-24]. Similarly, dynamic grain growth during deformation is thought to harden the AZ31 sheet during GBS creep [23]. However, further investigation is needed to link the deformation behavior of AZ31 to microstructure evolution during deformation.

The purpose of this study is to conduct gas-pressure bulge tests of two AZ31 sheet materials with different initial microstructures at 450°C. The deformation behavior of these materials, which is characterized through height and thickness measurements from the formed specimens, is related to grain size and cavitation observations obtained from metallography.

Experimental Procedure

The materials studied were two Mg AZ31 rolled sheet materials each with a thickness of 2 mm. One material was supplied in the H24 temper [27], which results in a partially-recrystallized microstructure. The other material was supplied in the O temper [27], which results in a fully-recrystallized microstructure. The compositions of both materials are provided in Table 1. These AZ31 materials were used in biaxial and plane-strain gas-pressure bulge forming experiments.

Table 1: Mg AZ31 sheet material compositions (wt%) are listed.

Temper	Al	Zn	Mn	Fe	Cu	Ni
H24	3.1	1.0	0.42	0.006	0.003	<0.003
O	3.2	0.77	0.27	0.004	0.003	<0.003
Temper	**Si**	**Ca**	**Be**	**Sr**	**Ce**	**Mg**
H24	<0.1	<0.01	<0.005	<0.005	<0.01	Bal.
O	<0.1	0.0007	<0.005	<0.005	<0.01	Bal.

Biaxial bulge tests were performed using a circular die with an inner diameter of 100 mm. This die shape produces a dome-shaped specimen, as shown in Figure 1(a), that experiences a nearly balanced biaxial stress state at the dome pole during deformation. Plane-strain bulge tests were performed using a rounded rectangular die with dimensions of 254 mm × 51 mm. This die shape results in a tent-shaped specimen, shown in Figure 1(b), that experiences a nearly plane-strain state along much of its length. For plane-strain forming, the rolling direction of the sheet was oriented parallel to the long axis of the forming die. The two halves of each die were contained within an air-circulated furnace which is preheated to 450°C prior to testing.

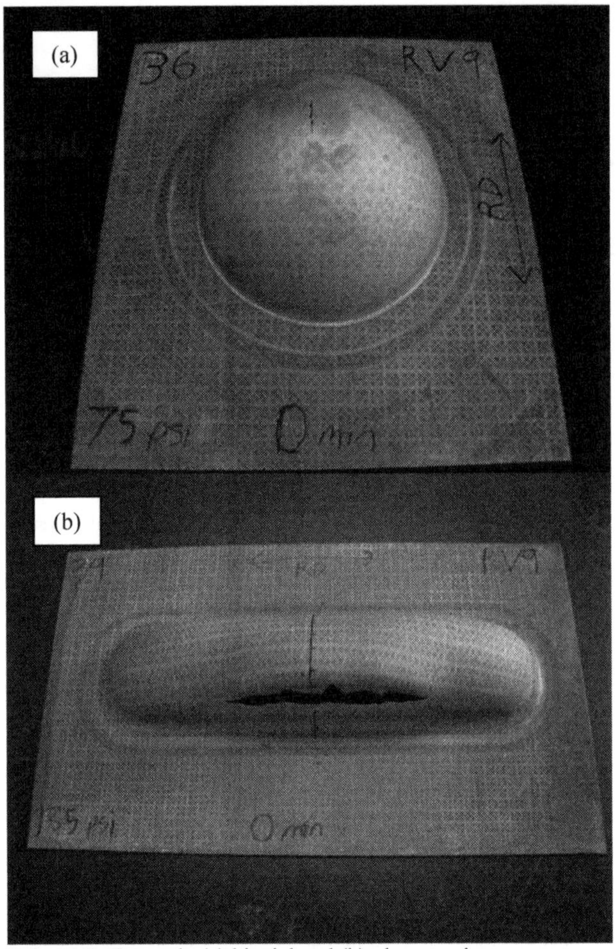

Figure 1: Example (a) biaxial and (b) plane-strain gas-pressure bulge specimens are shown.

Prior to each test, an AZ31 sheet blank sufficiently large to span the sealing beads in the forming dies was placed between the preheated die halves, which were then clamped together to provide a tight seal along the seal bead. Each blank was then preheat soaked for 0, 6, or 11 minutes at 450°C. Upon completion of preheat soaking, gas-pressure was rapidly applied to the lower surface of the blank, causing it to expand into the upper die cavity. The forming pressures used were 520 kPa (75 psi) for biaxial specimens and 930 kPa (135 psi) for plane-strain specimens. These forming pressures were chosen, after numerous experimental trials, to produce an average effective true-strain rate at the specimen pole (for the dome-shaped biaxial specimens) or peak (for the tent-shaped plane-strain specimens) of approximately 10^{-3} s^{-1}. During forming, temperature was monitored with a thermocouple attached to the die and maintained at 450°C. Initial bulge forming experiments were conducted until rupture. For these specimens only, the forming time is the time to rupture. Subsequent experiments were conducted for $1/3^{rd}$, $2/3^{rd}$, $5/6^{th}$ and $9/10^{th}$ of the average rupture time for each material and preheat soak time. Once either rupture or the desired forming time was reached, the specimen was removed from the dies and allowed to cool in air. For biaxial bulge specimens, the dome height and true (thickness) strain were measured at the dome pole. For plane-strain bulge specimens, the height and true strain were measured at the specimen peak. Specimen heights were measured using a stand built specifically for that task, and specimen thicknesses were measured using a magnetic sheet thickness gauge.

Samples for metallographic examination were excised from the gas-pressure bulge specimens to study grain growth and cavitation during deformation. Metallography was performed at locations with effective strains, $\bar{\varepsilon}$, of approximately 0.4, 0.6, 0.8, and 1.2, as calculated from

$$\bar{\varepsilon} = \frac{\sqrt{2}}{3}\left[(\varepsilon_1 - \varepsilon_2)^2 + (\varepsilon_2 - \varepsilon_3)^2 + (\varepsilon_3 - \varepsilon_1)^2\right]^{1/2}, \quad (2)$$

where ε_i are principal true strains [28]. At each location of interest, the local strain was measured along the longitudinal direction (same as the rolling direction), long-transverse direction (perpendicular to the rolling direction), and short-transverse direction (normal to the sheet) prior to sectioning. Strains normal to the sheet were calculated by measuring the change in thickness due to deformation. Strains within the plane of a sheet were measured with the aid of circle grids imprinted on the sheet prior to deformation. Metallographic specimens were sectioned from the formed bulge specimens at each location of interest and mounted in epoxy. After grinding and polishing, the specimen microstructures were revealed using an acetic-picral etchant (35 mL ethanol, 5 mL H_2O, 5 mL acetic acid, and 2.1 g picric acid). Grain sizes were measured along the rolling, long-transverse, and short-transverse directions using the lineal-intercept method [29]. An average grain size was then calculated by taking the geometric mean of the three directional grain size measurements.

Additional grain-size measurements were obtained from AZ31 sheet specimens annealed in salt. The temperature of the salt bath was maintained at 450°C and monitored using a K-type thermocouple. Specimens were annealed for 80, 360, or 960 s. After annealing was complete, the specimens were quenched in water and prepared for metallography as described in the previous paragraph. The average grain sizes in these specimens were also measured using the lineal-intercept method [29]. The grain sizes from these specimens represent the effects of static grain growth in AZ31 at 450°C.

Results

Height and Thickness Measurements

The specimens shown in Figures 1 exhibit rupture characteristics typical of the biaxial and plane-strain tests. Biaxial bulge specimens always ruptured at the dome pole along a line parallel to the rolling direction. The location of greatest thinning, and the largest strain in these specimens, is near the specimen pole. Rupture occurs as a result of damage accumulated at or near this location of greatest strain, as is shown Figure 1(a). Rupture of plane-strain specimens typically occurred along a line parallel to the specimen long axis and near the line of contact with the die wall. Flow localization occurs preferentially near the line of contact with the die for a plane-strain geometry, particularly when sliding contact is involved. Thus, the greatest specimen thinning was along this line, and rupture occurred here first as a result of damage accumulation. Once a rupture occurs in a plane-strain specimen, it tends to propagate along the line of greatest thinning in the side of the specimen, as is shown Figure 1(b). While rupture of the biaxial specimens does not significantly distort the specimen shape, rupture of the plane-strain specimens is typically violent enough to cause significant additional deformation. Thus, the final plane-strain specimen height is not representative of the formed height immediately prior to rupture. Because of this, all measurements of height for plane-strain specimens were made from tests halted prior to rupture.

Figure 2 shows the measured (a) dome height and (b) pole true strain for each biaxial specimen as a function of forming time, i.e., elapsed time under gas pressure. The measurements from the ruptured specimens shown in Figure 2 were used to calculate the average inflation rate and pole true strain at rupture for each material and preheat time, and these are provided in Table 2. Average inflation rate is defined as the dome height at rupture divided by the corresponding time to rupture. Table 2 shows that the preheat soak time has an effect on both inflation rate and pole strain at rupture in both materials. The inflation rate of both materials decreased as preheat soaking time increased from 0 to 6 min, though a further increase in preheat soaking time from 6 to 11 minutes had little additional effect on inflation rate. The pole strain at rupture increased as preheat soaking time increased over the range of times tested.

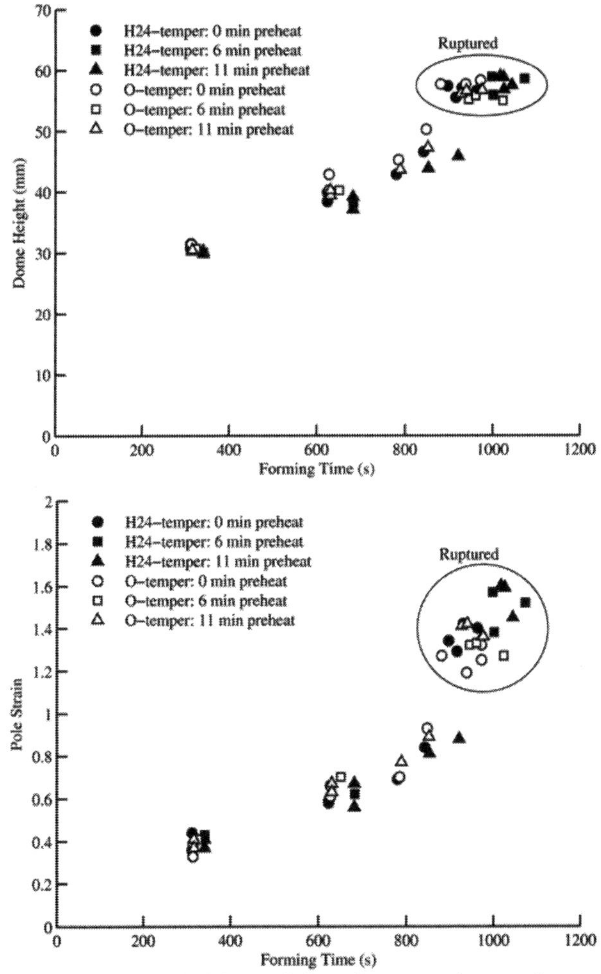

Figure 2: (a) Dome heights and (b) pole true strains are shown as functions of forming time for the gas-pressure biaxial bulge tests. Forming temperature was 450°C and gas pressure was 520 kPa (75 psi) for all tests. The average pole strain rate was approximately 10^{-3} s^{-1}. Specimens tested to rupture are labeled.

Table 2: Average inflation rate and pole strain at rupture are listed for biaxial bulge specimens tested to rupture. The forming temperature was 450°C and gas pressure was 520 kPa (75 psi) for all tests.

Temper	Preheat time (min)	Average inflation rate (mm/min)	Percent difference due to preheat soak
H24	0	3.67	-
	6	3.39	-7.6%
	11	3.39	-7.6%
O	0	3.70	-
	6	3.40	-8.1%
	11	3.55	-4.1%
Temper	**Preheat time (min)**	**Pole strain at rupture**	**Percent difference due to preheat soak**
H24	0	1.36	-
	6	1.49	9.6%
	11	1.56	14.7%
O	0	1.26	-
	6	1.31	4.0%
	11	1.39	10.3%

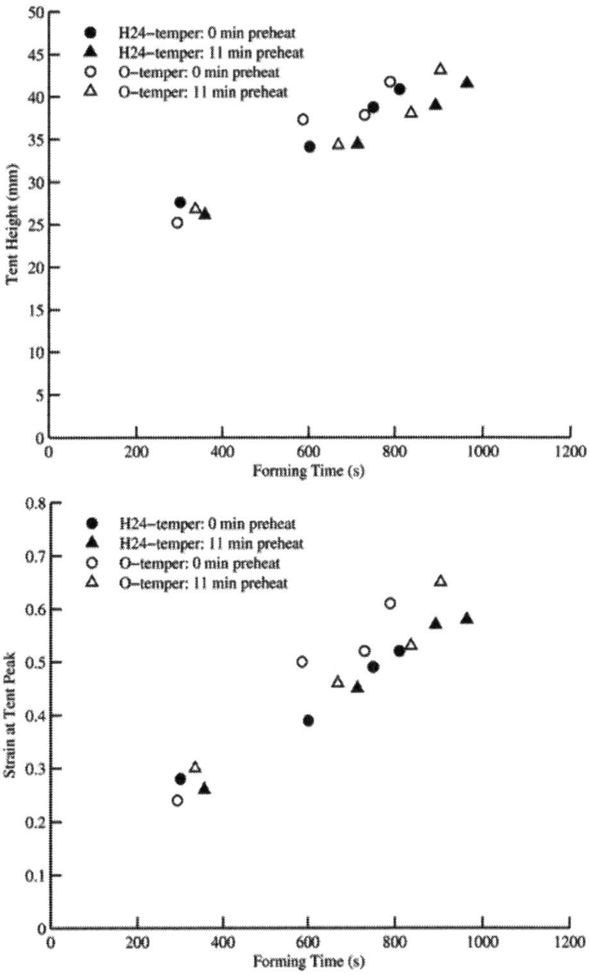

Figure 3: (a) Peak heights (tent heights) and (b) true strains at the peak are shown as a function of forming time for gas-pressure plane-strain bulge tests. Forming temperature was 450°C and gas pressure was 930 kPa (135 psi) for all tests. The average strain rate at the specimen peak was approximately 10^{-3} s^{-1}.

Figure 3 shows the (a) peak height (tent height) and (b) true strain at the peak (top) of plane-strain bulge specimens as a function of forming time. As preheat soak time increases, tent heights tend to decrease slightly for particular forming times. In addition, the peak strain tends to increase more quickly in the O-tempered material than in the H24-tempered material.

Grain Size Measurements

Grain sizes measured in salt-annealed specimens are plotted against annealing time in Figure 4. At an annealing time of 80 s, the grain size of the H24-tempered material is smaller than that of the O-tempered material. However, the grain size of the H24-tempered material grows larger than that of the O-tempered material as annealing time increases. The grain size of the O-tempered material is approximately constant as annealing time increases. The microstructure of the H24-tempered material is clearly less stable at 450°C than that of the O-tempered material.

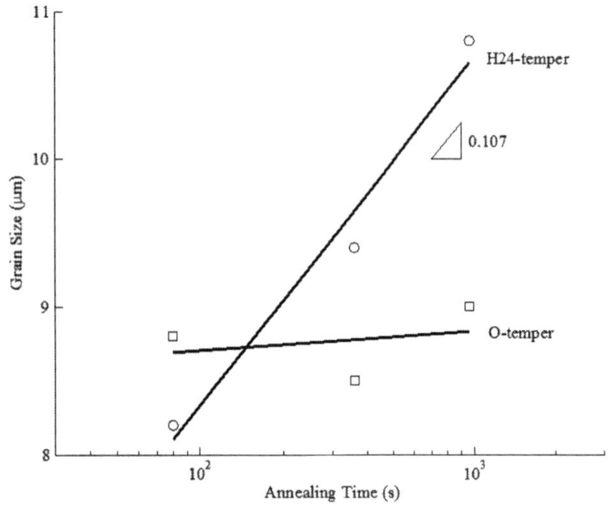

Figure 4: Lineal-intercept grain size measurements of annealed AZ31 specimens are plotted versus annealing time on dual logarithmic scales. Specimens were annealed at 450°C in a salt bath.

Figure 5 presents the average grain size measurements from the H24-tempered bulge specimens plotted against the corresponding effective uniaxial strain at the location of measurement, as calculated through Equation 2, regardless of forming time. Figure 6 presents average grain size measurements from the bulge specimens plotted against forming time, regardless of local strain. The outliers in each figure are from the same specimen and location, which was very close to the location of rupture. These outliers are attributed to recrystallization or abnormal grain growth beyond that from regions of lesser strain. Figure 5 indicates that grain size can be reasonably described as a single-valued function of true strain. However, Figure 6 shows that grain size can vary significantly between different locations with identical preheat soaking times and identical forming times. Thus, grain size is not a single-valued function of forming time. Figures 5 and 6 demonstrate that grain size evolution during forming is better described as a function of strain rather than as a function of time. As true strain increases, grain size increases for $\varepsilon < 0.6$, indicating that the AZ31 sheet materials exhibit dynamic grain growth at these strains. Dynamic grain growth likely causes the strain hardening observed in previous studies [23-24].

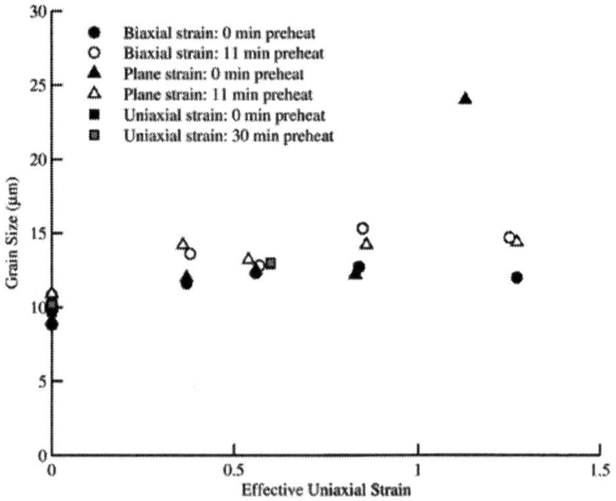

Figure 5: Average lineal-intercept grain size is shown for the H24-tempered material as a function of local effective uniaxial strain. Temperature was 450°C for all specimens.

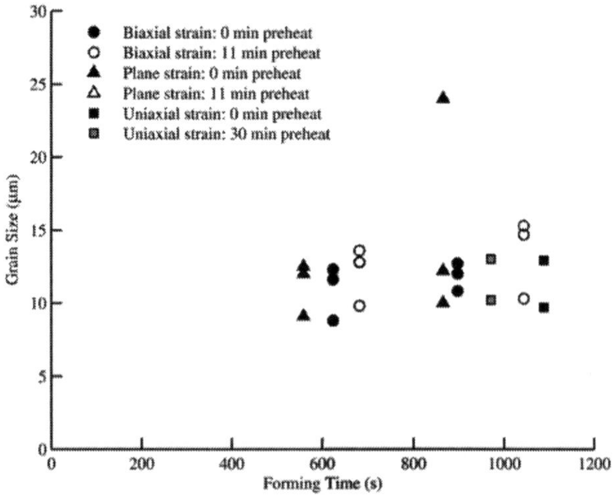

Figure 6: Average lineal-intercept grain size is shown for the H24-tempered material as a function of forming time. Temperature was 450°C for all specimens.

Discussion

The results of this study suggest that the decrease in forming rate during bulge tests as preheat soaking time increases is a result of static grain growth prior to deformation. As grain size increases, the GBS creep rate will decrease according to Equation 1, slowing the deformation rate. The GBS creep rate typically depends on the inverse of grain size to a power of two to three [25-26], making GBS creep rate quite sensitive to even slight grain growth. Figure 4 indicates that the H24-tempered material exhibits static grain growth at 450°C. Thus, this hypothesis reasonably explains why the inflation rate of the H24-tempered material slows as preheat soaking time increases; see Table 2. The data of Table 2 indicate that the inflation rate of the O-tempered material also slows as preheat soaking time increases. This occurs despite minimal static grain growth at 450°C; see Figure 4. Thus, an additional effect of preheat soaking may exist in the material. Preheat soaking may increase the susceptibility of the O-tempered material to dynamic grain growth in the earliest stages of deformation, and this would produce high initial strain hardening rates that should be observable in experiments. Unfortunately, the available data are not yet sufficient to fully test this hypothesis.

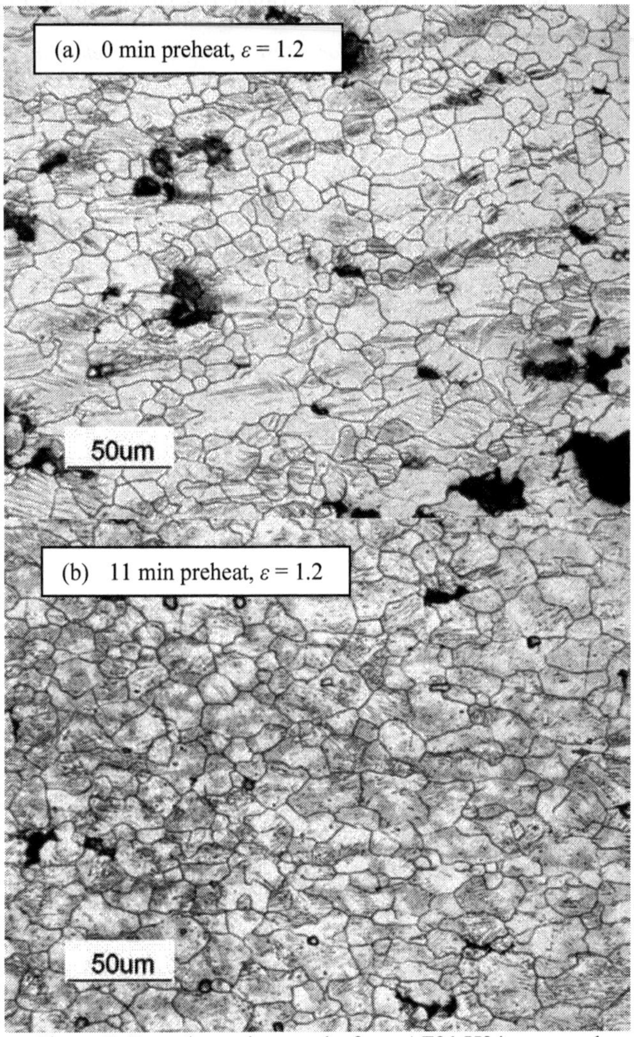

Figure 7: Two photomicrographs from AZ31 H24-tempered biaxial bulge specimens are shown. Both specimens were tested at 450°C and 520 kPa (75 psi). Both photomicrographs are of a region where $\varepsilon = 1.2$ with the rolling direction horizontal and short-transverse direction vertical. Photomicrograph (a) is from a specimen that was not preheat soaked prior to deformation. Photomicrograph (b) is from a specimen that was preheat soaked for 11 minutes prior to deformation. Cavitation damage is clearly less in the preheat soaked specimen.

The increase in pole strain at rupture with preheat soaking time in the biaxial specimens is the result of a reduction in the GBS creep contribution to deformation. Grain growth associated with preheat soaking will reduce the strain rate from GBS creep, thus decreasing the inflation rate; see Table 2. However, Chang *et al.* [30] observed that cavitation develops with strain much more quickly during GBS creep than during solute drag creep in AA5083 sheet, another fine-grained material. If cavitation in AZ31 is also most rapid during GBS creep, this can explain why preheat soaking time increases pole strain at rupture, while decreasing inflation rate. Although pole strain at rupture is increased by a reduction in GBS creep, the dome height is not increased because the greater role of dislocation-climb creep leads to faster localization of thinning at the pole. Figure 7 shows two photomicrographs of material from H24-tempered biaxial bulge specimens tested to an effective strain of 1.2. These photomicrographs show that cavitation does decrease as preheat soaking time increases from 0 to 11 minutes, supporting this theory.

There are two limitations to the grain size data of Figures 5 and 6. First, all of the grain size measurements were made from specimens deformed at approximately 10^{-3} s^{-1}. At this strain rate, both GBS creep and dislocation-climb creep contribute to deformation [22-23]. To separate the effects of GBS creep and dislocation creep on dynamic grain growth, grain size must be measured after deformation at slower strain rates (less than 3×10^{-4} s^{-1}), where GBS creep dominates deformation, and at faster strain rates (greater than 10^{-2} s^{-1}), where dislocation-climb creep dominates deformation [23-23]. Additionally, most of the grain size measurements in Figures 5 and 6 were made at locations of large strain. Further data are required to determine how grain size increases with strain and time at small strains. Note that Figure 5 suggests that strain state has little influence on average grain size. Thus, future studies of dynamic grain growth can be reasonably conducted using a single type of test geometry, such as a simple tensile test.

Conclusions

Gas-pressure bulge tests were conducted using two AZ31 sheet materials at 450°C and pressures selected to maintain true-strain rates of approximately 10^{-3} s^{-1}, and preheat soaking time prior to each test was varied. An increase in preheat soaking decreases the forming rate at a given constant gas pressure. It is proposed that this decrease in forming rate is a result of an increase in grain size caused by the preheat soaking treatment, which produces static grain growth. This increase in grain size decreases the rate of grain-boundary-sliding (GBS) creep, decreasing the forming rate when GBS creep is significant. An increase in preheat soaking time also increases the pole thickness strain prior to rupture in bulge test specimens at 450°C, but does not necessarily increase the maximum dome height achieved. Grain growth from preheat soaking reduces GBS creep, which reduces the rate of cavitation development and allows greater thinning at the dome pole prior to rupture. Dynamic grain growth was observed at 450°C in both biaxial and plane-strain bulge tests. This dynamic grain growth, which is primarily a function of strain rather than time at temperature, is likely responsible for strain hardening at slow strain rates [23], for which GBS creep is significant.

References

1. D. Eliezer, E. Aghion, and F.H. Froes, "Magnesium Science, Technology, and Applications," *Adv. Perform Mater.*, 5 (1998), 201-212.
2. M.M. Avedesian and H. Baker, *Magnesium and Magnesium Alloys* (Novelty, OH: ASM International, 1999), iv.
3. E. Aghion, B. Bronfin, and D. Eliezer, "The Role of the Magnesium Industry in Protecting the Environment," *J. Mater. Process. Tech.*, 117 (2001), 381-385.

4. B.L. Mordike and T. Ebert, "Magnesium: Properties-applications-potential," *Mater. Sci. Eng. A*, 302 (2001), 37-45.
5. M. Easton *et al.*, "Magnesium Alloy Applications in Automotive Structures," *JOM*, 60 (11) (2008), 57-62.
6. M.K. Kulekci, "Magnesium and Its Alloys Applications in Automotive Industry," *Int. J. Adv. Manuf. Technol.*, 39 (2008), 851-865.
7. L.H. Pomeroy, "Advantages of Light-Weight Reciprocating Parts," *Automotive Engineering*, XI (6) (1922), 508-519.
8. G.S. Cole and A.M. Sherman, "Light Weight Materials for Automotive Applications," *Mater. Char.*, 35 (1995), 3-9.
9. A.I. Taub, "Automotive Materials: Technology Trends and Challenges in the 21st Century," *MRS Bulletin*, 31 (2006), 336-343.
10. A.I. Taub *et al.*, "The Evolution of Technology for Materials Processing over the Last 50 Years: The Automotive Example," *JOM*, 59 (2) (2007), 48-57.
11. M.W. Toaz and E.J. Ripling, "Correlation of the Tensile Properties of Pure Magnesium and Four Commercial Alloys with Their Mode of Fracturing," *J. Metals*, 8 (1956), 936-946.
12. J.A. Yasi, L.G. Hector, Jr., and D.R. Trinkle, "First-Principles Data for Solid-Solution Strengthening of Magnesium: From Geometry and Chemistry to Properties," *Acta Mater.*, 58 (2010), 5704-5713.
13. J.A. Yasi, L.G. Hector, Jr., and D.R. Trinkle, "Prediction of Thermal Cross-Slip Stress in Magnesium Alloys from Direct First-Principles Data," *Acta Mater.*, 59 (2011), 5652-5660.
14. G.C. Cornfield and R.H. Johnson, "The Forming of Superplastic Sheet Metal," *Int. J. Mech. Sci.*, 12 (1970), 479-490.
15. A.K. Ghosh and C.H. Hamilton, "Influences of Material Parameters and Microstructure on Superplastic Forming," *Metall. Trans. A*, 13 (1982), 733-743.
16. F.I. Saunders *et al.*, *U.S. Patent 5974847*, 1999.
17. A.J. Barnes, "Superplastic Forming 40 Years and Still Growing," *J. Mater. Eng. Perform.*, 16 (2007), 440-454.
18. M.S. Rashid *et al.*, *U.S. Patent 6253588*, 2001.
19. J.G. Schroth, "General Motors' Quick Plastic Forming Process," *Advances in Superplasticity and Superplastic Forming*, ed. E.M. Taleff *et al.*, (Warrendale, PA: TMS, 2004), 9-20.
20. P.E. Krajewski and J.G. Schroth, "Overview of Quick Plastic Forming Technology," *Mater. Sci. Forum*, 551-552 (2007), 3-12.
21. G. Giuliano and S. Franchitti, "The Determination of Material Parameters from Superplastic Free-Bulging Tests at Constant Pressure," *Int. J. Mach. Tool Manu.*, 48 (2008), 1519-1522.
22. E.M. Taleff *et al.*, "Material Models for Simulation of Superplastic Mg Alloy Sheet Forming," *J. Mater. Eng. Perform.*, 19 (2010), 488-494.
23. P.A. Sherek *et al.*, "The Effects of Strain and Stress State in Hot Forming of Mg AZ31 Sheet," *Magnesium Technology 2012*, ed. S.N. Mathaudhu, (Warrendale, PA: TMS, 2012), 301-306.
24. P.A. Sherek, "Simulation and Experimental Investigation of Hot Gas-Pressure Forming for Light-Alloy Sheet Material" (Master's thesis, The University of Texas at Austin, 2009).
25. O.D. Sherby and P.M. Burke, "Mechanical Behavior of Crystalline Solids at Elevated Temperature," *Progr. Mater. Sci.*, 13 (1968), 325-390.
26. T.G. Nieh, J. Wadsworth, and O.D. Sherby, *Superplasticity in Metals and Ceramics* (Cambridge: Cambridge University Press, 1997), 43-72.
27. ASM International: *Metals Handbook, Vol. 2* (Novelty, OH: ASM International, 2002).
28. J. Hu, Z. Marciniak, and J. Duncan, *Mechanics of Sheet Metal Forming* (Oxford: Butterworth-Heinemann, 2002), 14-27.
29. ASTM E 112-96, *Standard Test Methods for Determining Average Grain Size* (West Conshohocken, PA, ASTM, 1996).
30. J.-K. Chang, E.M. Taleff, and P.E. Krajewski, "The Effect of Microstructure on Cavitation during Hot Deformation of a Fine-Grained Aluminum-Magnesium Alloy as Revealed through Three-Dimensional Characterization," *Metall. Mater. Trans. A*, (40) 2009, 3128-3137.

Magnesium Technology 2013

Corrosion I

EFFICIENCY OF A NEW HEXAVALENT CHROMIUM-FREE CHEMICAL PICKLING PROCESS BASED ON ORGANIC AND INORGANIC ACIDS ON MAGNESIUM ALLOYS Mg-Y-RE-Zr and Mg-Zn-RE-Zr

Hélène Ardelean[1], Antoine Seyeux[1], Sandrine Zanna[1], Philippe Marcus[1]
Sophie Pettier[2], Nathalie Le Pottier [2] and Daniel Lecuru[2]

[1]Laboratoire de Physico-Chimie des Surfaces, UMR CNRS-ENSCP 7045, Ecole Nationale Supérieure de Chimie de Paris, Chimie Paristech, 11 rue P. et M. Curie, 75005 Paris, France.

[2]Eurocopter, EADS, Aeroport International Marseille-Provence 13725 Marignane

Keywords: Pickling solutions; Magnesium alloys; Passive films; SEM-EDX, XPS; Profilometry; Pickling rate.

Abstract

The removal of surface contaminants as result of corrosion attack, inadequate or incorrectly applied surface protection coating (HAE, DOW 19) on Mg-Y-RE-Zr and Mg-Zn-RE-Zr alloys was investigated by a new chemical pickling process based on organic and inorganic acids with addition of anticorrosion compounds and wetting agents. Various combinations of acetic, citric, oxalic, tartric acids with nitric, sulphuric, phosphoric and hydrofluoric acids were investigated. The influence of the composition, pH and time of pickling, on the surface morphology, roughness, composition (of contaminant film and of alloy surface) were evaluated using Bruker Dektak 150 profilometer, Scanning electron microscopy and X-Ray Photoelectron Spectroscopy.
The experimental results show that the best pickling efficiency was obtained with a combination of tartaric acid, nitric acid and a zirconium fluride for removal of ~ 20 μm of the contaminated surface. After completely removal of the incorrectly coatings, the alloy surfaces obtained were remarkably cleaned with a low roughness (between 3 and 4 μm) and without any attack at the grain boundaries and surface, and without any surface contaminant. An anti-corrosive effect was obtained by the formation of a protective layer on the surface of the alloy during the pickling. This new tartaric + nitric acids based pickling is an interesting alternative to the environmentally undesirable Cr^{6+}- surface pickling.

Introduction

Magnesium alloys have an interesting future wherever weight saving is essential, including automotive, aerospace and communication fields. However, Mg alloys exhibit low corrosion and wear resistances, which limit their widespread use. Corrosion can be avoided or minimized by the formation of protective layers on the surface. The surface of Mg alloys is supposed to be cleaned prior to surface protection treatments in order to enhance the corrosion resistance and surface appearance. Mechanical and chemical cleaning methods are used depending on the application involved. Mechanical cleaning include grinding and rough polishing, dry and wet abrasive blast cleaning, wet brushing and wet (bowl/ barrel) abrading. Chemical cleaning is achieved by solvent degreasing, alkaline cleaning and acid cleaning. In our study we focus on acid chemical cleaning/ pickling process to remove the incorrectly protective coating (HAE, DOS 19). Nowadays, most of the surface material removal treatments are done with hexavalent chromium compounds. All compounds containing hexavalent chromium radicals are classified as carcinogenic substances under the current REACH regulation [1] and will be restricted. The search for their substitution is mandatory and prevails against any other risk reduction measures. In this work, we focused upon this major challenge of designing a new hexavalent chromium-free coating removal process based on organic and inorganic acids for magnesium alloys Mg-Y-RE-Zr (WE43) and Mg-Zn-RE-Zr (RZ5). The classically used chemical pickling/ removing process is achieved with chromic acid 180 - 220 g/L at room temperature [2]. Acid pickling is a useful method for the removal of bad or incorrectly applied surface protective layers (obtained by DOW 19 chemical conversion or by HAE anodization), contaminations such as oxide tarnish, heavy metal impurities, embedded sand and burned-in lubricants. Taking into account the chemical composition of the surface layers which are supposed to be removed, we have studied the effects of aqueous solutions of different mineral and organic pure acids: i) separately or ii) in combination, and iii) with addition of anti-corrosion components, wetting agents, on the material removal (etching rate), corrosive attack (surface roughness, appearance, morphology, phases) of Mg alloys. The choice of acids was made according to their potential in pickling by forming soluble salts with the main components of the layers to be removed in agreement with published solubility data [3]. So, we have selected: HNO_3 (nitric acid), H_3PO_4 (phosphoric acid), H_2SO_4 (sulphuric acid), HF (hydrofluoric acid) as inorganic acids and $C_4H_6O_6$ (tartaric acid), CH_3COOH (acetic acid), $C_2H_2O_4.2H_2O$ (oxalic acid), $C_6H_8O_7$ (citric acid) as organic acids. The removal of the bad and heterogeneous coatings with thickness varying between 1 and 24 μm must be done without any attack of the surface alloys. Thus, we have selected inorganic inhibitors such as: $ZrO(NO_3)_3$ for the protection against acid attack of the pickled areas during the pickling of the area having higher thickness of the coating. This protective effect is similar to that furnished by the Cr_2O_3 film formed on the alloy surface during the pickling with chromic acid. The results obtained are reported in this paper.

Experimental

Sample preparation

The WE43 and RZ5 magnesium alloys used in this study were commercial alloys with the chemical composition given in Table 1. Some specimens were treated by HAE anodizing (thickness between ~ 5 and 24 μm) or by DOW 19 chemical conversion (thickness between ~ 1 and 5 μm). The HAE coating produced is a mixed $MgO\text{-}Mg(OH)_2$ system with additional compounds. The DOW 19 chemical conversion coating produced contains mainly MgO, $Mg(OH)_2$ with small amount of Cr_2O_3. The specimens were cut into sets of samples of dimensions ~ 50 mm x 50 mm x 3mm using a shear.

Table 1.
Chemical composition (wt. %) of the WE43 magnesium alloy.

Alloy	Fe	Zn	Y	RE(Nd)	RE(Ce)	RE(La)	Zr	Mn	Mg
WE43	-	<0.2	4.0	3.0	-	-	0.5	0.15	Bal
RZ5	-	4.4	-	-	0.8	0.3	0.4-1.0	-	Bal

The choice of the composition of the chemical pickling bath is important. We have tested (at medium and higher concentrations, room temperature and pH values between 0.7 and 2.4):

- a set of inorganic acids: HNO_3, H_3PO_4, H_2SO_4, HF,
- a set of organic acids $C_4H_6O_6$, CH_3COOH, $C_2H_2O_4.2H_2O$, $C_6H_8O_7$,
- additives for the pH correction (NH_4OH, KOH),
- moderators/ surfactants/ dispersers such as: NH_4F, HMTA, Dodecylsulphate sodium salt ($C_{12}H_{25}NaO_4S$),
- inorganic inhibitors such as: $ZrO(NO_3)_2$, $Ce(NO_3)_3$.

The coating material removal process was performed by dipping the specimens in the pickling bath. After completely coating removal, the specimens were rinsed in deionized water and then dried in a stream air.

After pickling the specimens were submitted to inhibitive removal pickling rate evaluation and to surface characterization by Profilometry, Scanning electron microscopy and X-Ray Photoelectron Spectroscopy.
A Bruker Dektak 150 profilometer was used to evaluate the surface roughness (RA- parameter) and thickness of the material removed (d – parameter).
Cross-sectional morphology and chemical composition of the WE43 and RZ5 alloys coated (with HAE or DOW 19) and pickled were characterized by SEM analyses using a Hitachi S2500 apparatus operating at 20 keV, coupled with EDS analyses using an EDS-PGT_IMIX system.
Chemical composition characterization by X-ray Photoelectron Spectroscopy (XPS), was done with Thermo Electron Escalab 250 spectrometer, usin g an Al Kα X-ray source. The XPS spectra recorded are: Mg 1s, Zr 3d, Cr 3p, Nb 3d, Y 3d, Nd 3d, O 1s, C 1s, F 1s. The reference binding energies are obtained from analyzed reference samples or from published data [4-7].

The inhibitive removal pickling rate - R_{irp} was performed by weighing the specimens before and after dipping in the pickling bath using an analytical balance. The weight loss gave the inhibitive removal pickling rate as calculated following Eq. (1).

$$R_{irp}=Px10^4/\rho xSxT \quad (1)$$

with R_{irp} - the inhibitive removal pickling rate in μm/min. P is weight loss in g, ρ is the coating density in g/cm^3 (3), S is the pickled surface in cm^2, T is the dipping time in min, 10^4 is a unit conversion factor.

Results

The etching effectiveness of the coating material removal depends on the solubility conditions in aqueous solutions of the reaction products formed between the different inorganic acids and the compounds present in the HAE/ DOW 19 coatings and the addition elements and the impurities present in the coating/ Mg alloy interface. Nitric acid form soluble nitrates ($Mg(NO_3)_2$, $Y(NO_3)_3$, $Nd(NO_3)_3$, $Zr(NO_3)_4$, $Zn(NO_3)_2$, $Ce(NO_3)_3$, $Al(NO_3)_3$, $Fe(NO_3)_3$) [3] and hence it has the capability to remove coating/ material. Sulphuric acid forms soluble reaction products [3]. Phosphoric and hydrofluoric acids form insoluble or slightly soluble phosphates and fluorides, respectively [3].

The organic acids tested for this study include: tartaric acid, acetic acid, oxalic acid, citric acid. The tartaric acid forms soluble salts with the compounds present in the surface HAE/ DOW 19 coatings and the addition elements and the impurities present in the substrate. The acetic acid give rise to soluble salts [3] but the roughness was increased. The pickling in tartaric and acetic acids give rise to a greyish metallic appearance. Oxalic acid forms insoluble or slightly soluble salts [3] with the HAE/ DOW19 coating compounds and with Mg, Nd, Zn and Ce elements present in the alloys. A brownish spotted surface appearance was observed on the surface of the pickled alloys. The citric acid forms soluble salts but whitish and dusty deposition precipitates were observed on the pickled surfaces.

The final optimal selected pickling solution was chosen after preliminary tests looking at the surface appearance that is related to the amount of the surface material pickled and the surface roughness. After the completely removal of the coating by pickling in nitric acid, the surface of the alloys has the shiniest metallic appearance. In sulphuric acid the surface had a metallic appearance but the roughness was non uniform. In phosphoric and hydrofluoric acids dark greyish surface films were observed.

After the completely removal of the coating in tartaric acid, the surface of the alloys has the shiniest metallic appearance.

The mainly compounds of the optimal bath are: tartaric acid, nitric acid and hexamethylene tetramine.

effect on the surface roughness as can be see in Fig. 1. – 4. It was observed that the roughness increases slightly from ≈ 1.0 μm to ≈ 4.0 μm with ~ 20 μm of HAE material removal in the optimal bath. Moreover, SEM and EDS analyses carried out after HAE material removal (Fig. 5 and 6) show the absence of any intergranular attack on the pickled alloy surface. After HAE coating removal, we observe the presence in the surface of a thin film mainly containing ZrOx (Fig. 6). So, the absence of any attack on the pickled surface area can be attributed to the protective effect of $ZrO(NO_3)_3$ (added The amount of the material removed varies with the pickling composition bath. The inhibitive removal pickling rates evaluated for the optimal bath with Eq. 1.

Surface appearance and roughness after inhibitive removal pickling in optimal bath

The surface appearance is related to the amount of the surface material removed and to the surface roughness. The pickling has an

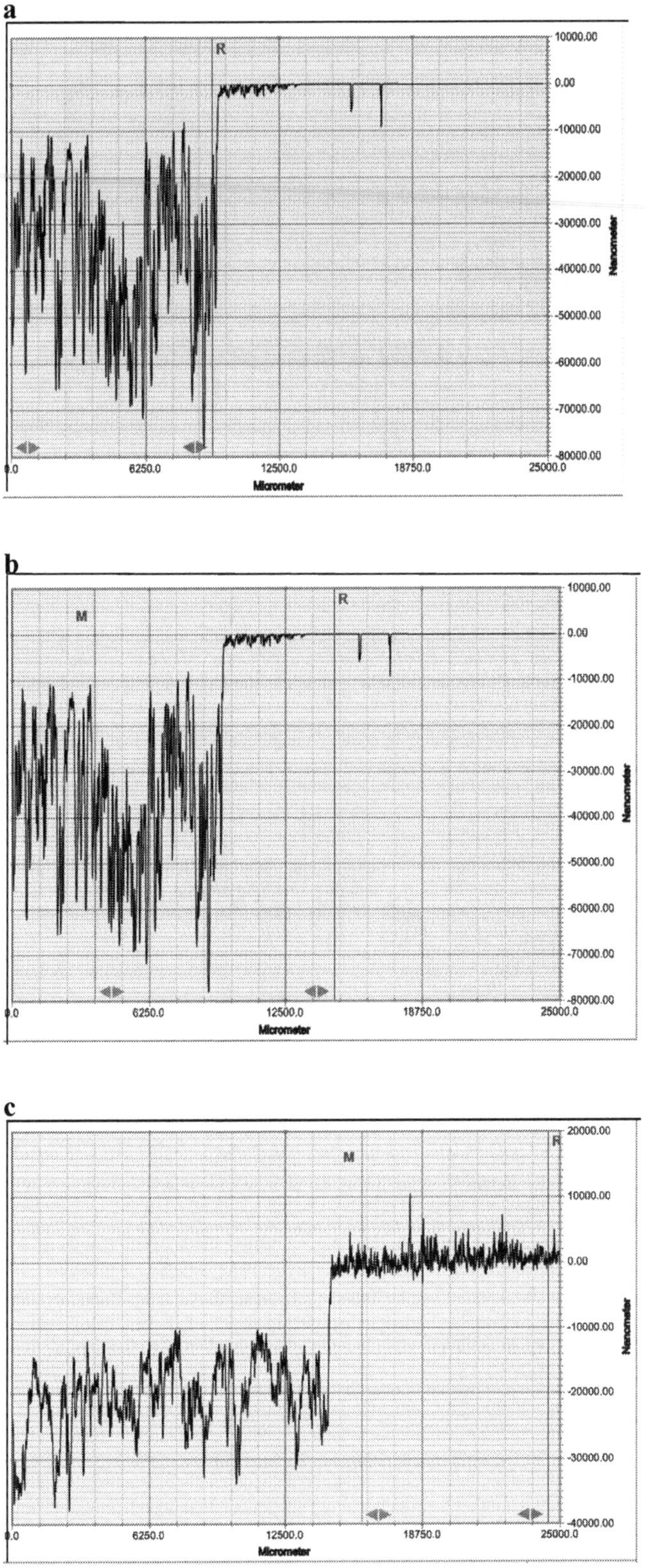

Fig. 1. Variation of the surface roughness with HAE coating removal (in OB-optimal pickling bath,) on the WE43 alloy - HAE coated; a) roughness on HAE coating: RA ≈ 1.0 μm. b) roughness on the WE43 alloy after removal of HAE coating: RA ≈ 4,2 μm. c) thickness of the coating removed: d ≈ 23.0 μm.

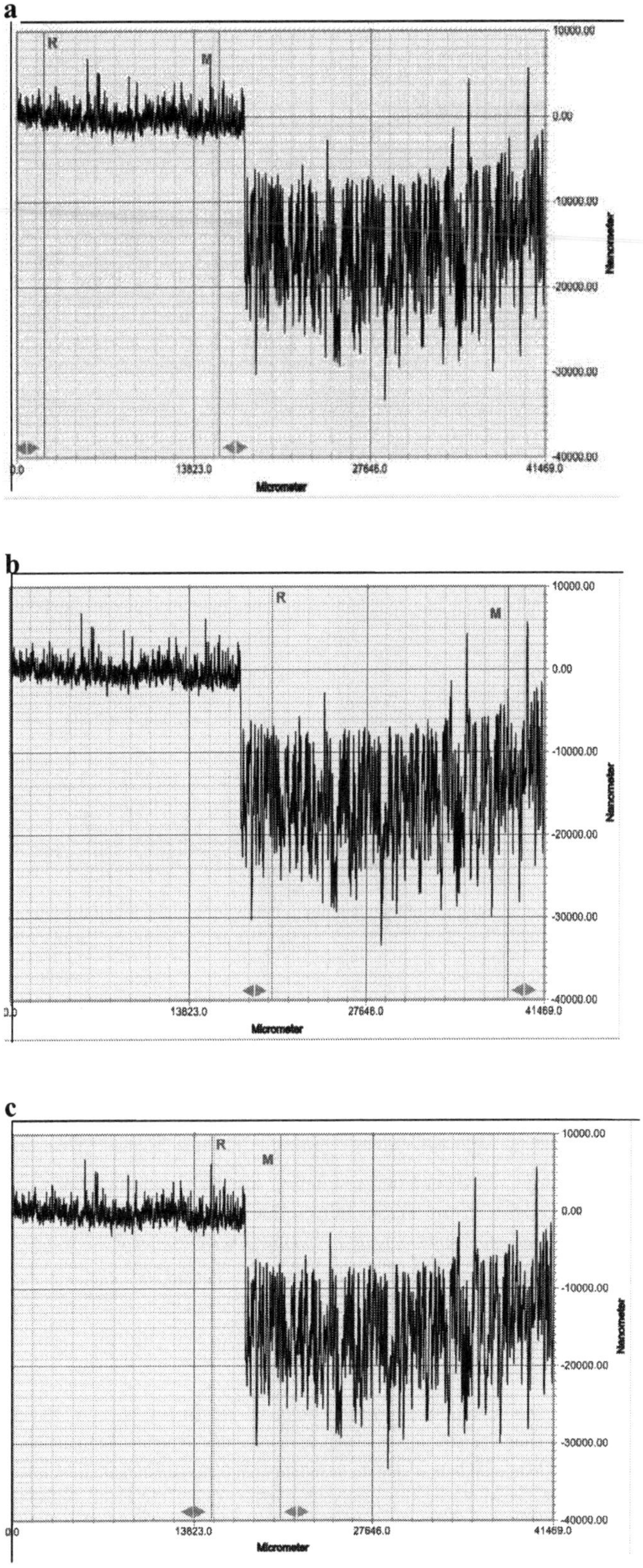

Fig. 2. Variation of the surface roughness with HAE coating removal (OB- optimal pickling bath) on the RZ5 alloy - HAE coated; a) roughness on HAE coating: RA ≈ 1.0 μm. b) roughness on the RZ5 alloy after removal of HAE coating: RA ≈ 3.2 μm. c) thickness of the HAE coating removed: d ≈ 5.0 μm.

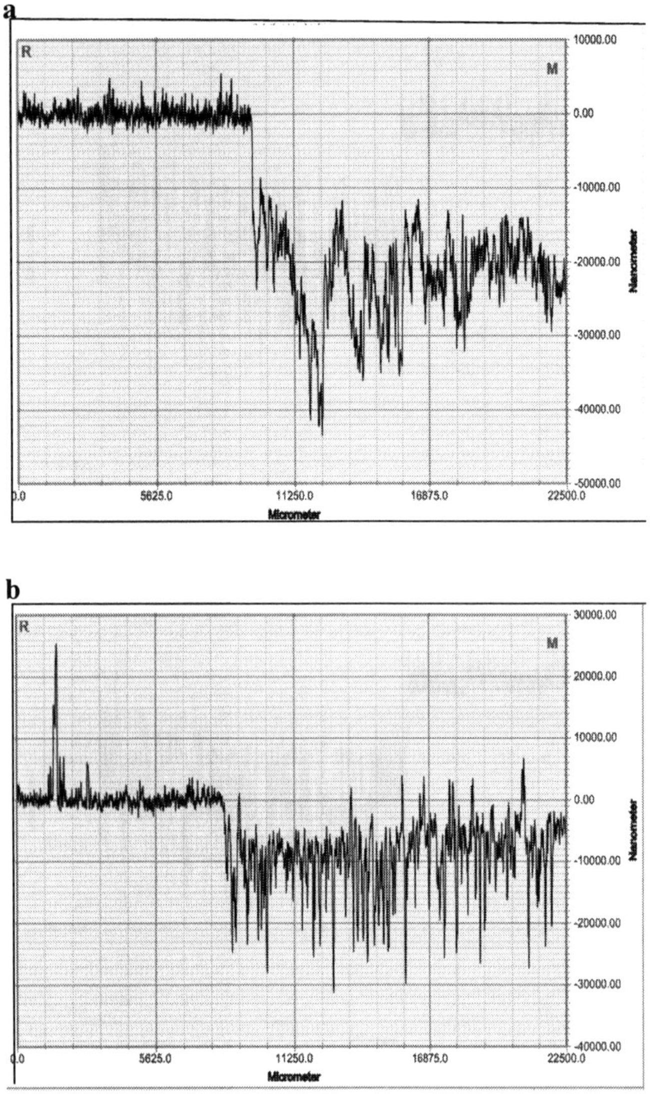

Fig. 3. Comparison of the variation of the surface roughness with HAE coating removal (in OB optimal pickling bath). a) on the WE43 alloy - HAE coated; roughness on HAE coating: RA ≈ 1.0 µm. roughness on the WE43 alloy after removal of HAE coating: RA ≈ 4.4 µm, thickness of the HAE coating removed d ≈ 16.5 µm. b) on the RZ5 alloy - HAE coated; roughness on HAE coating: RA ≈ 1.3 µm, roughness on the RZ5 alloy after removal of HAE coating: RA ≈ 4.1 µm, thickness of the HAE coating removed: d ≈ 9.2 µm.

effect on the surface roughness as can be see in Fig. 1. – 4. It was observed that the roughness increases slightly from ≈ 1.0 µm to ≈ 4.0 µm with ~ 20 µm of HAE material removal in the optimal bath. Moreover, SEM and EDS analyses carried out after HAE material removal (Fig. 5 and 6) show the absence of any intergranular attack on the pickled alloy surface. After HAE coating removal, we observe the presence in the surface of a thin film mainly containing ZrOx (Fig. 6). So, the absence of any attack on the pickled surface area can be attributed to the protective effect of $ZrO(NO_3)_3$ (added in the pickling bath), that forms a thin film of ZrOx on the pickled alloy surface.

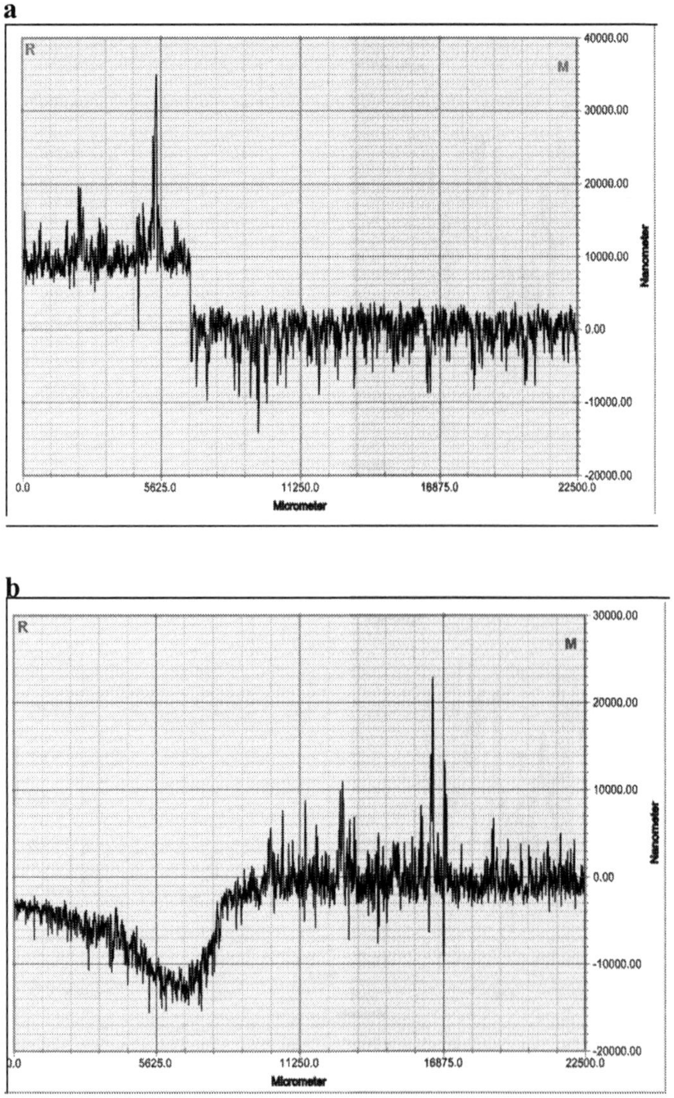

Fig. 4. Variation of the surface roughness with HAE coating removal (in chromic acid bath) on the WE43 alloy - HAE coated; a) HAE / pickled WE43 interface, roughness on HAE coating: RA ≈ 2,1 µm, roughness on the WE43 alloy after removal of HAE coating: RA ≈ 1.7 µm, thickness of the HAE coating removed: d ≈ 14.4 µm. b) HAE / polished WE43 interface, roughness on the WE43 alloy after mechanical polishing RA ≈ 1,0 µm, thickness of the HAE film at HAE / mechanical polished WE43 interface, d = 11.4 µm.

Composition characterization by XPS

The XPS spectra of WE43 after optimal bath (OB) pickling (Fig. 7.) show the presence of metallic Mg, of Mg, Zr and Y oxides and hydroxides, Mg and Zr fluorides and of low amount of addition element Nd oxides. We note the absence of P 2p pick at≈ 133 – 134 eV.

Surface morphology and elemental composition

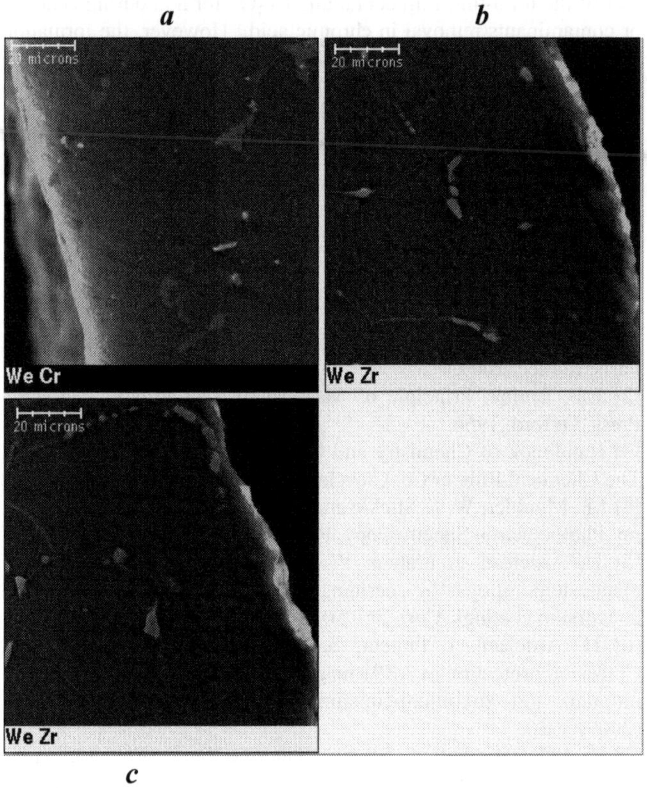

Fig. 5. SEM of WE43 alloy after HAE material removal in: a) chromic acid and b), c) optimal bath.

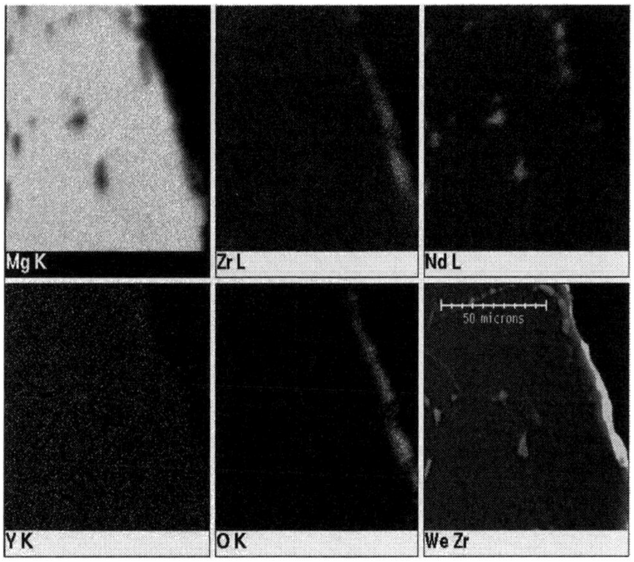

Fig. 6. Elemental composition (Mg, Zr, Nd, Y, O, SEI) by EDS analyses of WE43 alloy after HAE material removal in optimal bath.

The XPS spectra of WE43 after chromic pickling (Fig.8.) show the presence of metallic Mg, of Mg, Cr and Y oxides and hydroxides. We observed also the F 1s peak at$\approx$ 687 eV , of high intensity, attributed to the presence of MgF_2. We note the peak P 2p localised at $\approx$ 133 eV, however, its intensity is low. This peak is attributed to the presence of **phosphates** at the surface of the pickled alloy. We observed also small amounts of Nd oxides. We conclude that the removal of the HAE coating was incomplete by chromic acid pickling, contrary to OB pickling (no phosphates at the surface). Moreover, Nd and Zr are observed after removal of HAE coating in OB bath contrary to sample pickled in chromic acid. As previously observed by SEM-EDS, OB bath pickling leads to the formation of a protective layer (containing mainly ZrOxF compound) and to a better removal of the HAE coating with reference to the pickling in chromic acid.

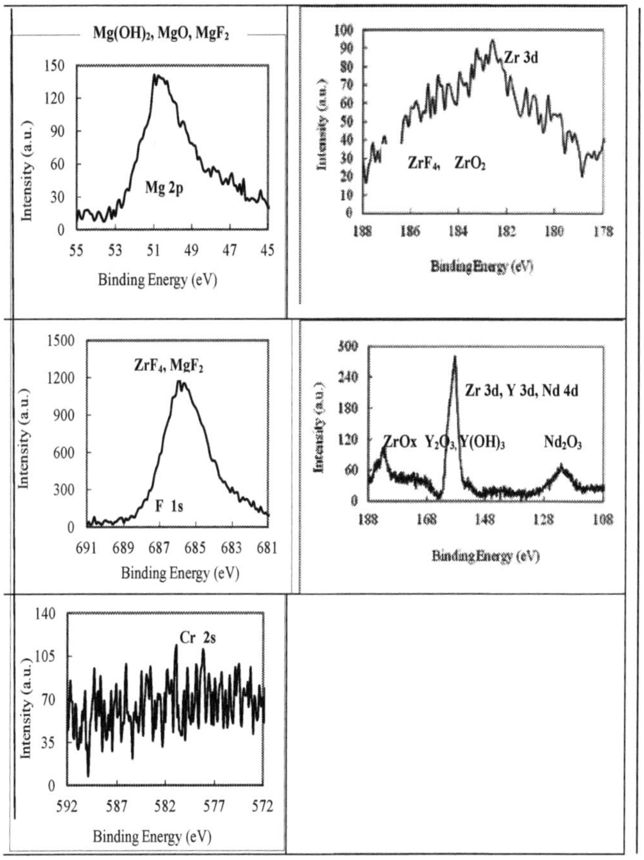

Fig. 7. XPS spectra (Mg 2p, **Zr 3d, F 1s**, Y 3d , Nd 4d, Cr 2s) on WE43 surface after OB pickling.

Conclusion

A new pickling process allowing the removal of HAE and DOW 19 coatings (with the thickness between 1 and 24 µm) and the cleaning of magnesium alloys was designed. The process is based on the chemical dissolution of the coatings into a optimal bath based on tartaric and nitric acids, surfactants (such as sodium lauryl sulphate, moderators (HMTA, NH_4F), and an additive (such as KOH) for the pH correction.

The addition of the inorganic inhibitors (such as $ZrO(NO_3)_3$) allows to obtain a good protection against corrosion of the pickled surface area, during the removal of thicker coating area. These inhibitors favor the formation, on the pickled surface, of a thin film containing mainly ZrO_xF compound which plays a key role in the anti-corrosion efficiency of the optimal bath. This effect is similar to

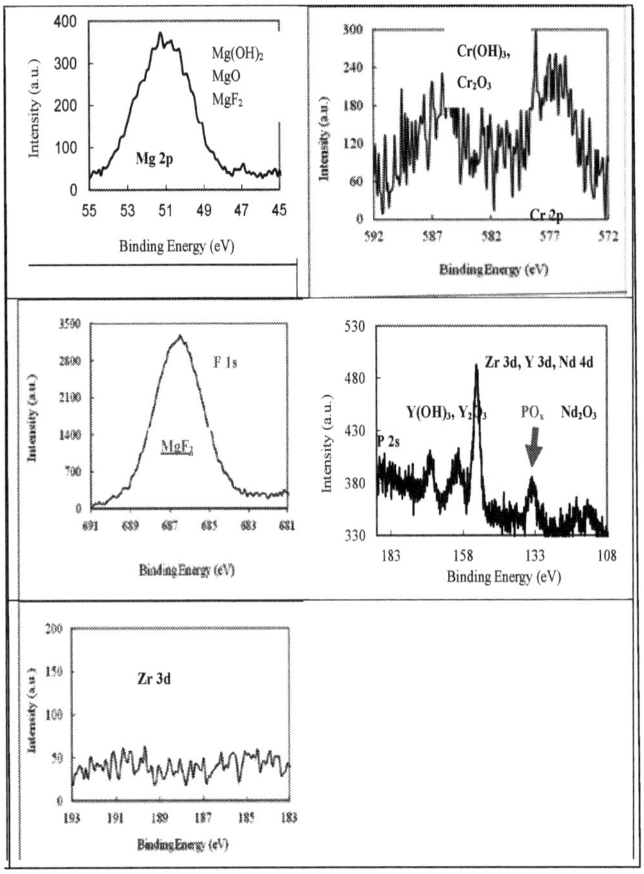

Fig. 8. XPS spectra (Mg 2p, Cr 2p, F 1s, Zr 3d, Y 3d , P 2p, Nd 4d) on WE43 surface after chromic acid pickling.

that of the ultra-thin film containing Cr_2O_3 formed during coatings or contaminants removal in chromic acid. However, the formation and the thickness of this film could be decreased by the decreasing of the inhibitor concentration and of the pH of the pickling bath.
The new green treatment provides an alternative to the pickling treatments based on chromic acid, used for Mg alloys (and other alloys).

References

[1] REACH S. Abdesslam, REACH et les composés du Chrome VI, 1ère Conférence Internationale « Le traitement des alliages légers », Paris, France, 2009.

[2] E.F. Emley, Principle of Magnesium Technology, Pergamon Press, Oxford, 1966.

[3] Handbook of Chemistry and Physics, 45th edition (1964-1965), The Chemical Rubber Co. Cleveland, Ohio.

[4] J.F. Moulder, W. F. Stickle and K. D. Bomben, Handbook of X-ray Photoelectron Spectroscopy, Perkin Elmer, Eden Praire (1992).

[5] H. Ardelean, I. Frateur, P. Marcus, Corrosion protection of magnesium alloys by cerium, zirconium and niobium-based conversion coatings, Corr. Sci. 50 (2008) 1907.

[6] H. Ardelean, I. Frateur, S. Zanna, A. Atrens, P. Marcus, Corrosion protection of AZ91 magnesium alloy by anodizing in niobium and zirconium-containing electrolytes, Corr. Sci., 51 (2009) 3030.

[7] H. Ardelean, Antoine Seyeux, Sandrine Zanna, Frédéric Prima, Isabelle Frateur, Philippe Marcus, Corrosion processes of Mg-Y-Nd-Zr alloys, submitted to Corr. Sci., (2012).

Magnesium Technology 2013
Edited by: Norbert Hort, Suveen N. Mathaudhu, Neale R. Neelameggham, and Martyn Alderman
TMS (The Minerals, Metals & Materials Society), 2013

GALVANIC CORROSION OF MG-ZR ALLOY AND STEEL OR GRAPHITE IN MINERAL BINDERS

David Lambertin, Adrien Rooses, Fabien Frizon
CEA, DEN, DTCD, SPDE, LP2C F-30207 Bagnols Sur Ceze, France

Abstract

The dismantling of UNGG nuclear reactor generates numerous nuclear wastes such as fuel decanning commonly composed of Mg-Zr alloy. A conditioning strategy consists in encapsulating these wastes into a hydraulic binder in a suitable state for storage. The eventual presence of steel and graphite accompanying the magnesium wastes could imply corrosion by galvanic coupling. This work is an experimental investigation of the galvanic coupling between Mg-Zr alloy and steel or graphite using ZRA electrochemical method in Portland cement or geopolymer pastes. The lowest corrosion activity of magnesium alloy while coupled to graphite or steel cathode has been observed in geopolymer pastes. Indeed, in this binder, an efficient corrosion protection of the magnesium alloy maintains the galvanic current very low during all the hardening process. In geopolymer paste, current densities of anodised Mg-Zr alloy is not dependent of the cathode/anode surface ratio in the range of 0.1 to 5 due to the dominance of the anode resistance.

Keywords: galvanic corrosion, Mg-Zr alloy, cement, geopolymer

Introduction

This study takes part of magnesium waste processing project of UP1 plant of Marcoule site in France. The dismantling of Uranium Natural Graphite Gas nuclear reactor generates a large volume of fuel cladding materials. Nuclear wastes such as Mg-Zr alloy fuel cladding, graphite and steel bars have been stored conjointly in tanks, and a solution for their processing is their immobilisation in a hydraulic binder [1]. Mg and its alloys act as anode in contact with other metals, and could suffer from a severe galvanic corrosion attack. Previous works [2] have evidenced the existence of intermediate Mg^+ which leads to the formation of Mg^{2+} and H_2. Therefore, in the case of magnesium alloy coupled to another metal such as steel, the hydrogen can be produced either by the water reduction or by AHE (Anodic Hydrogen Evolution) [3] on magnesium alloy (1).

$$Mg^+ + H_2O \rightarrow Mg^{2+} + OH^- + \frac{1}{2} H_2 \quad (1)$$

The knowledge of the corrosion activity while immobilizing the wastes is essential to avoid any damage of the storage container on the long run. If corrosion studies have been carried out on anodised magnesium in solutions [4-7], no data on the corrosion behaviour of anodised magnesium alloy in hydraulic binders such as Portland cement or geopolymer is available in the literature.
A theoretical and conceptual relation of the galvanic current i_{galv} between an anode and a cathode is proposed in the literature by:

$$i_{galv} = (E_c - E_a)/(R_c + R_a + R_s + R_m) \quad (2)$$

where $E_c - E_a$ is the electrical potential difference of the cathode and anode, R_c and R_a their resistances, R_s the resistance of the solution between the electrodes and R_m the resistance of the metallic path between the cathode and the anode. R_m is normally considered negligible.

In the case of Graphite|Mg and Steel|Mg couples, the electrical potential difference is usually high, whatever the media. However, the binder paste media evolve chemically and physically, due to the dissolution of particles and to the curing processes respectively. The level of the solution resistance R_s could therefore change drastically. Furthermore, aggression or protection phenomenon on the surface of the magnesium anode may also modify the values of R_a during the immersion in the binder pastes.

The purpose of this study is to compare and understand the galvanic corrosion behaviour of Mg-Zr alloy coupled with graphite or steel in different Portland cement and geopolymer pastes to propose a suitable embedding binder for these wastes.

Experimental

The two selected Portland Cements are CEM I 52.5 N CE CP2 White from Calcia Cruas factory (WPC) and CEM I 52.5 PM ES CP2 from Lafarge Le Teil factory (OPC). Their compositions are reported in Table 1. The cement pastes were obtained by mixing the cement powder with water, with a water to cement mass ratio of 0.40.

Table 1 – Compositions of Portland cements WPC and OPC.

Portland Cement type		WPC	OPC
Clinker mineral composition	C_3S	73.5 %	67.9 %
	C_2S	13.5 %	19.9 %
	C_3A	12.0 %	4.1 %
	C_4AF	1.0 %	7.9 %
Gypsum content %wt	$CaSO_4{\cdot}H_2O$	5.5%	2.7%

The geopolymer binders (Na-GEO) were activated in waterglass alkali solutions and formulated with a Al:Si:Na:H_2O molar ratio of 1:1.8:1:11.5. Waterglass activating solutions were prepared by dissolving sodium hydroxide pellets (Prolabo, Rectapur, 98 %), and amorphous silica provided by Rhodia (Tixosil 38) in Milli-Q water. Solutions were then stirred for 24 h. Metakaolin was purchased under the brand name of Pieri Premix MK from Grace Construction Products. The oxide composition of metakaolin determined by X-ray Fluorescence (XRF) can be found in Table 2. Geopolymer pastes were then prepared by mixing metakaolin and waterglass solution with a standard laboratory mixer and then directly used for electrochemical measurements.
Small pieces of a few square centimetres of non-radioactive Mg-Zr alloy fuel decanning from UNGG reactors stored for long time under ambient atmosphere were used as electrodes. The weight composition of this alloy is Mg 99.48 %, Zr 0.49 % and Mn 0.016%.

Table 2 – Oxide composition of metakaolin.

Oxide	SiO_2	Al_2O_3	Fe_2O_3	K_2O	TiO_2
% weight	54.4	38.4	1.3	0.6	1.6

Steel and graphite electrodes were respectively made of 2 mm wire of SAE 1020 steel and 6 mm pure graphite cylinder.

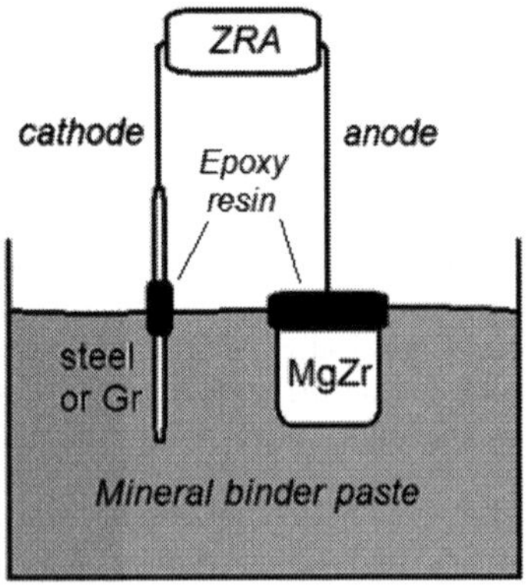

Fig. 1. Test arrangement diagram.

Zero-Resistance Ammeter test (ZRA) has been used to register the galvanic current density (J_{galv}) of coupled metals immersed in a mineral binder paste versus time. Galvanic current densities were calculated as a function of the magnesium surface. The experiments were performed with a Princeton Applied Research VersaSTAT 4 model. The electrodes distance has been kept equal to 2 cm as described in Fig 1. As binder pastes are changing media, the experiment duration has been fixed at 14 hours to cover the dissolution and curing steps.

Results and discussions

Galvanic corrosion of Mg-Zr alloy coupled to graphite and steel in Portland cement and geopolymer pastes.

Galvanic currents densities of Mg-Zr alloy coupled to graphite and steel immersed in WPC, OPC and Na-GEO are presented in Fig. 2. The surface ratio of graphite or steel to Mg-Zr (called S) has been kept equal to 0.1.

The current densities of magnesium alloy in the binder pastes present quite same evolutions either coupled to graphite or steel.

In OPC and WPC pastes, for both couples, anodised Mg-Zr alloy first displays an increase of current and then progressively decreases. Indeed, as anhydrous cement compounds are dissolving in the aqueous solution, the concentrations of aggressive soluble species increase and enhance metal corrosion. Theoretically, this can be explained by a decrease of both solution and anode resistances R_s and R_a while the electrical potential difference remains unchanged. When the point of saturation concentration of ionic constituents is reached, a maximum of magnesium current density is observed [8]. Then, the setting of the binder occurs and the consumptions of water and dissolved elements lead to a progressive decrease of the corrosion activity. Indeed, from then on, the solution resistance R_s is rising continuously. It has been noticed that the level of anodic current density is more important in OPC binder than in WPC binder. This difference can be explained by the effect on R_s and R_a of the cement constituents such as the clinker composition, the gypsum content and the use of different adjuvants part of cement making process.

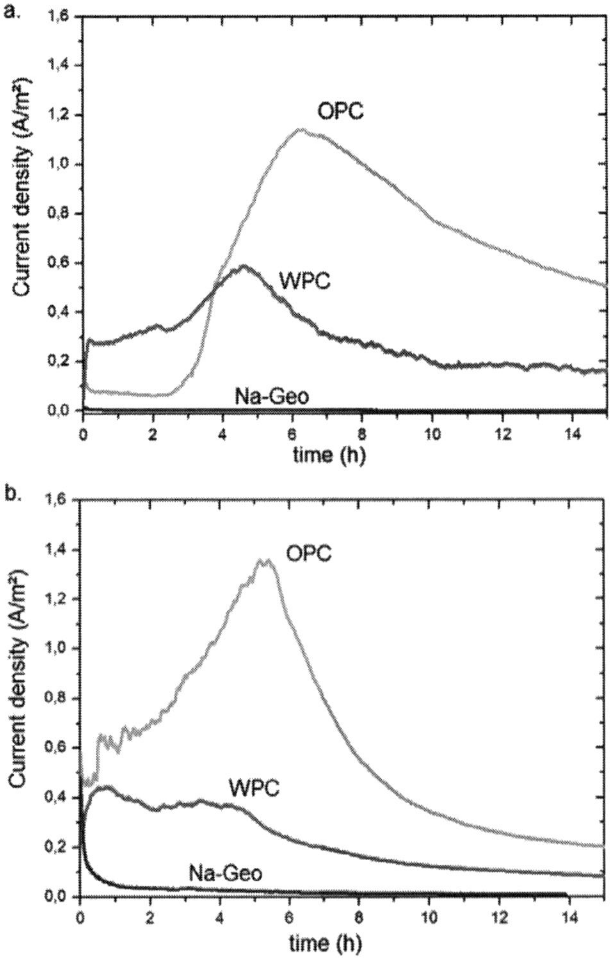

Fig. 2. Galvanic current densities of Mg-Zr alloy coupled to graphite (a) and steel (b) immersed in OPC, WPC and Na-Geo binder pastes for a $S_{cathode}/S_{Mg-Zr}$ ratio equal to 0.1.

In Na-geo paste, the magnesium current densities exhibit a totally different behaviour than in Portland cements. The anodic current evolution presents a sharp decrease as soon as the electrodes are immersed in the Na-geo paste. This important decrease of i_{galv} is a result of the formation of a efficient corrosion protection on the surface of the magnesium alloy which considerably amplifies R_a. The anodic protection could be a result of the formation of layers such as silicates [9], quickly after immersion in the geopolymer paste. For both cases, galvanic current densities remain very low over time and no not evolve during the material setting. This underlines that the R_s evolution is negligible towards the fast rise of R_a and that the latter leads by itself the electrochemical system. Nevertheless, the decrease seems to be faster with graphite than with steel cathode because of the difference of the cathode potentials E_c.

As a conclusion of the very low corrosion activity of anodised magnesium alloy in Na-Geo pastes, sodium geopolymer binders tend to be an appropriate matrix for the waste embedding in order to minimise the production of hydrogen gas related to corrosion.

Nevertheless, as the metal wastes could have been positioned randomly in the tanks, the contact surface ratio between magnesium alloy and the other metals could change significantly. The next part of this study will assess the influence of the surface ratio on the corrosion activity of Mg-Zr alloy coupled to steel.

Influence of the metal surfaces ratio on the corrosion of Mg-Zr alloy coupled to steel in geopolymer paste.

In a medium, galvanic current densities depend on the nature of the connected metals but it is also related to the cathode/anode surface ratio. Indeed, an increase of the cathode/anode surface ratio lead to a decrease of the values of R_c and consequently to a rise of i_{galv}. It is well known that the worse combination is to have a high area ratio of cathode to anode. Indeed, G. Song et al. [10] have evidenced the linear increase of the anodic current density of Steel|Mg couple while increasing cathode/anode area ratio in 5% NaCl aqueous solutions.

The current densities of Mg-Zr alloy coupled to steel in Na-Geo paste with different Steel/Mg-Zr surface ratios (*S*) are presented in Fig. 3. On the one hand, the curves of anodic current density for the four surface ratios from 0.1 to 5 remain very close. This points out that the galvanic current density of Steel|Mg-Zr in geopolymer pastes is not dependent on the surface ratio of steel to magnesium. On the other hand, for all surface ratios, anodic current densities are very low and unaffected by the material hardening as already seen in the previous part. This two observations illustrate not only the high values of the anode resistance R_a related to the magnesium surface protection, but also the negligibility of evolution of R_c whatever the cathode/anode surface ratio in the range of 0.1 to 5.

Conclusions

- Na-Geo paste is the binder that implies the lowest corrosion activity of magnesium alloy while coupled to graphite or steel cathode. The galvanic current remains very low during the hardening process thanks to the efficient corrosion protection once immersed in the geopolymer paste.
- In geopolymer paste, the current density of anodised Mg-Zr alloy is not dependent to the cathode/anode ratio in the range of 0.1 to 5 due to the dominance of the anode resistance R_a in the electrochemical system.
- Na-Geo paste appears to be a suitable encapsulation matrix for magnesium wastes

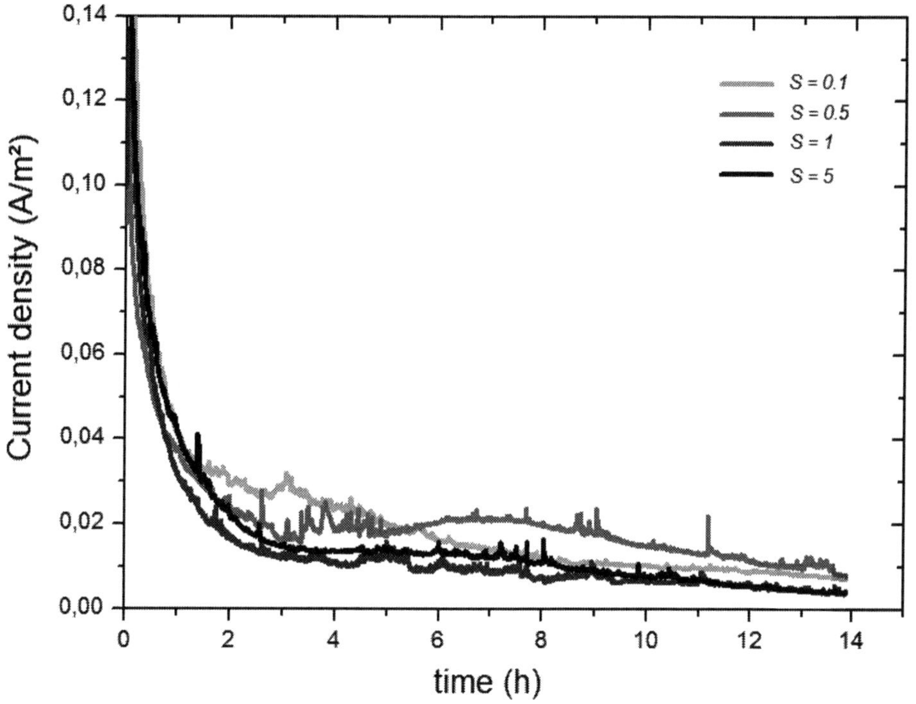

Fig. 3. Galvanic current densities of Mg-Zr alloy coupled to steel with cathode/anode surface ratios of 0.1, 0.5, 1 and 5 in Na-Geo paste

References

[1] D. Lambertin, F. Frizon, F. Bart, Surface and Coatings Technology, 206 (2012) 4567-4573.

[2] G.L. Song, A. Atrens, Advanced Engineering Materials, 1 (1999) 11-33.

[3] G.L. Song, in, Woodhead Publishing in Materials, USA, 2011.

[4] S. Bender, J. Goellner, A. Heyn, E. Boese, Mater. Corros., 58 (2007) 977-982.

[5] J.X. Jia, G.L. Song, A. Atrens, Corrosion Sci., 48 (2006) 2133-2153.

[6] D. Lambertin, A. Blachère, F. Frizon, F. Bart. W.H. Sillekens, S.R. Agnew, M.R. Neelameggham, M.O. Pekguleryuz (Eds.) Magnesium Technology 2011.

[7] Z.M. Shi, J.X. Jia, A. Atrens, Corrosion Sci., 60 (2012) 296-308.

[8] J.W. Bullard, H.M. Jennings, R.A. Livingston, A. Nonat, G.W. Scherer, J.S. Schweitzer, K.L. Scrivener, J.J. Thomas, Cement and Concrete Research, 41 (2011) 1208-1223.

[9] J.E. Gray, B. Luan, Journal of Alloys and Compounds, 336 (2002) 88-113.

[10] G. Song, B. Johannesson, S. Hapugoda, D. StJohn, Corrosion Sci., 46 (2004) 955-977.

Magnesium Technology 2013
Edited by: Norbert Hort, Suveen N. Mathaudhu, Neale R. Neelameggham, and Martyn Alderman
TMS (The Minerals, Metals & Materials Society), 2013

The influence of Mg-Zr master alloy microstructure on the corrosion of Mg

D.S. Gandel[a,b], M.A. Easton[a,b], M.A. Gibson[a,c], T. Abbott[a,d] and N. Birbilis[a,b]
[a]*CAST Cooperative Research Centre*
[b]*Department of Materials Engineering, Monash University, Clayton VIC. 3800, Australia*
[c]*CSIRO Process Science and Engineering, Clayton, VIC, 3168, Australia*
[d]*Magontec Limited, Sydney, NSW, 2000, Australia*

Keywords: Magnesium, Zirconium, Polarisation, Weight loss, Microstructure

Abstract

In this study, sixteen Mg-Zr alloys were produced to investigate the role of Zr on corrosion of Mg. Alloys were produced using two different commercial Mg-Zr master alloys commonly used for grain refining Mg, but which contain different Zr particle size distributions. It is seen that the master alloy with a smaller Zr particle size leads to an alloy containing more Zr in solid solution. The ratio of Zr in solid solution and in particle form was observed to have a marked effect on the corrosion of Mg.

Introduction

Zirconium (Zr) has a low solid solubility in magnesium (Mg) of 0.73 at.% and it is acknowledged that there are no intermetallic phases that form between Mg and Zr [1]. Zr is commonly and predominantly added to Mg in order to refine the grain size [2-4]. The reduction in grain size in Mg alloys provides greatly improved casting quality and increased mechanical properties [5-7]. As Mg has a Hall-Petch coefficient of 280-320 MPa/µm, the reduction of the grain size in Mg alloys significantly increases the strength of the alloy and allows a greater control over the alloy texture [6, 8, 9]. Thus, Zr is used in the commercial alloys WE43, ZE41 and ZK60. Moreover, newer Mg alloys, such as AM-SC1 and Elektron 21, have recently been developed which also contain Zr for the purpose of grain refinement [10-13]. Given the considerable present interest in Mg alloys a better understanding of the influence of Zr additions on the corrosion of Mg is necessary.

There are reported cases where the addition of Zr to Mg alloys has decreased the corrosion rate through the removal of Fe impurities by the Zr addition [4, 14-16]. Fe impurities in Mg are scavenged by Zr in the melt due to the Zr and Fe combining to form insoluble precipitates, usually Fe_2Zr, which can settle to the bottom of the melt crucible owing to a difference in density. Thus, Mg-Zr alloys are generally considered 'high purity', with commercial Mg alloys that incorporate Zr, usually containing under 50 ppm of Fe [14].

Whilst the removal of Fe by Zr is beneficial in reducing the corrosion rates of Mg alloys, the addition of Zr to the Mg matrix can cause its own corrosion related issues. It has been reported that when Zr is not homogeneously dispersed within the Mg matrix, the corrosion rate can increase compared to a more even distribution of smaller Zr particles [17]. Furthermore, excess Zr in Mg increases the amount of elemental Zr particles that form in the matrix, which is detrimental for corrosion [18]. These elemental Zr particles have been reported to cause micro-galvanic couples with the surrounding Mg matrix and the disruption of the protective oxide film on the alloy surface [19]. As such, further investigation into elucidating the role of Zr in corrosion of Mg is of considerable importance to improve both current and future commercial Mg alloys containing Zr.

In this study, the role of Zr additions on the increasing corrosion rate of Mg and the fundamental effects of different Zr particle sizes and size distributions, added from different Mg-Zr master alloys, is examined. The effect on the corrosion rates for given Mg-Zr alloy mixtures made from two different Mg-Zr master alloys, i.e. Microzir (formally AM-Cast) and Zirmax, are reported.

Experimental Methods

Alloy production and characterization

Sixteen Mg-Zr alloys were produced by adding different amounts of Mg-Zr master alloys to commercially pure Mg. The Mg-Zr master alloys used herein were Microzir, supplied by Magontec, and Zirmax supplied by Magnesium Elektron. Melting was carried out in a resistance furnace using AM-Cover® as a cover gas. The commercially pure Mg was initially melted in a steel crucible at 700°C to which small amounts of the master alloys were added to achieve specific Zr contents in the final castings. The melt was poured into a graphite coated cast iron mould and the ingots were allowed to air cool. Stirred samples were vigorously mixed immediately prior to casting; where as unstirred specimens were held stationary for 20 minutes prior to casting.

The compositions of the alloys were determined independently via inductively coupled plasma atomic emission spectroscopy (ICP-AES) (Spectrometer Services, Coburg, Australia). The precise composition of the experimental alloys is given in Table 1. The values for Zr content in solid solution and overall Zr content were measured by via an acid pre-treatment procedure prior to ICP-AES analysis as per Crawley [20]. The soluble Zr content was determined by dissolving the sample in a 10% HCl solution. The total Zr content was determined by dissolving the specimen in a 50% HCl – 6% HF solution.

Several alloys were examined via scanning electron microscopy. Prepared Mg-Zr alloy specimens and both Mg-Zr master alloys were polished to a 1µm diamond paste finish and then imaged using a JEOL 7001F SEM in back scattered electron (BSE) mode. The microscope was equipped with energy dispersive x-ray spectroscopy (EDX) (Oxford Instruments X-Max 80 detector).

Electrochemical and corrosion testing

Specimen surfaces were ground to a 2000 grit surface finish. A 3-electrode electrochemical flat-cell with an exposed sample area of 1 cm^2 was used in conjunction with a 0.1M NaCl electrolyte. A VMP 3Z potentiostat was used, with potentiodynamic polarisation conducted at 1 mV/s. Prior to polarisation, the samples were conditioned for ten minutes at open circuit to ascertain a stable potential. The polarisation curves were used to determine i_{corr} (via a Tafel-type fit) using EC-Lab software.

Tafel-like fits were executed by selecting a portion of the curve that commenced >50mV from E_{corr}, and i_{corr} was estimated from

the value where the fit intercepted the potential value of the true E_{corr}. Polarisation testing was also able to visually reveal comparative information related to the kinetics of both the anodic and cathodic reactions of the various Mg specimens. Each sample was tested five times and an average result was determined.
Additionally, weight loss was determined by exposing alloy samples via immersion in 0.1M NaCl for a period of 24 hours. Subsequent corrosion products were removed by a light scrubbing following a 3s immersion in 15% HNO_3. The mass loss was determined on three unique samples and an average result was determined and reported.

Results and Discussion

Alloy characterization

All alloys examined in this study have Zr additions below the levels expected for grain refinement to take place. Stirring the Mg melt prior to casting was found to increase the overall Zr content in binary Mg-Zr alloys (Table 1). However, while the unstirred alloys had a lower average total Zr content, they did have similar percentages of Zr in solid solution compared to the stirred alloys. The lower total Zr content in the unstirred alloys compared to the stirred alloys is most likely due to Zr particles contained in the unstirred Mg melt settling down to the bottom of the crucible during the extended holding time prior to casting. This would result in less Zr in suspension in the Mg melt at the point in time where the molten liquid is poured from the crucible to the mould while making the Mg-Zr ingot.
The ICP-AES compositional data (Table 1) confirm that the Mg-Zr alloys made from the Microzir master alloy have on average a much higher percentage of Zr in solid solution than the Mg-Zr alloys made from the Zirmax master alloy. The stirred alloys made with the Microzir master alloy averaged between 36% to 47% of the total Zr content in solid solution compared to the stirred alloys made with the Zirmax master alloy which only averaged between 10% to 25% of the total Zr content in solid solution. This difference between the Microzir and Zirmax samples does not appear to be caused by a stirring effect due to the similarities in percentage of Zr in solid solution in both the stirred and unstirred conditions. The difference in microstructure is caused by the different sizes of Zr particles introduced by the Mg-Zr master alloys (Figures 1 (A & B)) [21].
Both the Zirmax and Microzir Mg-Zr alloys (Figures 2 (A, & C)) contain Zr particles embedded in the Mg matrix. EDX mapping elucidates the difference in microstructure of the two alloy groups due to the difference in Zr particles size introduced by the different Mg-Zr master alloys. Apart from the greater number of Zr particles present, there is a large amount of Zr in solid solution in the Mg matrix of the Microzir sample (Figure 2 (B)). The Zirmax alloy (Figure 2 (D)) does not appear to have much Zr in solid solution, despite having a similar overall Zr content to the Microzir alloy. The finer particle sizes introduced by the Microzir master alloy has allowed a greater amount of Zr to go into solid solution rather than forming insoluble Zr particles in the Mg matrix [22]. Due to the very low Fe content in both of the Mg-Zr alloys, no discernable Fe peaks of note were seen while performing EDX testing on the selected specimens.
There is a noticeable difference in the Zr particle structure between the two Mg-Zr alloy groups. The Microzir alloys generally have smaller Zr particles that cluster together, compared to the Zirmax alloys that have larger and more isolated individual Zr particles. Again, the difference in particle size between the two master alloys is likely to be the cause for the different morphologies in the Zr particles present in the matrix. Vigorous stirring has been shown to minimise the presence of large Zr particle clusters, however, larger individual Zr particles need longer holding times to let the particles settle out of the Mg melt effectively [23]. The extended holding times required to remove the larger individual Zr particles have been shown to reduce the overall Zr content in the Mg alloy. The EDX map of the selected Microzir and Zirmax specimens corroborates the compositional analysis which indicate that the Microzir alloys have a much higher percentage of Zr in solid solution compared to the Zirmax alloys.

Assessment of Zr in solid solution

There is a higher average Zr content in the stirred Mg-Zr alloys compared to the unstirred alloys (Figure 3 (A & B)). This is due to the fact that more Zr particles are able to settle to the bottom of the crucible during processing of the unstirred samples, which were held motionless for an extended period of time prior to casting. It can be seen that there is a trend where the overall percentage of Zr in solid solution increases as the overall Zr content increases in the Microzir containing alloys (Figure 3 (A)). The Zirmax alloys, however, show the opposite effect, in that as the overall Zr content increases the percentage of Zr in solid solution decreases. Despite the reduced uptake of Zr in Mg overall, these trends are also seen in the unstirred Mg-Zr alloys (Figure 3 (B)).
The amount of Zr in solid solution is influenced by the difference in Zr particle size characteristic of the two master alloys. The Zirmax master alloy has a larger and more variable Zr particle size compared to the Microzir master alloy, which has a more even distribution of smaller Zr particles [22, 23]. The finer Zr particle size in the Microzir master alloy, with most Zr particles ranging between 1 and 5 μm in size, allows for more Zr to dissolve into solid solution. This finer distribution is highly desired due to the beneficial effects favouring the alloying effectiveness of the Mg-Zr master alloy and the ensuing grain-refining effect of Zr [4, 24-26]. Moreover, with more Zr in solid solution, less elemental Zr particles will be retained in the Mg matrix to contribute to corrosion related problems and increased corrosion rates.

Effect of Zr alloying additions on Mg corrosion rates

I_{corr} increases in the Zirmax containing alloys as the Zr content in solid solution increases (Figure 4 (A)). The Zirmax alloys, however, typically have only low levels of Zr in solid solution with increasing total Zr content, with only up to 0.012 wt.% Zr observed in solid solution at best. The Microzir alloy, with over 0.1 wt.% Zr in solid solution, displays an initial slight decrease in i_{corr} before increasing, however, the later increase does not exceed the initial measured i_{corr} value.
The i_{corr} values for the Zirmax alloys appear to be quite variable with very little increase in Zr content in solid solution (Figure 4 (B)). The Microzir alloys, with a slightly higher percentage of Zr in solid solution, do have some semblance of a decreasing corrosion rate. There is initially a drop in the measured i_{corr} value as the Zr content in solid solution is increased. However, the overall Zr content is still quite low (Table 1).
In the Microzir alloys, i_{corr} decreases, then increases slightly as the overall Zr content increases (Figure 4 (C)). The higher percentage of Zr in solid solution allows for higher overall Zr contents in Mg

before accelerating the corrosion rate. The Microzir-4 alloy contains five times as much Zr in solid solution as the Zirmax-4 alloy with a similar total Zr content. The greater amount of Zr present as elemental Zr particles in the Zirmax alloys thus contributes to increased i_{corr} values than the Microzir produced Mg-Zr alloys.

The stirred and unstirred samples do not have a similar measured i_{corr} trend. There is an overall decrease in i_{corr} for the unstirred Microzir alloys and a slight overall increase for the unstirred Zirmax alloys (Figure 4 (D)) compared to stirred alloys of a similar content. The greater percentage of Zr in solid solution at a lower overall Zr content in the unstirred Microzir alloys compared to the stirred alloys may account for this difference as a greater amount of Zr has gone into solid solution. The unstirred Zirmax alloys have a lower percentage of Zr in solid solution compared to the stirred alloys with a similar Zr content. The greater amount of Zr particles embedded in the matrix, rather than dissolving into solid solution, is the likely cause of the increase i_{corr} values measured.

Influence of Zr on the corrosion reaction kinetics of Mg

As Zr is introduced into Mg there is an increase in the anodic reaction kinetics whilst the cathodic reaction kinetics are only slightly increased for the Zirmax containing alloy. The changes in the reaction kinetics when Zr is added can be seen in Figure 5. The increase in anodic and cathodic reaction kinetics can be observed through the shifts in the anodic and cathodic branches of the polarisation curves compared to pure Mg. At the alloying content limit of the Mg alloys examined (Table 1), Zr is still below the solid solution limit in Mg according to the phase diagram [1], with a large percentage of the total Zr content in the Mg-Zr alloys in solid solution.

The data reveals that Zr is an anodic activator of Mg and increases the corrosion rate of Mg when added. This anodic activation effect is observed even at very low levels of Zr and does not seen to be influenced by the initial microstructure of the Mg-Zr master alloy used in the manufacturing of the Mg-Zr alloys. This data supports previous work that the initial addition of Zr into Mg increases the anodic kinetics before increasing the cathodic reaction kinetics via micro-galvanic coupling with large Zr particles embedded in the Mg matrix [27]. The increase in anodic kinetics is the key electrochemical reason for the increased corrosion rates seen in the Mg-Zr alloys produced herein. As such, the increase in the cathodic reaction kinetics in the Zirmax containing Mg-Zr alloy can be explained by the increased number of large Zr particles present in the microstructure due to less Zr dissolving into solid solution.

Comparison between Microzir and Zirmax alloys via long corrosion testing

There is a noticeable trend linking the overall Zr content and the mass loss from immersion results (Figure 6). The relationship shows that as the Zr content increases the mass loss also increases per unit time. The data is presented in its native units to avoid problems or errors from assuming uniform corrosion for the immersion weight loss data data.

There is a divergence in the weight loss values for the two Mg-Zr alloy groups. The Microzir alloys not only have a higher average weight loss than the Zirmax alloys but they also have a larger calculated standard error. As the two alloy groups have a similar overall Zr content (Table 1), the most likely cause for the increased weight loss observed in the Microzir alloys is the higher Zr content in solid solution.

General discussion

The increase in Zr content in Mg has a great influence on the anodic corrosion characteristics of Mg. As discussed previously, however, there is no large overall increase in i_{corr} with increasing Zr content in the Microzir alloys, Figure 6 shows that when there is a greater amount of Zr in solid solution there is an increase in the corrosion rate according to the longer term immersion weight loss tests.

The higher levels of Zr in solid solution lead to a greater anodic activating effect in the Mg matrix. This is an indicator that there is a greater anodic activation by the higher amount of Zr in solid solution increasing the long-term corrosion rates. With a higher Zr content in solid solution there should be a greater amount of generalised corrosion attack across the surface area of the alloy as less Zr particles would be present in the matrix to serve as corrosion initiation sites via micro-galvanic coupling.

Despite the desire for a greater content of Zr in solid solution, which is beneficial for grain refinement [27], even low levels of Zr appear to increase the overall corrosion of Mg. The difference in the size of the Zr particles introduced to Mg does appear affect the way in which Zr affects the long-term corrosion rate. Higher percentages of Zr in solid solution for a given total Zr content increase the long-term corrosion rates of Mg, where as lower percentages of Zr in solid solution for a similar Zr content will have a greater impact on the short-term electrochemical corrosion behaviour of Mg.

While other studies have suggested that the addition of Zr is favourable for corrosion resistance of Mg [14, 15], these studies have generally examined Mg-alloys that contain other elements, such as rare earths. As such, they do not reflect the direct interaction between only Zr and Mg. Any decrease in corrosion rates observed in such studies are likely from secondary interactions between Zr and ternary or quaternary elements or the changes in grain sizes - which in turn also impact the corrosion of the particular alloys studied.

Conclusions

1. For the same nominal composition, higher overall Zr contents were obtained through stirring of the Mg melt prior to casting due to settling of Zr in the furnace crucible with the long holding times associate with the unstirred alloys.
2. Stirring of the melt did not greatly affect the average percentage of Zr in solid solution for a given total Zr content.
3. Mg-Zr master alloys with a finer particle size used in the production of Mg-alloys will have an increased amount of Zr in solid solution in Mg-alloys for a given Zr content.
4. The corrosion current density, i_{corr}, increases more significantly with an increasing Zr content when there is a lower percentage of Zr in solid solution.
5. Zr in solid solution in Mg is an anodic activator, increasing the anodic reaction kinetics and thereby accelerating the generalised corrosion attack of the matrix.

6. Mass loss testing indicated that greater (and more highly variable) mass loss occurs with time in alloys which have a higher proportion of Zr in solid solution.

Acknowledgments

The CAST Co-operative Research Centre was established under, and is funded in part by, the Australian Governments Co-operative Research Centres Scheme. Andy Yob, Ming Sun and Sherly Simanjuntak are gratefully acknowledged for their assistance in preparing the alloys used in this research.

[1] H. Okamoto, Mg - Zr (Magnesium - Zirconium), Journal of Phase Equilibria and Diffusion, 28(3), (2007), 305-306.

[2] I.J. Polmear, Light Alloys. 3rd Edition (Ed). 1995, London, England: Arnold.

[3] E.F. Emley, Principles of Magnesium Technology. 1st edition (Ed). 1966, Manchester: Pergamon Press.

[4] M. Qian and D.H. StJohn, Grain nucleation and formation in Mg-Zr alloys, International Journal of Cast Metals Research, 22(1-4), (2009), 256-259.

[5] K.V. Kutniy, I.I. Papirov, M.A. Tikhonovsky, A.I. Pikalov, S.V. Sivtzov, L.A. Pirozhenko, V.S. Shokurov, and V.A. Shkuropatenko, Influence of grain size on mechanical and corrosion properties of magnesium alloy for medical implants, Material Wissenschaft Und Werkstofftechnik, 40(4), (2009), 242-246.

[6] C.D. Lee, Effect of grain size on the tensile properties of magnesium alloy, Materials Science and Engineering A, 459, (2007), 355-360.

[7] H.E. Friedrich and B.L. Mordike, Magnesium Technology. 2006, Berlin: Springer. 708.

[8] W. Yuan, S.K. Panigrahi, J.Q. Su, and R.S. Mishra, Influence of grain size and texture on Hall-Petch relationship for a magnesium alloy, Scripta Materialia, 65, (2011), 994-997.

[9] H. Han, S. Liu, L. Kang, and L. Liu, Refinement role of electromagnetic stirring and calcium in AZ91 Magnesium alloy, Journal of Wuhan University of technology - Materials Science Edition, 23(2), (2007), 194-197.

[10] P. Lyon, New magnesium alloy for aerospace and specialty applications, in: A.A. Luo(Ed) Magnesium Technology, (2004), TMS, pp. 311-315.

[11] T. Rzychon, J. Michalska, and A. Kielbus, Corrosion resistance of Mg-RE-Zr alloys, Journal of Achievments in Materials and Manufacturing Engineering, 21(1), (2007), 51-54.

[12] C.J. Bettles, M.A. Gibson, and S.M. Zhu, Microstructure and mechanical behaviour of an elevated temperature Mg-rare earth based alloy, Materials Science and Engineering A, 505, (2009), 6-12.

[13] A.C. Hanzi, F.H.D. Torre, A.S. Sologubenko, P. Gunde, R. Schmid-Fetzer, M. Kuehlein, J.F. Loffler, and P.J. Uggowitzer, Design strategy for microalloyed ultra-ductile magnesium alloys, Philosophical Magazine Letters, 89(6), (2009), 377-390.

[14] P. Cao, M. Qian, D.H. StJohn, and M.T. Frost, Uptake of iron and its effect on grain refinement of pure magnesium by zirconium, Materials Science and Technology, 20, (2003), 585-592.

[15] G. Song and D. StJohn, The effect of zirconium grain refinement on the corrosion behavior of magnesium-rare earth alloy MEZ, Journal of Light Metals, 2, (2002), 1-16.

[16] M. Qian, D.H. StJohn, and M.T. Frost, Zirconium alloying and grain refinement of magnesium alloys, in: H.I. Kaplan(Ed) Magnesium Technology 2003, (2003), The Minerals, Metals and Materials Society pp. 209-214.

[17] G. Ben-Hamu, D. Eliezer, K.S. Shin, and S. Cohen, The relation between microstructure and corrosion behaviour of Mg-Y-RE-Zr alloys, Journal of Alloys and Compounds, 341, (2007), 269-276.

[18] Z. Rong-chang, Z. Jin, H. Wei-Jiu, W. Dietzel, K.U. Kainer, C. Blawert, and K.E. Wei, Review of studies on corrosion of magnesium alloys, Transactions of Nonferrous Metals Society of China, (2006), 763-771.

[19] W.C. Neil, M. Forsyth, P.C. Howlett, C.R. Hutchinson, and B.R.W. Hinton, Corrosion of magnesium alloy ZE41 - the role of microstructural features, Corrosion Science, 51, (2009), 387-394.

[20] R.H.A. Crawley, Determination of soluble and insoluble zirconium in magnesium alloys, Analytica Chimica Acta, (1961), 281-284.

[21] M. Sun, G. Wu, M.A. Easton, D.H. StJohn, T. Abbott, and W. Ding, A comparison of the microstructure of three Mg-Zr master alloys and their grain refinement efficientcy, in: W.J. Poole and K.U. Kainer(Eds), Mg2012: 9th International Conference on Magnesium Alloys and their Applications, (2012, pp. 873-880.

[22] M. Qian, D.H. StJohn, and M.T. Frost, A new zirconium-rich master alloy for the grain refinement of magnesium alloys, in: K.U. Kainer(Ed) 6th International Conference Magnesium Alloys and their Applications, (2003), Wiley-VCH Verlag Gmbh & Co., pp. 706-712.

[23] M. Qian, L. Zheng, D. Graham, M.T. Frost, and D.H. StJohn, Settling of undissolved zirconium particles in pure magnesium melts, Journal of Light Metals, 1, (2001), 157-165.

[24] M. Qian, D.H. StJohn, M.T. Frost, and M.R. Barnett, Grain refinement of pure magnesium using rolled Zirmax master alloy (Mg-33.3Zr), in: H.I. Kaplan(Ed) Magnesium Technology 2003, (2003), TMS (The Minerals, Metals & Materials Society), pp. 215-220.

[25] M. Qian, D.H. StJohn, and M.T. Frost, Characteristic zirconium-rich coring substructures in Mg-Zr alloys, Scripta Materialia, 46, (2002), 649-654.

[26] M. Qian, D.H. StJohn, and M.T. Frost, Heterogeneous nuclei size in magnesium-zirconium alloys, Scripta Materialia, 50, (2004), 1115-1119.

[27] D.S. Gandel, M.A. Easton, M.A. Gibson, and N. Birbilis, Influence of Mn and Zr on the corrosion of Al-free Mg-alloys, Part 2: Impact of Mn and Zr on Mg-alloy electrochemistry, Corrosion, (Submitted).

List of tables and figures

Table 1: Composition of alloys produced in this study as tested via ICP-AES.

Figure 1: (A): BSE-SEM micrograph of Microzir master alloy, (B): BSE-SEM micrograph of Zirmax master alloy. **Figure 2**: (A): BSE-SEM micrograph of Mg-0.22Zr Microzir alloy, (B): EDX map of Zr content in Mg-0.22Zr Microzir alloy, (C): BSE-SEM micrograph of Mg-0.19Zr Zirmax alloy, (D): EDX map of Zr content in Mg-0.19Zr Zirmax alloy.

Figure 3: (A): Zr in solid solution vs. total Zr content for stirred Mg-Zr alloys (B): Zr in solid solution vs. total Zr content for unstirred Mg-Zr alloys.

Figure 4: (A): Zr content in solid solution (wt.%) vs. i_{corr} for stirred Mg-Zr alloys, (B): Zr content in solid solution (wt.%) vs. i_{corr} for unstirred Mg-Zr alloys, (C): Total Zr content (wt.%) vs. i_{corr} for stirred Mg-Zr alloys, (D): Total Zr content (wt.%) vs. i_{corr} for unstirred Mg-Zr alloys.

Figure 5: Electrochemical polarisation curves for Microzir and Zirmax containing Mg-Zr alloys compared with commercially pure Mg.

Figure 6: Weight loss vs. Zr content for Mg-Zr alloys made with Microzir and Zirmax master alloys.

Table 1: Composition of alloys produced in this study as tested via ICP-AES.

Sample ID	Mg wt.%	Zr wt.% (Total content)	Overall % of Zr in solid solution	Fe wt.%
Microzir-1 Stirred	~Bal	0.028	42.86	0.01
Microzir-1 Unstirred	~Bal	0.012	33.33	0.008
Microzir-2 Stirred	~Bal	0.061	36.07	0.01
Microzir-2 Unstirred	~Bal	0.014	42.86	0.008
Microzir-3 Stirred	~Bal	0.12	45	0.011
Microzir-3 Unstirred	~Bal	0.019	47.37	0.005
Microzir-4 Stirred	~Bal	0.22	47.27	0.014
Microzir-4 Unstirred	~Bal	0.037	51.35	0.006
Zirmax-1 Stirred	~Bal	0.028	25	0.006
Zirmax-1 Unstirred	~Bal	0.018	22.22	0.006
Zirmax-2 Stirred	~Bal	0.057	15.79	0.006
Zirmax-2 Unstirred	~Bal	0.025	16	0.005
Zirmax-3 Stirred	~Bal	0.12	10	0.001
Zirmax-3 Unstirred	~Bal	0.056	8.93	0.005
Zirmax-4 Stirred	~Bal	0.19	10.53	0.006
Zirmax-4 Unstirred	~Bal	0.075	8	0.013

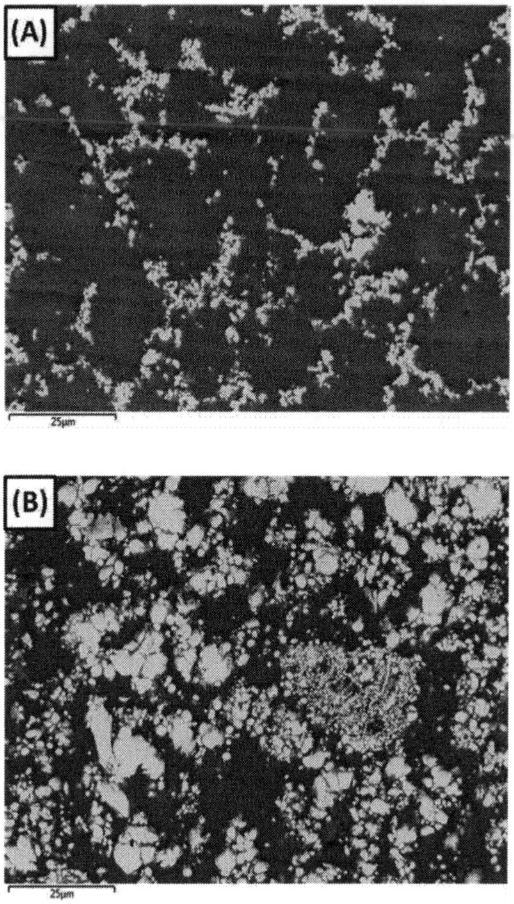

Figure 1: (A): BSE-SEM micrograph of Microzir master alloy, (B): BSE-SEM micrograph of Zirmax master alloy.

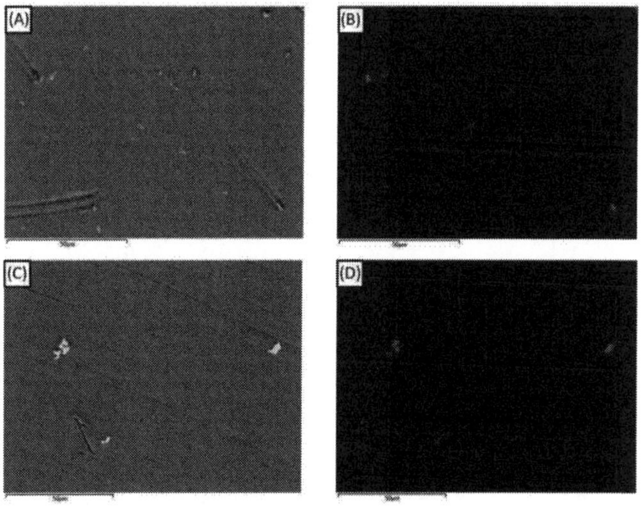

Figure 2: (A): BSE-SEM micrograph of Mg-0.22Zr Microzir alloy, (B): EDX map of Zr content in Mg-0.22Zr Microzir alloy, (C): BSE-SEM micrograph of Mg-0.19Zr Zirmax alloy, (D): EDX map of Zr content in Mg-0.19Zr Zirmax alloy.

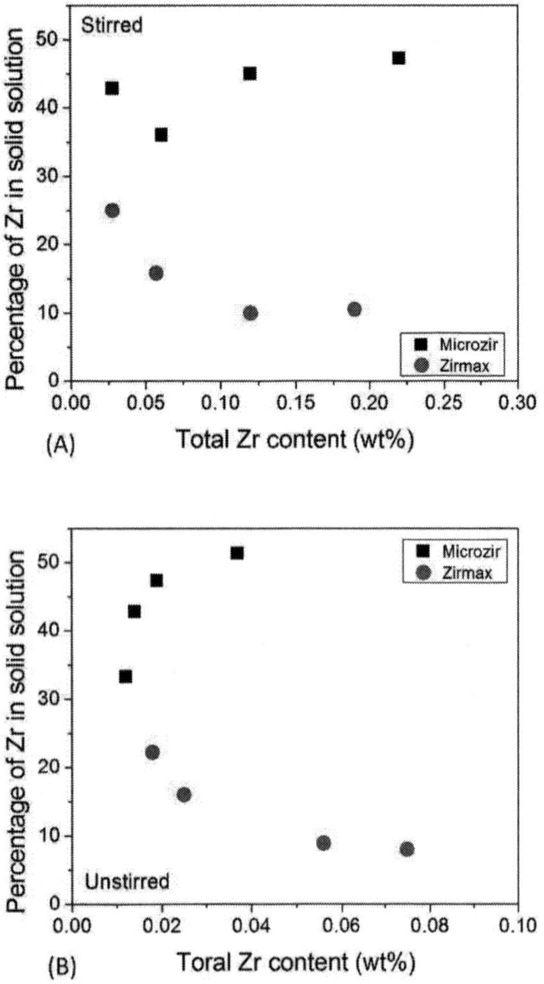

Figure 3: (A): Zr in solid solution vs. total Zr content for stirred Mg-Zr alloys (B): Zr in solid solution vs. total Zr content for unstirred Mg-Zr alloys.

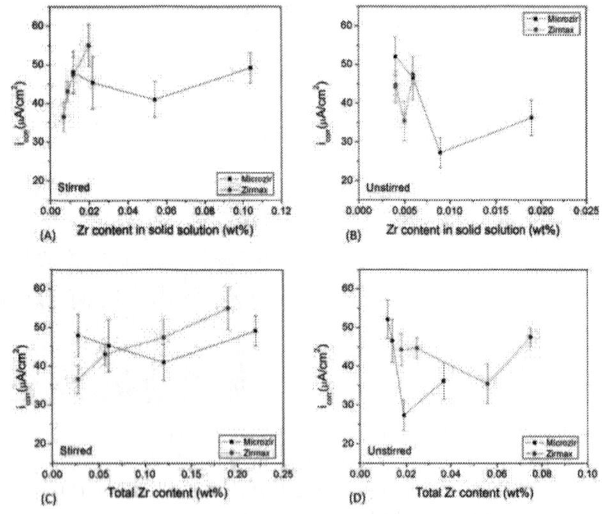

Figure 4: (A): Zr content in solid solution (wt.%) vs. i_{corr} for stirred Mg-Zr alloys, (B): Zr content in solid solution (wt.%) vs. i_{corr} for unstirred Mg-Zr alloys, (C): Total Zr content (wt.%) vs. i_{corr} for stirred Mg-Zr alloys, (D): Total Zr content (wt.%) vs. i_{corr} for unstirred Mg-Zr alloys.

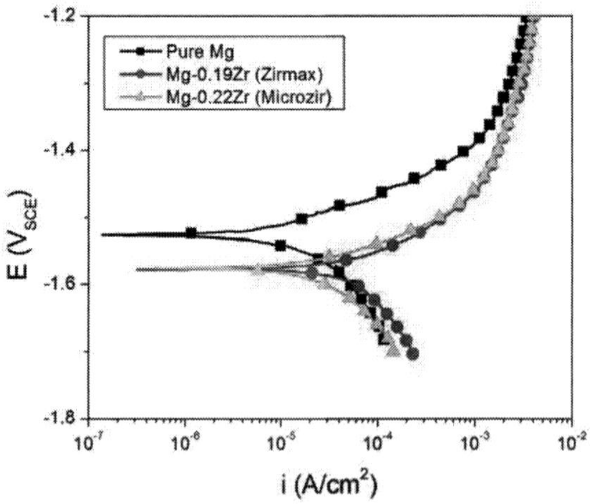

Figure 5: Electrochemical polarisation curves for Microzir and Zirmax containing Mg-Zr alloys compared with commercially pure Mg.

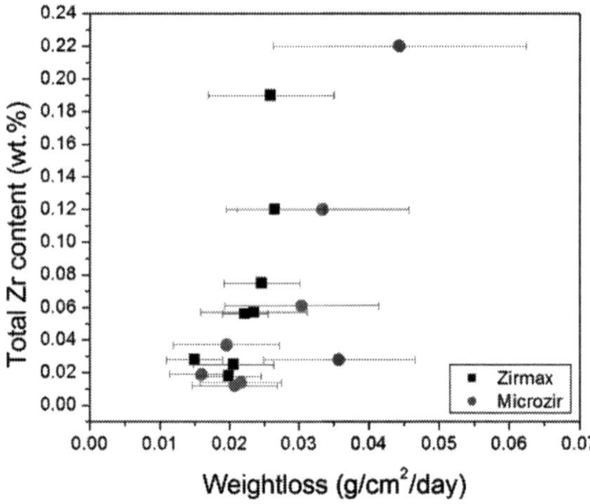

Figure 6: Weight loss vs. Zr content for Mg-Zr alloys made with Microzir and Zirmax master alloys.

Magnesium Technology 2013
Edited by: Norbert Hort, Suveen N. Mathaudhu, Neale R. Neelameggham, and Martyn Alderman
TMS (The Minerals, Metals & Materials Society), 2013

THE USE OF AC-DC-AC METHODS IN ASSESSING CORROSION RESISTANCE PERFORMANCE OF COATING SYSTEMS FOR MAGNESIUM ALLOYS

Robert C. McCune[1], Vinod Upadhyay,[2] Yar-Ming Wang[3], and Dante Battocchi[4]
[1]Robert C. McCune & Associates, LLC, West Bloomfield, MI, 48322, USA
[2]North Dakota State University, Fargo, ND, 58105 (currently: University of Akron, Akron, OH, 44325)
[3]General Motors Co., R&D Center, Warren, MI, 48090, USA (retired)
[4]North Dakota State University, Fargo, ND, 58105

Keywords: magnesium, coatings, corrosion

Abstract

The potential utility of AC-DC-AC electrochemical methods in comparative measures of corrosion-resisting coating system performance for magnesium alloys under consideration for the USAMP "Magnesium Front End Research and Development" project was previously shown in this forum [1]. Additional studies of this approach using statistically-designed experiments have been conducted with focus on alloy types, pretreatment, topcoat material and topcoat thickness as the variables. Additionally, sample coupons made for these designed experiments were also subjected to a typical automotive cyclic corrosion test cycle (SAE J2334) as well as ASTM B117 for comparison of relative performance. Results of these studies are presented along with advantages and limitations of the proposed methodology.

Introduction

The Magnesium Front End Research and Development Project ("MFERD"), to which this work contributes, was recently reviewed [2]. Briefly, the project is an international collaboration between the U.S. Department of Energy, the U.S. Automotive Materials Partnership, LLC (USAMP), Natural Resources – Canada, and the Ministry of Science and Technology of the People's Republic of China. The goal is to advance both the knowledge base and enabling manufacturing technologies necessary to engineer and produce magnesium-intensive automotive body substructures with markedly reduced weight, but with equivalent mechanical strength and durability of comparator baseline steel structures.

One of several research subject matters within the ongoing MFERD Project is corrosion prevention and surface treatment for magnesium components and assemblies, particularly as envisioned for automotive substructures. An ongoing objective of the "corrosion" task has therefore been the evaluation of various coating systems as might be incorporated in both the individual component pieces and the completely assembled structure. Figure 1 illustrates a model corrosion protection system for magnesium incorporating the following elements: a.) a cleaning or preparation step, which includes any metal cleaning and pickling or 'activation,' followed by b.) one of several possible processes to provide a "pretreatment" or an anchoring layer for c.) the topcoat which is presumed to be a polymeric layer or "paint." The pretreatment process would also apply to the use of structural adhesives for joining the component pieces, or total assembly to other portions of the automotive body structures. Various pretreatment processes for magnesium have been considered over the years [3] including conversion coating (e.g. chromating), anodizing or other electrically-based processes used to develop primarily inorganic layers with sufficient surface roughness to which the topcoat or adhesive can adhere. Considering the choices of base metal magnesium (e.g. die-casting, extrusion, formed sheet), pretreatment options, and possible topcoats or adhesives, the number of possible surface treating permutations is extraordinary. Prior work [4] offered an approach to coating systems corrosion evaluation, predicated on observable physical metrics such as scribe creepback (paint undercutting) and area fraction of accumulated surface corrosion product.

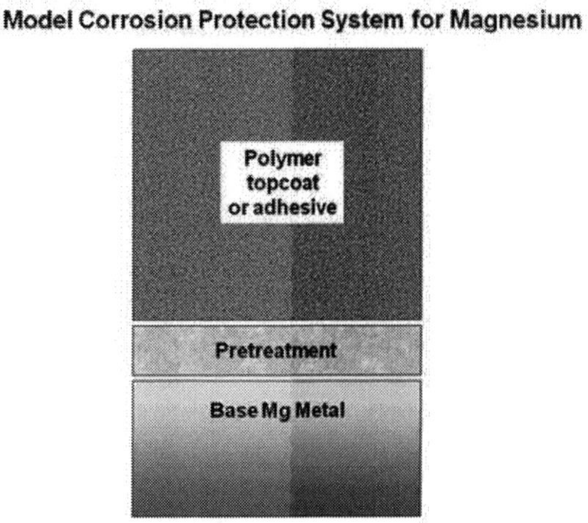

Figure 1. Model corrosion protection system proposed for magnesium components and front end subassembly.

The background for exploring the use of an electrochemical impedance spectroscopy (EIS) approach to the characterization of polymeric coatings on magnesium alloys and their breakdown was discussed in the earlier work [1]. A simple primer on EIS methods may be found online [5]. In the prior study, the EIS of polymer-coated metals was further extended via the use of intermittent cathodic excursions of approximately - 4 V DC (relative to the open circuit potential), for periods of 30 minutes, during a total measurement cycle of four hours. This "AC-DC-AC" approach to accelerated testing of insulating coatings has been attributed to Hollaender, et al. [6]. Cathodic polarization, particularly of ferrous materials with polymeric coatings, is a relatively common practice for assessing the resistance of the polymer layer to the localized pH transition accompanying cathodic reactions, which often generates molecular hydrogen resulting in increase of the local $(OH)^-$ concentration. This transition may be deleterious to the polymer-surface bonding. This phenomenon should certainly be considered for magnesium alloys wherein hydrogen evolution is usually the principal cathodic reaction associated with corrosion, and local stability for the magnesium corrosion product occurs in basic solutions at high pH (i.e. strong base conditions). Loss of adherence of the polymer topcoat would add to the progressive corrosion of the underlying magnesium.

The previous work [1] suggested the utility of the AC-DC-AC approach for assessing behavior of coating systems in relatively short periods when compared to field exposures and manufacturer's cyclic test procedures, often lasting many months. Should such an approach be both practical and verifiable, it could be useful in sorting through the vast array of potential surface process combinations available to protect magnesium alloys. The aim of the current work was to extend the initial study to a broader array of variables in a designed experiment approach, and to continue exploration of the method as a rapid assessment tool.

Experimental Procedures

Materials and Surface Treatments. The magnesium materials used in this work were AM60B low-pressure die castings in the form of 100 x 150 x 3mm plates, provided by ACT Test Panels, Inc. (Hillsdale, MI), and AZ31 H24 sheet of thickness 2 mm, and comparable size cut from larger sheets as produced by Luoyang Copper, of Luoyang, Henan, China. Several coupons of AZ31 sheet (Magnesium Elektron) of the "benchmark" process [7] used by Ford Motor Company in a production line for F-150 radiator surrounds remained from the earlier work [1]. Unfortunately, the "benchmark" process was unavailable for the current study, and a comparable, but different "baseline" process was used instead to emulate the benchmark.

Table 1. Materials and processes considered in this work.

Designation	Material(s)	Treatments
Benchmark Process	AZ31 2mm Magnesium Elektron	acetic acid de-ox Henkel Alodine® 5200 degas Protech ES-542-N49 epoxy powder coat (commercial line)
Baseline Process	AZ31 – Luoyang Copper AM60B ACT Panels	sulfuric acid de-ox Henkel Alodine® 5200 Protech ES-542-N49 lab over bake 10 min
Designed Experiment	AZ31 (Luoyang) AM60B ACT Panels	
		Pretreatt:Alodine® 5200, sulfuric acid de-ox or AHC MagPass®
		Topcoat: Protech® Epoxy Powder, lab bake or PPG P 6000 CX cathodic electrocoat
		Topcoat thickness: typical (75μm – powder, 25 μm e-coat) or approx. 2X typical thickness.
Commercial Process	AZ31 (Luoyang) AM60B (ACT)	Alodine® 5200 PPG 590-534 cathodic electrocoat

The "benchmark" process included alkaline cleaning, deoxidation ("activation") with acetic acid, pretreatment with Henkel Alodine® 5200 followed by a degassing (curing) cycle and application of an electrostatic epoxy powder topcoat (Protech ES-542-N49), with subsequent thermal fusion[7]. The "baseline" process also employed Alodine® 5200 and epoxy power coat, but applied in a laboratory vs. a production setting. A commercially-available coating process designed for use with aluminum and magnesium was also explored. This process included the Alodine® 5200 pretreatment (with sulfuric acid de-ox) and cathodic epoxy (PPG 590-534) electrocoat.

Table 1. summarizes the materials and nomenclature in this work. Table 2 details the layout of a 2^4 full factorial designed experiment based on magnesium alloy (AZ31 or AM60B), pretreatment process (Alodine® 5200 or AHC MagPass®), topcoat (Protech ES-542-N49 powder epoxy or PPG P-6000CX cathodic electrocoat) and coating thickness of "typical" (i.e. the usually specified coating thickness for the process) and 2X typical (produced by exposure time or current to achieve approximately twice the nominal thickness). Nominal thickness for electrocoat is usually less than 25 μm and for the powder coat 75 μm.

Table 2. Factors and levels for 2^4 full factorial design of experiments for polymeric coatings on magnesium.

Factor	(-) level	(+) level
Substrate	AZ31	AM60B
Pretreatment	Alodine ®5200	MagPass®
Topcoat	cathodic e-coat	epoxy powder coat
Thickness	nominal	2X nominal

Test Exposures and Measurement Protocols

Test Solutions. The original intent of the work was to focus on AC-DC-AC evaluation for the various materials and surface treatments when exposed to the SAE J2334 [8] electrolyte consisting of 0.5% NaCl, 0.1% $CaCl_2$ and 0.075% $NaHCO_3$ in water solution. The prior AC-DC-AC measurements [1] had been made using the 5% NaCl electrolyte according to ASTM B-117 [9]. Most manufacturer cyclic tests in use today avoid the higher NaCl concentration of ASTM B-117 in favor of lower salt concentrations as exemplified by SAE J2334, which may be more typical of actual environments to which the product is exposed. During the course of this work, however, it was found that the lower salt concentrations prolonged the degradation of coatings, to the extent that even with cathodic polarization employed to aggravate delamination of the paint layer, conditions did not permit comparisons of relative performance to be made in a reasonable time period. Later experiments employed the 5% NaCl ASTM B-117 electrolyte in hopes of securing some data in more reasonable time frames.

AC-DC-AC Protocol. EIS was employed to evaluate the barrier properties of coatings. Measurements were made using a two electrode set-up with the magnesium alloy substrate as the working electrode and a platinum mesh as the counter/reference electrode. The SAE J2334 salt solution was the test electrolyte used in all measurements. A Perspex™ cylinder with a surface area of 7.07 cm2 was mounted on the samples and was clamped with an O-ring insert to facilitate electrochemical measurements. A Gamry Instrument R 600 Potentiostat/Galvanostat/ZRA in conjunction with Gamry Framework Version 5.20/EIS 300 software was used for these experiments (Gamry Instruments, Inc., Willow Grove, PA). Impedance response corresponding to the applied frequency of 100 kHz to 0.01Hz was measured with an acquisition rate of 10 points per decade. A 10mV amplitude perturbation potential with respect to the open circuit potential was used. In order to induce coating breakdown, the DC cathodic bias was varied to promote degradation of the coating performance. This was substantially different from the original approach [1], wherein a constant – 4V vs. OCP was employed throughout the procedure. Each DC step was applied for 4 hours during an AC-DC-AC cycle. The number of hours (where indicated) is the total number of DC hours to which the samples were subjected. While such a practice revealed which coating configurations required a greater cathodic potentials to effect a breakdown, it did not permit easy comparisons among the treatments tested, since this in itself became a variable.

SAE J2334 Cyclic Exposure and Visual Inspection. In addition to the AC-DC-AC protocol described above, scribed panels were exposed to cyclic testing as prescribed in the SAE J2334 standard [8]. This is a cyclic test protocol that exposes panels to a 24 hour cycle consisting of 100% relative humidity at 50°C for 6 hours, salt fog exposure (using the electrolyte described previously) at ambient condition for 15 minutes, and drying at 60°C and 50% relative humidity for 17 hours and 45 minutes. Cyclic testing is preferred among the auto manufacturers and was sought as a validation or confirmation of coatings performance via the AC-DC-AC protocol.

ASTM B-117 Exposures. When it became apparent that the lower NaCl concentration of the SAE J2334 electrolyte was not producing distinguishable results, particularly for the scribe creepback measurements, a decision was made to employ the ASTM B-117 5% NaCl salt fog environment as a means to provide at least some indications of the relative performance for the designed experiment matrix of Table 2. AC methods remained employable for probing of the coating impedance, with now a comparison of "before" and "after" values of the low-frequency (e.g. 0.1Hz) impedance for a 50 day exposure to the B-117 environment as the response function for the designed experiment. In this case, improved performance was indicated by a lower degree of impedance loss from the initial value. For these coupons, there was no DC bias applied to accelerate paint layer delamination. Degradation of the coating by the environmental exposure was inferred instead from loss of low-frequency impedance as a consequence of exposure to the environment. No attempt was made to assign the impedance loss to any particular equivalent circuit element (e.g. moisture ingress into the polymer dielectric layer). B-117 environment exposure also permitted a range of scribe creepback data to be obtained for the experimental array.

Results and Discussion

AC-DC-AC Measurements with SAE J2334 Electrolyte. As mentioned previously, the need to continually increase the cathodic bias levels in an attempt to force the coating breakdown was generally unsuccessful, and moreover precluded direct comparison of the various coating configurations since differing bias levels and times had been employed. Figure 2 shows the trend for low-frequency impedance as a function of DC polarization time for the "benchmark" and "baseline" coatings, using the SAE J2334 electrolyte. The "benchmark" coating, (designated in this figure as AZ31+Al5200+PC) did not exhibit loss of impedance, despite polarizations to as much as -8V vs. OCP. The "baseline" process used to replicate the benchmark did show breakdown of the coating, both on AM60 and the AZ31 alloys. By this approach, the AZ31 actually exhibited a longer time to breakdown than had been witnessed in the prior study.

Whereas the "benchmark" process applied to AZ31 exhibited breakdown by cathodic polarization of -4 V vs. OCP when using 5% NaCl as the electrolyte, it could not be breached by as much as -8 V using the SAE J2334 electrolyte. Interestingly, however, the "benchmark" coating did exhibit appreciable scribe undercutting and creepback during exposure to the cyclic SAE J2334 testing, even though the polymer film and interface remained protective as inferred from AC-DC-AC testing. Figure 3 illustrates the degree of creepback witnessed for the "benchmark" processed coupon after 74 cycles of SAE J2334 and compared to the "baseline" coupons which suggested breakdown more readily by AC-DC-AC, but were more durable in scribe creepback.

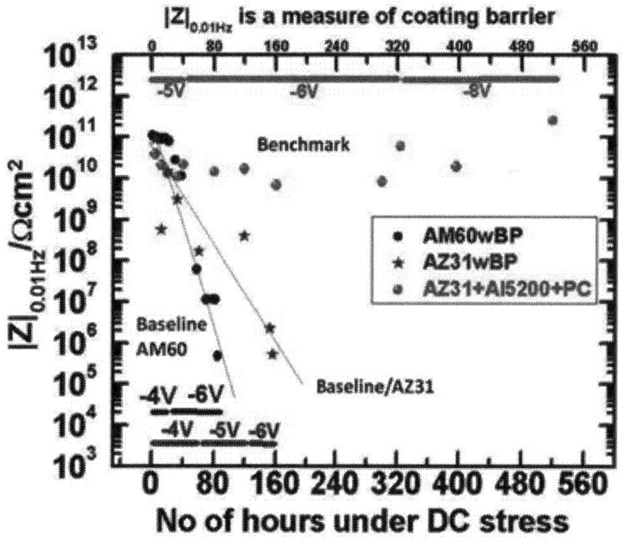

Figure 2. Loss of impedance (modulus) at low frequency (0.01Hz) in AC-DC-AC testing for hours of DC polarization at the levels shown. BP = "baseline process."

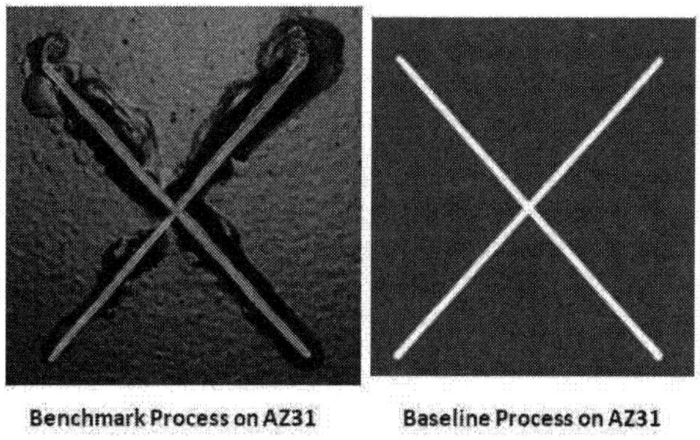

Figure 3. Scribe creepback for "benchmark" and "baseline" processed AZ31 after 74 cycles of SAE J2334.

The AC-DC-AC method probes internal changes in the coating layer and electrochemical reactions at the metal/pretreatment/polymer interface. For durable coatings, which are resistant to moisture and electrolyte ingress, the changes induced by cathodic DC polarization can be minimal. Conversely, an example of a coating which showed both a breakdown under AC-DC-AC conditions at 5 V of applied cathodic polarization, as well as breakdown in later cyclic testing (for AZ31) was the commercially-processed coating which employed Alodine®5200 as the pretreatment and PPG 590-534 cathodic electrocoat as the topcoat. Figure 4 shows a comparison of the breakdown behavior for this coating on both AZ31 and AM60 as compared to a more robust coating consisting of the AHC MagPass® pretreatment and Protech ES-542 N49 powder epoxy. Unfortunately there are no observational results at this time on cyclic performance of the latter combination.

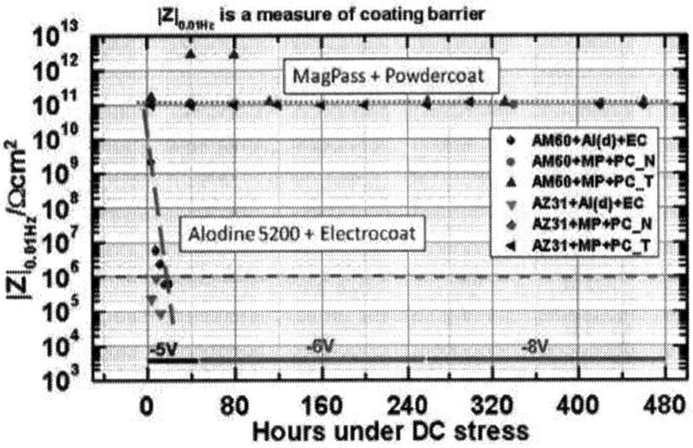

Figure 4. Impedance decay under DC stress behavior for two categories of coatings on AZ31 and AM60 – one a commercial process based on Alodine® 5200 with cathodic epoxy topcoat, and the other based on the AHC MagPass® pretreatment with powder epoxy topcoat.

Impedance Degradation with Salt Fog Exposure. It was indicated above that the AC-DC-AC methodology did not permit a ready comparison of the designed experiment matrix of substrates and coatings of Table 2 using the SAE J2334 electrolyte. This was due, in part, to the inability to produce breakdown of the particular coating configurations through cathodic polarization, even at substantial, and possibly even unreasonable potentials (e.g. in excess of -8V vs OCP). On the other hand, later experiments revealed loss of low-frequency impedance following exposures to the B-117 environment for periods of 50 days. Analysis of this data also indicated higher values for the initial layer impedance of all powder coats relative to the electrocoat of around an order of magnitude (i.e. 10^{10} for powder coats vs 10^9 ohm-cm^2 for electrocoat). It should be appreciated, however, that the low-frequency, high impedance regime is the most tenuous experimental space for precise measurement using AC methods.

Figure 5 illustrates a summary of the impedance loss (in orders of magnitude) for the initial sample, and for the same sample after 50 days of exposure to the B-117 environment. As mentioned, no cathodic polarization was employed. Table 3 illustrates the sample coding for the experiment set.

Analysis of this designed experiment set suggested, as had previously been advanced [4], that the choice of pretreatment was a significant factor in durability of the coating system. Figure 6 illustrates a "main effects" plot for this data using the statistical software Minitab®. In this particular layout, the response factor was "loss of impedance" such that lower values are preferable for more durable coating systems. For this particular designed experiment, analysis suggested that the MagPass pretreatment vs. the Alodine 5200 pretreatment had the largest effect on retention of the initial coating/system impedance, whereas the topcoat thickness had virtually no influence. Surprisingly, AZ31 sheet vs AM60 die casting showed slight improvement as did the electrocoat vs. powdercoat. Both of these results were unexpected based on prior experiments.

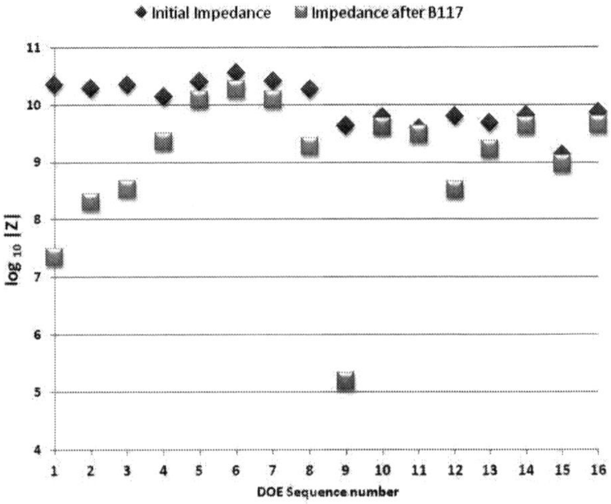

Figure 5. Summary of low-frequency initial and final impedance moduli for the designed experiment matrix of Table 2, before and after 50 days of exposure to the ASTM B-117 environment.

Table 3. Experimental nomenclature and layout for the 2^4 full factorial designed experiment of Table 2. AZ= AZ31, AM= AM60B, Al= Alodine 5200, MP= MagPass, P= powder coat, E = electrocoat, - = nominal thickness, + = 2X nominal.

No.	Base	Pretreat	Topcoat	Thickness
1	AZ	Al	P	+
2	AZ	Al	P	-
3	AM	Al	P	+
4	AM	Al	P	-
5	AM	MP	P	-
6	AM	MP	P	+
7	AZ	MP	P	+
8	AZ	MP	P	-
9	AM	Al	E	-
10	AM	Al	E	+
11	AM	MP	E	-
12	AM	MP	E	+
13	AZ	Al	E	-
14	AZ	Al	E	+
15	AZ	MP	E	-
16	AZ	MP	E	+

No.	Base	Pretreat	Topcoat	Thickness
1	AZ	Al	P	+
2	AZ	Al	P	-
3	AM	Al	P	+
4	AM	Al	P	-
5	AM	MP	P	-
6	AM	MP	P	+
7	AZ	MP	P	+
8	AZ	MP	P	-
9	AM	Al	E	-
10	AM	Al	E	+
11	AM	MP	E	-
12	AM	MP	E	+
13	AZ	Al	E	-
14	AZ	Al	E	+
15	AZ	MP	E	-
16	AZ	MP	E	+

AZ= AZ31, AM=AM60B, Al=Alodine5200
MP = MagPass, - = nominal thickness
+ = 2X nominal P=Powder, E= E-coat

For the 50 day B-117 scribe creepback, a rating scale according to ASTM D1654 (0 = poor,worst; 10 = zero creepback, best) was employed. These values were also treated statistically with Minitab®. Figure 7 shows this main effects plot supporting improvements with MagPass vs Alodine 5200, AZ31 vs AM60B, and thicker vs nominal coatings. In this layout, there was not a noticeable influence of powder coating over electrocoat.

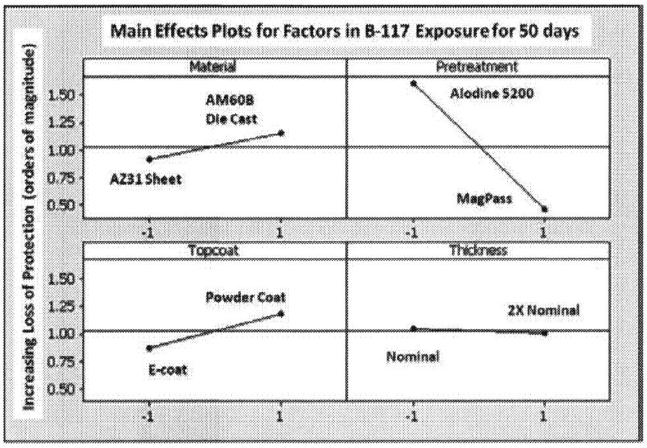

Figure 6. Main effects plot showing influence of the factors from Tables 2 and 3 on the loss of low-frequency impedance in orders of magnitude from the initial value to the final value after 50 days of exposure to the B-117 environment.

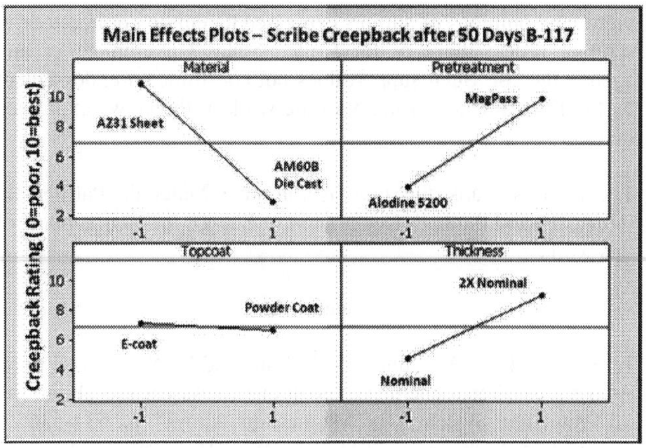

Figure 7. Main effects plot when scribe creepback rating (according to ASTM D1654) is used as the experiment response. In this scheme a higher number refers to increased resistance to film undercutting.

Summary and Conclusions

While the initial study of the AC-DC-AC approach for characterization of protective coatings on magnesium alloys was encouraging [1], the present work indicates that clearly more study is required, particularly with regard to suitable electrolytes for the measurement and also the concerns over possibly unreasonable values for cathodic polarization required to effect breakdown of the coating system – which in some instances was not achieved, even at potentials as high as -8 V vs. OCP. The one instance of a predicted rapid coating system breakdown of the commercial product at cathodic potential of -5 V vs. OCP (Fig. 4.) has been at least supported by parallel studies of the corrosion of coated assembled magnesium-intensive front-end substructures which are not detailed in this report. The use of the ASTM B-117 environment, which appeared to be needed here to effect any sort of useful measurements in a reasonable time period, is generally unattractive to automotive manufacturers as being much more severely aggressive than test environments normally employed for cyclic corrosion tests. The continuing work of the "MFERD" effort will likely also include extension to improved methods for coating system validation and comparison for magnesium alloys and magnesium-intensive subassemblies.

Acknowledgments

This material is based upon work supported by the Department of Energy National Energy Technology Laboratory under Award Number Nos.DE-FC05-95OR22363, DE-FC05-02OR22910, DE-FC26-02OR22910, or DE-0003583. This report was prepared as an account of work sponsored by an agency of the United States Government. Neither the United States Government nor any agency thereof, nor any of their employees, makes any warranty, express or implied, or assumes any legal liability or responsibility for the accuracy, completeness, or usefulness of any information, apparatus, product, or process disclosed, or represents that its use would not infringe privately owned rights. Reference herein to any specific commercial product, process, or service by trade name, trademark, manufacturer, or otherwise does not necessarily constitute or imply its endorsement, recommendation, or favoring by the United States Government or any agency thereof. The views and opinions of authors expressed herein do not necessarily state or reflect those of the United States Government or any agency thereof. Such support does not constitute an endorsement by the Department of Energy of the work or the views expressed herein.

The authors thank the U.S. Automotive Materials Partnership (USAMP) and the U.S. Department of Energy for their support of this work.

References

1. S. Song, R. C. McCune, Y.-M. Wang and W. Shen, "Use of an AC/DC/AC Electrochemical Technique to Assess the Durability of Protection Systems for Magnesium Alloys" p. 531-536 in Magnesium Technology 2011, eds. Wim H. Sillekens, Sean R. Agnew, Neale R. Neelameggham, Suveen N. Mathaudhu, TMS (The Minerals, Metals and Materials Society), Warrendale, PA, 2011.
2. A. A. Luo, J. F. Quinn, R. Verma, Y.-M. Wang, T.M. Lee, D. A. Wagner, J.H. Forsmark, X. Su, J. Zindel, M. Li, S.D. Logan. S. Bilkhu and R. C. McCune, "The USAMP Magnesium Research and Development Project," PP. 54-58 in Light Metal Age, April 2012.
3. J. E. Gray and B. Luan, "Protective coatings on magnesium and its alloys – a critical review," Journal of Alloys and Compounds 336, 88-113 (2002).
4. P. J. Blanchard, D. J. Hill, G. T. Bretz and R. C. McCune, "Evaluation of Corrosion Protection Methods for Magnesium Alloys in Automotive Applications," in Magnesium Technology 2005, Neale R. Neelameggham, Howard I. Kaplan, and Bob R. Powell, eds.,TMS (The Minerals, Metals & Materials Society), Warrendale, PA, 2005.
5. Gamry, Inc. : http://www.gamry.com/App_Notes/EIS_Primer/EIS_Primer.htm.
6. J. Hollaender, E. Ludwig, S. Hillebrand, " Assessing protective layers on metal packaging material by electrochemical impedance spectroscopy," Proceedings of the Fifth International Tinplate Conference, London, (1992) pp. 300-315.
7. J. S. Balzer, P. K. Dellock, M. H. Maj, G.S. Cole, D. Reed, T. Davis, T. Lawson and G. Simonds, "Structural Magnesium Front-End Support Assembly," Society of Automotive Engineers, Warrendale, PA, Paper 2003-01-0186 (2003).
8. Standard J2334 version 200312, "Automotive Cyclic Corrosion Test," Society of Automotive Engineers (SAE), Warrendale, PA (2003).
9. ASTM B117 - 11 "Standard Practice for Operating Salt Spray (Fog) Apparatus," ASTM International, 100 Barr Harbor Drive, PO Box C700, West Conshohocken, PA, 19428-2959 USA.
10. ASTM D-1654 - 08 "Standard Test Method for Evaluation of Painted or Coated Specimens Subjected to Corrosive Environments", ASTM International, 100 Barr Harbor Drive, PO Box C700, West Conshohocken, PA, 19428-2959 USA.

CORROSION BEHAVIOR OF CERIUM-BASED CONVERSION COATINGS ON MAGNESIUM ALLOYS EXPOSED TO AMBIENT CONDITIONS

Carlos E. Castano[1], Surender Maddela[1], Matthew J. O'Keefe[1]
[1] Materials Research Center, Department of Materials Science and Engineering, Missouri University of Science and Technology; 101 Straumanis-James Hall 401 W. 16th St.; Rolla, MO, 65409, USA

Keywords: Cerium-based Conversion Coatings, Corrosion of Magnesium Alloys, Ambient Exposure, Sunlight Exposure.

Abstract

Exposure of CeCCs on AZ31B and AZ91D Mg alloys to ambient sunlight, temperature, and humidity was done to determine the effect on corrosion resistance. It was found that the CeCCs changed from pale yellow to almost translucent after 24 hours of sunlight exposure. The effect of the solar electromagnetic radiation on the morphological, chemical and optical properties of these coatings was investigated using SEM and UV-Vis characterization techniques. In addition, the corrosion performance of CeCCs before and after ambient exposure was studied by ASTM B117 neutral spray testing and electrochemical polarization measurements. In general, the changes in appearance did not adversely affect the corrosion performance of the coatings.

Introduction

The development of environmentally benign coating systems is one of the most active areas of research in surface engineering. Chromium and fluorine-free surface treatments have been the main objectives in the research and development of new coating alternatives [1-5]. A global effort has led to the development of a large number of more environmentally friendly conversion coatings such as stannates, phosphates, zirconium and rare earth-based coatings [1,2,4-6]. In addition to replacement of the hexavalent chromium and fluorine baths, the use of lightweight materials is desirable for a number of reasons. For instance, magnesium alloys have been selected by the automotive industry for several lightweight applications [3]. Cerium-based conversion coatings have shown to improve the corrosion resistance of several magnesium alloys [6-9]. The effectiveness of the CeCCs on the corrosion protection has been highly dependent on process parameters. Detailed studies of the effect of process variables such as coating thickness, CeCC bath temperature, alkaline and acid surface pretreatments, and postreatments on the corrosion performance of CeCCs applied on AZ31B and AZ91D Mg alloys have been reported [8-12]. However, effects of several concomitant variables on the behavior of CeCCs are still unknown. In this study, the effect ambient sunlight of cerium-based conversion coatings applied to AZ31B and AZ91D Mg alloys was investigated.

Experimental Procedure

Coupons of AZ31B and AZ91D Mg alloys of 10 x 5 cm^2 of exposed surface area were mechanically polished using 180 grit abrasive silicon carbide papers, cleaned with isopropyl alcohol, then rinsed with deionized water, and finally dried at room temperature. The cleaned samples were pretreated in 1 wt% HNO_3 aqueous solution for 30 seconds followed by an alkaline cleaning in 5 wt% of $Na_2SiO_3 \cdot 5H_2O$ aqueous solution for 5 minutes at room temperature. The activated Mg alloys were then immersed in an aqueous solution of 4wt% of $CeCl_3 \cdot 7H_2O$ (99.9 %, Alfa Aesar), 6 vol% of hydrogen peroxide (Fisher Chemical, 30 vol%), and 0.25 wt% of organic gelatin (RDH, Rousselot) in deionized (DI) water for 60 seconds. Following deposition, the coated samples were postreated for 5 min at 85 °C in a 2.5 wt% NaH_2PO_4 aqueous solution.

Four samples for each Mg alloy were deposited in order to have reproducible results. Two of the samples for each alloy were used for salt spray testing and the other two for the other characterization techniques. Half of the total area in each sample was wrapped with aluminum foil to protect that region from sunlight exposure. The panels were then exposed to ambient conditions during 6 hours/day (9:00 am to 3:00 pm) for consecutive 4 days.

Morphological characterization was performed using the Dual Beam Helios NanoLab 600 in SEM mode. Ultraviolet-visible spectra analyses were measured using a Varian Cary 5 ultraviolet-visible-near-infrared spectrometer (UV-Vis-NIR) in the wavelength range of 310-750 nm. The corrosion performance of the coated panels was evaluated by ASTM B117 neutral salt fog testing and cyclic potentiodynamic polarization scans. The latter technique was carried out using a flat cell of 1.0 cm^2 of exposed area (model K0235, Princeton Applied Research) with a saturated calomel electrode (SCE) and platinum mesh counter electrode. A 0.6 wt.% sodium chloride + 0.6 wt.% ammonium sulfate solution in DI water was used as electrolyte. Open circuit potential (OCP) was monitored for 1500 seconds and the cyclic potentiodynamic polarization scans were conducted at 1 mV/s from -0.3 to +0.8 V and then decreased back to -0.25 V respect to OCP.

Results and Discussion

Optical images of cerium-based conversion coatings on AZ31B and AZ91D Mg alloy panels partially exposed to ambient conditions before and after 168 hours of salt spray testing are shown in Figure 1. A color change from pale yellow to almost translucent was observed in the sunlight exposed area of the CeCCs of both alloys compared with the Al foil protected area. Traditionally yellow to orange coatings are obtained when panels are coated through the method described in the experimental procedure. Since the optical properties can be explained by physical and/or chemical changes, the corrosion behavior of both exposed and unexposed areas was studied. The bottom part of Figure 1 shows the panels after salt spray testing. Although the color is different in both areas the corrosion performance is almost identical in each magnesium alloy. No differences in size or number of pits were observed in the different regions for each panel.

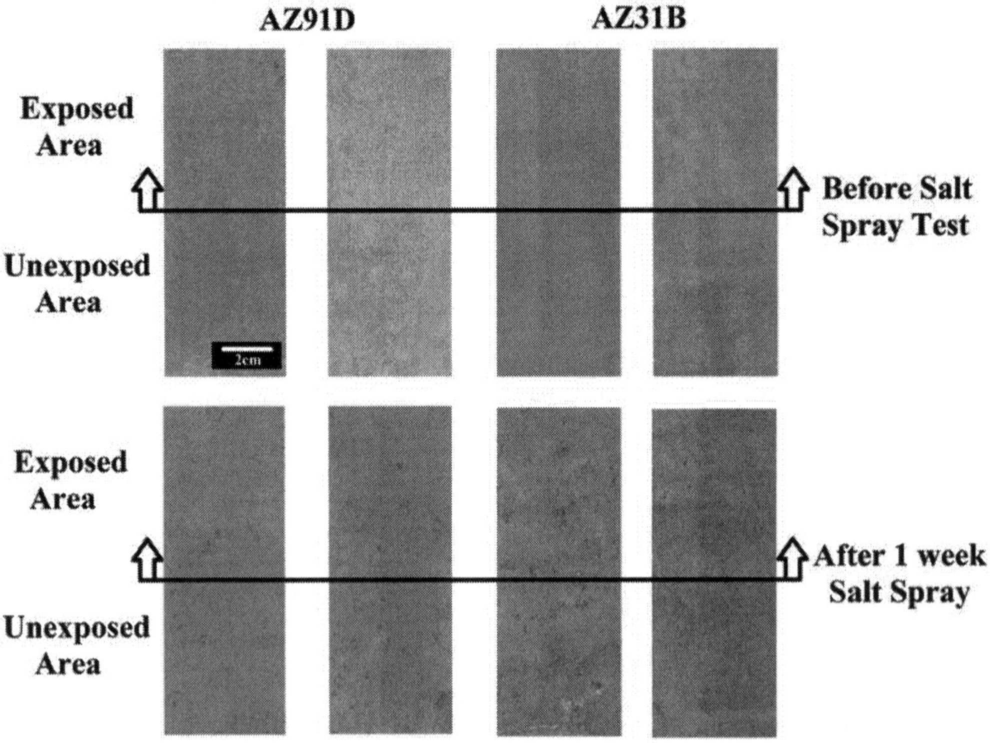

Figure 1. Optical images of cerium-based conversion coatings on AZ91D and AZ31B Mg alloys partially exposed to sunlight before and after 1 week of salt spray testing.

The surface morphologies of the cerium-based conversion coatings on AZ31B and AZ91D Mg alloys after 24 hours of ambient exposure are shown in Figure 2.

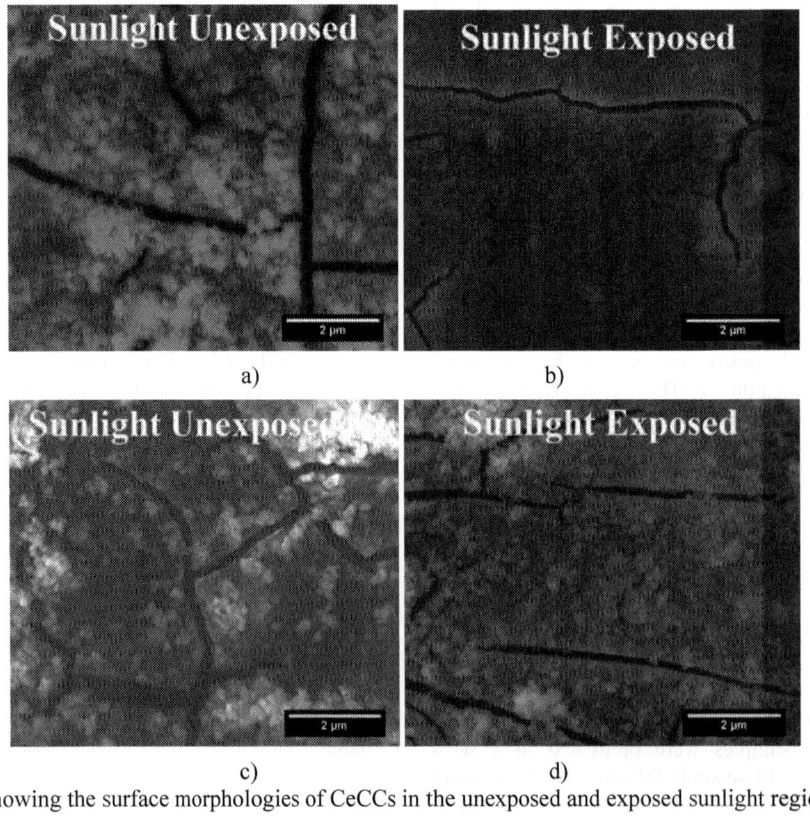

Figure 2. SEM images showing the surface morphologies of CeCCs in the unexposed and exposed sunlight regions on: a), b) AZ31B and c), d) for AZ91D.

Figures 2 a) and 2 c) illustrate the surface morphologies of the CeCC - AZ31B and the CeCC – AZ91D systems without sunlight exposure, respectively. The same uniform cracked surface morphology with small nodular agglomerates has been observed on both magnesium alloys. In addition, this morphology is very similar to the as-coated CeCCs reported previously on magnesium alloys [9-12]. Figures 2 b) and 2 c) reveal the surface structures of the CeCCs on AZ31B and AZ91D after being exposed to 24 hours of sunlight. These coatings showed a slight decrease in cracking and nodule size compared to the unexposed regions of coatings.

The optical reflectance spectra of CeCCs applied on AZ31B and AZ91D Mg alloys with and without being subjected to sunlight are shown in Figures 3 a) and b), respectively. The spectra of the CeCCs subjected to sunlight revealed an increase in the reflectivity at wavelengths between 400 and 500 nm. With this increase, the reflectance of the sunlight exposed samples is consistently higher across the visible range (400-700 nm), appearing lighter than the unexposed regions. Additionally, the spectra of exposed coatings are approximately with the same reflectance value in the visible range which makes the coatings appear white in color. In contrast, the unexposed region presents less reflectance values at wavelengths of 400-500 nm than the rest of the visible range, resulting in a yellowish color.

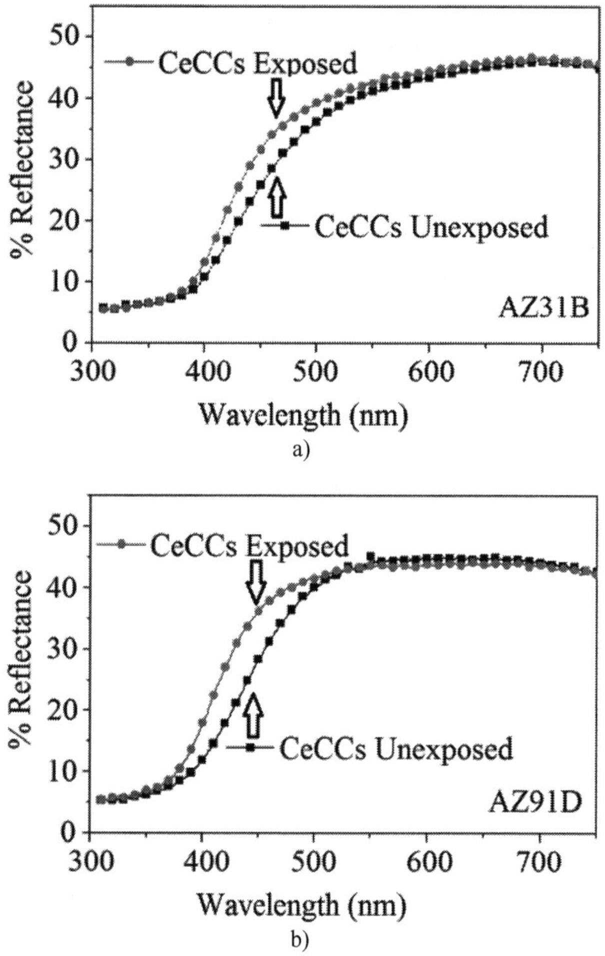

Figure 3. UV-Vis spectra of CeCCs after ambient exposure with and without sunlight exposure on: a) AZ31B and b) AZ91D.

Cyclic potentiodynamic scans of the CeCCs deposited onto AZ31B and AZ91D substrates were measured for ambient exposed regions with and without blocking the sunlight (Figure 4). The corrosion parameters calculated from cyclic potentiodynamic polarization curves are presented in Table I.

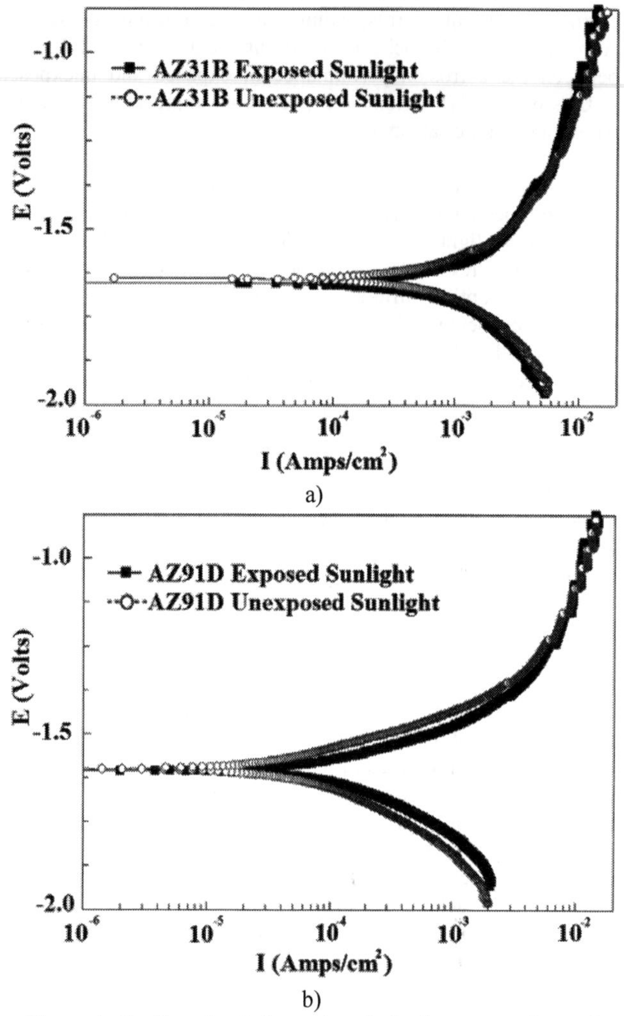

Figure 4. Cyclic potentiodynamic polarization comparison of cerium coated unexposed and exposed to sunlight samples on: a) AZ31B Mg alloy and b) AZ91D Mg alloy.

TABLE I. Corrosion parameters derived from the potentiodynamic polarization experiments.

Sample	i_{corr} µA/cm^2	E_{corr} mV$_{SCE}$
AZ31B Unexposed Sunlight	450	-1640
AZ31B Exposed Sunlight	470	-1660
AZ91D Unexposed Sunlight	41	-1600
AZ91D Exposed Sunlight	82	-1600

The Figure 4 a) suggests that corrosion performance of CeCCs with and without sunlight exposure is equivalent since both curves follow the same trend. The calculated corrosion potentials (E_{corr}) and corrosion current densities (i_{corr}) for both regions in AZ31B are basically the same value. The CeCCs on AZ91D exposed and unexposed to sunlight also showed similar corrosion performance. Nevertheless the unexposed region has slightly better corrosion resistance, Figure 4 b). This observation is based on the smaller corrosion current corresponding to the unexposed sample. Although from electrochemical measurements a slight difference between the corrosion performance of exposed and unexposed CeCCs on AZ91D regions is expected, no appreciable variations were found in the salt spray test.

From electrochemical results practically no difference in corrosion behavior was observed for exposed and unexposed regions to sunlight. This observation is consistent with one week in salt spray testing, where panels performed similar in both regions. The optical difference in regions of CeCCs on both Mg alloys after being subjected to direct sunlight has the same tendency; the UV-Vis results showed an increase in the reflectance values at wavelengths in the range of 400 – 500 nm. These results are in agreement with the visual observations and the change in color might be related with the amount of cerium species in each oxidation state. Yellow precipitates have been correlated with high amounts of Ce(IV) species; colorless or translucent precipitates have been identified with larger amounts of Ce(III) species [13]. The morphologies of samples have been studied; fewer cracks and smaller nodule sizes were observed on sunlight exposed panels against unexposed regions.

Conclusions

The exposure of CeCCs on AZ31B and AZ91D Mg alloys to ambient sunlight changed the appearance of the surface. The exposed regions presented different morphologies and reflectance curves but relatively equivalent corrosion properties. The different morphologies observed in ambient sunlight exposed panels do not give clues on why the color could change. Although the surfaces are slightly different after sunlight exposure on both Mg alloys the similar mud-cracking morphology does not suggest different protection performance. The differences found in the reflectance curves are the confirmation of the color change from pale yellow to almost uncolored after 24 hours of sunlight exposure. The decrease of Ce(IV) species on exposed CeCCs might be an explanation for color change, because Ce(IV) species has been related with the yellowish appearance of CeCCs. The studies of CeCCs before and after ambient exposure by ASTM B117 neutral spray testing and electrochemical polarization measurements showed similar results concluding that the sunlight exposed panels did not modify the corrosion properties.

Acknowledgements

The authors appreciate the technical assistance of Dr. Jay Switzer's group at Missouri University of Science and Technology for the UV-Vis measurements.

References

1. J.E. Gray and B. Luan, "Protective Coatings on Magnesium and its Alloys," *Journal of Alloys and Compounds*, 336 (2002), 88-103.
2. W. X. Zhang et al., "Electroless Ni-P Layer with a Chromium-free Pretreatment on AZ91D Magnesium Alloy," *Surface and Coatings Technology*, 201 (2007), 4594-4600.
3. M. K. Kulekci, "Magnesium and its Alloys Applications in Automotive Industry," *The International Journal of Advanced Manufacturing Technology,* 39 (9-10) (2008), 851–865.
4. H. Umehara, M. Takaya, and S. Terauchi, "Chrome-free Surface Treatments for Magnesium Alloys," *Surface and Coatings Technology*, 169-170 (2003), 666-669.
5. H. Ardelean, I. Frateur, and P. Marcus, "Corrosion Protection of Magnesium Alloys by Cerium, Zirconium and Niobium-based Conversion Coatings," *Corrosion Science*, 50 (2008), 1907-1918.
6. A.L. Rudd, C.B. Breslin, and F. Mansfeld, "The Corrosion Protection Afforded by Rare Earth Conversion Coatings Applied to Magnesium," *Corrosion Science,* 42 (2000), 275-288.
7. M. Dabalà et al., "Cerium-based Chemical Conversion Coatings on AZ63 Magnesium Alloy," *Surface and Coatings Technology*, 172 (2003), 227-232.
8. K. Brunelli et al., "Effect of HCl Pre-treatment on Corrosion Resistance of Cerium-based Conversion Coatings on Magnesium and Magnesium Alloys," *Corrosion Science*, 47 (4) (2005), 989-1000.
9. S. Maddela et al., "Influence of Surface Pretreatment on Coating Morphology and Corrosion Performance of Cerium-based Conversion Coatings on AZ91 D Alloy," *Corrosion*, 66 (11) (2010), 115006-1/115006-8.
10. C. E. Castano et al., "A Comparative Study on the Corrosion Resistance of Cerium-based Conversion Coatings on AZ91D and AZ31B Magnesium Alloys," *ECS Transactions*, 41 (15) (2012), 3-12.
11. C. E. Castano et al. "Effect of Thickness on the Morphology and Corrosion Behavior of Cerium-based Conversion Coatings on AZ31B Magnesium Alloy," *Magnesium Technology 2012*, ed. S.N. Mathaudhu et al. (Hoboken, NJ: TMS, John Wiley & Sons, Inc., 2012) 247-252.
12. C. E. Castano et al., "Effect of Deposition Temperature on the Properties of Cerium-based Conversion Coatings on AZ31 and AZ91 Magnesium Alloys," (Paper presented at the 9th International Conference on Magnesium Alloys and their Applications, Vancouver, British Columbia, Canada, 9 July 2012.
13. S. A. Hayes et al., "The Phase Stability of Cerium Species in Aqueous Systems: I. E-pH Diagram for the System," *Journals of the Electrochemical Society*, 149 (12) (2002), C623/C630.

Magnesium Technology 2013

Nanostructured Materials

Magnesium Technology 2013
Edited by: Norbert Hort, Suveen N. Mathaudhu, Neale R. Neelameggham, and Martyn Alderman
TMS (The Minerals, Metals & Materials Society), 2013

THERMAL STABILITY OF ULTRA-FINE GRAINED MAGNESIUM ALLOY PROCESSED BY EXTRUSION AND ECAP

Jitka Vrátná[1], Miloš Janeček[1]
[1]Charles University in Prague, Department of Physics of Materials, Ke Karlovu 5, CZ-12116 Czech Republic

Keywords: AZ31 magnesium alloy, Severe plastic deformation (SPD), Equal channel angular pressing (ECAP), Thermal stability

Abstract

The mechanical properties and thermal stability of ultra-fine grained (UFG) structure of magnesium alloy AZ31 during annealing was investigated. UFG specimens were prepared by a combined two-step severe plastic deformation process: the extrusion (EX) and the equal-channel angular pressing (ECAP). This combined process leads to microstructure refinement and enhanced microhardness. Specimens were annealed isochronally at temperatures 150 - 400 °C for 1 hour and isothermally at temperatures 170 - 210 °C. The evolution of microstructure and mechanical properties were studied by light and scanning electron microscopy and microhardness measurements. The coarsening of a fine-grained structure at higher temperatures was accompanied by a gradual decrease of the microhardness.

Introduction

Magnesium alloys are energy-efficient materials having the potential to replace some conventional structural materials, e.g. steel or aluminum alloys. Magnesium is a very light metal (its density is 1.74 g/cm^3) with relatively high specific mechanical properties which results in expanding use of magnesium-based materials in weight-critical applications [1].

The mechanical and other physical properties may be improved by refining the grain size to submicrometer or even nanometer level. It has been known for many decades, going back to the 1950s, that the structure of deformed metals can change with increasing plastic deformation such that random dislocation arrays can lower the energy of the system by "self-assembling" into "cells" or "subgrains" such that there is a high dislocation density in the cell walls and lower dislocation density within the cells [2].

A variety of special techniques are used for the production of bulk ultra-fine grained (UFG) materials, e.g. equal channel angular pressing (ECAP) [3], high pressure torsion (HPT) [4], accumulative roll-bonding (ARB) [5], twist extrusion [6] or multi-directional forging [7].

Among these techniques, which introduce the severe plastic deformation (SPD) in the material, ECAP is quite an easy and widely-used method of preparation of very fine grain structure. A combined two-step process involving an initial extrusion step and subsequent processing by ECAP (so called EX-ECAP) leads to the enhanced mechanical properties and homogeneous ultra-fine grain microstructure [8, 9].

However, the application of UFG materials is limited due to the structure stability at elevated temperatures. Thermal stability depends on many variables (e.g. stacking fault energy of the material, preparation process and its parameters or properties of grain boundaries) [10].

This work is therefore motivated by this fact and its main objective is to investigate thermal stability - microstructure and microhardness evolution during isochronal and isothermal annealing of the UFG AZ31 magnesium alloy prepared by extrusion and 4 passes of ECAP.

Experimental procedures

Commercial AZ31 magnesium alloy, with a nominal composition of Mg-3%Al-1%Zn in the initial as cast condition was used in this investigation. The material was first extruded at T = 350 °C with an extrusion ratio of ER = 22 using a 630 t direct extrusion press. Billets of the dimensions 10 × 10 × 120 mm were machined from the extruded bar. ECAP pressing was performed at 180°C to reduce grain growth during pressing [11] following route B_c at the speed of 50 mm/min for 4 passes.

Specimens for light (LM) and scanning electron microscopy (SEM) observation of the microstructure were taken from the middle part of the billets perpendicular to the pressing direction. Specimens were first mechanically grinded on watered abrasive papers, and then polished with polishing diamond suspension of grade 3, 1 and ¼ μm and alumina suspension of grade 0.05 μm. Using this procedure, flat samples for Vickers microhardness (load 100 g, 10 s) measurements with minimum surface scratches were obtained. Finally, the specimens′ surface was etched with picric acid for 1 second which enables observations in LM and SEM.

Results and discussion

The first set of specimens was annealed isochronally at temperatures 150 – 400 °C for 1 hour and quenched in water. Subsequently, the Vickers microhardness HV0.1 was measured at the specimens (at least 15 indents were made in each sample). The results are shown in Fig.1a. The first data point (HV0.1 = 86) corresponds to the initial non-annealed specimen. One can see that microhardness values after annealing at 150 and 170 °C differ within the statistical error only. However, the microhardness declines abruptly in the range of 170 - 230 °C and then continuously up to 400 °C.

Results of isothermal annealing at the temperatures 170, 190 and 210 °C are depicted in Fig.1b and 1c. The HV0.1 values decline more rapidly at higher temperatures after short annealing times. The final microhardness of the specimen annealed at 170 °C for 64 hours is approximately equal to 76, at 190 °C 72 and at 210 °C 67 HV0.1.

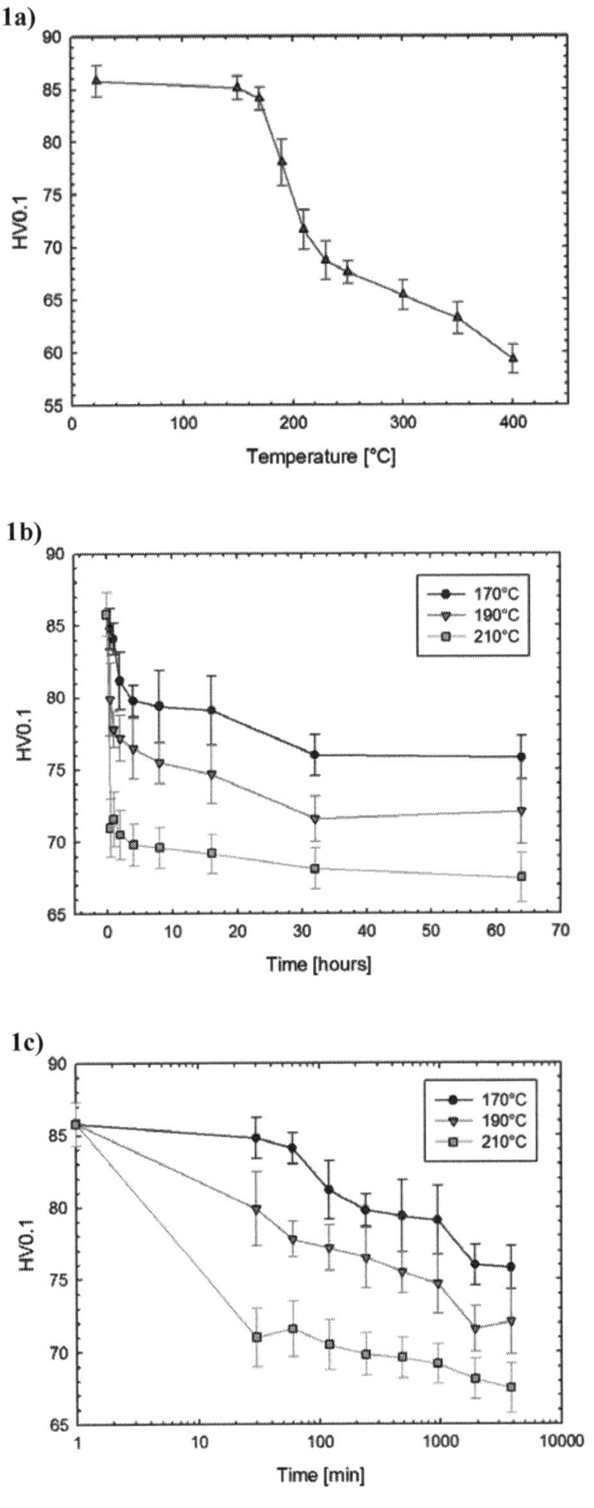

Figure 1: Measurements of thermal stability of AZ31 alloy after extrusion and 4 passes of ECAP: a) microhardness after isochronal annealing, b) and c) microhardness after isothermal annealing at temperatures 170, 190 and 210 °C in a linear and logarithmic scale, respectively.

Microstructure of the AZ31 alloy after extrusion and 4 passes of ECAP is nearly homogeneous with very fine grains of the average size 1 μm and few remaining larger grains (see Fig. 2 and details in Janeček et al [8]).

In order to characterize microstructure changes during annealing we will compare the final conditions of the isothermal annealing (after 64 hours) in the temperature range of 170 - 210 °C in which microhardness drop was found and therefore grain structure variations are expected to occur due to static recovery and grain growth. The grain structure after long term exposure (64 hours) in this temperature range will be characterized both by light microscopy using Nomarski contrast and by scanning electron microscopy using secondary electron signal.

The microstructure of sample annealed at 170 °C for 64 hours is shown in Figs. 3a (LM) and 4a, 5a (SEM). It is seen that in this condition the ultra-fine grained character of microstructure remains with the majority of very fine grains (approximately 80% of the volume). However, first larger grains of the sizes ~ 5 - 10 μm started to be formed.

The bimodal character of grain structure is also observed at long term annealing at 190 °C, see Figs. 3b (LM) and 4b, 5b (SEM) consisting of agglomerates of ultra-fine grains and large grains of the sizes ~ 5 – 15 μm. However, the fraction of corse grains significantly increases as compared to annealing at lower temperature. Approximately the same fraction of coarse and fine grains was found in this specimen.

Annealing at the highest temperature from the range of microhardness drop (210 °C) results in much more homogeneous and coarser grain structure as compared to previous conditions as depicted in Figs. 3c (LM) and 4c - 5c (SEM). Coarse grains of the average size of ~ 5 – 15 μm prevail in the microstructure. However, a certain fraction of fine grains still remains.

Fig. 3d (LM) and 4d - 5d (SEM) present the microstructure of the condition upon annealing at 400 °C for 1 hour, where the processes of recovery and grain growth are expected to be mostly developed due to very low values of microhardness. The microstructure in this condition is much more homogeneous in comparison with the isothermally annealed specimens. It consists of larger grains of the average sizes ~ 5 – 20 μm only.

Microstructure observations confirmed that different microstructures correspond well to the different microhardness values measured at isochronally and isothermally annealed specimens.

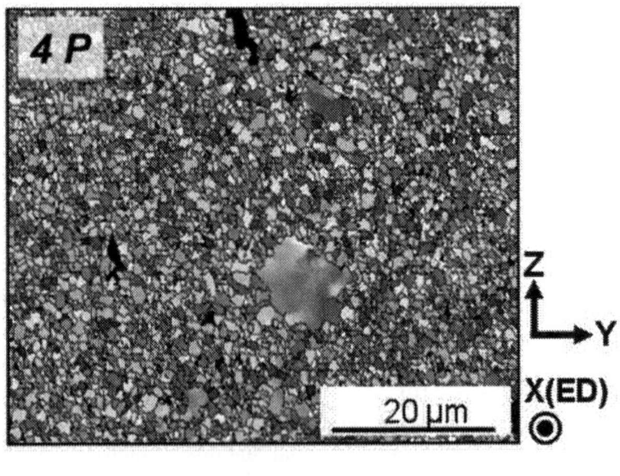

Figure 2: Microstructure of the AZ31 alloy after extrusion and 4 passes of ECAP (EBSD) [8].

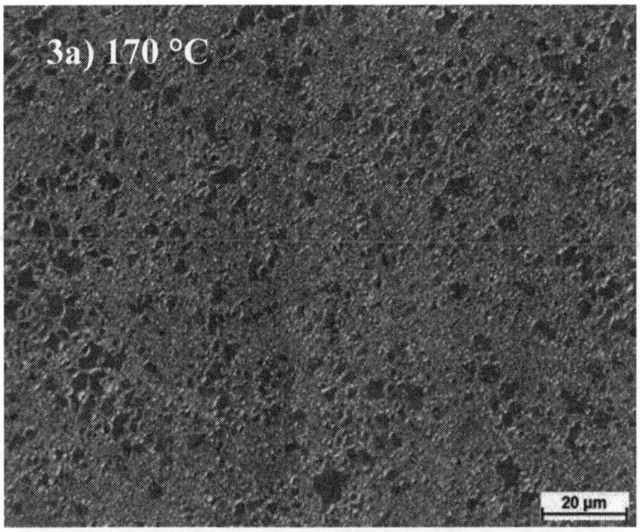

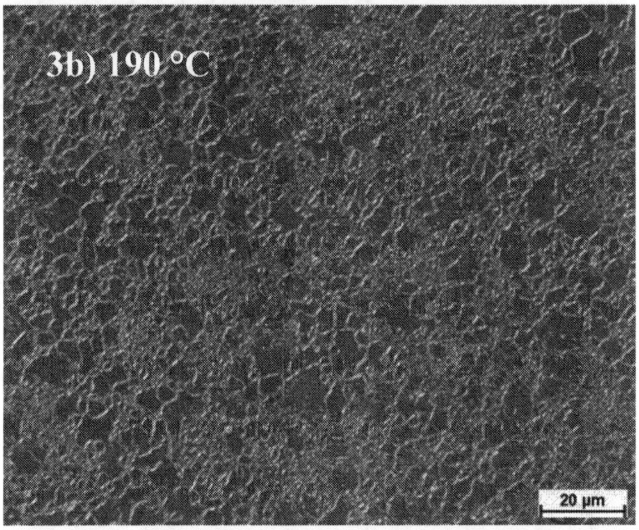

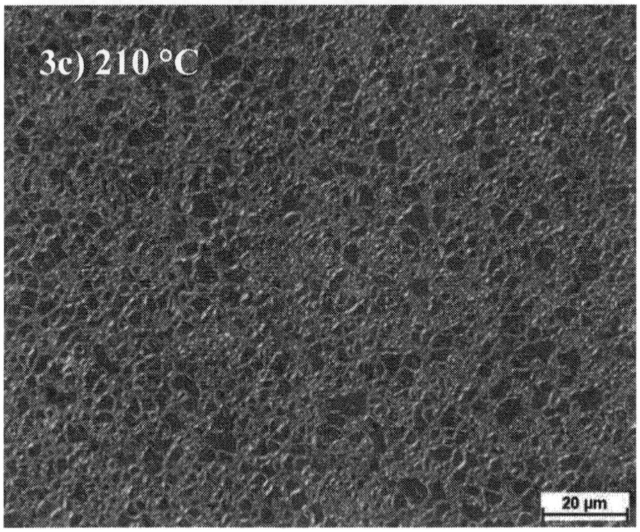

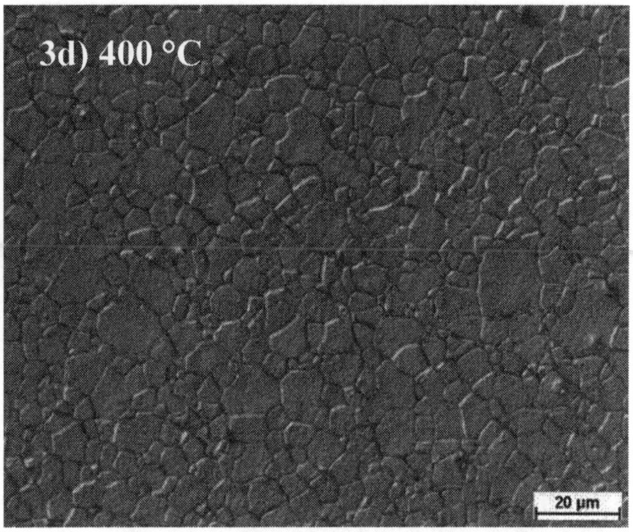

Figure 3: Microstructure of the AZ31 alloy after annealing: a) 64 hours at 170 °C, b) 64 hours at 190 °C, c) 64 hours at 210 °C and d) 1 hour at 400 °C. (Light microscopy, magnification 1000×, etchant picric acid, Nomarski contrast)

Figure 4: Microstructure of the AZ31 alloy after annealing: a) 64 hours at 170 °C, b) 64 hours at 190 °C, c) 64 hours at 210 °C and d) 1 hour at 400 °C. (Scanning electron microscopy, secondary electron signal, magnification 4000×, etchant picric acid)

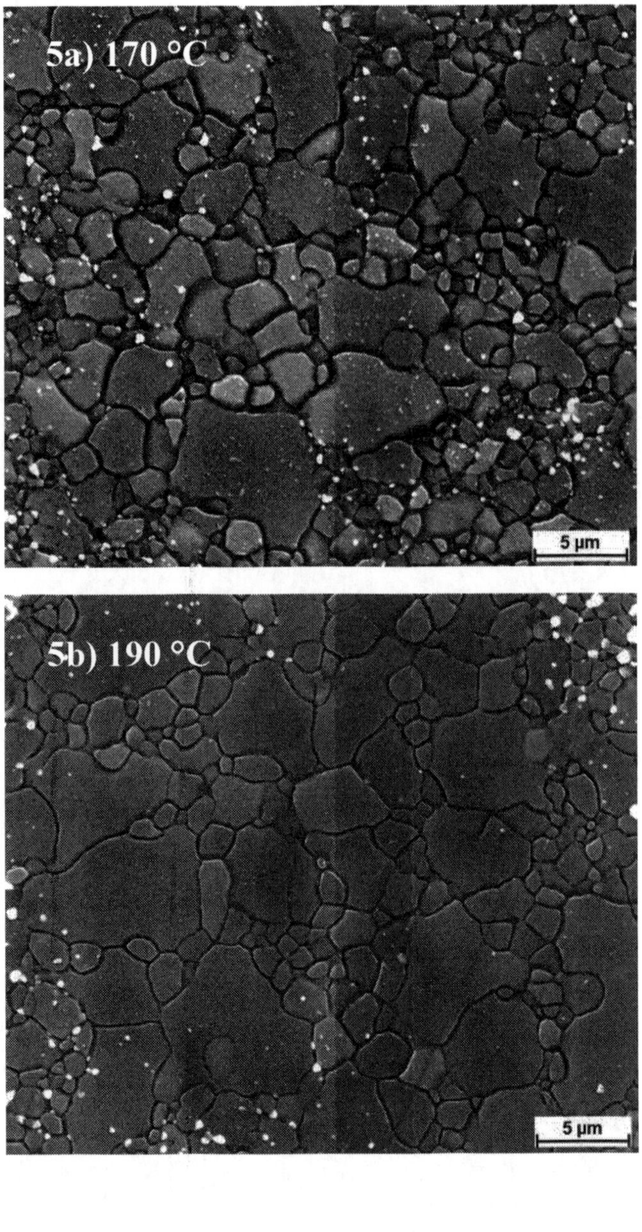

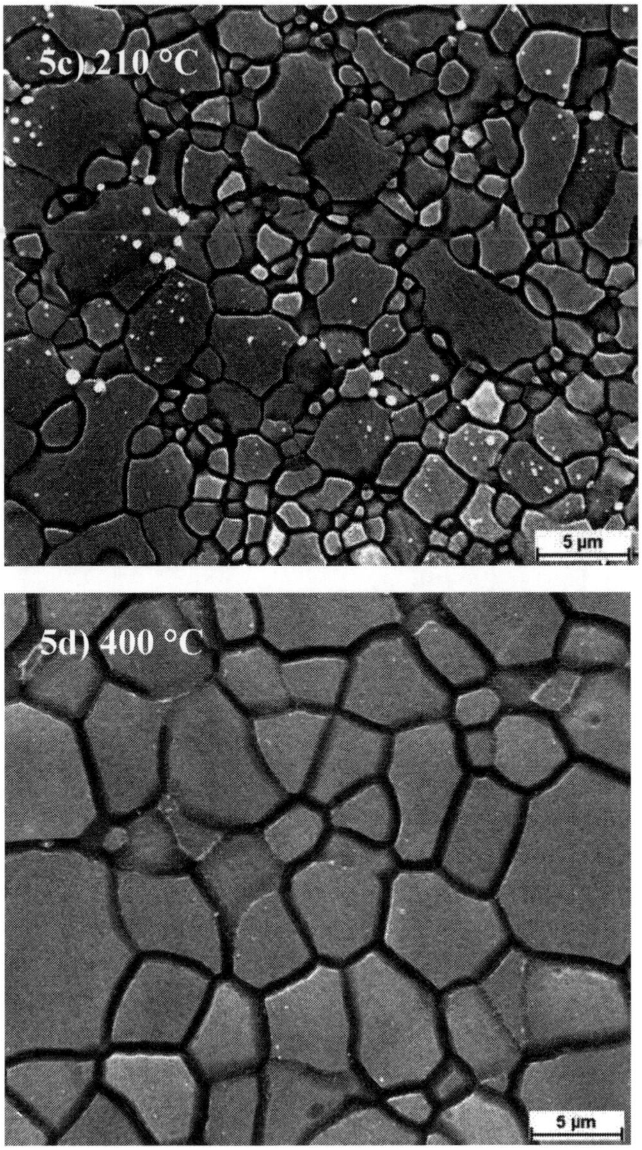

Figure 5: Microstructure of the AZ31 alloy after annealing: a) 64 hours at 170 °C, b) 64 hours at 190 °C, c) 64 hours at 210 °C and d) 1 hour at 400 °C. (Scanning electron microscopy, secondary electron signal, magnification 8000×, etchant picric acid)

Conclusions

The microstructure and microhardness of ultra-fine grained (UFG) magnesium alloy AZ31 and its evolution with annealing temperature and time were investigated. The following conclusions may be drawn from this work:

- The isochronal annealing leads to decrease of the microhardness (from 86 to 59 HV0.1) and significant coarsening of the initial UFG microstructure.

- The isothermal annealing measured at 170, 190 and 210 °C results in microhardness decrease to the values of 76, 72 and 67 HV0.1 and formation of inhomogeneous microstructure.

Acknowledgements

This work was financially supported by the Czech Grant Agency (GACR) under the Grant 106/09/0482. J. Vrátná acknowledges finacial support by GAUK 530712/2012 and SVV 2012-265303.

References

[1] M. Gupta, N.M.L. Sharon, *Magnesium, Magnesium Alloys, and Magnesium Composites*, (New Jersey, John Wiley &Sons, Inc., 2011), 1-3.

[2] C.C. Koch. "Nanostructured Materials: An Overview," Bulk Nanostructured Materials", ed. M.J. Zehetbauer and Y.T Zhu (Weinheim, Wiley-VCH Verlag GmbH & Co. KGaA, 2009), 9.

[3] R.Z. Valiev, T.G. Langdon, "Principles of equal-channel angular pressing as a processing tool for grain refinement" *Progress in Materials Science* (2006) 51, 881–981.

[4] A.P. Zhilyaev, T.G. Langdon "Using high-pressure torsion for metal processing: Fundamentals and applications" *Progress in Materials Science* (2008) 53, 893-979.

[5] Y. Saito, H. Utsunomiya, N. Tsuji, T. Sakai, "Novel ultra-high straining process for bulk materials-development of the accumulative roll-bonding (ARB) process" *Acta Mater* (1999) 47, 579-583.

[6] N.V. Varyutkhin, Y. Beygelzimer, S. Synkov, D. Orlov, "Application of Twist Extrusion" *Mater Sci Forum* (2006) 503-504, 335-340.

[7] X. Yang, Y. Okabe, H. Miura, T. Sakai, "Effect of pass strain and temperature on recrystallisation in magnesium alloy AZ31 after interrupted cold deformation" *J Mat Sci* (2012) 47, 2823-2830.

[8] M. Janeček, S. Yi, R. Král, J. Vrátná, K.U. Kainer, "Texture and microstructure evolution in ultrafine-grained AZ31 processed by EX-ECAP" *J Mater Sci* (2010) 45, 4665-4671.

[9] M. Janeček, J. Čížek, J. Gubicza, J. Vrátná, "Microstructure and dislocation density evolutions in MgAlZn alloy processed by severe plastic deformation" *J Mater Sci* (2012) 47, 7860-7869.

[10] J. Gubicza, N.Q. Chinh, S.V. Dobatkin, E. Khosravi, T.G Langdon, "Stability of ultrafine-grained microstructure in fcc metals processed by severe plastic deformation" *Key Engineering Materials* (2011) 465, 195-198.

[11] R.B. Figueiredo, T.G. Langdon, "Principles of grain refinement and superplastic flow in magnesium alloys processed by ECAP" *Mater Sci Eng A* (2009) 501, 105.

Magnesium Technology 2013

Corrosion II

Magnesium Technology 2013
Edited by: Norbert Hort, Suveen N. Mathaudhu, Neale R. Neelameggham, and Martyn Alderman
TMS (The Minerals, Metals & Materials Society), 2013

FORMATION OF VANADATE CONVERSION COATING ON AZ31 MAGNESIUM ALLOY

S. A. Salman[1, 2], K. Kuroda[1], M. Okido[1]
[1]*EcoTopia Science Institute, Nagoya University, Furo−cho, Chikusa, Nagoya 464−8603, Japan*
[2]*Graduate School of Engineering, Al-Azhar University, Nasr City, Cairo 11371, Egypt*

Keywords: Mg alloy, Anodizing, Vanadate, Magnesium, conversion coating.

Abstract

Magnesium alloys have high strength-to-weight ratios, excellent castability, machinability, weldability, and thermal stability, and good damping capacity. Therefore, Magnesium alloys are recognized as alternatives to Al alloys and steel to reduce the weight of structural materials. However, a major obstacle to the widespread use of magnesium alloys is its poor corrosion resistance, particularly in wet environments. Therefore, further surface treatment of magnesium and its alloy is important in meeting several industrial specifications. In the present investigation, a chromate-free, corrosion-resistant conversion coating using vanadium based solution was applied to AZ31 magnesium alloy. Conversion coating was carried out in vanadium based solution. The effect of vanadium concentration and treatment time on the coating film was characterized by SEM, EDX and XRD. The corrosion characteristics of anodic films were evaluated using potentiodynamic polarization test in deaerated NaCl solution. Corrosion resistance property was improved with conversion coating treatment.

Introduction

Magnesium is the lightest of all structural metal materials in practical use, with a density equivalent to two-third the density of Al, one-third that of Zn and one quarter that of steel. It also has a good electrical/heat conductivity, damping properties and machining abilities. These properties make magnesium alloys an attractive target in a number of applications, especially in fields where weight reduction is critical or there are particular technical requirements. However, a major obstacle to the widespread use of magnesium alloys is its poor corrosion resistance: magnesium alloys are highly susceptible to corrosion attack, particularly in wet environments. Therefore, further surface treatment of magnesium and its alloy is important in meeting several industrial specifications. The common coating processes are electrochemical plating, chemical conversion, anodizing, physical vapor deposition, and electroless nickel plating.

Of these, conversion coatings are economical, easy to use and offer a variety of valuable properties to extend the service life and improve the performance characteristics of the magnesium alloy parts being treated. Chromate conversion coating is widely used in industry because of its desirable properties: it provides excellent corrosion protection through the corrosion-inhibiting properties of the hexavalent chromium present in the film and through the physical barrier of the film itself. However, changing environmental regulations and pollution prevention requirements have led to a significant push to find a new alternative to the use of poisonous Cr^{6+}, therefore, developing an environmentally friendly surface treatment has attracted a great deal of research interest in recent years. Several non-chromate solutions are thus developed, such as rare earth salt [1, 2, 3, 4, 5], stannate [6], phosphate [7] and phosphate/permanganate solutions [8, 9]. Vanadate was also suggested as pretreatment candidates aiming to self-healing conversion coating [10] Guan et al. developed a new type of inorganic conversion coating formed on aluminum alloy substrates by dipping the substrate in aqueous vanadate-based solutions at ambient temperatures [11]. The anticorrosion properties was improved with vanadate contained coatings, which showed that in near-neutral solutions, vanadate increases the pitting potential, and decreases the oxygen reduction reaction rate on AA2024-T3 Al alloy. Sodium vanadate was added to solution contains phosphate and fluoride ions with other active corrosion inhibitor in order to form an improved chromate-free corrosion resistant conversion coating for magnesium alloy [12]. The addition of sodium vanadate is a preferred choice because it improves the humidity resistance of the conversion coating over a wide range of concentrations and enables use of a 50% shorter coating cycle. Vanadium solution is usually used as the corrosion resistant inhibitor for the paint or pigment systems. However, only few works are reported on the application of vanadium-based conversion coatings on magnesium alloys. The vanadate-based conversion coating can significantly improve the corrosion resistance of the AZ61 magnesium alloy [13]. Vanadate conversion coating was formed on the Mg–Li–Al–Ce alloy in $NH_4VO_3 + K_3(Fe(CN)_6)$ solutions. Uniform coating with regular distributed pores was formed the anticorrosion property was improved with his conversion coating [14]

In the present research, the formation of conversion coatings on AZ31 magnesium alloy in vanadium based solution was investigated. The effect of time and concentration of vanadium on the anti-corrosion property of the coating films were evaluated using the potentiodynamic polarization curves. Furthermore, the surface morphologies and phase structure were detected using SEM, EDX, XPS and XRD.

Experimental

Specimens

Commercially available AZ31 Mg alloys were used as the substrate. The chemical composition of the alloy is listed in Table 1. The surface of the alloy was polished up to #2000 emery paper followed by 0.05 μm alumina powders. The specimens were carefully cleaned with water, rinsed with acetone and dried under air. All of the experiment specimens were mounted using polytetrafluoroethylene (PTFE) resin tape, leaving 1 cm^2 surface area.

Table 1 Chemical composition of AZ31 Mg alloy (mass%)

Al	Zn	Mn	Si	Cu	Ni	Fe	Mg
3.0	1.0	0.43	0.01	<0.01	0.001	<0.003	Bal.

Conversion coating

Anodizing was performed at room temperature for 300 - 1800 s in the aqueous electrolyte of Vanadyl Sulfate at various concentration range 5 – 25 g/l. After the treatment, the specimens were carefully rinsed using distilled water and dried under air before analysis.

Morphology and Structure of Anodic Film

The morphology and microstructure of the anodic films were observed with a Hitachi S-800 scanning electron microscope (SEM), the phase structure and the composition of the coating films were identified with X-ray diffraction (XRD) and Energy Dispersive X-ray Spectrometry (EDS).

Corrosion Measurements

The potentiodynamic polarization test was carried out using a Solartron 1285 potentiostat from Solartron Analytical, Farnborough, United Kingdom. The measurements were controlled by Scribner Associates Corrware electrochemical experiment software. The anodic and cathodic polarization curves were measured in 0.017 M NaCl and 0.1 M Na_2SO_4 solution at 298 K with a scanning rate of 1 mVs^{-1}.

Results and Discussion

Figure 1 shows the open-circuit potential (OCP) as a function of time throughout the conversion coating treatment of AZ31 magnesium alloy in 5 g/l $VOSO_4.xH_2O$ for 600s. The OCP increased from approximately -1.72 V upon immersion in vanadium-based solution to -1.5 V after 600 s.

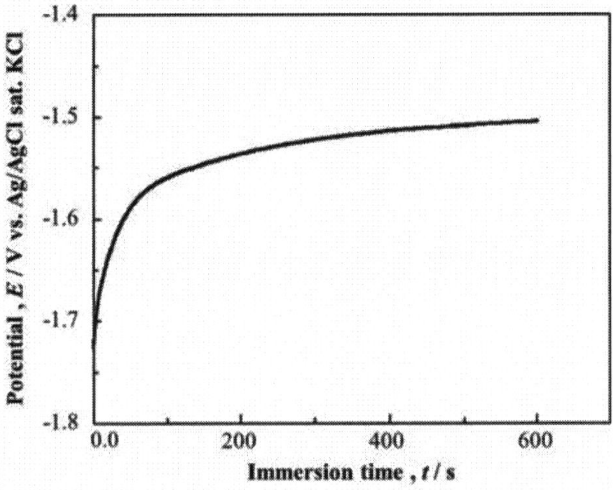

Figure 1. The open-circuit potential-time transient curve throughout conversion coating treatment of AZ31 magnesium alloy in 5 g/l $VOSO_4.xH_2O$ for 600s.

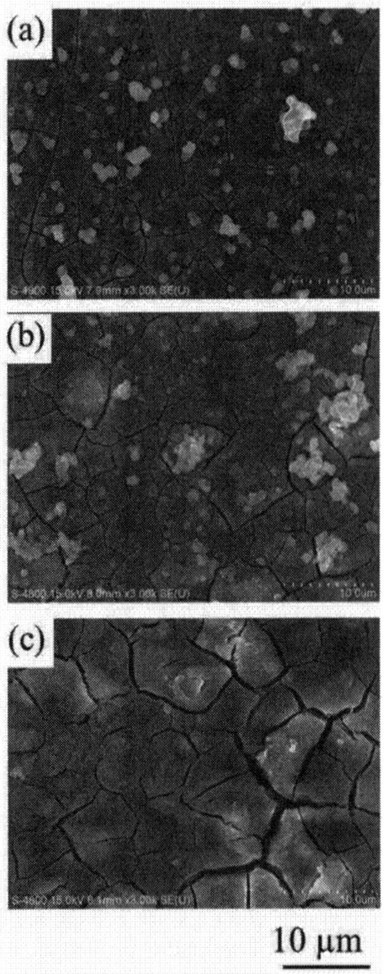

Figure 2. SEM images of conversion coating films formed on AZ31 Mg alloy in vanadium-based solution with various $VOSO_4.xH_2O$ concentration (a) 10g/l, (b) 15 g/l, and (c) 25 g/l for 15 min.

Figure 2 shows the SEM images of conversion coating films formed on AZ31 Mg alloy in vanadium-based solution with various concentrations. The coating film has a brown color to the naked eye. The coating revealed the presence of several micro-cracks, the wide of these micro-cracks increases with vanadium concentration increasing in the treatment solution. The maximum size of these micro-cracks was 0.35, 0.64, and 1.0 μm for the film formed in the solution with concentration of 10g/l, 15 g/l, and 25 g/l respectively.

Potentiodynamic polarization curves of AZ31 Mg alloys before and after treatments in vanadium-based solution were measured in 0.017 M NaCl and 0.1 M Na_2SO_4 solution at 298 K with a scanning rate of 1 mVs^{-1}. It was noted that the corrosion potentials (E_{Corr}) were significantly increased with treatment in vanadium solution. Figure 3 shows that the film formed in the solution containing 25 g/l and 15g/l $VOSO_4.xH_2O$ has nearly the same corrosion potential of -1.5 V. The film formed in the solution containing 10 g/l has the largest corrosion potential. The pitting potential (E_p) of the coating film formed at 10g/l has the largest potential of -1.21 V compared to non-treated specimen -1.35 V and those treated in 15 and 25g/l $VOSO_4.xH_2O$ -1.30 and -1.34 respectively. This indicate that the film formed in 10g/l has better anticorrosion property and these results agree with the SEM image results as this film has a very narrow micro-cracks compared with those formed in the solution with concentration of, 15 g/l and 25 g/l $VOSO_4.xH_2O$. So, the concentration of 10 g/l was used to determine the effect of treatment time.

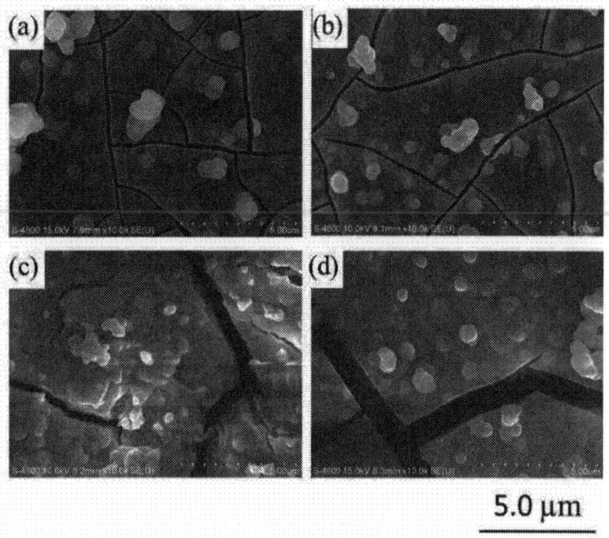

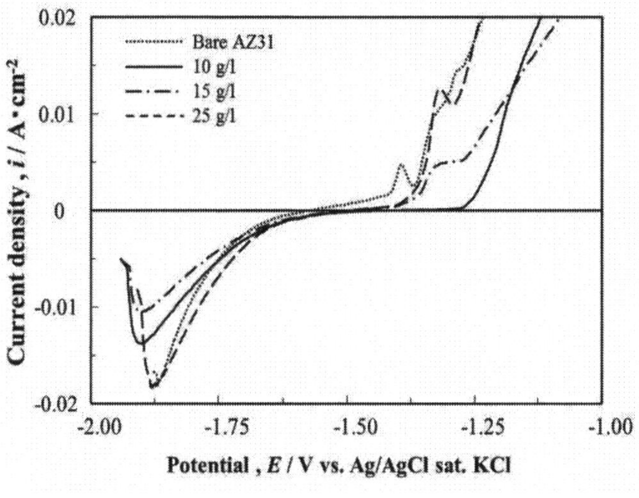

$VOSO_4.xH_2O$

Figure 4 shows the SEM images of the conversion coating films formed on AZ31 Mg alloy throughout the immersion in 10g/l $VOSO_4.xH_2O$ solution for (a) 15 min, (b) 30 min, (c) 60 min, and (d) 120 min. The figure shows that the micro-cracks have propagated throughout the coating surface. The films formed mainly from a compact white precipitates from magnesium and vanadium oxides. Only a few unattached white precipitates were located on the surface of the coatings. The maximum size of these micro-cracks was 0.09, 0.09, 0.71 and 1.19 μm for the film formed in the solution with concentration of 10g/l for 15 min, 30 min, 60 min, and 120 min respectively. The results clearly show that the film formed at 30 min. has smooth surface with minor micro-cracks.

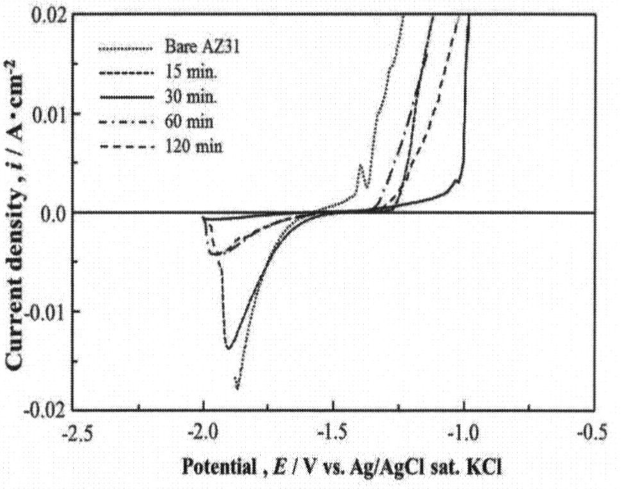

Figure 5 shows the potentiodynamic polarization curves of AZ31 Mg alloy after treatment in various times in vanadate solution. The pitting potential of the films formed in the solution containing 10 g/l $VOSO_4.xH_2O$ for various treatment times were better than bare AZ31. However, the film formed at 30 min. has the noblest pitting potential of 1.03 V. Therefore, the EDS analysis was carried out on the coating film formed in 10g/l $VOSO_4.xH_2O$ for 30 min to reveal the elemental presence and distribution on the surface. Figure 6(a) show the SEM image of the conversion coating film. The EDS elemental mapping Fig 6(b, c, d) clearly shows the distribution of Mg, O, and V respectively in the coatings. The Vanadium elemental mapping confirms their presence in the coating film as shown in Fig. 6(c). Magnesium and oxygen elemental mapping reveal that the film contains MgO and/or $Mg(OH)_2$. Furthermore, results shows that the white color precipitates which observed on the surface are mainly compose of Vanadium and Oxygen which reveal to formation of Vanadium oxide and/or hydroxides. Further investigation is needed to determine the exact mechanism of the conversion coating.

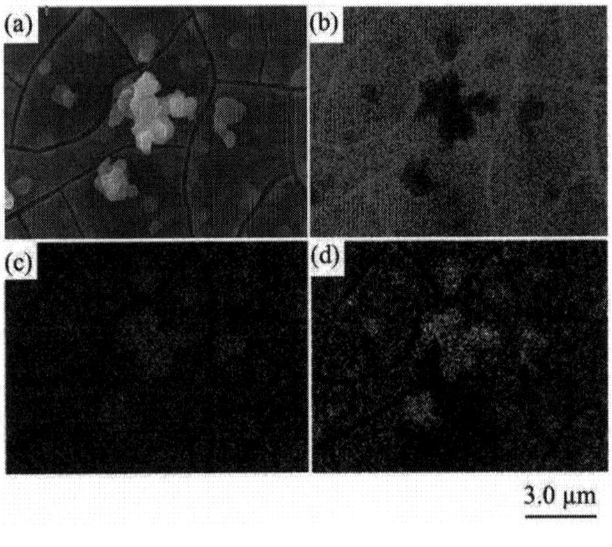

Conclusions

Conversion coating was performed on AZ31 Mg alloy in various concentration of $VOSO_4.xH_2O$. The effect of treatment time on the microstructure and anticorrosion property of the films were examined and. The best anticorrosion property was obtained with treatment in 10g/l $VOSO_4.xH_2O$ for 30 min. this film has a smooth surface with minor micro-cracks compared to other films. The results show that vanadium oxides and/or hydroxides were included in the coating films.

Acknowledgment

The authors gratefully acknowledge the financial support of Aichi Center for Industry and Science Technology and the Ministry of the Education, Culture, Sports, Science and Technology, Japan. We wish also to acknowledge Professor Ryoichi Ichino of Ecotopia Science Institute., Nagoya University for valuable discussion during this research.

References

1. A.L. Rudd, C.B. Breslin, F. Mansfeld, "The corrosion protection afforded by rare earth conversion coatings applied to magnesium," *Corrosion Science,* 42 (2000), 275-288.
2. C.S. Lin and S.K. Fang, "Formation of Cerium Conversion Coatings on AZ31 Magnesium Alloys," *Journal of the Electrochemical Society*, 152 (2) (2005), B54-B59.
3. A.L. Rudd, C.B. Breslin, and F. Mansfeld, "The corrosion protection afforded by rare earth conversion coatings applied to magnesium" *Corrosion Science*, 42 (2) (2000), 275-288.
4. M. Dabalà, K. Brunelli, E. Napolitani, and M. Magrini, "Cerium-based chemical conversion coating on AZ63 magnesium alloy," *Surface & Coatings Technology*, 172 (2003), 227-232.
5. S.A. Salman, R. Ichino, and M Okido, "Development of Cerium-based Conversion Coating on ZA31 Magnesium Alloy," *Chemistry Letters*, 36 (8) (2007), 1024-2025.
6. F. Zucchi, A. Frignani, V. Grassi, G. Trabanelli, C. Monticelli, "Stannate and permanganate conversion coatings on AZ31 magnesium alloy," *Corrosion Science*, 49 (2007), 4542-4552.
7. W. Zhou, D. Shan, E. Han, W. Ke, "Structure and formation mechanism of phosphate conversion coating on die-cast AZ91D magnesium alloy," *Corrosion Science,* 50 (2008), 329-337.
8. K.Z. Chong, T.S. Shih, "Conversion-coating treatment for magnesium alloys by a permanganate–phosphate solution," Materials Chemistry and Physics, 80 (2003), 191-200.
9. M. Mosiałek, G. Mordarski, P. Nowak, W. Simka, G. Nawrat, M. Hanke, R.P. Socha, J. Michalska, "Phosphate–permanganate conversion coatings on the AZ81 magnesium alloy: SEM, EIS and XPS studies," *Surface & Coatings Technology*, 206 (2011), 51-62.
10. Swapan K. Ghosh, Self-healing Materials: Fundamentals, Design Strategies and Applications, (Weinheim: Germany. WILEY-VCH Verlag GmbH & Co., 2008), 114.

11. H. Guan and R.G. Buchheit, "Corrosion Protection of Aluminum Alloy 2014-T3 by Vanadate Conversion Coatings," *Corrosion Science Section, Corrosion.* 60 (3) (2004), 285- 296.
12. Owen M. Briles, M. Jaworowski, M.A. Kryzma, X. Tang, "Corrosion resistant, chromate-free conversion coating for magnesium alloys," US Patent, US (2003), 20030150525.
13. K.H. Yang, M.D. Ger, W.H. Hwu, Y. Sung, Y.C. Liu, "Study of vanadium-based chemical conversion coating on the corrosion resistance of magnesium alloy," *Materials Chemistry and Physics,* 101 (2007), 480-485.
14. Y. Ma, N. Li, D. Li, M. Zhang, X. Huang, "Characteristics and corrosion studies of vanadate conversion coating formed on Mg–14 wt% Li–1 wt% Al–0.1 wt% Ce alloy," *Applied Surface Science*, In Press, Corrected Proof, Available online 23 July 2012.

Magnesium Technology 2013

Joining and Friction Processing

Magnesium Technology 2013
Edited by: Norbert Hort, Suveen N. Mathaudhu, Neale R. Neelameggham, and Martyn Alderman
TMS (The Minerals, Metals & Materials Society), 2013

Microstructure modification and performance improvement of Mg-RE alloys by friction stir processing

Yujuan Wu[1, 2], Liming Peng [1, 2*], Feiyan Zheng[1, 2], Xuewen Li[1, 2], Dejiang Li[1,2*], Wenjiang Ding [1, 2]
[1] *National Engineering Research Center of Light Alloy Net Forming, Shanghai Jiao Tong University, 800 Dongchuan Road, Shanghai 200240, PR China*
[2] *The State Key Laboratory of Metal Matrix Composites, Shanghai Jiao Tong University, 800 Dongchuan Road, Shanghai 200240, PR China*

Keywords: Friction stir processing, Mg–Gd–Zn–Zr, Mg-Gd-Ag-Zr, Mg-Nd-Zn-Zr, microstructure modification, performance improvement, LPSO structure

Abstract

Friction stir processing (FSP) is a severe plastic deformation (SPD) processing, which is very useful to refine grain size and secondary phase as well as change the texture of metal materials. Many FSP research were focused on aluminum alloys, while there are few reports on FSP of magnesium alloys, esp. on precipitation-hardening Mg-RE alloys. This paper overviewed the microstructures and mechanical properties of several FSPed Mg-RE alloys, such as Mg-Gd-Zn-Zr, Mg-Gd-Ag-Zr, and Mg-Nd-Zn-Zr with or without long period stacking ordering (LPSO) structure. The effects of processing parameters, such as rotation rate and traversing speed, on microstructure and mechanical properties were evaluated. It shows that FSP can effectively lend to performance improvement by microstructure modification, including obtaining remarkable finer and more homogenized grains, changing distribution and volume percentage of secondary phase etc.

Introduction

Magnesium (Mg) and its alloys, being the lightest structural metals, have remarkable potential applications in the aerospace, aircraft, and automotive industries for weight reduction in recent years [1]. However, the low strength, low ductility and poor creep resistance at increased temperatures restrict their wide applications. Recently, Mg alloys containing rare earth (RE) elements (especially heavy RE) have widely been developed and systematically investigated as promising materials with high strength and high creep resistance [2-4], which meet the demand for high temperature applications. Among them, Mg–RE–Zn/Cu/Al/Ni alloys (RE= Y, Dy, Ho, Er, Gd, Tb, Tm) alloys have attracted much attention, due to solution strengthening, aging strengthening [2-4] and long period stacking ordered (LPSO) structures strengthening and toughening [9-18].

At present, Mg–RE alloy parts are produced mainly by casting. However, coarser grains and eutectic at grain boundaries limit their mechanical properties and application, especially at higher temperatures. Therefore, it is need to modify the microstructures to enhance the mechanical properties, broadening its applications. Conventionally, heat treatment and various plastic deformation methods including severe plastic deformation (SPD) such as equal channel angular pressing (ECAP) [5] are applied to modify the microstructure of the Mg castings. However, the heat treatment for both the as-cast parts and pre-deformation parts at high temperature for lengthy time not only is time and cost consuming but also results in coarse grains as a result of their poor formability.

Friction stir processing (FSP), as one of SPD, is developed as a relatively novel metal working technique as based on the basic principle of friction stir welding (FSW) [19,20]. FSP causes SPD, material mixing and thermal exposure, so that it reduces intermetallic compounds size, grain size, and elimination of porosity through forging pressure, due to the combined influence of the tool tilt angle and shoulder diameter. At present, FSP has demonstrated as an effective process for producing the fine-grained heterogeneous bulk metal materials or surface composites, resulting from modifying the microstructure of metallic parent materials (PMs) and reinforced particles [19- 28].

In the past few years, several studies had been conducted to understand the effect of FSP on the microstructure and properties of Mg–Al–Zn (AZ) [21] and Mg–Al–Mn (AM) alloys [22,23]. It was reported that FSP resulted in remarkable grain refinement and significant breakup and dissolution of the coarse, network-like secondary eutectic phase–β-$Mg_{17}Al_{12}$, distributed at grain boundaries, which improved significantly the tensile properties of the casting, Recently, there is study on FSP for Mg-RE alloys, for example, WE43 [24], Mg-Gd-Y-Zr [27], Mg-Y-Zn [26], Mg-Nd-Zn-Zr [28] and Elektron 21 (ASTM EV31A) [29]. However, there are no reports on FSP for Mg-Gd-Ag-Zr [2,3] Mg-Gd-Zn-Zr [6-15] alloys.

Therefore, in this study, Mg-Gd-Zn-Zr, Mg-Gd-Ag-Zr, and Mg-Nd-Zn-Zr alloys ingots with or without LPSO structure were prepared by conventional ingot metallurgy (I/M). Then, FSP was used to these alloys for preparing fine-grained structure. The effects of FSP on the microstructure and mechanical properties of the alloy were investigated.

Experimental procedures

The alloys studied in the paper were GQ132K (Mg-13.1Gd-1.8Ag-0.48Zr, wt.%) , GZ132K (Mg-12.7Gd-1.9Zn- 0.52Zr, wt.%) and NZ20K (Mg-2.0Nd-0.3Zn-0.45Zr, wt.%). Alloys ingots were prepared by conventional ingot metallurgy (I/M). Pure Mg was

* Corresponding author at: Tel.: +86 21 54742911; fax: +86 21 34202794.
E-mail address:plm616@sjtu.edu.cn (Liming Peng); lidejiang@sjtu.edu.cn (Dejiang Li)

melted in an electric resistance furnace with a mild steel crucible under a protective gas (0.3% SF_6 and 99.7% CO_2). Pure Zn/Ag metal and Mg–25wt.%Gd / Mg-25 wt.% Nd were melted at about 973 K. Mg–30 wt. %Zr was added into the melt at about 1053 K. Finally, the melt was poured at about 1023 K into the mild steel mold preheated to 473 K. GZ132K and GQ132K alloys were cast into plates. NZ20K alloy were cast into cylinders. The alloy ingots were cooled in air, then cut into small specimens. The solid solution heat treatment for NZ20K and GQ132K alloys were carried out at 813K for 4h, and 723K for 12h, respectively, in a SX2-8-10-type high temperature heat treatment furnace, and immediately quenched in cold water. Then, the solution-treated ingot of NZ20K alloy was extruded to plate with extrusion ratio of 6 and temperature of 648 K.

The as-cast and as-extruded plates were machined to plates with thickness of 7 mm for FSP. A steel tool with a shoulder 22 mm in diameter, and a threaded cylindrical pin with 8 mm in diameter and 6 mm in length were used for the FSP experiment. A tool rotation rate of 800 rpm, and a traversing speed of 200 mm/min and the tool tilt angle of 2.8° were adopted. The aging treatment for as-FSPed GQ132K alloy was carried out in a furnace oil bath with constant temperature of 473K, and immediately quenched in cold water.

Tensile specimens with a gauge length of 10 mm, a width of 3 mm and a thickness of 1.5 mm were machined. For as-FSPed alloys, they were parallel to the FSP direction with the gauge being completely within the nugget zone. Four tensile specimens were measured in each zone. Tensile tests at room temperature were conducted on a Zwick T1-FR020TN A50 universal testing machine at a strain rate of 1.67×10^{-3} s^{-1}. Moreover, Vickers hardness tests were measured in the cross-sectional as-FSPed and aged samples by a HV-30 Vickers hardness tester with a 5 Kg load for 30 s.

Phase analysis was carried out by X-ray diffraction (Rogaku D/max 2550V) with angle from 10° to 90°, step size of 0.02° and step time of 15s and database is MDI Jade 5.0. To characterize the as-cast, as-extruded and as-FSPed microstructures for these three alloys, specimens etched by 4% HNO_3 and 96% ethanol were observed by optical microscopy (OM, Zeiss Axio Observer A1) and scanning electron microscopy (SEM, JEOL-7600F) at 10–20 KV equipped with an Oxford energy disperse X-ray spectrometer (EDS). For further study the microstructures and structures for the alloy, transmission electron microscopy (TEM, JEOL-2010) observations were carried out. The thin foils were thinned by a twin jet electropolishing in a solution of 5% $HClO_4$ and 95% ethanol under conditions of 20 mA, 75 V and 233 K and then by low energy beam ion thinning with the incidence angle of 4° and the voltage of 3.5 V for 0.5 h. Then, TEM observations were performed at 200 KV equipped with EDS.

Results and disscussion

Pre-FSPed microstructures

The microstructure of as-cast GQ132K alloy Fig. 1 shows OM and SEMmages of as-cast GQ132K alloy. The average grain size in the as-cast GQ132K alloy is about 55 μm. It shows that it consists of the α-Mg solid solution, the eutectic. Interestingly, there are two contrasts secondary phases in the eutectic, marked B and C in Fig. 1 (b).

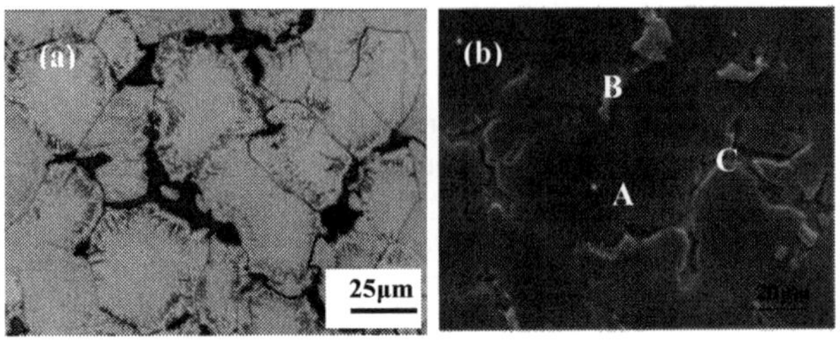

Fig. 1 (a) OM image, (b) SEM image of as-cast GZ132K alloy

By analysis of the EDS in the SEM mode, the compositions of Gd and Ag elements are different between the two phases. Among them, the B phase has higher Ag and lower Gd than C. However, both of them were established to Mg_5(Gd,Ag) phase, having a fcc structure with lattice constant of a = 2.22 nm, which will be reported in detail elsewhere.

In conclusion, the as-cast GZ132K alloy mainly consists of the α-Mg solid solution and β-Mg_5(Gd,Ag) phase. Interestingly, no LPSO structure is formed in the Mg-Gd-Ag-Zr alloy.

The microstructure of as-cast GZ132K alloy Fig. 2 shows OM, SEM, TEM images and corresponding selected-area electron diffraction (SAED) patterns of as-cast GZ132K alloy. The average grain size in the as-cast GZ132K alloy is about 60 μm. Fig. 2 (a) and (b) show that it consists of the α-Mg solid solution, the eutectic and lamellae [black structure within α-Mg matrix in Fig.2 (a)]. Fig.2 (c) and (d) show that the lamellae has a 14H-LPSO structure. There is coherent relationship, i.e., $(001)_{2H\text{-}Mg}$// $(0014)_{14H\text{-}LPSO}$ and $[001]_{2H\text{-}Mg}$// $[001]_{14H\text{-}LPSO}$. In addition, the lattice constants of the 2H-Mg and the 14H-type LPSO structure are estimated to be a = 0.325 nm, c = 0.520 nm, and a = 0.325 nm, c = 3.722 nm, respectively Furthermore, average chemical compositions of the 14H-type LPSO structure is Mg–8.10±1.0 at.% Zn–11.05±1.0 at.% Gd, by analysis of the EDS in the SEM mode [12]. Fig. 1(e)-(h) show that the secondary phase is the β-phase has a face-center cubic (fcc) structure with lattice constant of a = 0.719 nm [12].

In conclusion, the as-cast GZ132K alloy mainly consists of the α -Mg solid solution, the fine-lamellae with the 14H-type LPSO structure with matrix and the eutectic, in which the secondary phase is the β-phase [$(Mg,Zn)_3Gd$]. Therefore, the β-phase in the Mg-Gd-Zn-Zr alloy is different from that in Mg-Gd-Ag-Zr alloy in composition and lattice constant. There is almost same RE element and the third element contents (wt.%), compared to Mg-Gd-Ag-Zr alloy. However, no LPSO structure being formed in Mg-Gd-Ag-Zr alloy is possible to the causes of the different Zn an d Ag atom radius, solid solubility in Mg matrix, mixing enthalpy of the RE–X pair and the effect on stacking energy etc.

The microstructures of as-extruded NZ20K alloys Fig. 3 shows X-ray diffraction patterns of the as-extruded and as-FSPed NZ20K samples, respectively. It indicates that the microstructure of the as-extruded NZ20K alloy consists of α-Mg and precipitated phases $Mg_{12}Nd$.

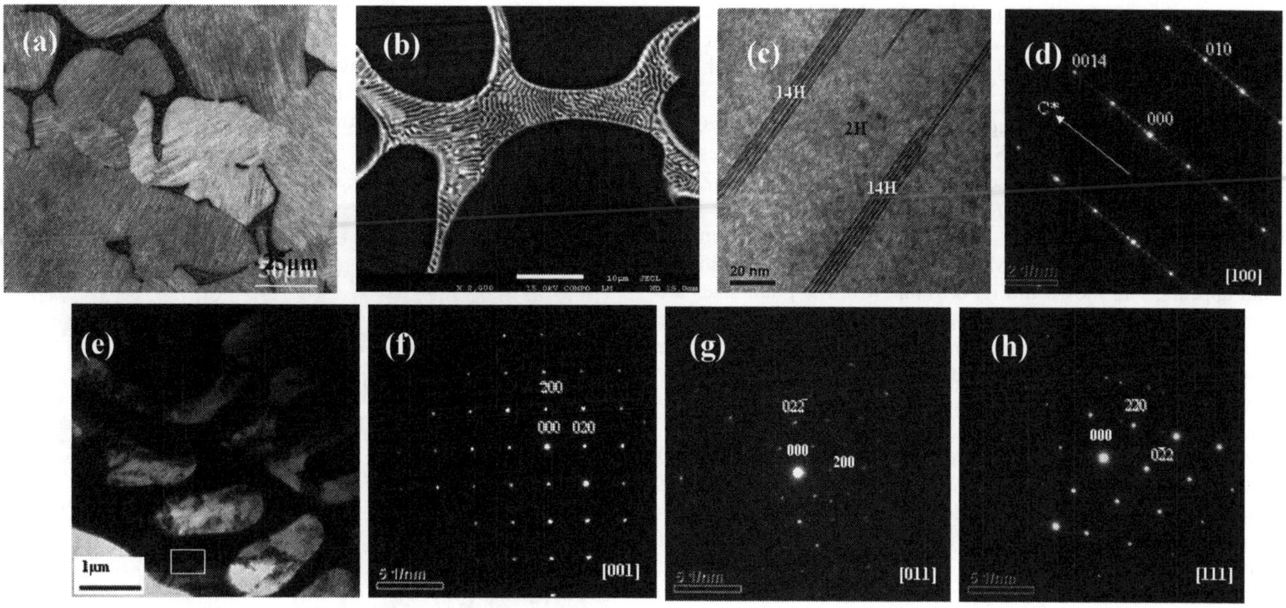

Fig. 2 (a) OM image, (b) SEM image, (c) Bright-field (BF) TEM image and (d) SAED patterns of the lamellar 14H-LPSO structure with electron beam // $[100]_{\alpha}$; (e) BF TEM image, (f)-(h) corresponding selected-area electron diffraction (SAED) patterns with the electron beams parallel to [001], [011] and [111] zones from TEM study for the secondary phase [$(Mg,Zn)_3Gd$]in eutectic in as-cast GZ132K alloy, respectively

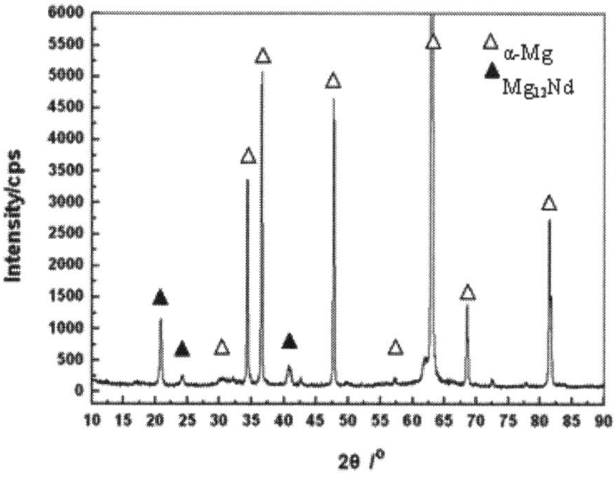

Fig. 3 X-ray diffraction pattern of as-extruded NZ20K alloy

Fig. 4 shows OM and SEM images of as-cast NZ20K alloy. The average grain size in the as-extruded alloy is about 21 μm. it indicates that the microstructure of the as-extruded NZ20K alloy consists of α-Mg and precipitated phases $Mg_{12}Nd$, combined with X-ray diffraction patterns analysis [30]. The $Mg_{12}Nd$ phase is formed by dynamic precipitation during hot-extrusion of the billet, as shown in Fig. 3(b), a large amount of grainy $Mg_{12}Nd$ distributes within the matrix, only a small amount of $Mg_{12}Nd$ distributes in the grain boundary.

Post-FSPed microstructures

The microstructure of as-FSPed GQ132K alloy Fig. 5 shows typical OM images of overall cross-section, TMAZ, and SZ for the as-FSPed GQ132K alloy. No defects in the joint, for example, tunnel, were detected in the sample, indicating the sound FSP parameter was achieved for the GQ132K alloy (rotation rate of 800 rpm, and a traversing speed of 200 mm/min). It shows distinct 4 regions: stir zone (SZ), thermo-mechanically affected zone (TMAZ), heat-affected zone (HAZ) and parent material (PM) after FSP in the alloy.

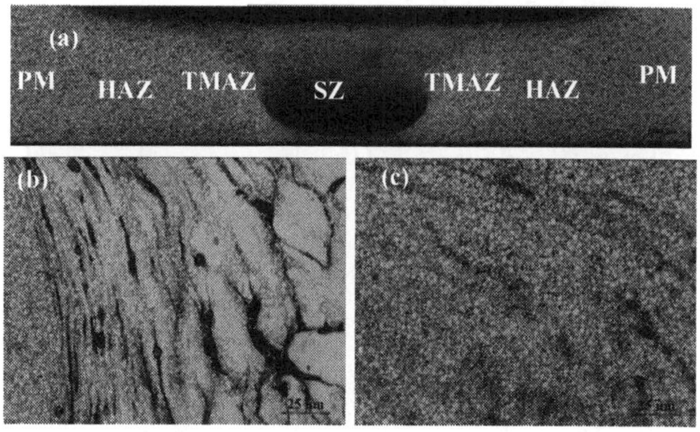

Fig. 5 Typical OM images of (a) overall cross-section, (b) TMAZ, and (c) SZ for the as-FSPed GQ132K alloy

After FSP, the primitive grains in SZ and TMAZ experience severe plastic deformation and dynamic recovery, while, the coarse grains have been broken. It is characterized by fine and equiaxed α-Mg grains in the SZ resulting from continuous dynamic recrystallization (CDRX) [Fig. 5(c)] , as compared with that in the PM [Fig. 1(a)].The grain size significantly decreases, and the average grain size is refined to about 1-4 μm from 55μm for the as-cast alloy [Fig. 5(c)]. While, it indicates that dynamic recrystallization (DRX) occurs in TMAZ during FSP; majority of primitive grains experience SPD and are elongated and bent. Moreover, it indicates that the β-Mg_5(Gd,Ag) phase is severely elongated and bent in TMAZ, and it is broken and part of it is dissolved into matrix resulting in the volume percentage and size decreasing in SZ.

The microstructure of as-FSPed GZ132K alloy Fig. 6 shows OM and SEM images of the as-FSPed GZ132K alloy. It indicates that the as-cast alloy experiences severe plastic deformation in SZ and TMAZ. The β-phase [$(Mg,Zn)_3Gd$] and grains with lamellar 14H-LPSO structure are severely bent Correspondingly, the lamellar 14H-LPSO structure with α-Mg matrix occurs bend and kinking band appears, , as shown in Fig. 6(b) and (d). The grain sizes are refined and β- phase are brokenup and part of the particles are dissolved into matrix, compared to as- cast (PM) alloy. In SZ, the plastic deformation degree is increased, grain sizes are obviously refined from 60 μm to 1-3 μm, and the lamellar 14H-LPSO structure with matrix is much bent, as shown in Fig. 6(c) and (e), . Moreover, it shows that many lamellae with the 14H-type LPSO structure are formed within the fine-grains in SZ and TMAZ during FSP, like other plastic deformation processes such as extrusion [14], as shown in Fig. 6(b) - (e), which detailed content will be reported elsewhere.

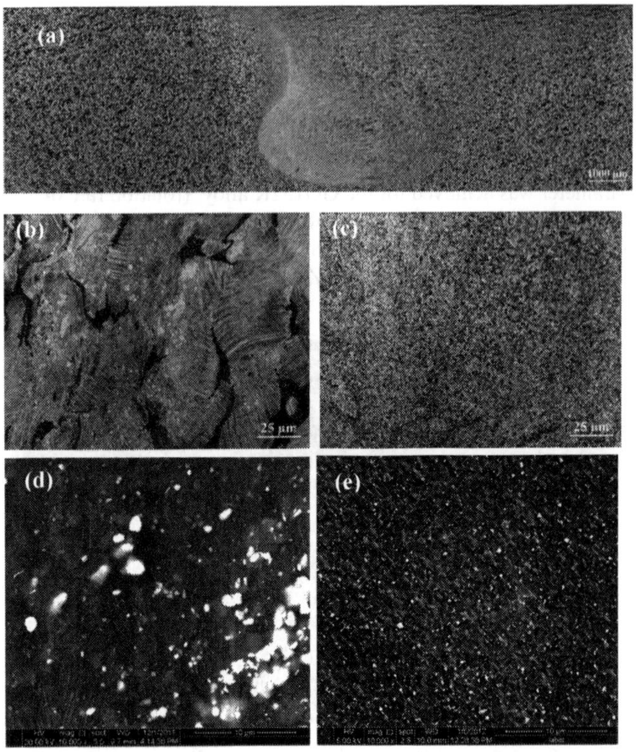

Fig. 6 Typical OM images of (a) overall cross-section, (b) TMAZ, and (c) SZ; SEM images of (d) thermo-mechanically affected zone (TMAZ) an (e) stir zone (SZ) for the as-FSPed GZ132K alloy

The microstructure of as-FSPed NZ20K alloy Fig. 7 shows typical OM images of the as-FSPed NZ20K alloy. It also shows distinct 4 regions like GQ132K alloy. It shows that the grain size significantly decreases in the SZ, as compared with that in the PM [Fig. 3(a)]. The average grain size in the as-extruded alloy is about 21 μm. After FSP, the microstructure of the SZ is characterized by fine and equiaxed α-Mg grains [Fig. 7(c)]. The average grain size in the SZ is measured to be 3-5μm. The primitive grains in SZ experience severe plastic deformation and dynamic recovery, while, the coarse grains have been broken. Therefore, it indicates that DRX occurs during FSP, but the time is too short for recrystallization grain to grow up, leading to significant grain refinement.

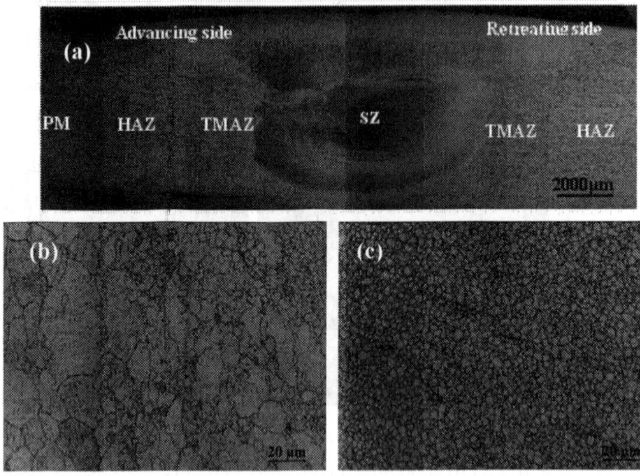

Fig. 7 Typical OM images of (a) overall cross-section, (b) TMAZ, and (c) SZ for the as-FSPed NZ20K alloy

TMAZ is found in the close vicinity of SZ, and also subjected to the thermal cycle and mechanical stress. DRX occurs in the part of zone and fine equiaxial grains nucleate and grew up, then, form the mixed zone of bulky grain and fine grain, as shown in Fig. 7 (b). The elongated and DRXed grain structures are characterized in the TMAZ and average grain size of TMAZ is 4.5-21μm.

Mechanical properties

Fig. 8 shows room-temperature tensile properties of the as-cast, solution-treated and as-FSPed GQ132K alloys. The as-cast alloy exhibits a yield tensile strength (YS) of 169 MPa, and ultimate tensile strength (UTS) of 241 MPa, and an elongation (δ) of 3.3%. After solution treatment at 773 K for 12 h, UTS and δ increase to 295 MPa, and 17.5 h, while, YS decreases little to 151MPa, mainly due to dissolution of coarse β-Mg_5(Gd,Ag) phase in grain boundaries which will be given elsewhere in the future. Interestingly, the tensile properties are obviously improved after FSP for the as-cast alloy without solution treatment, i.e., UTS, YS and δ increase to 344 MPa, 283 MPa and 20.2% from 241 MPa, 169 MPa and 3.3%. The reasons come from the dissolution, refinement and uniform distribution of β-Mg_5(Gd,Ag) phase, and grain refinement, otherwise, solution treatment would commonly increase the grain sizes for the as-cast alloy.

It can be concluded that FSP can effectively improve mechanical properties, by modify microstructure of Mg-RE alloys including refining grains and breaking and dissolution of particles.

Fig. 9 shows aging strengthening curves of PM and SZ for as-FSPed GQ132K alloy at 473K. The primary Vickers hardness of PM and SZ is 86.1 HV and 110.4 HV which increases 28% compared to PM. During aging, it is divided into 3 stages: under-aging, peak-aging and over-aging. They increase firstly, and arrive peak curves of 114.8 HV for 32 h, and 138.6 HV for 18h for

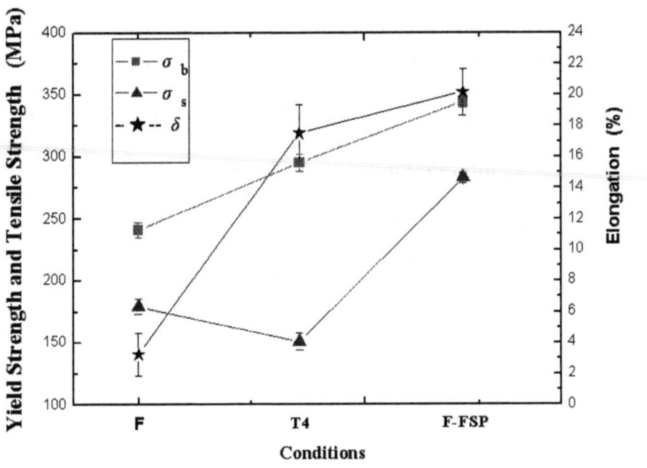

Fig. 8 Room-temperature tensile properties of the as-cast, solution-treated and as-FSPed GQ132K alloys

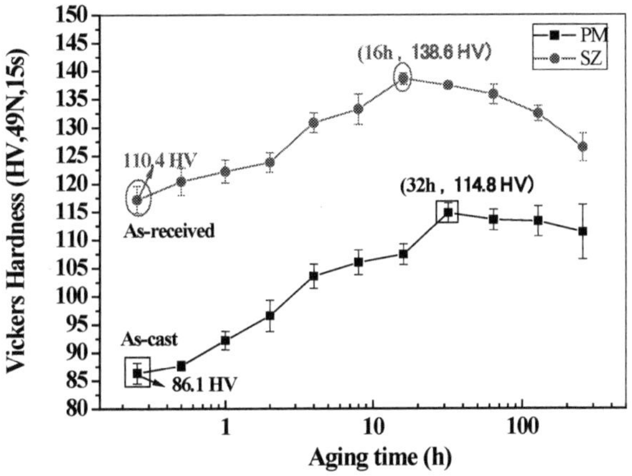

Fig. 9 Aging strengthening curves of PM and SZ for as-FSPed GQ132K alloy at 473K

PM and SZ, respectively. It indicates that the peak time for SZ

shortens than PM and the peak hardness increases, which saving cost and time of aging and improving mechanical properties. The obvious aging strengthening for the aged GQ132K alloy is attributes to complex precipitation effect of ellipsoidal non-basal phase (β') and lamellar basal phase [3,4] in Mg-Gd-Ag(-Zr) sery alloy.

Conclusions

The conclusions are summarized as follows:

(1) It shows distinct 4 regions: stir zone (SZ), thermo-mechanically affected zone (TMAZ), heat-affected zone (HAZ) and parent material (PM) after FSP in the Mg-Gd-Zn-Zr, Mg-Gd-Ag-Zr, and Mg-Nd-Zn-Zr alloys.

(2) Continuous dynamic recrystallization (CDRX) occurs in SZ of, while, dynamic recrystallization (DRX) occurs in TMAZ during FSP for the Mg-RE alloys.

(3) To the as-cast GQ132K, as-cast GZ132K and as-extruded NZ20K alloys, the grain sizes are significantly refined from to 3-5μm, respectively, in SZ of FSP.

(4) The secondary eutectic phase in grain boundaries including β-Mg_5(Gd,Ag) phase in GQ132K, β-$(Mg,Zn)_3$Gd phase in GZ132K alloys are severely deformed including being elongated and bent in TMAZ, and they are broken and part of them are dissolved into matrix resulting in the volume percentage and size decrease in SZ.

(5) In as-FSPed GZ132K alloy, the lamellar 14H-LPSO structure with primary α-Mg matrix occurs bend and kinking band appears in TMAZ. The lamellar 14H-LPSO structure with primary α-Mg matrix.

(6) It shows that many lamellae with the 14H-type LPSO structure are formed within the fine-grains in SZ and TMAZ during FSP.

(7) FSP can effectively improve mechanical properties, by modify microstructure of Mg-RE alloys including refining grains and breaking and dissolution of particles. For as-cast GQ132K, UTS, YS and δ increase to 344 MPa, 283 MPa and 20.2% from 241 MPa, 169 MPa and 3.3%. The Vickers hardness in SZ is improved from 86.1 HV in PM.

(8) The GQ132K exhibits obvious aging strengthening effect. For as-FSPed alloy, the peak time for SZ (16h) shortens than PM (32h) and the peak hardness increases from 114.8 HV to 138.6 HV.

Acknowledgements

The authors are grateful for the support f from National Natural Science Foundation of China (No.50971089), Shanghai Phospherus Program Project (No. 11QH1401200), China Postdoctoral Science Foundation (No.20090460615) and China Postdoctoral Science Special Foundation (201003267).

References

[1] A. Wendt, K. Weiss, A. Ben-Dov, et al., "Magnesium castings in aeronautics applications-Special requirements," *Magnesium Technology 2005*, (2005), 269-273.

[2] S.M. He, X.Q. Zeng, L.M. Peng et al., "Precipitation in a Mg–10Gd–3Y–0.4Zr(wt.%) alloy during isothermal aging at 250℃," *J. Alloy Comp.*, 421 (2006), 309-313.

[3] X. Gao, J.F. Nie, "Enhanced precipitation-hardening in Mg–Gd alloys containing Ag and Zn," *Scripta Mater.*, 58 (2008), 619-622.

[4] K. Yamada, H. Hoshikawa, S. Maki, "Enhanced age-hardening and formation of plate precipitates in Mg－Gd－Ag alloys," *Scripta Materialia*, 61 (2009): 636-639.

[5] B. Chen, D. L. Lin, X.Q. Zeng et al., "Microstructure and mechanical properties of ultrafine grained $Mg_{97}Y_2Zn_1$ alloy processed by Equal Channel Angular Pressing," *J. Alloy Comp.*, 440 (1-2) (2007), 94-100.

[6] Y. Kawamura, T. Morisaka, M. Yamasaki, "Structure and mechanical properties of rapidly solidified $Mg_{97}Zn_1RE_2$ Alloys," *Mater. Sci. Forum*, 419-422 (2003), 751-756.

[7] K. Amiya, T.Ohsuna, A. Inoue, "Long-period hexagonal structures in melt-spun $Mg_{97}Ln_2Y_1$ (Ln = Lanthanide metal) alloys," *Mater. Trans.*, 44 (10) (2003), 2151-2156.

[8] M. Yamasaki, T. Anan, S. Yoshimoto et al., "Mechanical properties of warm-extruded Mg–Zn–Gd alloy with coherent 14H long periodic stacking bordered structure precipitate," *Scripta Mater.*, 53 (2005), 799-803.

[9] Y. Kawamura, M. Yamasaki, "Formation and mechanical properties of $Mg_{97}Zn_1RE_2$ alloys with long-period stacking ordered structure," *Mater. Trans.*, 48 (11) (2007), 2986-2992.

[10] M. Yamasaki, M. Sasaki, M. Nishijima et al., "Formation of 14H long period stacking ordered structure and profuse stacking faults in Mg–Zn–Gd alloys during isothermal aging at high temperature," *Acta Mater.*, 55 (2007), 6798-6805.

[11] J.P.Li, Z. Yang, T. Liu et al., "Microstructures of extruded Mg–12Gd–1Zn–0.5Zr and Mg–12Gd–4Y–1Zn–0.5Zr alloys," *Scripta Mater.*, 56 (2007), 137-140.

[12] Y.J. Wu, D.L. Lin, X.Q. Zeng et al., "Formation of a lamellar 14H-type long period stacking ordered structure in an as-cast Mg-Gd-Zn-Zr alloy," *J. Mater. Sci.*, 44(2009):1607-1612.

[13] Y.J. Wu, X.Q. Zeng, D.L. Lin et al., " The microstructure evolution with lamellar 14H-type LPSO structure in an $Mg_{96.5}Gd_{2.5}Zn_1$ alloy during solid solution heat treatment at 773K," *J. Alloys Compd.*, 477 (2009):193-197.

[14] Y.J. Wu, L.M. Peng, D.L. Lin, et al., "A high-strength extruded Mg-Gd-Zn-Zr alloy with superplasticity," *J Mater. Res.*, 24(12)(2009): 3596-3602.

[15] W.J. Ding, Y.J. Wu, L.M. Peng, et al., " Formation of 14H-type long period stacking ordered structure in the as-cast and solid-solution-treated Mg-Gd-Zn-Zr alloys," *J Mater. Res.*, 24(5) (2009):1842-1854.

[16] H. Yokobayashi, K. Kishida, H. Inui, "Enrichment of Gd and Al atoms in the quadruple close packed planes and their in-plane long-range ordering in the long period stacking-ordered phase in the Mg-Al-Gd system, " *Acta Materialia* , 59(2011) : 7287-7299.

[17] P. P′erez , S. Gonz′alez, G. Garc′es, et al., "High-strength extruded $Mg_{96}Ni_2Y_1RE_1$ alloy exhibiting superplastic behaviour," *Mater. Sci. Eng. A,* 485 (2008), 194-199.

[18] Y. Kawamura, T. Kasahara, S. Izumi, et al., "Elevated temperature $Mg_{97}Y_2Cu_1$ alloy with long period ordered structure," *Scripta Mater.*, 2006, 55: 453-456

[19] R.S. Mishra, Z.Y. Ma, "Friction stir welding and processing, Materials Science and Engineering," *Materials Science and Engineering R,* 50(2005):1-78.

[20] R.S. Mishra, M.W. Mahoney, S.X. McFadden, et al., "High strain rate superplasticity in a friction stir processed 7075 Al alloy," *Scripta Mater.*, 42(2000):163-168.

[21] C.I. Chang, X.H. Du, J.C. Huang, "Achieving ultrafine grain size in Mg-Al-Zn alloy by friction stir processing," *Scripta Materialia*, 57(2007):209-212.

[22] R. Johnson, "Friction stir welding of magnesium alloys," *Mater. Sci. Forum*, 419-422(2003), 365-370.

[23] W.B. Lee, Y.M. Yeon, S.B. Jung, "Joint properties of friction stir welded AZ31B-H24 magnesium alloy, " Mater. Sci. Technol., 19(6)(2003), 785-790.

[24] M. Tsujikawa, S.W. Chung, M. Tanaka, et al., " High-strengthening of Mg - 5.5 mass % Y-4.3 mass % Zn cast alloy by friction stir processing," *Mater. Trans.*, 46 (12) (2005), 3081-3084.

[25] Q. YANG, B.L. XIAO, Z.Y. MA, "Influence of Process Parameters on Microstructure and Mechanical Properties of friction-Stir-Processed Mg-Gd-Y-Zr Casting," *Metallurgical and Materials transactions A*, 43A(2012), 2094-2109.

[26] S. W. Chung, M. Tsujikawa, T. Morishige, "Mechanical Properties on the Friction Stir Processed Cast Mg-1at.%Zn-2at.%Y Alloy," *Magnesium alloy 2007*,(2007), 299-305.

[27] B.L. Xiao, Q. Yang, J. Yang, "Enhanced mechanical properties of Mg-Gd-Y-Zr casting via friction stir processing," *Journal of Alloys and Compounds*, 509 (2011), 2879-2884.

[28] Z.L. Ning, F.Y. Cao, H.H Liu, et al., "Microstructure Analysis of Mg-2.54Nd-0.26Zn-0.32Zr Alloy, " R*are Metal Materials and Engineering*, 38(1)(2009):1997-2000.

[29] T. A. FREENEY, R.S. MISHRA, "Effect of Friction Stir Processing on Microstructure and Mechanical Properties of a Cast-Magnesium–Rare Earth Alloy," *Metallurgical and Materials transactions A*, 43A (2010), 73-84.

[30] F.Y. Zheng, Y.J. Wu, X.W. Li, "Microstructure and mechanical property evolution of Mg-2.0Nd-0.3Zn-1.0Zr alloy by friction stir processing," *Metallurgical and Materials transactions*, 2012, submitted.

Magnesium Technology 2013
Edited by: Norbert Hort, Suveen N. Mathaudhu, Neale R. Neelameggham, and Martyn Alderman
TMS (The Minerals, Metals & Materials Society), 2013

A Multi-stage Approach for Predicting Fatigue Damage in Friction Stir Spot Welded Joints of Mg AZ31 Alloy

H.M. Rao[1] and J.B. Jordon[1]
[1]The University of Alabama, Dept. of Mechanical Engineering, Tuscaloosa, AL, 35487

Keywords: Friction Stir Spot Welding, Mg AZ31, Multi Stage Fatigue Modeling

Abstract

In this work, we propose a model for predicting fatigue damage in friction stir spot welded (FSSW) joints made of Mg AZ31 alloy. In this modeling approach, an attempt is made to capture failure mechanisms due to the influence of variation in welding parameters including tool plunge depth, tool rotation speed, and tool pin diameter. As such, the fatigue model presented here is a deterministic approach, where fatigue lifetimes are estimated based on specific geometrical and microstructural information. In particular, the model addresses the observed variation in failure mechanisms commonly observed in Mg FSSW coupons under a range of applied loading. Further, a distinction is made between fatigue crack incubation, microstructural small and physically small fatigue crack growth, and finally long crack growth of the coupon. The fatigue model presented here showed good correlation for fatigue lifetimes for variation in welding conditions.

Introduction

In order to meet and exceed federal standards regarding fuel economy, the automotive industry is exploring lightweight designs. Lightweight metals, such as magnesium alloys, provide substantial weight savings over traditional structure metals, i.e. steels. However, industrial joining techniques for magnesium alloys are not well established compared to more established metals and this is especially true for spot joining. One possible alternative spot joining technology is friction stir spot welding. Friction stir spot welding (FSSW) is a variant of friction stir welding (FSW) and is an attractive spot welding technique due to its solid state process and its low energy requirements.

FSSW is a thermo-mechanical process for spot lap-joing of sheet [1,2] and Fig. 1 shows a schematic of the process. The rotating tool comprises a probe pin that is plunged into two sheets of metal to be joined. The probe pin typically penetrates the upper sheet completely and then passes into the bottom sheet to various depths depending on process conditions. Here, the downward force and rotational speed of the tool generates localized friction as the pin interacts with the upper sheet. The heat generated by friction softens the sheet materials adjacent to the tool and deforms plastically to form a solid state bond between the surface of upper and bottom sheet [3]. A longer dwell period typically provides the driving force for the upwards displacement of lower sheet material which promotes good bonding [2]. Since the dwell time in a manufacturing setting is relatively short, e.g., 2 to 5 seconds, the tool rotation speed and plunging motion largely determines the heat generation [4], joint formation and weld mechanical properties [5]. In general, FSSW is free of defects commonly associated with fusion welding, largely because the temperature attained during FSSW is less than the melting point of base materials. The other advantages of FSSW include no need of cooling agent, filler materials, or post weld treatment. In addition, FSSW is more energy efficient and clean when compared to other types of spot welding techniques.

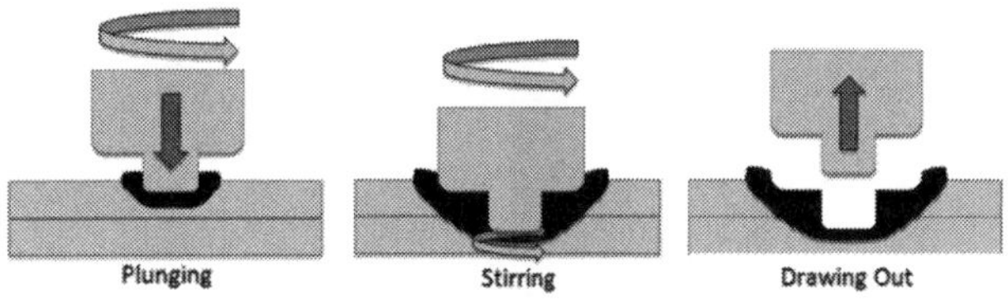

Figure 1. A schematic illustrating the friction stir spot welding process.

While fundamental relationships governing the static strength of FSSW in magnesium alloys have been investigated, only a few studies exist regarding fatigue behavior and performance in Mg alloys [8-10]. In fact, Rao and Jordon [10] analyzed the influence of microstructural and geometrical features on fatigue lifetimes of FSSW on AZ31 Mg alloy. Fractography analysis conducted in this study suggested that the effective top sheet thickness largely determined the failure mode, which in turn influenced the final number of cycles to failure. While the height of the interfacial hook was greater in the process with better fatigue performance, it was the larger effective top sheet thickness that promoted crack propagation modes more favorable to greater fatigue resistance. Furthermore, scanning electron microscopy (SEM) of striation spacing confirmed differences in crack growth rates.

The current state of fatigue modeling of spot welds is successful in correlating the number of cycles to failure for various types of joining methods. The most common approaches use some variation of linear elastic fracture mechanics (LEFM), where the variation in methods is on the determination of the stress intensity solution. However, a deterministic approach to correlate fatigue lifetimes based on specific microstructure and geometrical features of a specific welding process does not currently exist. As such, the purpose of this paper is to establish a microstructure sensitive fatigue model for FSSW joints in Mg alloys. The premise of this model is to capture the classical defined stages of fatigue: incubation; microstructurally small/ physically small crack growth; and long crack growth. To the best of the authors' knowledge, this type of approach has not been applied for FSSW or other similar spot welds.

Materials and Experiments

In order to provide the foundation for the develop of a multi-stage model for FSSW in Mg alloys, results from a recent study [10] and are summarized here. Two sets of FSSW coupons, made using 2.0 mm thick AZ31 Mg alloy sheets, were welded in an overlap configuration. For process #1, the individual sheet dimensions were: length of 100 mm, width of 35mm, and a welded overlap area of 35 x 35 mm. For process #2, the individual sheet dimensions were: length of 100 mm, width of 38 mm and a welded overlap area of 38 x 38 mm. For process #1, the FSSW tool was made from standard tool steel (H13) material with a shoulder diameter of 12mm and a cylindrical pin having a diameter of 5 mm, pin length of 3.2 mm, and left hand threads (M5), as shown in Fig. 2a. For process #2, the FSSW tool is identical to the tool used in process #1; however the tool had a triangular pin as shown in Fig. 2b. Both tools had a 10° concave shoulder as shown in Fig. 2c.

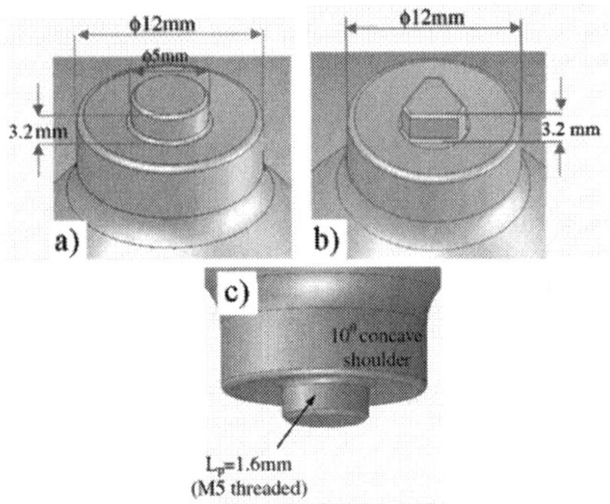

Figure 2. Schematic of the FSSW tool geometry employed for (a) process #1 and (b) process #2.

Table I. lists the process conditions used to join the two sets of lap-shear coupons employed in this study. Figure 3 shows representative lap-shear samples from both process #1 and #2. For the fatigue tests, the lap-shear coupons were cyclically tested in an MTS 810 servo-hydraulic load frame under load control with a sinusoidal waveform at three different load ratios (0.1, 0.3, and 0.7). The fatigue tests were tested at frequencies that ranged from 5–30 Hz. While most of the tests were conducted at 10 Hz, higher frequencies were employed for the low load amplitudes to expedite failure. Steel shims were employed in the test frame to prevent additional bending forces. The grip-to-grip distance employed for each coupon was 110 mm.

Table I. FSSW welding process parameters used in current study.

Welding Parameters	Process #1	Process #2
Tool rotation (RPM)	1000	750
Tool Plunge Speed (mm/min)	20	20
Shoulder Plunge Depth (mm)	0.5	0.1
Dwell Time (sec)	2.5	2.5

For weld characterization, samples were sectioned through the center of the weld nugget coupon and parallel to the loading direction. Samples were cold mounted in epoxy and then ground and polished. Using a Keyence VHX-1000 digital optical microscope, key features of the weld were characterized including size and shape of the interfacial hook, the bond width, and effective sheet thickness. Representative fatigued coupons from process #1 and process #2 were also observed under the Keyence microscope to identify the failure modes at different cyclic load levels.

Interfacial Hooks

In typical FSSW joints, the faying surface oftentimes exhibits a hook-like shape as shown in Fig. 3. This hooking of the faying surface is a result of trapped oxide films are that displaced upward due to the plastic flow of the material resulting from the downward plunge of the pin into the bottom sheet. In fact, the degree of oxide distribution at the faying surface is greatly influenced by the tool geometry and tool rotation speed.

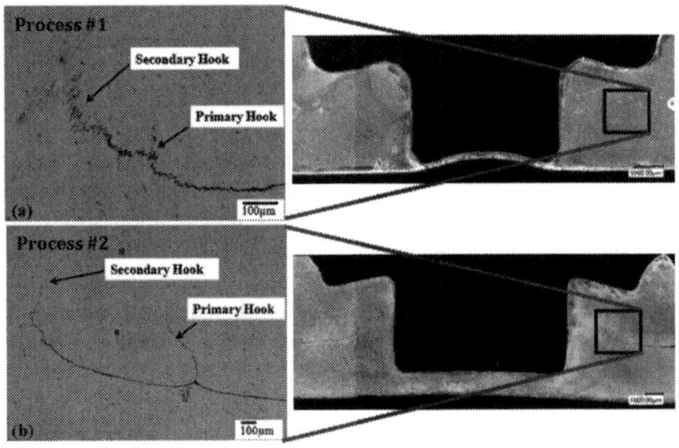

Figure 3. Magnified views of the primary and secondary hooks formed in (a) process #1 and (b) process #2.

Consequently, the link between the sheet thickness, T (see Fig. 4), and hook height are directly related to the shoulder plunge depth achieved during FSSW. If the tip-to-tip distance of the primary interfacial hooks is used to define the bond width, D2 (refer Fig.4), process #2 had a slightly larger bond width. Dimension nomenclature of the various weld geometry measured are indicated in Fig. 4 and the quantified results are shown in Table II.

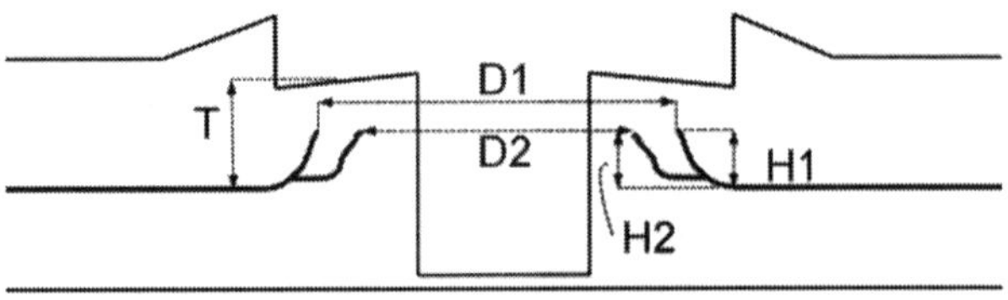

Figure 4. Schematic of key geometrical features of FSSW coupon with dimension nomenclature. (Refer to Table II for dimensions)

Table II. Summary of dimensions of the FSSW features for process #1 and process #2 (refer Figure 5. for nomenclature).

Nomenclature	Process #1 (mm)	Process #2 (mm)
D1	10.8	11.0
D2	10.6	9.5
H1	0.15	0.63
H2	0.28	1.04
T	1.20	1.90

Optical Fractography

Figure 5 shows an overview of representative fractured FSSW coupons from both process #1 and #2. Several macroscale observations were made regarding the failure modes. It can be seen that the dominant crack originated from the interfacial hooks in coupons from the two processes. The mode of failure varied from nugget pullout, top sheet pullout, nugget shear failure and bottom sheet pullout for different loading conditions in the FSSW coupons of the two processes. Table III summarizes the failure modes for the FSSW Mg AZ31 coupons subjected to cyclic loading for both process #1 and process #2 based on the optical fractography conducted in this study.

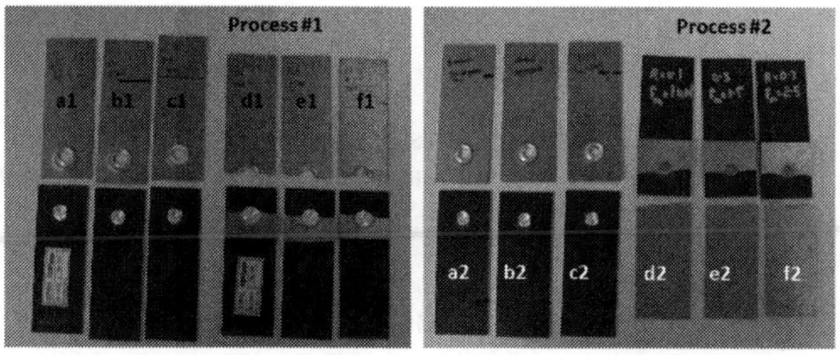

Figure 5. Representative fractured coupons tested at a range of maximum cyclic loads of 1-3 kN at R=0.1, 0.3, and 0.7 (Refer to Table III for a summary of failure modes).

Table III. Summary of failure modes in coupons of process #1 and process #2 (ID correlates with specimens shown in Figure. 5).

Process#1: Mode A (Low Cycle Fatigue Regime)	Dominant crack grew from primary hook into the top sheet, then halfway around the outer circumference of the nugget, finally failing by nugget pullout.
Process#1: Mode B (High Cycle Fatigue Regime)	Dominant crack grew from primary hook into the top sheet, halfway around the outer nugget circumference and then along the width of the top sheet.
Process#2: Mode C (Low Cycle Fatigue Regime)	Dominant crack grew across the weld nugget and parallel to loading direction in mode II propagation.
Process#2: Mode D (High Cycle Fatigue Regime)	The Dominant crack grew from root of the secondary hook into the bottom sheet, propagated around the outer circumference of the nugget region and then along the width of the bottom sheet.

Multi Stage Fatigue Modeling

The multistage fatigue (MSF) model is a microstructure-sensitive model that considers the various experimentally observed stages of fatigue damage evolution such as crack incubation, microstructurally small crack (MSC), physically small crack (PSC), and long crack growth [11]. While the model was originally developed for a cast A356 Al alloy, it has been modified to extend its application to other materials and other processing methods (e.g. processing of wrought materials) [12]. The MSF model is based on three distinct stages of fatigue damage, where total fatigue life N_{Total} is given by

$$N_{Total} = N_{Inc} + N_{MSC/PSC} + N_{LC} \tag{1}$$

where N_{Inc} is the number of cycles to incubate a crack near or at a micro-notch that includes the nucleation of crack-like damage and early crack propagation through the zone of the micronotch's root influence [11]. The $N_{MSC/PSC}$ term is the number of cycles required for propagation of a microstructurally small crack (MSC) and physically small crack (PSC), and N_{LC} is the number of cycles required for long crack propagation.

Crack incubation involves nucleation plus small crack growth at the micronotch of the order of ½ D, where D is the micronotch size. Small crack growth (MSC) comprises propagation of microstructural cracks with lengths that fall within $a_i < a < k\,MS$, where MS is a characteristic length scale (defined as the smallest grain size) of interaction with microstructural (MS) features, and k is a multiplier where $1 \leq k \leq 3$ [11, 12]. The PSC range consists of crack propagation with a crack length in the interval of $k\,MS < a <\sim 10\ MS$). Depending on the micronotch and texture of the matrix, the PSC regime may extend from 300 μm to ~2-3 mm in length.

Crack Incubation Regime

The fatigue damage incubation life, N_{inc} is characterized by the cyclic damage at the micronotch root and takes on a modified Coffin-Manson law form [11] in the microscale as follows:

$$C_{inc} N_{inc}^{\alpha} = \beta = \frac{\Delta\gamma_{max}^{p*}}{2} \tag{2}$$

where β is the nonlocal damage parameter around a micronotch, and C_{inc} and α represent the linear coefficient and exponent, respectively in a modified Coffin-Mason approach for the fatigue crack incubation life. The numerical value of the exponent α was selected to be in the range of the macroscopic Coffin-Mason law [11, 12]. The coefficient in the Coffin-Manson equation, $C_{inc} = c_n + z(c_m - c_n)$, is modified [11] to include a coefficient for nucleation of small cracks at micronotch in the high cycle regime (c_n) and a coefficient for the low cycle regime (c_m). We further note that $z = (1/0.7)\langle \frac{l}{D} - 0.3\rangle$ is a localization multiplier [11] that is nonzero below the microplasticity percolation limit. Just beyond the percolation limit, the localization multiplier transitions to unity as the plastic shear strain localizes interdendritically [11]. Above that point, incubation is negligible because of the high strain level that localizes at the micronotch [11]. Note that the ratio $\frac{l}{D} = \sqrt{\frac{A_{plastic\ zone}}{A_{notchroot}}}$ is defined as the square root of the ratio of the plastic zone over the micronotch root area. We further note that $\frac{l}{D}$ ratio is typically determined through micro-mechanical simulations. In this study, however, no such simulations were available, and as such, C_{inc} is assumed to be constant for both the high cycle and low cycle regime. The β parameter is related to the local average maximum plastic shear strain amplitude, $\frac{\Delta\gamma_{max}^{p*}}{2}$, and is estimated by the following relations:

$$\beta = \frac{\Delta\gamma_{max}^{p*}}{2} = Y100[\varepsilon_a - 0.00025(1-R)]^X \tag{3}$$

Here, ε_a is the remote applied strain amplitude, the parameter Y [12] is correlated as $Y = y_1 + (1+R)y_2$, where R is the load ratio, and y_1, y_2, X are model constants. For completely reversed loading cases, $Y = y_1$.

Small Crack Growth

Similar to modeling efforts in wrought aluminum alloys [12], we combine the mathematical expressions for the MSC and PSC regimes into a single mathematical form. Crack growth in the MSC/PSC is governed by the range of crack tip displacement, ΔCTD, which is proportional to the crack length, and the nth power of the applied stress amplitude, $\sigma^n{}_a$, in the high cycle fatigue (HCF) regime and to the macroscopic plastic shear strain range, $\frac{\Delta\gamma^P_{max}}{2}$, in the low cycle fatigue (LCF), and it is given by the following,

$$\left(\frac{da}{dN}\right)_{MSC} = \chi(\Delta CTD - \Delta CTD_{th}), \tag{4}$$

$$\Delta CTD = C_{II}\left(\frac{GS}{GS_0}\right)^{\omega}\left[\frac{U\Delta\hat{\sigma}}{S_{ut}}\right]^{\zeta} a_i + C_I\left(\frac{GS}{GS_0}\right)^{\omega}\left(\frac{\Delta\gamma^P_{max}}{2}\right)^2 \tag{5}$$

Here, χ is a constant for a given microstructure, typically less than unity and usually taken as 0.32 for several different alloys [12], and a_i is the initial crack length. The C_I and C_{II}, and ζ are material dependent parameters which capture the microstructural effects on MSC growth [11, 12]. The threshold value for crack tip displacement was defined on the order of the Burger's vector for the Mg-rich matrix [14], $\Delta CTD_{th} = 3.2 * 10^{-4}\mu m$. The term $\Delta\hat{\sigma}$ is the combination of the uniaxial effective stress amplitude, $\bar{\sigma}_a = \sqrt{\frac{3}{2}\frac{\Delta\sigma'_{ij}}{2}\frac{\Delta\sigma'_{ij}}{2}}$, and the maximum principal stress range, $\Delta\sigma_1$, and is given as $\Delta\hat{\sigma} = 2\theta\bar{\sigma}_a + (1-\theta)\Delta\sigma_1$, with $0 \le \theta \le 1$ as the path dependent loading parameter after [12], where θ is a weighting parameter such that $\theta = 0$ gives the von Mises stress state, and $\theta = 1$ gives the maximum for principal stress state. The parameter U is employed to capture the load ratio effects and is defined as $U = \frac{1}{1-R}$ [12]. In order to capture the effect of grain size on small crack growth, we employ the ratio of grain size to the reference size $\left(\frac{GS}{GS_0}\right)^{\omega}$ for each of the three materials in this study, where GS_0 is the reference grain size, GS is the specific grain size, and ω is a material parameter [12].

Long Crack Growth

Linear elastic fracture mechanics is used to model long crack growth in the FSSW coupons. The general expression of the number of cycles were calculated using the relation

$$\left(\frac{da}{dN}\right)_{LC} = C((\Delta K)^M - (\Delta K_{th})^M) \tag{6}$$

where for Mg AZ31, $M \sim 3.68$ and $C \sim 3 \times 10^{-10}$ [15] and the intrinsic threshold is given by $\Delta K_{th} \sim 0.91$ MPa$\sqrt{m}$ for Mg AZ31 [16], as determined from the threshold for long cracks with grains in longitudinal direction. We assume the long crack grows once the MSC/PSC reaches the free surface of the coupon and/or a threshold length of 2mm. Since final separation of the joint in the high cycle fatigue (HCF) regime for both processes occurs through the sheet and not the weld, we assume the following stress intensity solution [17]:

$$K_{eff} = \sigma\sqrt{\pi a}.F\left(\frac{a}{b}\right) \tag{7}$$

where σ is the applied stress, a is the crack length at full separation, b is half the width of the specimen. For process#1 and process#2 in the low cycle fatigue (LCF) regime, the coupons failed due to shear overload in the nugget. Based on work of Jordon et al. [9] for FSSW joints in Mg AZ31 alloy, the following relation is used to estimate the crack growth rate in LCF:

$$K_I = \frac{P}{r^{1.5}}\left[0.314\left(\frac{2r}{t}\right)^{0.397}\right] \tag{8}$$

$$K_{II} = \frac{P}{r^{1.5}}\left[0.282 + 0.162\left(\frac{2r}{t}\right)^{0.710}\right] \tag{9}$$

$$K_{eff} = \sqrt{(K_I{}^2 + K_{II}{}^2)} \tag{10}$$

Where K_I and K_{II} are global stress intensity solutions, P is the load (N), r is the radius of the weld and t is the thickness of the sheet. Table IV (Appendix) lists the MSF model parameters employed in this study.

Modeling Results and Discussion

The MSF model used in this current study for the FSSW joints is an adaptation from the model developed for base materials. The modifications and assumptions were made to closely relate conditions in the FSSW joints that were experimentally observed in [10]. Figure 6 shows the correlation of the MSF model to the experimental load-life results for the FSSW Mg AZ31 alloy for the two sets of processes. It is important to note that the model correlated to the distinct failure modes as listed in Table III. Furthermore, the MSF employed the three distinct stages of fatigue damage: incubation, small crack (MSC/PSC)) and long crack growth (LC) to capture the long crack growth.

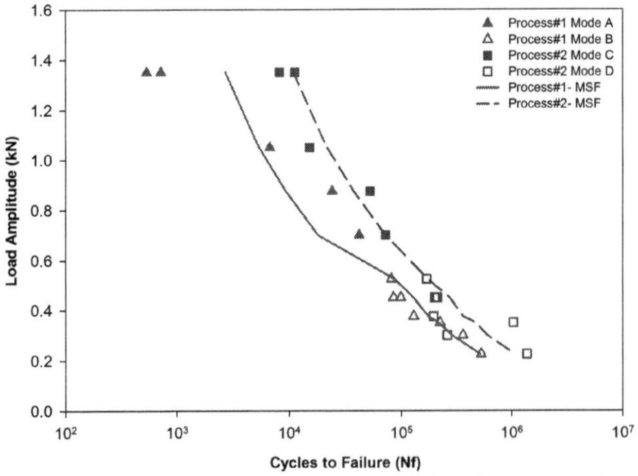

Figure 6. A comparison of experimental fatigue life results with crack growth model prediction of fatigue life of FSSW Mg AZ31 alloy.

The primary inputs to the model include the load amplitude, load ratio, weld geometry such as the hook height and weld radius. The remaining constants are based on assumptions and micromechanical simulations and other accepted values pertaining to microstructure used prior MSF work [11-13]. However, we note that future work is needed to validate some of the model parameters, in particularly the incubation and small crack growth parameters. In fact, as mentioned earlier, micromechanical simulations are planned to validate the parameters employed in this study. However, the results of the MSF correlation to the experimental results suggest that the model framework is well adapted to predicting the fatigue damage in FSSW joints.

Figure 7 and 8 show the contribution of incubation, small crack growth and long crack growth for process #1 and process #2, respectively. As shown in Fig. 7 and Fig. 8 the MSC/PSC is the dominant stage in both the processes in the LCF conditions, whereas the crack incubation cycle is negligible in both of the process for LCF and HCF. This trend correlates with the experimentally observed conditions of crack propagation dominated damage [10] as compared to some base materials where total fatigue is dominated by incubation [11]. In FSSW, the notch root at the faying surface promotes a short incubate life and thus the majority of total life is spent in internal crack propagation.

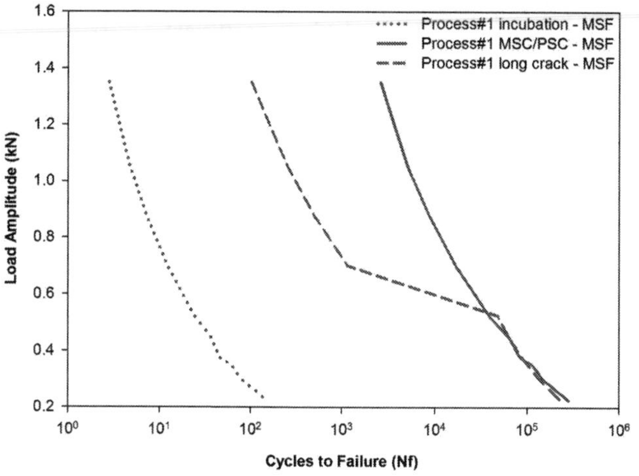

Figure 7. Crack life cycles as predicted by the fatigue model in process#1 under different loading conditions.

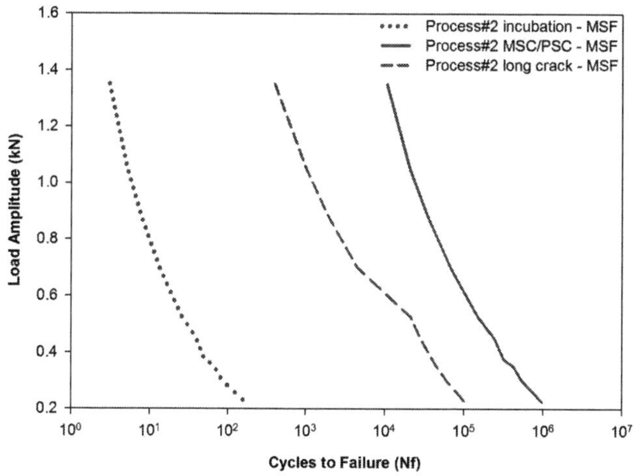

Figure 8. Crack life cycles as predicted by the fatigue model in process#1 under different loading conditions.

Experimentally, we observed that the MSC/PSC propagated through the primary interfacial hook in coupons from process #1, which is in thermomechanically affected zone of the weldment. This region of the weldment is characterized by larger grains which has the least resistance to crack growth. In coupons from process #2, the crack propagated through the root of the secondary hook and traversed to the bottom sheet in the stir zone of the weldment which is characterized by finer recrystalized grains [9, 10]. This crack growth regime contributed to ~ 94% of the total life of the coupons from process #1in LCF and ~ 50% of the total life in HCF. In coupons from process #2, the MSC/PSC contributed to ~ 95% of the total life in LCF and ~ 89% in HCF. This trend indicates that in LCF, the microstructure of the weldment does not contribute much to crack growth resistance and final failure. However, in HCF, the microstructure of the weldment is more crucial in resisting the crack growth and thus as has more of an effect on the final number of cycles as compared to the LCF regime.

Once the MSC/PSC reached the surface of the top sheet in process #1 and bottom sheet in process #2 in HCF, the crack grew along the width of the respective sheet. This growth along the width of the sheets is assumed to be the long crack regime for the fatigue modeling purpose. The long crack growth in process #1 contributed to about ~ 50% of the total life and ~ 11% of the total life in coupons from process #2 in the HCF as predicted by the MSF model. In process #1, the long crack negotiated through the nugget surface and along the width of the top sheet. The grain structure of the nugget and top sheet may contribute to the inhabitation of the growth of the crack. As in coupons from process #2, the long crack propagated in the bottom sheet away from the weld nugget which is not exposed to high physical deformation by the welding process. Hence the grains in this region of the coupon are not recrystalized and do not inhibit crack growth. For coupons in process #1 and process #2 in LCF, the long crack regime contributed to ~ 6% of the total life in process#1 and ~ 11% of the total life in coupons from process #2. Under high loads, the shear stresses are large enough for the long crack to grow through the weld nugget. The MSC/PSC and long crack together constitute to the final shear failure of the nugget in coupons from both the process.

Conclusions

In this paper, we present the formulation of a multi-stage fatigue (MSF) model for friction stir spot welding in Mg alloy. This model not previously applied to joints, was adapted to the lap-shear spot welded coupon to capture the incubation, small crack, and long crack growth stages observed experimentally. In general the MSF model showed good correlation to experimental results. In particular, the model correlated the differences in process conditions employed in fabricating the lap-shear coupons and their effect on the number of cycles to failure. This is a significant development since the results presented here suggest that the MSF model can be used to help elucidate the effect of a particular welding process and loading condition on fatigue lifetimes and help design future friction stir welds based on fatigue requirements.

References

1. S.M. Chowdhurya, D.L. Chena, S.D. Bholea, and X. Caob, "Tensile Properties of a Friction Stir Welded Magnesium Alloy: Effect of Pin Tool Thread Orientation and Weld Pitch," *Materials Science and Engineering A*, 527 (2010), 6064–6075.

2. Y.H. Yin, N. Sunb, T.H. North, and S.S. Hu, "Hook Formation and Mechanical Properties in AZ31 Friction Stir Spot Welds," *Journal of Materials Processing Technology*, 210 (2010), 2062–2070.

3. D.A. Wang, and C.H. Chen, "Fatigue Lives of Friction Stir Spot Welds in Aluminum 6061-T6 Sheets," *Journal of Materials Processing Technology*, 209 (2009), 367-375.

4. A. Gerlich, P. Su, and T.H. North, "Tool Penetration During Friction Stir Spot Welding of Al and Mg Alloys," *Journal of Materials Science*, 40 (2005), 6473–6481.

5. Y.H. Yin, A. Ikuta, and T.H. North, "Microstructural Features and Mechanical Properties of AM60 and AZ31 Friction Stir Spot Welds," *Materials and Design,* 31 (2010), 4764–4776.

6. N. Sun, Y.H. Yin, A.P. Gerlich, and T.H. North, "Tool Design and Stir Zone Grain Size in AZ31 Friction Stir Spot Welds," *Science and Technology of Welding and Joining,* 14 (2009), 747-752.

7. Y.H. Yin, N. Sun, T.H. North, and S.S. Hu, "Influence of Tool Design on Mechanical Properties of AZ31 Friction Stir Spot Welds," *Science and Technology of Welding and Joining,* 15 (2010), 81-86.

8. P.K. Mallick, and L. Agarwal, "Fatigue of Spot Friction Welded Joints of Mg-Mg, Al-Al and Al-Mg Alloys," (Paper presented at Society of Automotive Engineers, Warrendale, PA, 2009), 01-0024.

9. J.B. Jordon, M.F. Horstemeyer, S.R. Daniewicz, H. Badarinarayan, and J. Grantham, "Fatigue Characterization and Modeling of Friction Stir Spot Welds in Magnesium AZ31 Alloy," *Journal of Engineering Materials and Technology*, 132 (2010), 041008-1-10.

10. H.M. Rao, J.B. Jordon, H.Badarinarayan, M.F. Hostemeyer "Effect of weld structure on fatigue life of fssw in Mg AZ31 alloys," (Paper presented at 141st Annual Meeting of TMS, Orlando, FL, March 15th 2012), 3.

11. D.L. McDowell, Ken Gall, M.F. Horstemeyer, and J. Fan. "Microstructure-Based Fatigue Modeling of Cast A356-T6 Alloy," *Engineering Fracture Mechanics*, Vol. 70, pp. 49-80, 2003.

12. Y. Xue, D.L. McDowell, M.F. Horstemeyer, M.H. Dale, and J.B. Jordon. "Microstructure-based Multisatge Fatigue Modeling of Aluminum Alloy 7075-T651," *Engineering Fracture Mechanics*, Vol. 74, Issue 17, pp. 2810-2823, 2007.

13. J.B. Jordon, J. Gibson, M.F. Horstemeyer, H. El Kadiri, J. Baird, and A.A. Luo. "Effect of Twinning, Slip, and Inclusions on the Fatigue Anisotropy of Extrusion-Textured AZ61 Magnesium Alloy," *Materials Science and Engineering A*, Vol. 528, pp. 6860– 6871, 2011.

14. S. Groh, E.B. Marin, M.F. Horstemeyer,and D.J. Bammann. "Dislocation Motion in Magnesium: A Study by Molecular Statistics and Molecular Dynamics, Modelling and Simulation," *Materials Science and Engineering*, Vol. 17, pp. 1-15, 2009.

15. Sabrina Alam Khan, Yukio Miyashita, Yoshiharu Mutoh, Zainuddin Bin Sajuri, "Influence of Mn content on mechanical properties and fatigue behavior of extruded Mg alloys," *Materials Science and Engineering A 420 (2006) 315–32.*

16. Zainuddin Bin Sajuri, Yukio Miyashita, Yasunobu Hosokai, Yoshiharu Mutoh, "Effects of Mn content and texture on fatigue properties of as-cast and extruded AZ61 magnesium alloys," *International Journal of Mechanical Sciences 48 (2006) 198–209*

17. Hiroshi Tada, Paul Paris and George Irwin, *The Stress Analysis of Cracks Handbook* (New York, NY: ASME, 2000), 40.

Appendix

Table IV. Microstructure-sensitive fatigue modeling parameters for AZ31 magnesium alloy

Constant	Parameters	Description
C_{inc}	9	Linear coefficient (Eq. 2)
α	-0.65	Ductility exponent in Modified Coffin Manson Law (Eq. 2)
X	3	Exponent in remote strain to local plastic shear strain (Eq. 3)
E	155 (MPa)	Elastic Limit
K'	494 (MPa)	Cyclic strain strength coefficient
N'	0.14	Cyclic strain hardening exponent
χ	0.28	Crack growth rate constant (Eq. 4)
C_{I}	0.00007	HCF constant in small crack growth (Eq. 5)
C_{II}	0.002	LCF constant in small crack growth (Eq. 5)
ω	2	Omega (Eq. 5)
ζ	1.2	Exponent in Small crack growth (Eq. 5)
t	2 (mm)	Sheet thickness (Eq.8 & 9)

Magnesium Technology 2013

Wrought Materials II

Magnesium Technology 2013
Edited by: Norbert Hort, Suveen N. Mathaudhu, Neale R. Neelameggham, and Martyn Alderman
TMS (The Minerals, Metals & Materials Society), 2013

RECRYSTALLIZATION BEHAVIOR OF A MgAlCa ALLOY DURING THERMOMECHANICAL PROCESSING AND SUBSEQUENT HEAT TREATMENT

V.M. Miller, T.M. Pollock

Materials Department; University of California, Santa Barbara, CA 93106-5050

Keywords: AXJ810, Texture, Fine Grain, Thixomolding, Thermomechanical Processing, Recrystallization

Abstract

Microstructure refinement via static and dynamic recrystallization during thermomechanical processing is crucial to the development of wrought magnesium product with desirable combinations of strength and ductility. Microstructural evolution in thixomolded AXJ810 is investigated during warm rolling and subsequent annealing at 300°C and 450°C using scanning electron microscopy and electron backscatter diffraction. The rolled sheets exhibited a partially recrystallized microstructure with a strong basal texture. Significant grain refinement was observed during both static and dynamic recrystallization. Retarded grain growth and strong texture randomization were observed only in regions containing intermetallic particles.

Introduction

As the lowest density structural metal, magnesium and its alloys are highly attractive engineering materials for lightweighting. However, Mg alloys tend to have low formability at room temperature, rendering conventional deformation processing unsuitable and hindering widespread implementation. It has been demonstrated that microstructures with a fine grain size and a low degree of preferred crystallographic orientation ("texture"), can produce ductile behavior in magnesium alloys at room temperature [1-3].

Microstructural refinement and changes in texture are often achieved in magnesium alloys through recrystallization during hot deformation and subsequent annealing. One promising approach is for producing these desirable microstructural features is thixomolding and thermomechanical processing (TTMP). Previous research in Mg-Al-Zn alloys has produced TTMP sheet material with yield strengths above 250MPa and elongations above 14% [4-5].

In the present study, the TTMP process is applied to a Mg-Al-Ca-Sr alloy. The additions of calcium and strontium to Mg-Al alloys has been shown to be an effective means to improve creep resistance at a lower cost than rare earth additions [6-8]. Recent work on this class of alloys demonstrates that the alloy studied, AXJ810, is composed of three major phases upon solidification: α-Mg, β-$Mg_{17}Al_{12}$, and the $(Mg, Al)_2Ca$ Laves phase with a dihexagonal C36 structure [9]. At lower temperatures, the C36 phase will transform to the Al_2Ca C15 phase; however, this transformation is kinetically slow [10]. Therefore, the C36/C15 transformation is not examined further in this study.

Experimental

Thixomolded 3mm thick plates of AXJ810, composition shown in Table I, were warm rolled in one pass at an approximate temperature of 250°C to a 50% reduction in thickness. The sheets were then subjected a flattening treatment at 270°C for two minutes. Additional heat treatments of varied duration below the β-solvus at 300°C and above the β-solvus at 450°C. were then conducted.

Table I. Composition (wt%) of AXJ810 alloy

Al	Ca	Sr	Mn	Mg
8.0	1.02	0.31	0.20	bal.

Microstructural characterization was carried out via scanning electron microscopy (SEM), energy dispersive spectroscopy (EDS). Texture and grain size measurements were made using electron backscatter diffraction (EBSD) measurements from a minimum of 10,000 grains. An FEI DB235 microscope equipped with an EDAX Orientation Imaging Microscopy system was used for SEM, EDS, and EBSD measurements. Additional microstructural observation was carried out using an FEI Tecnai T20 transmission electron microscope (TEM). Material was characterized in the as-molded, as-TTMP, and post-TTMP heat treated states. SEM samples were mechanically polished to 1μm diamond paste, then etched for ten seconds in a 1:10 solution of o-phosphoric acid in methanol. Samples for EBSD and TEM were mechanically polished, then ion milled using the Gatan Precision Ion Polishing System. Fraction recrystallized was estimated as the fraction of grains with the grain reference orientation deviation of less than 1°.

Results

As-Molded Microstructure

In the as-molded condition, the AXJ810 microstructure consists of consists of large particles that were not melted during the molding process surrounded by a matrix of equiaxed fine grains, as shown in Figure 1. The grain boundaries are decorated by the intermetallic phases, β-$Mg_{17}Al_{12}$ and the C36 phase. The average grain size, excluding the unmelted particles, is 6.5μm, while the diameter of the unmelted particles ranges from approximately 20μm to greater than 60μm. Significant amounts of porosity are also observed.

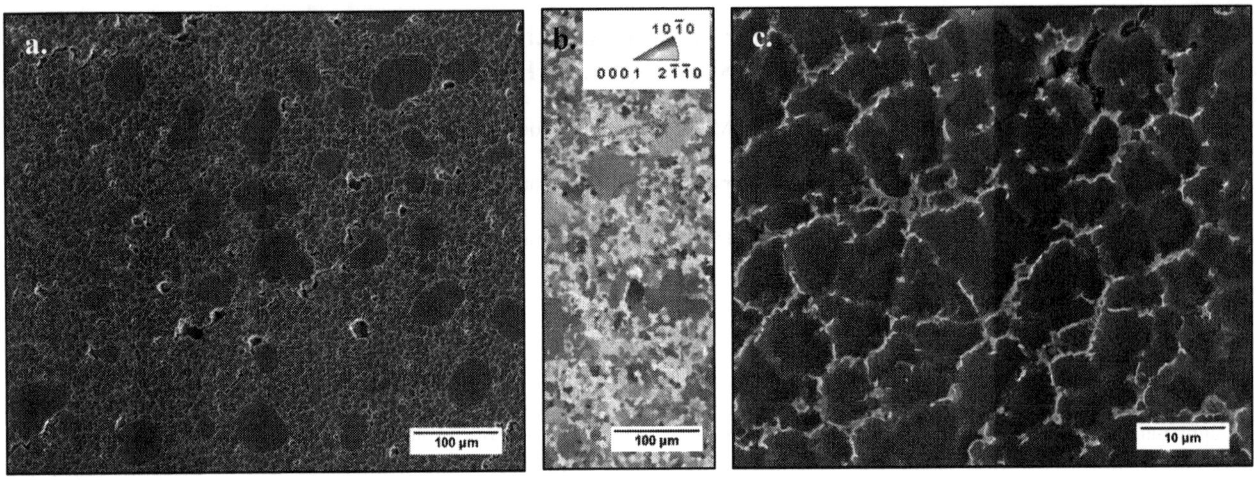

Figure 1. As-molded microstructure in secondary electron imaging (a, c) and an inverse pole figure map (b).

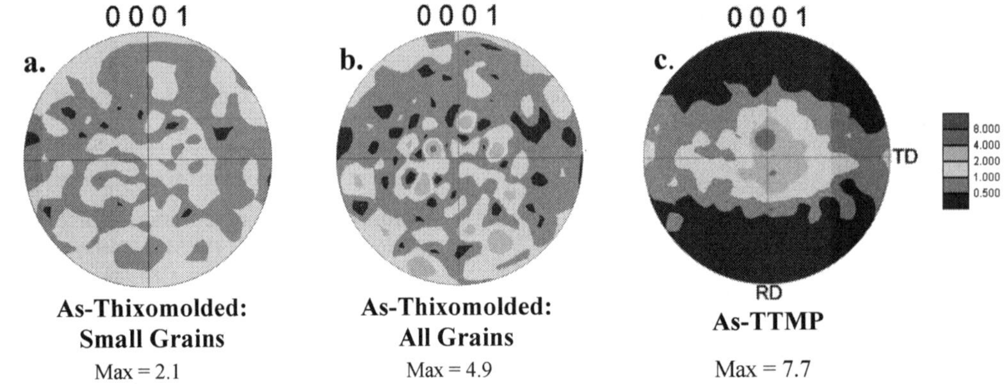

Figure 2. As-molded microstructure in secondary electron imaging (a, c) and an inverse pole figure map (b).

Texture analysis of the as-molded material shows some weak texture due to sampling bias caused by the presence of the large unmelted particles, as shown in Figures 2a and b. Exclusion of these particles reveals a nearly random orientation distribution, with a peak intensity of 2.1 multiples of random distribution (MRD).

As-TTMP Microstructure

After the TTMP process, the microstructure is partially recrystallized, and additional β–$Mg_{17}Al_{12}$ has precipitated, as shown in Figure 3. The average grain size measured via EBSD is 1.5µm. The as-TTMP material also possesses a strong, doubly peaked basal texture, as shown in Figure 4. The measured peak intensity of 7.7 MRD, though this is likely understated due to the sampling bias toward recrystallized grains.

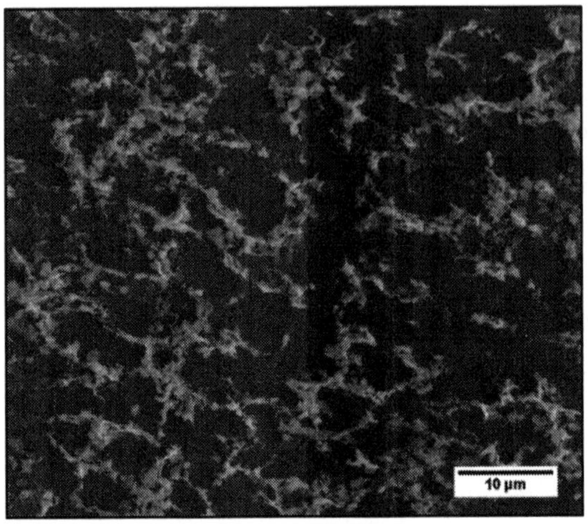

Figure 3. As-TTMP microstructure.

300°C Heat Treatments

Post-TTMP heat treatment at 300°C results in >80% fraction recrystallized within the first five minutes of annealing, and increased to >90% recrystallized after 24 hours. Minimal dissolution of the β phase is observed at short annealing times, though it partially dissolves during the longer heat treatments, as shown in Figure 4. The recrystallized grain size observed after 5 minutes is 4.9µm, and only a slight increase to 5.1µm is observed after 24 hours annealing, as summarized in Table II. The larger grains from unmelted particles are not observed. After recrystallization, a major reduction in texture is observed, with a peak intensity of 3.1 MRD after 5 minutes and 3.3 MRD after 24 hours, as shown in Figure 4a and b.

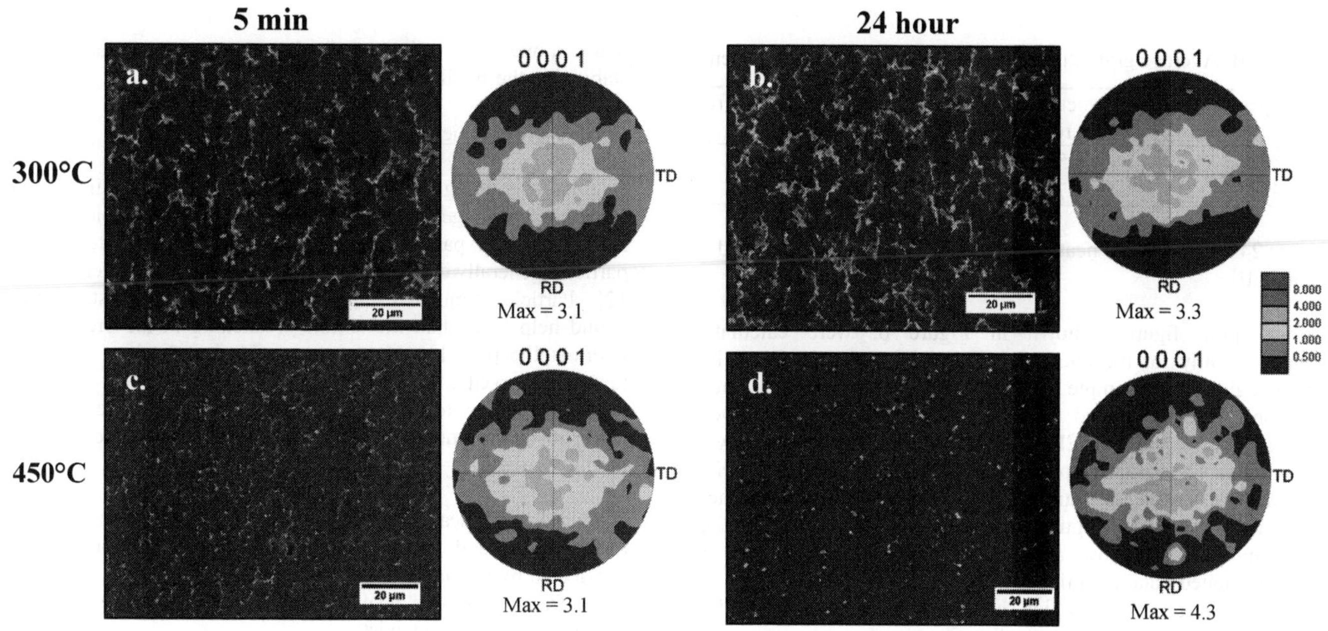

Figure 4. Secondary electron micrographs and EBSD (0001) pole figures for the 5 minute (a, c) and 24 hour (b, d) heat treatments at 300°C (a, b) and 450°C (c, d).

450°C Heat Treatments

Heat treatment at 450°C also produces a >80% recrystallized microstructure within five minutes, with >90% recrystallized after 24 hours, as shown in Figure 4c and d. However, unlike at 300°C, nearly complete dissolution of the β phase was achieved after less than 5 minutes. The recrystallized grain size observed after 5 minutes is 4.6μm, which increases to 7.1μm after 24 hours. The peak texture intensity after 5 minutes is 3.1 MRD, but this increases to 4.3 MRD after 24 hours.

Table II. Average grain diameter (d) after 5 minute and 24 hour heat treatments

Temperature	d after 5 minutes	d after 24 hours
300°C	4.0μm	4.9μm
450°C	4.6μm	7.1μm

Intermetallic-Free Regions

In material subjected to longer heat treatments at both temperatures, regions with a grain size larger than average were observed. Inverse pole figure maps and EDS maps, as shown in Figure 5, were used to confirm that these regions correspond to areas of low alloy content created by the unmelted grains in the original thixomolded material. The Al/Ca overlay EDS map clearly denotes the locations of intermetallic particles. A region free of intermetallics is present in the lower portion of the image, corresponding with the coarse-grained region in the IPF map.

Grain size and texture analysis was conducted independently on the intermetallic-containing and intermetallic-free regions. Note that the sample size for the intermetallic-free regions (approximately 500 grains) was significantly smaller than that of the intermetallic-containing regions due to the lower fraction

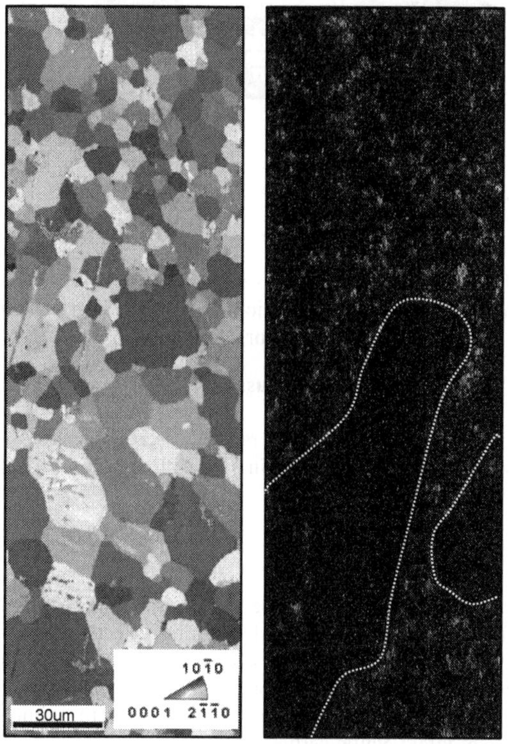

Figure 5. Inverse pole figure map (left) and Al/Ca EDS map (right) of material heat treated at 450°C for 24 hours demonstrating that the coarse-grained region is largely free of intermetallic particles, outlined in white on the EDS map.

present in the material. After 5 minute heat treatments, neither temperature showed any statistically significant difference in grain size in the intermetallic free regions. The average grain size

Table III. Average grain diameter (d) after 24 hour heat treatments

Temperature	d (intermetallic-containing region)	d (intermetallic-free region)
300°C	5.1μm	13.8μm
450°C	7.1μm	15.4μm

after 24 hours at each heat treatment temperatures is presented in Table III.

Basal pole figures, shown in Figure 6, were calculated independently for the intermetallic-containing and intermetallic-free regions of a sample heat treated at 450°C for 24h. These regions were partitioned using a grain size criterion and were visually correlated with EDS maps to ensure accuracy. However, this grain size criterion prevents a similar analysis from being conducted on material subjected to a shorter post-TTMP heat treatment. The pole figure calculated from the intermetallic-containing regions shows a lower value of peak intensity than the un-partitioned data. Additionally, the pole figure representing the intermetallic-containing region shows a more significant spread of the basal poles in the transverse direction.

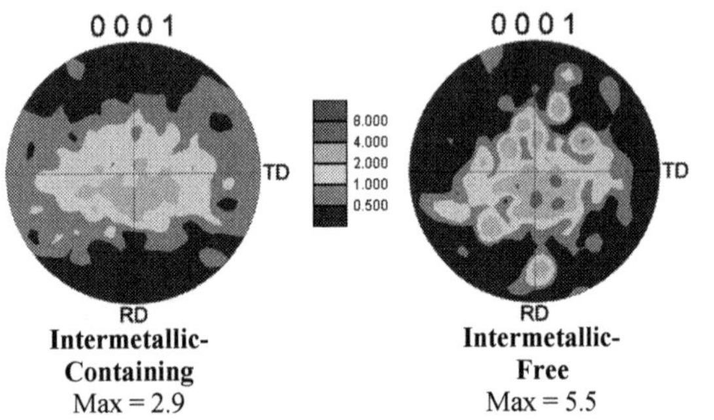

Figure 6. Basal pole figures of the intermetallic-free and intermetallic-containing regions in a sample heat treated at 450° C for 24 hours.

Discussion

At both heat treatment temperatures, recrystallization occurs rapidly and produces a fine grain size with significant degree of texture randomization.

Grain Growth

Severe grain growth was not observed in any of the heat treatment conditions in this study. Particle pinning effects appear to be effective in restricting the grain size, as evidenced by comparison of the regions with and without intermetallic particles. Grain growth appears to be significantly more severe without the presence of intermetallic particles.

Additionally, only small differences in the extent of grain growth was observed between the two heat treatment temperatures, despite the complete dissolution of the β phase at 450°C. The extent of coarsening observed at both temperatures is low in comparison to the extent of grain coarsening in the intermetallic-free regions. The similarity in the extent of grain coarsening regardless of the presence of the β phase indicates that the C36 phase alone is effective at pinning the grain boundaries and stabilizing the small grain size.

Texture Development

Regions containing intermetallic particles show dramatic texture weakening at longer heat treatment times. This is atypical; static recrystallization, particularly without the presence of intermetallic particles generally does not result in major texture weakening [11-12]. Particle stimulated nucleation (PSN) of recrystallization would help to explain the texture weakening in the presence of intermetallic particles. However, considering the low degree of texture observed after the 5 minute heat treatments, additional work is necessary to verify whether the stronger texture observed in the absence of intermetallics is a result of oriented nucleation or oriented growth.

At the shortest heat treatment duration, 5 minutes, the fraction recrystallized and the recrystallization texture appeared to be unaffected by the heat treatment temperature and corresponding β phase fraction. Shorter heat treatment durations are necessary to discern any effects that exist.

Conclusions

In the AXJ810 alloy, the TTMP and subsequent heat treatment process is highly effective for producing material with a low degree of texture and a fine grain size, suggesting it will have good room temperature formability. Comparison of regions with and without intermetallic particles demonstrates that the particles play a critical role in texture randomization and stabilization of the fine grain structure during recrystallization and grain growth. The presence of the thermally-stable C36 phase is effective at restricting grain growth to temperatures above the β solvus, allowing for enhanced microstructural control during heat treatment.

Acknowledgements

The authors would like to thank nanoMAG, LLC for providing the material used in this study.

References

1. J. Koike *et al.*, "The activity of non-basal slip systems and dynamic recovery at room temperature in fine-grained AZ31B magnesium alloys," *Acta Materialia*, 51 (7) (2003), 2055-2065.

2. R. Gehrmann, M. Frommert, and G. Gottstein, "Texture effects on plastic deformation of magnesium," *Materials Science and Engineering: A*, 395, (1-2) (2005), 338-349.

3. M. Barnett, D.L. Atwell, and A. Beer, "Grain Size in Mg Alloys: Recrystallization and Mechanical Consequences," *Materials Science Forum*, 558-559 (2007), 433-440.

4. T.D. Berman *et al.*, "Microstructure Evolution in AZ61 During TTMP and Subsequent Annealing Treatments," *Magnesium Technology*, eds. W. Sillekens *et al.* (Wiley, 2011), 599-603.

5. T.D. Berman *et al.*, "Microstructure Modification and Deformation Behavior of Fine Grained AZ61L Sheet Produced by

Thixomolding and Thermomechanical Processing," *Magnesium Technology 2012*, eds. S.N. Mathaudhu *et al.* (Wiley, 2012), 339-344.

6. M.O. Pekguleryuz and A.A. Kaya, "Creep Resistant Magnesium Alloys for Powertrain Applications," *Advanced Engineering Materials*, 5 (12) (2003), 866-878.

7. A.A. Luo, B.R. Powell, and M.P. Balogh, "Creep and Microstructure of Magnesium-Aluminum-Calcium Based Alloys," *Metallurgical and Materials Transactions A*, 33 (3) (2002), 567-574.

8. K. Hirai *et al.*, "Effects of Ca and Sr addition on mechanical properties of a cast AZ91 magnesium alloy at room and elevated temperature," *Materials Science and Engineering: A*, 403 (1-2) (2005), 276-280.

9. A. Suzuki *et al.*, "Phase equilibria in the Mg-Al-Ca ternary system at 773 and 673K," *Metallurgical and Materials Transactions A*, 37 (12) 2006, 975-983.

10. A. Suzuki *et al.*, "Structure and transition of eutectic $(Mg, Al)_2Ca$ Laves phase in die-cast Mg-Al-Ca base alloy," *Scripta Materialia*, 51 (10) (2004), 1005-1010.

11. X. Yang, H. Miura, and T. Sakai, "Recrystallization Behaviour of Fine-Grained Magnesium Alloy After Hot Deformation," *Transactions of Nonferrous Metals Society of China*, 31 (2007), 1139-1142.

12. R.K. Nadella, I. Samajdar, and G. Gottstein, "Static Recrystallisation and Textural Changes in Warm Rolled Pure Magnesium," *Magnesium: Proceedings of the 6th International Conference on Magnesium Alloys and Their Applications* (2005), 1052-1057.

Magnesium Technology 2013
Edited by: Norbert Hort, Suveen N. Mathaudhu, Neale R. Neelameggham, and Martyn Alderman
TMS (The Minerals, Metals & Materials Society), 2013

THE INFLUENCE OF DEFORMATION MECHANISMS ON RUPTURE OF AZ31B MAGNESIUM ALLOY SHEET AT ELEVATED TEMPERATURES

Aravindha R. Antoniswamy[1], Alexander J. Carpenter[2], Jon T. Carter[3], Louis G. Hector, Jr.[3], and Eric M. Taleff[1, 2]
[1]The University of Texas at Austin; Materials Science and Engineering; 1 University Station C2200; Austin, TX 78712-0292, USA
[2]The University of Texas at Austin; Mechanical Engineering; 1 University Station C2200; Austin, TX 78712-0292, USA
[3]General Motors Research and Development Center; 30500 Mound Rd.; Warren, MI 48090-9055, USA

Keywords: AZ31B, bulge forming, rupture, deformation mechanism, cavity interlinkage, necking, anisotropy

Abstract

Gas-pressure bulge tests were conducted on Mg alloy AZ31B wrought sheet until rupture at temperatures from 250 to 450°C. The rupture orientation was observed to change with forming pressure, which controls the forming strain rate, at 350 to 450°C. This phenomenon is a result of associated changes in the mechanisms of plastic deformation. At slow strain rates ($\leq 3\times10^{-2}$ s^{-1}), cavity interlinkage associated with grain boundary sliding (GBS) creep induced rupture along the sheet rolling direction (RD). At fast strain rates ($\geq 3\times10^{-2}$ s^{-1}), flow localization (necking) associated with dislocation-climb-controlled (DC) creep induced rupture along the long-transverse direction (LTD), a result of mild planar anisotropy. Biaxial bulge specimens tested at 250 to 300°C ruptured explosively, hence preventing any further analysis.

Introduction

Wrought magnesium alloy sheets have high strength- and stiffness-to-weight-ratios compared to traditional steel and aluminum alloys [1]. Increased use of Mg can provide a significant reduction in the overall mass of a vehicle [2]. Vehicle light-weighting offers several performance advantages, which include improved fuel economy and reduced greenhouse gas emissions [3-6]. The extended range facilitated by the use of lightweight materials, such as Mg, is also beneficial for vehicles with alternative propulsion systems based upon fuel cells or batteries, for example.

Because of limited room-temperature ductility, forming of magnesium alloy sheet at elevated temperatures, where ductility is quite high, is of interest. Potential commercial hot-forming processes include gas-pressure forming, such as with superplastic forming and quick-plastic forming techniques [7, 8], and hot-die stamping. This study is primarily interested with the former. For gas-pressure forming, a sheet metal blank is clamped between two hot dies, and the sheet is formed into a cavity in one die using gas pressure [9]. The performance of a sheet material in such a forming process can be studied in the laboratory using the gas-pressure bulge test, for which the sheet is formed into an open cylindrical cavity. The stress states that occur in the sheet during gas-pressure bulge tests are characteristic of those that occur during hot gas-pressure forming of automotive components in manufacturing practice [10]. The die geometry in these tests can be tailored to form the sheet along a chosen strain path. For example, a cylindrical die is used to produce balanced biaxial tension at the bulge apex. Biaxial bulge testing is one type of test that can be used in constructing forming limit diagrams (FLDs). These tests can produce a nearly constant strain rate at the dome pole while avoiding any significant frictional effects, as the forming bulge does not contact the die cavity wall beyond a region of relief near the sealing bead [11]. Carpenter et.al. used biaxial bulge forming experiments to validate physics-based material models that predict the forming of AZ31B sheet at 450°C [12].

Cavity interlinkage and necking are the two mechanisms that control rupture in uniaxial tension tests of AZ31B sheet at elevated temperatures [13]. At 350 to 450°C, AZ31B deforms primarily by grain boundary sliding (GBS) creep at slow strain rates and by five-power dislocation-climb (DC) creep at fast strain rates [14]. Cavity interlinkage is associated with GBS creep [15] and, necking is associated with DC creep in AZ31B sheet.

The rupture behavior during biaxial bulge testing of AZ31B at 250 to 450°C is analyzed in this study. Metallography and EBSD studies are used to relate the change in rupture direction from along the sheet rolling direction (RD) to the sheet transverse direction (LTD) at 350 to 450°C to the underlying physics of deformation.

Experimental Procedure

Material

The material of this investigation is a 2-mm-thick AZ31B rolled alloy sheet supplied in the H24 temper [16]. The as-received microstructure is highly deformed, as shown in Figure 1. Hence, the average grain size could not be measured in the as-received condition. The composition of the alloy, as provided by the supplier, is given in Table I.

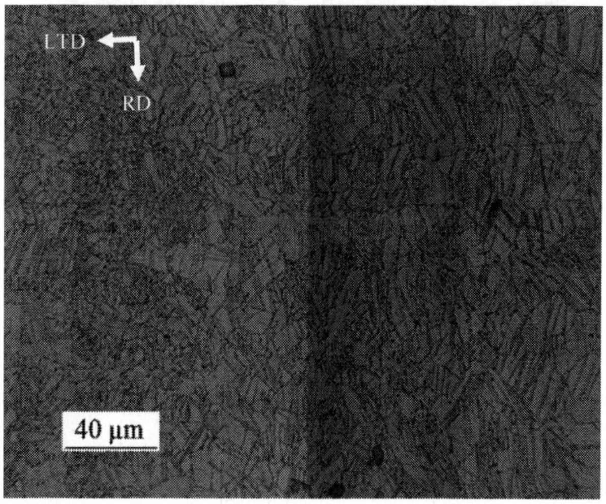

Figure 1. The as-received microstructure of AZ31B.

Table I. Composition of the AZ31B sheet material in wt. pct.

Al	Zn	Mn	Fe	Cu	Ni
3.1	1.0	0.42	0.006	0.003	<0.003
Si	**Ca**	**Be**	**Sr**	**Ce**	**Mg**
<0.1	<0.01	<0.005	<0.005	<0.01	Bal.

Gas-Pressure Biaxial Bulge Testing

Gas-pressure biaxial bulge testing was conducted using a custom instrument. This instrument forms a sheet blank at elevated temperature by gas pressure into an open cylindrical die with a 100-mm inner diameter to produce dome shapes. The dome pole experiences approximately balanced-biaxial tension. The stress state gradually changes down the slope of the dome toward the sealing bead, where a nearly plane-strain state exists. Prior to testing, a convective furnace heats the die set to the desired test temperature. A K-type thermocouple attached to each die monitors temperature. Once the dies reach the desired temperature, an AZ31B sheet blank is inserted, and the dies are then clamped together to form a sealing bead with the blank. The blank is held in the dies for a time sufficient to reach the desired test temperature. A prescribed gas pressure is then applied to form the blank into the die cavity. All tests were conducted until rupture, as determined by the leakage of gas pressure, as in Ref [11]. Figure 2 shows (a) the die setup used for biaxial bulge testing and (b) a tested specimen. Rupture typically occurred at the dome pole of the biaxial bulge specimens. Thickness strains were measured near the rupture. Average true-strain rates were then calculated by dividing rupture time into true strain near rupture. After rupture, the die halves are unclamped and the specimen is removed from the apparatus and allowed to air cool.

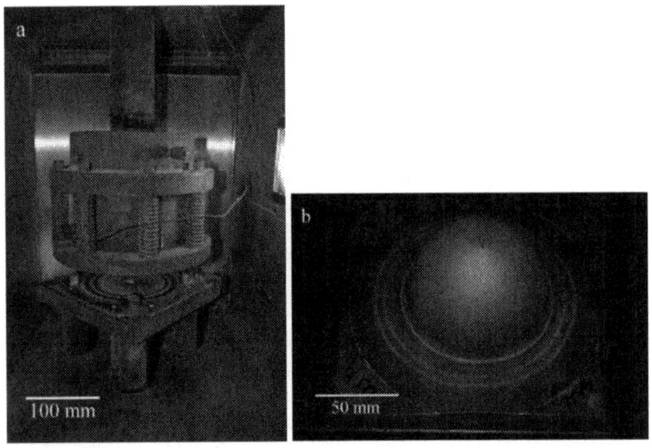

Figure 2. (a) The biaxial bulge testing dies and (b) a tested specimen are shown

Metallography

In order to examine the cross-section near rupture, tested biaxial bulge domes were sectioned perpendicular to the rupture orientation, and the sections were mounted in epoxy for metallographic examination. A specimen that ruptured along the RD was mounted to view the LTD-Short transverse direction (STD) plane, and a specimen that ruptured along LTD was mounted to view the RD-STD plane. All specimens were ground, polished, and then etched in an acetic-picric etchant (4.2 g picric acid, 10 ml water, 10 ml acetic acid and 70 ml ethanol) to reveal grain boundaries. Optical photomicrographs were obtained from the etched specimens using a Nikon EPIPHOT microscope equipped with a digital camera.

Electron Back Scattered Diffraction (EBSD)

For EBSD analysis, the annealed AZ31B alloy sheet was polished using a sequence of oil-based diamond suspensions of 9, 3, and 1 µm, respectively. This was followed by fine-polishing using colloidal silica. Finally, a 1 to 2 s etch using a solution of 60 ml ethanol, 20 ml distilled water, 15 ml acetic acid, and 5 ml nitric acid was applied. The EBSD data were collected in a Philips/FEI XL30 SEM at a voltage of 20 kV. A pole figure was calculated from the EBSD data using the MTEX open source toolbox for Matlab [17].

Results and Discussion

AZ31B alloy sheet was tested at 250 to 450°C at various gas-pressures, producing average true-strain rates measured near rupture to range from 10^{-3} to 10^{-1} s^{-1}. Three types of rupture were observed during these tests: (a) rupture along the RD, (b) rupture along the LTD, and (c) explosive rupture. Table II lists the rupture characters for both AZ31B under different testing conditions. The strain rates in Table II are given in ranges over which a particular rupture character was observed. From Table II, it can be concluded that, for temperatures from 350 to 450°C, AZ31B ruptured along the RD at slow rates (≤ 3×10^{-2} s^{-1}) and along the LTD at fast rates (≥ 3×10^{-2} s^{-1}). Multiple tests were conducted to evaluate the reproducibility of these rupture orientations. By changing the sheet orientation with respect to the die, it was confirmed that rupture direction was not caused by the orientation of the sheet in the dies.

Table II. The rupture characteristics of biaxial-bulge specimens are tabulated for different forming conditions

Temperature (°C)	Average true strain rate at pole (s^{-1})	Rupture orientation
450	10^{-3} to 3×10^{-2}	RD
	3×10^{-2} to 10^{-1}	LTD
350	10^{-3} to 3×10^{-2}	RD
	3×10^{-2} to 10^{-1}	LTD
300	10^{-3} to 2×10^{-2}	RD
	2×10^{-2} to 5×10^{-2}	Explosive
250	10^{-3} to 3×10^{-2}	Explosive

To investigate this rupture behavior, the biaxial-bulge specimens tested at 450°C and at strain rates of 10^{-3} s^{-1} and 10^{-1} s^{-1} were studied further. Explosive ruptures, as shown in Figure 3, prevented any such analysis of biaxial bulge specimens tested at 250 to 300°C. Figure 4 shows ruptured domes and photomicrographs of rupture cross-sections for these two specimens. The specimen tested at 10^{-3} s^{-1} ruptured primarily by cavitation, while the specimen tested at 10^{-1} s^{-1} ruptured primarily by necking. Grain-boundary-sliding (GBS) creep and dislocation-climb (DC) creep were identified as the dominant creep mechanisms, respectively, for the deformation of these specimens at 450°C [9].

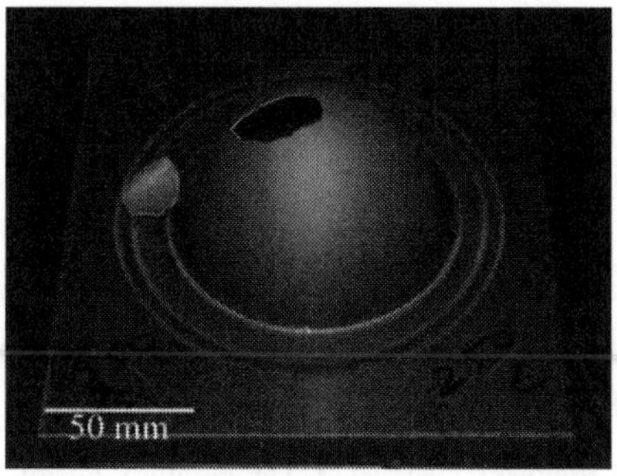

Figure 3. Explosive rupture in an AZ31B biaxial bulge specimen tested at 250°C and 2×10^{-3} s^{-1} is shown.

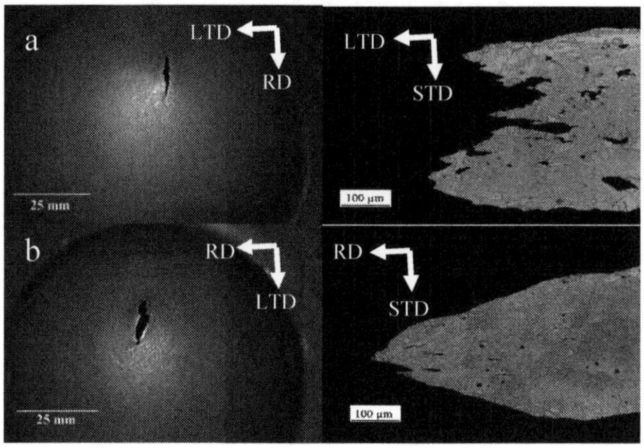

Figure 4. Ruptured domes and photomicrographs from the cross-sections of the ruptures are shown for AZ31B alloy sheet tested at 450°C and at (a) 10^{-3} s^{-1} and (b) 10^{-1} s^{-1}.

GBS creep dominates at slow strain rates, and DC creep dominates at the fast strain rates. Rupture by cavitation is consistent with deformation by GBS creep, which inhibits necking through a low stress exponent (approximately two) and promotes cavitation. Rupture by necking is consistent with deformation by DC creep, which is associated with a higher stress exponent (approximately 5). This high stress exponent provides less resistance to necking.

Rupture by cavity interlinkage is most likely to occur along the RD because the cavity nucleation sites, primarily second-phase particles and inclusions, tend to reside in stringers oriented along the RD. Thus, the biaxial bulge specimens failed by cavitation along the RD at slow strain rates $\lesssim (3\times10^{-2}$ $s^{-1})$. The ruptures at fast strain rates occurred in necked regions. It is hypothesized that a small amount of planar anisotropy in the sheet leads to plane-strain necks aligned along the LTD at fast strain rates ($\gtrsim 3\times10^{-2}$ s^{-1}). Although the planar anisotropy is not significant enough to affect the performance of material constitutive models that predict deformation [10], it might be sufficient to induce flow localization along a preferred direction.

Figure 5 shows stress-strain curves for three different specimen orientations, described as angle of the tensile axis from the sheet rolling direction (RD=0°, LTD=90° and 45°) in the AZ31B sheet at 450°C and at strain rates of 3×10^{-2} s^{-1} and 10^{-1} s^{-1}. These data are from tests previously analyzed in Ref [9]. The flow stress of the LTD (90°) specimens is slightly greater than that of the RD (0°) specimens. This indicates that the LTD is slightly stronger (harder) than the RD. The flow stress along the LTD is slightly higher than along the RD, necking will prefer deformation along the RD instead of the LTD, promoting plan-strain necking along a line parallel with the LTD (lowest-strain direction). This supports the hypothesis of planar anisotropy that affects flow localization.

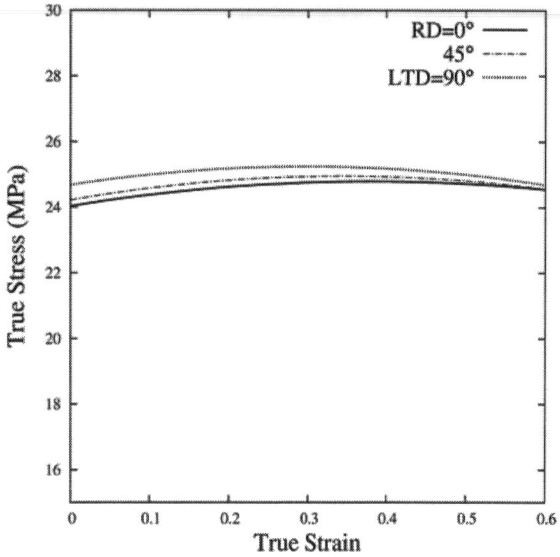

(a)

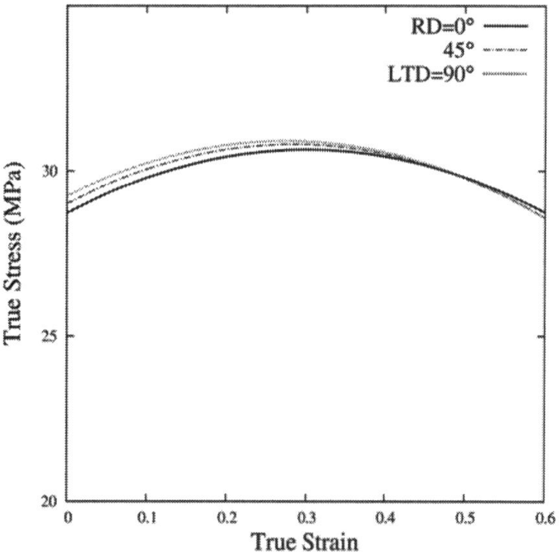

(b)

Figure 5. Stress-strain curves are shown for AZ31 alloy sheet at 450°C at (a) 3×10^{-2} s^{-1} and (b) 10^{-1} s^{-1}. The orientation of the tensile direction with respect to the sheet rolling direction (RD) is specified in degrees (RD = 0°and LTD = 90°) [9].

The reason behind this anisotropy can be directly related to the crystallographic texture, as previously explained by Agnew *et al.* [18]. Figure 6 shows the <0001> pole figure obtained from AZ31B alloy sheet annealed and recrystallized at 450°C.

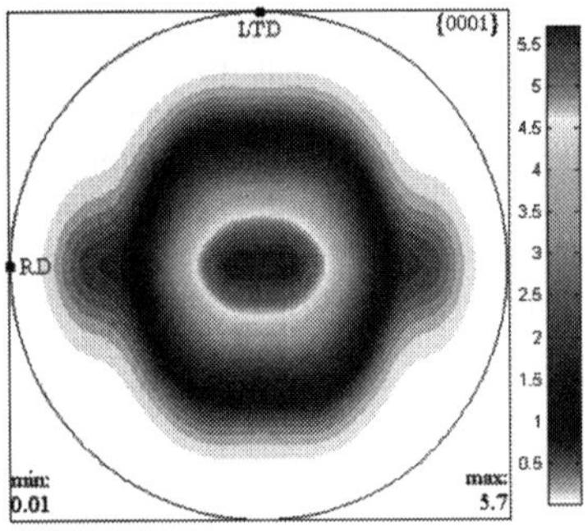

Figure 6. The <0001> pole figure of AZ31B alloy sheet annealed and recrystallized at 450°C for 30 minutes is shown.

Other than the basal fiber texture, the basal poles are spread slightly along the RD, which is consistent with the expected sheet texture noted in Ref [19]. This texture causes the LTD to be the harder in-plane direction and the RD to be the softer in-plane direction during deformation by slip, even though the difference may be slight. Dislocation-climb-controlled creep is the active deformation mechanism at the fast strain rates of interest ($\gtrsim 3\times10^{-2}$ s^{-1}) at 450°C [9]. During the final stages of forming, deformation preferentially occurs along the RD, which causes plane-strain necking and the resulting rupture to be oriented along the LTD, as shown schematically in Figure 7.

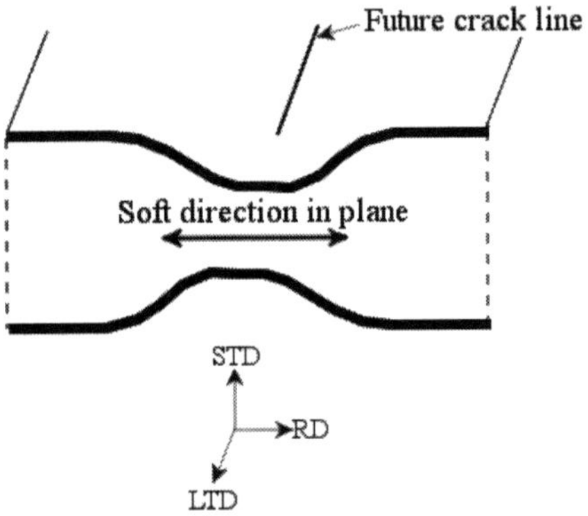

Figure 7. The effect of planar anisotropy on rupture orientation is shown. Deformation during necking occurs preferentially along the soft in-plane direction (the RD in AZ31B).

Similar texture [18], deformation [14] and rupture [13] mechanisms observed in AZ31B down to 350°C suggest that this hypothesis explains the rupture behavior from 350 to 450°C. The rupture orientation in AZ31B at 350 to 450°C helps identify the dominant rupture mechanism, among cavitation and necking, at a chosen strain rate.

Conclusions

Biaxial bulge testing was conducted for AZ31B alloy sheet at temperatures from 250 to 450°C. The rupture orientation at 350 to 450°C is controlled by the active mechanisms of plastic deformation. At slow strain rates ($\leq 3\times10^{-2}$ s^{-1}), cavity interlinkage associated with grain boundary sliding (GBS) creep is the dominant rupture mechanism, as shown in cross sections close to ruptures. This cavity interlinkage induces rupture along the rolling direction (RD). At fast strain rates ($\gtrsim 3\times10^{-2}$ s^{-1}), flow localization (necking) associated with dislocation-climb (DC) creep is the dominant rupture mechanism. Planar anisotropy from the texture of the sheet causes necking to induce rupture along the long-transverse direction (LTD). This change in rupture orientation can be used as a method to identify the dominant rupture mechanism for AZ31B at a chosen strain rate and temperatures from 350 to 450°C. Explosive rupture prevented the analysis of AZ31B tested at 250 to 300°C.

References

1. A.A. Luo, "Magnesium: Current and Potential Automotive Applications," *JOM*, 54 (2002), pp. 42-48.

2. P.E. Krajewski, "Elevated Temperature Forming of Sheet Magnesium Alloys," *SAE Technical Paper*, (2001), Paper Number. 2001-01-3104.

3. L.H. Pomeroy, "Advantages of Lightweight Reciprocating Parts," *SAE Technical Paper,* (1922), Paper Number. 220044.

4. G.S. Cole, A.M. Sherman, "Light Weight Materials for Automotive Applications," *Material Characterization,* 35 (1995), pp. 3-9.

5. A.I. Taub, "Automotive Materials: Technology Trends and Challenges in 21st Century," *MRS Bulletin,* 31 (2006), pp. 336-34

6. A.I. Taub, P.E. Krajewski, A.A. Lou, J.N. Owens, "The Evolution of Technology for Materials Processing over the Last 50 Years: The Automotive Example," *JOM,* 59 (2007), pp. 48-57.

7. J.G. Schroth, "General Motors' Quick Plastic Forming Process," *Advances in Superplasticity and Superplastic Forming*, E.M. Taleff, P.A. Friedman, P.E. Krajewski, R.S. Mishra, and J.G. Schroth, eds., TMS, Warrendale, PA, 2004, pp. 9-20.

8. P.E. Krajewski, J.G. Schroth, "Overview of Quick Plastic Forming Technology," *Mat. Sci. Forum*, 2007, vol. 551-552, pp. 3-12.

9. P.A. Sherek, A.J. Carpenter, L.G. Hector, Jr., P.E. Krajewski, J.T. Carter, J. Lasceski, E.M. Taleff, "The Effects of Strain and Stress State in Hot Forming of Mg AZ31 Alloy Sheet," *Magnesium Technology*, S.M. Mathaudhu, W.H. Sillekens, N.R. Neelameggham , N.Hort, eds., TMS (2012), pp. 301-306.

10. G. Giuliano, S. Franchitti, "The Determination of Material Parameters from Superplastic Free-Bulging Tests at Constant Pressure," *International Journal of Machine Tools and Manufacture,* 48 (2008), pp. 1519-1522.

11. F. Abu-Farha, R. Verma, L.G. Hector Jr., "High Temperature Composite Forming Limit Diagrams of Four Magnesium AZ31B Sheets Obtained by Pneumatic Stretching," *Journal of Materials Processing Technology*, 212 (2012), pp. 1414-1429.

12. A.J. Carpenter, "Physics-Based Material Constitutive Models for the Simulation of High-Temperature Forming of

Magnesium Alloy AZ31," PhD Thesis, University of Texas at Austin, (2012).

13. E. Hsu, J. Szpunar, R. Verma, "Effect of Temperature and Strain Rate on Formability of AZ31 Magnesium Sheet alloy," *SAE Technical Paper*, (2006), Paper Number. 2006-01-0258.

14. E.M. Taleff, L.G. Hector, Jr., R. Verma, P.E. Krajewski, J.K. Chang, "Material Models for Simulation of Superplastic Mg Alloy Sheet Forming," *Journal of Materials Engineering and Performance,* 9 (2010), pp. 488-494.

15. Yin D.L., "Superplasticity and cavitation in AZ31 Mg Alloy at elevated temperatures," *Materials Letters*, 59 (2005), pp. 1714-1718.

16. ASM International, *Metals Handbook. Vol. 2: Properties and Selection: Nonferrous Alloys and Pure Metals*, (Materials Park, OH, 2002).

17. F. Bachmann, R. Hielscher, H. Schaeben, "Texture Analysis with MTEX – Free and Open Source Software Toolbox," *Solid State Phenomena*, 160 (2010), pp. 63-68.

18. S.R. Agnew, O. Duygulu, "Plastic Anisotropy and the Role of Non-basal Slip in Magnesium Alloy AZ31B," *International Journal of Plasticity*, (2005), pp. 1161-1193.

19. U.F. Kocks, C.N. Tome, H.-R. Wenk (Eds.), *Texture and Anisotropy: Preferred Orientations in Polycrystals and their Effect on Materials Properties,* (Cambridge, New York 2000), pp. 204-206.

Magnesium Technology 2013
Edited by: Norbert Hort, Suveen N. Mathaudhu, Neale R. Neelameggham, and Martyn Alderman
TMS (The Minerals, Metals & Materials Society), 2013

Effect of the Extrusion Conditions on the Microstructure and Mechanical Properties of Indirect extruded Mg-Zn-Y Alloy with LPSO Phase

Jonghyun Kim[1] and Yoshihito Kawamura[2]
[1]Magnesium Research Center, Kumamoto University; 2-39-1, Kurokami, Kumamoto, 860-8555, JAPAN
[2]Department of Materials Science, Kumamoto University; 2-39-1, Kurokami, Kumamoto, 860-8555, JAPAN

Keywords: Indirect extrusion, Mg alloy, LPSO, Mechanical properties

Abstract

Mg alloys are the lightest commercial structural alloys and have excellent specific strength and stiffness characteristics. Due to its good castability, the die casting process has been established for the fabrication of various automobile components such as instrumental panels and seat frames. In particular, the application of Mg alloys to a conventional direct extrusion process is not considered to be cost-effective mainly due to their low extrudability.

The indirect extrusion process is free of friction between the billet and the container; hence, the extrusion pressure as well as surface cracking can be reduced considerably compared to the direct extrusion process, leading to its implementation at lower temperatures and higher speeds. In this study, Mg-Zn-Y alloy was subjected to the direct and indirect extrusion processes and the effects of the processing conditions on the microstructure and mechanical properties were investigated.

Introduction

Magnesium alloys, as one of the lightest structural metallic materials, have recently attained an increasing interest for use in computer, mobile phone and automobile industries because of their low density, high specific strength and stiffness, and good damping properties [1-3]. It is generally regarded that the main obstacle that must be overcome in Mg extrusions is their low extrusion speed, which leads to high production costs. Normally, high-strength Mg alloys are extrudable only at speeds of 0.5–2.5 m/min [4-5].

The direct extrusion process is the most widely used method of producing complex shapes in the non-ferrous metals industry. The major disadvantage of this process is that metal flow is inhomogenous, and consequently, the heat generated results in localised heating. This leads to undesirable mechanical and metallurgical features such as surface tearing and surface recrystallisation in extruded rods [6].

Conversely, in the indirect extrusion process, there is no friction between the billet and the container. This can result in significant reductions in extrusion pressure and also surface cracking, compared to the direct extrusion process, thus enabling its implementation at lower temperatures and higher speeds [7].

In this study, Mg-Zn-Y alloy with LPSO phase was underwent direct and indirect extrusion, and the effects of the processing conditions (metal flow rate) on the microstructure and mechanical properties were studied.

Experimental procedure

The investigated alloy, with a nominal composition of $Mg_{97}Zn_1Y_2$, was prepared by high-frequency induction melting in an Ar atmosphere, and then cast into a cylindrical steel mold (31 and 44 mm in diameter for direct and indirect extrusion, respectively).

Direct extrusion was performed using billets of 70 mm height and 29 mm length, whilst indirect extrusion utilized a billets of 120 mm height and 42 mm length. Extrusion was performed at 648K using an extrusion ratio of 10 and a ram speed of 1.0 - 9.9 mm/s.

The phase structures of as-cast and extruded alloys were investigated through X-ray diffractometry (XRD), optical microscopy (OM), scanning electron microscopy (SEM) and transmission electron microscopy (TEM). The extrusion texture and grain size of the as-extruded alloys were measured via electron backscatter diffraction (EBSD) using a field emission scanning electron microscope (FE-SEM). The mechanical properties of the investigated alloys were determined using tensile test specimens with a gage section of ϕ 2.5 mm × 15 mm.

Results and discussion

Fig. 1 shows the dependence of tensile properties on metal flow rate of the $Mg_{97}Zn_1Y_2$ alloy after extrusion at (a) room temperature and (b) 523K. The constitutive equation for the extruded $Mg_{97}Zn_1Y_2$ alloy followed by metal flow rate was determined to be:

$$\dot{\varepsilon} = \left\{ \frac{6V_e \ln R \times \tan\theta}{D(1-1/R\sqrt{R}} \right\}$$

where R is the extrusion ratio, V_e is the extrusion speed, θ is the half angle of the dies, and D is the diameter of billet.

At both room and elevated temperature (523 K), from 0.28 - 1.99 s^{-1}, the tensile yield strengths of the direct extruded alloys sharply decreased with increasing metal flow rate, while the elongation increased. The tensile yield strength of the indirect extruded alloys also decreased with increasing metal flow rate until 1.55 s^{-1}, after which the tensile yield strength were slightly decreased. At room temperature and 0.28 s^{-1}, the yield strength of direct and indirect extruded alloy were 342 MPa and 275 MPa, respectively.

Figure 2 shows microstructures of direct ((a) and (c)) and indirect ((b) and (d)) alloys extruded with a metal flow rate of 0.28 ((a) and (b)) and 1.99 s^{-1} ((c) and (d)). The direction of extrusion is parallel to the horizontal direction in these as well as all subsequent microstructural figures presented in this paper. In In each micrograph in Figure 2, three areas can be seen: the dynamically recrystallized (DRXed) α-Mg fine-grain region, the hot-worked α-Mg coarse-grain region, and the LPSO phase region.

To quantitatively describe the change in the dispersion of the LPSO phase, the dispersion level, D_L (μm^{-1}), is defined in this study by the following expression: [8]

$$D_L = \frac{\sum N_v}{\sum L_v}$$

where L_v (μm) is the length of a vertical line segment L drawn perpendicular to the extrusion direction within the micrographs in Fig. 2, and N_v is the number of LPSO phase regions intersected by segment L. The total length of segments, $\sum L_v$, is greater than 5 mm.

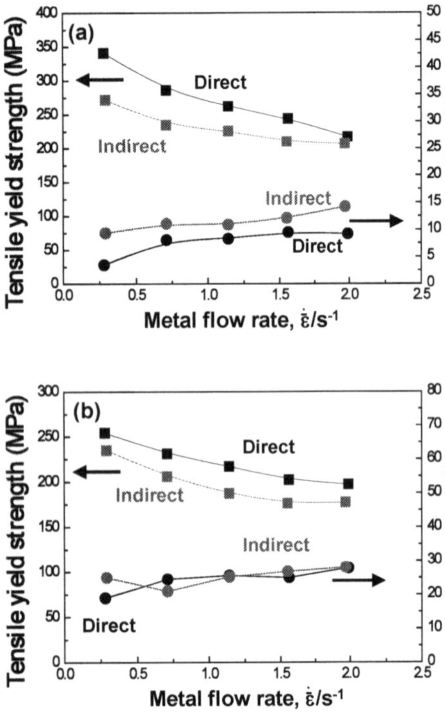

Figure. 1 Dependence of mechanical properties of direct and indirect extruded $Mg_{97}Zn_1Y_2$ alloys on the metal flow rate, at(a) room and (b) elevated temperature (523 K).

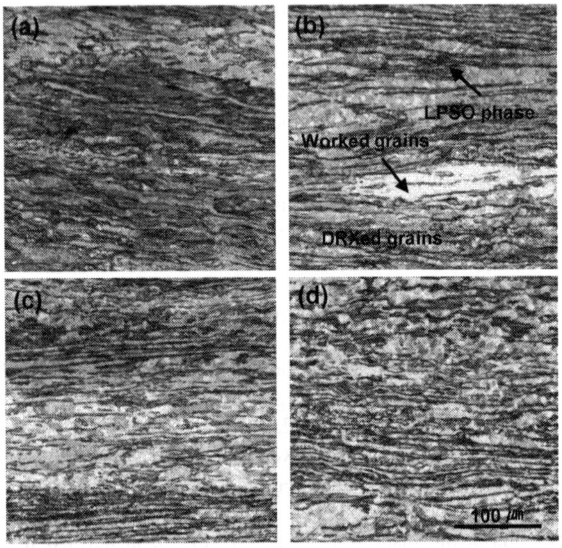

Figure. 2. Optical micrographs of direct and indirect extruded $Mg_{97}Zn_1Y_2$ alloys at low and high metal flow rates: (a) direct and $0.28s^{-1}$, (b) indirect and $0.28s^{-1}$, (c) direct and $1.99s^{-1}$, and (d) indirect and $1.99s^{-1}$

Fig. 3 shows the relationship between the dispersion of LPSO phase in the direct and indirect extruded $Mg_{97}Zn_1Y_2$ alloys and the metal flow rate. The dispersion and refinement of the fiber-shaped LPSO phase for indirect extruded $Mg_{97}Zn_1Y_2$ alloys are higher than those of the direct extruded Mg alloys.

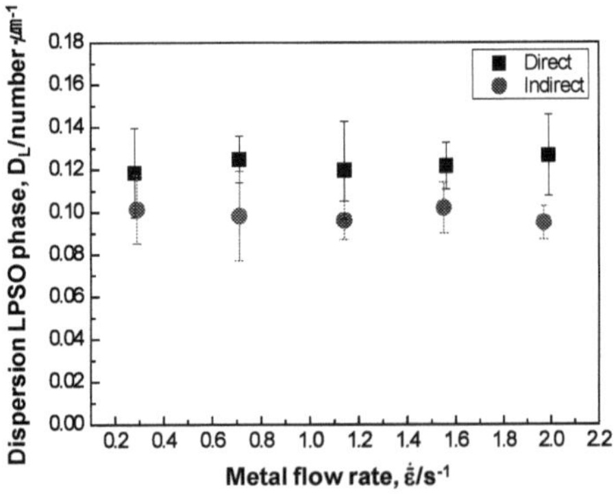

Figure 3. Relationship between dispersion LPSO phase of the direct and indirect extruded $Mg_{97}Zn_1Y_2$ alloys and metal flow rate

Figs. 5 and 6 show inverse pole figure (IPF) maps of longitudinal cross-sections of direct and indirect extruded $Mg_{97}Zn_1Y_2$ alloys, at metal flow rates of 0.28 and $1.99s^{-1}$, respectively.. The $Mg_{97}Zn_1Y_2$ alloy shows a bimodal α-Mg grain structure composed of fine recrystallized grains of approximately 1-3 μm in size, and coarse unrecrystallized parent grains with grain sizes greater than 10 μm. It is noteworthy that dynamic recrystallization occurs in the α-Mg matrix in the vicinity of LPSO phase.. Comparing both metal flow rates, the volume fraction of the DRXed region is larger for the slower rate ($0.28s^{-1}$), while that of the worked grains region is larger for the faster rate ($1.99s^{-1}$)..

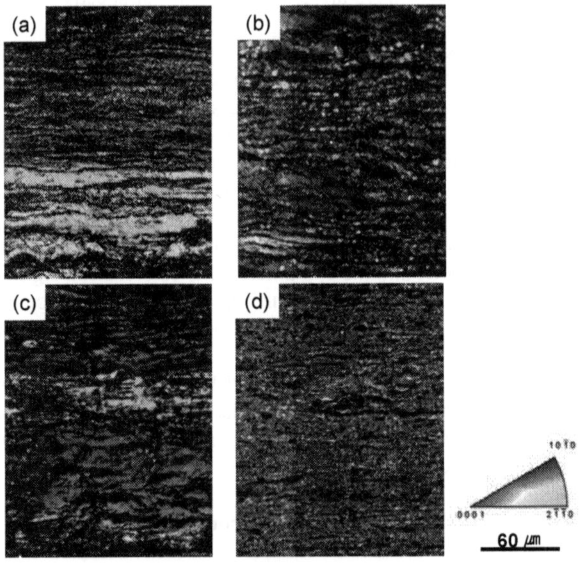

Figure. 4. IPF maps obtained from a longitudinal cross-section of direct and indirect extruded $Mg_{97}Zn_1Y_2$ alloy at a metal flow rate

of 0.28 s^{-1}: (a) direct, center region (b) indirect, center region (c) direct, edge region and (d) indirect, edge region

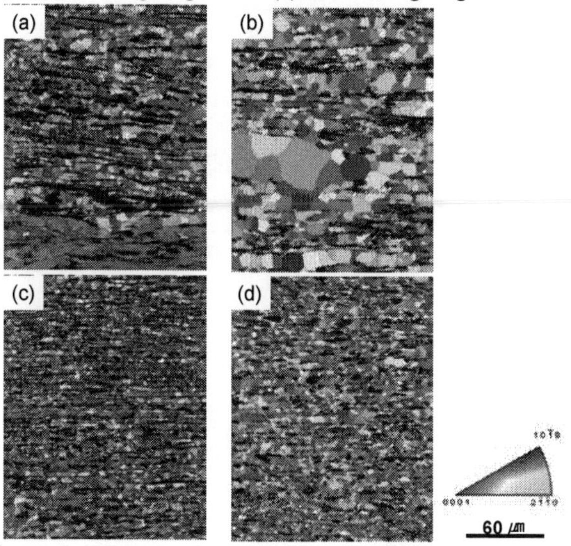

Figure. 5. IPF maps obtained from a longitudinal cross-section of direct and indirect extruded $Mg_{97}Zn_1Y_2$ alloy at a metal flow rate of 1.99 s^{-1}: (a) direct, center region (b) indirect, center region (c) direct, edge region and (d) indirect, edge region

Figure 6. shows the DRXed α-Mg grain size at center and edge regions of direct and indirect extruded $Mg_{97}Zn_1Y_2$ alloys. The grain size in both direct and indirect extruded $Mg_{97Z}n_1Y_2$ alloys increase with increasing metal flow rate. The grain sizes of the indirect extruded $Mg_{97}Zn_1Y_2$ alloys in center and edge regions of extruded samples are both larger than those in the respective regions in the direct extruded Mg alloys.

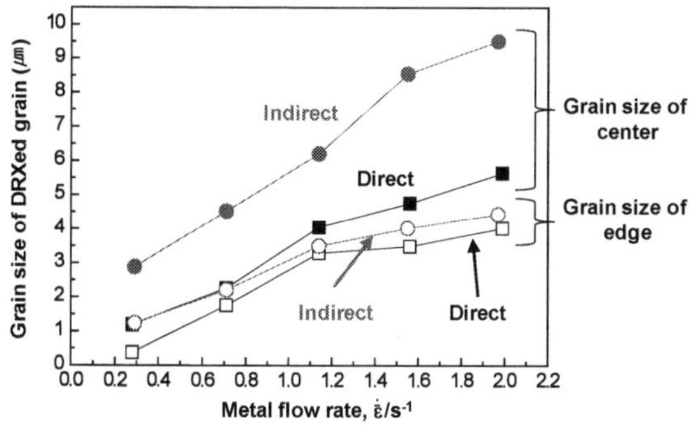

Figure. 6. Grain size of center and edge regions in direct and indirect extruded $Mg_{97}Zn_1Y_2$ alloys as a function of metal flow rate.

Conclusions

(1) The tensile yield strength of both direct and indirect extruded alloys decrease with increasing metal flow rate. At a given metal flow rate, the tensile yield strength of direct extruded $Mg_{97}Zn_1Y_2$ alloys is higher than that of the indirect extruded alloys.

(2) The dispersion and refinement of the fiber-shaped LPSO phase regions in indirect extruded $Mg_{97}Zn_1Y_2$ alloys are both higher than those of the direct extruded Mg alloys.

(3) The grain sizes of the indirect extruded alloy in the edge and center regions are both larger than that those in the respective regions in the direct extruded alloys.

References

1. A. A. Luo, "Recent magnesium alloy development for automotive powertrain applications," *Materials Science Forum*, 419-422 (2003) 57-66.
2. Magnesium & Magnesium alloys, ed. M. M. Avedesian & H. Baker: ASM International, Materials Park, OH (1999), 3-6.
3. A. Luo, M.O. Pekguleryuz, "Cast magnesium alloys for elevated temperature applications," *Journal of Materials Science*, 29 (1994) 5259-5271.
4. Y. Kawamura and M. Yamasaki, "Formation and Mechanical Properties of $Mg_{97}Zn_1RE_2$ Alloys with Long Period Stacking Ordered Structure," *Materials Transactions*. 48 (2007) 2986-2992.
5. S. Yoshimoto, M. Yamasaki and Y. Kawamura, "Microstructure and Mechanical Properties of Extruded Mg-Zn-Y Alloys with 14H Long Period Ordered Structure," *Materials Transactions*. 47 (2006) 959-965.
6. Atwell, D.L., Barnett, M.R., "Extrudability of magnesium alloys," *Magnesium Technology*, (2007) 145-150.
7. T. Sheppard and S.J. Paterson, "Some observations on metal flow and the development of structure during the direct and indirect extrusion of commercial purity aluminum," *Journal of Mechanical Working Technology*, 7 (1982) 39-56.
8. Hirano M, Yamasaki M, Hagihara K, Higashida K, Kawamura Y, "Effect of Extrusion Parameters on Mechanical Properties of Mg97Zn1Y2 Alloys at Room and Elevated Temperatures," *Materials Transactions*. 51 (2010) 1640-1647.

Magnesium Technology 2013
Edited by: Norbert Hort, Suveen N. Mathaudhu, Neale R. Neelameggham, and Martyn Alderman
TMS (The Minerals, Metals & Materials Society), 2013

INVESTIGATION OF MECHANICAL PROPERTIES AND DEFORMATION BEHAVIOR OF CaO ADDED Mg-6Zn-1.2Y SHEETS

Hyun Kyu Lim, Young-Ok Yoon, Shae K. Kim
Korea Institute of Industrial Technology; 156, Gaetbeol-ro, Yeonsu-gu; Incheon, 406-840, KOREA

Keywords: Mg-Zn-Y alloy, CaO addition, Hot rolling, Mechanical properties, Deformation behavior

Abstract

Recently, CaO added Mg-Al alloy sheets exhibited improved strength and elongation due to the effects of dispersion hardening and grain-refining by intermetallic compounds such as Al_2Ca and $(Mg,Al)_2Ca$ and improved melt cleanliness. However, there is little study on these effects in Mg-Zn alloy system. In this study, as-cast ingots were prepared by melting Mg, Zn, Mg-25Y master alloy and Mg-3CaO master alloy in an electric resistance furnace under SF_6+CO_2 atmosphere. Sheet specimens were prepared by hot rolling process after indirect extrusion. The mechanical properties of CaO added Mg-6Zn-1.2Y specimens varied significantly with annealing process while the variation of mechanical properties of Mg-6Zn-1.2Y specimen was small. In addition, the texture of specimens with CaO (Ca in ICP analysis) was different from that of specimen without CaO. Therefore, we discussed the effect of microstructure and texture on mechanical properties and deformation behavior of CaO added Mg-6Zn-1.2Y sheet.

Introduction

There are some limitation to apply magnesium alloys to structural parts, although magnesium alloys have attractive properties such as low density, high specific strength (ratio of strength to density), excellent castability, good machinability and so on [1]. Therefore, many researches on the development of magnesium alloys with not only high strength, but also high ductility have been occurred to overcome the drawbacks of magnesium alloys.

One of these researches is on ECO-Mg alloy. The concept of ECO-Mg alloy was to simply introduce CaO particles as an ingredient into conventional magnesium alloys. The surface oxide layer of molten magnesium alloy changed from porous oxide film to dense and complex oxide films. Therefore, they have been possible (1) non-SF6 process, (2) Be elimination, (3) improved melt cleanliness, (4) ensured original process adjustability for casting, forming, joining as well as surface treatment, (5) improved mechanical property by grain refinement and internal soundness, (6) ensured safety during manufacturing and application by raising oxidation and ignition resistances of machined chips and products, and (7) improved recyclability [2]. In the case of extruded alloys, the CaO added magnesium alloys had good surface quality and the surface roughness decreased with CaO contents [3, 4]. Moreover, the as-rolled CaO added AZ31 alloy exhibited higher strength than as-rolled conventional AZ31 alloy due to the effects of dispersion hardening and grain-refining by intermetallic compounds such as Al_2Ca and $(Mg, Al)_2Ca$ and improved melt cleanliness [5]. However, there is little study on these effects in Mg-Zn alloy system.

Quasicrystals are isotropic and posse specially ordered lattice structure called the quasiperiodic lattice structure [6]. When an alloy possesses quasicrystals as a second phase, they are stable against coarsening at high temperatures due to the low interfacial energy of quasicrystals [7], providing the improved bonding properties in the I-phase/matrix interface. Therefore, to increase both strength and ductility, many researches on magnesium alloys reinforced with quasicrystalline icosahedral phase (I-phase) particles have been performed because magnesium alloys reinforced with I-phase particles, especially Mg-Zn-Y alloys, have been reported to exhibit a good combination of elongation and strength [8-10].

This study has been carried out to investigate microstructures and mechanical properties of CaO added Mg-6Zn-1.2Y sheets with annealing process and understand the effect of the addition of CaO in Mg-Zn-Y alloys.

Experimental procedure

As-cast ingots were prepared by melting commercial purity Mg, Zn, Mg-25Y master alloy and Mg-3CaO master alloy in an electric resistance furnace under SF_6+CO_2 atmosphere, followed by pouring the melt into the steel mold of cylindrical cavity with the dimension of 50 mm in diameter and 200 mm in height. Chemical compositions of as-cast alloys were analyzed by ICP-mass spectrometer and the results were listed in Table 1.

The ingots preheated at 300~320 °C for 60 min. were extruded with a reduction ratio of 20:1 and the cross-sectional area of the extruded strips was 25*4 mm^2. The extruded strips preheated at 350 °C for 30 min. were rolled with a reduction ratio of 10 % to 28 % per pass to final thickness of 1.3 mm. After hot rolling, sheets were annealed at 250~350 °C for 15~60 min. Uniaxial tensile tests were carried out on dog-bone specimens (specimen gauge length of 25 mm in length, 6 mm in width, and 1.3 mm in thickness) machined from annealed sheets at room temperature with initial strain rate of 10^{-3} /s.

Table 1. ICP analysis results of as-cast ingots

Alloy No.	Description		Chemical composition (wt.%)			
			Mg	Zn	Y	Ca
1	ZW61	Nominal	92.8	6	1.2	
		Analyzed	Bal.	5.93	1.16	
2	ZWO6103	Nominal	92.5	6	1.2	0.3
		Analyzed	Bal.	6.09	1.11	0.33
3	ZWO6107	Nominal	92.1	6	1.2	0.7
		Analyzed	Bal.	6.20	1.13	0.72

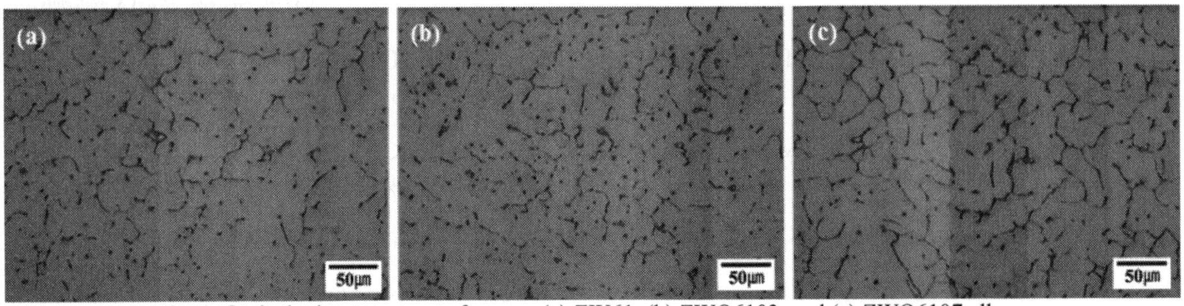

Fig. 1. Optical microstructures of as-cast (a) ZW61, (b) ZWO6103, and (c) ZWO6107 alloys

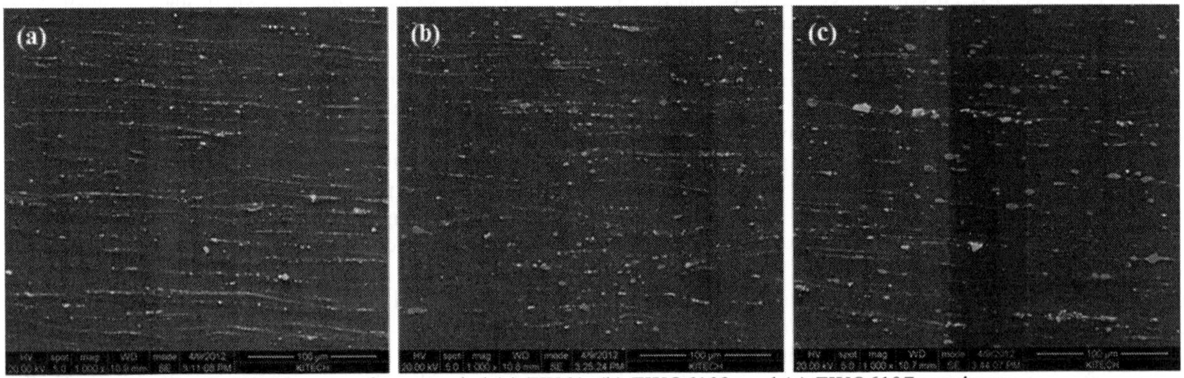

Fig. 2. SEM images of as-extruded (a) ZW61, (b) ZWO6103, and (c) ZWO6107 specimens

For microstructure observation, specimens for an optical microscope were etched with acetic picral etchant or 2 % nital etchant and for an EBSD were ion-milled.

Table 2. Electronegativity of elements [12]

Element	Mg	Zn	Y	Ca
Electronegativity	1.31	1.65	1.22	1.00

Results and Discussion

Table 1 showed the results of chemical composition analysis by ICP-mass spectrometer. Although Ca was detected by ICP in alloys, we used "O" as code letter to be distinguished from alloys with metallic Calcium whose code letter is "X".

The microstructures of as-cast alloys were shown in Fig. 1. All alloys were consisted of α-Mg and the secondary solidified phases. According to the EDS analysis results, in CaO added alloys, e.g. ZWO6103 and ZWO6107 alloys, there are two kinds of secondary phases. One consisted of Mg, Zn, and Y, which might be I-phase and the other consisted of Mg, Zn, and Ca, which might be τ_2 ($Ca_2Mg_5Zn_{13}$) phase. Furthermore, with CaO content, the volume fraction of τ_2 phase increased and α-Mg and τ_2 phases composed eutectic pockets. Generally, the greater difference between electronegativity of the elements the more stable compound formation between them [11]. The electronegativity of Mg, Zn, Y, and Ca were listed in Table 2. In Table 2, the electronegativity of Ca is the lowest among elements. Therefore, with increasing CaO or Ca content, the τ_2 phase might be formed easily.

Fig. 2 showed the SEM images of as-extruded specimens. The broken secondary phase particles formed during extrusion from grain boundary networks of as-cast alloys and were distributed as streamlines in parallel to extrusion direction. From the results of EDS mapping, coarser τ_2 phase existed in ZWO6107 strip than ZWO6103 strip. That means with increasing CaO content, the volume fraction of τ_2 phase increased and τ_2 phase is harder than I-phase.

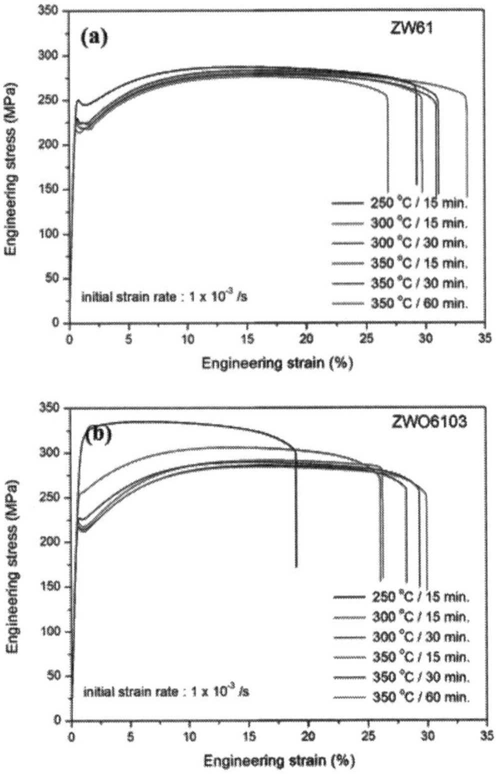

Fig. 3. Engineering stress-strain curves of (a) ZW61 and (b) ZWO6103 sheets

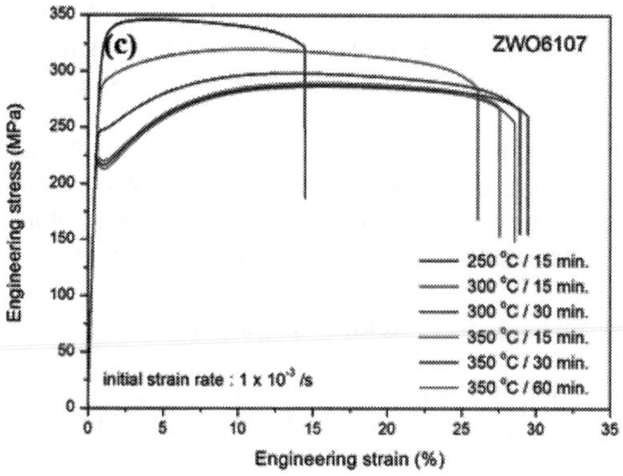

Fig. 3. (*continued*) Engineering stress-strain curves of (c) ZWO6107 sheets

The engineering stress-strain curves of annealed sheets were shown in Fig. 3. In the case of annealed ZW61 sheets, the variation strengths were small in all annealing condition. However, the strengths of annealed ZWO6103 and ZWO6107 sheets varied significantly with annealing times and temperatures. This phenomenon could be explained that the thermal stability of CaO added Mg-6Zn-1.2Y sheet was higher than that of ZW61 alloy sheet. Also, some CaO added specimens which were annealed over 300 °C exhibited the yield phenomenon while the yield phenomenon of ZW61 specimens could be observed in all annealing conditions. J. Koike et al. had reported that the yield phenomenon occurred by the activation of non-basal slip system due to the compatibility stresses in fine-grained microstructures [13]. That is, yield phenomenon could be occurred in fine grains with no or little dislocation. Though CaO added specimens annealed at 250 °C had fine-grained microstructures, there might be many dislocations in grains. Thus deformation mechanism of CaO added specimens annealed at 250 °C was different from that of specimens annealed over 300 °C.

In addition, when specimens were fully annealed (in this case, annealing temperature was 350 °C), ZWO6103 and ZWO6107 sheets exhibited hither ultimate tensile strengths than ZW61 did, although tensile yield strengths of CaO added alloys specimens were slightly lower than those of specimens without CaO. Overall, annealed ZW61 specimens had larger elongation than CaO added specimens due to the dispersion of fine particles.

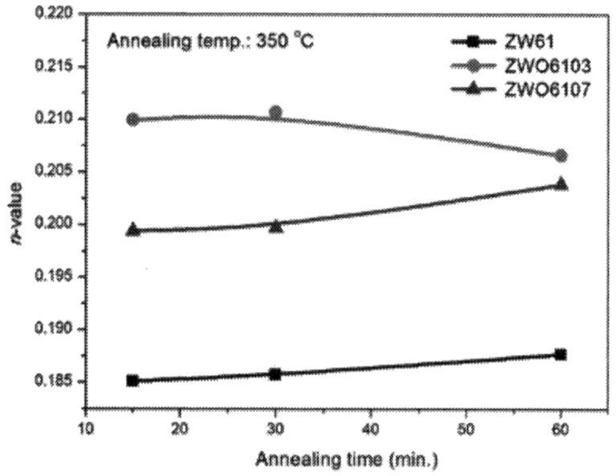

Fig. 4. Variations of strain-hardening exponents (n-value) with annealing time

Fig. 4 showed the variations of strain-hardening exponents (n-value) with annealing time at 350 °C of the tested specimens, which were measured in the engineering strain range from 5 % to 10 %. As shown in Fig. 4, the n-values of ZWO6103 and ZWO6107 sheets were higher than those of ZW61 sheets. When comparing the n-values of ZWO6103 and ZWO6107 specimens, ZWO6103 specimens exhibited higher n-values than ZWO6107 did. Generally, the strong work-hardening in materials containing non-deforming particles could be deduced from the cutting and bowing process of dislocations. The effective obstacle spacing for a second dislocation to be extruded between the particles was reduced due to the dislocation loops formed by the passage of the first dislocation. Thus, the stress required to generate a second or further loop was increased [14]. In this study, the addition of CaO in Mg-6Zn-1.2Y alloy might reduce the effective obstacle spacing, but when adding 0.7 wt.% CaO, it could be thought that the effective obstacle spacing might increase due to the large τ_2 phase particles.

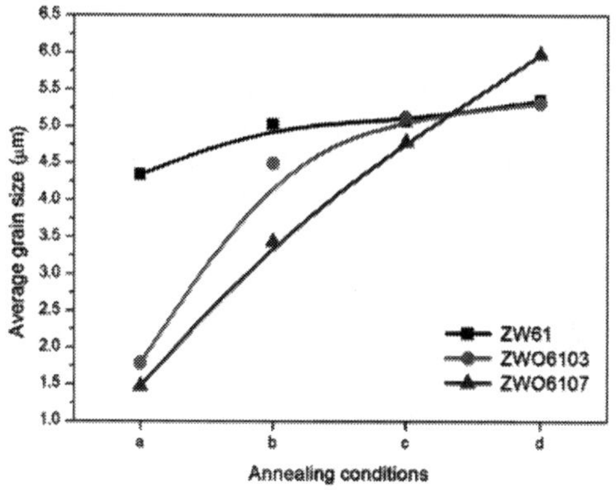

Fig. 5. Comparison of average grain size with different annealing process (a: 250 °C / 15 min., b: 300 °C / 15 min., c: 350 °C / 30 min., and d: 350 °C / 15 min.)

To find the relationship between grain size and yield strength, gain sizes of some tested alloys were measured by EBSD. Fig. 5 showed the average gain sizes of specimens with different annealing process. The variation of grain size of ZW61 specimens was small while the grain sizes of ZWO6103 and ZWO6107 specimens increased with annealing condition. These results corresponded with previous tensile test results.

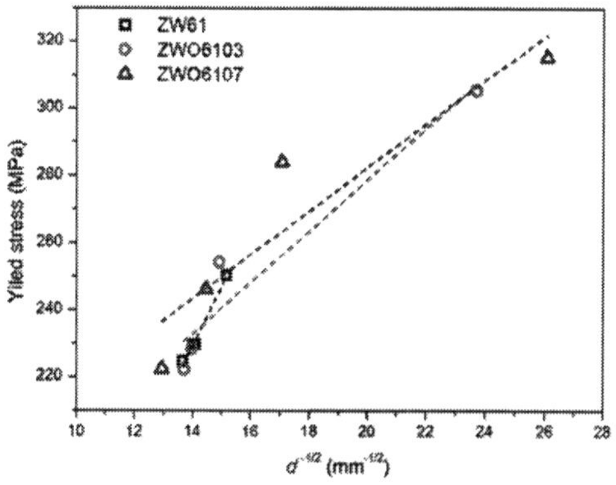

Fig. 6. Relation between yield stress and the inverse square root of the average grain size

Fig. 6 showed the relationship of yield strengths against the inverse square root of the average grain sizes in tested alloys.

Table 3 listed the intercept stress, σ_o and the slope of tested specimens. From data of Fig. 6 and Table 3, it could be noticed that with adding CaO, σ_o increased but k_y decreased. In Hall-Petch relationship, σ_o referred to the average yield strength of a single grain and was determined by all the operative strengthening mechanisms except grain boundary strengthening. And k_y is a complex parameter that determines the effectiveness of grain boundaries in raising yield strength [15]. Therefore, it could be thought that the addition of CaO in Mg-6Zn-1.2Y alloy made average yield strength of a single grain higher but the effectiveness of grain boundaries in raising yield strength lower.

Table 3. Hall-Petch parameters of tested specimens

Alloy	Intercept stress, σ_o (MPa)	Slope, k_y (MPa mm$^{1/2}$)
ZW61	-16.4±17.2	17.6±1.2
ZWO6103	125.7±25.3	7.7±1.5
ZWO6107	152.4±32.9	6.5±1.8

Fig. 7 showed the results of the texture analysis using EBSD obtained from tested specimens annealed at 300 °C for 30 min. Egg-shaped distribution of {0002} pole was observed in annealed ZW61 specimen, which indicated that the strong basal plane texture had been developed during rolling and annealing processes. However, in the case of CaO added specimens, {0002} pole tilted toward the transverse direction (Fig. 7 (b) and (c)). Moreover, the texture intensity decreased from 10.512 (Fig. 7 (a)) to 6.689 (Fig. 7 (c)) with increasing CaO content. This change of texture by adding CaO might contribute to the little reduction of elongation for annealed sheet specimens with high strength.

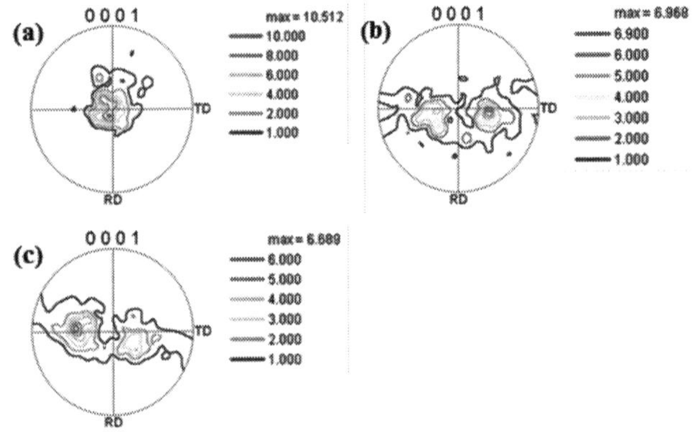

Fig. 7. {0002} Pole figure of tested specimens annealed at 300 °C for 30 min.; (a) ZW61, (b) ZWO6103, and (c) ZWO6107

Conclusion

In the present study, mechanical properties and deformation behavior of CaO added Mg-6Zn-1.2Y sheets were investigated. The conclusions are as follows:

(1) τ_2 ($Ca_2Mg_5Zn_{13}$) phase was formed by addition of CaO in Mg-6Zn-1.2Y alloy.

(2) As-cast structures of icosahedral and τ_2 phases were destroyed by extrusion process and the size of broken τ_2 phase particles was larger than that of icosahedral phase.

(3) In annealed conditions, CaO added Mg-6Zn-1.2Y sheets exhibited higher ultimate tensile strength than ZW61 sheet, but elongation decreased due to the coarser τ_2 phase particles by adding CaO.

(4) The addition of CaO in Mg-6Zn-1.2Y alloy made average yield strength of a single grain higher but the effectiveness of grain boundaries in raising yield strength lower.

(5) Also, the addition of CaO in Mg-6Zn-1.2Y alloy made texture different from ZW61 alloy. And texture change might contribute to the little reduction of elongation for annealed sheet specimens with high strength.

Acknowledgments

The authors are grateful to Prof. D.H. Kim and Mr. Y.K. Kim in Yonsei University and Prof. Y.B. Park and Mr. T.Y. Gwak in Sunchon National University for hot rolling and EBSD experiments.

Reference

[1] M.M. Avedesian, H. Baker, eds., *Magnesium and Magnesium Alloys*, ASM International, Metals Park, 1999

[2] F. Czerwinski ed., *Magnesium Alloys - Design, Processing and Properties*, InTech, 2011

[3] J.K. Lee, Y.O. Yoon, S.K. Kim, "Development of CaO Added Wrought Mg Alloy for Cleaner Production", *Magnesium Technology 2006* (2006) 185-189

[4] S.K. Kim, J.K. Lee, Y.O. Yoon, H.H. Jo, "Development of AZ31 Mg alloy wrought process route without protective gas", *Journal of Materials Processing Technology*, 187-188 (2007) 757-760

[5] D.I. Jang, Y.O. Yoon, D.U. Kim, S.B. Jung, S.K. Kim, "Behavior of CaO in hot-rolled AZ31 Mg alloys", *Proceedings of 6th International Conference on Materials Processing for Properties and Performance (MP3-2007)*, Beijing, September 2007, China

[6] C. Janot, *Quasicrystals*, Oxford, Clarenson Press, 1994

[7] J. M. Dubois, P. Plaindoux, E. Belin-Ferre, N. Tamura, D. J. Sordelet, *Proceedings of the 6th international Conference on Quasicrystals*, World Scientific, Singapore, 1997

[8] D.H. Bae, S.H. Kim, D.H. Kim, W.T. Kim, "Deformation behavior of Mg-Zn-Y alloys reinforced by icosahedral quasicrystalline particles", *Acta Materialia*, 50 (2002) 2343-2356

[9] J.Y. Lee, D.-H. Kim, H.K. Lim, D.H. Kim, "Effects of Zn / Y ratio on microstructure and mechanical properties of Mg-Zn-Y alloys", *Materials Letters*, 59 (2005) 3801-3805

[10] J.Y. Lee, H.K. Lim, D.-H. Kim, W.T. Kim, D.H. Kim, "Effect of volume fraction of qusicrystal on the mechanical properties of quasicrystal-reinforced Mg-Zn-Y alloys", *Materials Science and Engineering A*, 449-451 (2007) 987-990

[11] L.L. Rokhlin, *Magnesium Alloys Containing Rare Earth Metals: Structure and Properties*, Talor & Francis, New York, 2003

[12] http://chemed.chem.wisc.edu/chempaths/GenChem-Textbook/Electronegativity-861.html

[13] J. Koike, T. Kobayashi, T. Mukai, H. Watanabe, M. Suzuki, K. Maruyama, K. Higashi, "The activity of non-basal slip systems and dynamic recovery at room temperature in fine-grained AZ31B magnesium alloys", *Acta Materialia*, 51 (2003) 2055-2065

[14] T.H. Courtney, *Mechanical Behavior of Materials 2nd ed.*, McGraw-Hill, Now York, 2000

[15] C.R. Barrett, W.D. Nix, A.S. Tetelman, *The Principles of Engineering Materials*, Prentice-Hall, New Jersy, 1973

Magnesium Technology 2013

Phase Formation

Magnesium Technology 2013
Edited by: Norbert Hort, Suveen N. Mathaudhu, Neale R. Neelameggham, and Martyn Alderman
TMS (The Minerals, Metals & Materials Society), 2013

BOUNDS TO HARDENING BY SOLID SOLUTION, PRECIPITATION AND SHORT RANGE ORDER IN Mg BINARY ALLOYS

C.H. Cáceres*, Saeideh Abaspour
ARC Centre of Excellence for Design in Light Metals
Materials Engineering, School of Engineering
The University of Queensland, QLD 4072, Australia

Keywords: Miedema's coordinates, Mg alloys, precipitation hardening; short range order

Abstract

Miedema's coordinates are used to rank 4 model binary alloys considering the respective values of enthalpy of formation and the tendency to developing random solid solution, precipitation, short range order (SRO) and intermetallic compounds. The terminal solid solubility generally increases whereas the tendency to order decreases with decreasing heat of formation, and hardening by near-random solid solution and/or precipitation is expected to be dominant for solutes with low tendency to order, such as Al. For solutes with an intermediate tendency to order, hence solubility, such as Zn, or to form compounds, such as Gd, SRO is predicted to dominate the hardening. For solutes whose very large heat of formation leads to very high melting point intermetallics forming congruently, such as Sb, the terminal solid solubility is too low for any solute based hardening to be feasible. Implicancies for alloy design and selection regarding solute or precipitation hardening, SRO and creep resistance are discussed.

Introduction

Magnesium's locus on the group IIA manifests in highly attractive pairing with most other metals, i.e., the interaction energy of mixing is strongly negative. A strongly negative interaction energy normally [1] leads to a high melting point intermetallic bounded by two eutectics, of the type illustrated by Fig. 1-a, and to which many binary Mg alloys closely conform to. Cases in point are, in order of increasingly negative interaction energy, Mg-Al, Mg-Zn and Mg-Sb alloys. In the opposite case, i.e., strongly positive energy, limited solubility and a single eutectic is observed, as in Fig. 1-b. An example for this case is Mg-Th.

When the interaction energy is negative, the ensuing tendency to pairing between unlike atoms in a solution of X atoms in Mg introduces an ordering effect in an otherwise random solid solution, with pre-eminence of pairs Mg-X in detriment of X-X or Mg-Mg pairs [2]. In a dilute alloy the net result is the development of short range order (SRO), whereas in the concentrated ones it leads to compound formation. In some alloys SRO is observed already in the liquid, leading to highly viscous melts, and a propensity to form metallic glasses [1] (pp. 81–82). SRO has been confirmed by X-ray diffraction in a number of Mg-based solid solutions, namely: Mg-Sn, Mg-Gd, Mg-In and Mg-Er [3,4], whereas indirect evidence has been presented for Mg-Zn [5] and Mg-Y [6,7]. SRO has also been predicted from first principles and experimentally confirmed in liquid Mg-Zn [8]. The formation of SRO in a solid solution leads to increased hardening well above that of random solid solution [5,9,10]. For a metal like Mg whose alloys are reluctant to develop precipitation hardening [11-13], the exploitation of SRO as a strengthening mechanism is an important alternative.

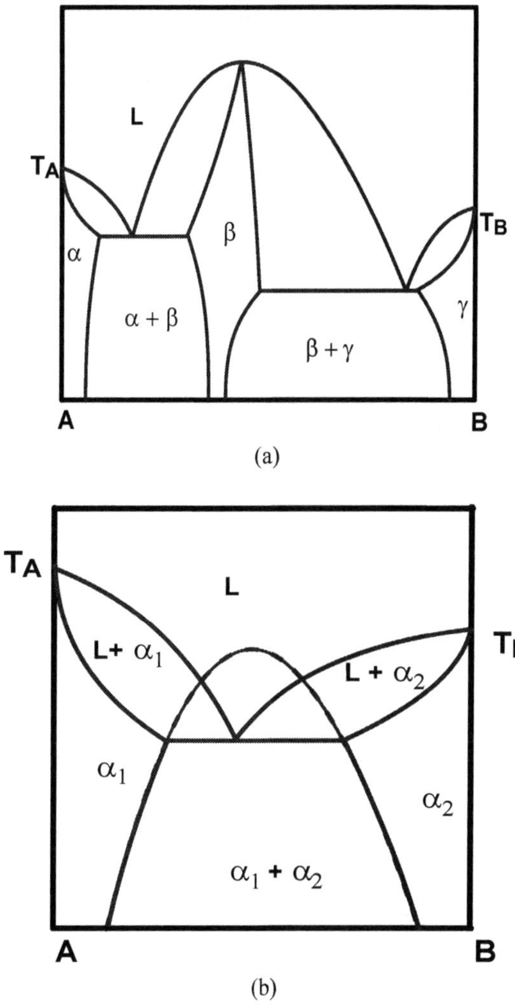

Figure 1: The phase diagrams for (a) highly negative and (b) highly positive interaction energy between solute and solvent.

* Corresponding author, c.caceres@uq.edu.au, Fax:+ 61-7-33467015.

Predicting the presence of SRO is immediate from a phase diagram like that of Figure 1-a, but not all phase diagrams of Mg alloys are so amenable of interpretation even when the interaction energy with the solute is manifestly negative. This is the case of Mg-Gd, Mg-Y or Mg-Nd alloys, amongst others, which are characterized by a sequence of intermetallic compounds, some forming from the solid solution and some forming congruently from the liquid. These intermetallics are often associated with particular size ratios between host and solute, i.e., the formation of intermetallic compounds at given concentrations normally takes pre-eminence; the same applies for particular electron concentrations [2,14]. That is, electronegativity is not the only factor at play.

The sign and magnitude of the enthalpy of mixing can be used to sort out alloys of a given host, Mg in this case, predicting, among other things, the relative solubility. These predictions can be made regardless of the details of the alloy's phase diagram, i.e., regardless of whether it conforms to those of Figure 1 or exhibits a complex sequence of compounds. This sorting method, developed on the principles laid out in the 1970's by Miedema *et al.* [15-18], will be used in this work to try and anticipate which alloy systems are more amenable of developing either strong solid solution, precipitation, short range order or, in some extreme cases, no solute-based hardening at all. Four model alloys will be used for the analysis: Mg-Al, Mg-Zn, Mg-Sb and Mg-Gd. The phase diagrams of the first three systems closely conform to that of Fig. 1-a, whereas the Mg-Gd is meant to represent the many intermetallic compound forming alloy systems.

Miedema's cellular model

Miedema *et al.*'s [15-19] phenomenological approach, in its simpler form involves only two parameters, and is valid for alloys of the transition metals with divalent metals such as Mg and Zn. It combines a work function, ϕ^*, (closely related to Pauling's electronegativity value), and the electron density at the boundary of the Wigner–Seitz (WS) cell, $n_{WS}^{1/3}$. The approach enables calculating the enthalpy of formation of dilute solutions in the liquid state, predicting the sign and magnitude of the interaction, hence the relative tendency to form either ordered, random or immiscible solutions. *Ab-initio* and Calphad calculations normally reproduce closely Miedema *et al.*'s, providing the approach with strong independent support. A limitation often pointed out is its fundamentally isotropic nature, i.e., the method applies most accurately to metals in the liquid state, where size related elastic energy effects can be neglected. Further developments in the *ab-initio* methods enabled overcoming some of these limitations (e.g., see Zhang *et al.* [20]).

The enthalpy of formation [17], ΔH, of a given alloy is the balance between a negative contribution from the difference between the host's and solute's work functions, and a positive contribution stemming from the need to smooth any discontinuity of the electron density at the boundary of the WS cell between solute and solvent. The difference in electronegativity is of course the main driving force either to form a solution or an intermetallic compound by charge transfer.

Using values from [18,21] (see Table 1 for specific values for Al, Zn, Gd and Sb), a diagram using Miedema's coordinates was created for Mg and a number of solutes in Figure 2. The two perpendicular lines through Mg sort out solutes as follows: elements sitting right on the lines have no enthalpy of mixing with Mg; the upper (or "north) and lower, (or "south") sectors identify solutes with a negative energy of mixing; those lying on the east or west sectors have a positive energy of mixing.

Solutes located right on the lines are expected exhibit complete miscibility, those on the north or south sections alloy readily with the host and show a tendency to develop phase diagrams similar to that of Fig. 1-a, whereas diagrams for solutes on the east-west sections should form simple eutectics akin to that of Fig. 1-b. The correctness of these predictions can be easily verified for the solutes considered in this work, as well as for many others.

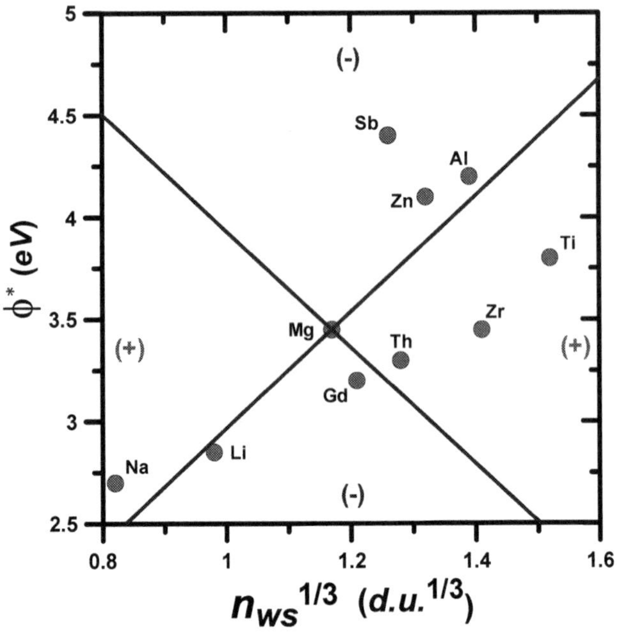

Figure 2: Miedema plot for Mg with the work function, ϕ^*, and the electronic density, $n_{WS}^{1/3}$, as coordinates (see Table 1).

Increasing values of ϕ^* are associated with a tendency to develop SRO and an attendant decrease in the terminal solubility [14]. The later follows as well from the Hume-Rothery's rule involving electronegativity. This is illustrated by Figure 3, where Al, with the smallest ΔH-value, has the largest terminal solubility, whereas Sb, with the largest one, exhibits virtually no solubility.

Solid solution hardening

In Figure 4 the room temperatures strength of Mg-Al, Mg-Zn and Mg-Gd solid solutions is compared using data from the literature. The respective rate of the solid solution hardening correlates well with the alloys' ΔH value (read from Figure 3), i.e., a rather weak hardening rate, consistent with a near random solid solution [7,10], is observed for Al, whereas the much stronger hardening in Mg-Zn and Mg-Gd is consistent with a well-developed SRO [5], expected from their increasingly negative ΔH-value.

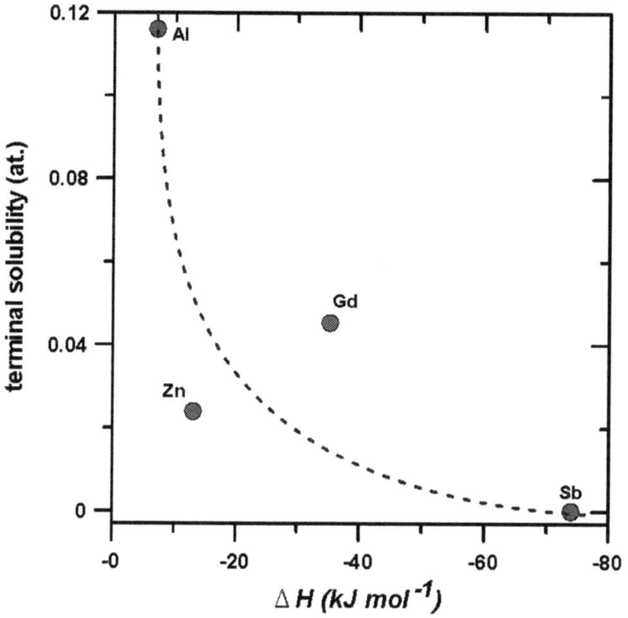

Figure 3: The terminal solid solubility as a function the enthalpy of formation for the solutes studied.

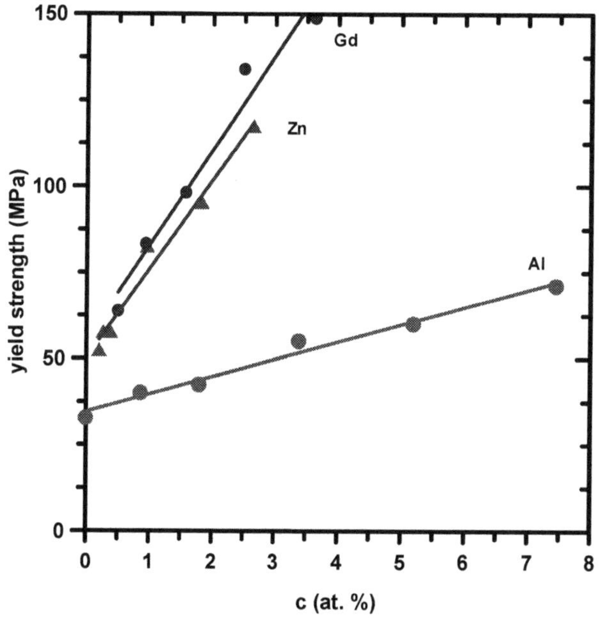

Figure 4: The strength of Mg-based solid solutions. Mg-Zn data from [22]; Mg-Al from [10] and Mg-Gd from [7].

Ranking of alloys

A straightforward way of ranking the strength of Mg alloys within the north-south sectors of Figure 4, i.e., those with negative interaction energy, and, more importantly, regardless of whether the respective phase diagrams conform to that of Figure 1-a, can be easily developed based on the potential strength of the SRO. For a solute concentration c , the strength of the SRO, $\tau_{SRO,}$ can be assumed to obey the relationship of the form [5]:

$$\tau_{SRO} \sim [\Delta H^* c(1\text{-}c)]^2. \qquad (1)$$

SRO is a typical athermal hardening mechanism, and alloys which develop it can be expected to be more creep resistant than those exhibiting only random solid solution [23,24]. Thus, to make the ranking valid for predicting creep strength, and in order to prevent any precipitation hardening effects that might arise when testing a supersaturated solid solution at high temperature, the analysis was done for the respective solute solubilities at ~200°C. These c-values are listed in Table 2.

Table 1: Miedema's coordinates values for Figures 2, 3 and 5 .

Solute	Al	Zn	Gd	Sb
ϕ^*	4.2	4.1	3.2	4.4
$n_{WS}^{1/3}$	1.39	1.32	1.21	1.26
ΔH	-7	-13	-35.3	-74

Table 2: Approximate solute solubilities (at.%) at 200°C for the alloys studied.

Solute	Al	Zn	Gd	Sb
c @ 200°C	2.5	1.5	1.0	0.05

The ranking of the alloys as per Eq. 1 is shown in Figure 5. Accordingly, the Mg-Al alloy is expected to exhibit the lowest strengthening by SRO, in this case due to its low ΔH-value. Zn and Gd containing alloys should exhibit increased strength due to the larger ΔH despite the somewhat reduced c. At the other end, for Mg-Sb the c-value drops to nearly zero, i.e., the alloy should exhibit no solute based strengthening despite its largΔ H. The latter is obviously an upper bound, as Mg-Al is a lower bound. That is, other Mg alloy systems within the north-south sectors of Figure 2 can be expected to fit in between these two bounds, with some systems exhibiting SRO strengths matching or surpassing that of Mg-Gd.

The alloys' behavior at room temperature depicted in Figure 4 is closely consistent with the ranking of Figure 5. Regarding the creep behavior, Figure 5 suggests that Mg-Gd should exhibit a well-defined athermal behavior due to its strong SRO, whereas Mg-Al should not, and Mg-Zn should lie in between. Creep data confirming these predictions are presented in another session in this symposium [25].

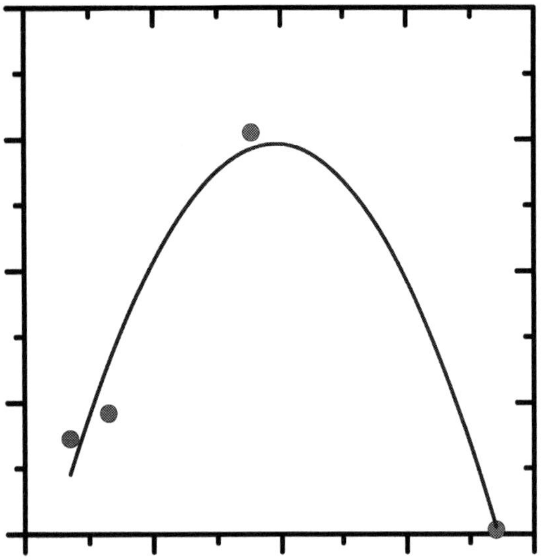

Figure 5: The predicted strength of the SRO (Eq. 1) *vs.* the heat of formation, for the alloys studied.

Figures 3 and 5 allow making predictions concerning the feasibility of precipitation hardening. A strong tendency to order, as opposed to clustering, is expected to interfere with the homogeneous nucleation of small and finely distributed precipitates upon ageing of the supersaturated solid solution [1]. The feasibility of developing strong precipitation hardening is thus expected to be opposite to the tendency to develop SRO. That is, alloys to the left of the maximum in Figure 5 should be increasingly prone to developing precipitation hardening. The classical ageing experiments on Mg-Al and Mg-Zn by Clark [11-13] support this assertion, i. e., of the alloys considered here Mg-Al exhibits the strongest (albeit mild by Al alloys' standards) response to ageing at intermediate temperatures.

Conclusions

Miedema *et al.*'s phenomenological approach has been used to rank the potential solute effects on the strength of four model Mg binary alloys considering the alloys' enthalpy of mixing.

A maximum in the random solid solutions and precipitation hardening is predicted for solutes with low heat of formation, such as Al.

Solutes with an intermediate heat of formation, such as Zn or Gd, are expected to exhibit strong SRO. These alloys are predicted to exhibit maximum strength at high temperature as a result of an extended athermal stress range.

For extreme values of heat of formation, e.g., Mg-Sb alloys, the very limited terminal solubility implies negligible solute-based hardening effects.

References

[1] P. Haasen: *Physical Metallurgy* (Cambridge University Press, Cambridge, 1996).

[2] W. Hume-Rothery: *Atomic Theory* (Institute of Metals, London, 1946).

[3] D.S. Gencheva, A.A. Katsnel'son, L.L. Rokhlin, V.M. Silonov, and F.A. Khavadzha: *Fiz. metal. metalloved.* 1981, vol. 51, pp. 788-793,

[4] L.A. Safronova, A.A. Katsnel'son, S.V. Sveshnikov, and Y.M. L'Vov: *Fiz. metal. metalloved.* 1977, vol. 43, pp. 76-80,

[5] C.H. Cáceres and A. Blake: *Phys. Stat. Sol.* 2002, vol. 194 (a), pp. 147-158,

[6] L. Gao, R.S. Chen, and E.H. Han: *J. Alloys Comp* 2009, vol. 472, pp. 234-240,

[7] L. Gao, R.S. Chen, and E.H. Han: *J. Alloys Comp* 2009, vol. 481, pp. 379-384,

[8] N. Jha and A.K. Mishra: *J. Alloys Comp* 2001, vol. 329, pp. 224-229,

[9] A. Akhtar and E. Teghtsoonian: *Metall. Trans. A* 1971, vol. 2, pp. 2757-2763,

[10] C.H. Cáceres and D.M. Rovera: *J. Light Metals* 2001, vol. 1/3, pp. 151-156,

[11] J.B. Clark: *Acta Metall.* 1965, vol. 13, pp. 1281-1289,

[12] J.B. Clark: *Acta Metall.* 1968, vol. 16, pp. 141-152,

[13] J.B. Clark: *Trans. JIM* 1968, vol. 9 Sup., pp. 354-355,

[14] Y.M. Zhang, J.R.G. Evans, and Y. Shoufeng: *The J. of Crystallization, Phys. and Chem.* 2010, vol. 1, pp. 103-119,

[15] A.R. Miedema, F.R. de Boer, and R. Boom: *Calphad* 1977, vol. 1, pp. 341-359, 10.1016/0364-5916(77)90011-6.

[16] A.R. Miedema: *J. of the Less Common Metals* 1973, vol. 32, pp. 117-136, 10.1016/0022-5088(73)90078-7.

[17] A.R. Miedema, R. Boom, and F.R. De Boer: *J. of the Less Common Metals* 1975, vol. 41, pp. 283-298, 10.1016/0022-5088(75)90034-x.

[18] F.R. de Boer, R. Boom, W.C.M. Mattens, A.R. Miedema, and A.K. Niessen: *Cohesion in Metals: Transition Metal Alloys* (North-Holland, Amsterdam, 1989).

[19] K.C. Russell; in: *Kinetic processes in advanced alloys*; www.dtic.mil/dtic/tr/fulltext/u2/a212049.pdf ,1987.

[20] R.F. Zhang, S.H. Sheng, and B.X. Liu: *Chemical Physics Letters* 2007, vol. 442, pp. 511,

[21] R.F. Zhang: *Miedema calculator*, http://zrftum.wordpress.com/; Accessed on: Sept. 9th, 2012.

[22] A.H. Blake and C.H. Cáceres: *Mater. Sci. Engng. A* 2008, vol. 483-484, pp. 161-163,

[23] D. Kuhlmann-Wilsdorf: *Metall. Mater. Trans.* 2004, vol. 35A, pp. 369-418,

[24] A. Seeger, in: J.C. Fisher, W.G. Johnston, R. Thomson and T. Vreeland (Ed.) *Dislocations and mechanical properties of crystals*, Chapman and Hall, London, 1957), pp. 243-329.

[25] S. Abaspour and C.H. Caceres, in: *TMS 2013*, 2013, This volume.

Magnesium Technology 2013
Edited by: Norbert Hort, Suveen N. Mathaudhu, Neale R. Neelameggham, and Martyn Alderman
TMS (The Minerals, Metals & Materials Society), 2013

A new magnesium alloy system: TEXAS

Björn Wiese[1], Chamini Mendis[1], Carsten Blawert[1], Eric Nyberg[2], Karl Ulrich Kainer[1], Norbert Hort[1]
[1]MagIC-Magnesium Innovation Center, Helmholtz-Zentrum Geesthacht, Max-Planck-Straße 1, D-21502 Geesthacht, Germany
[2]Pacific Northwest National Laboratory, 902 Battelle Boulevard, Richland, WA, USA

Keywords: magnesium alloy system, MgCaSn phase, microstructure, mechanical properties, corrosion

Abstract

A new TEXAS alloy system (Mg-Sn-Nd-Ca-Al-Si) is presented in order to extend the range of applications for magnesium alloys. The alloy has been produced by permanent mould direct chill casting, a process that provides a homogenous distribution of alloying elements throughout the entire casting. This work presents microstructural features and a new Mg-Sn-Ca phase with the morphology of hexagonal platelets. Additionally mechanical properties and the corrosion behaviour of TEXAS alloys are presented in as cast and heat treated conditions.

Introduction

Magnesium (Mg) alloys have low density and are used in automobile and aerospace applications where lightweight is of importance. Mg-Sn-Ca (TX) alloys has been developed as creep resistant alloys [1]. In this system Sn forms solid solution with Mg to provide corrosion resistance while Ca forms intermetallic particles in the matrix. The alloys based on this system, have tensile yield strengths similar to AZ31 alloy, commercially available Mg alloy with good combination of strength and corrosion resistance.

The new magnesium alloys, TEXAS, are based on the Mg-Sn-Nd-Ca-Al-Si system and are expected to extend the range of the applications possible for the TX system [1, 2, 3, 4, 5, 6, 7, 8, 9]. The first steps in extending the TX system were based on investigations into TXAS, TXA and TXS form [10]. Formation new phases, the increase the mechanical properties at the room temperature and at elevated temperatures was expected with Si and Al addition [11]. Studies on the influence of rare earth elements such as Ce, Zr and Gd, on the Mg3Sn2Ca (TX32) alloys were already investigated [12, 13, 14, 15].

This paper is the first step to investigate microstructure, mechanical properties and corrosion resistance of this TEXAS alloy system. Here we report the microstructure characterisation of the as cast alloys. The mechanical properties and corrosion resistance of the as cast (F) and the T4 heat treated alloys are also reported in this contribution.

Experimental Procedure

Pure Mg (99.99 %, Hydro Magnesium), pure Sn (99.995 %, ChemPur), pure Nd (99.5 %, Griem Ad- vanced Material co., Ltd), pure Al (99.99 %), pure Si (99.9 %, ChemPur) and pure Ca (99 %, Alfa Aesar) melted in a mild steel crucible to produce alloys with 3.0 wt% Sn, 0.3 wt% Nd, 0.3 wt% Al, 0.3 wt% Si and varying amounts of Ca from 0.5 wt% to 2.0 wt%. All alloys have been prepared by permanent mould direct chill casting, because this is a special method of casting [16]. Alloys were melted in flowing Ar+3 % SF6 in an induction furnace (Nabertherm, Germany) at a temperature of 720 °C and Sn, Nd, Ca, Al and Si were added to the magnesium melt. The melt was kept for five minutes without stirring and was poured into a permanent steel mold (55 x 110 x 270 mm) which was preheated to 350 °C. The melt was kept for 15 min at 680 °C. The cast ingots were cooled down to room temperature by water quenching as reported elsewhere [16]. Table 1 list the nominal composition of the alloys investigated. The specimens were solution heat treated at 500 °C for 6 h and cooled down to room temperature by water quenching (T4 condition).

Table 1: Nominal composition of the TEXAS alloys (wt%)

alloy	Compositions (wt%)					
	Sn	Nd	Ca	Al	Si	Mg
TEXAS A	3	1.5	2.0	0.3	0.3	bal
TEXAS B	3	1.5	1.0	0.3	0.3	bal
TEXAS C	3	1.5	0.5	0.3	0.3	bal

The specimens for scanning electron microscopy (SEM) were prepared by grinding with SiC paper to 2500 grit, and then polished with, 3 µm diamond paste followed by a mixture of 1 µm diamond suspension and OPS™ anhydrous suspension. Microstructures were investigated using a ZEISS, ULTRAT M 55 SEM equipped with an EDAX energy-dispersive X-ray spectrometer (EDXS) and local compositions determined by point analysis. The specimens of 5 x 20 x 30 mm were ground down to a 2500 grit SiC paper, polished with 1 µm OPST M anhydrous suspension and then cleaned with ethanol and dried are used for X-ray diffraction (XRD). The phase identification was carried out with a X'Pert PRO PANalytical and HighScorePlus XRD operating at 40 kV and 40 mA with Cu Kα (wavelength λ = 0.15406 nm) radiation. The measurements were conducted by scanning (2Θ) from 15° to 100° with a step size of 0.025°. A acquisition time of 2 s per step was used.

Tensile tests were conducted in accordance to DIN EN 10002 and DIN 50106 at room temperature using a Zwick 050 testing machine (Zwick GmbH & Co., KG). An initial strain rate of 0.001 s^{-1} in was used. The Vickers hardness was measured using a EMCO-TEST the M1C 010, indenter with a 5 kg load and the value reported is the average from 10 measurements per sample. The specimens for hardness measurement were ground to a 2500 grit SiC finish.

Potentio-dynamic polarisation measurements were performed in 0.5 % NaCl solution. A starting pH value of 7 was adjusted with

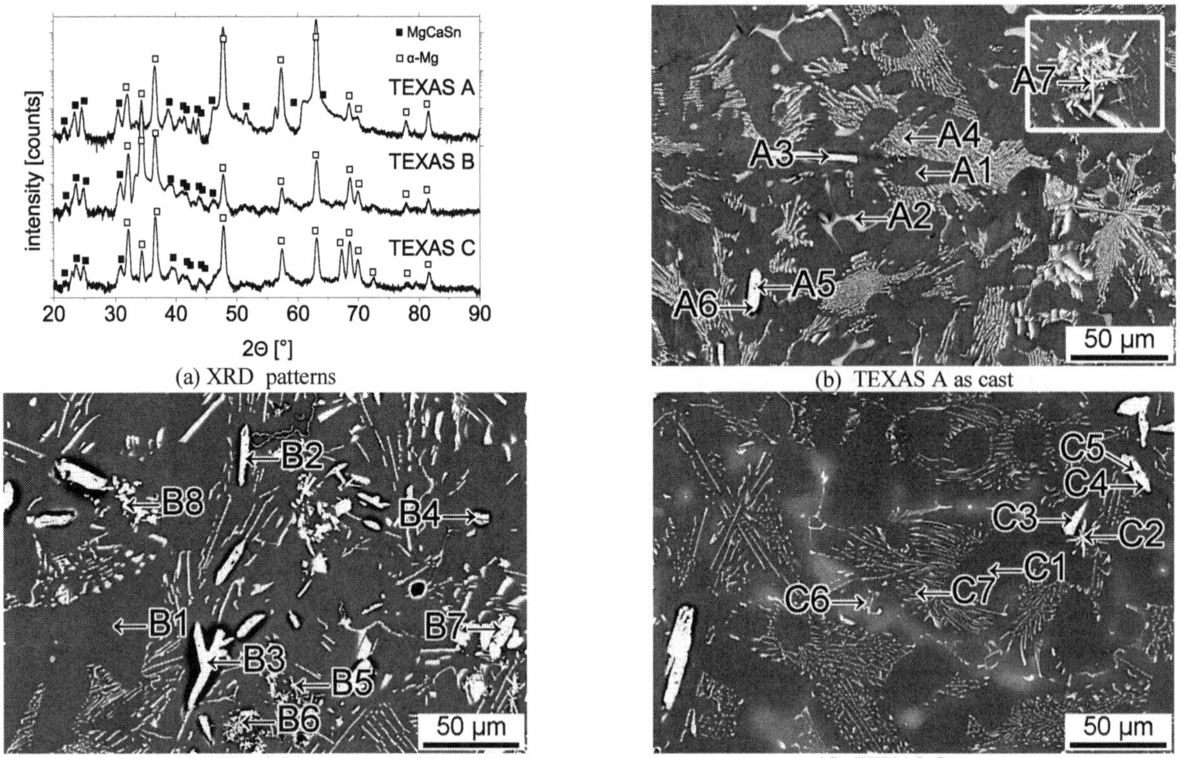

(a) XRD patterns (b) TEXAS A as cast (c) TEXAS B as cast (d) TEXAS C as cast

Figure 1: (a) XRD patterns and (b,c,d) SEM Backscattered Electrons (BSE) microstructure of TEXAS alloys.

Table 2: EDX of striking points from TEXAS A, B and C to the Figure 1.

	Compositions at%									
Alloys	points	Mg	Al	Si	Sn	Ca	Nd	Mg/(Sn+Ca)	Si/Nd	Sn/Ca
TEXAS A	A1	99.7	0.0	0.0	0.1	0.1	0.0	-	-	-
	A2	94.7	0.0	0.9	1.9	2.4	0.2	-	-	0.8
	A3	86.7	2.1	0.0	0.1	10.9	0.2	7.9		0.0
	A4	49.1	0.0	4.8	22.5	21.8	1.8	1.1	2.7	1.0
	A5	67.0	0.4	5.0	13.3	12.9	1.3	2.6	3.7	1.0
	A6	41.0	0.0	5.5	25.8	24.3	3.4	0.8	1.6	1.1
	A7	23.9	0.0	34.7	0.8	1.1	39.1	-	0.9	-
TEXAS B	B1	99.7	0.0	0.0	0.2	0.1	0.0	-	-	-
	B2	42.2	0.1	7.9	22.7	25.4	1.7	0.9	4.6	0.9
	B3	54.9	0.1	6.0	17.9	19.8	1.4	1.5	4.3	0.9
	B4	35.9	0.0	8.1	25.1	29.0	1.9	0.7	4.2	0.9
	B5	68.3	0.1	15.6	0.1	0.3	15.6	-	1.0	-
	B6	65.6	0.0	17.4	0.2	0.4	16.5	-	1.1	-
	B7	25.9	1.4	32.1	1.7	0.1	38.9	-	0.8	-
	B8	35.8	0.0	31.0	0.6	0.6	32.0	-	1.0	-
TEXAS C	C1	99.7	0.0	0.1	0.2	0.1	0.0	-	-	-
	C2	37.4	0.0	29.0	1.8	5.1	26.8	-	1.1	-
	C3	41.3	0.0	11.8	19.8	24.9	2.2	0.9	5.3	0.8
	C4	35.6	0.0	11.6	21.6	28.5	2.8	0.7	4.2	0.8
	C5	58.7	0.1	4.7	15.7	20.2	0.7	1.6	6.8	0.8
	C6	76.9	0.0	20.1	2.8	0.2	0.0	-	-	14.8
	C7	92.2	0.0	2.2	2.2	3.4	0.2	16.8	14.4	0.6

NaOH. The measurements were performed in a standard three electrode set-up with the alloy specimens as working electrodes (1.54 cm2 area in contact with the electrolyte) in conjunction with an Ag/AgCl reference electrode and a platinum counter electrode. After 30 min the free corrosion potential were recorded, the polarisation scan was measured with a scan rate of 0.2 mV s^{-1}. The test completed at a corrosion current of 10 mA. The corrosion rate was determined from the cathodic component of the corrosion curve.

Results and Discussion

Microstructure and Phase Characterization

The XRD patterns of as cast TEXAS A, B and C are shown in Figure 1(a). The XRD shows the presence of Mg and MgCaSn phase (orthorhombic phase) and no phase containing Nd or Si were detected. The investigations by SEM and EDX showed in addition to a phase containing Mg, Ca and Sn number of phases containing Mg, Ca, Si, Nd and another phase with Nd and Si. These phases are not found in the XRD measurement as the volume fraction of these phases are likely to be lower than the detection limit of XRD and need to be investigated by TEM. The SEM microstructures of the three TEXAS alloys A, B and C shown in Figures 1 with EDX measurement are listed in Table 2.

Dendritic microstructures were observed in all three alloys, with hexagonal platelets of MgCaSn phase similar to that observed in the TX alloys [1, 2]. The hexagonal shape of these precipitates varies with Ca content of the alloy and they show a more irregular hexagonal shape reduction of Ca. The platelets (Figure 1(b) to 1(d)) show dark points in the centre and the EDX measurements give a higher Mg composition at the centre compared to the areas away from these points (Point A5 and C5 in the centre and A6 and C4 in the platelets). The Mg-rich areas in the centre of plates are expected to act as a nucleus for the formation of the MgCaSn phase. The composition of the platelets are similar to that reported previously [1, 2, 3, 4] with a ratio of Sn to Ca of 0.8 to 1.0 in the ternary phase CaMgSn which decreases with the decreases in Ca concentration. However, the eutectic areas showed a comparable ratio of Sn to Ca. The stoichiometric ratio between Sn:Ca changed in this alloy system in TEXAS A 1:2, in TEXAS B 1:1 and in TEXAS C 2:1. The TEXAS A and B alloys can bind the entire Ca with Sn in the MgCaSn phases. In TEXAS C Mg_2Sn and in TEXAS A Mg_2Ca are also likely to form as in the TX system [6]. The XRD patterns do not shows the Mg_2Sn and Mg_2Ca phases Figure 1(a).

In TX alloys with the addition of RE no RE containing particles were detected [13, 14, 15], In this case Nd and as well Si was formed in the platelets but such phases are yet to be characterised. The Nd and Si leads to needle-shaped precipitates as shown by SEM-EDX Figure 1 on the points A7, B5, B6 and C2. The ratio between Nd and Si is approximately 1 in all alloys. The detection of the Nd-Si precipitates by XRD was not possible.

Another phase that is observed in the TEXAS C contained of Mg and Si (point C6). The formation of Mg-Si phase appears to be suppressed by the Ca concentration in TEXAS A and B but with the reduction of Ca content Mg-Si phase also forms. Additions of Ca changed the microstructure and caused a morphological change the characteristic shape of Mg_2Si Chinese Script particles from sharp needle like particles to more curved particles. The modification mechanism of Ca is polygonal type Mg_2Si particles nucleating from $CaSi_2$ particles has been reported previously [17]. Again the XRD analysis do not show the Mg_2Si phase.

Mechanical Properties

Table 3 list the hardness of the three TEXAS alloys in the as cast (F) and the T4 heat treated conditions. The TEXAS A alloy with the highest Ca concentration shows the highest hardness value in each condition. The alloys TEXAS B and C with the lower Ca concentration show similar to hardness values in the as cast and the T4 heat treated conditions. The heat T4 treatment did not reduce the hardness of any TEXAS alloys and any reductions were within the standard deviation of the measurements. The reduction in the Ca concentration from 2 to 1 wt % reduced the hardness, as less of Ca-rich phases are present in the microstructure.

Table 3: Mechanical properties like TYS, UTS, A and hardness for the TEXAS alloys in as cast (F) and T4 heat treated conditions.

		TEXAS		
	cond.	A	B	C
TYS [MPa]	F	79.69 ± 3.60	66.78 ± 1.56	70.52 ± 0.18
	T4	62.39 ± 1.62	56.03 ± 1.39	61.78 ± 3.80
UTS [MPa]	F	113.64 ± 3.12	108.46 ± 13.29	125.56 ± 3.77
	T4	99.42 ± 4.48	104.45 ± 7.15	117.36 ± 2.30
A [%]	F	1.16 ± 0.12	1.45 ± 0.54	2.59 ± 0.21
	T4	1.54 ± 0.28	2.03 ± 0.37	2.61 ± 0.57
hardness [HV]	F	47.91 ± 3.04	41.69 ± 3.81	41.78 ± 3.18
	T4	46.29 ± 3.72	39.52 ± 2.86	37.59 ± 4.07

The Table 3 and the Figures show the tensile properties of alloys tested in the as cast (F) and the T4 heat treated conditions. In comparison with the as cast condition, the T4 condition show a decrease in the tensile yield strength (TYS) and ultimate tensile strength (UTS) for all alloys but with an increase in the elongation (A) for TEXAS A and B. The decrement of the UTS does not show such a clear trend after T4. The T4 heat treatment has the biggest influence on the TYS a decrease 80 to 62 MPa in TEXAS A. TEXAS B shows the highest increase of the elongation (A) after T4. The highest ductility was observed in the T4 heat treated TEXAS C alloy. The Ca content has a major influence on the strength where higher the Ca content higher the TYS.

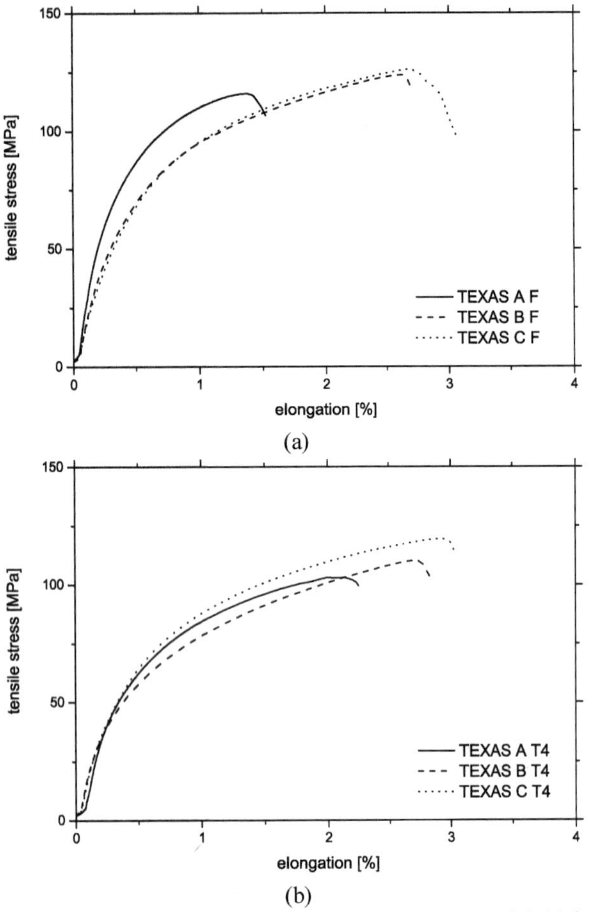

Figure 2: Load-deformation curves of TEXAS A, B and C (a) in as cast (F) and (b) T4 heat treated conditions.

Corrosion Properties

The results from the potentio-dynamic polarisation measurements are listed in Table 4. All three alloys are similar and show filiform corrosion on the surface and deep holes in the shape of hexagonal platelets. These particles probably act as a galvanic site and are the starting points for corrosion. The EDX measurements showed that this platelets are phases rich of Sn, Ca and Mg and the corrosion rate is dependent on the Ca and Sn concentration. The cast TEXAS A, with the highest Ca concentration, shows the maximum corrosion rate.

After the T4 heat treatment, the corrosion at TEXAS A is apparently lower, as the microstructure has been homogenized and some of the Mg,Ca,Sn phase has gone into solution. The corrosion rate of TEXAS B after T4 heat treatment is similar to that of the as cast condition. However, the corrosion rate increases after T4 heat treatment of the TEXAS C alloys, likely due to the precipitates, probably Sn phase is dissolved in preference to the Ca, Sn containing phase.

Table 4: Polarisation measurements for the TEXAS alloys.

TEXAS	cond.	Free corr Potential [mV]	Icorr [mA]	Corrosion-Rate [mm/a]
A	F	-1564.0 ± 1.0	0.253 ± 0.036	3.75 ± 0.53
	T4	-1568.3 ± 1.2	0.179 ± 0.019	2.66 ± 0.28
B	F	-1552.3 ± 0.6	0.132 ± 0.034	1.96 ± 0.51
	T4	-1555.7 ± 0.6	0.122 ± 0.029	1.82 ± 0.42
C	F	-1577.0 ± 20.0	0.121 ± 0.017	1.80 ± 0.25
	T4	-1575.3 ± 4.9	0.158 ± 0.006	2.34 ± 0.09

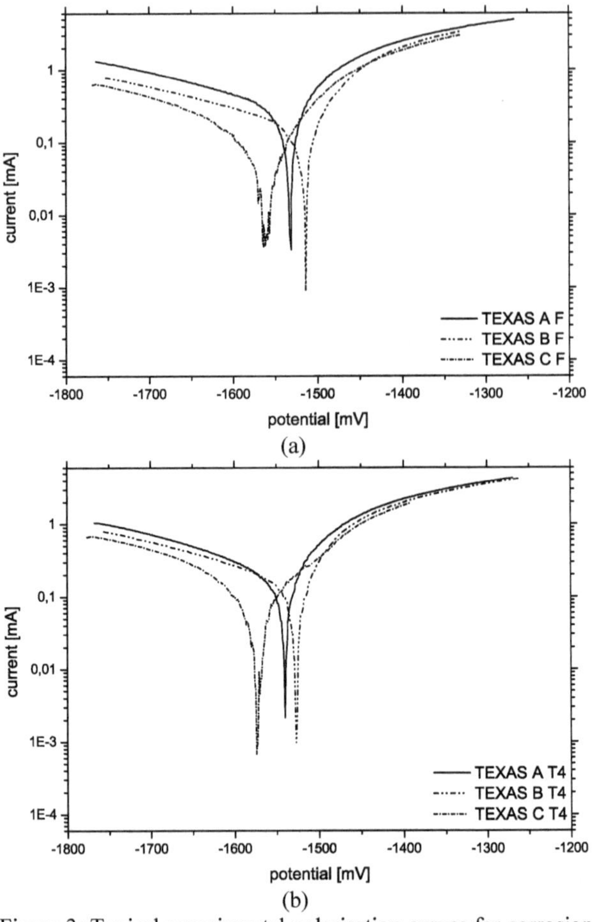

Figure 3: Typical experimental polarisation curves for corrosion resistance of TEXAS A increases, TEXAS B remains similar to the as cast condition and TEXAS C decreases.

Conclusions

1) The variation of the Ca content of between 0.5 and 2.0 wt% has a strong influence on the microstructure of the as cast TEXAS alloys. The hexagonal shape of the MgCaSn platelets can be seen clearly with increasing Ca content. With the lowest Ca content of 0.5 wt% was Mg-Si phase was observed. All three alloys show needle-shaped Nd-Si precipitates but are not observed with XRD.

2) The increase in Ca content reduces the ductility in the TEXAS alloys and leads to an increase in tensile yield strength. The ductility of TEXAS alloy system rather low, with maximum elongation of 2.6 % measured for the TEXAS C containing lowest content of Ca.

3) The corrosion resistance of the three alloys are different in the as cast and the T4 heat treated conditions. With a lower Ca concentration in the TEXAS alloys show a decrease in the corrosion rate in the as cast condition. After the T4 heat treatment the corrosion resistance of TEXAS A increases, TEXAS B remains similar to the as cast condition and TEXAS C decreases.

References

[1] T. Abu Leil et al. "Microstructure, corrosion and creep of as-cast magnesium alloys," In R.S. Beals, M.O. Pekguleryuz, R. Neelameggham, and A.A. Luo, editors, *TMS Magnesium Technology* (2007).

[2] N. Hort et al. "Microstructural Investigations of the Mg-Sn-xCa System," *Advanced Engineering Materials*, 8 (2006), 359–364.

[3] A. Kozlov et al. "Phase equilibria, thermodynamics and solidification microstructures of Mg-Sn-Ca alloys, Part 1: Experimental investigation and thermodynamic modeling of the ternary Mg-Sn-Ca system," *Intermetallics*, 16 (2008), 299–315.

[4] A. Kozlov et al. "Phase equilibria, thermodynamics and solidification microstructures of Mg-Sn-Ca alloys, Part 2: Prediction of phase formation in Mg-rich Mg-Sn-Ca cast alloys," *Intermetallics*, 16 (2008), 316–321.

[5] K.P. Rao et al. "Hot workability characteristics of cast and homogenized Mg-3Sn-1Ca alloy," *Journal of Materials Processing Technology*, 201 (2008), 359–363.

[6] T. Abu Leil et al. "Microstructure and corrosion behavior of Mg-Sn-Ca alloys after extrusion," *Transactions of Nonferrous Metals Society of China*, 19 (2009), 40–44.

[7] Y.V.R.K. Prasad et al. "Hot working parameters and mechanisms in as-cast Mg-3Sn-1Ca alloy," *Materials Letters*, 62 (2008), 4207–4209.

[8] M. Yang, L. Cheng, and F. Pan. "Comparison of as-cast microstructure, tensile and creep properties for Mg-3Sn-1Ca and Mg-3Sn-2Ca magnesium alloys," *Transactions of Nonferrous Metals Society of China*, 20 (2010), 584–589.

[9] G.H. Hasani and R. Mahmudi. "Tensile properties of hot rolled Mg-3Sn-1Ca alloy sheets at elevated temperatures," *Materials and Design*, 32 (2011), 3736–3741.

[10] K.P. Rao et al. "Effect of Minor Additions of Al and Si on the Mechanical Properties of Cast Mg-3Sn-2Ca Alloys in Low Temperature Range," *Materials Science Forum*, 654-656 (2010), 635–638.

[11] N. Hort, Y. Huang, and K. U. Kainer. "Intermetallics in Magnesium Alloys," *Advanced Engineering Materials*, 8 (2006), 235–240.

[12] B. Shi, R. Chen, and W. Ke. "Effect of element Gd on phase constituent and mechanical property of Mg-5Sn-1Ca alloy," *Transactions of Nonferrous Metals Society of China*, 20 (2010), 341–345.

[13] M. Yang et al. "Effects of zirconium addition on as-cast microstructure and mechanical properties of Mg-3Sn-2Ca magnesium alloy," *Materials and Design*, 32 (2011), 1967–1973.

[14] M. Yang, Y. Ma, and F. Pan. "Effects of little Ce addition on as-cast microstructure and creep properties of Mg-3Sn-2Ca magnesium alloy," *Transactions of Nonferrous Metals Society of China*, 19 (2009), 1087–1092.

Magnesium Technology 2013
Edited by: Norbert Hort, Suveen N. Mathaudhu, Neale R. Neelameggham, and Martyn Alderman
TMS (The Minerals, Metals & Materials Society), 2013

Effect of Sn additions on the age hardening response, microstructures and corrosion resistance of Mg-0.8Ca (wt%) alloys

C.L. Mendis, D. Tolnai, C. Blawert and N. Hort

MagIC -Magnesium Innovation Centre, Helmholtz Zentrum Geesthacht,
Institute for Materials, Max Planck Strasse 1 Geesthacht 21502 , Germany

Keywords: Mg-Ca-Sn alloys, Microstructures, Age hardening

Abstract

The as-cast Mg-3Sn-2Ca alloy shows a good creep resistance that is better than that of AE42 alloy. The high content of alloying additions in these alloys results in low ductility due to the presence of primary intermetallic particles. The effect of reduced amount of Ca and Sn, below the maximum solid-solubility limit of each alloy additive, was studied to determine the potential of the Mg-0.8Ca-*x*Sn alloys in structural applications. The microstructure of the alloys resulting from different processing routes was characterized and the corrosion resistance of the alloys measured to determine the best combination of alloy composition for further alloy development.

Introduction

Magnesium alloys have been considered in many applications where a combination of light weight and high specific strength is of importance. However, there is a lack of low cost, high strength magnesium alloys that may be used in structural applications where elevated temperature stability is required for temperatures up to 200°C. Mg-3Sn-2Ca (wt%) alloy has been recently developed as a possible high strength creep resistant alloy that has comparable creep resistance to AE42 alloys [1,2]. However the ductility to failure and the corrosion resistance of Mg-3Sn-2Ca alloy is not sufficient for many applications [2]. The high content of alloying additions found in this alloy cause formation of high volume fraction of primary intermetallic particles, which result in low ductility and corrosion resistance observed in the previous investigations [1,2]. The reduction in the volume fraction of intermetallic particles found in the alloys is expected to enhance the corrosion resistance but at a cost of reducing tensile yield strength. For improved ductility and corrosion resistance a magnesium alloy that contains lower content of alloying additions is very attractive. Precipitation hardening does not significantly reduce the ductility of heat treated, extruded or hot rolled magnesium alloys significantly [3]. This concept may be used to enhance the strength of magnesium alloys without significant loss in yield strength while retaining ductility of the alloy due to lower total solute concentration. The precipitation hardened Mg-Ca-Zn alloys with low solute contents has been shown to have a good combination of yield strength and ductility by Gao *et al.* [4] compared to a heat treated AZ91 alloy while the creep resistance of the alloy is relatively high. It is noted that the grain boundaries of the alloys investigated do not contain intermetallic particles that provide galvanic sites, which promote corrosion in many alloys strengthened by grain boundary intermetallic particles. It has been also shown for die cast AXJ530 alloy the creep resistance may be enhanced with precipitation of fine scale particles following the T6 heat treatment at 250°C [5].

Recently, a number of publications have shown that the ternary additions of Al [6], Zn [7-9] and In [10] have enhanced the age hardening response at 200°C for Mg-0.5Ca alloy (wt%). Each of these alloys was strengthened via a high density of very fine scale precipitates distributed uniformly through the microstructure. These precipitates are only few atomic layers in thickness and form on the basal planes for the Mg-Ca-Al [6] and Mg-Ca-Zn [9] alloys but form on the prismatic planes for the Mg-Ca-In alloys [6,10]. The refinement of precipitate distribution has been attributed to the negative enthalpies of mixing between the two alloying additions which has been proposed to enhance the nucleation of precipitates as compared with the binary alloy [11]. The enthalpy of mixing between Sn and Ca in the binary Sn-Ca system is reported to be negative [12] and thus expected the fine distribution of precipitates in the Mg-Ca-Sn alloys to enhance the precipitation hardening response of the alloy as compared with the binary Mg-Ca alloy.

In this paper we report the findings from a preliminary investigation on the as-cast and heat treated microstructures of Mg-0.8Ca-*x*Sn (wt%) alloys with varying additions of Sn. The age hardening response of the Mg-0.8Ca-*x*Sn (wt%) alloys was also evaluated at 200°C.

Experimental procedure

Pure Mg (99.99% Hydro Magnesium), was melted in a mild steel crucible and pure Sn (99.995% Chempur) and pure Ca (99% Alfa Aersar) was added to the melt at 700°C. The melt was held at ~700°C for 5 min stirred and then cast into a hardened steel finger mould pre-heated to 250°C. The nominal compositions of the alloys prepared are reported in Table 1. The samples of Mg-0.8Ca-*x*Sn (wt%) was solution heat treated at 525°C for 5h in flowing Ar + SF_6 to prevent oxidation of the alloys. The samples were then cold water quenched to ensure dissolution of alloying additions. The samples for age hardening experiments were heat treated at 200°C for various times immediately following the solution heat treatment to ensure maximum amount of solute is available for precipitation. The age hardening response was measured after different ageing times at 200°C by measuring the Vickers hardness on samples ground to a 2500 grit SiC paper finish using an EMCOTEST M1C10 Vickers hardness tester with a load of 5kg. The hardness values reported are an average of 10 measurements for each of the testing conditions.

The samples were ground with SiC paper to 2500 grit finish and then polished in 3μm diamond suspension and a mixture of 1μm diamond suspension and OPS suspension for optical microscopy and scanning electron microscopy (SEM). SEM investigations were conducted with a Carl Zeiss ULTRA™ 55 SEM equipped with an EDAX energy dispersive X-ray spectrometer.

The intermetallic phases found in the as-cast and the solution heat treated samples were characterized at the BW2 beam line of the Hamburger Synchrotronstrahlungslabor (HASYLAB) at the Deutsches Elektronen-Synchrotron (DESY). The high brilliance of the beam allows the detection of the low volume fraction intermetallic phases. The beam energy was set to 100 keV. During the experiment the diffraction patterns were recorded with a 10s acquisition time by a mar555 flatpanel with an area of 3070×2560 pixels with a pixel size of 140 μm².

Table 1 Nominal compositions of the alloys investigated

Alloy	Composition (wt%)		
	Mg	Ca	Sn
Mg-0.8Ca-	balance	0.8	-
Mg-0.8Ca-1.2Sn	balance	0.8	1.2
Mg-0.8Ca-2.3Sn	balance	0.8	2.3
Mg-0.8Ca-3.5Sn	balance	0.8	3.5

Preliminary analyses of precipitate microstructures were conducted in a Phillips CM200 transmission electron microscope operating at 200 kV. The samples for TEM analysis were prepared by electric discharge machining 3 mm discs from alloy specimens that were ~ 200 mm in thickness and twin jet electropolishing using a solution of 1.5% perchloric acid in methanol at a polishing temperature of ~-45°C with a current of ~0.8-1.2mA and a voltage of 50V. Upon perforation the specimens were washed in ethanol to remove any trace of acid.

Potentio-dynamic polarization tests for the as-cast Mg-0.8Ca-*x*Sn alloys were conducted in 0.5%NaCl solution with starting pH of 7. Measurements were conducted with standard 3 electrode cell (330ml) method where the sample was the working electrode while the counter electrode was platinum mesh. The reference electrode used in this experiment was Ag/AgCl. The free corrosion potential was measured after 30mins and the polarization curve was measured at a scan rate of $0.2mVs^{-1}$ and the corrosion rate was determined from the corrosion current at the intersection of the cathodic Tafel slope with the vertical through the corrosion potential.

Results and discussion

The as-cast and the solution heat treated microstructures

The optical micrographs of the as-cast and the solution treated Mg-0.8Ca-*x*Sn alloys are shown in Figure 1. The as-cast microstructures, Fig 1 (a-c) show the presence of a semi continuous to continuous network of primary intermetallic particles at the grain boundaries and in the inter-dendritic regions. The solution treated samples of Mg-0.8Ca-*x*Sn alloys, Fig 1 (d-f) showed a significant reduction in the volume fraction of intermetallic particles found at the grain boundaries. Additionally, solution heat treatment removed the dendritic nature of the microstructure and caused some grain growth in the alloys investigated. The change in the microstructure from the as-cast samples to the solution heat treated samples may be clearly observed in SEM microstructures shown in Fig 2. The microstructures of Mg-0.8Ca-3.5Sn alloy is shown here as an example which shows the changes to contrast along the dendrites showing changes in the composition of the alloy, Fig 2 (a), in the as-cast condition which is not evident in the solution heat treated alloy of the same composition, Fig. 2(b).

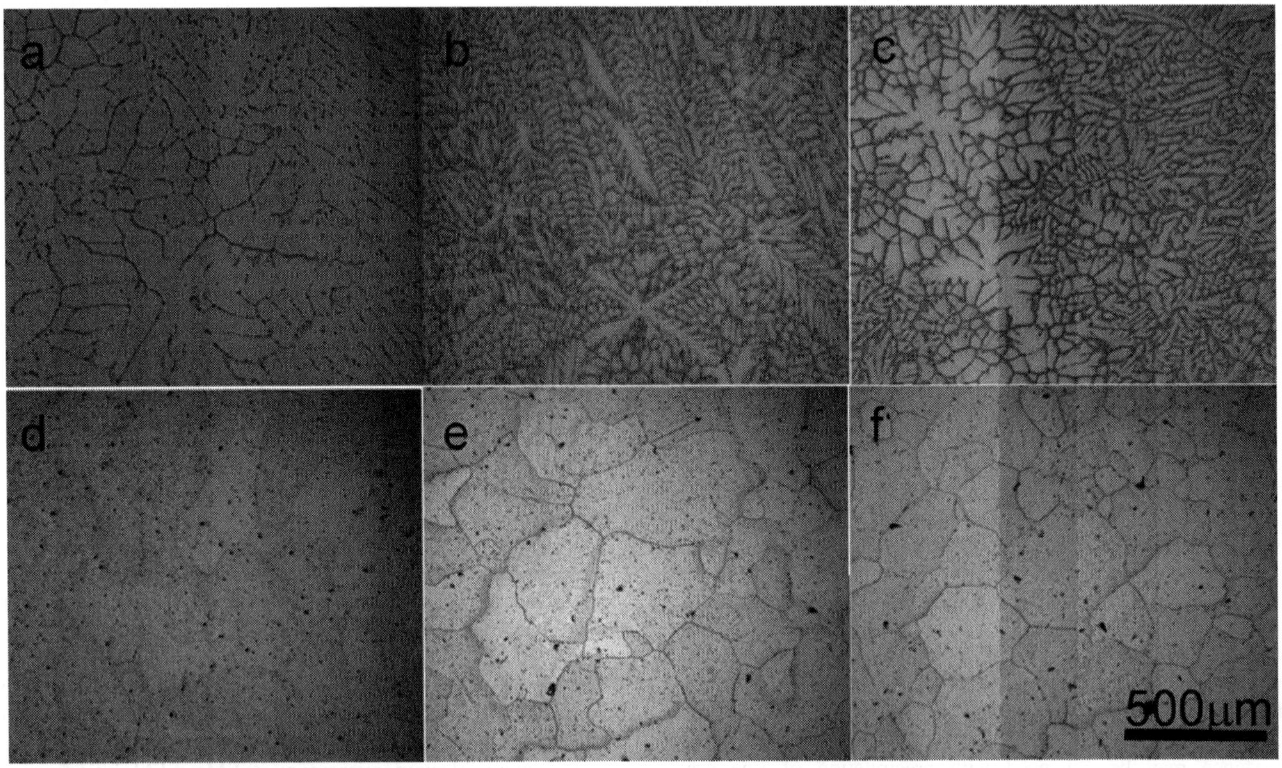

Figure 1. Optical micrographs typical of (a-c) the as-cast and (d-f) the solution heat treated 525°C/5h alloys. (a,d) Mg-0.8Ca-1.2Sn (b,e) Mg-0.8Ca-2.3Sn and (c,f) Mg-0.8Ca-3.5Sn

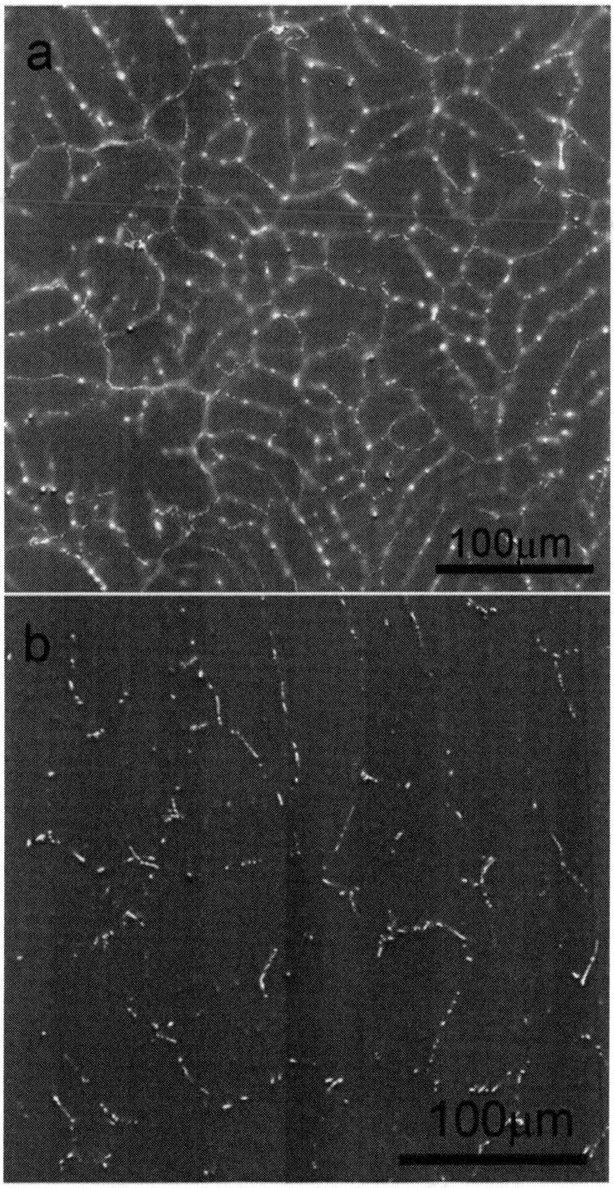

Figure 2 SEM micrographs showing (a) the as-cast and (b) the solution heat treated at 525°C/5h microstructures of Mg-0.8Ca-3.6Sn alloy.

Two dimensional diffraction patterns were recorded during synchrotron diffraction measurements. The Debye-Scherrer rings were integrated through 360° to obtain the line profiles shown in Fig 3. These line profiles show that in the as-cast samples, Fig 3 (a) the main peaks are due to magnesium while some additional peaks due to intermetallic phases could also be observed. In the Mg-0.8Ca alloy some very small reflections due to Mg_2Ca may be observed but the main peaks due to Mg_2Ca phase are overlapping with the magnesium peaks in the line profile. In Mg-0.8Ca-*x*Sn alloys the MgSnCa phase (orthorhombic crystal structure with a = 0.786nm b = 4.66nm c = 8.74nm [13]) was observed in addition to the peaks due to magnesium. After the solution heat treatment, the peaks due to the intermetallic phases have largely disappeared from the line profiles and only peaks due to the magnesium may be observed, Fig 3 (b). It is noted that the magnesium peaks had shifted slightly from the position for the as-cast alloy of the same composition. The shift in the peaks may be associated with the incorporation of solute elements into the matrix during solution heat treatment.

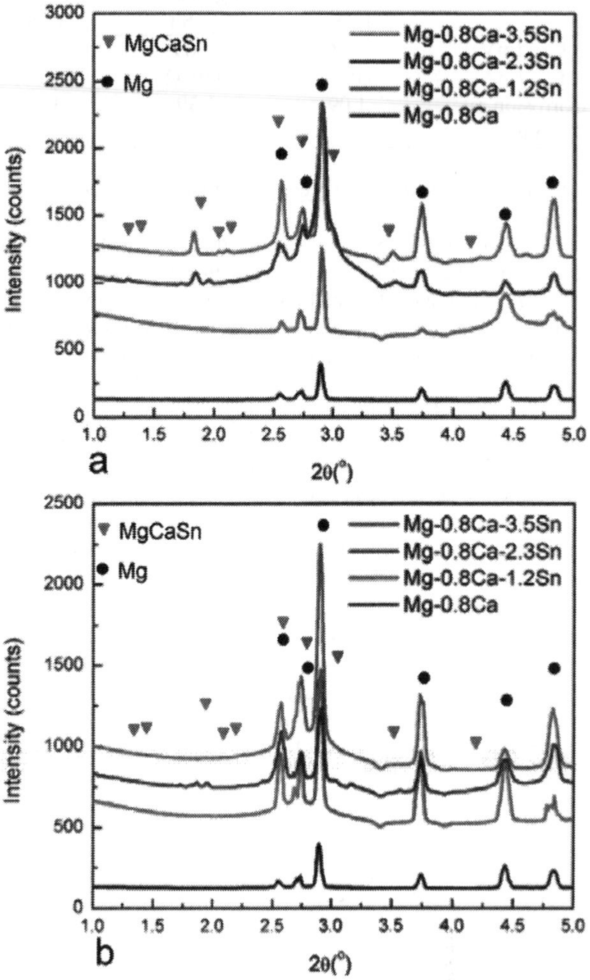

Figure 3 XRD line profiles resulting from the azimuthal integration of the diffraction patterns recorded from Mg-0.8Ca-*x*Sn and Mg-0.8Ca alloys in (a) the as-cast and (b) solution heat treated 525°C/5h conditions (Wavelength $\lambda = 0.0124$ nm)

<u>Corrosion resistance of the as-cast microstructures</u>

The pertinent results from the potentio-dynamic polarization curves are listed in Table 2. The free corrosion potential and the I_{corr} measured increases with increased Sn concentration. It is obvious that the nobler Sn and Sn-based intermetallics are responsible for the observed increase in the free corrosion potential. Furthermore the intermetallics can cause internal galvanic corrosion thus the increase in the corrosion rate is consistent with the increase in the volume fraction of the intermetallic phases. The Mg-0.8Ca-1.2Sn alloy shows with a 1.1mm/year the lowest rate of corrosion and the Mg-0.8Ca-3.5Sn with a 1.5mm/year the highest The corrosion curves, not shown in this contribution, show that passivation of the corroded samples was observed. This is especially evident in the Mg-0.8Ca-1.2Sn alloy. This suggests that the Mg-0.8Ca-*x*Sn alloys show good potential as a structural magnesium alloy.

Table 2 Pertinent features from the potentio-dynamic corrosion curves of the as-cast Mg-0.8Ca-*x*Sn alloys.

	Alloy		
	Mg-0.8Ca-1.2Sn	Mg-0.8Ca-2.3Sn	Mg-0.8Ca-3.5Sn
E_{corr} (mV)	1756	1669	1651
I_{corr} (mA)	0.0477	0.0592	0.0683
Corrosion rate mm/y	1.09	1.39	1.47

<u>Age hardening response and precipitate microstructures</u>

The age hardening response of the Mg-0.8Ca-*x*Sn alloys at 200°C are shown in Fig 4 (a). The age hardening response of the Mg-0.8Ca alloy is relatively small and a hardness increment of only 9VH was observed. The addition of Sn to the Mg-0.8Ca resulted in an increase in the maximum hardness observed by two times while it also decrease the time to peak hardness from ~16h in the binary alloy to ~10h in the ternary alloys. The maximum hardness increment was observed for the Mg-0.8Ca-3.5Sn alloy after ageing for 10h. The TEM micrographs illustrating the precipitate microstructure of the peak aged Mg-0.8Ca-3.5Sn alloy is shown in Fig. 4 (b). The conventional TEM micrographs show that the precipitate particles form on the basal planes of magnesium with a particle diameter of ~20 ±5 nm, but the particle thickness cannot be measured accurately from conventional TEM images. The diffraction pattern recorded $<10\bar{1}0>_{Mg}$, Fig 4 (b insert), show additional diffraction contrast due to solid-state precipitates. The diffused line like contrast associated with the precipitates suggests that these particles are very thin in the thickness direction as compared with the diameter of the precipitates. The diffraction contrast due to the precipitate particles in Mg-0.8Ca-Sn alloy is similar to that observed for the Mg-0.5Ca-1.2Zn [7-9] and Mg-0.5Ca-0.5Al [6] alloys in previous investigations. The diffraction patterns and precipitate morphology suggest that the precipitates observed in the Mg-0.8Ca-Sn are of the same morphology as the previous investigations.

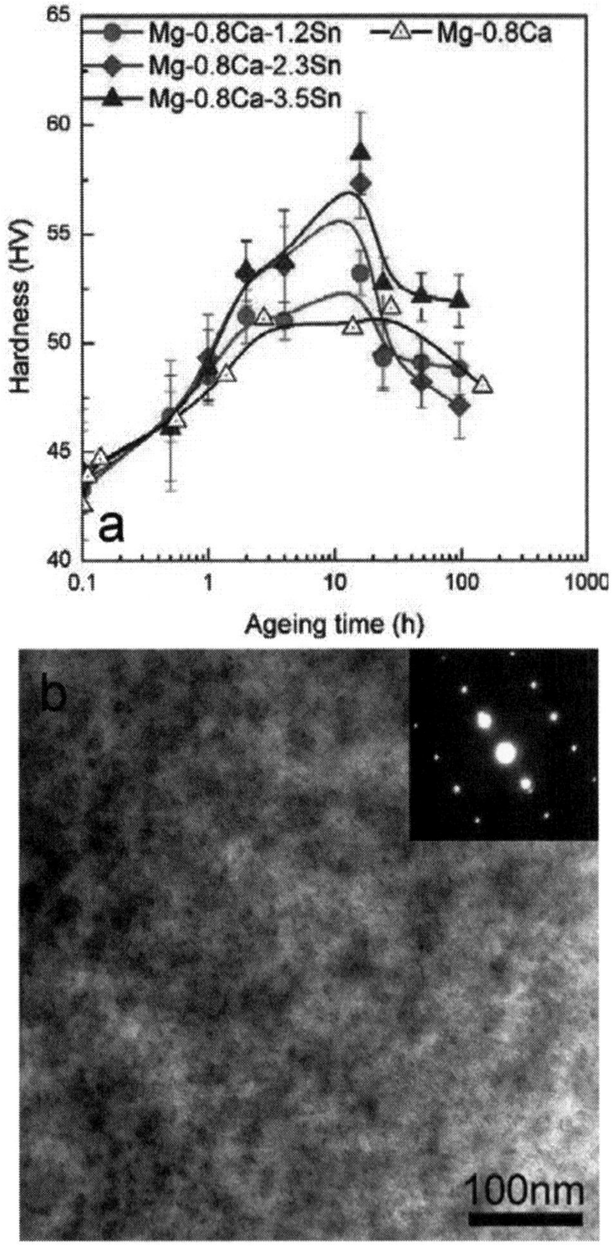

Figure 4 (a) the age hardening response of Mg-0.8Ca-*x*Sn alloys at 200°C. (b) The precipitate microstrcuture of Mg-0.8Ca-3.5Sn alloy aged at 200°C forl0h to peak hardness. Electron beam is approximately parallel to $<10\bar{1}0>_{Mg}$.

Conclusions

The Sn addition to Mg-0.8Ca alloy resulted in the formation of MgCaSn phase at the grain boundaries in the as-cast condition which was dissolved into the magnesium matrix by solution heat treatment in all alloys wiht the exception of the Mg-0.8Ca-3.5Sn alloy.

A relatively low corrosion rate was observed for the Mg-0.8CXSn alloys in the as-cast condition with the Mg-0.8Ca1.2Sn alloy showing the lowest corrosion rate, thus the best resistance to corrosion.

The age hardening experiments at 200°C showed that the hardness of Mg-0.8Ca-*x*Sn alloys increased with the addition of Sn compared to the binary Mg-0.8Ca alloy. The maximum hardening response was from the Mg-0.8Ca-3.5Sn alloy.

The hardening increment was due to a uniform distribution of very thin basal plate precipitates forming in the peak-aged alloy thought to be similar to those observed in Mg-Ca-Zn and Mg-Ca-Al alloys previously under heat treatment conditions.

Acknowledgements

The authors would like to thank the Deutsches Elektronen-Synchrotron the provision of synchrotron radiation facilities in the framework of the proposal II-20100225.

References

1. N. Hort *et al.* Microstructural Investigations of the Mg-Sn-*x*Ca system *Advanced Eng. Mater* 8 (2006) 359-364

2. T.Abu Leil *et al* Microstructure corrosion and creep of as-cast magnesium alloys In *Magnesium Technology 2007* (Ed R. Beals M.O. Pekguleryux R. Neelameggeham and A.A. Luo) TMS Materials Park Ohio (2007)

3. K. Hono *et al* Towards the development of heat-treatable high-strength wrought Mg alloys *Scripta Mater.* 63 (2010) 710-15.

4. X. Gao *et al* Precipitation-hardened Mg–Ca–Zn alloys with superior creep resistance *Scripta Mater* 53 (2005) 1321-26.

5. A. Suzuki *et al* Precipitation strengthening of a Mg-Al-Ca–Based AXJ530Die-cast Alloy *Metall. Mater. Trans. A.* 39A (2008) 696-702.

6. J. Jayamani *et al* Enhanced precipitation hardening of Mg-Ca alloy by Al additions *Scripta Mater* 63 (2010) 831-34

7. J.F. Nie and B.C. Muddle Precipitation hardening of Mg-Ca-(Zn) alloys 37 (1997) 1475-81

8. J.C. Oh *et al* TEM and 3DAP characterization of an age-hardened Mg–Ca–Zn alloy *Scripta Mater* 53 (2005)675-679

9. K. Oh-ishi *et al* Age hardening response of Mg-0.3at.%Ca alloys with different Zn contents *Mater Sci Eng A* 526 (2009)177-84

10. C.L. Mendis *et al* Precipitation of prismatic plates in Mg-0.3Ca alloys with In additions *Scripta Mater.* 64 (2011) 137-40

11. C.L. Mendis *et al* Microalloying effect on the precipitation processes of Mg-Ca alloys *Metall. Mater Trans A* 43 (2012) 3978-87

12. F. R. de Boer *et al* Cohesion in Metals : Transition Metal alloys (Cohesion and Structure) (1989) Elsevier New York

13. Villars, P. and Calvert, L.D., Pearson's Handbook of Crystallographic Data for Intermetallic Phases. 2nd ed (1991) ASM International, Materials Park Ohio.

Magnesium Technology 2013
Edited by: Norbert Hort, Suveen N. Mathaudhu, Neale R. Neelameggham, and Martyn Alderman
TMS (The Minerals, Metals & Materials Society), 2013

Thermodynamics of phase formation in Mg-La-Ce-Nd alloys

Rainer Schmid-Fetzer[1], Joachim Gröbner[1], Artem Kozlov[1], Milan Hampl[1],
Mark A. Easton[2], Suming Zhu[2], Mark A. Gibson[3] and Jian-Feng Nie[2]

[1]Institute of Metallurgy, Clausthal University of Technology, Robert-Koch-Str. 42, D-38678 Clausthal-Zellerfeld, Germany
[2]CAST CRC, Department of Materials Engineering, Monash University, Victoria 3800, Australia
[3]CAST CRC, CSIRO Process Science & Engineering, Clayton, Victoria 3169, Australia

Keywords: Simulation, Rare earth, Phase Diagrams, Thermodynamics, Solidification paths

Abstract

Experimentally validated thermodynamic descriptions have been developed for the ternary Mg-La-Ce, Mg-La-Nd, and Mg-Ce-Nd systems by selecting key alloys in both systems and analyzing the phase formation in both the as-cast and heat treated state by SEM/EDS and DSC. These results were combined to form the validated thermodynamic Calphad-type description for quaternary Mg-La-Ce-Nd alloys. It is shown that for these light rare earth elements (La, Ce, Nd) the intermetallic phases with Mg exhibit significant mutual solid solubility in the ternary systems, extending into the quaternary alloy system. This is reflected by considering the shared crystal structures in the thermodynamic modeling. Simulated solidification paths of three Mg-La-Ce-Nd alloys with different La:Ce:Nd ratios and a common total content of 5 wt.% rare earth (RE) metals are evaluated using computational thermodynamics. Unexpected and distinctly different solidification behavior of these three alloys is revealed. The sequence La→Ce→Nd in the periodic table is not at all reflected in a monotonous solidification behavior. The demonstrated individual impact of each of these elements forbids treating the RE additions as a mere wt.% sum of RE elements.

Introduction

In recent years there has been considerable interest in the addition of RE metals to magnesium and its alloys to improve the mechanical properties, high temperature creep resistance in particular. The RE elements are generally added to Mg alloys in the form of mischmetal, a mixture of various RE metals. Commonly used mischmetal consists mainly of the light RE metals, i.e. La (25-34%), Ce (48-55%), Nd (11-17%) and small amounts of Pr and Sm. It has been realized that the individual RE elements may exert distinctly different effects on the mechanical properties of Mg alloys [1-3].

The purpose of this work is to present a concise comparison of the key features of the three ternary subsystems at the Mg-corner of the Mg-La-Ce-Nd quaternary phase diagram and to apply our validated knowledge of this system in solidification and heat treatment simulations of quaternary Mg-La-Ce-Nd alloys with different ratios of RE additions.

Phase diagrams of ternary Mg-RE-RE systems

The first ternary system of Mg with two light rare earths elements investigated by the present authors was the Mg-Ce-Nd system [1]. Easton et al. [2] have elaborated the crystallographic and thermodynamic effects on phase selection in binary magnesium-rare earth (Ce or Nd) alloys. Subsequently the addition of La was investigated in the two ternary Mg-Ce-La and Mg-La-Nd systems by Gröbner et al. [3].

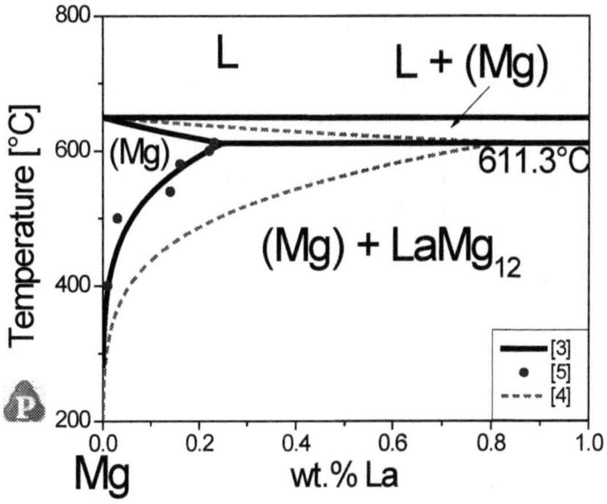

Figure 1: The Mg-rich part of the Mg-La binary phase diagram based on our recent assessment [3] compares well with the experimental data for the solvus of (Mg) by Rokhlin [4] and significantly differs from the calculation by Guo and Du [5].

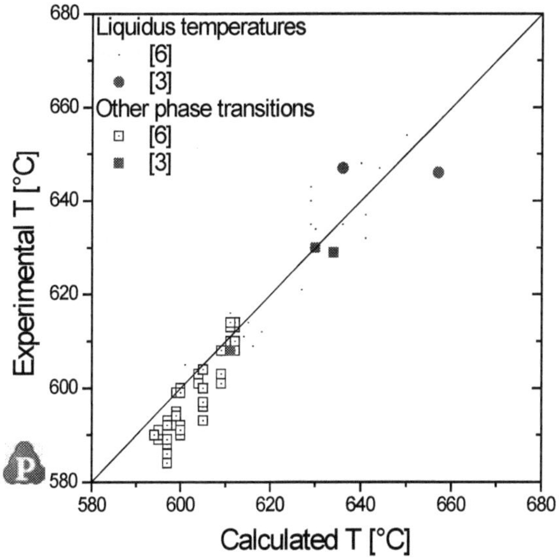

Figure 2: Comparison of all experimentally observed thermal signals in the Mg-La-Ce system [6, 3] with temperatures from the thermodynamic calculation of [3].

The binary phase diagram Mg-La was also re-assessed in that work [3] and its Mg-rich part is shown in Fig. 1. The focus is on the solid solubility of La in (Mg). Our calculated solvus data agree with the experimental data Rokhlin [4] whereas the calculated data by Guo and Du [5] are about four times larger at the eutectic temperature of 611.3°C.

For ternary Mg-La-Ce alloys the experimental thermal signals for samples on six vertical sections at constant 95, 90, 85, 80, 75 and 70 wt.% Mg were reported by Rokhlin and Bochvar [6] and additional DSC experiments were performed for selected key alloys in our previous work [3]. Figure 2 shows the comparison between the experimental thermal signal temperatures and those calculated for all these alloys from the thermodynamic description by Gröbner et al. [3]. Essentially, the agreement is better than ±10 K which may be considered satisfactory in view of the experimental challenges for the Mg-RE systems.

The calculated Mg-rich parts of the liquidus projections of the three ternary subsystems at the Mg-corner of the Mg-La-Ce-Nd quaternary are shown in Figs. 3(a–c), based on our recently obtained data [1, 3]. A large extension of the primary crystallization field of (Mg) up to 15 wt.% RE or more is noted, way beyond the typical alloying additions of RE in applied magnesium alloys. The Mg-richest intermetallic phase in all those ternary systems is $REMg_{12}$, denoting the solid solution phase $(La,Ce,Nd)Mg_{12}$. Its liquidus surface intersects that of (Mg) in the monovariant equilibrium Liquid + (Mg) + $REMg_{12}$, except for alloys very close to the Mg-Nd binary edge where the RE_5Mg_{41} phase appears, forming the monovariant equilibrium Liquid + (Mg) + RE_5Mg_{41}. In most of the Mg-rich ternary alloys depicted in Figs. 3(a–c) $REMg_{12}$ is expected to solidify as a stable second phase after primary solidification of (Mg). The exception being alloys with a high Nd/(La+Ce) ratio where RE_5Mg_{41} is expected to solidify as a stable second phase. The problem of retarded nucleation of the stable Nd_5Mg_{41} phase in binary hypoeutectic Mg-Nd cast alloys, which tend to contain $NdMg_{12}$ at quite slow cooling rates or $NdMg_3$ if the cooling rate is particularly rapid, has been solved recently [2]. It will be addressed for a quaternary Nd-rich alloy later in this study.

In Figs. 4(a-c) the calculated Mg-rich parts of the isothermal phase diagram sections at 500°C of the three ternary systems are compared. The phases REMg (structure type cP2-CsCl) and $REMg_3$ (cF16-BiF_3) exhibit complete solid solution ranges in all three ternaries. Moving closer to the Mg-corner the intermetallics RE_5Mg_{41} (tI92-Ce_5Mg_{41}), RE_2Mg_{17} (hP38-$Ni_{17}Th_2$), and $REMg_{12}$ (tI26-$ThMn_{12}$) may exhibit limited or complete solid solution ranges as clearly indicated in Figs. 4(a-c). The intricacies of modeling such intermetallic solution phase in a wider perspective of the extensive thermodynamic Mg alloy database are recently detailed elsewhere [7].

The single-phase region of (Mg) is highlighted by the red colored regions in the ternary phase diagrams at 500°C, but just visible as a dot in Fig. 4(a) and as a narrow line in Figs. 4(b) and (c). That is due to the small solid solubilities in the binary systems at 500°C: 0.06 wt.% La, 0.2 wt.% Ce, and 2.2 wt.% Nd in the binary (Mg) phases, respectively.

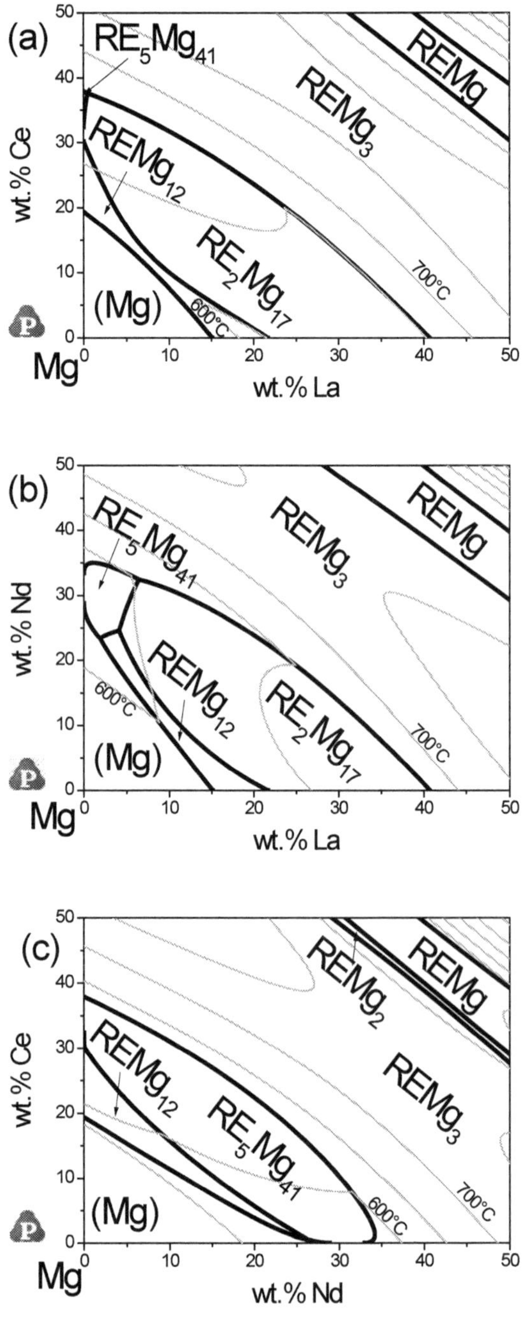

Figure 3. Calculated Mg-rich part of the liquidus projections of three ternary Mg-RE-RE systems: (a) Mg-La-Ce, (b) Mg-La-Nd, (c) Mg-Ce-Nd.

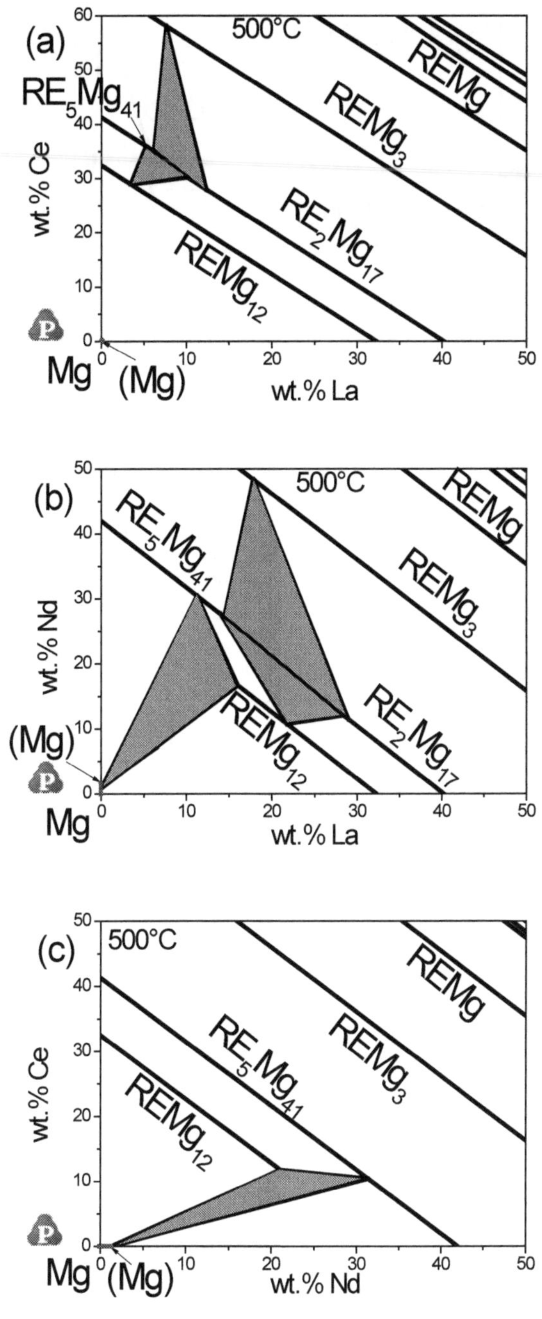

Figure 4. Calculated Mg-rich part of the isothermal sections at 500°C of three ternary Mg-RE-RE systems: (a) Mg-La-Ce, (b) Mg-La-Nd, (c) Mg-Ce-Nd. Triangular three-phase regions are gray-shaded and the single-phase (Mg) solid solution range is highlighted by red color.

Solidification and heat treatment simulation of quaternary Mg-La-Ce-Nd alloys with different ratios of RE additions

Based on our validated assessments of the ternary subsystems and the consistent description of the multicomponent intermetallic solution phases the phase formation in quaternary Mg-La-Ce-Nd alloys can be reliably predicted by thermodynamic calculations. This is exemplified on three alloys with a total addition of 5 wt.% RE but different ratios of La:Ce:Nd. The three alloys investigated are denoted as "3-1-1", "1-3-1", and "1-1-3" with the compositions 3La-1Ce-1Nd (wt.%), 1La-3Ce-1Nd (wt.%), and 1La-1Ce-3Nd (wt.%), respectively. Their location in the quaternary system is sketched in Fig. 5. The intention is to study the impact of variations in the RE composition of alloying additions such as mischmetal.

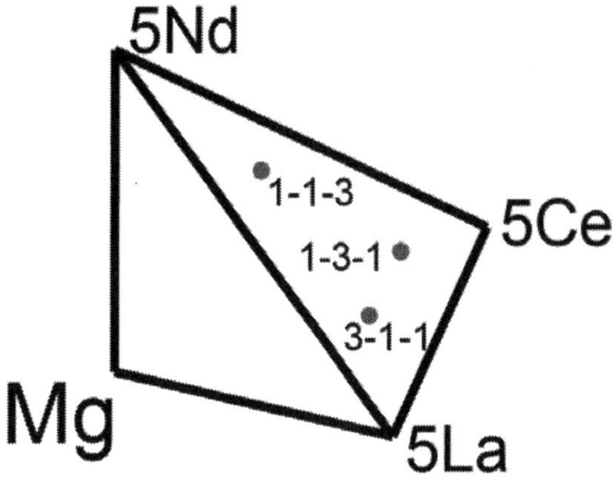

Figure 5: Perspective sketch of the Mg-rich corner of the Mg-Ce-La-Nd quaternary system sectioned at the 95 wt.% Mg plane, showing the locations of the three investigated alloys: Alloy 3-1-1 with 3La-1Ce-1Nd (wt.%), alloy 1-3-1 with 1La-3Ce-1Nd (wt.%), and alloy 1-1-3 with 1La-1Ce-3Nd (wt.%).

All thermodynamic calculations are performed using the Pandat software [8] and the new and significantly upgraded PanMg2012 database [7, 9] (www.computherm.com).

The constitution of the as-cast alloys is approximately simulated using Scheil solidification conditions and the developing phase amounts are given in Figs. 6(a-c). The final amount of intermetallic phases at the Scheil solidus, considered to constitute the frozen-in microstructure, is denoted in the diagrams, e.g. 9.2% of $REMg_{12}$ phase and 1.5% of RE_5Mg_{41} phase in alloy 3-1-1. The at.% scale based on moles of atoms of all phases is used (and not the phase weight fraction) since it is often a better approximation to the volume fraction of the phases in the microstructure.

The complex compositions of these $(La,Ce,Nd)Mg_{12}$ and $(La,Ce,Nd)_5Mg_{41}$ phases, also resulting from the calculations, are not reported here but may be envisaged from the extended solubility ranges in the ternary subsystems shown for 500°C in Figs. 4(a-c). The total amount of intermetallic phases reported in Figs. 6(a-c) decreases from 10.7%→9.5%→7. 8% if the majority RE component changes from La→Ce→Nd. In the same sequence the dominating $REMg_{12}$ phase is replaced by the RE_5Mg_{41} phase. That interdependence could only be revealed by the quantitative thermodynamic simulation considering all the quaternary solubilities in the intermetallics and in the (Mg) phase as well as the quaternary liquidus properly.

The sequence La→Ce→Nd reflects the increasing order of these elements in the periodic table. One might, thus, be tempted to assume a monotonous change in the solidification behavior, as already suggested by the monotonously decreasing

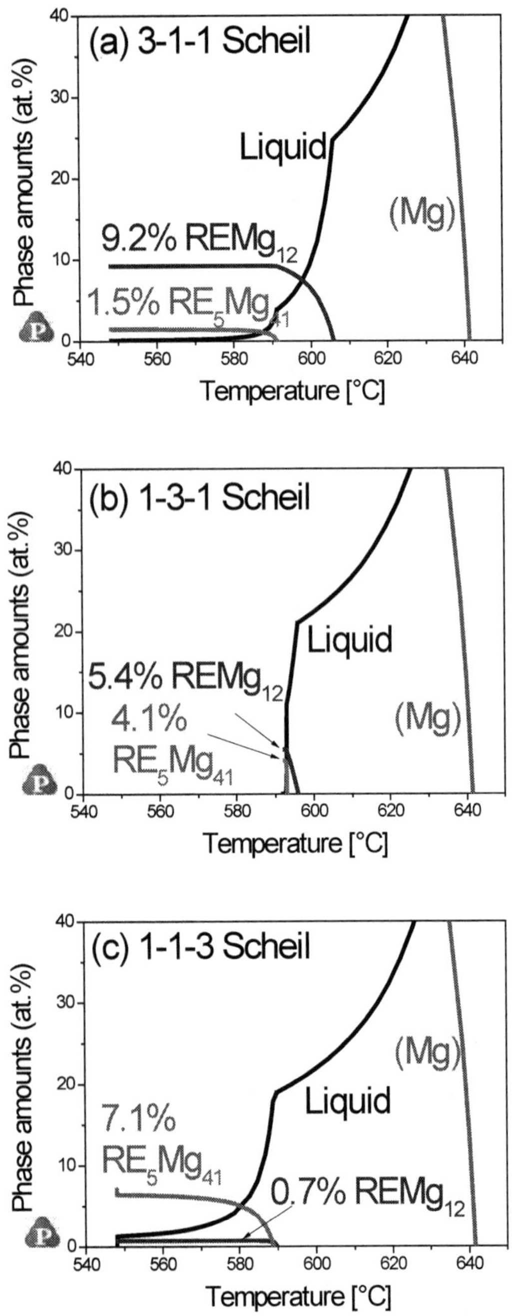

Figure 6. Calculated phase amounts during Scheil solidification simulation of the three Mg-La-Ce-Nd alloys detailed in Figure 5: (a) Alloy 3-1-1, (b) Alloy 1-3-1, (c) Alloy 1-1-3.

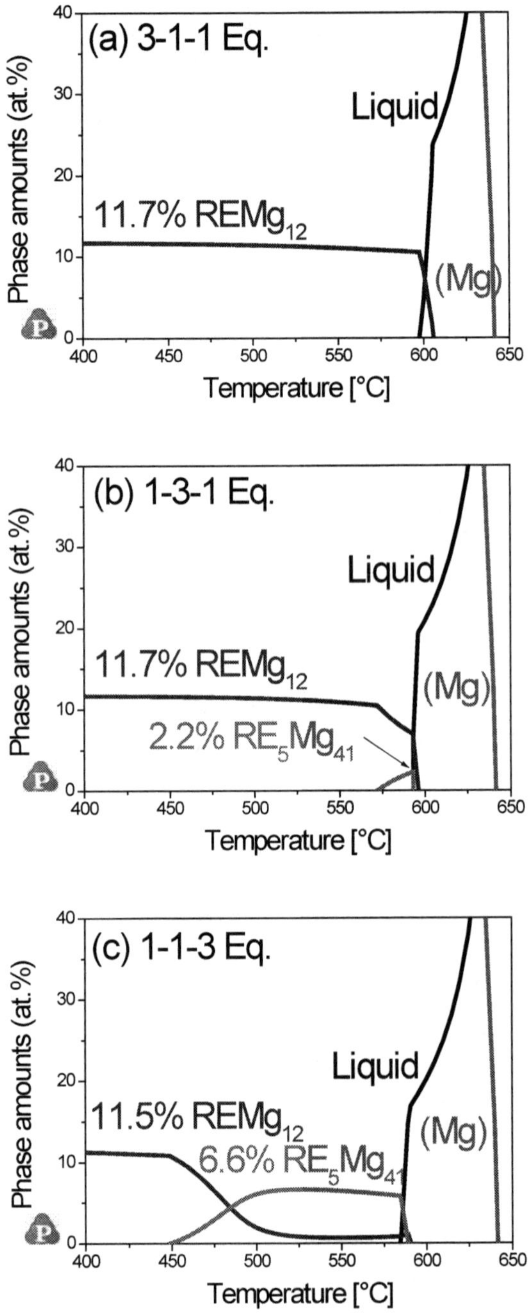

Figure 7. Calculated phase amounts at equilibrium for the three Mg-La-Ce-Nd alloys detailed in Figure 5: (a) Alloy 3-1-1, (b) Alloy 1-3-1, (c) Alloy 1-1-3.

total amount of intermetallic phases. That is, however, not the case at all. The Scheil solidus of the "middle" alloy 1-3-1 is much higher, 592°C, compared to the alloys 3-1-1 and 1-1-3, both with 548°C. That gives alloy 1-3-1 a much narrower freezing range since the liquidus temperatures of all three alloys are virtually the same, at 641.5°C. The reason is that alloy 1-3-1 terminates solidification under Scheil conditions with a truly quaternary liquid composition of Mg-0.7La-15.3Ce-4.8Nd (wt.%) and still precipitating according to the monovariant reaction

$$\text{Liquid} \rightarrow (\text{Mg}) + \text{REMg}_{12} + \text{RE}_5\text{Mg}_{41}$$

occurring between 593 and 592°C for this alloy. That is, coincidentally, very close to the equilibrium solidus of this alloy 1-3-1 at 593°C with the same reaction. By stark contrast both alloys 3-1-1 and 1-1-3 terminate Scheil solidification with the binary liquid composition of Mg-28.8Nd (wt.%), which is attained from quaternary liquid compositions while running down the polythermal reaction

$$\text{Liquid} \rightarrow (\text{Mg}) + \text{RE}_5\text{Mg}_{41}$$

from about 578°C down to the binary Mg-Nd eutectic at 548°C. This startling and distinct behavior could not have been envisaged without the tool of computational thermodynamics, even if one had known the graphical presentation of the ternary phase diagrams.

It is noted that in all cases the $REMg_{12}$ phase is the first intermetallic to precipitate in all alloys. The temperature step until the RE_5Mg_{41} phase appears decreases as seen from Figs. 6(a-c).

For comparison the results from equilibrium solidification simulation (Lever rule) for these three alloys are given in Figs. 7(a-c). Again, the $REMg_{12}$ phase is the first intermetallic to precipitate in all alloys. Termination of solidification is somewhat different though, occurring in the following reactions under equilibrium for alloys "3-1-1", "1-3-1", and "1-1-3", respectively:

597°C Liquid → (Mg) + $REMg_{12}$

593°C Liquid → (Mg) + $REMg_{12}$ + RE_5Mg_{41}

584°C Liquid + $REMg_{12}$ → (Mg) + RE_5Mg_{41} .

The subsequent solid state reaction

(Mg) + RE_5Mg_{41} → $REMg_{12}$

occurs in the latter two alloys, and below 450°C $REMg_{12}$ is the only stable intermetallic phase in all three alloys. The maximum intermetallic phase amounts occurring at any temperature is denoted in the diagrams in Figs. 7(a-c). A significant heat treatment response with changing phase amounts is expected. That is seen by comparing the initial phase amounts in the as-cast structures, denoted in Figs. 6(a-c) to the ones expected after long-term heat treatment at selected temperatures in Figs. 7(a-c).

Impact of metastable constrained solidification

It has been shown recently that in Mg-Nd and Mg-Ce-Nd alloys the formation of the stable Nd-rich RE_5Mg_{41} in the as-cast state is easily overrun and that at small supercooling the competing $REMg_{12}$ phase may form instead [1] and that the Nd-rich RE_5Mg_{41} phase develops at high heat treatment temperatures and long enough times. Based on thermodynamic and crystallographic matching calculations, the phase selection is well explained by the competition between the driving force (reduction in bulk Gibbs energy) and the nucleation energy barrier (increment of surface energy) [2].

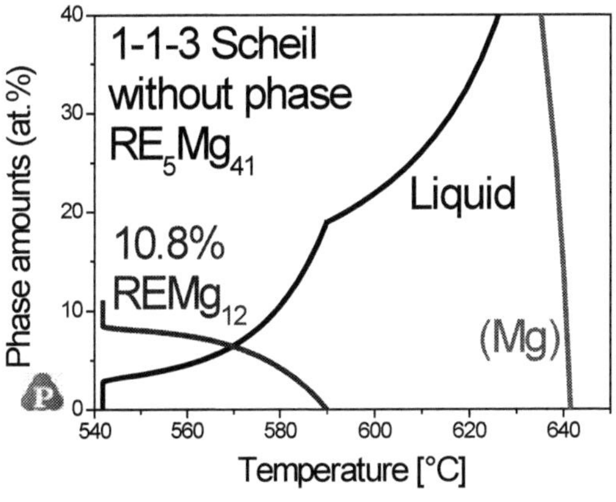

Figure 8. Calculated phase amounts during Scheil solidification simulation of alloy 1-1-3 suppressing the phase RE_5Mg_{41}, compare to Fig. 6(c).

That knowledge is applied to the Nd-richest of the three quaternary alloys, 1-1-3, by simulating such a metastable as-cast state in a Scheil calculation with the RE_5Mg_{41} phase being suspended. The phase selection is constrained by this exclusion; otherwise the Scheil conditions apply which, of course, always generate a metastable constitution in view of the blocked solid state diffusion. The resulting phase amounts are given in Fig. 8. The stable and metastable solidification paths are of course identical until the RE_5Mg_{41} phase precipitates at 589°C, compare with Fig. 6(c). In the stable scenario 1.1% of residual liquid solidify in the stable binary Mg-Nd eutectic:

548°C Liquid → (Mg) + Nd_5Mg_{41} (stable)

In the metastable scenario, Fig. 8, the suppression of the RE_5Mg_{41} phase does not prevent the liquid composition from segregating to the binary Mg-Nd edge, where 2.7% of residual liquid solidify in the metastable binary Mg-Nd eutectic:

541.8°C Liquid → (Mg) + $NdMg_{12}$ (metastable)

Subject to a sufficiently long heat treatment, attaining the equilibrium state, the phases will transform to the constitution shown in Fig. 7(c), irrespective of the conditions assumed for the as-cast state.

Terminal freezing range (TFR)

The key temperatures during the Scheil solidification simulations are listed for the three quaternary Mg-La-Ce-Nd alloys at preset values of fraction solid, f_s, in Table I. In addition to the "Scheil solidus" at f_s = 100% two additional key temperatures are given at the fractions solid of f_s = 88% and f_s = 98%. The temperature difference between these points had been defined as the *terminal freezing range* (TFR) [10]. The idea put forward with this definition is that a small TFR should be beneficial in order to avoid hot tearing. It has been detailed that the value of ΔT-88/98, describing the partial freezing range of the "almost" last 10% of solidifying liquid, is more relevant than ΔT-90/100. That is because the exact termination of solidification, f_s = 100%, cannot generally be expected to be predicted precisely from the limiting Scheil model since the path of the last 2% of liquid may be prone to some artifacts [10]. One might argue about the values of f_s = 88% and f_s = 98% selected here, however, they are located approximately in the region where interdendritic feeding, burst and solid feeding become dominant, triggering casting defects [11]. The concept of TFR [10] has found some interest and was successfully applied to the hot tearing susceptibility of Mg-Zn alloys [12] and the development of high-strength and ductile Mg–Zn–Zr–Ca–Mn(–Yb) alloys [13].

Table I. Temperatures attained during Scheil solidification simulation at preset values of fraction solid, f_s, and the terminal freezing range, **ΔT-88/98,** for the three quaternary Mg-La-Ce-Nd alloys detailed in Figure 5. Their liquidus temperatures vary only between 641.4 and 641.5°C.

Alloy	**Temperature at preset f_s [°C]**			**ΔT-88/98 [K]**
	f_s = 88 %	**f_s = 98 %**	**f_s = 100%**	
3-1-1	601	589	548	12
1-3-1	593.3	592.8	592.0	0.5
1-1-3	587	564	548	23
1-1-3 *	582	542	542	20

* Constrained phase selection, without RE5Mg41

It is another startling result that these values of ΔT-88/98, given in Table I, are drastically different for the three quaternary alloys with all the same total RE addition of 5 wt.%. And these values do not even vary monotonously in the sequence La→Ce→Nd of the major RE addition. Again, the "middle" alloy 1-3-1 shows an extremely small value of ΔT-88/98 = 0.5 K. One might thus expect the Ce-rich alloy to perform better during casting compared to the La- and Nd-rich ones.

References

1. J. Gröbner, A. Kozlov, R. Schmid-Fetzer, M. A. Easton, S. Zhu, M. A. Gibson, and J.-F. Nie, "Thermodynamic analysis of as-cast and heat treated microstructures of Mg-Ce-Nd alloys," *Acta Mater.*, 59 (2011) 613-622.
2. M.A. Easton, M.A. Gibson, D. Qiu, S.M. Zhu, J. Gröbner, R. Schmid-Fetzer, and J.F. Nie, M. Zhang, "The role of Crystallography and Thermodynamics on Phase Selection in Binary Magnesium-Rare Earth (Ce or Nd) Alloys," *Acta Mater.*, 60 (2012) 4420-4430.
3. J. Gröbner, M. Hampl, R. Schmid-Fetzer, M.A. Easton, S.M. Zhu, M.A. Gibson, and J.F.Nie, "Phase Analysis of Mg-La-Nd and Mg-La-Ce Alloys," *Intermetallics*, 28 (2012) 92–101.
4. L.L. Rokhlin, "Magnesium Alloys Containing Rare Earth Metals, " Taylor & Francis, London, 2003.
5. C. Guo and Z. Du, "Thermodynamic assessment of the La–Mg system," *J. Alloys Comp.* 385 (2004) 109–113.
6. L.L. Rokhlin, N.R. Bochvar, "Phase Diagrams of the Mg-La-Ce System," *Izv. Akad. Nauk SSSR, Metally*, (2) 193-197 (1977).
7. R. Schmid-Fetzer, J. Gröbner, "Thermodynamic database for Mg alloys — progress in multicomponent modeling," *Metals*, 2 (2012) 377-398. Special Issue "Magnesium Technology", open access.
8. W. Cao, S. Chen, F. Zhang, K. Wu, Y. Yang, Y. Chang, R. Schmid-Fetzer, W. A. Oates, "PANDAT Software with PanEngine, PanOptimizer and PanPrecipitation for Materials Property Simulation of Multi-Component Systems," *Calphad*, 33 (2009) 328-342.
9. J. Gröbner, R. Schmid-Fetzer, "Key issues in a thermodynamic Mg alloy database," *Metall. Mater. Trans. A*, (2012) accepted
10. M.B. Djurdjevic, R. Schmid-Fetzer, "Thermodynamic Calculation as a Tool for Thixoforming Alloy and Process Development," *Mater. Sci. Engin.* A, 417 (2006) 24-33.
11. A.K. Dahle, D.H. StJohn, "Rheological behaviour of the mushy zone and its effect on the formation of casting defects during solidification," Acta Mater., 47 (1998) 31–41.
12. P. Gunde, A. Schiffl, P.J. Uggowitzer, "Influence of yttrium additions on the hot tearing susceptibility of magnesium–zinc alloys," *Mater. Sci. Engin.* A, 527 (2010) 7074–7079.
13. P. Gunde, A.C. Hänzi, A.S. Sologubenko, P.J. Uggowitzer, "High-strength magnesium alloys for degradable implant applications," *Mater. Sci. Engin.* A, 528 (2011) 1047–1054.

Magnesium Technology 2013
Edited by: Norbert Hort, Suveen N. Mathaudhu, Neale R. Neelameggham, and Martyn Alderman
TMS (The Minerals, Metals & Materials Society), 2013

PHASE STABILITY INVESTIGATION OF THE MG-ZN-SM SYSTEM

Xiangyu Xia[1], Amirreza Sanaty Zadeh[1], Chuan Zhang[2], Xiaoqin Zeng[3], Donald Stone[1], Alan.A. Luo[4]
[1] Materials Science Program; University of Wisconsin, Madison; 1509, University Ave; Madison, WI, 53706, USA
[2] Computherm LLC, 437 S. Yellowstone Dr, Suite217, Madison, WI, 53719, USA
[3] Department of Materials Science and Engineering; Shanghai Jiaotong University; 800, Dongchuan Road, Shanghai, 200240, China
[4] Chemical and Materials Systems Laboratory; General Motors Research and Development Center; 30500 Mound Road, Warren, MI, 48090, USA

Keywords: Mg-Zn-Sm system; intermetallics; isothermal section

Abstract

Phase equilibrium of the Mg-Zn-Sm ternary system at the Mg-rich region at 400°C was investigated. One ternary phase Z was confirmed, with a composition formula of $Mg_3Zn_{10}Sm$ and crystal structure of $P6_3/mmc$. Zn was found to have large solubility in both Mg_3Sm (~35.6 wt.%) and $Mg_{41}Sm_5$ (~4 wt.%) phases. An isothermal section of 400°C in the Mg-rich corner was constructed, which provides important understanding of phase relationships in the Mg-Zn-Sm system.

Introduction

Magnesium alloys have been increasingly used in the automotive and aerospace industries for light weighting and energy saving, because of their low density, high specific stiffness, excellent castability, and other favorable attributes [1-4]. However, the relatively poor mechanical properties like strength and creep resistance have limited their applications. The addition of Zn has been reported to significantly enhance strength and creep resistance in Mg alloys by both solid solution strengthening and age hardening. However, Zn also brings brittleness and coarse grains [5], so a third strengthening element is always considered in Mg-Zn alloys. Mg-Zn-RE (rare earth) alloys have attracted much interest recently. RE's solid solution strengthening in α(Mg) and precipitation strengthening from the intermetallics could provide strong influence on Mg alloy mechanical properties[6,7].

Samarium (Sm), as one of the cheapest RE elements, attracts significant interest. Although there have been some reported studies on the mechanical properties of Mg-Zn-Sm alloys [8,9,10], the strengthening mechanisms, intermetallics and phase diagrams are yet to be fully understood. Drits, et al [11] studied the phase relationships in this system, as well as liquidus projection and a partial isothermal section at Mg corner, as shown in Fig.1. Three ternary phases X ($(Mg_{1.39}Zn_{0.62})_3Sm_2$), Y ($Mg_{9.54}Zn_{5.21}Sm$), Z ($Mg_3Zn_6Sm$) were also identified. However, their study was not sufficient for further applications for the following reasons: 1) The homogeneity ranges of the three ternary phases were not clearly specified; 2) Despite the clear crystal structure information for X (FCC, a=6.994Å) and Z (HCP, a=14.62 Å, c=8.78 Å) phase, such information for Y phase was not available; 3) These three phases differ significantly in composition and crystal structure from those investigated in [12-14], which are named as μ_3 ($Mg_{13}Zn_{30}Sm_3$), μ_5 ($Mg_{15}Zn_{40.5}Sm_{6.5}$) and μ_7 ($Mg_{58.1}Zn_{155.3}Sm_{26.6}$), all with hexagonal crystal structure. Thus, a detailed investigation into ternary phases and phase relationships in this system is necessary.

For this purpose, three typical Mg-Zn-Sm alloys (their compositions are listed in Table. 1) were identified for experimental investigation based on the work of Drits et al [11]. As shown in Fig.1, the alloys selected in this study are located in the three-phase region, which will provide us with more phase-equilibrium information and less experimental work. In this study, phase equilibria of these three alloys at 400°C were investigated.

Experimental

The Mg-Zn-Sm alloy samples were prepared using commercially pure Mg, Zn (>99.9wt%) and Mg-20wt.%Sm master alloy ingots. Melting process was under protection of SF_6/CO_2 and the samples were cut into pieces of around 100g. The chemical composition of each alloy was then analyzed using Inductively Coupled Plasma Mass Spectrometry (ICP-MS) and listed in Table.1. Afterwards, samples were wrapped with tantalum foils respectively, sealed in a quartz tube and back-filled with high-purity argon atmosphere (>99.999%), and then annealed at 400°C for one month. Phase identification and lattice parameter measurements were performed by X-ray diffraction (XRD) using General Area Diffraction Detector System (GADDS), with Cu Kα radiation (40kV, 50 mA), as well as Transmission Electron Microscopes (TEM, Tecnai T12). SEM (EVO-50) was employed to study the microstructures. The compositions of each element within the phases in equilibrium were determined using Electron Probe Microscope Analysis (EPMA) (CAMECA SX100).

Table.1. Chemical composition of selected Mg-Zn-Sm alloys and EPMA measured compositions of each phase in equilibrium from this study (the chemical composition is measured by ICP)

Alloy #	Overall Compositions (wt.%)	Phases	Mg (wt.%)	Zn (wt.%)	Sm (wt.%)
1	Mg-23.35Zn-5.73Sm	α(Mg)	90.2	9.7	0.1
		Z	12.43	70.16	17.76
2	Mg-1.17Zn-	α(Mg)	97.19	0.36	1.70

	13.01Sm	$Mg_{41}Sm_5$	54.71	3.88	40.64
3	Mg-8.16Zn-13.38Sm	α(Mg)	97.17	0.8	1.10
		Mg_3Sm	12.36	35.59	50.13

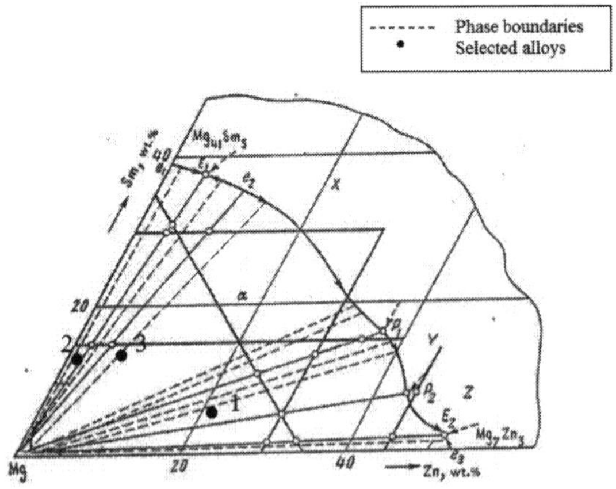

Fig.1. Projection of crystallization surfaces, and phase regions at 300°C of Magnesium corner of system Mg-Sm-Zn [11]

Results and Discussion

A typical microstructure of alloy #1 Mg-23.35Zn-5.73Sm is shown in Fig.2.a. Two phases are in equilibrium: white dendritic phase and dark matrix. The XRD results shown in Fig.3 indicate that the α (Mg) and Z phase coexist. The compositions of either phase determined from EPMA are listed in Table.1. Based on the above phase identification results, the dark matrix is α(Mg), in which Zn has a solubility of around 9.7 wt%; the light Z phase has a composition of Mg-72Zn-16.5Sm and a formula of $Mg_3Zn_{10}Sm$. The XRD pattern corresponds well with the hexagonal phase μ_3, with the lattice constants a=14.6Å and c=8.71Å.

It should be noted that though μ_3, μ_5, μ_7 [12-14] are three individual phases, they are similar in many perspectives: same crystal structures ($P6_3/mmc$), same "c" lattice constant (8.8Å), similar composition (Mg: Zn: Sm≈3:10:1) and the only difference is the "a" lattice parameter (14.6Å/23.5Å/33.6Å). For our purpose of constructing Mg-Zn-Sm phase diagram around Mg-rich corner and further assisting alloy optimization, it would be difficult and unnecessary to distinguish these three phases; thus we take the average composition for Z phase, while the crystal structure resembles that of μ_3.

The BSE image of alloy #2 Mg-1.17Zn-13.01Sm is presented in Fig.2.b. The dark phase is identified to be α(Mg), with 1.7% Sm and negligible Zn. The light phase corresponds well to $Mg_{41}Sm_5$, with a solubility of Zn of around 4%, as listed in Table.2. The existence of these two phases is supported by the XRD spectrum as seen in Fig.3. Compared with $Mg_{41}Nd_5$ [16], $Mg_{41}Sm_5$ has the same crystal structure (I4/m), similar lattice constants ($Mg_{41}Sm_5$: a=14.8Å, c=10.36Å; $Mg_{41}Nd_5$: a=14.7Å, c=10.40Å) and similar Zn solubility (~2-3%), thus it is reasonable to infer that Zn's substitution mechanism would be the same in both phases, which means Zn substitute Mg atoms in $Mg_{41}Sm_5$.

The BSE image in Fig.2.c shows the two phase equilibrium in alloy #3 Mg-8.16Zn-13.38Sm. The dark matrix is α(Mg) and the eutectic white is identified to be Mg-35.6Zn-50Sm, i.e. the X phase [11]. The formula is calculated as $Mg_{34.1}Zn_{36.6}Sm_{22}$. The eutectic structure indicates an eutectic reaction L = X + Mg, as described in Drits'[11] and the eutectic dark phase is eutectic Mg. The presence of the two phases is confirmed by the XRD pattern in Fig.3.

Fig. 4 shows the bright field image of the X phase and the selected area diffraction patterns from two different zone axes. The crystal structure of the X-phase is confirmed to be FCC, as it was found from the XRD patterns.

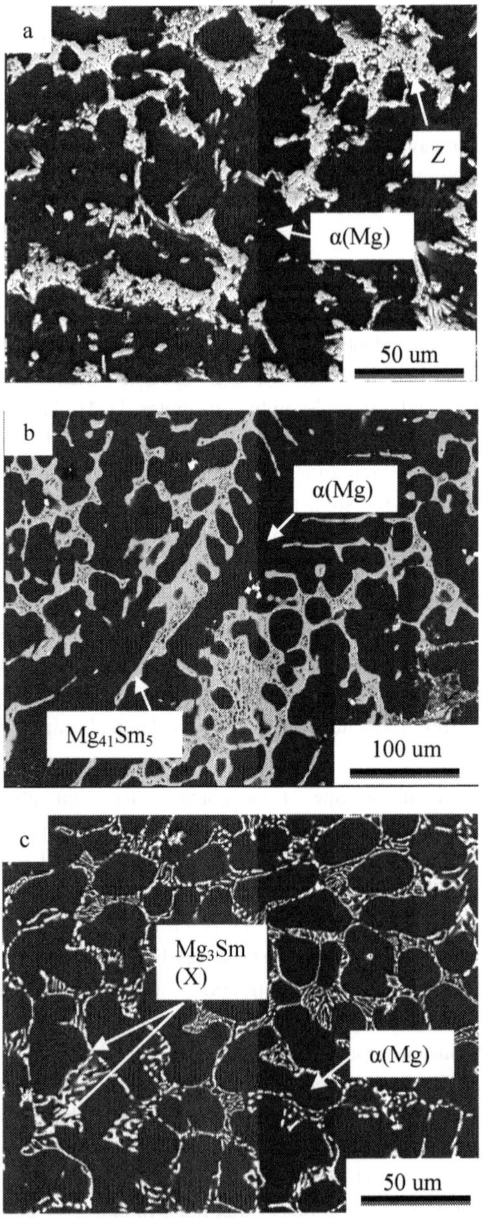

Fig.2. BSE images of Annealed Alloys ((a):#1;(b):#2; (c):#3)

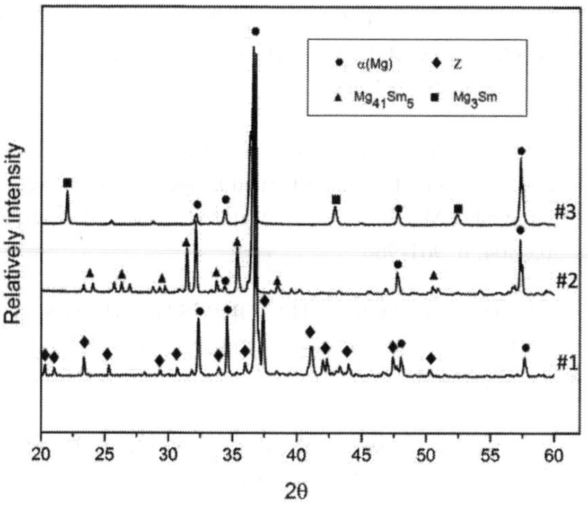

Fig.3. XRD Patterns of Annealed Alloys

The lattice parameter, calculated from the diffraction patterns for X-phase is between 6.64-6.94 Å. These values are consistent with the calculated values from the XRD patterns, i.e. 6.97Å. The lower value for lattice parameter, compared to that of Mg_3Sm (a=7.731Å) [17], is probably caused by magnesium atoms, with larger atomic radius, being randomly replaced with smaller zinc atoms in the structure. Therefore, while the X-phase was proposed to be ($(Mg_{1.39}Zn_{0.62})_3Sm_2$) in an early study [11], the XRD and TEM results in this paper, along with the fact that the composition of Mg and Zn together is 3 times as that of the Sm, suggest that this phase is instead the Mg_3Sm binary phase with a large solubility of zinc atoms in the structure.

Based on the above results, a schematic isothermal section of Mg-Zn-Sm system in Mg-rich corner at 400°C is determined, as shown in Fig.5.

In alloy#1, only two phases exist, indicating that #1 is either on the α(Mg)-Z tie-line, or the amount of Mg_7Zn_3 or Mg_3Sm is too small to be detected. In either situation, the tie-line between α(Mg) and Z phase would be close to #1's composition. The limited homogeneity is represented by the dashed circles.

In #2, $Mg_{41}Sm_5$ contains around 4% of Zn. Thus its homogeneity range expands towards Zn corner and #2 lies within the two phase region. #3 lies in the α(Mg)-Mg_3Sm region. The α(Mg)-Mg_3Sm-$Mg_{41}Sm_5$ ternary region can be inferred from phase rule, shown as dashed lines in Fig.5.

The ternary region α(Mg) + Mg_3Sm + Z was not observed from present experimental results. However its boundary can be estimated from compositions of Z and Mg_3Sm, and labeled as dashed lines in Fig.5. Further study with more selected alloys under different annealing temperatures is undergoing to clarify phase equilibria at the Mg-rich corner of the Mg-Zn-Sm ternary system. .

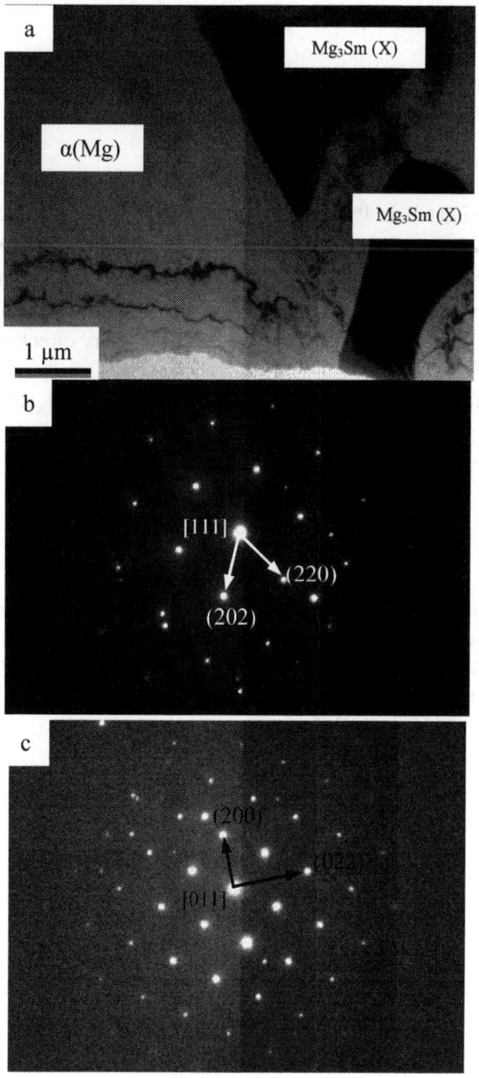

Fig.4. (a). TEM BF image of alloy #3 Mg-8.16Zn-13.38Sm; (b). TEM SAD pattern of Mg_3Sm (X-phase) with zone axis B = [111]; (c). TEM SAD pattern of Mg3Sm (X-phase) with zone axis B = [011].

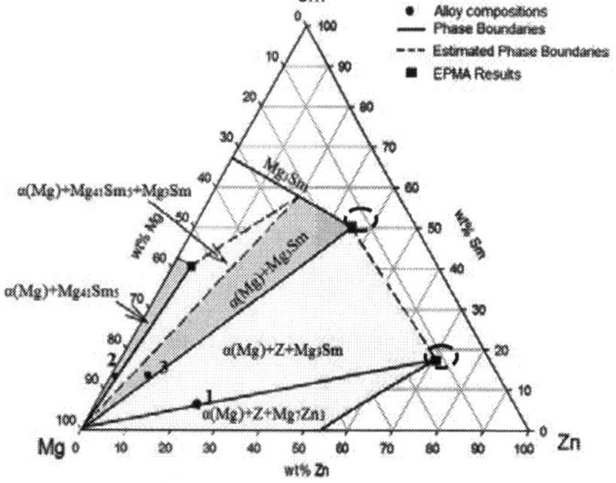

Fig.5. Isothermal section of Mg-Zn-Sm system in Mg corner at 400°C

Conclusions

1. The ternary phase Z was confirmed, with a modified formula of $Mg_3Zn_{10}Sm$ and hexagonal structure.

2. The solubility of Zn in binary phase $Mg_{41}Sm_5$ in 400°C is estimated as 4%. The crystal structure remains the same.

3. The X-phase identified by Drits [11] is experimentally confirmed to be the binary Mg_3Sm phase with the solubility of Zn around 35.6% along the Mg-Zn binary direction.

4. Three binary phase equilibriums α(Mg)+Z, α(Mg)+$Mg_{41}Sm_5$, α(Mg)+Mg_3Sm were identified and partial isothermal section of Mg-Zn-Sm at 400°C was constructed.

Acknowledgements

The authors wish to thank Anil Sachdev, Rick Waldo and Misle Tessema from General Motors Global R&D for helpful discussion and technical support, Shanghai Jiaotong University for sample preparation and the financial support from the National Science Foundation (No. 1005762).

References

1. M. Bamberger, G. Dehm, "Trends in the Development of New Mg Alloys", Annu. Rev. Mater. Res. 2008. 38, p. 505-33.

2. A.A.Luo, "Recent magnesium alloy development for elevated temperature applications", International materials reviews, 2004, P. 13-30.

3. M. O. Pekguleryuz, A.A. Kaya. In: K.U. Kainer, editor. Magnesium alloys and their Applications. Germany, Wiley-VCH Verlag GmbH, (2003), p. 74.

4. A. A. Luo, R. K. Mishra, etc, "Magnesium Alloy Development for Automotive Applications", Materials Science Forum, 2012, 706-709, p. 69-82.

5. C.J. Boehlert, "The tensile and creep behavior of Mg-Zn Alloys with and without Y and Zr as ternary elements", J Mater Sci, 2007, P. 3675-3684.

6. L. L. Rokhlin, Magnesium Alloys Containing Rare Earth Metals: Structure and Properties. CRC Press, 2003

7. I. P. Moreno, T. K. Nandy, etc, "Microstructure and Creep Behavior of a Die Cast Mgnesium-Rare Earth alloy", Magnesium Technology 2002, p. 111-116.

8. Rokhlin L. L, "PHASE DIAGRAM AND MECHANICAL PROPERTIES OF Mg-Sm ALLOYS", Russian Metallurgy, 1979, 4, p. 165-167.

9. L. L. Rokhlin, S.V. Dobatkin, etc, "Effect of Severe Plastic Deformation on the Structure and Properties of the Age-Hardenable Mg-Sm Alloys", Materials Science Forum, 2006, 503-504, p. 961-966.

10. N. P. Abrukina, "MICROHARDNESS OF PHASES IN THE SYSTEMS Mg-Sm-Y and Mg-Sm-Zn", Russian Metallurgy, 1987, 2, p. 210-212.

11. M. E. Drits, L. L. Rokhlin, etc, "PHASE EQUILIBRIUM IN THE Mg-Sm-Zn SYSTEM", Russian Metallurgy, 1985, 6, p. 194-200.

12. Kazumasa Sugiyama, etc, "Crystal structure of μ_7-MgZnSm", Journal of Alloys and Compounds, 1999, 285, p. 172-178.

13. E. Abe, H. Takakura, etc, "Hexagonal superstructures in the Zn-Mg-rare-earth alloys", Journal of Alloys and Compounds, 1999, 283, p. 169-172.

14. K. Sugiyama, K. Yasuda, etc, "The structure of hexagonal phases in Mg-Zn-RE (RE=Sm and Gd) alloys", Z. Kristallogy, 1998, 213, p. 537-543.

15. Pandat 8.0 – Phase Diagram Calculation Software for Multi-component Systems, Computherm LLC, 437 S. Yellowstone Dr, Suite217, Madison, WI, 53719, USA, 2008.

16. Guo. C, etc, Mg-Nd Phase Diagram, ASM Alloy Phase Diagrams Center, P. Villars, editor-in-chief; H. Okamoto and K. Cenzual, section editors; http://www1.asminternational.org/AsmEnterprise/APD, ASM International, Materials Park, OH, 2006

17. Okamoto H, etc, Mg-Sm Phase Diagram, ASM Alloy Phase Diagrams Center, P. Villars, editor-in-chief; H. Okamoto and K. Cenzual, section editors; http://www1.asminternational.org/AsmEnterprise/APD, ASM International, Materials Park, OH, 2006

Magnesium Technology 2013
Edited by: Norbert Hort, Suveen N. Mathaudhu, Neale R. Neelameggham, and Martyn Alderman
TMS (The Minerals, Metals & Materials Society), 2013

In situ synchrotron diffraction of the solidification of Mg-RE alloys

D. Tolnai, C.L. Mendis, A. Stark, G. Szakács, B. Wiese, K.U. Kainer, N. Hort

Helmholtz-Zentrum Geesthacht, Institute of Materials Research, Max-Planck Str. 1, D – 21502 Geesthacht, Germany

Keywords: Magnesium Rare Earth alloys, Solidification, In situ synchrotron diffraction

Abstract

Mg-RE alloys have a good potential to be used in bio-applications as degradable implants. Their macroscopic characteristics are strongly dependant on the microstructure, which can be tailored through the alloy composition and the solidification parameters. *In situ* synchrotron diffraction provides a unique tool to follow the phase formation, possible meta-stable and stable phase transformations and grain growth during cooling. In the present study Mg alloys containing Gd, Y and Nd were investigated to characterize the solidification phenomenon during cooling from 680°C to room temperature. Samples were melted and solidified in a Bähr 805 dilatometer modified for *in situ* synchrotron measurements. The molten samples were contained in steel crucibles, and the temperature was controlled by type S thermocouples during cooling. The results provide an experimental validation of thermodynamic calculations; and provide experimental input for refining the thermodynamic models, which contribute to the better understanding of the microstructure evolution to control desirable macroscopic characteristics.

Introduction

Based on their promising mechanical and corrosion properties Mg-RE based alloys are attractive for structural [1] and as well for biomaterial applications [2]. Y, Gd and Nd are important alloying elements for these types of materials. Gd has a maximum solubility of 23.49 wt% at 548 °C [3] and offers both solid solution strengthening and precipitation hardening. The Mg_5Gd, phase forms at the grain boundaries during casting of Mg-Gd alloys [3]. These particles have a size in the micrometer range. During solution heat treatment they dissolve until the maximum solubility at the given temperature. A T6 heat treatment leads to the precipitation of fine particles. The selected area diffraction investigations using transmission electron microscopy (TEM) indicate that these fine precipitates are the meta-stable phases β' and β" after peak hardness at 250°C. The β' phase is homogeneously distributed over the matrix, while β" is only observed heterogeneously at peak hardness. The β" phase is isomorphous with Mg and the precipitates are coherent. The lattice parameter of β' is twice that of Mg and has a c-based centred orthorhombic structure (a = 0.641 nm, b = 2.223 nm, c = 0.521 nm) [2].

Nd has a maximum solubility of ~ 3 wt % at 548 °C therefore the main strengthening mechanism is precipitation hardening by the $Mg_{12}Nd$ phase mainly at the grain boundaries. During solution heat treatment they spheroidize, while the T6 treatment leads to the precipitation of the β' Mg_3Nd phase. The β' has a fcc structure with a lattice parameter a ≈ 0.375 nm [4]. Mg-Y-Nd based alloys, such as WE43 and WE54 form an important class of cast magnesium alloys used in elevated temperature applications with superior mechanical properties at both room and elevated temperatures [5]. The WE54 and WE43 alloys are conventionally used in the T6 condition with the precipitation hardening response as well as the sequence and crystal structure of the resultant phases investigated before [6-9]. The intermetallic phases found in the as-cast Mg-5Y-2Nd (wt%) and Mg-4Y-2Nd (wt%) alloys have been characterized, with transmission electron microcopy to be a ternary phase with a face centered cubic crystal structure with a = 2.223 nm and a composition of $Mg_{14}Nd_2Y$ [10] or $Mg_{15}Nd_2Y$ [9].

However, no detailed studies on the microstructural evolution during solidification from the melt has been conducted to date even though majority of these alloys are used as heat treated sand or permanent mould castings and, as such, their microstructure is determined during solidification. The macroscopic mechanical properties of multiphase materials depend strongly on the chemical composition, volume fraction and spatial distribution of the internal phases [11] which are determined during solidification. Additionally, the cast microstructures influence structure-property relationships of the alloys during subsequent thermo-mechanical processing. Therefore the understanding of the sequence of formation and evolution of the internal phases during solidification is a prerequisite to achieve control of the microstructure of these alloys.

The thermodynamic databases and binary phase diagrams [3] provide information on the presumed solidification path. Based on this data, the resultant microstructure can be modeled [12]. The majority of the commercial Mg alloys are not binary, but complex alloys with more, than one alloying addition. For these materials the information in the simulation databases is limited. Thus the simulation of the experimentally observed microstructure and modeling of the solidification sequence is not feasible due to the lack of available data. To investigate solidification phenomenon experimentally, traditionally ex situ measurements were performed. The cooling is interrupted at different temperatures i.e. at different stages of the microstructure development. The quenched samples can be investigated *ex situ* by conventional metallographic methods in order to characterize the phases and to quantify microstructural parameters [13-15]. The advantage of this method is its relative simplicity, while its disadvantage is that it lacks time-temperature resolution of the investigated solidification process. The continuous development of the X-ray acquisition systems at synchrotron sources provides a unique tool to characterize the phase formation and evolution during the solidification *in situ*. Recording the diffraction patterns while cooling the system down, allows to identify the internal phases, to determine their solidification sequence [16,17] and to use these results as experimental validation of the existing thermodynamic databases and in certain cases contribute to the completion of them.

The present study reports results of *in situ* synchrotron diffraction studies conducted during solidification of Mg-RE alloys, and their comparison with the existing thermodynamic data on this system.

Experimental methods

The alloys Mg-15Gd and Mg-4Y-3Nd (wt %) were chosen for the experiments, as they are of particular interest in structural and in biomaterial applications. The microstructures of the alloys were examined with a Carl Zeiss Gemini Ultra 55 scanning electron microscope (SEM) operating at 15kV equipped with an EDAX energy dispersive X-ray spectrometer (EDXS).

For the diffraction experiments small samples taken from the materials in as-cast condition; were encapsulated in steel crucibles. The measurements were carried out at the HZG beamline HARWI II of the Hamburger Synchrotronstrahlungs-labor (HASYLAB) at the Deutsches Elektronen-Synchrotron (DESY). The experiments were performed inside the coil of a DIL 805A/D dilatometer (Bähr-Thermoanalyse GmbH, Hüllhorst, Germany). The experimental setup is shown in Figure 1.

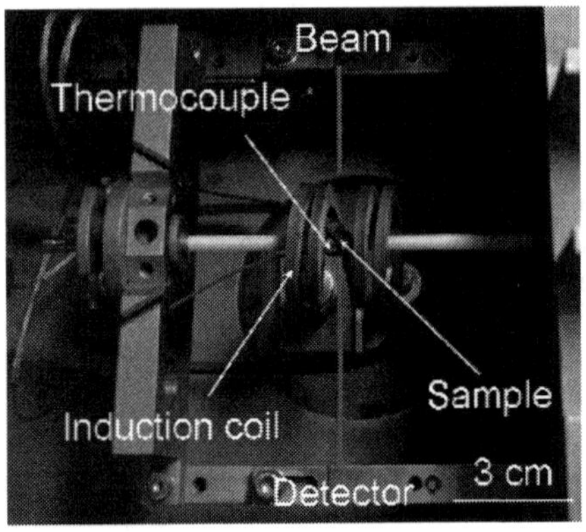

Figure 1. Experimental setup of the *in situ* solidification experiment inside the chamber of a Bähr 805A/D dilatometer modified for *in situ* synchrotron measurements. The windows on the sides are covered by Kapton foil, while the coil is opened, for the beam to pass through the sample.

During the tests the samples were held in sealed stainless steel crucibles in order to hold the molten metal and to isolate it from the surrounding atmosphere. The dilatometer has been modified to meet the requirements of *in situ* synchrotron measurements. The two windows on the sides covered by Kapton foils, prevent interference with the X-ray beam. The windings of the induction coil are opened so the beam only passes through the sample [18]. The temperature was regulated by a type S thermocouple welded on the crucible containing the sample. The heating and cooling was performed in Ar flow.

To melt the samples the system was heated up to 680 °C, held for 5 min was and then cooled to 180 °C at a cooling rate of 5 K/min for the Mg15Gd samples and of 10 K/min for the Mg4Y3Nd samples, respectively.
The *in situ* diffraction experiments were conducted in transmission geometry using a beam with a cross section of 1 × 1 mm^2. In order to penetrate the sample inside the crucibles, high-energy X-rays were used with a photon energy of 100 keV, corresponding to a wavelength of λ = 0.0124 nm. During the experiment the Debye-Scherrer diffraction rings were recorded in every 25 s (~ 2.5 K and ~ 5 K) by a mar555 flatpanel detector (Marresearch GmbH, Norderstedt, Germany) with an area of 3070×2560 pixels with a pixel size of 140 μm^2. Conventional diffraction patterns (line diagrams) were achieved by filtering the detector image to avoid the effect of hot pixels, followed by an azimuthal integration of the Debye-Scherrer rings.

Results

Microstructure

The secondary electron micrograph of Mg15Gd in as-cast condition is shown in Figure 2.

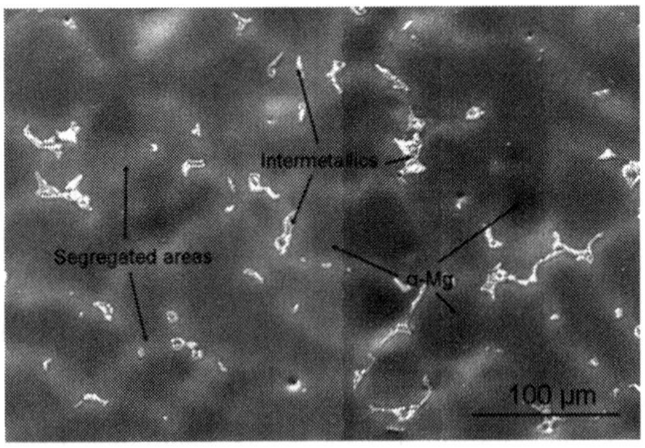

Figure 2 Secondary electron micrograph of the investigated Mg-15Gd alloy in as-cast condition.

The microstructure consists of α-Mg dendrites, segregated regions and intermetallic particles identified as Mg_5Gd using EDXS. The secondary electron micrograph of Mg-4Y-3Nd in as-cast condition is shown in Figure 3.

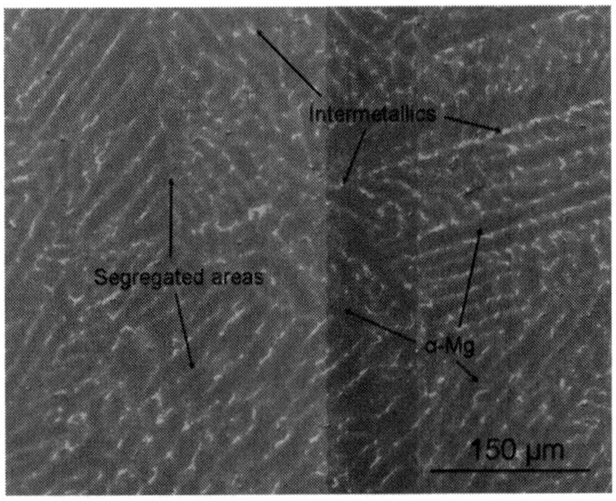

Figure 3 Secondary electron micrograph of the investigated Mg-4Y-3Nd alloy in as-cast condition.

The alloy contains α-Mg dendrites, segregated regions and intermetallic particles with Y and Nd, identified as; $Mg_{24}Y_5$-, $Mg_{12}Nd$-binary phases and a ternary phase $Mg_{14}Y_4Nd$ using EDXS.

In situ synchrotron diffraction

The line profiles resulting from the azimuthal integration of the 2D diffraction patterns acquired during solidification are shown in Figure 4 for Mg15 Gd and in Figure 5 for Mg4Y3Nd, respectively.

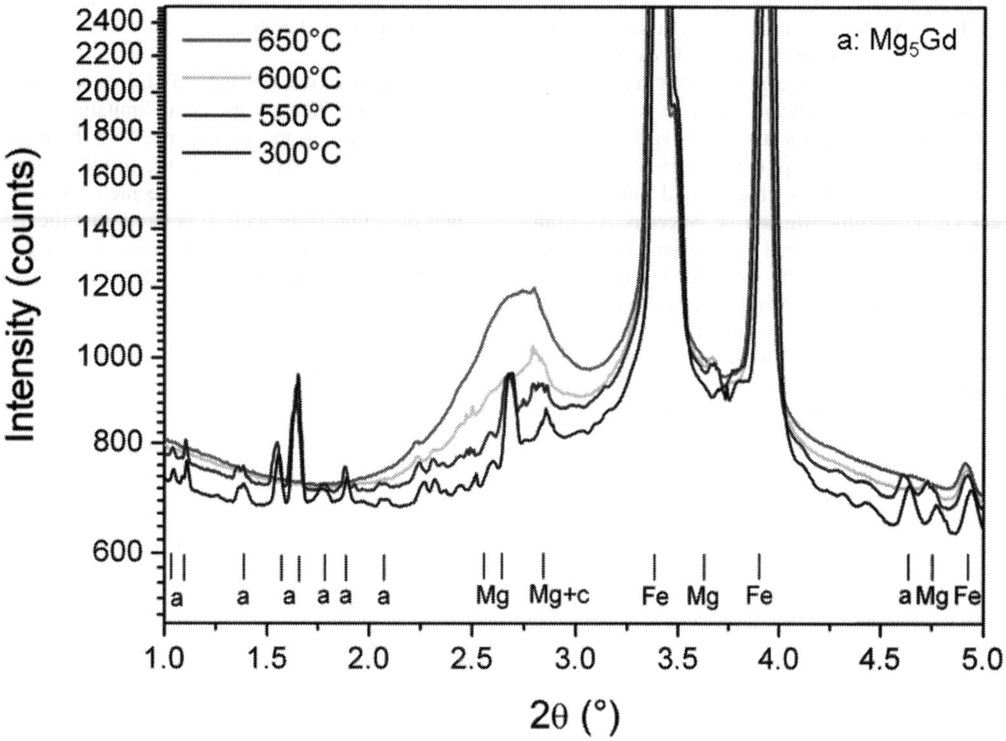

Figure 4 Sequence of XRD line diagrams obtained by the integration of the diffraction patterns recorded during solidification of Mg15Gd. (Wavelength $\lambda = 0.0124$ nm)

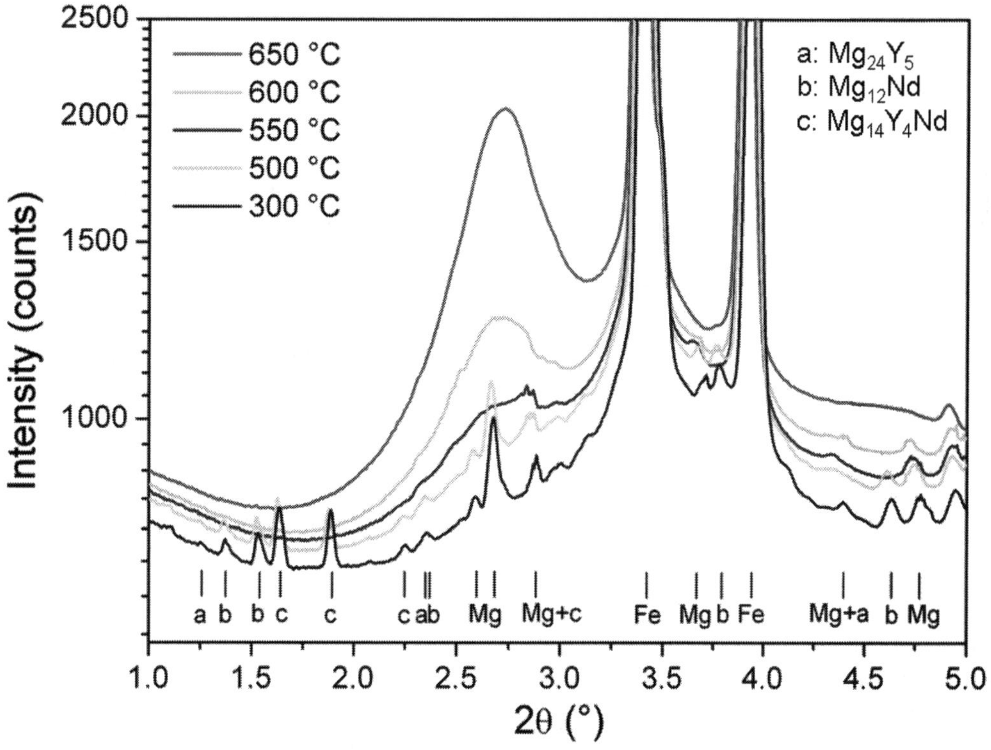

Figure 5 Sequence of XRD line diagrams obtained by the integration of the diffraction patterns recorded during solidification of Mg4Y3Nd. (Wavelength $\lambda = 0.0124$ nm)

From indexing the peaks in the line profiles the solidification sequence of the alloys can be determined with a temperature resolution of ~ 3 °C (5 °C in the case of Mg-4Y-3Nd). When the samples are molten only a diffuse background originating from the melt and the peaks of the steel crucibles are observed. In case of the Mg-15Gd (Figure 4) the solidification starts with the α-Mg grains at a measured temperature of 630°C. At a measured temperature of 550°C the formation of the Mg_5Gd intermetallic phase and the peaks of crystalline Mg can be detected. As from 550°C the cooling advances no additional peaks appear on the line profiles. In the case of Mg4Y3Nd the solidification also starts with the initiation and growth of the α-Mg dendrites at a measured temperature of 625 °C. As the cooling advances the dendrites are followed by the Nd containing intermetallic phases. The binary $Mg_{12}Nd$ and the ternary $Mg_{14}Y_4Nd$ phases are first observed at 545 °C. Finally, solidification of $Mg_{24}Y_5$ can be observed at 320 °C.

Discussion

The Mg-Gd binary phase diagram is shown in Figure 6.

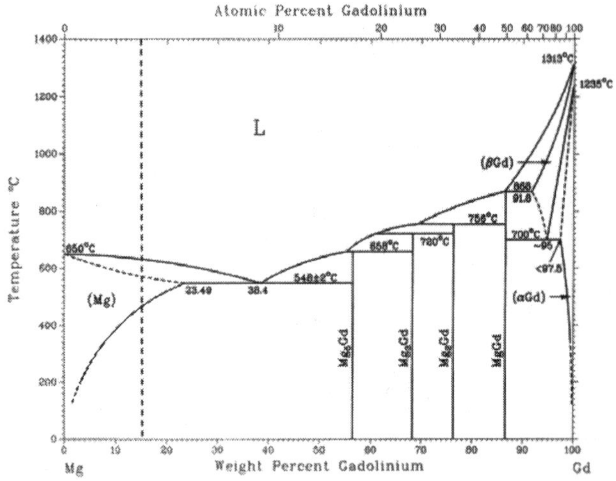

Figure 6. Mg-Gd binary phase diagram [3].

Comparing the phase diagram of Mg-Gd with the results from the *in situ* solidification tests, it can be seen that the experimental results correlate with the thermodynamic data from the literature. During cooling the system reaches its liquidus temperature at 627 °C and the formation of α-Mg grains start. On the diffraction pattern they can be barely detected as they are not in a fixed position but moving in the melt. As the cooling proceeds and the grains grow, their detection becomes more feasible. After further cooling; the system reaches its eutectic temperature (548 °C) where the solidification of Mg_5Gd takes place.

The thermodynamic simulation of the solidification of Mg-Y-Nd system was carried out with the software Pandat 8.1. The resulting section of the ternary phase diagram at an Nd content of 3 wt% is shown in Figure 7. The vertical line corresponds to a Y content of 4 wt%. According to the thermodynamic database the α-Mg dendrites initiate at 639 °C. They are followed by the $Mg_{12}Nd$ phase at 538 °C, while the solidification ends with the $Mg_{24}Y_5$ phase at 316 °C. The formation of the ternary $Mg_{14}Y_4Nd$ phase is not predicted by the thermodynamic software.

The determination of the solidification sequence by *in situ* synchrotron diffraction has some limitations. A compromise must be made between the cooling rate and the acquisition time. The crystallographic orientation of the solidifying grains plays a vital role. As these grains float in the melt without fixed orientation their detection is difficult. It is likely that the α-Mg grains can be detected properly for the first time as the system reaches the dendrites coherency point, the temperature where a continuous dendrite skeleton is built up.

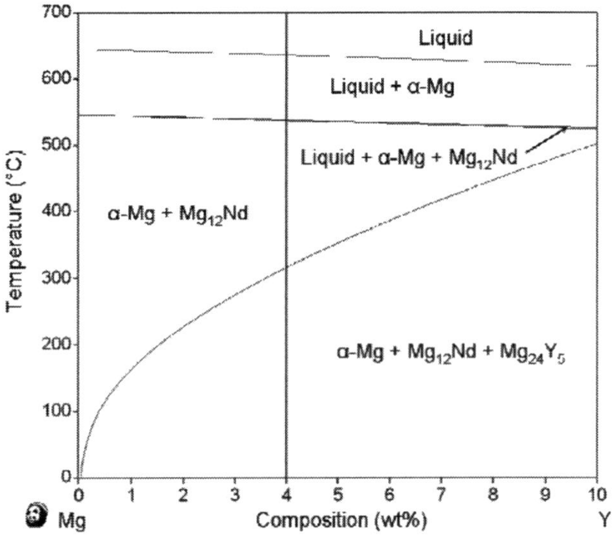

Figure 7 Section of the ternary Mg-Y-Nd phase diagram at a Nd content of 3 wt%, calculated by Pandat 8.1.

Despite these limitations, the onset temperatures, obtained experimentally by *in situ* diffraction and that predicted theoretically from the thermodynamic calculations, for the solidification of different phases correlate reasonably well. The ternary intermetallic phase $Mg_{14}Y_4Nd$ was only shown experimentally and not predicted by the thermodynamic simulations due to limited information available.

Conclusions

In situ solidification experiments were performed with a Bähr 805 A/D dilatometer, on Mg-15Gd and Mg-4Y-3Nd alloys by synchrotron diffraction, for the first time. From the results the following conclusions can be drawn:

- In the case of Mg-15Gd, first α-Mg solidifies at a measured temperature of 530 °C, followed by the solidification of α-Mg/Mg_5Gd at a measured temperature of 550 °C.
- In the case of Mg-4Y-3Nd the solidification starts with α-Mg which solidifies at a temperature of 625 °C followed by the solidification of the intermetallic particles; $Mg_{12}Nd$ and $Mg_{14}Y_4Nd$ at 545 °C and finally the $Mg_{24}Y_5$ phase at a temperature of 320 °C.

- The results for both of the alloys are in correlation with the data from thermodynamic databases. Furthermore *in situ* synchrotron diffraction reveals the presence of the ternary $Mg_{14}Y_4Nd$ phase and the onset of its solidification at 545 °C which could not be predicted by the simulations.

Acknowledgements

The authors would like to thank the Deutsches Elektronen-Synchrotron the provision of synchrotron radiation facilities in the framework of the proposal II-20100225.

References

[1] L.L. Rokhlin, *Magnesium Alloys Containing Rare Earth Metals, Structure and Properties*, Taylor & Francis, (2003).

[2] N. Hort et al., „Magnesium alloys as implant materials – Principles of property design for Mg–RE alloys" *Acta Biomaterialia* 6 (2010) 1714-1725.

[3] T.B. Massalski, *Binary Alloy Phase Diagrams,* ASM International Materials Park (1990).

[4] L.Y. Wei, G.L. Dunlop and H. Westengen, „Age hardening and precipitation in a cast magnesium - Rare-earth alloy" *Journal of Materials Science* 31 (1996) 387-397.

[5] G. Barucca et al., „Formation and evolution of the hardening precipitates in a Mg–Y–Nd alloy" *Acta Materialia* 59 (2011) 4151-4158.

[6] C. Antion et al., „Hardening precipitation in a Mg–4Y–3RE alloy" *Acta Materialia* 51(2003) 5335-5348.

[7] J.F. Nie and B.C. Muddle, „Characterisation of strengthening precipitate phases in a Mg–Y–Nd alloy" *Acta Materialia* 48 (2000) 1691-1703.

[8] P.J. Apps et al., „Phase compositions in magnesium-rare earth alloys containing yttrium, gadolinium or dysprosium" *Scripta Materialia* 48 (2003) 1023-1027.

[9] P.J. Apps et al., „Precipitation reactions in Magnesium-rare earth alloys containing Yttrium, Gadolinium or Dysprosium" *Scripta Materialia* 48 (2003) 475-479.

[10] H. Karimzadeh Ph.D Thesis Manchester University United Kingdom (1985).

[11] G. Requena et al., „The effect of the connectivity of rigid phases on strength of Al-Si alloys" *Advanced Engineering Materials* 13 (2011) 674-684.

[12] W.J. Boettinger et al., „Phase-field simulation of solidification *Annual Review of Materials Science,* 32 (2002) 163-194.

[13] Y. L. Liu, S and B. Kang, „The solidification process of Al-Mg-Si alloys *Journal of Materials Science* 32 (1997) 1443-1447.

[14] H. Jones, „Some effects of solidification kinetics on microstructure formation in aluminium-base alloys" *Materials Science and Engineering A* 413-414 (2005) 165-173.

[15] N. Iqba let al., „Nucleation kinetics during the solidification of aluminum alloys" *Journal of Non-Crystalline Solids* 353 (2007) 3640-3643.

[16] S.S. Babu et al., „Time-resolved X-ray diffraction investigation of primary weld solidification in Fe-C-Al-Mn steel welds" *Acta Materialia* 50 (2002) 4763-4781.

[17] O. Shuleshova et al., „*In situ* observations of solidification in Ti–Al–Ta alloys by synchrotron radiation" *Intermetallics* 19 (2011) 688-692.

[18] P. Staron et al., „In situ experiments with synchrotron high-energy X-rays and neutrons" *Advanced Engineering Materials* 13 (2011) 685-689.

Magnesium Technology 2013
Edited by: Norbert Hort, Suveen N. Mathaudhu, Neale R. Neelameggham, and Martyn Alderman
TMS (The Minerals, Metals & Materials Society), 2013

NUCLEATION KINETICS OF THE γ-PHASE IN A BINARY MG-AL ALLOY

Mehdi Lalpoor[1,2], J.S. Dzwonczyk[1,2], N. Hort[3], S.E. Offerman[2]
[1]Materials innovation institute, Mekelweg 2, 2628 CD Delft, The Netherlands
[2]Delft University of Technology, Department of Materials Science and Engineering, Mekelweg 2, 2628 CD Delft, The Netherlands
[3]Magnesium Innovation Center, Helmholtz Zentrum Geesthacht, Max-Planck-Straße 1
21502 Geesthacht, Germany

Keywords: Mg-Al alloys, precipitation, Nucleation kinetics

Abstract

The binary magnesium-aluminum system has been studied extensively the last decades. Surprisingly, the kinetics of nucleation of the γ-precipitates ($Mg_{17}Al_{12}$) in the α-matrix have not been determined experimentally in detail. The main goal of this research is to investigate the precipitation kinetics of the γ-phase in a Mg-Al alloy during annealing, and to understand the relation between the precipitate volume fraction and the mechanical properties of the material. Comprehensive micro-structural analysis was undertaken to reveal the number density, volume fraction, and morphological changes of the γ-precipitates after solution heat treatment (homogenization) and during subsequent annealing below the solvus temperature. As a result, the incubation time and steady state nucleation rate of γ-precipitates in the α-matrix were determined by fitting the experimental data to the classical nucleation theory.

Introduction

The burst in the global population in the next few decades has made new concerns regarding the materials consumption and environmental issues[1]. Replacement of regular alloys such as aluminum and steel in transportation means by lighter ones such as magnesium helps reduce the fuel consumption and CO_2 emission. Magnesium is a light metal which is readily available and has excellent castability [2], but its low yield strength, yield strength anisotropy and poor room temperature formability [3-5]have resulted in the development of Mg-alloys chiefly as cast products. The high strength of cast products obtained by solid solution and precipitation hardening is mainly due to the addition of rare earth metals (RE's) [6]. The similar electrochemical potential of RE's to Mg helps improve the corrosion resistance, and the high melting point precipitates formed by them increase the creep resistance, strength and ductility of Mg-alloys [7].However, the rare earth alloying elements that provide magnesium alloys with high strength, high ductility, creep and corrosion resistance should be avoided in the near future due to their high price and risk of supply disruption[8]. Therefore, in order to gain high strength and high ductility in Mg-alloys in the absence of RE elements other strengthening mechanisms must be applied while using abundantly available elements. Severe plastic deformation may be used to modify the texture by rearrangement of the distribution of basal planes [9] in order to improve the ductility, since the critical resolved shear stress (CRSS) for basal plane slip in Mg is ~100 times lower than that for non-basal plane slip. The grain size reduction is also an efficient strengthening mechanism in Mg alloys because the Hall-Petch coefficient for magnesium is reported to be approximately 0.7 MP $m^{-1/2}$ for coarse grains (30-87 μm) and 0.13 MPa $m^{-1/2}$ for fine grains (17-30 μm) [10], which is significantly higher compared to aluminum alloys (0.07 MPa $m^{-1/2}$) [11, 12]. Precipitation hardening, which is widely practiced in aluminum alloys, would be another potential strengthening mechanism, however, not so efficient for magnesium alloys due to their low precipitation hardening response. It has been long known that age-hardening response of Mg alloys is poor compared with many age-hardenable aluminum alloys.

Among the abundantly available elements, Al has the highest solubility in Mg at elevated temperatures (12.6 wt.% at 437 °C), which falls to about 2% at room temperature [13]. In spite of this, Al fails to act as an efficient precipitation hardenable element in Mg. The main reason for the poor age-hardening response of Mg-Al based alloys has been attributed to the coarseness and unfavorable orientation of the continuous precipitates with respect to the basal plane of Mg, which makes them inefficient obstacles to dislocation movement [14-16]. The age hardening response of magnesium alloys may be enhanced by promoting precipitates on the most effective planes for strengthening. However, the promotion of prismatic or pyramidal precipitates in magnesium alloys is difficult since the factors that contribute to the choice of precipitates morphology are not fully understood [17]. The number density of precipitates may be enhanced by increasing the number of potential nucleation sites, which would result in increased precipitation hardening. While such technical solutions are practically applied to stimulate the aging response of Mg-Al alloys, a thorough understanding of the nucleation mechanism sounds essential. The overall kinetics of precipitation of some Mg-Al alloys have been investigated [18, 19], however the nucleation kinetics of such alloys are not studied yet. In this research work the nucleation kinetics of a Mg-7 wt.% Al alloy are studied with the aim to understand the slow precipitation response of such alloys during heat treatment. Using the classical nucleation theory, the incubation time and steady state nucleation rate are assessed for the first time.

Experimental procedure

The material used in this project was a Mg-7 wt.% Al alloy produced by permanent mold casting in the form of a 100 mm diameter cylinder. The as-cast microstructure consisted of coarse γ-$Mg_{17}Al_{12}$ constituents in an α-Mg matrix. After cutting, a big piece of alloy was encapsulated in a quartz tube filled with argon and sealed in order to prevent oxidation at elevated temperatures. The solution heat treatment was performed at 425 °C for 160 hours and the XRF as well as EPMA experiments proved that Al is homogenously distributed in the matrix. After the end of solution heat treatment the tube was broken and the sample was quenched in water. The precipitation hardening was performed at 260 °C for various times namely, 10 and 30 min, 1, 2, 3, 5, 10, 12, 15 and 24 h. The samples were enclosed in steel envelopes flushed with argon for protection against oxidation and they were immediately quenched in water afterwards. In order to reveal the microstructure and quantify the number density as well as volume fraction of the precipitates, the samples were mechanically ground/polished and then etched in 1% Nital solution for 1 minute. Due to the importance of statistical reliability of data, the scanning electron microscopy (SEM) was used to be able to cover a large area including multiple grains. Image analysis was performed on the back scattered electron images to determine the number density and volume fraction of the precipitates. The areas covered per image were 16×21 μm^2 for samples annealed for short times, 61×81 μm^2 for 5 h to 12 h annealing, and 93×123 μm^2 for samples annealed longer than 12 h. 30 images were analyzed to determine the number density and 20 images for volume fraction of the precipitates, and the average values are reported here.

The micro-hardness tests were performed using the Struers-Dura Scan 70 machine with an indentation load of 100 g. 40 indentations were done for each case in grids of 4×10 points 200 μm away from each other, and the average values were calculated. Thus, an area of 1.1 mm^2 was covered in order to assure the statistical reliability of the data.

Results and discussion

Fig. 1 shows the evolution of microstructure in the Mg-7 wt% Al alloy after various heat treatment conditions. As can be seen in Fig. 1a, some precipitates have formed right after quenching of the solution heat-treated sample. Our observations indicated that there is no noticeable change in the number as well as volume fraction of the γ-precipitates during the early stages of precipitation hardening treatment, i.e. from t = 0 min to t = 3 h. It is only after 5 hours annealing that the precipitates start showing up on the microstructure (Fig. 1b). From this moment on, the volume fraction and number density of the precipitates increase drastically up to 12 h annealing (Fig. 1c). After 12 h annealing however, the number density of the precipitates start to fall sharply due to coarsening while the volume fraction remains constant (Fig. 1d). The precipitates have a lath shape and grow both in thickness and length with increasing annealing time as observed from the SEM images.

Fig. 2a shows the changes in the volume fraction of γ-precipitates with aging time at 260 °C. As explained before, the volume fraction of the precipitates starts increasing only after 5 h (18,000 s) annealing and gains its equilibrium value predicted by the phase diagram (4.96%) after 12 h (43,200 s). By applying the lever rule (Eq. 1) to the volume fractions measured through image analysis, the aluminum weight fraction of the matrix as a function of annealing time was calculated [20] (Fig. 2b).

$$X_{Al}^{\alpha-Mg}(t) = \frac{X_0 - f^{\gamma}(t) \cdot {}^{e}X_{Al}^{\gamma}}{1 - f^{\gamma}(t)} \qquad (1)$$

Where, the $f^{\gamma}(t)$ is the volume fraction of the precipitates, $X_0 = 0.07$ and ${}^{e}X_{Al}^{\gamma} = 0.44$. The driving force for nucleation was calculated by Thermo-Calc for various aluminum concentrations of the matrix starting from a supersaturated solid solution up to the end of transformation where the equilibrium aluminum concentration corresponding to 260 °C is reached (~ 4.68 wt.%). Knowing the changes in aluminum concentration of the matrix with time (Fig. 2b), the time dependence of the driving force for nucleation was determined correspondingly and the results are presented in Fig. 2c. As can be seen, the driving force is initially high (where the aluminum concentration of matrix is high) and it decreases as the precipitation proceeds.

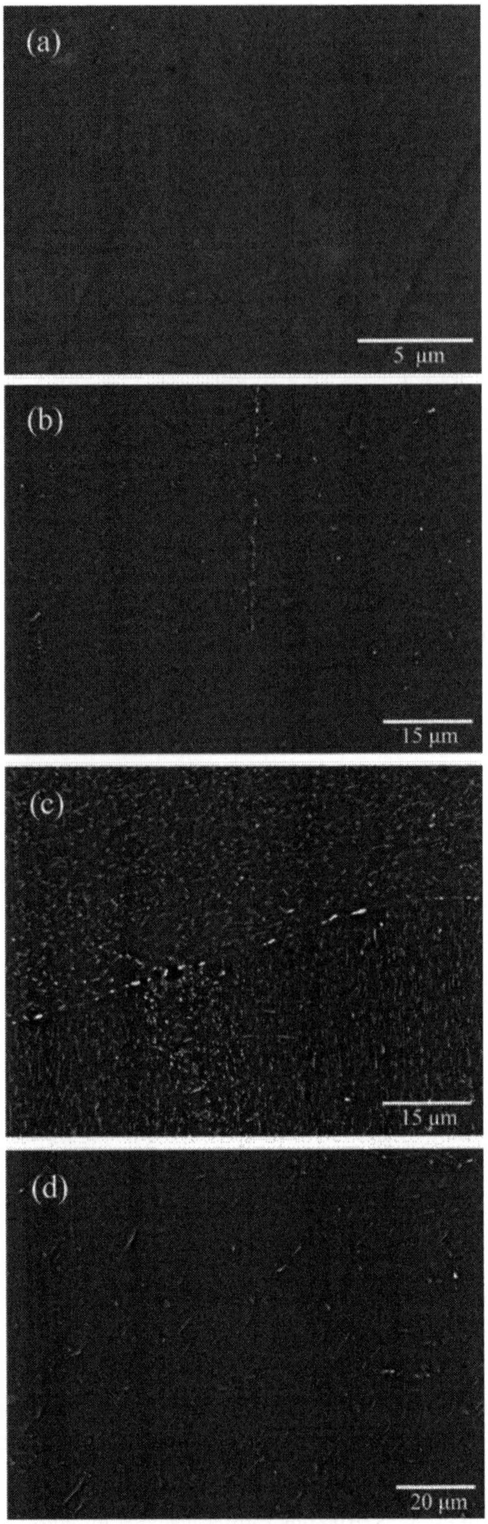

Figure 1. Evolution of microstructure of the Mg-7 wt.% Al alloy under various heat treatment conditions: a) solution heat treated at 425 °C and quenched, b, c and d) precipitation hardened at 260 °C for 5, 12 and 15 hours respectively (mind the different magnifications).

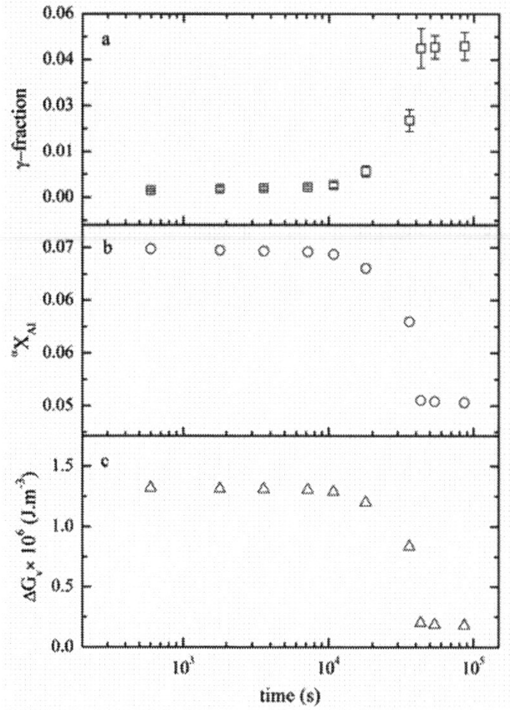

Figure 2. The results of quantitative metallography and thermodynamic calculations: a) volume fraction of the γ-precipitates ($Mg_{17}Al_{12}$) measured from SEM images, b) the aluminum weight fraction of the α-Mg matrix versus annealing time calculated using Eq. 1, and c) the driving force for nucleation; ΔG_v, versus annealing time (calculated by Thermo-Calc).

Fig. 3a shows the number density of precipitates measured form the SEM images. The number density remains very low up to 5 h (18,000 s) and then increases sharply until 12h (43,200 s) annealing after which it drastically falls due to the coarsening of the precipitates. The number densities of the precipitates in this alloy are an order of magnitude smaller compared to the ones reported for AZ91 at 250 °C [14]. The higher concentration of the precipitates in AZ91 may be on the one hand attributed to the higher Al content of the matrix which enhances the chemical driving force of nucleation. The addition of Zn on the other hand, reduces the solid solubility of aluminum in magnesium and increases the amount of precipitate phase formed upon aging [14]. In Fig. 3b, the nucleation rate has been calculated per untransformed fraction of the matrix (1-f^{γ}). The large errors at small annealing times are due to the small time steps and tend to decrease with increasing the aging time. As can be seen, the nucleation passes through a relatively long incubation period after which the rate increases dramatically up to 12 h (43,200 s). Afterwards, due to very rapid coarsening the nucleation rate turns to negative, which is not presented in the graph. In order to quantify the nucleation kinetics, the classical nucleation theory was applied, as our observations indicated that the nucleation mechanism is homogenous and occurs predominantly within the grains.

Having fitted the classical nucleation theory (Eq. 2) to the experimental results gained in this study, it turned out that the results barely fit to the initial increasing exponential part of the curve ($\exp\left(-\frac{\tau}{t}\right)$), and the other part which comprises the decrease of the nucleation rate due to driving force reduction ($\exp\left(-\frac{\psi/\left(\Delta G^*\right)^2+Q_d}{kT}\right)$) does not appear owing to the immediate coarsening that follows the nucleation rate peak.

$$\dot{N}_u=\dot{N}_{u,ss}\exp\left(-\frac{\tau}{t}\right)=N\cdot\beta^*\cdot Z\cdot\exp\left(-\frac{\psi/\left(\Delta G^*\right)^2+Q_d}{kT}\right)\cdot\exp\left(-\frac{\tau}{t}\right)\qquad(2)$$

In Eq. 2, N is the density of the potential nucleation sites, β^* is the frequency factor, or rate at which single solute atoms are added to the critical nucleus, Z is the Zeldovichnon-equilibrium factor, ψ contains information on the interfacial energy, ΔG^* is the free energy of activation for the formation of the critical nucleus, Q_d the activation energy for diffusion of Mg in Al, k Boltzmann constant, T absolute temperature, τ incubation time, and t the annealing time.
The fact that the entire classical nucleation equation could not be fitted to the experimental results (shown in Fig. 2b) inhibited us from direct assessment of ψ, and surface energy of the precipitates accordingly. However, by fitting the $\dot{N}_u=\dot{N}_{u,ss}\exp(-\tau/t)$ curve to the experimental data (Fig. 3b), one gets the steady state nucleation rate $\dot{N}_{u,ss}$=1.6×10^{14} ± 2.7×10^{13} m^{-3}s^{-1} and the incubation timeτ = 74210.1 ± 6438.1 s as the fitting parameters. Selecting lower aging temperatures may help prevent rapid coarsening, but the longer incubation periods expected at such low temperatures makes the aging time required to reach the steady state nucleation rather long [14].

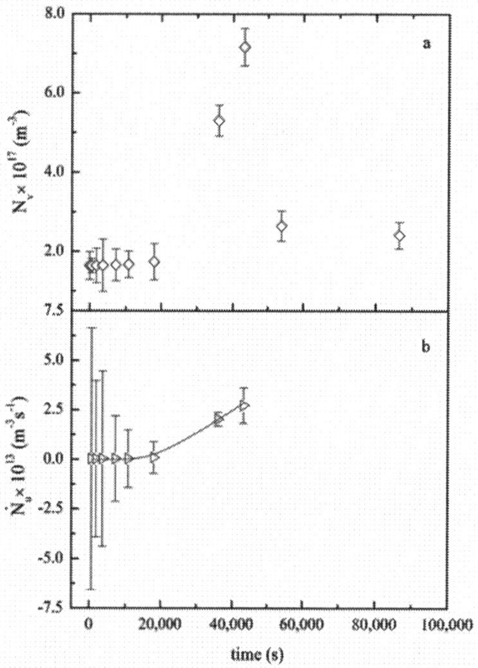

Figure 3. Evolution of precipitates number density and nucleation rate in the course of precipitation hardening: a) the number of precipitates per unit volume plotted versus annealing time in the Mg-7 wt.% Al alloy, b) the nucleation rate up to 12 hours annealing (due to rapid coarsening after 12 h, the nucleation rate is not plotted). The solid line is a fit of $\dot{N}_u=\dot{N}_{u,ss}\exp(-\tau/t)$.

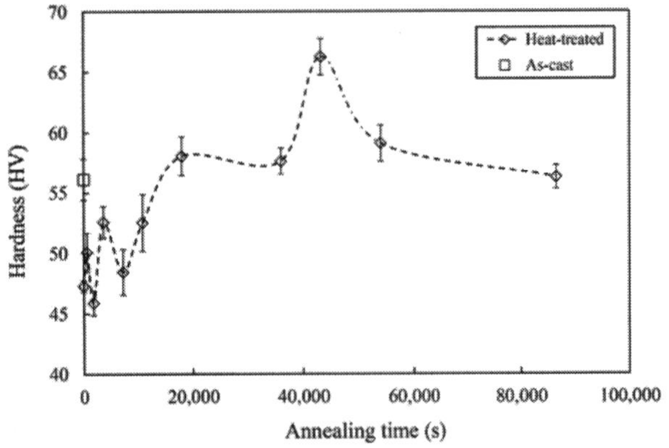

Figure 4. The results of micro-hardness tests on Mg-7 wt.% Al alloy in as-cast and after various heat treatment conditions.

The age hardening response of the Mg-7 wt.% Al can be seen from Fig. 4. The hardness of the alloy in the as-cast condition is 56.1 ± 1.7 HV, but after solution heat treatment and quenching it falls to 47.3 ± 0.5 HV. The hardness values fluctuate from t = 0 up to 3 h (10,800 s) annealing and then reach a plateau at 5 h (18,000 s). The hardness peak (66.2 ± 1.5 HV) is reached after 12 h (43,200 s), which is in agreement with our number density and volume fraction measurements. After 12 h annealing when coarsening becomes dominant, the hardness falls and eventually reaches 56.3 ± 1.0 HV at t = 24 h (86,400 s). The peak hardness value is only 18% higher than the as-cast condition and 40% higher than the solution heat treated state indicating that under normal conditions the aging response of the material is rather low. Another interesting observation in this study was the stimulated nucleation of precipitates on the deformation bands formed during the sample preparation prior to aging treatment (Fig. 5; the precipitates formed on deformation bands are not included in the number density and volume fraction measurements). In the corresponding sample (aged at 260 °C for 5 h), while the number density of the precipitate is slightly higher compared to the sample aged for 3 h, the deformation bands have acted as potential nucleation sites and apparently reduced the energy barrier for the nucleation. This observation strengthens the idea that the plastic deformation prior to aging may be used as an effective tool for improvement of the aging response in magnesium alloys.

Figure 5. Microstructure of the Mg-7 wt.% Al alloy after 5 h annealing at 260 °C. The role of deformation bands in stimulating the nucleation is indicated by arrows.

Conclusions

The nucleation of the $Mg_{17}Al_{12}$ precipitates in the Mg-7 wt% Al alloy remains in the transient state and coarsening occurs before reaching the steady state regime, which makes the direct assessment of the precipitate's interfacial energy from the entire nucleation curve impossible. While the nucleation rate is rather low during the incubation period, any external parameter may stimulate the nucleation. The deformation bands accidentally formed during cutting of the samples before annealing acted as efficient nucleation sites. This observation is a proof that deformation prior to precipitation hardening may help improve the aging response of Mg-alloys.

Acknowledgments

This research was carried out under the project number M42.5.10395 in the framework of the Research Program of the Materials innovation institute M2i (www.m2i.nl).

References

[1] R. Dobbs, J. Oppenheim, F. Thompson, M. Brinkman, M. Zornes, in, McKinsey Global Institute, 2011.
[2] K. Hono, C.L. Mendis, T.T. Sasaki, K. Oh-ishi, Scripta Materialia 63 (2010) 710–715.
[3] I.J. Polmear, Light Alloys; From Traditional Alloys to Nanocrystals, Elsevier Butterworth-Heinemann, Oxford, 2006.
[4] J. Koike, T. Kobayashi, T. Mukai, H. Watanabe, M. Suzuki, K. Maruyama, K. Higashi, Acta Materialia, 51 (2003) 2055–2065.
[5] J. Bohlen, P. Dobron, E.M. Garcia, F. Chmelik, P. Lukac, D. Letzig, K.U. Kainer, Advanced Engineering Materials, 8 (2006) 422-427.
[6] I.J. Polmear, Materials Science and Technology January 1994 Vol. 10 1, 10 (1994) 1-16.
[7] S.K. Das, L.A. Davis, Materials Science and Engineering, 98 (1988) 1-12.
[8] S.W. Xu, K. Oh-ishi, S. Kamado, F. Uchida, T. Homma, K. Hono, Scripta Materialia 65 (2011) 269–272.
[9] W.J. Kim, H.G. Jeong, H.T. Jeong, Scripta Materialia 61 (2009) 1040–1043.
[10] J. Koike, Metallurgical and Materials Transactions A, 36 (2005) 1689-1696.
[11] T. Mukai, K. Higashi, Scripta Materialia, 44 (2001) 1493–1496.
[12] J.W. Wyrzykowski, M.W. Grabski, Philosophical Magazine A, 53 (1986) 505-520.
[13] H. Okamoto, Desk handbook phase diagrams for binary alloys, ASM International, Materials Park, OH 44073-0002, 2000.
[14] S. Celotto, Acta Materialia, 48 (2000) 1775-1787.
[15] A.F. Crawley, B. Lagowski, Metallurgical Transactions 5(1974) 949-951.
[16] A.F. Crawley, K.S. Milliken, Acta Metallurgica, 22 (1974) 557-562.
[17] C.L. Mendis, K. Oh-ishi, Y. Kawamura, T. Honma, S. Kamado, K. Hono, Acta Materialia, 57 (2009) 749–760.
[18] D. Duly, Y. Brechet, B. Chenal, Acta Metall. Mater. , 40 (1992) 2289-2300.
[19] B.A. Esgandari, H. Mehrjoo, B. Nami, S.M. Miresmaeili, Materials Science and Engineering A 528 (2011) 5018–5024.
[20] S.E. Offerman, H. Strandlund, N.H.v. Dijk, J. Sietsma, E.M. Lauridsen, L. Margulies, H.F. Poulsen, J. Ågren, S.v.d. Zwaag, Materials Science Forum 550 (2007) 357-362.

Magnesium Technology 2013

Solidification

Magnesium Technology 2013
Edited by: Norbert Hort, Suveen N. Mathaudhu, Neale R. Neelameggham, and Martyn Alderman
TMS (The Minerals, Metals & Materials Society), 2013

COMPUTATIONAL MULTI-SCALE MODELING OF THE MICROSTRUCTURE AND SEGREGATION OF CAST Mg ALLOYS AT LOW SUPERHEAT

Laurentiu Nastac and Nagy El-Kaddah
Department of Metallurgical Engineering, University of Alabama, Tuscaloosa, Alabama 35487 USA

Keywords: Casting Mg alloys, Microstructure evolution, Stochastic mesoscopic modeling

Abstract

It is well known that casting at low superheat has a strong influence on the solidification structures of the cast alloy. Recent studies on casting magnesium AZ alloys at low superheat using the Magnetic Suspension Melting (MSM) process have shown that the cast alloy exhibit a fine globular grain structure, and the grain size depend on the cooling rate. This paper describes a stochastic mesoscopic model for predicting the grain structure and segregation in cast alloys at low superheat. This model was applied to predict the globular solidification morphology and solute redistribution of Al in cast Mg AZ31B alloy at different cooling rates. The predictions were found to be in good agreement with the observed grain structure and Al segregation. This makes the model a very useful tool for optimizing the solidification structure of cast magnesium alloys.

Introduction

It is well established that the grain size, and solute segregation and precipitation of intermetallic phases within the grain have significant influence on mechanical properties of the cast alloys [1]. For typical cooling rates of sand and die castings -- in the range of 0.1-100°C/s -- the grain structure is dendritic, with segregation of the alloying elements in the matrix between the secondary dendrite arms [2, 3]. The detrimental effect of segregation on mechanical properties is generally minimized by reducing the grain size of as-cast alloy through either grain refinement and/or casting at high cooling rates [2-4].

Much of research efforts to minimize the effects of segregation and intermetallic precipitates on the mechanical properties have been focused on grain refinement of the alloys. The addition of various grain refiners such as carbon and zirconium on the final grain structure has been extensively investigated [5-7]. These studies showed that grain refiners significantly reduce the grain size, and retain the dendritic solidification morphology of cast alloy which is generally undesirable as it cause entrapment of aluminum rich secondary-α phase between secondary dendrite arms [7,8].

Casting at low superheat is another method for grain refinement of cast alloys. This method not only reduces the grain size, but also changes the solidification morphology from a dendritic to a globular structure. The idea behind this technique is to suppress the formation of the columnar zone by reducing temperature gradient at the solidification front and to significantly reduce the growth rate to achieve plane front solidification. Casting at low superheat was successfully applied to grain refine superalloys [9-11], Al-Li 2090 and 8090 alloys [12,13] and more recently to Mg AZ31B alloy [14-16]. Superalloys studies were carried out using Microcast X process [9], while grain refinement of Al and Mg alloys was investigated using the Magnetic Suspension Melting (MSM) process [17]. The average grain size for unidirectionally-solidified Al-Li 2090 and 8090 alloys using a stainless steel bottom chill mold was about 30 μm, while that for Mg AZ31B alloy was approximately 80 μm. Increasing the cooling rate by using a copper bottom chill mold was found to further decrease the grain size of the Mg alloy [16]. These studies have also shown that segregation of secondary and intermetallic phases did not occur within the grain, but rather only at the grain boundaries.

Recently, simulation of microstructure evolution during solidification has become a powerful tool for predicting the grain structure and segregation in cast alloys. Phase Field [18], Monte Carlo [19], Cellular Automaton (CA) [19-21], Stochastic Modeling [19-22], and Voronoi Tessellation [23] are the most common methods for simulating microstructure evolution during solidification. This paper describes a stochastic globular solidification model for cast alloys at low superheat and critically examines the model by comparing theoretical predictions of grain structure and segregation of cast Mg AZ31B alloy at low superheat against experimental measurements.

The stochastic globular solidification model

This section presents the formulation for a stochastic model for globular solidification. This model is based on previously developed models for dendritic solidification [19-22]. The presented globular solidification model takes into account the nucleation and growth of hexagonal crystals.

Model formulation

A stochastic mesoscopic model for grain structure evolution includes nucleation and growth kinetics, as well as the growth anisotropy and grain selection mechanisms. The geometry of the solidified material is first enmeshed in macro-volume elements for which energy, mass, and momentum transport can be calculated by a classical deterministic methods. Then, each macro-volume is subdivided into a number of cubic micro-volume elements that have a state index associated with them. In the time-step calculation, the average temperature of the specimen is calculated from an energy balance. When the average temperature is lower than the equilibrium (or nucleation) temperature, the nucleation and growth of grains begin. The structure of the stochastic micro-model consists of a regular network of microcells that resembles the macro-volume element.

The mesoscopic model is characterized by: (a) geometry of the microcell, (b) state of the microcell, (c) neighborhood configuration, and (d) transition rules that determine the state of the microcell. The geometry of microcell is chosen to be cubic in 3-D and square in 2-D calculations. Each microcell can have only two possible states: either liquid or solid. The ratio between the number of solid microcells and the total number of microcells represents the fraction of solid within the macro-volume element. The neighborhood configurations are based on the first- and second-order nearest neighbor microcell. The probabilistic selection of these configurations is based upon the crystallographic anisotropy of grain growth. More details about the selection of the neighborhood configurations are shown in references [19, 21]. Solidification behavior depends to a great extent on the transition rules. In the present model, the change of state of the microcell from liquid to solid is initiated either by either nucleation or growth kinetics. The models for the description of nucleation and growth kinetics in the microcells are initiated based on the local solidification start times, t_L, which are computed by interpolating the local solidification start times of the macro-volume elements.

Growth anisotropy and probabilistic selection of neighbourhood configuration

At the beginning of the simulation, all the microcells are liquid, and their state index is set to zero. As nucleation proceeds, some microcells become solid, and their index is changed to an integer larger than zero, *n*. The microcells in contact with the mold wall are identified with a different reference index, *m*. The index is transferred from the parent microcell to adjacent microcells, as they become solid through growth. For the case of dendritic solidification of AZ31B alloys, the integer takes into account the preferential growth of hexagonal crystals in the prism direction. The anisotropic characteristic in the prism direction certainly results in 6-fold preferred growth directions with angles between the primary trunks of dendrites of 60°. For graphical representation each integer has a color associated with it, and each microcell is a pixel on the computer screen. The crystallographic orientation of the new grains is chosen randomly among 255 orientation classes, which are the first 255 colors used for graphical representation.

In 2-D calculations, the probability, $dp(\theta)$, that a newly nucleated grain has a principal growth direction in the range $(\theta, \theta + d\theta)$ is given by:

$$dp(\theta) = \frac{3}{\pi} d\theta \qquad (1)$$

where $3/\pi$ takes into account the six-fold symmetry of the hexagonal crystal, *i.e.*, the integral of $dp(\theta)$ from $-\pi/6$ to $\pi/6$ is equal to unity.

Nucleation

The number of grains, δN_e, that nucleate in the volume of the liquid and at the surface of the mold during one micro-time step, δt, are calculated by using nucleation site distribution, $N_e(\delta t)$. These distributions can be calculated by assuming some experimental approaches similar to those already used in deterministic models of solidification. The instantaneous nucleation model presented in [24] was used in this work to calculate the nucleation site distribution. Other nucleation models are described in references [21, 25]. Assuming no grain movement in the liquid, the grain density at any given location, *N*, can be expressed as a function of local cooling rate:

$$\delta N_e = C_0 + C_1 \dot{T} + C_2 \dot{T}^2 \qquad (2)$$

where C_0, C_1, and C_2 are the nucleation parameters determined from experiments and $\dot{T}$ is the local cooling rate at the beginning of solidification.

The probability, dp_e, for a microcell located in the bulk of the liquid or at the surface of the mold to nucleate during the micro-time step, δt, are calculated as

$$dp_e = \frac{\delta N_e}{N^v_{\mu VE} - N^s_{\mu VE}} \qquad (3)$$

where $N^v_{\mu VE}$ and $N^s_{\mu VE}$ are the number of microcells in the bulk and at the metal-mold interface, respectively. During each time-step calculation both metal-mold interface and bulk microcells are scanned and a random number, *rand*, is generated for each of them ($0 \leq rand \leq 1$). The nucleation of a microcell that is still liquid will occur only if $rand \leq dp_e$.

Growth

It is assumed that a nucleus formed at a particular location will grow based upon a growth kinetics model and a neighborhood configuration rule previously described. As nucleation proceeds, the microcell becomes active (n or $m > 0$) and can grow over a distance, $d_{\mu VE}(t)$, that is given by the following equation:

$$d_{\mu VE}(t) = \int_{t_0}^{t} V(\Delta T) dt \qquad (4)$$

where t_0 is the initial time, t is the actual time, and V is the growth velocity.

The initial time is defined as either the nucleation time, if growth is initiated through a nucleus, or the capturing time, if growth is initiated by capturing another microcell. At the capturing time, t_c, the initial microcell on which the growth was initiated captures the nearest neighbors based on a selected neighborhood configuration. For the case of equiaxed growth, this occurs when $d_{\mu VE}(t_c)$ is equal to the distance

$$a_\theta = a\sqrt{tan^2\theta + 1} \qquad (5)$$

where a is the size of the microcell for the stochastic network and θ takes values from $-\pi/6$ to $\pi/6$. At the capturing time, the initial microcell becomes solid and its state index is transferred to the captured neighbors.

In order to reflect the probabilistic nature of grain extension, the nearest neighbors are trapped by the active microcells only when a randomly generated number, *rand*, is smaller than the capture probability, p_g, defined as follows:

$$p_g = \frac{d_{\mu VE}(t)}{a_\theta} \qquad (6)$$

Further, the same procedure is used until all microcells become solid.

The growth kinetics model presented in [19] is applied for equiaxed solidification, where the growth velocity is described by

$$V = \mu\, \Delta T^2 \qquad (7)$$

where $\mu = \frac{2\sigma^*}{\Gamma}\left[\frac{m_L(k-1)C_L^*}{D_L} + \frac{\rho L}{K_L}\right]^{-1}$

D_L is the liquid diffusivity, Γ is the Gibbs-Thomson coefficient, k is the partition ratio, m_L is the liquidus slope, ρ is the density, L is the latent heat of fusion, K_L is the liquid thermal conductivity, C_L^* is the liquid interface concentration, and the stability constant $\sigma^* = 1/4\pi^2$.

The melt undercooling for the system under consideration can be calculated based on the following definition/assumption:

$$\Delta T = T_L + m_L\left(C_L^* - C_o\right) - T_b \qquad (8)$$

where T_L is the equilibrium liquidus temperature, $\langle C_L\rangle^L$ is the intrinsic volume-averaged liquid concentration, and T_b is the bulk temperature defined as the average temperature in the volume element, C_L^*, is calculated with the segregation model described in [19, 22].

Comparison of the simulated and observed structure of cast Mg AZ31B

In this section, we shall provide a critical assessment of the stochastic model by comparing the theoretically predicted grain size and Al segregation of cast Mg AZ31B alloy at low superheat against experimental measurements at two different cooling rates. The cast Mg alloy was prepared using the MSM process, which is capable of casting metal at superheat as low as 5°C. Casting of the metal was carried out in bottom-chilled mold shown in Figure 1. The mold was made of a ceramic tube with a metal block in the bottom. A battery of thermocouples was used to measure the cooling rates along the ingot. In this work, stainless steel and copper chill blocks were used to investigate the effect of solidification rate on cast structure. The details of the experimental MSM system and operating conditions are available in [14 - 16].

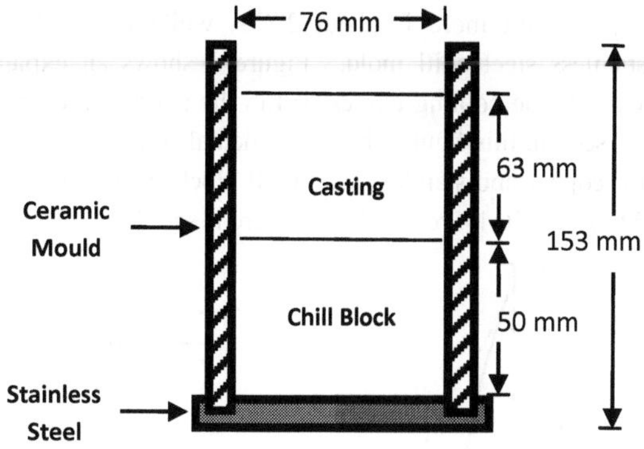

Fig 1. Sketch of the casting mold

The simulation of the microstructure in the bottom-chill mold shown in figure 1 was carried out assuming zero heat flux at the top and side walls. In these computations, the pouring temperature of the Mg alloy was 635°C and the initial temperature of the chill block was 20°C. Table 1 presents the data used in the microstructure modeling of Mg AZ31B alloy.

Table 1. Data Used in Microstructure Modeling of AZ31B

Solidification Kinetics Property	AZ31B
Liquid diffusivity, D_L [$m^2\ s^{-1}$]	$3x10^{-9}$
Solid diffusivity, D_S [$m^2\ s^{-1}$]	$1x10^{-12}$
Liquidus slope [K wt. $\%^{-1}$] [14]	$m_L = -6.6$
Initial Al concentration [wt. %]	$C_o = 3.14$
Al partition coefficient [14]	$k = 0.39$
Eutectic [wt. %]	32.3
Gibbs-Thomson coefficient, G [K m]	0.9×10^{-7}
Equiaxed nucleation parameters dN_e [nuclei m^{-3}]	$C_0 = 5x10^{12}$; $C_1 = 1x10^{12}$; $C_2 = 0$

Figure 2 shows the measured cooling curves of the cast alloy in the mold for copper and stainless steel chill blocks. This figure shows that the cooling rate of the Mg alloy in the copper chill mold is much faster than that in the stainless steel chill mold. The Initial cooling rate in the copper chill mold is about 2°C/s, while it is 1°C/s for stainless steel chill mold. Figure 3 shows an expanded view of the cooling curves within the solidification range. As seen in this figure, the experimental solidification times for copper and stainless steel chill blocks were found to be 24.7s and 36.1s, respectively. Here it should be mentioned that these solidification times were in reasonable agreement with model predictions.

Figures 4 and 5 show a comparison between the theoretically predicted and experimentally measured grain structure of cast Mg AZ31B alloy for low and high cooling rates, respectively. As seen in these figures, the model reasonably predicts the mean grain size and size distribution over the range of the cooling rate investigated. It also qualitatively predicts the expected decrease of the grain size with increasing the cooling rate.

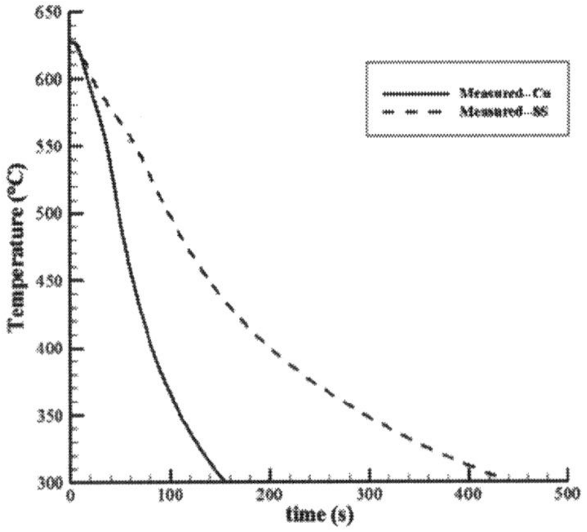

Fig 2. Cooling curves for copper and stainless steel chill blocks

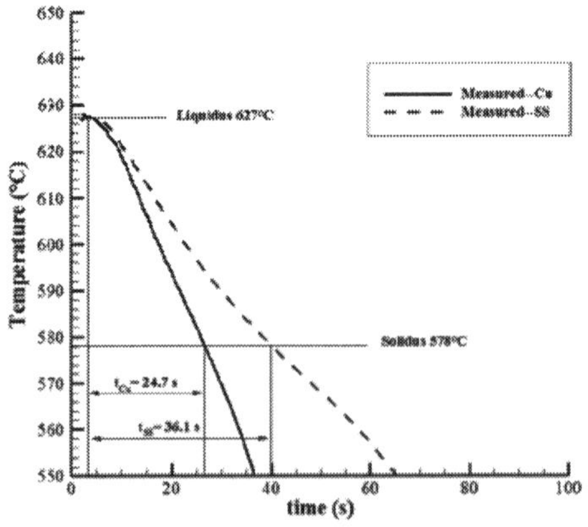

Fig 3. Cooling curves for solidification region.

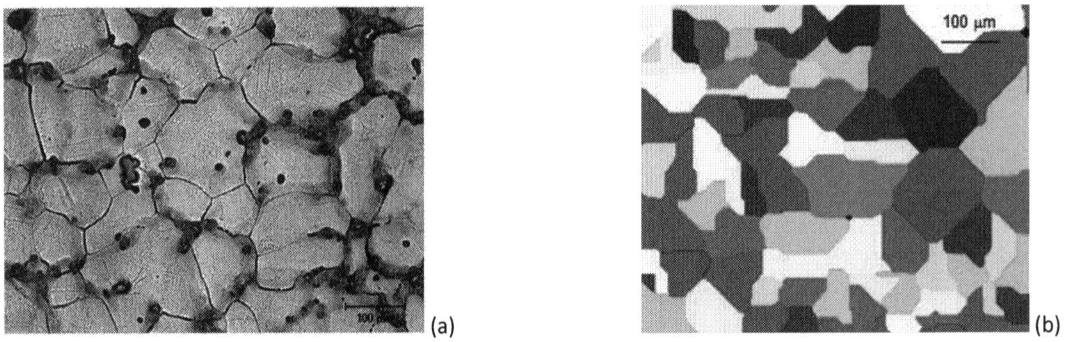

Fig 4. Comparison between (a) experimental and (b) simulated grain structure for stainless steel bottom chill mold

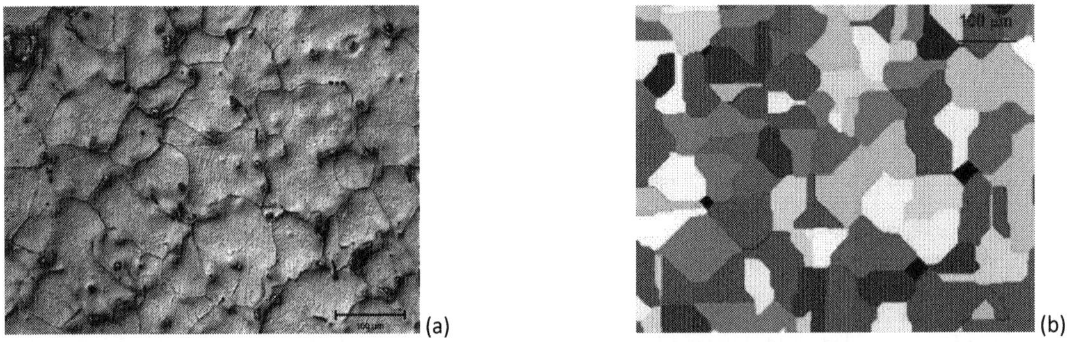

Fig 5. Comparison between (a) experimental and (b) simulated grain structure for stainless steel bottom chill mold

Figure 6 shows the EDS maps of Al segregation for the two cooling rates investigated. The light areas correspond to Al-rich secondary-α phase, while the more bright spots are the $Mg_{17}Al_{12}$ intermetallic. These phases are essentially found at the grain boundaries, thus confirming globular solidification morphology. These figures also show that the concentration of aluminum in the matrix is much lower than that at the grain boundaries, and the cooling rate has little effect on concentration profile across the grain. The measured and predicted Al segregation profile at the end of solidification within 2 globular grains is illustrated in figure 7. The excellent agreement between the measured and predicted magnitude and profile of Al distribution validates the segregation formulation of the model for globular solidification.

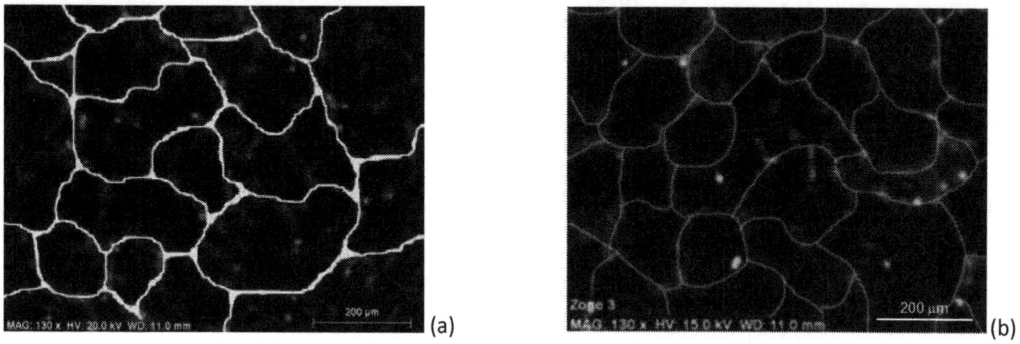

Fig 6. EDS Maps of Al solute distribution in MSM cast Mg AZ31B at 8°C superheat for (a) stainless steel and (b) copper chill block

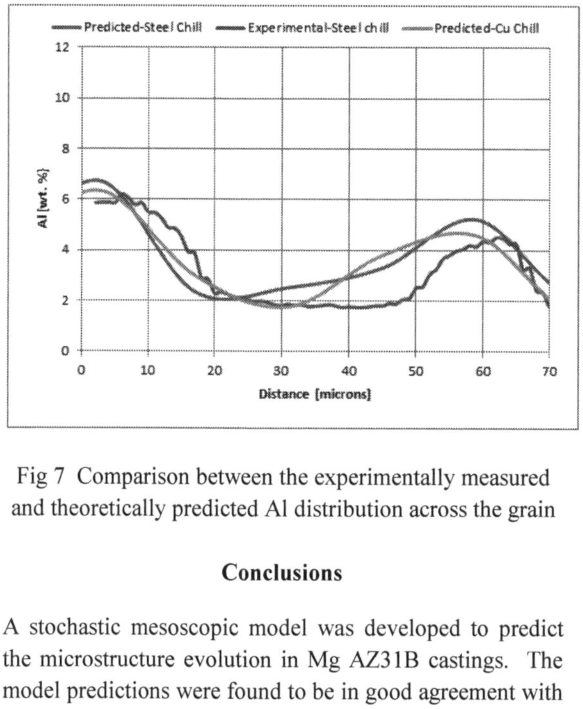

Fig 7 Comparison between the experimentally measured and theoretically predicted Al distribution across the grain

Conclusions

A stochastic mesoscopic model was developed to predict the microstructure evolution in Mg AZ31B castings. The model predictions were found to be in good agreement with the experimental measurements in terms of the microsegregation and the average grain size of the cast alloy. This suggests that the developed stochastic mesoscopic model can be used to assist in controlling the microstructure of cast alloys with globular morphologies.

Acknowledgment

The authors graciously thank the National Science Foundation for funding this investigation under Grant number CMMI-0856320.

References

1. N. Li, R. Osborne, B. Cox, B., and D. Penrod, "Magnesium Engine Cradle: The USCAR Structure Cast Magnesium Development Project: ASE paper," 337 (2005).
2. J.P. Thomson, S. Xu, M. Sadayappan, P.D. Newcombe, L. Millette, and M. Sahoo, "Low Pressure Casting of Magnesium Alloys AZ91 and AM50," AFS Transactions, 1-10 (2004).
3. S.G. Lee, G.R. Patel, and A.M. Gokhale, "Characterization of the effects of process parameters on macrosegregation in a high-pressure die-cast Magnesium alloy," Materials Characterization, 55, 219-224 (2005).
4. Y. Chino, T. Furuta, M. Hakamada, and M. Mabuchi, "Influence of distribution of oxide contaminants on fatigue behavior in AZ31 Mg alloy recycled by solid-state processing," Materials Science and Engineering: A, 424, 355-360 (2006).
5. D. H. StJohn, M. Qian, M. A. Easton, and P. Cao, Native Grain Refinement of Magnesium Alloys: Scripta Materialia 53 (2005) 841-844.
6. P. Cao, M. Qian, and D. H. StJohn, "Effect of iron on grain refinement of high-purity Mg–Al alloys," Scripta Materialia, V51 (2004) p.125-129

7. G. Han, X. Liu, and H. Ding, "Grain refinement of AZ31 magnesium alloy by new Al-Ti-C master alloys," Transactions of Nonferrous Metals Society of China, V19 (2009) p.1057-1064
8. D.G. Leo Prakash, Doris Regener, "Quantitative characterization of $Mg_{17}Al_{12}$ phase and grain size in HPDC AZ91 Magnesium Alloy," Journal of Alloys and Compounds, V461 (2007) p. 139-146
9. J.R. Brinegar, *et al.*, U.S. patent, No. 4832112 (1989).
10. Y. Ma, et al., Journal of Materials Processing Technology 137, (2003) p 35–39.
11. L. Liu, *et al.*, Materials Science and Engineering A 394 (2005) p1–8.
12. J. Reynolds, M. Shamsuzzoha and N. El-Kaddah, Characterization of Al-Li Castings Produced by the Magnetic Suspension Melting Process: in "Proceedings of the 4th Decennial International Conference on Solidification Processing", edited by J. Beech and H. Jones (Department of Engineering Materials, University of Sheffield, Sheffield, 1997) p. 45-48.
13. C. Mahato, M. Shamsuzzoha and N. El-Kaddah, Solidification Morphology and Structure of Cast Al-Li 2090 Alloy at Low Superheats: in "Solidification of Aluminum Alloys", edited by M. G. Chu, D. A. Granger and Q. Han (TMS, Warrendale, OH, 2004) p. 321-328.
14. N. Rimkus, L. Vihtelic, and N. El-Kaddah, "Grain and Microstructure of Cast Mg AZ31-B Alloy Produced by the Magnetic Suspension Melting Process," *Proceedings of* Light Weight Materials for Vehicles and Components Symposium, (TMS, Warrendale, OH, 2010) p.753-761.
15. N. Rimkus, M.L. Weaver, and N. El-Kaddah, "Microstructure and Mechanical Behavior of Cast Mg AZ31B Alloy Produced by the Magnetic Suspension Melting Process," *Proceedings of Magnesium Technology 2011,* (TMS, Warrendale, OH, 2011) p. 335-343.
16. G. Poole, N. Rimkus, A. Murphy, P. Boehmcke, and N. El-Kaddah, "Effect of Solidification Rate on Microstructure of Cast Mg Alloys at Low Superheat," *Proceedings of Magnesium Technology 2012*, (TMS, Warrendale, OH, 2012) p. 161-164.
17. N. El-Kaddah, T.S. Piwonka, and J.T. Berry, U.S. patent, No. 5033948 (1991).
18. J.A. Warren, and W.J. Boettinger, *Acta Materialia*, **43** 2 (1995) p. 689.
19. L. Nastac, *Modeling and Simulation of Microstructure Evolution in Solidifying Alloys* Springer-New York (2004): http://www.springer.com/materials/special+types/book/978-1-4020-7831-6
20. D.M. Stefanescu, *Science and Engineering of Casting Solidification* Springer (2002) p. 145-148.
21. L. Nastac, and D.M. Stefanescu, *Modelling and Simulation in Materials Science and Engineering* **5** (1997) p. 391-420.
22. L. Nastac, *Acta Materialia* **17** (1999) p. 4253.
23. S. Vernede, and M. Rappaz, *Acta Materialia* **55** (2007) p.1703.
24. D.M. Stefanescu, G. Upadhya, and D. Bandyopadhyay, "Heat Transfer-Solidification Kinetics Modeling of Solidification of Castings", *Metallurgical and Materials Trans. A* **21** (1990) p 997.
25. W. Olfield, "A Quantitative Approach to Casting Solidification: Freezing of Cast Iron", *ASM Trans.* **59** (1996) p 945.

Magnesium Technology 2013
Edited by: Norbert Hort, Suveen N. Mathaudhu, Neale R. Neelameggham, and Martyn Alderman
TMS (The Minerals, Metals & Materials Society), 2013

Effect of Casting Defects Distribution on the Beginning of Tensile Fracture in Semi-solid Injected Magnesium AZ91D Alloy

Yuichiro Murakami[1], Kenji Miwa[2], Naoyuki Kanetake[3], Shuji Tada[1]
[1]Materials Research Institute for Sustainable Development, National Institute of Advanced Industrial Science and Technology (AIST); 2266-98, Anagahora, Shimo-Shidami, Moriyama-ku; Nagoya, Aichi, 463-8560, Japan
[2]Aichi Science and Technology Foundation; 1-157-1, Onda-cho; Kariya, Aichi, 448-0013, Japan
[3]Nagoya University; Furo-cho, Chikusa-ku; Nagoya, Aichi, 464-8603, Japan

Keywords: Semi-soid injection process, Magnesium alloy, Tensile fracture, Casting defects, X-ray CT

Abstract

Semi-solid process is useful for magnesium alloys because processing temperatures lower than conventional casting processes result in decreased combustibility. Additionally it can decrease casting defects by the increased viscosity and decreased solidification shrinkage. In this study, casting defects of semi-solid injected AZ91D specimens were observed by X-ray CT tomography and tensile test was carried out. Thus, relations between casting defects and fracture starting point were investigated. As a result, the specimens were not always fractured at the site of the largest defect; meanwhile the defects situated near the surface or perpendicularly elongated to the tension axis exerted a potent influence on fracture.

Introduction

In recent years, global warming and global climate change have recently become an international concern. Hence the development of environmental protection programs and green technology is becoming considerably important. To solve these issues, the reduction of carbon dioxide emissions and improvement of fuel efficiency are urgent issues for the transportation machine. In particular, the reduction of vehicle weight is highly effective means of improving fuel efficiency and it has a positive impact for reduction of CO_2 emissions from transport. The expanded use of light metals such as aluminum and magnesium is considered one of the most effective methods for reducing vehicle weight. Magnesium in particular is considered to be a promising lightweight structural material because it has excellent specific strength, and has the lowest density among applicable metallic materials.

The die-casting process is among the most commonly used methods for forming magnesium alloy products with complex configurations. However, products molded by die-casting exhibit low engineering performance due to the existence of inherent defect such as porosity, hot cracks and oxide inclusions. Additionally, die-casting of magnesium alloys requires the use of cover gas because magnesium alloys are easy combustible in the liquid state.

The semi-solid forming process is a near net shape variant of die casting, which is carried out at the solid-liquid co-existence phase, which was first developed in the early 1970s [1,2]. Recently, this process has been applied to fabricate high-quality aluminum alloy [3-5] or magnesium alloy products [6-8]. The semi-solid forming process uses semi-solid slurry that has a higher viscosity and smaller liquid fraction than liquid metal; hence this process is expected to reduce casting defects in the final components. Additionally this process is considered useful for magnesium alloys because processing temperatures lower than conventional casting processes result in decreased combustibility of the magnesium alloy.

We have developed a new type of semi-solid injection process that allows magnesium alloys form in high material yield [9 - 12]. In this process, magnesium billets are heated to the semi-solid temperature range in an injection cylinder. Thus, the semi-solid magnesium alloy is not exposed to air, so this process requires no cover gas. Also particular billet is used in the semi-solid injection process in general because the billet cannot have fluidity even if the commercial billet is heated to the semi-solid temperature. On the other hand, the fluidity of the semi-solid slurry can be improved by applying shear stress [13]. In our new process, shear stress is applied to the slurry before injecting into the mold. Therefore commercial magnesium billets can be used.

On the other hand, the semi-solid products differ in character from conventional castings. For example, the distribution of the primary Mg particles was not uniform in the thickness direction [11], and mechanical strength was affected by not only casting defects but also fraction solid [12]. In this study, plate specimens were made by using a testing machine implementing the proposed semi-solid injection method, after which tension test was carried out. The effects of distribution of casting defects on the fracture behavior were investigated.

Experimental procedure

The semi-solid samples were prepared by using the vertical injection system shown in Fig. 1. The injection cylinder is able to heat the magnesium alloy billets to the semi-solid temperature, as well as to maintain this temperature. The injection cylinder in the apparatus was constantly filled with billet to a height of 320 mm under the nozzle, and the billet in the uppermost part of the cylinder was heated to a given semi-solid temperature. The semi-solid billet can then be injected into a permanent mold by a piston. The injection cylinder has an inner diameter of 25 mm, an outer diameter of 60 mm and a length of 54 mm; the nozzle is 3 mm in diameter. The injection cylinder is equipped with six heaters on the surface of the outer wall; these heaters can control the temperature of the billet precisely. The six heaters are controlled independently on the basis of measurements from six thermocouples (H1 to H6), each of which is inserted at the location of a heater. The temperature of the billet in the uppermost part of the injection cylinder (from the nozzle to a height of 130 mm) was controlled within ±1 °C by setting the temperature of H1 (from the nozzle to a height of 21 mm) and H2 (from the nozzle to a height of 151 mm). The temperature of the billet in the uppermost part of the injection cylinder (from the nozzle to 130 mm height) was set to the temperature in the semi-solid range, namely, 591 °C or 586 °C. The fraction solid f_s was 0.3 or 0.4, respectively, at each of these temperatures. On the other hand, the

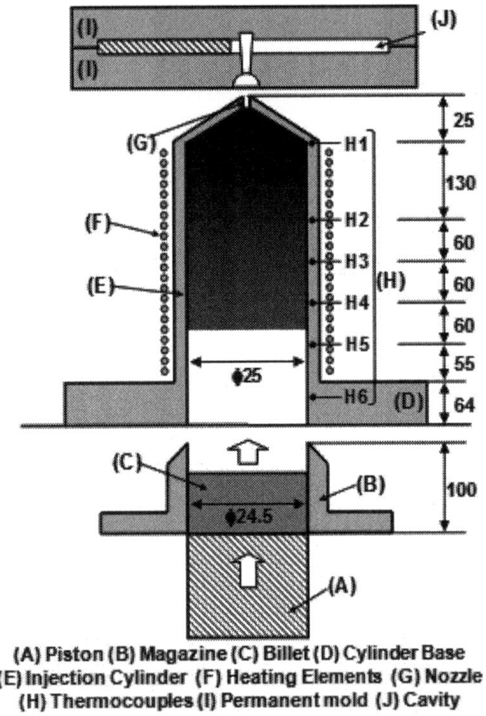

Fig. 1 Schematic representation of semi-solid forming apparatus

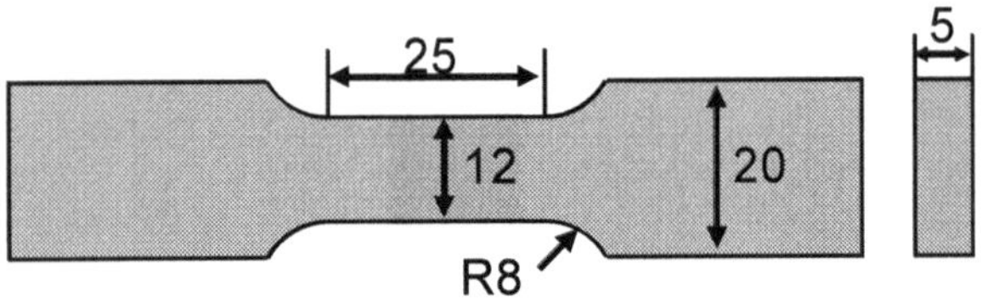

Fig. 2 Schematic image of tensile test piece

temperature of the billet closer to the bottom was decreased by controlling the temperature of heaters H3 to H6, and billet at the bottom was in a solid state. There is not much risk of combustion in a system with such design since billet in the semi-solid state is not exposed to air. Therefore, this experiment was carried out in air, without the use of a cover gas or inert gas.

The permanent mold in the experiment was a plate cavity of 20 mm in width, 100 mm in length and 5 mm in thickness. The slurry was injected horizontally since the mold was mounted vertically, and the mold clamping force was set to 274.4 kN. The amount of injected slurry per cycle was about 30 g, which was equivalent to the amount of material from the nozzle to a height of 53 mm. The injection speed V, which was calculated from the average piston speed, was set to 400 mm/s. The flow speed of the slurry at the nozzle was 27.8 m/s at each of these injection speeds.

The microstructures of these specimens were observed by optical microscopy. The samples were molded in epoxy resin and polished by grinding with SiC paper, followed by polishing with diamond paste, then, the specimens were etched in a solution of 75 ml ethylene glycol, 1 ml nitric acid and 24 ml distilled water. The plate specimens were mechanically processed into tensile test pieces shown as Fig. 2. Each tensile test piece was 100 mm in length, 20 mm in width 25 mm in gage length, 5 mm in thickness and 8 mm in shoulder radius. In addition, the surface of the test piece was as cast. Tensile tests were carried out at the room temperature by applying tension at a speed of 1.0 mm/s to 4 samples per set of casting conditions.

Also, distribution of casting defects in the specimens were analyzed on an X-ray computerized tomography (CT) scanner before and after tensile test. The rendered images were created from cross-sectional images obtained with 3D rendering software (VGStudio Max 2.0, Volume Graphics Inc.). In addition, the resolution of X-ray CT was 46 μm, allowing for detection of casting defects greater than 8 voxels.

Results and discussion

Microstructure of specimen and distribution of α-Mg particles

Fig. 3 shows a typical microstructure at the plane perpendicular to the slurry flow direction. The observation point was 50 mm from the gate. These micrographs show that the specimen has a homogeneous microstructure with a uniform dispersion of primary α-Mg in the matrix (eutectic α-Mg and β-$Mg_{17}Al_{12}$). When the slurry was injected, the solid phase consisted of the α-Mg particles with the matrix as the liquid phase. The distribution of the solid particles in the thickness direction was nonuniform. The solid particles have a tendency to concentrate in the center of the plate for all conditions. Additionally, the density of the solid particles at the surface was low, and, especially at a fraction solid of 0.3, the surface was nearly devoid of solid particles.

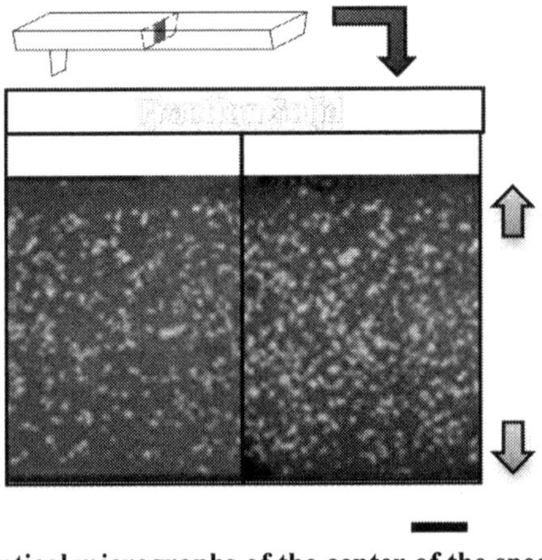

Fig. 3 Optical micrographs of the center of the specimen in the plane perpendicular to the flow direction.

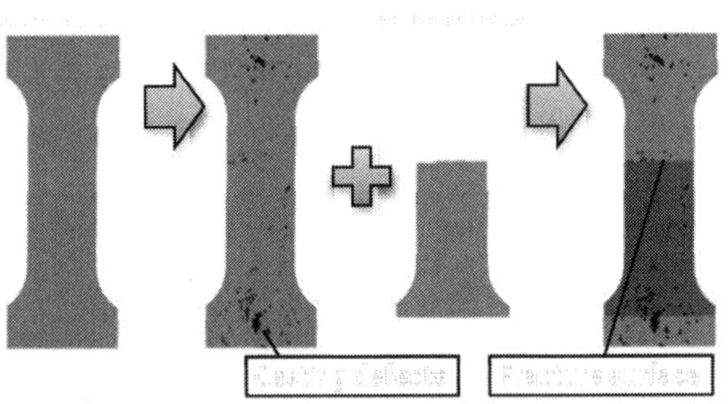

Fig. 4 Schematic diagram of the method for detection of the fracture surface.

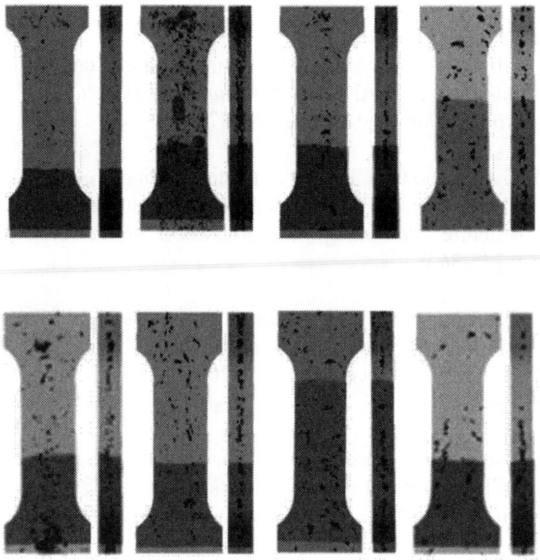

Fig. 5 Results of fracture surface detection and relation between distribution of casting defects and fracture surface.

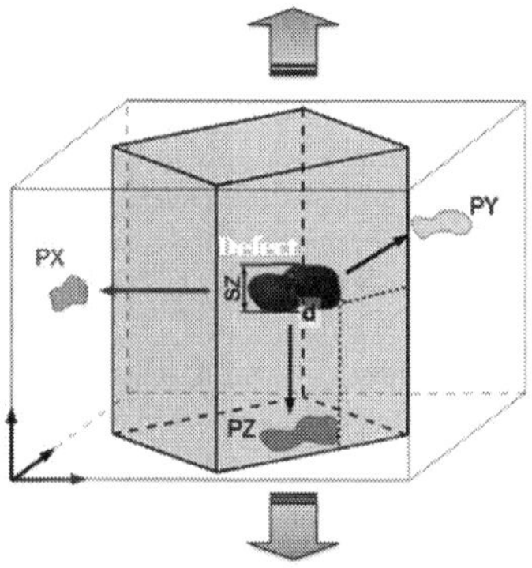

Fig. 6 Schematic view of the parameters for calculating indexes.

Fracture surface detection by X-ray CT

Detection of the fracture surface was carried out in the following way. First, the samples were scanned by X-ray CT before and after tensile test, as shown in Fig. 4. A 3D image of the sample before the tensile test was rendered by X-ray CT data, and casting defects in the sample were detected by contrast difference in the rendering image. Next, a rendering image of the sample after the tensile test was made from X-ray CT data. After that, this image was superimposed on the rendering image before the tensile test, thus the fracture surface was appeared. Additionally, casting defects involved in the fracture could be detected.

Fig. 5 shows results of fracture surface detection and relation between distribution of casting defects and fracture surface. Weiler et al. [14] showed the areal fraction of defects appeared to be the major factor in the fracture of the tensile test at magnesium high pressure die-cast samples. In this study, some samples fractured at the largest point of the areal fraction of defects (e.g., sample (a)-2). On the other hand, many samples were not always

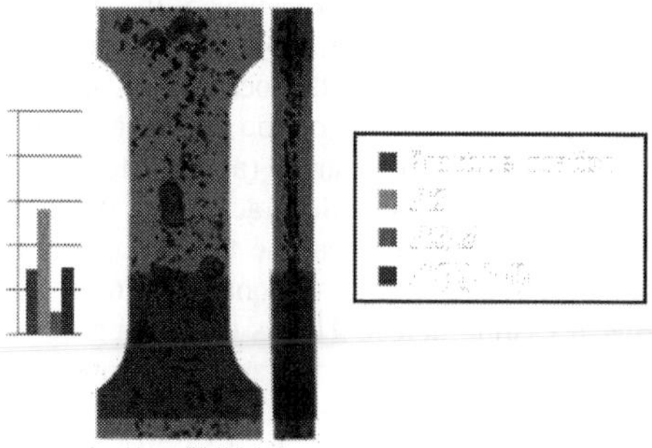

Fig. 7 Comparison of the sample (a)-2 with the maximum point of calculated indexes PZ, PZ/d, and PZ/(d・F).

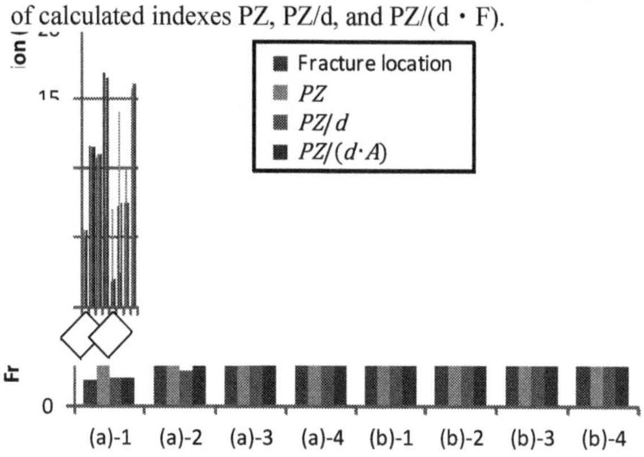

Fig. 8 Relation between the fracture location of samples and the maximum point of calculated indexes *PZ*, *PZ/d*, and *PZ/(d·F)*. The fracture surface location was measured as a distance from the bottom of the gage.

fractured at the largest point of cross sectional areal fraction of defects.

Relation between fracture surface and casting defects

From preceding section, the cross sectional area *PZ*, distance from the surface *d* and shape of casting defects are thought to have effects on fracture. There, we calculated two indexes to clarify the factor of fracture of semi-solid materials. Schematic view of the parameters for calculating indexes was shown in Fig. 6. First one was *PZ/d*, where *PZ* is cross sectional area of casting defects projected on the plane perpendicular to the tensile axis, and *d* is distance from surface. This index includes the effect of distance from the surface, then this index becomes greater if the defect is located near the surface. Second one was *PZ/(d·F)*, where *F* is described as follows.

$$F_x = SZ/\left(2\sqrt{\frac{PX}{\pi}}\right), F_y = SZ/\left(2\sqrt{\frac{PY}{\pi}}\right)$$

where, *SZ* is the length of the tensile axis direction of a defect, *PX* and *PY* are projected areas of perpendicular to the tensile axis. Then, if $F_x \geqq F_y$, $F = F_y$, else if $F_x < F_y$, $F = F_z$. If a defect has oblate shape and expands to perpendicular to the tensile axis, this parameter becomes small. Thus, the index *PZ/(d·F)* includes the effect of both the location and size effect of casting defect. Fig.7 shows the comparison of sample (a)-2 with the maximum point of calculated indexes PZ, PZ/d, and PZ/(d・F). By this comparison, it is made clear that this sample was fractured at the largest PZ/(d・F).

Fig. 8 shows comparison between the fracture location of samples and the maximum point of calculated indexes. The vertical axis in Fig. 8 shows the location of the fracture which was measured as a percentage of the entire gage length from the bottom of the gage. Only sample (a)-4 fractured at the largest *PZ*. On the other hand, the samples fractured at the largest *PZ*/ *d* were 6 of 8. This result means that the fracture of semi-solid metal is affected by not only cross sectional area of casting defects but also site of casting defect. That is the casting defect located near surface has the large impact on fracture. Furthermore, all samples but (b)-3 were fractured at the largest $PZ/(d\cdot F)$. Therefore, the casting defects with flat perpendicular to the tensile axis shape are thought to have larger impact on fracture.

These different results from die-casting [15] are thought to be after effects of not uniform structure. Such as the primary α-Mg particles have a tendency to concentrate in the center of the plate and the density of the primary α-Mg particles at the surface was low. In the semi-solid magnesium AZ91D, a hardness of the matrix (eutectic α-Mg and $\beta\text{-}Mg_{17}Al_{12}$) is greater than primary α-Mg particles. The defect near the surface is thought to be located at the high strength matrix. Thus the lack of matrix which can shoulder high stress is thought to be reason that the casting defect located near surface has the large impact on fracture. Then the casting defects with the shape like flat perpendicular to the tensile axis have the large impact on fracture. This is considered the effect of stress concentration on the sharp edge of defects perpendicular to the tensile axis.

Conclusions

To investigate the effect of the distribution of casting defects on the fracture, X-ray CT scanning and tensile test of the AZ91D magnesium specimens made by the semi-solid injection testing machine were performed. The fracture surface was appeared by the X-ray CT image then the casting defects involved in the fracture could be detected. In this study, the samples were not always fractured at the site of the largest cross sectional areal fraction of casting defects.

Therefore, the two indexes were calculated. One was *PZ*/ *d* which includes the effect of location, another was $PZ/(d\cdot F)$ which also includes the effect of shape of casting defects. By comparison between the fracture location of samples and calculated indexes, the fracture of semi-solid metal product is appeared to be affected by not only cross sectional area of casting defects but also site and shape of casting defect. That is the casting defect located near surface has the large impact on fracture. Furthermore the casting defects with the shape like flat perpendicular to the tensile axis are thought to more largely affect fracture.

References

1. M.C Flemings and R. Meharabian, "Casting Semi-solid Metals," *AFS Transactions*, 102 (1973), 81-88.

2 . M.C Flemings, "Solidification processing," *Metallurgical and Materials Transaction*, 5B (1974), 2121-2134

3. W.G. Cho and C.G. Kang, "Mechanical properties and their microstructure evaluation in the thixoforming process of semi-solid aluminum alloys," *Journal of Materials Processing Technology*, 105 (2000), 269-277

4. S. Nafisi and R. Ghomashchi, "Grain refining of conventional and semi-solid A356 Al–Si alloy," *Journal of Materials Processing Technology*, 174 (2006), 371-383

5. H.K. Jung and C.G. Kang, "Induction heating process of an Al–Si aluminum alloy for semi-solid die casting and its resulting microstructure," *Journal of Materials Processing Technology*, 174 (2006), 355-364

6. T. Tsukeda, K. Saito, "Mechanical Properties of AM -Series Magnesium Alloy Made by Injection Molding Process" Journal of Japan Foundry Engineering Society, 70 (1998), 697-701 (in Japanese)

7. Z. Koren, H. Rosenson, E.M Gutman, Ya. B. Unigovski, A Eliezer, "Development of semi-solid casting for AZ91 and AM50 magnesium alloys", *Journal of Light Metals*, 2 (2002) 81-87

8 . F. Czerwinski, "Size evolution of the unmelted phase during injection molding of semi-solid magnesium alloys", *Scripta Materialia*, 48 (2003), 327-331

9. N. Omura, Y. Murakami, M.G. LI, T. Tamura, K. Miwa, "Effect of Volume Fraction Solid and Injection Speed on Mechanical Properties in New Type Semi-solid Injection Process," *Solid State Phenomena*, 141-143 (2008) 761-766

10. Y. Murakami, N. Omura, M.G. Li, T. Tamura, S. Tada, K. Miwa "Microstructures and Casting Defects of Magnesium Alloy Made by a New Type of Semi-solid Injection Process," *Magnesium Technology 2011*, (2011) 107-112

11. Y. Murakami, N. Omura, M. Li, T. Tamura, K. Miwa, "Effect of Injection Speed on Microstructure of AZ91D Magnesium Alloy in Semi-solid Injection Process", *Materials Transactions*, 53 (2012), 1094-1099

12 . Murakami, K. Miwa, N. Omura, S. Tada, "Effects of Injection Speed and Fraction Solid on Tensile Strength in Semi-solid Injection Molding of AZ91D Magnesium Alloy", *Materials Transactions*, 53 (2012), in press

13. Y. Murakami, K. Miwa, M, Kito, T. Honda, N. Kanetake, S. Tada, "Effect of Injection Conditions in Semi-solid Injection Process on Fluidity of AC4CH Aluminum Alloy", *Journal of Japan Foundry Engineering Society*, 84 (2012), in press

14. J.P. Weiler, J.T. Wood, R.J. Klassen, E. Maire, R. Berkmortel, G. Wang, "Relationship between internal porosity and fracture strength of die-cast magnesium AM60B alloy", *Materials Science and Engineering A*, 395 (2005), 315-322

15. J.P. Weiler, J.T. Wood, R.J. Klassen, E. Maire, R. Berkmortel, G. Wang, "Relationship between internal porosity and fracture strength of die-cast magnesium AM60B alloy", *Materials Science and Engineering A*, 395 (2005), 315-322

Magnesium Technology 2013
Edited by: Norbert Hort, Suveen N. Mathaudhu, Neale R. Neelameggham, and Martyn Alderman
TMS (The Minerals, Metals & Materials Society), 2013

Effect of inoculation method of refiner on the grain refinement of AZ91 alloy

Jun Ho Bae, Young Min Kim, Chang Dong Yim, Ha-Sik Kim and Bong Sun You

Light Metal Group, Korea Institute of Materials Science, Changwon City, Gyeongnam 641-831, Republic of Korea

Keywords: Magnesium alloy, grain refinement, grain refiner, inoculation method

Abstract

Grain refinement involving inoculation of a grain refiner was investigated for application to a commercial Mg alloy. Cylindrical pellets composed of a mixture of $MnCO_3$ and carbon powder with several different ratios were plunged into a commercial AZ91 alloy. The results show that the addition of the mixed pellets can more efficiently refine AZ91 compared to the individual use of carbon powder or $MnCO_3$. In particular, the addition of mixed pellets with a 3:7 ratio between $MnCO_3$ and carbon provides an excellent refining efficiency, decreasing grain size from 460μm to 52μm of the AZ91 alloy. This is attributed to inoculation of $MnCO_3$ and carbon, which brings about the formation of heterogeneous nucleants such as MgO, Al_4C_3, and Al_8Mn_5 before solidification of α-Mg and also to melt agitation by the release of CO_2 gas

Introduction

Grain refinement is an effective way to improve the mechanical properties of magnesium alloys. In general, refined microstructure in as-cast components leads to a more uniform distribution of solute elements, secondary phases and microporosity on a fine scale. It also provides superior extrudability and rollability, excellent resistance to hot tearing, good surface finish and other beneficial attributes, leading to considerable cost reduction in the production of wrought magnesium parts [1-4].

Many grain refining methods including Zr addition, super-heating, Elfinal or ferric chloride inoculation, carbon inoculation, direct addition of nucleant particles and so on have been developed for Mg-Al alloy systems [1-10]. Among them, the addition of carbon containing agents to the melt is known as the most effective way for grain refinement of aluminum containing magnesium base alloys because they offer some important practical advantages such as lower operating temperature, large melt volume, and less fading with long holding time [1, 2].

The addition of carbon-containing agents such as C_2Cl_6, SiC, Al_4C_3, and carbon powder have been reported to successfully refine Mg-Al based alloys [11-14]. Among them, C_2Cl_6 is the most useful grain refiner for commercial application; however, it is widely prohibited because it can cause serious environmental problems due to the release of toxic gas during refining. Other refining technologies also have accompanying problems, and the detailed refining mechanisms have not yet been clearly established. Accordingly, a reliable, effective and eco-friendly grain refiner should be developed.

We recently reported in a previous study [15] that manganese carbonate has excellent grain refining efficiency in Mg-9wt%Al alloys. Its refining effect is attributed to the inoculation of manganese carbonate, which brings about the formation of effective heterogeneous nucleant particles such as MgO, Al_4C_3 and Al_8Mn_5 in the melt as well as to melt agitation of carbon dioxide gas generated by the decomposition of manganese carbonate. However, direct inoculation of carbonate powder raises serious concerns regarding the safety of the magnesium foundry because carbonate powder is accompanied with an explosive reaction and the melt to be boiled. In addition, most of the reaction can occurs at the surface of the melt, which affects the efficiency of grain refinement. The present study investigates grain refinement of a commercial magnesium alloy by inoculation of a grain refiner of a pressed pellet comprising manganese carbonate and carbon powders in varying ratio. The microstructures and refining performance for an AZ91 alloy were studied.

Experimental Procedure

Two types of refiners, carbon and manganese carbonate powders were used for grain refinement of a commercial AZ91 alloy. The average size of the carbon and manganese carbonate powders ranged from 3 to 5μm. The powders were mixed for 30 min with zirconia balls using a milling machine with varying ratios (Table 1). Next, the mixture of powders was pressed and formed to a cylindrical pellet shape.

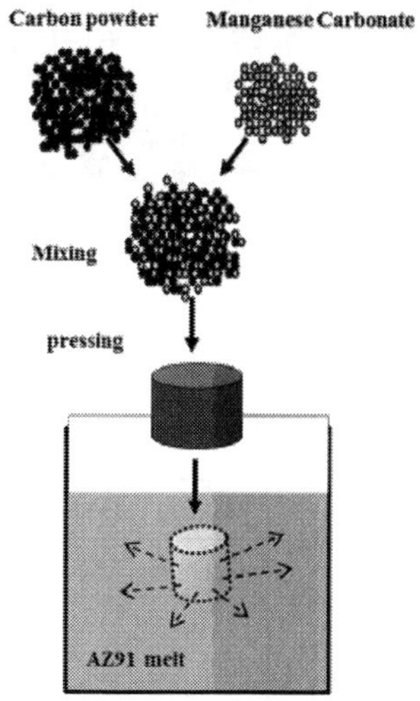

Figure 1. Schematic drawing of grain refining process using mixed pellets.

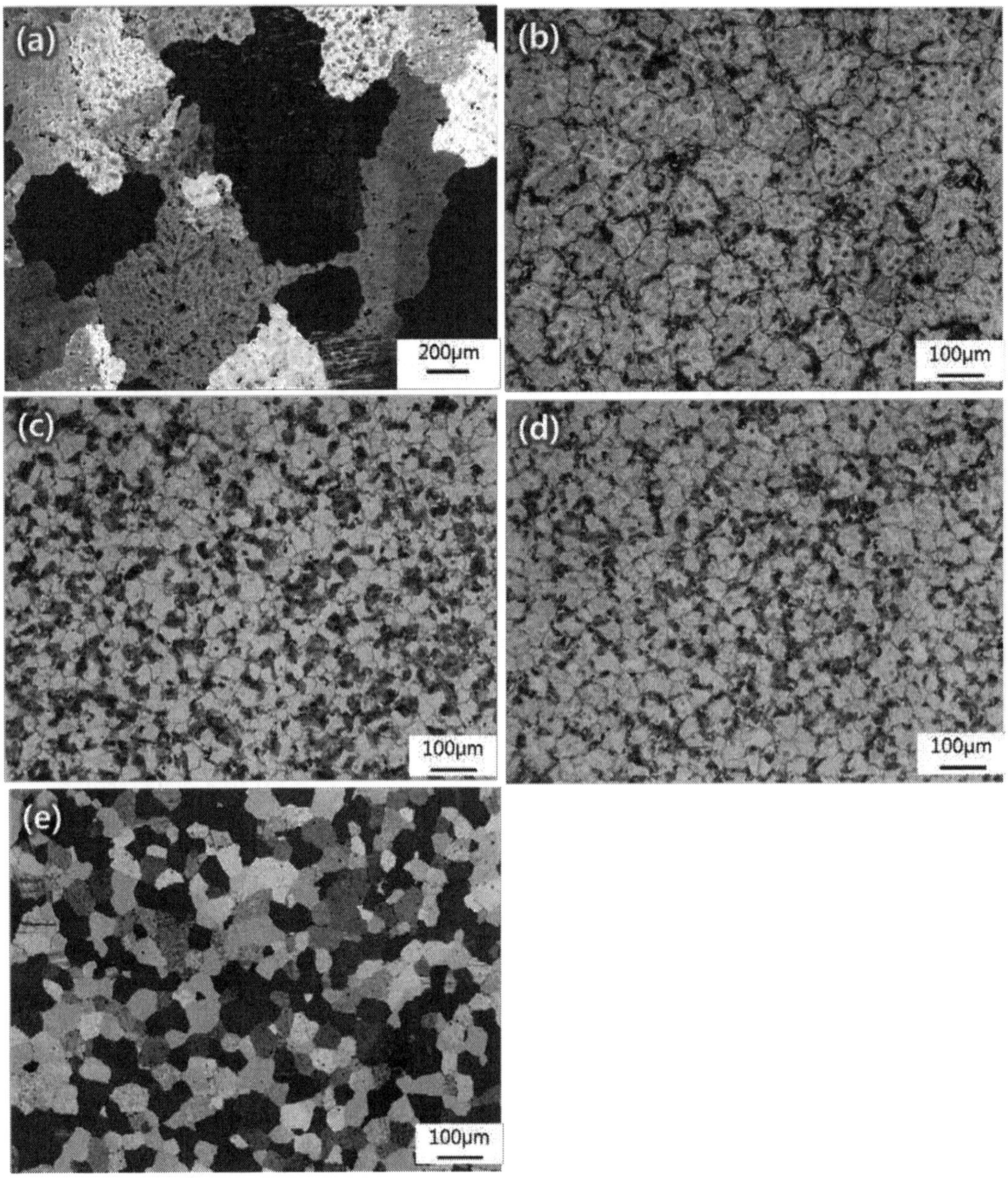

Figure 2. Microstructural changes of AZ91 alloys (a) without addition of the grain refiner and (b, c, d, e) with the addition of mixed pellet according to mixing ratio- (2, 3, 4, 5), respectively.

Table 1. Pellet mixing ratio of carbon and manganese carbonate

Number	Mixing ratio (wt%)	
	Carbon	$MnCO_3$
1	100	-
2	10	90
3	30	70
4	50	50
5	-	100

Commercial AZ91 alloys supplied by DSM in Israel were prepared. They were melted in an electrical resistance furnace using a mild steel crucible under the protection of a mixed gas of CO_2 and SF_6. The mixed and pressed pellets, in an amount of 1wt% were plunged into the melt at 740°C. The melt was held at 740°C for 30 minutes and then poured into a permanent mould which was coated with BN and preheated to 200°C. During holding at the melting temperature, the melts were periodically stirred using steel rod. The schematic drawing presented in Figure 1 shows the sequential process of grain refinement of the magnesium alloy including the fabrication of the mixed pellets of refiners. Metallographic samples were cut from the same position, *i.e.* at 20mm from the bottom of the castings, and prepared according to a standard procedure. Samples for optical microscopy were etched with a solution of 1 ml acetic acid, 50 ml of water and 150 ml of ethyl alcohol. Grain size was measured by the linear intercept method from micrographs taken using polarized light in and optical microscope.

Results and Discussion

Figure 2 shows the changes in the microstructure and grain sizes in grain refined as-cast AZ91 alloys using pellets with different ratios of manganese carbonate and carbon, in comparison to the un-refined AZ91 alloy. The average grain size of the AZ91 alloy without refining treatment was 460±40μm. The grain size decreased in all cases of treated grain refinement, notably, there was a significant decrease from 460±40μm to 52±3μm with the addition of mixed pellets having a 3:7 ratio between manganese carbonate and carbon powder, respectively.

The addition of more than 50% manganese carbonate, however, does not lead to further grain refinement as shown in Figure 3. In the case of using pellets composed 100% carbon, the grain refining ability was less (380±15μm) than that obtained with direct inoculation of carbon powder. Strongly pressed carbon powder cannot spread easily in the magnesium melt due to the poor wettability of the carbon pellet surface and the metal melt, and the powder has a tendency to agglomerate in the melt even with stirring. Pressed pellets composed solely of carbon powder thus have a detrimental effect on grain refinement of the Mg alloy. However, with the addition of manganese carbonate to the pellet, the grain size significantly decreased under the same casting conditions.

When the mixed pellet is plunged into the magnesium melt, manganese carbonate decomposes with the release of carbon dioxide gas and the melt is agitated. This facilitates the diffusion of the carbon powder in the magnesium melt at the same time. However, further addition of 50% manganese carbonate does not lead to a reduction of grain size, and the use of 100% manganese carbonate results in grain size of 70±2μm which is larger than that (under 60μm) obtained when using a 1:1 ratio of mixed refiners.

Normally, manganese carbonate decomposes with carbon dioxide at 200°C to give manganese oxide (MnO).

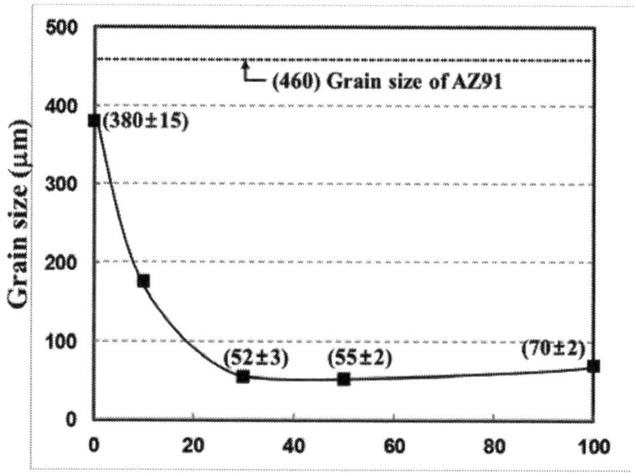

Figure 3. Changes in average grain size of AZ91 alloy with the addition of pellets mixed with different ratios.

However, when manganese carbonate is added into a commercial AZ91 alloy melt at 740°C, magnesium oxide (MgO) is formed by the following reaction instead of MnO (1):

$$89.9\,Mg + 9.0Al + 0.8Zn + 0.3Mn + 1MnCO_3 \rightarrow \text{Liquid} + MgO + Al_4C_3 \text{ (at 1013K)} \quad (1)$$

It is observed that there is no CO_2 gas at the equilibrium condition as the the excess Mg can consume CO_2 gas completely. At the same time, an Al_4C_3 phase can be formed by the reaction of carbon dioxide gas and aluminum atoms dissolved in the magnesium melt. Figure 4 shows the results of the thermodynamic calculation showing the solidification behavior of the AZ91 alloy with the addition of 1wt% $MnCO_3$.

In the actual process, carbon dioxide gas bubbles rise to the surface of the melt, and there is not sufficient time to react completely of carbon dioxide gas and aluminum, magnesium atoms dissolved in magnesium melt. The efficiency of refinement cannot be controlled easily, as the actual amounts of Al_4C_3 and MgO which are effective nucleants in this system are lower than the predicted values, as seen in Figure 4. The addition of carbon to manganese carbonate compensates for the loss of C from the released CO_2 gas without reaction with aluminum atoms.

According to previous reports, MgO, Al_4C_3 and Al_8Mn_5 phases can have beneficial effects on grain refinement of Al contained magnesium alloys. Al_4C_3 is known to be an excellent grain refiner due to its thermodynamic stability at the melting temperature of magnesium and crystallographic similarity with α-Mg; however, there have been no decisive experimental findings supporting this [13-19].

The fact that carbon addition leads to the formation of globular type $Al_8(Mn,Fe)_5$ particles causing the grain refinement in Mg-Al base alloys instead of Al_4C_3 particles is possibly conflicted with the Al_4C_3 nuclei hypothesis commonly accepted by many researchers. Although Al_4C_3 particles are known to provide more effective heterogeneous nucleation sites than Al_8Mn_5 particles in terms of crystallographic features [20], such a comparison of which one is better or worse is indeed meaningless in the melt where solute atoms can easily diffuse to long distance unlike solid-solid transformation. As far as crystallographic coherency

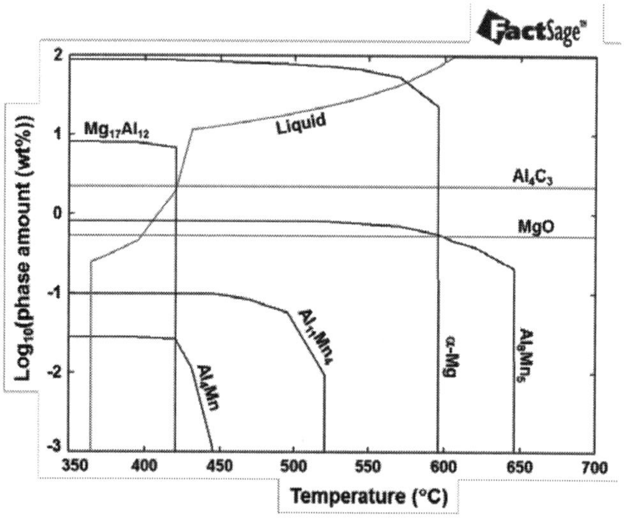

Figure 4. Thermodynamic calculation showing the sequence of microstructural evolution during solidification of manganese carbonate and carbon added in AZ91 alloy.

between magnesium matrix and heterogeneous nuclei is somewhat satisfied, other factors such as affinity between them, particle size and so on become more important. The existence of other particles such as Mg-rich Mg-Al-C ternary carbide in the melt must be not ignored. Although thermodynamic data for magnesium carbide are not sufficient yet, Viala [21] and Feldhoff [22] already reported that Al_2MgC_2 carbides were found in the interfaces between carbon fibers and magnesium alloys containing from 0.6 to 19%Al.

Experimental results by Kim et al show that carbon inoculation through the addition of 0.6wt% C_2Cl_6 lead to considerable grain refinement of commercial AZ91 with 0.25wt% Mn and Mg-9Al-0.3Mn alloys [23]. This indicates that the presence of Mn in Mg-Al alloy is necessary to acquire considerable grain refinement by carbon inoculation. Kim et al proposed a duplex nucleation theory, outlined as follows: (1) the formation of Al_4C_3 (R-3m, a=3.335Å, c=24.967Å) particles after carbon addition; (2) an Al_8Mn_5 (R3m, a=12.64Å, c=15.855Å) layer is formed on the surfaces of Al_4C_3 particles; and (3) α-Mg ($P6_3/mmc$, a=3.209Å, c=5.211Å) phases nucleate on the surfaces of Al_8Mn_5 particles. This concept was originally proposed for magnesium alloys by Mahoney *et. al.* [8] and has been known as a kind of grain refining mechanism in aluminum alloys [24, 25]. Similarly, Schumacher and Greer [25] also suggested that the TiB_2 particles were surrounded by Ti_3Al layer which in turn was surrounded by α-Al. That was same phenomenon as shown in this research.

Based on the above results in this research, therefore, the grain refining mechanism by carbon addition can be suggested as followed; after carbon inoculation, carbides containing aluminum are formed in magnesium melt and Al and Mn solute atoms in the magnesium melt segregate to the melt-carbides interfaces, and then globular type Al_8Mn_5 particles are nucleated on the surfaces of the carbides and enclose them before the formation of α-Mg phase as temperature decreases. Finally α-Mg phases are nucleated on the surfaces of Al_8Mn_5 particles. This is supported by a previous finding that the direct addition of fine Al_4C_3 powders into an AZ91 alloy melt led to the formation of polygonal Al_8Mn_5 particles, but Al_4C_3 particles could not be found. When 1.0%Al_4C_3 powders with average particle size of 13.7μm were directly plunged into the AZ91 alloy melt at 740°C, all the particles observed except β phases were globular type $Al_8(Mn,Fe)_5$ particles as shown in Figure 5, while Al_4C_3 particle was not found even in the sample poured just after addition of Al_4C_3 powders (Figure 5(a)). It implies that Al_4C_3 plays a role of heterogeneous nucleation site for $Al_8(Mn,Fe)_5$ particle instead of α-Mg phase [26].

The addition of $MnCO_3$ in the magnesium melt provides an additional advantage, the melt agitation effect of carbon dioxide gas generated with agitation of the melt. Although the effect of

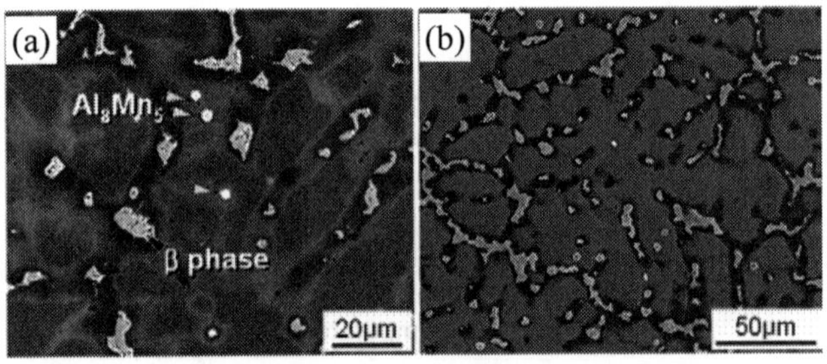

Fig. 5. SEM micrographs of the samples held for (a) 0min and (b) 30min after adding 1.0%Al_4C_3 powders into AZ91 alloy at 740°C. [26]

melt agitation on grain refinement of Mg-Al based alloys in not yet fully understood, it has been reported that melt agitation by gas bubbling leads to a grain refinement. Besides, decomposition of $MnCO_3$ provides additional Mn elements, and also contributes to the formation of additional Al_8Mn_5 particles leading to a positive grain refining effect.

Conclusions

In the present research, the grain refining efficiency of pressed pellets composed of manganese carbonate and carbon powder for a commercial AZ91 alloy were investigated. Mixed pellets with a ratio of 3:7 of manganese carbonate and carbon powder showed excellent grain refining efficiency in the AZ91 alloy, and a decrease of up to 52μm in the average grain size was obtained with the addition of 1wt% mixed pellet. The addition of pellets composed solely of carbon has detrimental effects as compared to direct inoculation or other methods. Mixed pellets with manganese carbonate and carbon powder bring about the formation of effective heterogeneous nucleant particles such as MgO, Al_4C_3 and Al_8Mn_5 in the melt as well as to melt agitation of carbon dioxide gas generated by the decomposition of manganese carbonate. These factors contribute to the diffusion of the carbon powder in the melt. The pellets also provide additional Mn which is the source element for the formation of Al_8Mn_5 nucleant. Therefore, mixed pellets with manganese carbonate and carbon powder offer several effective heterogeneous nucleant and refining mechanism, and are expected to provide the best refining performance among grain refiners developed thus far.

References

1. C.E. Nelson, "Grain Size Behavior in Magnesium Alloy", *Trans American Foundrymen's Soc.*, 56 (1948), 1-23.
2. Y.C. Lee, A.K. Dahle and D.H. StJohn, "The Role of Solute in Grain Refinement of Magnesium", *Metal. Mater. Trans.*, 31A (2000), 2895-2906.
3. E.F. Emley, *Principles of Magnesium Technology* (Oxford: Pergamon Press, 1966), 200-211.
4. N. Tiner, "Superheating of Magnesium Alloys", *Metals Technology*, 12 (1945), 1-19.
5. A.K. Dahle, et al., "Development of the As-cast Microstructure in Magnesium-Aluminium Alloys", *J. Light Metal.*, 1 (2001), 61-72.
6. D.H. StJohn, et al., "Grain Refinement of Magnesium Alloy", *Metall. Mater. Tran.*, 36A (2005), 1669-1678.
7. J.A. Davis, L.W. Eastwood and J.C. Dehaven, "Grain Refinement in Magnesium Casting Alloys", *Trans. American Foundrymen's Soc.*, 53 (1945), 352-362.
8. C.H. Mahoney, A.L. Tarr and P.E. LeGrand, "A Study of Factors Influencing Grain Size in Magnesium Alloys and a Carbon Inoculation Method for Grain Refinement", *Metals Technology*, 12 (1945), 1-20.
9. V. B. Kurfman, "Nucleation Catalysis by Carbon Additions to Magnesium Alloys", *Transactions of the Metallurgical Society of AIME*, 221 (1961), 540-546.
10. M. Bamberger, "Structural Refinement of Cast Magnesium Alloys", *Mater. Sci. Tech*, 17 (2001), 15-24.
11. Q. Jin, et al., "Grain Refining Mechanism of a Carbon Addition Method in a Mg-Al Magnesium Alloy", *Scripta Mater.*, 49 (2003), 1129-1132.

12. Q. Jin, et al., "Reply to Comments on Grain Refining Mechanism of a Carbon Addition Method in a Mg-Al Magnesium Alloy", *Scripta Mater.*, 52 (2005), 421-423.
13. L. Lu, A.K. Dahle and D.H. StJohn, "Heterogeneous Nucleation of Mg–Al Alloys", *Scripta Materialia,* 54 (2006), 2197–2201.
14. L. Lu, A.K. Dahle and D.H. StJohn, "Grain Refinement Efficiency and Mechanism of Aluminum Carbide in Mg-Al Alloys", *Scripta Mater.*, 53 (2005), 517-522.
15. Young Min Kim, et al., "Grain refinement of Mg-Al cast alloy by the addition of manganese carbonate", *J. alloys and compounds*, 490 (2010) 695-699
16. M.A. Easton, A. Schiffl, J.Y. Yao and H. Kaufmann, "Grain Refinement of Mg-Al(-Mn) Alloys by SiC Additions", *Scripta Mater.*, 55 (2006), 379-382.
17. M. Qian and P. Cao, "Discussions on Grain Refinement of Magnesium Alloys by Carbon Inoculation", *Scripta Materialia,* 52 (2005), 415–419.
18. P. Cao, M. Qian, D.H. StJohn, "Native Grain Refinement of Magnesium Alloys", *Scripta Materialia*, 53 (2005), 841–844.
19. T. Motegi, "Grain-Refining Mechanisms of Superheat-Treatment and Carbon Addition to Mg–Al–Zn Alloys", *Mater. Sci. and Eng. A*, 413-414 (2005), 408-411.
20. M.X. Zhang, et al, "Crystallography of Grain Refinement in Mg–Al Based Alloys", *Acta Materialia*, 53 (2005), 3261–3270
21. J.C. Viala, G. Claveyrolas, F. Bosselet and J. Bouix, "The Chemical Behavior of Carbon Fibres in Magnesium base Mg-Al Alloys", *J. Mater. Sci.*, 35 (2000), 1813-1825.
22. A. Feldhoff, E. Pippel and J. Woltersdorf, "Structure and Composition of Ternary Carbides in Carbon-Fibre Reinfored Mg-Al Alloys", *Philosophical Magazine A*, 79 (1999), 1263-1277.
23. Young Min Kim, Chang Dong Yim and Bong Sun You, "Grain refining mechanism in Mg-Al base alloys with carbon addition", *Scripta Materialia,* 57 (2007) 691-694
24. P.S. Mohanty and J.E. Gruzleski, "Mechanism of Grain Refinement in Aluminum", *Acta Metall. Mater.*, 43 (1995), 2001-2012.
25. P. Schumacher and A.L. Greer, "New Studies of Nucleation Mechanisms in Aluminium Alloys: Implications for Grain Refinement Practice", *Mater. Sci. Technol.*, 14 (1998), 394-404.
26. Young Min Kim, et al, "The Role of Carbon for Grain Refinement in Mg-Al Base Alloys", *TMS, Magnesium Tech.* (2007) 121- 126

Magnesium Technology 2013

Nanocomposites and Metal Matrix Composites

Magnesium Technology 2013
Edited by: Norbert Hort, Suveen N. Mathaudhu, Neale R. Neelameggham, and Martyn Alderman
TMS (The Minerals, Metals & Materials Society), 2013

NANOPARTICLE ADDITION TO ENHANCE THE MECHANICAL RESPONSE OF MAGNESIUM ALLOYS INCLUDING NANOSCALE DEFORMATION MECHANISMS

Muralidharan Paramsothy*, Manoj Gupta
Department of Mechanical Engineering, National University of Singapore
9 Engineering Drive 1, Singapore 117576

Keywords: Nanoparticles, Magnesium alloys, Mechanical properties, NENID, Nanoscale deformation mechanisms.

Abstract

In this study, various magnesium alloy nanocomposites derived from AZ (Aluminium-Zinc) or ZK (Zinc-Zirconium) series matrices and containing Al_2O_3, Si_3N_4, TiC or carbon nanotube (CNT) nanoparticle reinforcement (representative oxide, nitride, carbide or carbon nanoparticle reinforcement, respectively) were fabricated using solidification processing followed by hot extrusion. The main aim here was to simultaneously enhance tensile strength and ductility of each alloy using nanoparticles. The magnesium-oxygen strong affinity and magnesium-carbon weak affinity (comparison of extremes in affinity) are both well known in the context of magnesium composite processing. However, an approach to possibly quantify this affinity in magnesium nanocomposite processing is not clear. In this study accordingly, Nanoscale Electro Negative Interface Density or NENID quantifies the nanoparticle-alloy matrix interfacial area per unit volume in the magnesium alloy nanocomposite taking into consideration the electronegativity of the nanoparticle reinforcement. The beneficial (as well as comparative) effect of the nanoparticles on each alloy is discussed in this article. Regarding the mechanical performance of the nanocomposites, it is important to understand the experimentally observed nanoparticle-matrix interactions during plastic deformation (nanoscale deformation mechanisms). Little is known in this area based on direct observations for metal matrix nanocomposites. Here, relevant multiple nanoscale phenomena includes the emanation of high strain zones (HSZs) from nanoparticle surfaces.

Introduction

In comparing nanocomposites to microcomposites, the particle reinforcement is at nanoscale where physically confined yet beneficial effects are imparted to the host matrix. The function of nanoparticles in a metallic matrix is related to: (a) nanoparticle-matrix reactivity and (b) nanoparticle distribution in the matrix. Both factors can strongly affect the mechanical (crystallographic structure related) or functional (electronic structure related) properties of the composite. Magnesium alloys are an easily available lightweight and energy saving metallic matrix. The AZ (Aluminium-Zinc) series of magnesium alloys are characterized by: (a) low cost, (b) ease of handling, (c) good strength and ductility and (d) resistance to atmospheric corrosion [1]. The ZK (Zinc-Zirconium) series of magnesium alloys are characterized by: (a) high strength and ductility after T5 aging, (b) good creep resistance, (c) poor arc weldability due to hot-shortness cracking and (d) excellent resistance weldability. These qualities enable the common use of AZ and ZK series magnesium alloys [1]. Regarding metal matrix nanocomposites as abundant interface area systems, the ways to reduce chemical reactivity at the nanoparticle-metal matrix interface during solidification processing are extremely limited. It is necessary to take into consideration the: (1) Pauling electronegativity of the nanoparticle (El_{np}) and (2) electronegativity difference between the nanoparticle and the alloy matrix ($El_{np} - El_{alloy}$) to understand the chemical reactivity at the nanoparticle-alloy matrix interface. Electron attraction leading to chemical reaction increases with El_{np} while chemical reactivity at the nanoparticle-alloy matrix interface also increases with ($El_{np} - El_{alloy}$). Much of the existing representative research literature on solidification processed magnesium alloy nanocomposites indicates that: (a) good nanoparticle distribution can be achieved in the magnesium matrix and (b) better mechanical properties can be achieved due to the addition of nanoparticles [2-6]. However, it is important to understand the experimentally observed nanoparticle-matrix interactions during plastic deformation regarding the mechanical performance of a nanocomposite. Considering metal matrix nanocomposites, little is known in this area based on direct observations. The interaction between carbon nanotubes (CNTs) and aluminium based matrix during deformation has been briefly documented [7-9]. Here, the aluminium/CNT nanocomposites were deformed to 2% and transmission electron microscopy (TEM) revealed regions of severe deformation (based on high electron contrast) near the CNT tips in the aluminium based matrix. Dislocation accumulation at the aluminium/CNT interface was also observed in the 2% deformed nanocomposites. Comparatively, there are more reports based on modelling and simulation as compared to direct experimental observations regarding deformation mechanisms in metal matrix nanocomposites [10, 11].

Accordingly, this study is aimed at understanding the beneficial (as well as comparative) effect of representative oxide, nitride, carbide and carbon nanotube (CNT) nanoparticles on AZ and/or ZK series magnesium alloys taking Nanoscale Electro Negative Interface Density or NENID as well as the experimentally observed nanoparticle-matrix interactions during physical deformation into consideration. NENID quantifies the nanoparticle-alloy matrix affinity based on interfacial area per unit volume in the alloy nanocomposite taking into consideration the electronegativity of the nanoparticle reinforcement. Disintegrated melt deposition (DMD) [12] followed by hot extrusion was used to synthesize the magnesium alloy nanocomposites.

Experimental Procedures

The materials used in this study are listed in Table I. The matrix materials used were: (1) AZ31, (2) AA5083 for metallurgically upgrading AZ31 (addition of 3 wt.% relative to AZ31 weight), (3) AZ61, (4) AA1050 for metallurgically upgrading AZ61 (addition of 2 wt.% relative to AZ31 weight), (5) AZ91 for metallurgically upgrading AZ31 (nominal Al content of AZ31 increased by 3 wt.%.) and (6) ZK60A. All alloy pieces were sectioned to smaller pieces. All oxide and scale surfaces were removed using machining. All surfaces were washed with ethanol after machining. Reinforcement materials used were: (1) Al_2O_3

Table I. Materials selected for this study

Material	Available description and/or composition (wt%)	Particle size (nm)	Source
Alloys			
AZ31	nominally 2.40-3.60 wt.% Al, 0.50-1.50 wt.% Zn, 0.15-1.00 wt.% Mn, 0.10 wt.% Si max, 0.10 wt.% Cu max, 0.03 wt.% Ni max, 0.30.wt% others max, balance Mg	-	Alfa Aesar, Massachusetts, USA
AA5083	nominally 4.0-4.9 wt.% Mg, 0.40 wt.% Si, 0.40 wt.% Fe, 0.10 wt.% Cu, 0.40-1.00 wt.% Mn, 0.05-0.25 wt.% Cr, 0.25 wt.% Zn, 0.15 wt.% Ti, 0.15 wt.% others, balance Al	-	Yan San Metals Pte. Ltd., Singapore
AZ61	nominally 5.8-7.2 wt.% Al, 0.15-0.50 wt.% Mn, 0.04-1.50 wt.% Zn, 0.10 wt.% Si max, 0.05 wt.% Cu max, 0.005 wt.% Ni max, 0.005 wt.% Fe max, 0.30 wt.% others max, balance Mg	-	Tokyo Magnesium Co. Ltd., Yokohama, Japan
AA1050	nominally > 99.5 wt.% Al	-	Alfa Aesar, Massachusetts, USA
AZ91	nominally 8.30-9.70 wt.% Al, 0.35-1.00 wt.% Zn, 0.15-0.50 wt.% Mn, 0.10 wt.% Si, 0.030 wt.% Cu, 0.005 wt.% Fe, 0.002 wt.% Ni, 0.02 wt.% others, balance Mg	-	Tokyo Magnesium Co. Ltd., Yokohama, Japan
ZK60A	nominally 4.80-6.20 wt.% Zn, 0.45 wt.% Zr min, 0.30 wt.% others max, balance Mg	-	Tokyo Magnesium Co. Ltd., Yokohama, Japan
Nanoparticles			
Al_2O_3	γ-alumina	30-70	Baikowski, Japan
CNT	vapor grown, 94.7 % purity, up to 100 aspect ratio (Al-Saleh and Sundararaj 2009)	40-70	Nanostructured & Amorphous Materials Inc, Texas, USA
TiC	98+ % purity	30-50	Nanostructured & Amorphous Materials Inc, Texas, USA
Si_3N_4	98.5+ % purity	15-30	Nanostructured & Amorphous Materials Inc, Texas, USA

Please note that AZ31, AZ61, AZ91 and ZK60A are magnesium alloys while AA5083 and AA1050 are aluminium alloys.

nanoparticles, (2) carbon nanotube (CNT) nanoparticles, (3) TiC nanoparticles and (4) Si_3N_4 nanoparticles. Monolithic AZ31, AZ31/AA5083 (nominal Al content of AZ31 increased by 3 wt.%), AZ31/AZ91 hybrid alloy (nominal Al content of AZ31 increased by 3 wt.%), AZ81 (nominal Al content of AZ61 increased by 2 wt.% using AA1050) and ZK60A were each cast using the DMD method [12, 13]. This involved heating all alloy pieces to 750°C in an inert Ar gas atmosphere in a graphite crucible using a resistance heating furnace. The crucible was equipped with an arrangement for bottom pouring. Upon reaching the superheat temperature, the molten slurry was stirred for 2.5min at 460rpm using a twin blade (pitch 45°) mild steel impeller to facilitate the uniform distribution of heat. The impeller was coated with Zirtex 25 (86%ZrO_2, 8.8%Y_2O_3, 3.6%SiO_2, 1.2%K_2O and Na_2O, and 0.3% trace inorganics) to avoid iron contamination of the molten metal. The melt was then released through a 10mm diameter orifice at the base of the crucible. The melt was disintegrated by two jets of argon gas oriented normal to the melt stream located 265mm from the melt pouring point. The argon gas flow rate was maintained at 25lpm. The disintegrated melt slurry was subsequently deposited onto a metallic substrate located 500mm from the disintegration point. An ingot of 40mm diameter was obtained following the deposition stage. To form the corresponding alloy nanocomposites (1.0-1.5 vol.% reinforcement content), Al_2O_3, Si_3N_4, TiC or CNT nanoparticle powder was isolated by wrapping in Al foil of minimal weight (< 0.50 wt.% with respect to the total matrix weight) and arranged on top of the alloy pieces, with all other DMD parameters unchanged. All billets were machined to 35mm diameter and hot extruded using 20.25:1 extrusion ratio on a 150 ton hydraulic press. The extrusion temperature was 350 °C. The billets were held at 400 °C for 60 min in a furnace prior to extrusion. Colloidal graphite was used as a lubricant. Rods of 8 mm were obtained.

Prior to tensile or compressive deformation, microstructural characterization studies were conducted on metallographically polished monolithic and nanocomposite extruded samples to determine grain characteristics. Hitachi S4300 Field-Emission SEM was used. Image analysis using Scion software was carried out to determine the grain characteristics. Thin foils were prepared from the nanocomposite extruded samples using disc punch-out and ion-milling for (regarding localized effects): (a) nanoparticle distribution as well as (b) nanoparticle-matrix reactivity observation (based on selected area electron diffraction (SAED)) using transmission electron microscopy (JEOL JEM 3010 TEM with 300keV accelerating voltage). Regarding SAED, nanopowder samples (reinforcement dispersed in ethanol) were also prepared by droplet application onto holey carbon film mounted on copper grids followed by drying. All k-space measurements (k) from SAED patterns were manually obtained and converted to crystallographic lattice d-spacings (d) based on $d = 2\pi / k$. Goniometer XRD studies were conducted using CuK_α radiation (λ = 1.5406 Å) with a scan speed of 2 °/min in an automated Shimadzu LAB-X XRD-6000 diffractometer to determine the intermetallic phase(s) present and dominant textures in the transverse and longitudinal (extrusion) directions (regarding globalised effects). All d-spacings from SAED and goniometer XRD analysis were matched with corresponding d-spacings in the JCPDS database available in the Shimadzu LAB-X XRD-6000 diffractometer operating software to determine all phases present.

Table II. Nanoparticle size, interfacial area and electronegativity characteristics

Nanoparticle	Average size (nm) [a]	Average interfacial area x 10^3, IA (cm^2/cc) [b]	Average electronegativity, E [c]	$E - E_{matrix}$ [d]	NENID x 10^3 (cm^2/cc) [e]
CNT(X)	55.0 (40-70)	7.3	2.55	1.22	19 (15-26)
CNT(Y)	55.0 (40-70)	10.9	2.55	1.22	28 (22-38)
Al_2O_3	50.0 (30-70)	18.0	2.71	1.38	49 (35-81)
TiC	40.0 (30-50)	22.5	2.05	0.72	46 (37-62)
Si_3N_4	22.5 (15-30)	40.0	2.55	1.22	102 (77-153)

[a] Brackets indicate the available size (diameter) range of each nanoparticle.

[b] Nanoparticle-matrix interfacial area per cc in 1.5vol% nanocomposite in all cases except CNT(X) which is for 1.0 vol% nanocomposites. Nanoparticles assumed to be spherical (except in the case of CNT) and uniformly distributed in the matrix.

[c] Electronegativities of Al, O, Ti, C, Si and N are 1.61, 3.44, 1.54, 2.55, 1.90 and 3.04, respectively. E is the atomic (stoichiometric) average.

[d] Electronegativities of Mg, Al, Zn and Zr are 1.31, 1.61, 1.65 and 1.33, respectively. E_{matrix} is the weighted average based on the alloy matrix composition. In the case of the AZ31, AZ31/AA5083, AZ31/AZ91, AZ81 and ZK60A matrix compositions, $1.32 < E_{matrix} < 1.34$. $E_{matrix} \sim 1.33$.

[e] Nanoscale Electro Negative Interface Density or NENID = IA x E. Brackets indicate the range of calculated values based on the available size (diameter) range of each nanoparticle.

After tensile or compressive deformation, thin foils were prepared from the deformed nanocomposite extruded samples using disc punch-out and ion-milling for (regarding localized effects) high strain zone (HSZ) observation using a JEOL JEM 3010 transmission electron microscope.

Smooth bar tensile properties of the monolithic and nanocomposite extruded samples were determined based on ASTM E8M-05. Round tension test samples of 5 mm diameter and 25 mm gauge length were subjected to tension using an MTS 810 machine equipped with an axial extensometer with a crosshead speed set at 0.254 mm/min. 6-9 specimens were tested per formulation.

Results and Discussion

Synthesis of monolithic alloys and corresponding nanocomposites

Synthesis of monolithic and nanocomposite materials, the final form being extruded rods, was successfully accomplished with: (a) no detectable metal oxidation and (b) no detectable reaction between graphite crucible and melts. The inert atmosphere used during DMD was effective in preventing oxidation of the Mg melt. No stable carbides of Mg formed due to reaction with the graphite crucible.

Effect of nanoparticle electronegativity on tensile behavior

Nanoparticle size, interfacial area and electronegativity characteristics are listed in Table II. E is the stoichiometric average electronegativity of the nanoparticle. E_{matrix} is the weighted average electronegativity of the alloy matrix. In this study, $E_{matrix} \sim 1.33$. The AZ (aluminium-zinc) and ZK (zinc-zirconium) series of magnesium alloys are major representations of mainly aluminium containing and mainly zinc containing magnesium alloys (respectively) which collectively cover a significant portion of the entire range of solidification processed commercial magnesium alloys. ($E - E_{matrix}$) is the difference in electronegativity between the nanoparticle and alloy matrix. Nanoscale Electro Negative Interface Density or NENID is the product of IA (nanoparticle-matrix interfacial area per unit volume) and E. The range of calculated values of NENID is based on the available size (diameter) range of each nanoparticle. The NENID ranges of the different chemical types of nanoparticles listed (in brackets) in Table II were of an overlapping nature. The intention of this investigation was to provide as fair as possible representation to: (1) each chemical type of nanoparticle listed (oxide, nitride, carbide and pure carbon) and (2) the entire range of solidification processed commercial magnesium alloys as significantly covered by mainly aluminium containing and mainly zinc containing magnesium alloys.

The overall results of ambient temperature tensile testing of the extruded materials are shown in Table III. Among the tensile properties investigated, failure strain increment varied significantly with NENID as indicated by the simply fitted linear trend lines across the somewhat scattered data points for all plots in Figure 1 (increments are with respect to the corresponding mechanical property of the monolithic alloy). The scatter in data points was expected given the difference in chemical compositions among the various magnesium alloy matrices involved in this study. In the alloy nanocomposites, no nanoparticle-alloy matrix reaction products having more than 2% by volume were detected using X-ray diffraction analysis. Accordingly, it was expected that if nanoparticle-alloy matrix reaction products were formed, their size and volume fractions would be small, making detection by advanced methods difficult (or non-convincing at the least). Compared to the Al_2O_3, TiC and CNT nanocomposites (under non-equilibrium conditions), it was possible that the Si_3N_4 nanocomposites experienced more nanoparticle-alloy matrix interfacial reactions when: (1) NENID in the Si_3N_4 nanocomposites was up to 10 times that of the other nanocomposites (see Table II) and (2) ($E - E_{matrix}$) in the Si_3N_4 nanocomposite was sufficiently high at 1.22 (compared to 1.38 for the Al_2O_3 nanocomposites). This was the possible cause for the tensile failure strain increment of the Si_3N_4 nanocomposites to be

Table III. Results of tensile testing of magnesium alloys and related nanocomposites

Material	NENID x 10^3 (cm^2/cc)	0.2%TYS (MPa)	UTS (MPa)	Tensile failure strain (%)
AZ31	-	172 ± 15	263 ± 12	10.4 ± 3.9
AZ31/1.0vol%CNT	19	190 ± 13 (+10)	307 ± 10 (+17)	17.5 ± 2.6 (+68)
AZ31/1.5vol%Al_2O_3	49	204 ± 8 (+19)	317 ± 5 (+21)	22.2 ± 2.4 (+113)
AZ31/AA5083	-	203 ± 4	310 ± 4	8.7 ± 1.8
AZ31/AA5083/1.0vol%CNT	19	221 ± 4 (+9)	321 ± 1 (+4)	12.0 ± 1.0 (+38)
AZ31/AZ91	-	207 ± 4	316 ± 6	8.0 ± 0.1
AZ31/AZ91/1.5vol%Al_2O_3	49	232 ± 13 (+12)	339 ± 10 (+7)	15.9 ± 0.5 (+99)
AZ31/AZ91/1.5vol%TiC	46	236 ± 8 (+14)	337 ± 7 (+7)	14.5 ± 0.7 (+81)
AZ31/AZ91/1.5vol%Si_3N_4	102	232 ± 2 (+12)	331 ± 2 (+5)	13.1 ± 0.5 (+64)
AZ81	-	227 ± 3	321 ± 2	8.7 ± 0.9
AZ81/1.5vol%CNT	28	209 ± 9 (-8)	328 ± 4 (+2)	13.7 ± 2.2 (+57)
AZ81/1.5vol%Al_2O_3	49	212 ± 12 (-7)	339 ± 11 (+6)	13.1 ± 1.0 (+51)
AZ81/1.5vol%Si_3N_4	102	229 ± 9 (+1)	328 ± 7 (+2)	10.7 ± 0.9 (+23)
ZK60A	-	182 ± 4	271 ± 1	6.7 ± 1.0
ZK60A/1.0vol%CNT	19	180 ± 6 (-1)	295 ± 8 (+9)	15.0 ± 0.7 (+124)
ZK60A/1.5vol%TiC	46	184 ± 2 (+1)	309 ± 3 (+14)	11.6 ± 1.4 (+73)
ZK60A/1.5vol%Si_3N_4	102	198 ± 6 (+9)	313 ± 4 (+16)	12.2 ± 0.8 (+82)

( ) Brackets indicate % change with respect to corresponding result of monolithic alloy.

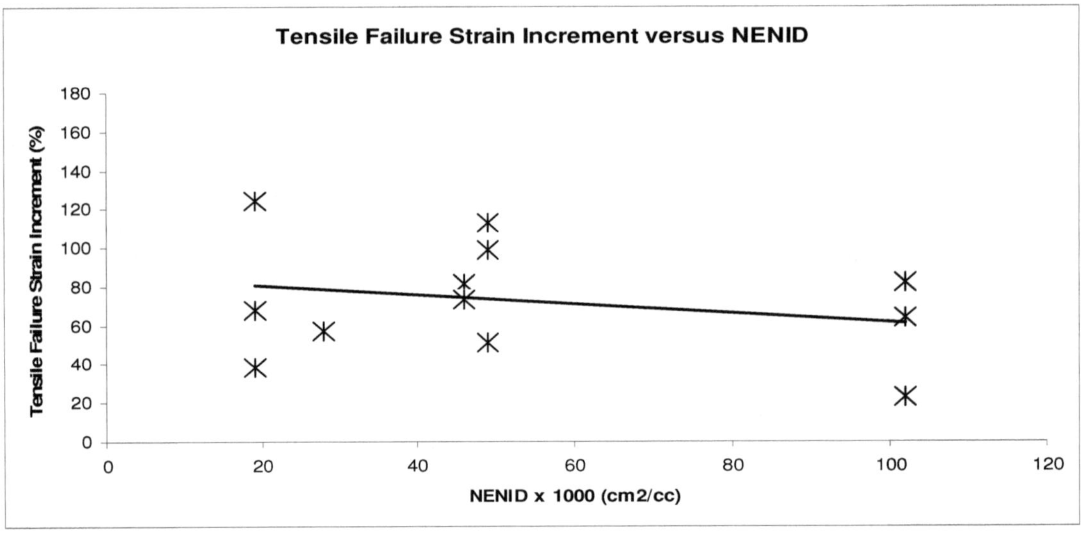

Figure 1. Significant variation of tensile failure strain increment with NENID as indicated by the simply fitted linear trend lines across the somewhat scattered data points for all plots. The scatter in data points was expected given the difference in chemical compositions among the various magnesium alloy matrices involved in this study. Please also note the overlapping NENID ranges of the different chemical types of nanoparticles listed (in brackets) in Table II.

closest to the bottom of the whole range exhibited by all the nanocomposites. Additionally, representative TEM SAED analysis for the Si_3N_4 nanocomposites is shown in Figure 2. Generally, TEM SAED occasionally indicated only for the Si_3N_4 nanocomposites that the Mg_3N_2 reaction product was formed. Also, the Si_3N_4 nanoparticle was originally amorphous in the as-supplied form but adopted a crystalline structure in the nanocomposite. In all other nanocomposites, no reaction product was *reliably* detected even by using TEM SAED. It was precisely for this reason (as previously mentioned) that it was expected that

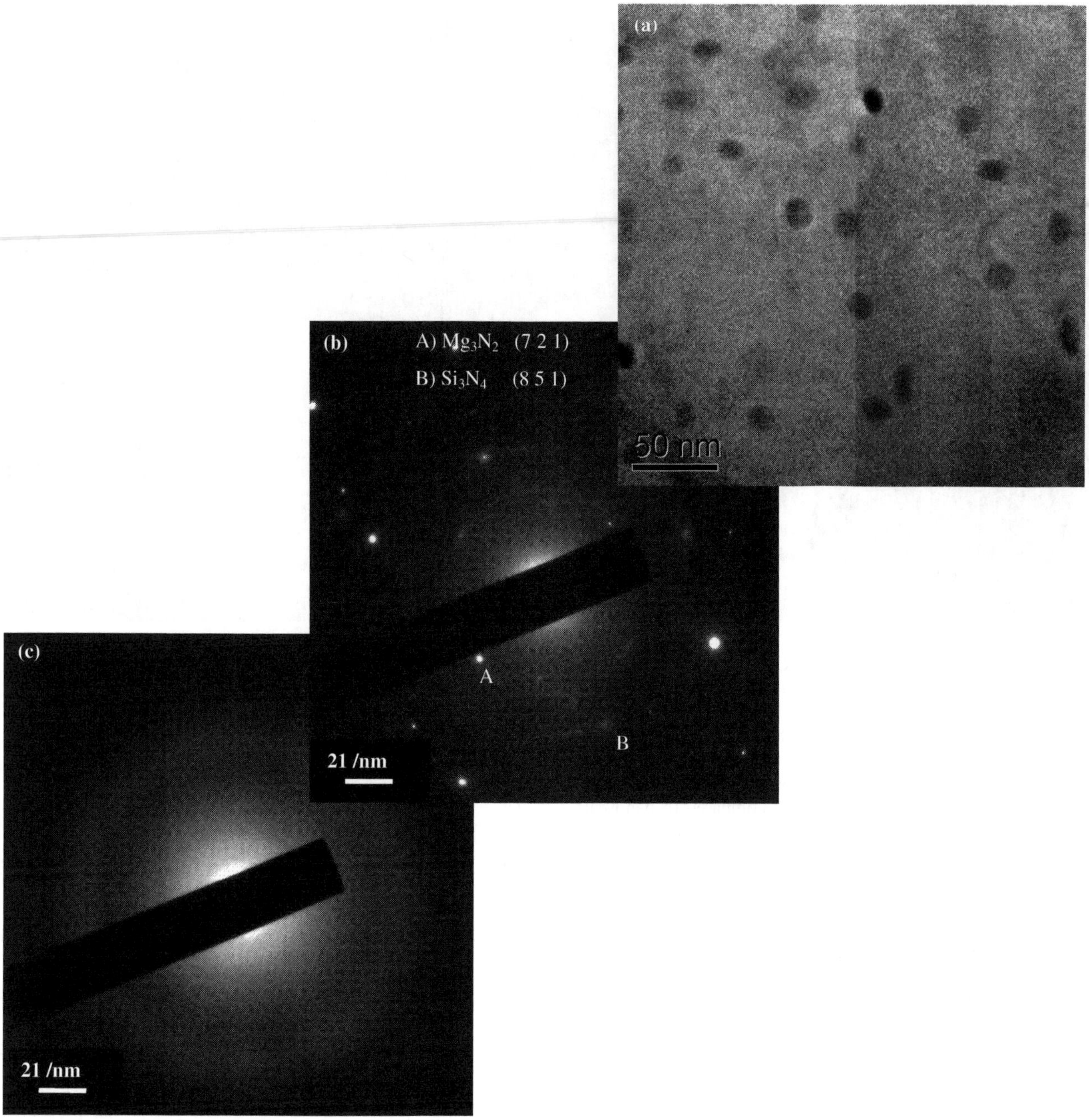

Figure 2. Representative TEM micrographs showing the: (a) presence of individual nanoparticles in each nanocomposite, (b) SAED pattern of Si_3N_4 nanocomposites and (c) SAED pattern of Si_3N_4 nanoparticles. Phases present but not labelled in the SAED pattern (in (b)) include Mg, Mg-Al and Mg-Zn phases only.

if nanoparticle-alloy matrix reaction products were formed, their size and volume fractions would be small, making detection by advanced methods difficult or non-convincing at the least.

Nanoscale Deformation Mechanisms

In each tensile deformed nanocomposite, the significant formation of high strain zones (HSZs) comprising dislocation slip was observed as shown in Figure 3. The uneven localizations of strain within the HSZ lead to: (a) severe electron contrast and (b) relatively less focus during TEM imaging. Figure 3a shows the emanation of the HSZ from a nanoparticle of approximately 100nm diameter. This direct observation of particle surface induced HSZ formation indicates the nanoscale dual role of fine particles in the magnesium matrix towards simultaneous strengthening and ductility enhancement. Strain hardening and dislocation slip within the HSZ contribute to strengthening and ductility enhancement, respectively. The HSZ formation from the particle surface also leads to shielding of the particle (and its immediate vicinity) from advancing nanoscale cracks, this being a tensile toughening mechanism. In the case of CNT nanocomposites, HSZs were formed to a significantly lower extent in the deformed nanocomposite compared to the near-spherical (and significantly lower aspect ratio) nanoparticle

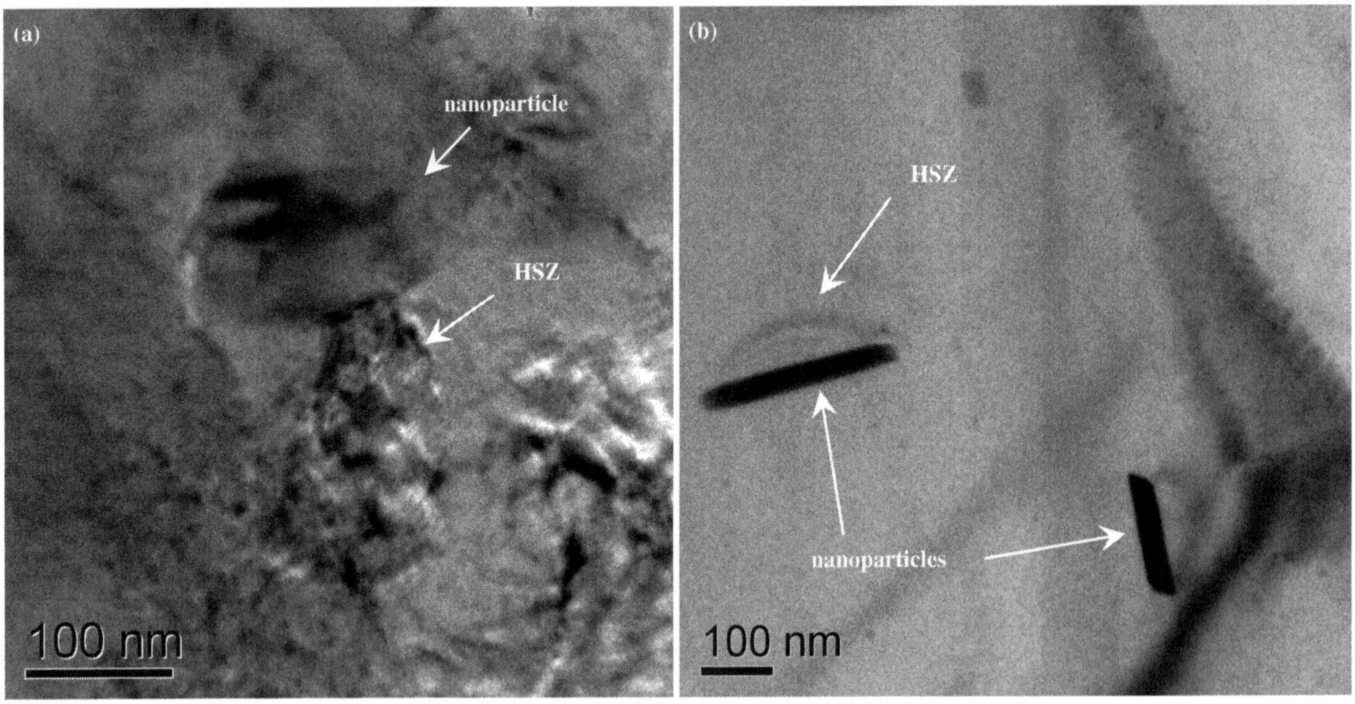

Figure 3. (a) Near-spherical nanoparticle induced HSZ formation and (b) higher aspect ratio nanoparticle induced relatively smaller bow-shaped HSZ formation in the deformed magnesium alloy matrix.

nanocomposite counterparts. Figure 3b shows the bow-shaped and relatively smaller HSZ formed around the CNT. This indicates the lower inclination of CNTs in the deformed magnesium matrix towards nanoscale ductility enhancement compared to nanoscale strength enhancement due to their increased aspect ratio at nanoscale.

Conclusions

1. The tensile failure strain increment of the Si_3N_4 nanocomposites were each closest to the bottom of the whole range exhibited by all the nanocomposites.
2. Compared to the Al_2O_3, TiC and CNT nanocomposites (under non-equilibrium conditions), it was possible that the Si_3N_4 nanocomposites experienced more nanoparticle-alloy matrix interfacial reactions when: (1) NENID in the Si_3N_4 nanocomposites was up to 10 times that of the other nanocomposites and (2) ($E - E_{matrix}$) in the Si_3N_4 nanocomposite was sufficiently high at 1.22 (compared to 1.38 for the Al_2O_3 nanocomposites).
3. The direct observation of particle surface induced HSZ formation indicates the nanoscale dual role of fine particles in the magnesium matrix towards simultaneous strengthening and ductility enhancement. Strain hardening and dislocation slip within the HSZ contribute to strengthening and ductility enhancement, respectively.
4. When the nanoparticle in the deformed magnesium alloy matrix has a higher aspect ratio (as in the case of CNT nanoparticles), it is less inclined towards nanoscale ductility enhancement compared to nanoscale strength enhancement.

References

[1] M.M. Avedesian, H. Baker, *ASM Specialty Handbook: Magnesium and Magnesium Alloys* (Ohio, OH: ASM International®, 1999) 12-25.

[2] M. Paramsothy, S.F. Hassan, N. Srikanth, M. Gupta, "Enhancing tensile/compressive response of magnesium alloy AZ31 by integrating with Al_2O_3 nanoparticles," *Mater. Sci. Eng. A*, 527 (2009), 162-168.

[3] M. Paramsothy, S.F. Hassan, N. Srikanth, M. Gupta, "Simultaneous enhancement of tensile/compressive strength and ductility of magnesium alloy AZ31 using carbon nanotubes," *J. Nanosci. Nanotechnol.*, 10 (2010), 956-964.

[4] M. Paramsothy, J. Chan, R. Kwok, M. Gupta, "Enhanced mechanical response of hybrid alloy AZ31 / AZ91 based on the addition of Si_3N_4 nanoparticles," *Mater. Sci. Eng. A*, 528 (2011), 6545-6551.

[5] M. Paramsothy, J. Chan, R. Kwok, M. Gupta, "TiC nanoparticle addition to enhance the mechanical response of hybrid magnesium alloy," *J. Nanotechnol.*, (2011), DOI: 10.1155/2011/401574.

[6] M. De Cicco, H. Konishi, G. Cao, H.S. Choi, L.S. Turng, J.H. Perepezko, S. Kou, R. Lakes, X. Li, "Strong, ductile magnesium zinc nanocomposites," *Metall. Mater. Trans. A*, 40 (2009), 3038-3045.

[7] H.J. Choi, J.H. Shin, B.H. Min, D.H. Bae, "Deformation behavior of Al–Si alloy based nanocomposites reinforced with carbon nanotubes," *Comp. Part A*, 41 (2010), 327-329.

[8] H.J. Choi, J.H. Shin, D.H. Bae, "Grain size effect on the strengthening behavior of aluminum-based composites containing multi-walled carbon nanotubes," *Comp. Sci. Technol.*, 71 (2011), 1699-1705.

[9] H.J. Choi, J.H. Shin, B.H. Min, J.S. Park, D.H. Bae, "Reinforcing effects of carbon nanotubes in structural aluminum matrix nanocomposites," *J. Mater. Res.*, 24(8) (2009), 2610-2616.

[10] H.T. Liu, L.Z. Sun, "Multi-scale modeling of elastoplastic deformation and strengthening mechanisms in aluminum-based amorphous nanocomposites," *Acta Mater.*, 53 (2005), 2693-2701.

[11] D.K. Ward, W.A. Curtin, Y. Qi, "Mechanical behavior of aluminum–silicon nanocomposites: A molecular dynamics study," *Acta Mater.*, 54 (2006), 4441-4451.

[12] L.M. Tham, M. Gupta, L. Cheng, "Influence of processing parameters during disintegrated melt deposition processing on near net shape synthesis of aluminium based metal matrix composites," *Mater. Sci. Technol.*, 15 (1999), 1139-1146.

Magnesium Technology 2013
Edited by: Norbert Hort, Suveen N. Mathaudhu, Neale R. Neelameggham, and Martyn Alderman
TMS (The Minerals, Metals & Materials Society), 2013

Properties of Extruded Disintegrable Metal Composites

Bobby Salinas[1], Zhiyue Xu[1], John Welch[1]
[1]Baker Hughes Incorporated; 14990 Yorktown Plaza Drive; Houston, TX, 77040, USA

Keywords: Extrusion, Metal composite, degradable material, completion tools

Abstract

Recently, a lightweight, disintegrable material with high strength and high ductility has been successfully used in oilfield tool applications to save well operation time and cost. The material is a powder metallurgy composite consisting of a metal matrix and mechanical/chemical reinforcements. Warm extrusion was used to produce rods and tubes of the material from cold pressed green preform. Disintegration rates of the extruded materials with different compositions range from 0 – 300 mg/cm^2/hr in 3% KCl at 93.3 °C with a strength of > 60 ksi (414 MPa) and ductility of 7 – 15%. This paper presents the microstructure, disintegration, and mechanical properties of the extruded composite. The results prove that disintegration properties can be controlled by altering the chemical composition of the reinforcements while maintaining good strength and ductility.

Introduction

Disintegrable metal matrix composite, DMMC, is a unique powder metallurgy composite where a base powder material is coated or modified via milling/CVD with customizable metallic or ceramic chemistries. Nano/micron level chemistries and thickness ultimately decide the balance between material strength and its dissolution in different wellbore fluids. The ultimate goal of this material development is to provide a disintegrable, high strength material system that can be used in interventionless downhole tools for completing different formation conditions. Completion devices such as fracturing balls, discs, plugs, and perforating gun bodies are often used for downhole fluid and pressure control.

Current polymer material must be milled away, flowed back or otherwise removed before production. Severe deformation of currently used materials that prevent flowback have been reported, requiring costly intervention operations to either remove or replace the tools and resulting in higher operational inefficiency. Using DMMC eliminates this possibility. The material is inherently designed for in-situ disintegration in downhole environments at a controlled customizable rate.

Conventional degradable materials have a well-known tradeoff between disintegration rate and mechanical strength, which limits their utility. The need for a high-strength disintegradable system is desirable; however, no system can provide the strength needed for this application [1 - 4]. An increased dissolution rate often lowers the mechanical strength of the material below an acceptable threshold for many downhole applications. DMMC provides the option to set the corrosion rate while maintaining its structural integrity in wellbore conditions.

Extrusion Process of DMMC

Extrusion is necessary to create a cost effective and post processing alternative without losing mechanical and disintegration properties. This can be accomplished by extruding green preform billets without losing the unique material properties of DMMC and creating objects of a fixed cross-sectional profile. The two advantages of this process are: 1. the production rate can be increased by extruding large preform billets 2. Higher strength can be achieved through work hardening of the material.

The press used for extrusion is illustrated in **Fig. 1**. A pre-heated preform billet, **Fig. 2**, is loaded into a container and pressed through a hot die. The starting billets were 4 in. (10.2 cm) and 9 in. (22.9 cm) OD and pushed through various tube and rod dies.

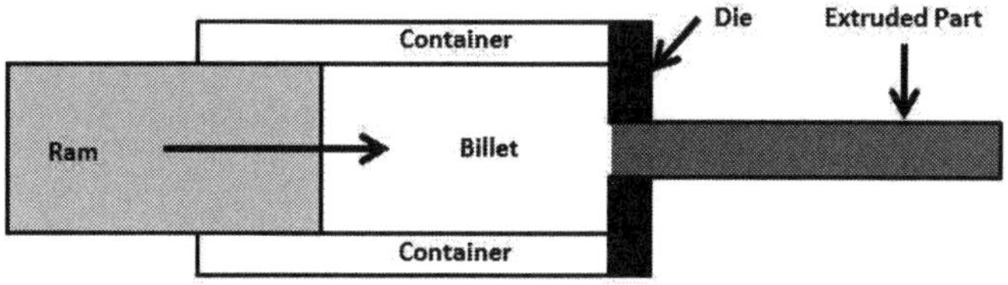

Fig. 1. Extrusion press diagram used for DMMC extrusion.

Fig. 2. Green preform of (a) powder, (b) billets.

Extrusion involved using critical pre-heating and process temperature controls to maximize results. The billets, dies, and containers were pre-heated to elevated temperatures for processing DMMC to the shape desired. Once the billet reached the optimum temperature, it was transferred to a container and pressed through the die with ram pressure. The ram pressure exerts pressure on the body which is moving through a fluid medium. The drag force exerted on the body if defined as (1), where P is the pressure, ρ is the density of the media, and v is the velocity of the body.

$$P = \rho v^2 \quad (1)$$

Table I lists the materials extruded at various reduction ratios. Each material family is composed of nanostructured coating or reinforcement materials.

Table I. DMMC materials extruded through various rod and tube dies.

ID	Starting Billet OD, cm	Final OD (ID), cm	Ratio
S1	10.80	2.54	18.1
S2	10.80	5.08	4.5
S3	20.32	7.62	7.1
S4	20.32	10.01	4.1
S5	20.32	7.62 (6.35)	23.7
S6	20.32	10.01	4.1
S7	20.32	12.07	2.8
S8	20.32	8.89 (6.35)	10.9
S9	20.32	13.34 (10.80)	6.9

The extrusion ratio is the cross-sectional area of the billet divided by the cross-sectional area of the final extruded part:

$$ER = \frac{A}{A'} \quad (2)$$

Extruded Parts

DMMC materials were extruded into different forms, including solid rods and tubes. Photos of extruded S5 and S8 tubes are shown in **Fig. 3**. The original billet was 8 in. (20.32 cm) OD with final tube OD of 3 in. (7.62 cm) (a) and 3.5 in. (8.89 cm) (b) with 0.25 in. (0.635 cm) and 0.5 in. (1.27 cm) wall thickness, respectively. **Fig. 4** shows both a 5.25 in. (13.34 cm) OD tube with 0.5 in. (1.27 cm) wall thickness (a), and 4.75 in. (12.07 cm) OD extruded rod (b).

Fig. 3. Extruded tubes (a) 7.62 cm OD, 0.635 cm wall thickness, (b) 8.89 cm OD, 1.27 cm wall thickness

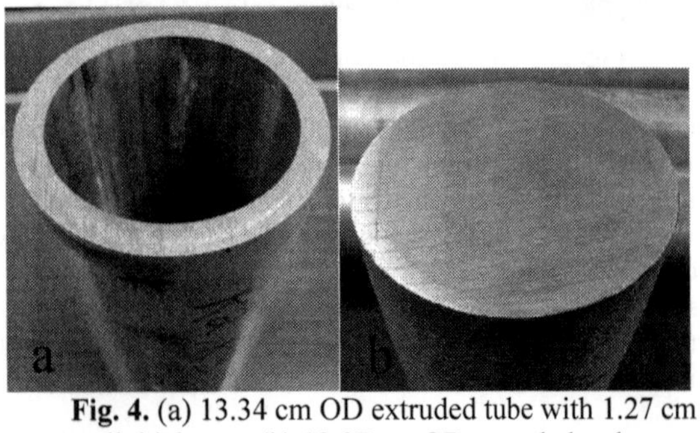

Fig. 4. (a) 13.34 cm OD extruded tube with 1.27 cm wall thickness, (b) 12.07 cm OD extruded rod.

Extruded rods of 1 in. (2.54 cm) OD were also produced (see **Fig. 5**) from a 4.25 in. (10.80 cm) OD starting billet. Test coupons for disintegration rates and compressive strength testing were machined according to ASTM E9.

Fig. 5. 2.54 cm extruded rods.

Properties and Microstructure of Extruded DMMC

Disintegration, mechanical, and microstructure properties provide the unique characteristics of extruding DMMC. Billet pre-heating temperature, extrusion rate, and die temperature play significant roles in providing the required properties needed for downhole use. Improper heating can drastically alter overall DMMC properties. To test these properties, 0.5 in. (1.27 cm) OD x 1.0 in. (2.54 cm) L coupons were machined from the final extruded parts and tested for the previously mentioned properties.

Disintegration properties

Rate of disintegration (ROD) testing was completed using a dynamic test cell in the presence of 3% KCl at 93.3°C. The sample was fully submerged in brine for a specific amount of time, and periodically checked for dimensional change and weight. The results are presented in Table II as mg/cm^2/hr, where the difference in weight is divided by the difference in area and total testing time:

$$ROD = \frac{\Delta wt}{\Delta A \times t} \quad (3)$$

Table II. Disintegration rates of various extruded rods and tubes.

ID	ROD (mg/cm^2/hr)
S1	0.649
S2	0.37
S3	2.24
S4	2.76
S5	6.365
S6	1.33
S7	58.78
S8	21.09
S9	15.12

Disintegration rates can be controlled using reinforcement and coating chemistry while maintaining uniform disintegration, as shown in **Fig. 6 (b).** In one case, rates as high as 300 mg/cm^2/hr were achieved. The tested coupons appeared dark grey with surface pitting when compared to **Fig. 6 (a).**

Fig. 6. Example of sample coupons of extruded parts (a) prior to testing and (b) after exposure to brine at 93.3°C .

Mechanical testing. The extruded product was tested for mechanical properties as shown in Table III (use table I as reference to extrusion type). 2.54 cm OD (S1) rods averaged 67.5 ksi (465 MPa) UCS with 7.1% strain while the 5.08 cm OD (S2) rod resulted in 63 ksi (434 MPa) UCS and 6.9% strain. The slight difference in UCS can be attributed to the lower extrusion ratio shown in table I, 18.1 versus 4.5.

Likewise, S3 and S4 billets were extruded into 7.62 cm and 10.01 cm OD rods, each showed average UCS of 62 ksi (428 MPa) with 11.4% and 13.3% strain, respectively. In this case, the slight difference in ratio, 7.1 versus 4.1, didn't affect the UCS.

Table III. Mechanical data of various DMMC rods and tubes.

ID	UCS, KSI (Mpa)	Strain % at Max UCS
S1	67.45 (465)	7.12
S2	63 (434)	6.93
S3	62.3 (430)	11.42
S4	62 (428)	13.32
S5	69.1 (476)	13.25
S6	53.7 (370)	13.23
S7	63.2 (436)	13.84
S8	54.8 (378)	10.76
S9	61.8 (426)	13.98

The ultimate compression strength (UCS) was plotted against the disintegration rate for rods and tubes (**Fig. 7** and **Fig. 8**, respectively). The disintegration can be controlled for a given strength.

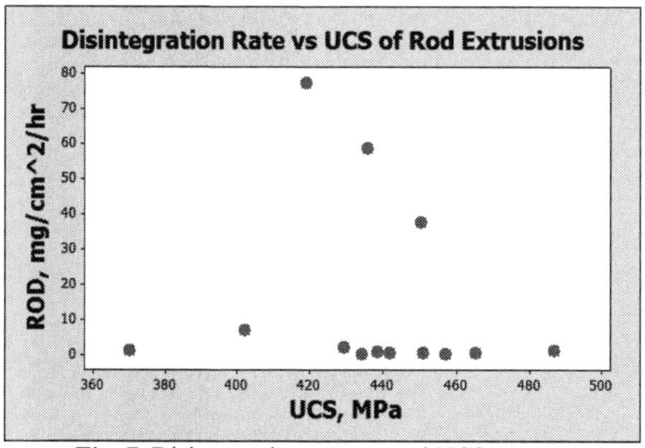

Fig. 7. Disintegration rate vs. rod UCS.

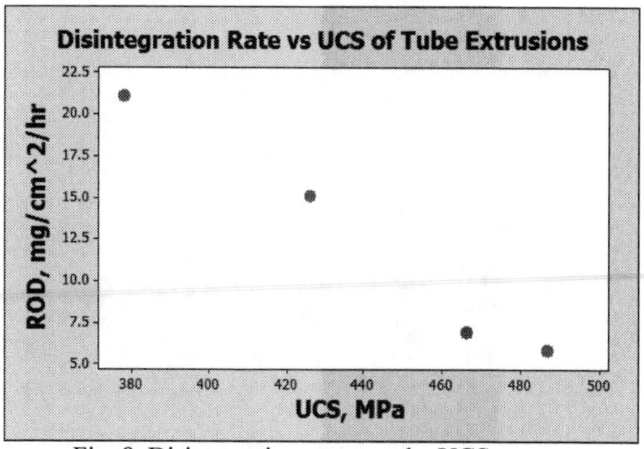

Fig. 8. Disintegration rate vs. tube UCS.

In **Fig. 9**, the UCS is plotted versus the extrusion ratio. The chart shows rods as solid dots and tubes are triangles. The UCS and extrusion ratios are proportional. Increased strength can be achieved with higher extrusion ratios.

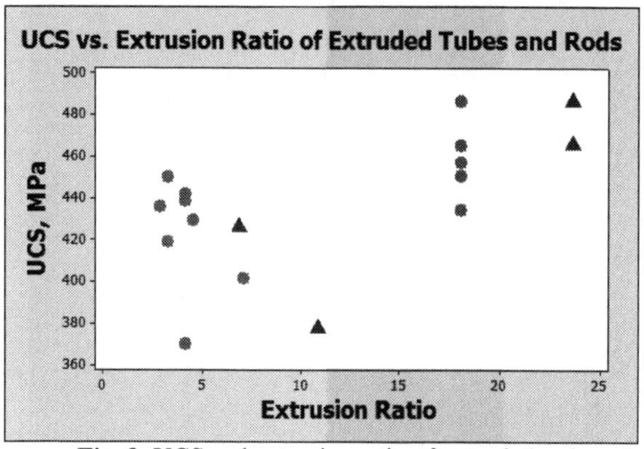

Fig. 9. UCS and extrusion ratio of extruded rods and tubes.

Microstructures.
As listed in Table I, rods and tubes were extruded into various ODs/IDs from forged and CIP billets. The microstructure of forged S2, **Fig. 10**, has a uniform structure, and the individual grain boundaries are clearly visible around the base particle. Once extruded, **Fig. 11**, the grain boundary, is spread along the direction of extrusion, and is depicted by the arrow.

Fig. 12 shows a micrograph image of as-extruded S4 extruded from a preform billet, as in **Fig. 2**. The extrusion direction is denoted with the arrow, but has a less visible grain boundary. The boundary difference is because of the pre-manufacturing conditions chosen for each powder. All materials extruded exhibited similar microstructure profiles relative to their prepping conditions. The increased surface area is clearly visible in the extruded profiles, and contributes to a slight increase in the disintegration rate, compared to as-forged.

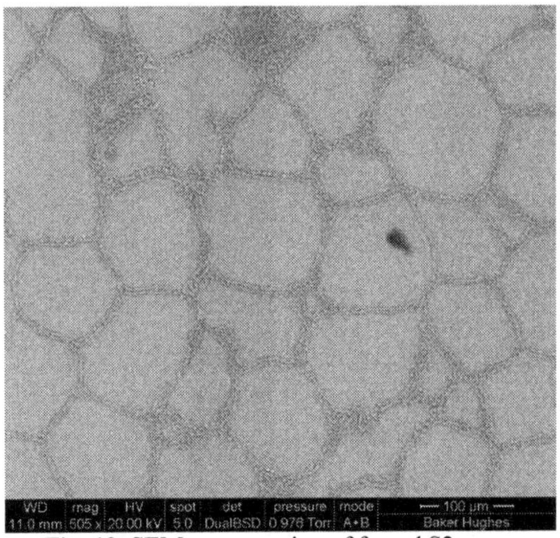

Fig. 10. SEM cross-section of forged S2.

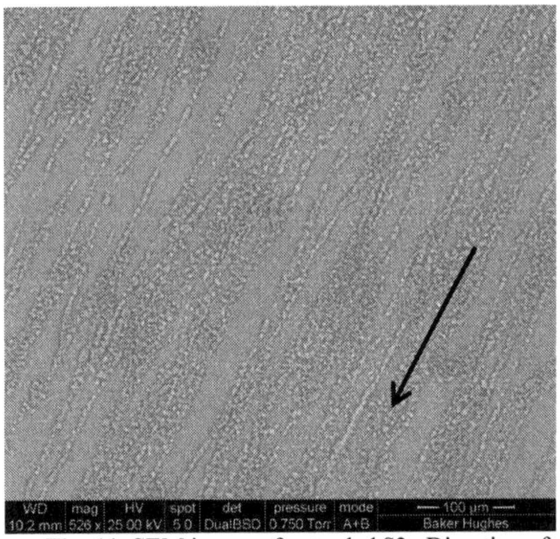

Fig. 11. SEM image of extruded S2. Direction of extrusion is depicted by arrow.

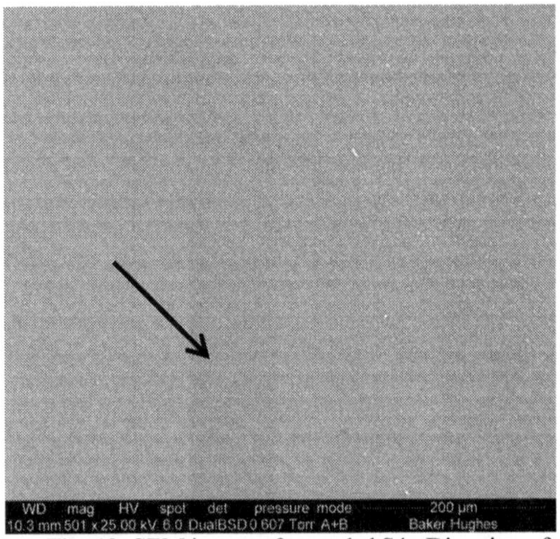

Fig. 12. SEM image of extruded S4. Direction of extrusion is depicted by arrow.

Conclusion

Disintegrable metal matrix composites with unique properties and microstructures are ideal structural materials for making interventionless downhole tools or components. Extrusion of preform DMMC billets, to tubes and rods, has been successfully completed. Extrusion from the preform billet shows we can achieve ROD and UCS comparable to as-forged. Disintegration rates can be controlled from 0 – 300 mg/cm^2/hr, in 3% KCl at 93.3 °C while maintaining >60 ksi (414 MPa) strength.

Acknowledgements

The authors wish to thank Baker Hughes for the permission to publish this work.

References

1. B.J. Salinas, Z. Xu, G. Agrawal, and B. Richard, "Controlled Electrolytic Metallics – An Interventionless Nanostructured Platform," Proceeding of SPE International Oilfield Nanotechnology Conference and Exhibition, Noordwijk, The Netherlands, 12 – 14 June 2012, SPE 153428.

2. Z. Xu, G. Agrawal, and B.J. Salinas, "Smart Nanostructured Materials Deliver High Reliability Completion Tools for Gas Shale Fracturing," Proceeding of the SPE Annual Technical Conference and Exhibition, Denver, Colorado, 30 October – 2 November 2011, SPE 146586.

3. Z. Zhang, Z, Xu, and B. Salinas, "High Strength Nanostructured Materials and Their Oil Field Applications," Proceeding of the SPE International Oilfield Nanotechnology Conference and Exhibition, Noordwijk, The Netherlands, 12 – 14 June 2012, SPE 157092.

4. Z. Xu, B. Richard, B. Salinas, and Z. Zhang, "Self-Disintegrating Materials and Their Field Applications," Presentation at the World Oil Shale Energy Technology Conference, Houston, Texas, 23 August 2012.

Magnesium Technology 2013
Edited by: Norbert Hort, Suveen N. Mathaudhu, Neale R. Neelameggham, and Martyn Alderman
TMS (The Minerals, Metals & Materials Society), 2013

EFFECT OF FIBER VOLUME FRACTIONS ON CORROSION RESISTANCE OF MG AM60 ALLOY-BASED COMPOSITES IN NACL SOLUTIONS

Xuezhi Zhang [1,a], Xiaoping Niu [2], Henry Hu [1,b]

[1]Department of Mechanical, Automotive and Materials Engineering
University of Windsor, 401 Sunset Ave.
Windsor, Ontario, Canada N9B 3P4
[2]Promatek Research Centre, Cosma International, Brampton, Ontario, Canada L6T 5R3,
E-mail: [a]zhang11w@uwindsor.ca, [b]huh@uwindsor.ca

Keywords: Magnesium composites, Corrosion behaviors, Squeeze casting, Fiber volume fractions

Abstract

Magnesium-based MMCs with their high stiffness-to-weight ratio bring a great interest to automotive and aerospace industries. Besides the enhanced mechanical properties of the MMCs, corrosion resistance is often a concern for applications being used in harsh environment. In this study, Magnesium alloy AM60 matrix composite reinforced with 7 vol%, 11 vol% and 22 vol% Al_2O_3 fibers were squeeze casted. The corrosion behavior in aqueous solutions containing 3.5% NaCl was investigated in comparison with magnesium alloy AM60. The scanning electron microscopy (SEM) was employed for the microstructure analysis of the composites and matrix alloy AM60 before and after corrosion test. The results show that the addition of Al_2O_3 fiber deteriorated the corrosion resistance of the matrix and the corrosion rate increases as the fiber volume fraction increased. The formation of the interfaces between the reinforcement and the matrix should be the primary mechanism for the corrosion to easily take place.

Introduction

To further improve the performance of lightweight materials in automotive and aerospace industries, metal matrix composites (MMCs) are introduced and developed. Magnesium metal matrix composites (Mg MMCs) become popular for applications in automotive industries due to their high strength and high stiffness with lower density. Alumina (Al_2O_3) fiber is one of the popular reinforcing materials for Mg MMCs. It has been reported that Mg/Al_2O_3 composites can be potentially applied for applications in power transmission housing, rotating shafts and also aircraft engines [1, 3].

Preform-squeeze casting technique is considered to be a desirable and cost effective process among the available techniques for fabricating Mg MMCs. The volume fraction and the distribution of Al_2O_3 fibers can be easily controlled during fabricating preforms. The volume fractions and the uniformity of the fiber have significant influence on the physical and mechanical properties of the composites [4-6].

Even though the addition of the reinforcement can improve the physical and mechanical properties of the composites, it also can deteriorate the corrosion resistance of the composites 7]. In this study, the influence of Al_2O_3 fiber volume fractions on the corrosion behavior of Mg AM60/Al_2O_3 composites by preform-squeeze casting was carried out in aqueous solutions containing 3.5% of NaCl by electrochemical experimentation and microstructure analysis.

Experimental Procedure

The base magnesium alloy selected for this study was the conventional magnesium alloy AM60. The chemical composition of this alloy is shown in Table 1[2]. Al_2O_3 short fibers with their cost effective and possess adequate properties were employed as the raw materials for preparation of preforms. The thermophysical properties of the matrix alloy AM60 and the Al_2O_3 fiber are listed in Table 2[8]. Both of the unreinforced AM60 and the composites with fiber volume fractions of 7%, 11% and 22% were cast under the same conditions. The details of fabricating preforms and composites can be found in reference [9].

Figure 1: A preform with fiber volume fraction of 9%.

Table 1: Chemical composition of the investigated alloy (mass fraction, %) [2]

Alloy	Al	Mn	Si	Fe	Mg
AM60	5.93	0.18	<0.02	0.013	Bal.

Table 2: Thermophysical properties of Al2O3 short fibers and matrix alloy AM60 [8]

Materials	**Al2O3 fiber**	**AM60**
Elasticity modulus (GPa)	200	35-44
Density (kg/m^3)	3400	1740
Diameter (μm)	4.0	
Heat expansion coefficient (10^{-6}/K)	6	45
Specific heat (J/kg K)	1000	1250
Thermal conductivity (W/ m K)	5	85

During casting process, protective gas mixture of sulfur hexafluoride, SF_6 and carbon dioxide, CO_2 in which CO_2 act as the carrier gas, was applied to protect the melt from any excessive oxidation or possible burning. All tools were preheated to 150 °C on the top of a box furnace for at least 20 minutes. The upper and lower molds were preheated to 300 °C. Before placing the preform into mold, the preform was preheated to 750 °C. Then, the molten matrix alloy AM60 with temperature of 760 °C was infiltrated into the preheated preform under an applied pressure of 90 MPa. The pressure was hold for 25 seconds. The heater for the mold was quickly turned off after the pressure withdrawal in order to cool the mold as soon as possible. After solidification, a cylindrical composite coupon was ejected. All of the 7, 11, 22 and 35 vol% composites were fabricated in the same procedure. Figure 2 illustrates the fabrication process of the fiber-reinforced composite by using the squeeze casting technique.

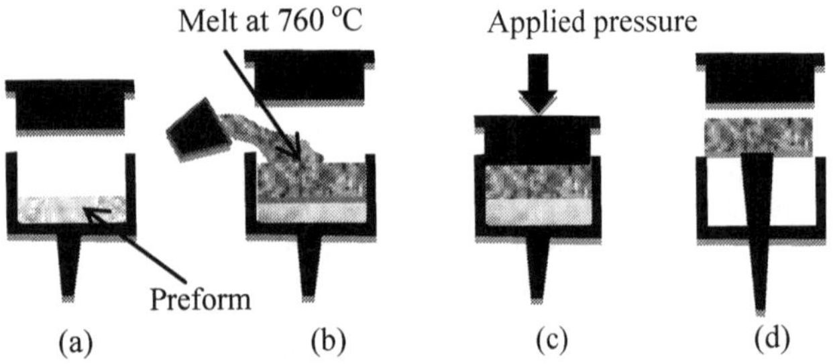

Figure 2: Schematic diagram of the squeeze casting procedure a) placing preform into the mold, b) pouring melt into the mold, c) applying pressure and d) composite is produced.

The rectangular samples (20x20x5mm) were cut form the middle part of the cast. Figure 3 shows a sample squeeze cast composite. All samples were polished prior the corrosion tests. All samples were ground by using silicon carbide papers with grades 240, 600 and 2500 grits. Then the samples were cleaned in acetone, rinsed with deionized water and dried before the potentiodynamic polarization.

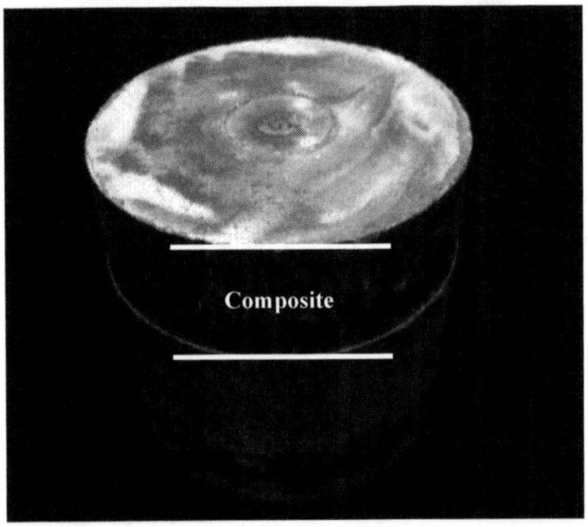

Figure 3: A squeeze cast magnesium matrix composite (AM60/ Al_2O_3).

Electrochemical Experimentation

The equipment used for the corrosion tests were EC-LAB SP-150 electrochemical apparatus with corrosion analysis EC-LAB software. A three-electrode cell was used for all tests. The prepared samples were set to be the working electrode, Ag/AgCl/sat'd KCL electrode as a reference electrode and Pt metal electrode as counter-electrode. For all of the experiment, 3.5% NaCl solution was prepared (salt mixing with deionized water). At the beginning of the test, samples were immerging into the salt solution to allow the open circuit potential to settle to a constant value. Potentiodynamic polarization scans were conducted at a rate of 10mv/s form –0.5v versus open circuit potential in a more noble direction up to 0.5v versus the reference electrode.

Microstructure Analysis

A Buehler optical image analyzer 2002 system was used for determining the primary characteristics of the composite. The detailed features were characterized at higher magnification by JSM-5800LV SEM, which had a maximum resolution of 100 nm in a backscattered mode and maximum useful magnification of 30,000X. Before placing the samples into the SEM, they were coated with Au and a copper tape was used on the surfaces to enhance the sample conductivity to eliminate the surface charging.

Results and Discussion

Microstructure

The distribution of the reinforcement has significant influence on the properties of the composites. To ensure the fibers were randomly and uniformly distribute in the composites, the samples were etched at difference time periods to bring out the fibers from the matrix alloy and to reveal their distribution at different depths of the composites. Figure 4 shows the composites surfaces of the composite with the reinforcement of 22% Al_2O_3 fibers which were etched for 10s, 50s and 110s.

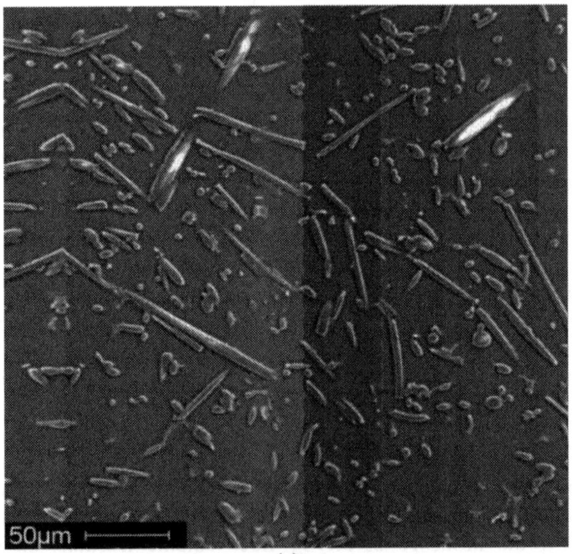

(a)

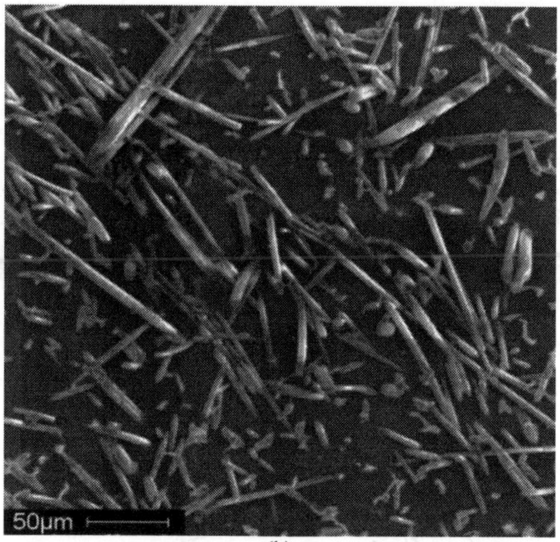

(b)

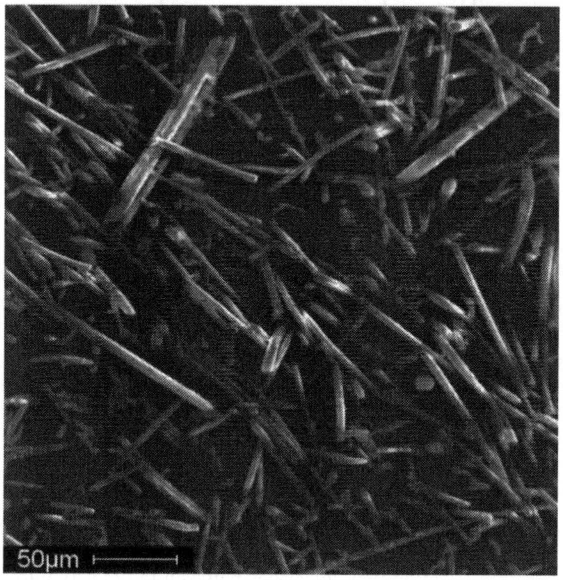

(c)

Figure 4: SEM micrographs showing the 22% fiber reinforced composite samples etched by (a) 10s, (b) 50s and (c) 110s.

As shown in figure 4, the fibers were distributed in a random and isotropic orientation and no agglomeration observed. This uniformity of the fiber distribution evidenced great contribution of the infiltration process during squeeze casting. The previous study found that the porosity has significant influence on the corrsion current values as well as the corrosion resistance [9]. Thus, the good infiltration could effectively improve the corrosion resistance from minimizing the formation of the porosity during squeeze casting.

An evident difference in grain size was found, between the unreinforced alloy AM60 and the composites with 7% and 11% fiber volume fractions. The change in grain size implied that the addition of fibers led to a finer grain structure of the composites. Figure 5 reveal the grain structures of the unreinforced alloy AM60 and the composites reinforced with 7% and 11% fiber. The grain size measurement for 22% fiber reinforced composite was not included in this work, since the grain boundaries were mostly covered by the reinforcing fibers and it was difficult to investigate the grain structure and measure the grain size quantitatively, as shown in Figure 5 (d).

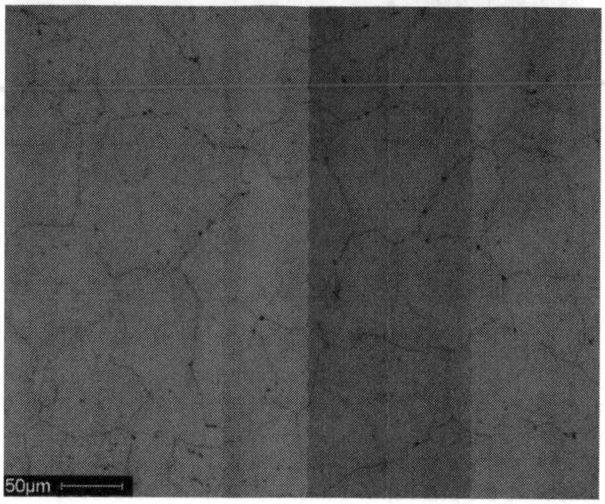

(a)

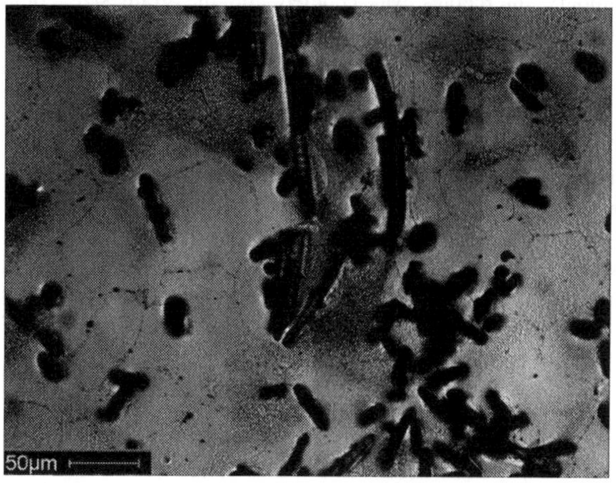

(b)

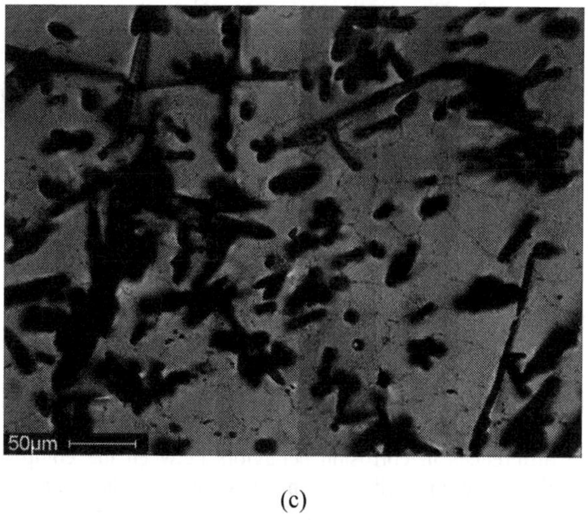

(c)

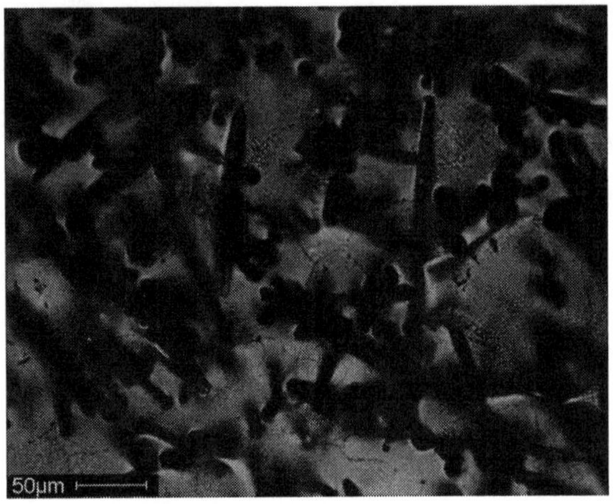

(d)

Figure 5: Optical micrographs showing grain structures of (a) unreinforced AM60, (b) 7%, (c) 11% and d) 22% fiber reinforced composites, repectively (all samples are under T4 condition).

To measure the grain size, a line was drawn across the mirograph and its length was noted. Then, the stage micrometer was used to determine the real length of the line. The number of the grain boundary interceptes was counted and the grain size could be determined by applying the following equation.

$$d = \frac{l}{n} \quad \text{(Eq. 1)}$$

Where, d is the grain size, l is the length of the line and n is the number of grain boundary intercepts.

As can be seen from Figure 6, the grain size of the composites decreased after the addition of the fibers. The grain size decreased from 67 μm for matrix alloy AM60 to 36 μm for 11% fiber reinforced composite.

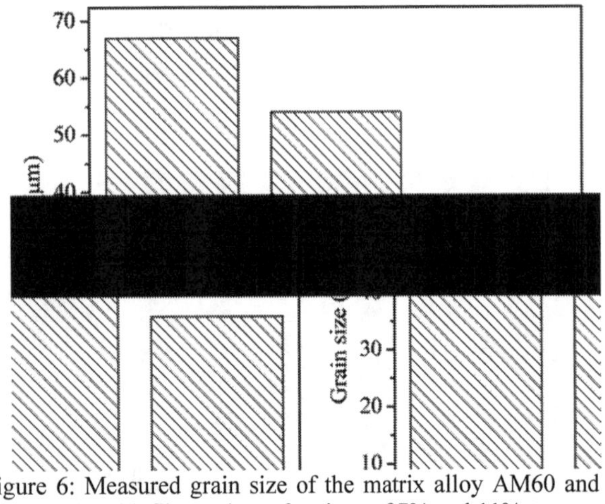

Figure 6: Measured grain size of the matrix alloy AM60 and its composites with fiber volume fractions of 7% and 11%.

Potentiodynamic Polarization

The differences in corrosion behaviors between the composites with the variation in fiber volume fractions and the matrix alloy AM60 are illustrated in Figure 7. The current density (i_{corr}) and polarization resistance (R_p) obtained by Tafel calculations are listed in Table 3. The polarization curves of the composites shifted to higher current densities from comparing the results between the composites and the matrix alloy. As fiber volume fraction increased, the current density rose. On the other hand, the R_p decreased by the addition of Al_2O_3 fibers and the Rp values of the composites were much lower than that of matrix alloy, as shown in Figure 5. By examining the values of the corrosion resistances listed in Table 3, the R_p of the 7% composite (2.301 kΩ cm^2) decreased by 73%, and the R_p of the 22% composite (1.061 kΩ cm^2) further decreased by almost four times in comparison with the matrix alloy (3.995 kΩ cm^2).

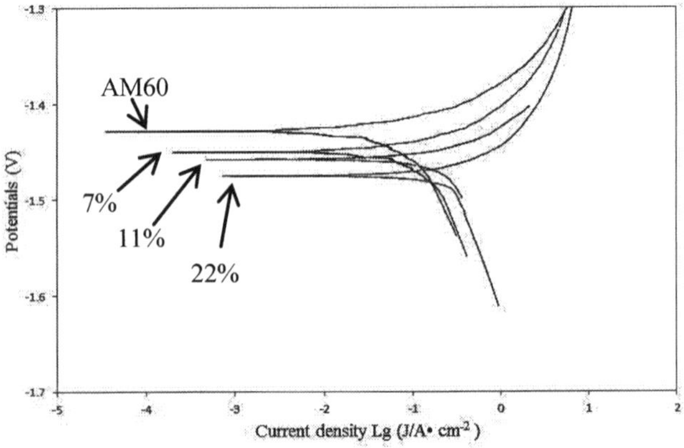

Figure 7: Potentiodynamic polarization curves for the unreinforced alloy AM60 and its composites with fiber volume fractions of 7%, 11% and 22%.

The previous corrosion studies on pure magnesium was carried out in the same 3.5% NaCl solution [10]. It has found that the grain refinement decrease the corrosion resistance of magnesium and the it was attributed to the higher density of grain boundaries and other defects from the grain refinement. However, in the case of fiber reinforced composite, the high density of grain boundaries might not be the main mechanism of decreasing in corrsion resistance. The introduction of the fibers in the composite generated new interfaces between the matrix alloy and Al_2O_3 fibers. As fiber volume fraction increased, a great deal of new interfaces generated in the composite. As a result, the presence of the new interfaces break the continuity of the matrix and form preferential spots for corrosion attack, and consequently the corrosion resistance of the composite decreased.

Table 3: Potentiodynamic polarization parameters of AM60 and its composites with fiber volume fractions of 7%, 11% and 22%

Sample	β_a (mV/dec)	β_c (mV/dec)	I_{corr} (μA/cm^2)	R_p (kΩ cm^2)
AM60	53.1	263.4	4.8	3.995
7%	31.5	189.1	5.7	2.301
11%	41.9	567.1	10.5	1.583
22%	32.3	245.3	11.6	1.061

Conclusions

As 7%, 11% and 22% fibers were added into the matrix alloy, uniform and random fiber distribution was obtain which guarantied the good infiltration during squeeze casting. The addition of Al_2O_3 fibers led to a significant grain refinement, which were 36 µm for 22% composite in comparison with 67 µm for matrix alloy AM60. The electrochemical testing results showed that the presence of Al_2O_3 fibers deteriorated the corrosion resistance of the composites. The decrease in corrosion resistance should attribute to the new interfaces generated as the fiber reinforcement added into the matrix alloy.

Acknowledgements

The authors would like to thank the Natural Sciences and Engineering Research Council of Canada, and University of Windsor for supporting this work.

References

[1] Q. Zhang, *Development of Hybrid Magnesium-based Composites, Advances in Light Weight Material- Aluminum, Casting Materials, Magnesium Technologies,* SAE World Congress, 2010, P107-114.

[2] H. Hu, *Squeeze Casting of Magnesium Alloys and Their Composites,* Journal of Material Science, 1998, P1579-1589.

[3] A. Banerji, *Ultra-mild Wear of Al2O3 Fiber and Particle Reinforced Magnesium Matrix Composites,* Advanced Materials Research, 2012, P503-508.

[4] Y.S. Yang, K. Feng and W. X. Zhao, *Fabrication and Properties of Al2O3 Short Fiber Reinforced Aluminum Matrix Composites*, Advanced Materials Research, 2011, P1277-1280.

[5] J. S. S. Babu, K. P. Nair, G. Unnikrishnan, C. G, Kang and H. H. Kim, *Fabrication and Properties of Magnesium (AM50)-based Hybrid Composites with Graphite Nanofiber and Alumina Short Fiber*, Journal of Composite Materials, 2010, P971-987

[6] M. Mizumoto, T. Qhgai and A. Kagawa, *Effect of Alumina fibers on Fabrication process and Characteristics of Alumina Fiber Reinforced Aluminum Alloy Composites,* Material Science Forum, 2010, P956-960.

[7] Q. Zhang, H. Hu, *Effect of Fiber Reinforcement on Corrosion Resistance of Mg AM60 Alloy-based Composites in NaCl Solutions,*TMS, 2011, P469-474.

[8] T. W. Clyne, M. G. Bader, G. R. Cappleman, P, A. Hubert, *The Use of a δ-Alumina Fiber for Metal Matrix Composites,* Jounal of Material Science, 1985, P85-96.

[9] Q. Zhang, *Development of Hybrid Mg-based Composites*, Master thesis, University of Windsor, Windsor, Ontario, Canada, 2010.

[10] E. Ghali, *Corrosion Resistance of Aluminum and Magnesium Alloy: Understanding, Perfromance, and Testing*, Wiley, 2010, P537

[11] D. Song, A. Ma, J. Jiang, P. Lin, D. Yang, J. Fan, *Corrosion behavior of Equal-Channel-Angular-Pressed Pure Magnesium in NaCl Aqueous Solution,* Corrosion Science, 2010, P481-490.

Magnesium Technology 2013
Edited by: Norbert Hort, Suveen N. Mathaudhu, Neale R. Neelameggham, and Martyn Alderman
TMS (The Minerals, Metals & Materials Society), 2013

SYNTHESIS OF DISINTEGRABLE METAL COMPOSITE FOR OILFIELD APPLICATIONS

Zhihui Zhang, Bobby Salinas, Caleb Newman and Zhiyue Xu
Baker Hughes Incorporated; 14990 Yorktown Plaza Drive; Houston, TX, 77040, US

Keywords: Composite, Disintegrable, Corrosion, Hydraulic Fracturing

Abstract

Lightweight metal composites were traditionally developed for weight-critical applications. Recently, significant efforts have been made to improve the corrosion performance. In this study we report the synthesis of a novel type of high-strength metal composites with enhanced corrosion rate for use in self-disintegratable tools in oil and gas wells. The composites were fabricated through a powder metallurgy procedure by consolidating reactive metal powders that were coated with nanoscale metallic and/or ceramic coatings. The interaction between the metal matrix and coating was studied using X-ray diffraction, differential scanning calorimetry, and electron microscopy. The composites exhibit simultaneous high strength (up to 460 MPa) and two orders of magnitude increase in the corrosion rate (i.e., 250 $mg/cm^2/hr$) under saline water. The corrosion behavior and associated field applications are described.

Introduction

The need for lightweight metal composites has recently generated renewed interest, given their potential to improve energy efficiency and limit carbon dioxide emissions- particularly in light of recent combinations with nanotechnology, which may offer them unusual physical, mechanical and other functional properties. Many inquiries/investigations have been directed toward the development of magnesium composites reinforced with nanoparticulates such as SiC, Y_2O_3, Al_2O_3, and carbon nanocubes because of their intrinsic low density and superior mechanical properties (for purposes such as weight savings and/or better design of weight distribution and balance in the aerospace, automobile and sporting goods industries). In most of these applications, however, corrosion prevention and mitigation represents a challenge, limiting their wide-spread use. In addition, because of the biocompatibility of magnesium and its closeness of Young's modulus to that of the human bone, magnesium alloys have been considered a promising material for biodegradable implant application (to thwart the need to re-open an implant after healing). In this case, a controlled manner of corrosion is preferred. Controlled corrosion of certain structural components and parts also represents tremendous technical and economic advantages in oil and gas well completions. In recent decades, production of natural gas from entrapped low permeability shale formations (shale gas) has grown rapidly from virtually nothing in 2000 to more than 10 billion cubic feet per day in 2010. Shale gas supply is expected to continue increasing, and account for more than 50 percent of the total US natural gas production by the 2030s. To extract natural gas economically from the low permeability shale plays, the wellbore must be drilled horizontally and completed with multistage hydraulic fracturing along the shale formations (ranging from ten to hundreds of meters in thickness) to extend the production zone. One-trip deployments offer an efficient multistage fracturing system by using setting balls and ball seat design to activate one sleeve at a time, thus allowing the entire lateral to be fractured stage by stage in the horizontal wellbores. If the balls and ball seats must be removed mechanically from the well after fracturing to clear the pathway for production, this means re-entering the wellbores with costly tools and decreasing system efficiency. One solution utilizing the wellbore fluid to flow the setting balls back may alleviate the problem, but the density requirement for flowback leaves these lightweight materials, such as phenolic, weak and prone to severe plastic deformations. If the formation fluid pressure is very low, it can be difficult to get the frac balls off seat or to flow back to the surface. The balls can also be obstructed by the residual frac/formation sand when production velocity is not high enough. Therefore, the superior solution is to develop strong and reliable materials so they are able to perform their functions during fracturing, the tool then chemically disintegrates itself under the wellbore fluids after the fracturing is completed [1]. Although materials possessing individual high strength, light weight or high disintegration rates have been well documented, integrating these three functions into one material still represents a technical challenge. This paper presents a novel composite, disintegrable metal matrix composite (DMMC), to meet such demand. The base materials were selected with three criteria to meet the interventionless downhole application requirement: (1) light weight, (2) controllable disintegration rate when engineered with corrosion enhancer constituents, (3) achievable high strength by microstructural and/or process engineering. Light weight is necessary for the setting balls to flow with wellbore fluid during deployment. A controllable disintegration rate requires the DMMC material to not only exhibit a wide range of disintegration rates for different applications, but also corrode slowly before a critical service time (for the system to complete its function) and then disintegrate rapidly following the service. In general the setting ball should disintegrate enough to become unseated in hours. During deployment, the ball can travel with a flow speed as high as 160 km/hr. Therefore, the DMCC material must also exhibit high strength and toughness.

Experimental Procedures

Reactive metallic powders were coated with metallic and Al_2O_3 coatings with a thickness in the range of 30~1000 nm using chemical vapor deposition (CVD). Table 1 lists the materials and sample designations in this study. The coated powders were then consolidated using a powder hot forging process with a pressure of 400 to 600 MPa at a processing temperature between 250 and 550 °C.

The microstructure was characterized using X-ray diffraction (XRD), differential scanning calorimetry (DSC), scanning electron microscopy (SEM) and transmission electron microscopy (TEM). The XRD was conducted on a Scintag X1 diffractometer using CuKα radiation. The DSC was carried out using a TA Instruments Q10 differential scanning calorimeter. The SEM was conducted in a FEI Quanta 600 scanning electron microscope. The TEM was conducted in a JEOL JEM2100F transmission electron microscope.

The corrosion rate was evaluated by weight loss in a 3% KCl solution at 100 °C, a modified corrosion immersion test method to simulate the downhole fluid environment. Cylinder test coupons were machined to a dimension of 12.7 mm in diameter and 25.4 mm in length. For each data point, the corrosion rate, or disintegration rate, was determined from the average of three coupons. Each sample coupon was measured prior to and after the test and the weight loss rate of corrosion/disintegration (disintegration rate, $mg/cm^2/hr$) was calculated with a time span of 24 hrs.

The mechanical properties were evaluated with compressive tests. Cylinder specimens were cut to a dimension 12.7-mm diameter and 25.4-mm length, according to ASTM E9 standard. The mechanical tests were conducted on a MTS electromechanical test machine with a load cell capacity of 100kN.

Table 1

Sample	Coating	
	Structure	Total thickness, nm
A	Base/M/Bond	1250
B	Base/M/Bond	1300
C	Base/M/Bond	1400
D	Base/Al_2O_3/Bond	2000

Results and Discussion

The DMMC matrix powder was produced via inert gas atomization with an average particle size of ~100 µm. Figure 1 shows a cross-section view of the DMMC powder following CVD coating (i.e. sample C). The powder particle was first coated with a metallic layer (designated as M hereafter) that has a standard electrode potential higher than Zn to form galvanic cell reactions when exposed to wellbore fluid. A bond coat was subsequently applied on top of the M layer to prevent contamination and facilitate consolidation. A series of different thicknesses were selected to study their effect on disintegration rates and mechanical strength (i.e. two key criteria for the ball deployment under the wellbore condition), as listed in Table 1. An Al_2O_3 layer substitute for the M layer was also selected for comparison purpose to differentiate the effect of M layer on the mechanical strength and disintegration rate. The chemical reactivity between the matrix and bond layer was accessed by using DSC scanning the CVD-coated powder with a heating rate of 10 °C/min, as shown in Figure 2. A moderate reaction between the bond coat and matrix powder is favored to form a metallurgical bond during the powder consolidation. The endothermic reaction at approximately 439 °C is expected from the phase diagram between the matrix alloy and bond coat. The sample with M layer shows a much lower reaction peak than the sample with Al_2O_3 layer. The DSC result is consistent with the microscopic examination showing that the M layer is almost fully continuous, whereas the Al_2O_3 layer is subjected to certain microporosity.

Despite different thickness and chemistry, the CVD coating was able to well bond to the matrix powder and undergo powder consolidation. As shown in Figure 3, the CVD coating formed a continuous network in the consolidated bulk composites, with no cracking or splitting observed between the matrix and M layer (Figure 3a) or Al_2O_3 layer (Figure 3b).

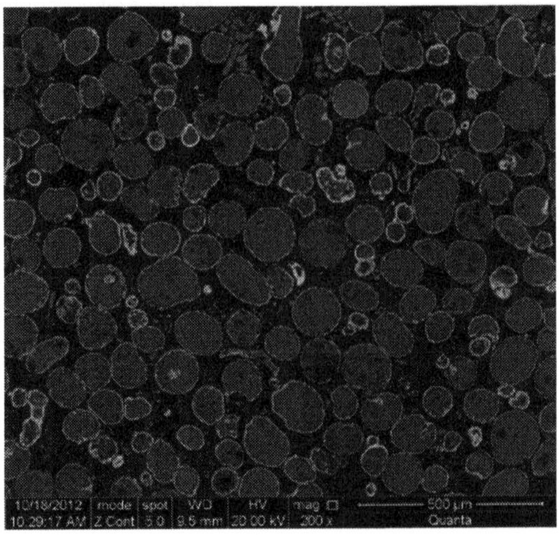

Figure 1. CVD coated DMMC powder.

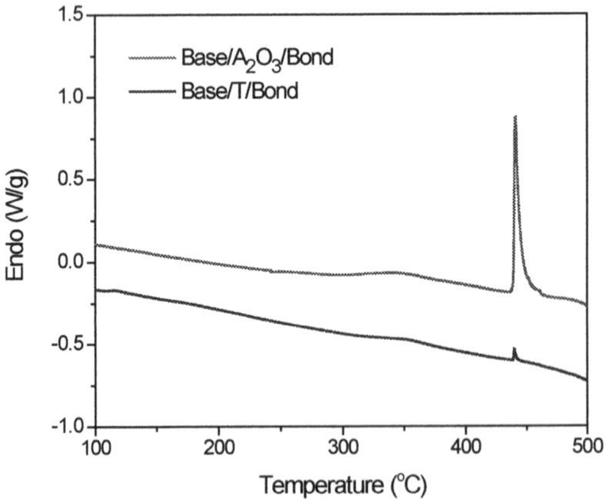

Figure 2. DSC continous heating curves of DMMC powders at a heating rate of 10 °C/min.

(a) 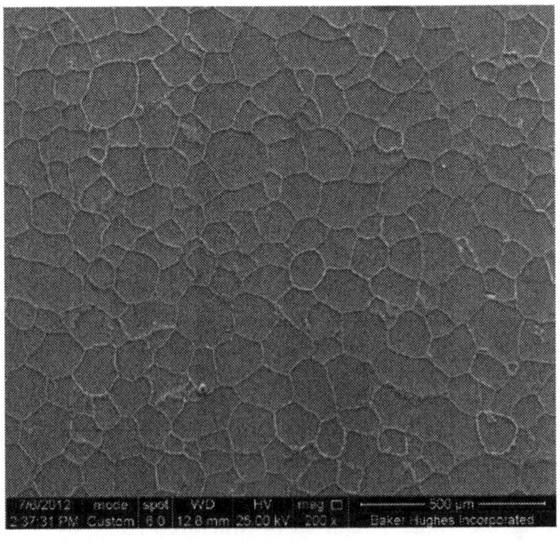

(b) 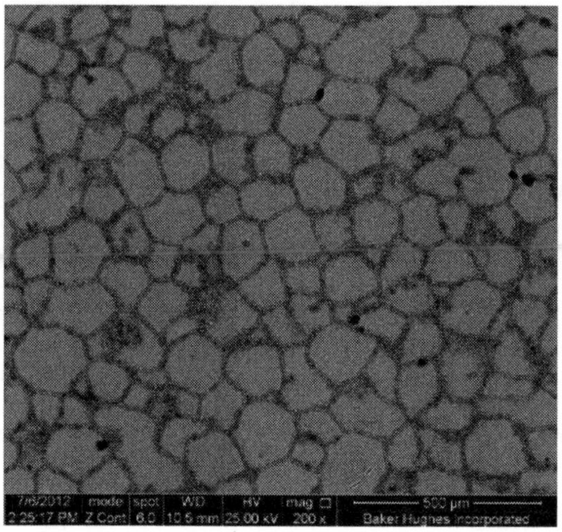

Figure 3. SEM micrographs of bulk DMMC composites: (a) Sample A, (b) Sample D.

Figure 4 shows a TEM bright field image taken from sample D. The interface between the matrix and CVD coatings (marked by a thick arrow) remained intact without interfacial debonding or delamination. Precipitates on the order of 200 nm were present in the matrix. XRD analysis (not shown here) indicated the majority of secondary phases were indexed to be $Mg_{17}Al_{12}$. The presence of $Mg_{17}Al_{12}$ was also identified in other samples by XRD.

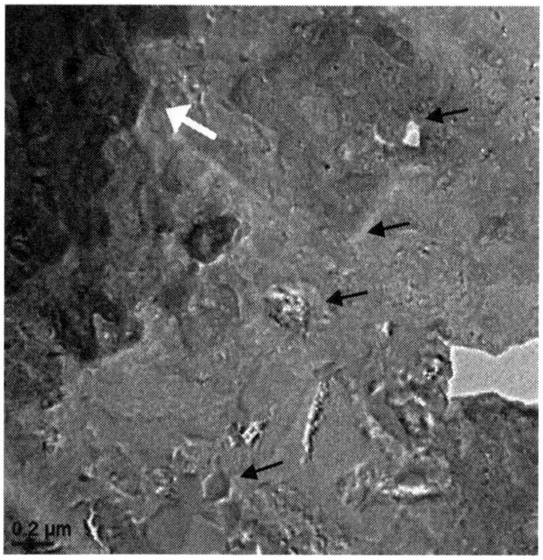

Figure 4. TEM bright field image from sample D.

The mechanical properties and rate of disintegration are shown in Figure 5(a) and (b), respectively. The disintegration rate for the first few hours (i.e. ~4 hours) are slightly lower and then become flat to nearly a constant up to 120 hours. An initial lower corrosion rate is an advantage to ensure the service functionality before the material is ready to dispose itself. In this study, the reported disintegration rate was measured with an average rate spanning 24 hrs. During application, a rate of 50 to 100 mg/cm^2/hr, or a diameter reduction rate up to 0.01 to 0.03 inch/hr is usually required to fit the hydraulic fracking operations where ball dismount time should match the shut-in time before production flows can be established. The dismounted ball could then be opted for either flowing back or disintegration by itself under the wellbore fluid, ensuring the elimination of any production pathway blockage. During the self-disposing application, the ultimate compressive strength is often cited as the criterion to determine the maximum pressure it can hold from the wellbore fluid. The compressive strength, as well as the strain to failure obtained from the DMMC composites is shown in Figure 5(a). The strain to failure is dependent on the chemistry of the coating and other factors such as processing defects; as a result, the compressive strength did not show a clear trend when compared to the coating thickness. This suggests an analysis at fixed plastic strain. Considering that moderate plastic deformation is required in the actual application, we selected an offset strain of 2% in this study to understand the strengthening mechanisms operating in the network DMMC (which was also plotted in Figure 5(b). The strengthening contributions in the DMMC were examined in order of (1) solid solution strengthening, (2) grain boundary strengthening, (3) Orowan strengthening, (4) internal stress strengthening, and (5) work hardening.

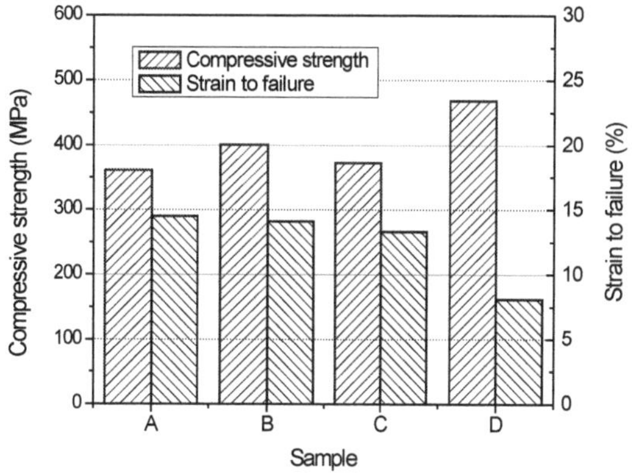

(a)

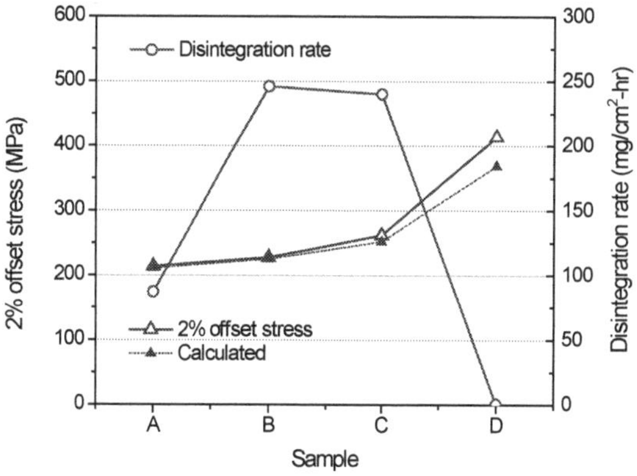

(b)

Figure 5. (a) Compressive strength and strain to failure of DMMC composites, (b) 2% offset stress and disintegration rate as a function of coating thickness and chemistry.

In dilute Mg-Al alloys, the yield stress increment as a result of dissolving solute atoms can be written as [2]:

$$\sigma_{ss} = M \cdot B \cdot c^n \quad (1)$$

where M is the Taylor factor and is usually taken to be 4 to 6.5. The term c is the solute concentration and n = 1/2 ~ 2/3. The term B is the solid solution hardening rate can be determined experimentally. From ref.[2], B = 21.2 MPa $(at.)^{-1/2}$ and M = 5.6 with n = 1/2. Here, the solute concentration c was estimated with an EDX line scan, as shown in Figure 6.

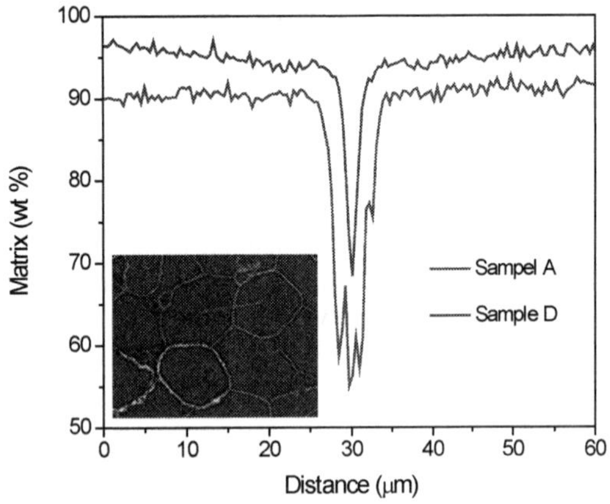

Figure 6. EDX line scan across the CVD coating

Grain boundary strengthening (or Hall-Petch strengthening) represents a strength advantage with a powder metallurgy processing technique over the conventional ingot metallurgy due to the rapid cooling process involved in the feedstock powder production. The yield stress as a function of grain size can be expressed as:

$$\sigma_{H-P} = \sigma_0 + k_y d^{-1/2} \quad (2)$$

where σ_0 = 11 MPa is the yield stress of coarse-grained counterpart, and k_y = 0.21 MPa $m^{1/2}$ is the Hall-Petch slope. The term d is the grain size and is determined to be ~40 to 50 μm by the line intersection method.

The Orowan strengthening results from the interaction of fine precipitates and dislocation in the matrix. The Orowan strengthening can be formulated as [3]:

$$\sigma_{OR} = M \frac{0.4Gb}{\pi(1-\nu)^{1/2}} \frac{In(\overline{d}_p / b)}{\overline{\lambda}_p} \quad (3)$$

where M is the Taylor factor (taken as 5.6), G is the shear modulus (17.4 GPa), b is the Burgers vector (0.32 nm), ν is the Poisson's ratio (0.29). $\overline{d_p}$ and $\overline{\lambda_p}$ are the average particle size and interparticle spacing. The value of $\overline{d_p}$ and $\overline{\lambda_p}$ was measured to be 200 nm and 500 nm, respectively, as shown in Figure 4.

The rate of work-hardening in these samples is dependent on factors such as solute atom concentration and material thermomechanical history; hence, it is generally difficult to formulate. Owing to the fact that all these samples underwent an identical processing procedure, only the effect of solute concentration is considered here using interpolation:

$$\sigma_{W-H} = \Delta\sigma(c_0) + \frac{\Delta\sigma(c_1) - \Delta\sigma(c_0)}{c_1 - c_0} \cdot (c - c_0) \quad (4)$$

where $\Delta\sigma(c_0)$ = 41.2 MPa and $\Delta\sigma(c_1)$ = 82.7. MPa are the work hardening rates for the pure metal and 10 wt% solid solution, respectively, which are determined experimentally from consolidated samples without the CVD coating.

Plastic deformation of a composite produces internal stress which impedes the flow by a stress proportional to the volume fraction of the inclusions and to the applied plastic strain. An estimate of the contribution from network phases was be made with the methods of Brown and Clarke [4]. The mean stress in the matrix due to inclusion is written as:

$$\sigma_{IS} = 4\gamma D G f \varepsilon_p \quad (5)$$

where γ = 1/2(1-ν) is an inclusion shape accommodation factor and ν the Poisson's ratio. $D = G^*/[G^* - \gamma(G^* - G)]$, where G and G^* are the shear modulus of the matrix and inclusion, respectively. The term D depends on the geometry of the inclusion and the ratio of the shear modulus of the inclusion and the shear modulus of the matrix. ε_p is the plastic strain and f is the volume fraction of the inclusion.

The predicted contributions to the flow stress (equations (1)-(5)) are summarized in Table 2. The total flow stress was calculated assuming simple additivity of each individual component. Table 2 indicates an agreement between the predictions and the experimental measurements. The theoretical estimate is on the conservative side, especially for sample D (when the network becomes more rigid due to Al_2O_3). This discrepancy becomes exacerbated if the root mean square law is assumed for each individual contribution. It should bear in mind that the simplified mean stress approach (i.e. equation (5)) ignores the continuous networking effect, which is expected to affect the deformation mode. Nevertheless, the equation (5) shows the CVD coating can significantly influence the material deformation behavior in addition to the disintegration rate.

Table 2

Sample	σ_{SS}	σ_{H-P}	σ_{OR}	σ_{W-H}	σ_{IS}	Cal.	Exp.
A	25.2	58.8	52.3	59.9	10.3	210.7	213.6
B	24.6	59.9	50.7	60.8	25.7	224.8	227.4
C	24.0	61.0	51.6	62.0	51.4	252.1	261.8
D	35.8	63.5	54.2	82.7	132.9	368.6	413.4

In summary, lightweight DMMC composites were successfully engineered with high disintegration rates up to 400 $mg/cm^2/hr$, and compressive strengths up to 460 MPa. A combination of mechanical strength and high disintegration rates enable the DMMC material to be a candidate for many interventionless downhole applications. The strengthening contributions were examined to shed light on ways of achieving even higher strength (>700 MPa) with high disintegration rates for future DMMC material development.

Acknowledgment

This research was sponsored by Baker Hughes and accomplished under the Type-1 research program.

References

1. Z. Xu, G. Agrawal, and B.J. Salinas, "Smart Nanostructured Materials Deliver High Reliability Completion Tools for Gas Shale Fracturing" (Paper presented at the SPE Annual Technical Conference and Exhibition 2011, Denver, Colorado), SPE 146586
2. C. H. Cáceres and D.M. Rovera, "Solid solution strengthening in concentrated Mg-Al alloys", Journal of Light Metals, 1(3) (2001), 151-156.
3. B. Q. Han et al., "Mechanical properties of an ultrafine-grained Al-7.5 pct Mg alloy", Metallurgical and Materials Transactions A, 34(3) (2003), 603-613.
4. L. M. Brown and D.R. Clarke, "Work hardening due to internal stresses in composite materials", Acta Metallurgica, 23(7) (1975), 821-830.

Magnesium Technology 2013

Wrought Materials III

Magnesium Technology 2013
Edited by: Norbert Hort, Suveen N. Mathaudhu, Neale R. Neelameggham, and Martyn Alderman
TMS (The Minerals, Metals & Materials Society), 2013

EFFECT OF PRECIPITATION ON DYNAMIC RECRYSTALLIZED GRAIN SIZE IN A MAGNESIUM ALLOY

Abu Syed Humaun Kabir, Jing Su, In-Ho Jung and Stephen Yue

Department of Mining and Materials Engineering, McGill University
3610 University Street, Montreal, Quebec, Canada H3A 0C5

Keywords: Hot compression, Precipitation, Dynamic recrystallization

Abstract

Magnesium, the lightest structural metal has specific tensile strength and rigidity superior to iron and aluminum showing a rising demand for use in the automotive industries to lighten the structure. But, magnesium and its alloys have limited ductility and poor formability at room temperature due to an insufficient number of operative slip and twinning systems associated with its hexagonal close packed structure. A fine grain structure and random texture can improve the ductility or formability. The aim of this work is to investigate the effect of precipitates on microstructure evolution during hot deformation of Mg-Al-Sn alloy. Mg-Al-Sn alloy has been designed using thermodynamic modeling software, FactSage, based on forming precipitates at deformation temperature of 300 °C. The formation of precipitates, mostly Mg_2Sn, during the dynamic recrystallization process may slow down the grain growth resulting in finer grain size and grain size homogenization.

Introduction

Magnesium is the highest specific strength structural metal and is the most potential candidate for the automotive industry to reduce the weight of the vehicle. It is 36 % lighter than aluminum and 78 % lighter than steel [1]. In comparison to aluminum and steel, very limited wrought magnesium alloys are commercially available [2]. One big reason behind this is, the low formability of Mg alloys at lower temperature (<200 °C) due to its hexagonal close pack (HCP) structure [3] and a preferred crystallographic orientation (texture) that develops in wrought alloys [4, 5]. This leads to a fabrication process with high temperature to increase the formability resulting in, increase in price of magnesium alloy products. Numerous investigations are ongoing in searching for suitable Mg alloy systems with superior properties.

It is well established that mechanical properties of polycrystalline metals at room temperature can be improved by grain refinement according to the Hall-Petch relationship [6]. Also, it has been reported that, the Hall-Petch relationship is four times more significant in magnesium mechanical properties than in aluminum, indicating the importance of grain refinement in magnesium alloys [7]. Grain refinement in hot worked magnesium alloys are usually done through a recrystallization process. Recrystallization is a restoration and usually thermally activated process by which relatively strain free grains are formed within the deformed structure consuming the deformed grains, resulting in, refined grain size [8]. Recrystallization during the hot deformation process is termed as dynamic recrystallization (DRX). Dynamic recrystallization is attributed to magnesium alloy because of its low stacking fault energy (~ 78 mJ m^{-2}) and limited slip systems when a critical deformation condition is reached [9]. DRX generally starts at the parent or old grain boundaries. New grains are subsequently nucleated at the old grain boundaries and make a "necklace structure" by the band of recrystallized grains and eventually the material becomes fully recrystallized. When the time (strain rate) and temperature is not sufficient enough for the complete recrystallization process, partial dynamic recrystallization is observed [10].

Precipitation strengthening is defined by the precipitation of a second phase, which is different in composition than the matrix [10]. In Mg–Sn binary phase diagram, the maximum solid solubility of tin in magnesium is relatively high with 14.85 wt% at the eutectic transformation temperature of 561 °C, which drops off sharply with the temperature. At 200 °C the solubility limit is 0.45 wt% and almost zero at room temperature [11]. This fast decrease in solubility with decrease in temperature makes tin suitable for precipitation in the form of Mg_2Sn. The Mg_2Sn phase precipitates mainly at grain boundaries and may take part in the strengthening effect (increase of the flow stress) by restraining the dislocation movement [12]. More uniform and fine precipitates are formed with plastic deformation, which can generate high energy sites like dislocations, twins and deformation bands. These defects accelerate the diffusion of solute atoms enhancing the subsequent aging kinetics and are the preferential location for the solute atoms to gather due to the minimum of energy needed [13].

Jonas and Weiss [14] reported for micro-alloyed steel that strain induced precipitates prevents the further motion of grain boundary by pinning effect and retards the dynamic recrystallization resulting in a fine grain structure. They also mentioned that the size of the precipitate is an important factor. Smaller precipitates create stronger retardation to dynamic recrystallization. Retardation continues as far as the precipitates size is no larger than the recrystallized nuclei. In general, for the hot working process, the dynamic recrystallization grain size depends on the deformation amount (strain), strain rate, and deformation temperature [15].

In AZ31, the leading alloy commercially available as wrought products, the precipitate forms at lower temperature (~200 °C) than the industrial deformation temperature range of 250-400 °C. So, there is no or little effect of second phase particles on the recrystallized grain size as they are mostly dissolved in the matrix during high temperature deformation [16]. But it is possible to design alloys where high temperature stable precipitates such as Mg_2Sn forms during the hot deformation temperature range of 250-350 °C. It is expected that this stable precipitates will retard the dynamic recrystallization process by the grain boundary pinning [17] or particle simulated nucleation

(PSN) process [10] resulting in reduced recrystallized grain size, which literally means the increase of ductility and strength at room temperature.

The goal of this research is to design alloy where it is possible to form precipitates during high temperature deformation and to investigate the possibility of recrystallized grain size reduction and homogenization by the pinning effect of the precipitates at the grain boundary.

Experimental Procedures

Thermodynamics software FactSage [18] was used to design a candidate Mg alloy with the alloying elements Al and Sn. Tin has been added in the magnesium-aluminum alloy to form thermally stable Mg_2Sn precipitate at deformation temperature. The alloy was then produced with commercially pure Mg, Al and Sn melted using an induction furnace (20 kW, 5 kHz) in a graphite crucible and cast in a copper mold preheated to 250 °C. The alloy addition and casting were done at 690 °C and 730 °C, respectively in a protective environment created by the mixture of CO_2 (99.5 %) and SF_6 (0.5 %) gases. The cast ingots were homogenized for 24 hours at 440 °C to fully dissolve the precipitates. The temperature for homogenization was chosen using the FactSage generated equilibrium diagrams.

Compression samples of 11.4 mm in height and 7.6 mm in diameter (h/d ratio of 1.5) were machined from the homogenized bar. The uniaxial compression was performed using a computer controlled servo-hydraulic materials testing system (MTS) with a 100 kN capacity, coupled with a radiant furnace. Isothermal compression testing was performed at 300 °C with strain rates of 1.0, 0.1, 0.01, and $0.001s^{-1}$. In this study the deformation amount was fixed to 90 % (strain of 0.9). The specimens were held at the deformation temperature for 10 minutes prior to deformation to ensure a uniform distribution of temperature throughout the sample and water quenched immediately after the deformation to preserve the dynamically formed microstructure.

Specimens for metallurgical examination were cut along the compression axis. Polished specimens were etched in a fresh solution of 10 ml acetic acid, 4.2 gm picric acid, 10 ml water and 70 ml alcohol for 3-8 seconds. Microstructural examination was carried out using Nikon Epiphot 200 optical microscope and Clemex image analysis software, with grain size measured by linear intercept method. A Philips XL30 FEG SEM with energy dispersive X-ray spectroscopy (EDS) was used to examine the microstructure and to characterize the second phase particles. Microindentation hardness was measured by Clark CM-100AT machine with a load of 50 gram and a dwell time of 10 seconds.

Results and Discussion

Mg-3 wt% Al-3 wt% Sn (AT33) alloy has been designed using thermodynamic modeling software FactSage in the target of forming precipitate at the deformation temperature of 300 °C. Figure 1 shows the equilibrium phase fraction obtained from FactSage indicating the types and amount of precipitates formed during the equilibrium cooling. Second phase particles Mg_2Sn and $Mg_{17}Al_{12}$ are starting to precipitate out from the Mg matrix just below 380 °C and 210 °C, respectively. Certain amount (80 °C in this study) of undercooling was well thought-out in the alloy design to generate the necessary driving force for the nucleation of second phase particle Mg_2Sn.

Figure 2-a shows the as-cast optical microstructures of the alloy. The microstructure consists of primary α-Mg, and secondary particles between dendrite arms. The average grain size was measured as ~120 μm. The precipitates are coarse having the size around 2-5 μm. EDS analysis confirms that the precipitates are mainly composed of Mg, Al, and Sn.

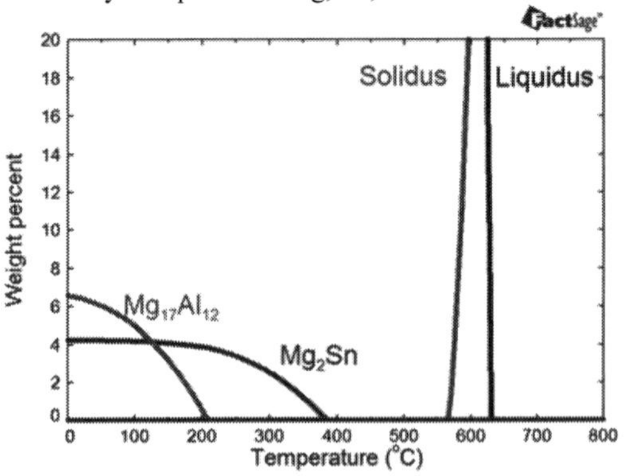

Figure 1: Equilibrium phase fraction of Mg-3Al-3Sn alloy predicted by FactSage.

No precipitate has been found after homogenization for 24 h at 440 °C, as shown by the microstructure in Figure 2-b. The average grain size of the homogenized sample displayed almost no grain growth during the homogenization treatment.

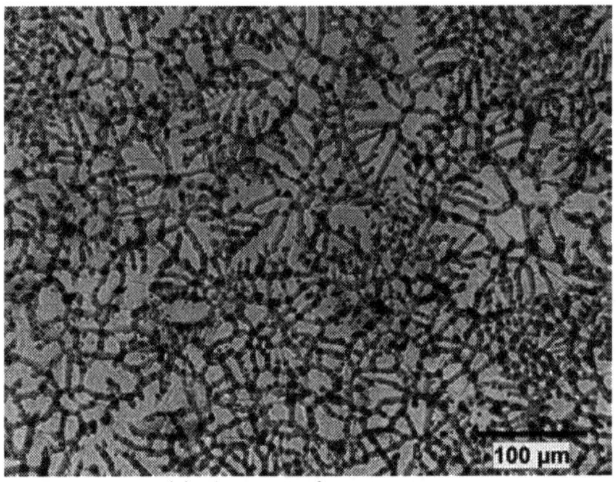

(a) As-cast microstructure

(b) Homogenized at 440 °C for 24 hours to dissolve precipitates

Figure 2: As cast and homogenized microstructures for alloy AT33

Figure 3 shows the variation of flow behavior at 300 °C for different strain rates. The typical true stress and true strain curve obtained at a strain of 0.9 exhibits work hardening, a single peak at low strain (0.1-0.3), after which the flow stresses decreases consistently followed by flow softening. This softening behavior is attributed to dynamic recrystallization phenomena [10]. The effect of strain-rate on the flow behaviour is significant and is shown in figure 4. Peak stress, flow stress, and yield stress (0.2% proof) decrease with decrease in strain rate. This is reasonable with regard to the time dependence of DRX. The decrease in flow stress with decreasing strain rate is due to the increasing ease of dynamic recrystallization and in part to the activation of additional slip systems. Lower strain rates provide more time for boundary migration and accelerate the nucleation and growth processes of dynamic recrystallization [19].

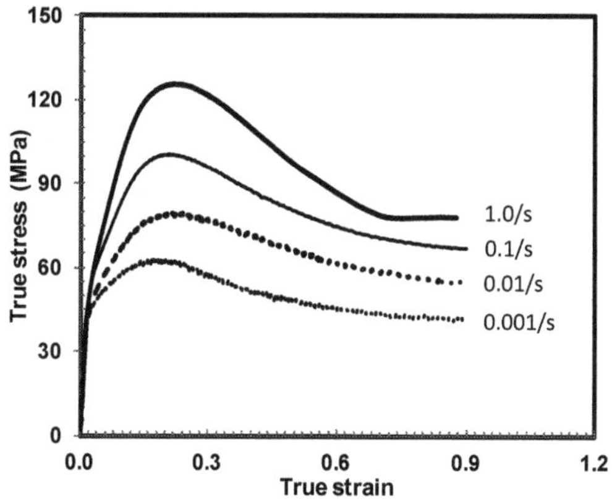

Figure 3: Flow behavior at 300 °C for different strain rates

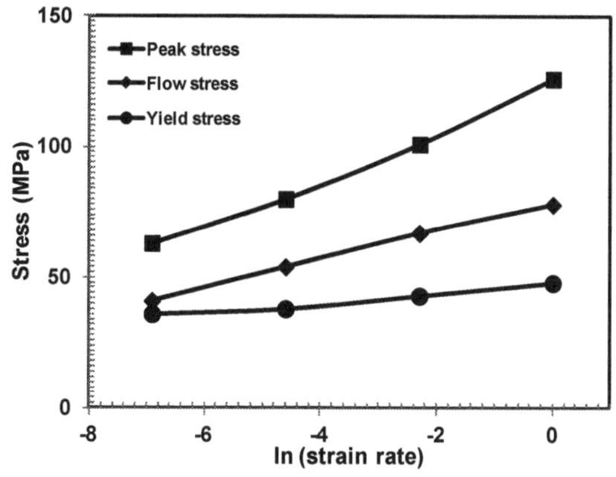

Figure 4: Effect of strain rate on peak stress, flow stress, and yield stress

Figure 5 shows the microstructures of the alloy deformed at 300 °C and a strain of 0.9 subjected to various strain rates. A necklace type structure was observed with recrystallized grains distributed along pre-existing boundaries. The microstructures are characterized by dynamic recrystallization and grains are significantly refined. The recrystallized grains evolved dynamically as the static recrystallization was prevented by quenching the samples immediately after deformation.

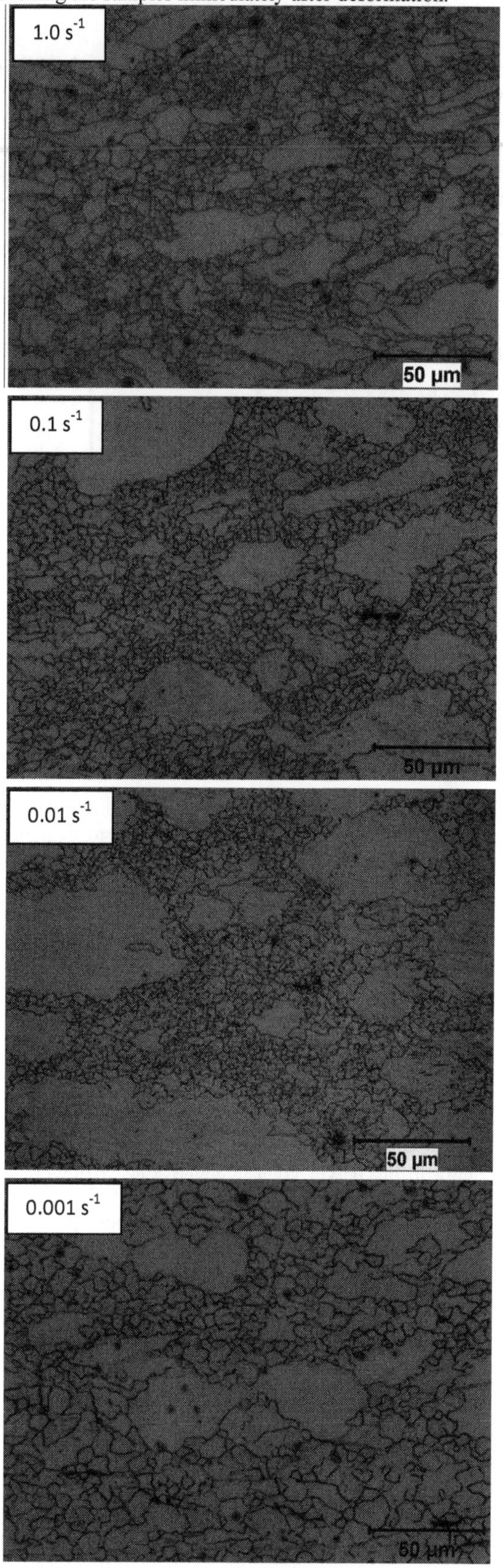

Figure 5: Deformed microstructure for different strain rates

Figure 6 shows the effect of strain rate on DRX grain size. The DRX grain size dependency on strain rate is related to the available time, with a lower strain rate ($0.001 s^{-1}$) equating to a longer time of deformation and more grain growth during the deformation. The time difference between the maximum and minimum strain rate is 15 minutes for a strain of 0.9 in this study. DRX grain size increases about 2 μm in total for the lowest strain rate. This increase can be reasoned by the time availability for grain growth in lowest strain rate as well as the size of the precipitates which is discussed in the next section.

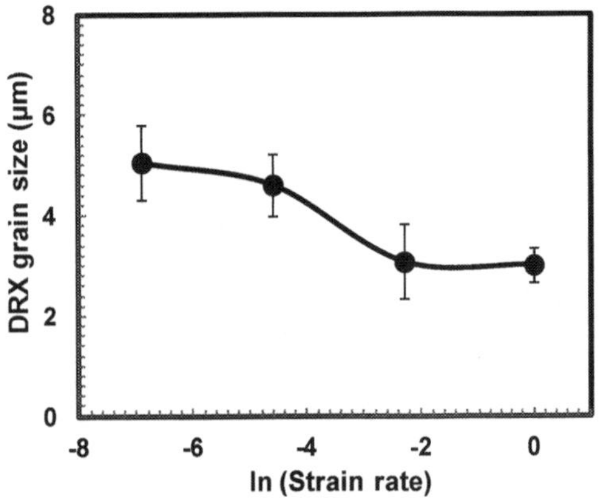

Figure 6: Effect of strain rates on DRX grain size at 300 °C

As mentioned earlier, the samples were held 10 minutes at the deformation temperature prior to deformation to achieve the uniform temperature throughout the whole sample. Quenching from this stage confirms the presence of Mg_2Sn precipitates in the parent grain boundaries. These precipitates are coarser compared to those forms during deformation.

Figure 7 shows the distribution of precipitates at different strain rates in the deformed alloy. Strain induced precipitates of 70-300 nm in size are observed mainly in the DRXed areas along or close to the DRX grain boundaries. It is distinguishable from the figure that both the size and the volume fraction of the precipitate increase with decreasing strain rate. For a strain rate of $1.0 s^{-1}$ (deformation time 0.9 sec), only the prior deformation precipitates in the parent grain boundaries are visible at lower magnification. But at higher magnification, very fine dispersion of precipitates (~70 nm) is visible close to those parent grain boundaries. Decrease in strain rate increases the time for more precipitates to nucleate and the existing one to grow. Thus, at the strain rate of $0.001 s^{-1}$, where the total deformation time is 15 minutes, more volume fraction of precipitates are visible as large as ~300 nm in size. The fine dispersion of precipitates may restrain grain growth by pinning the boundaries and dislocations. The increasing size of the precipitates for low strain rates may not be as effective as the fine precipitates are, for the pinning effect, resulting in the increases of DRX grain size as shown in figure 6.

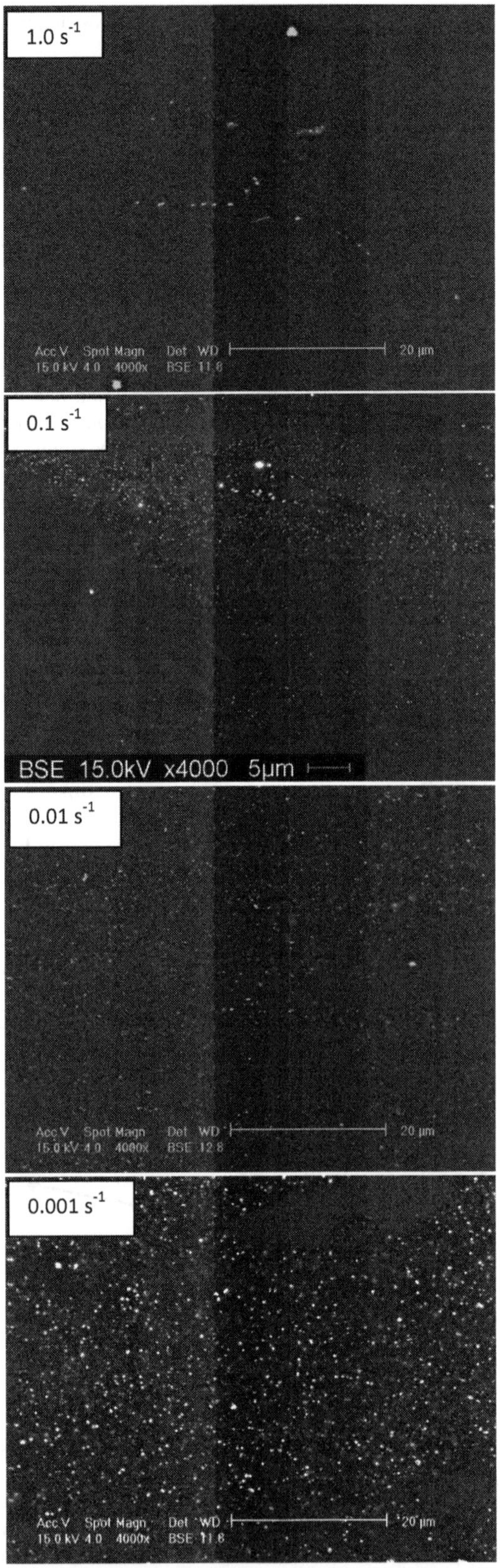

Figure 7: Distribution of Mg_2Sn particles with strain rates

The EDS study (Figure 8) confirms that the precipitates are mostly composed of Mg and Sn, which is consistent with the

Mg_2Sn precipitate predicted by the thermodynamic calculation results from FactSage at 300 °C. To study the room temperature strength, micro-indentation hardness experiment was performed. Figure 9 shows the effect of strain rate on the micro-indentation hardness at different strain rates. Hardness increases with decreasing the strain rate. Dotted line in the figure indicates the hardness of the homogenised sample. The increase in hardness in deformed samples than the homogenized one is reasoned by the presence of Mg_2Sn precipitates. Mg_2Sn is a hard (~119 HV) and brittle phase which contributes to strengthen the matrix [12]. The size and volume fraction of the Mg_2Sn phase increases with decreasing strain rate. This explains the continuous increase in micro-hardness of the alloy with decreasing strain rate.

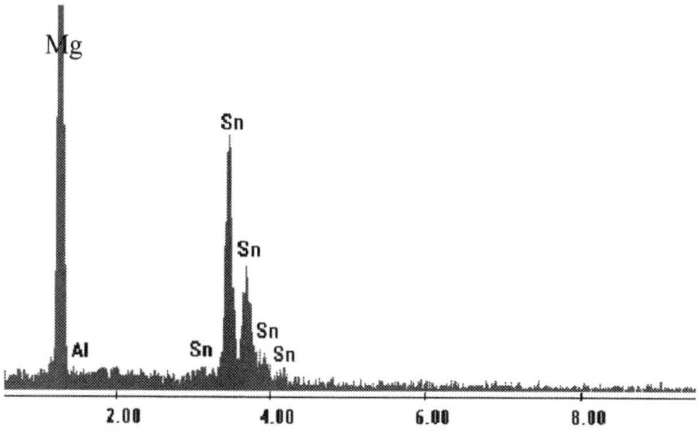

Figure 8: EDS qualitative analysis of the precipitate

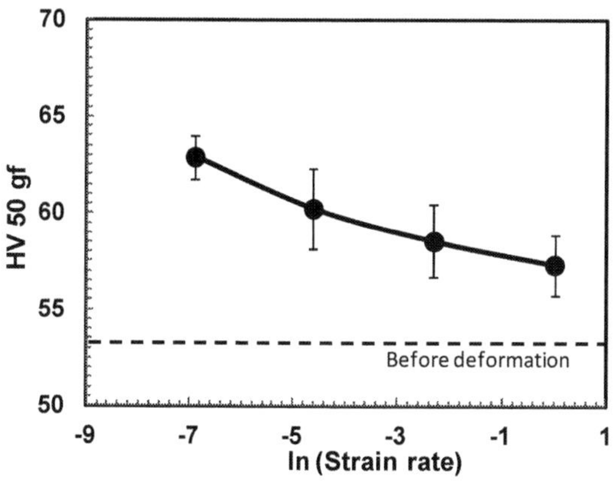

Figure 9: Effect of strain rate on micro-indentation hardness

Summary

The aim of this work was to investigate the grain boundary pinning effect by precipitates to reduce DRX grain size. So far, the result indicates that the precipitates are mostly along or close to the dynamically formed grain boundaries, which may be an indication of the precipitation pinning. The following results can be summarized:

1. The experimental result supports the FactSage calculation in terms of homogenization temperature and precipitation prediction.

2. Deformed microstructure is characterised by dynamic recrystallization and the grain size is significantly refined.

3. DRX grain size increases with decreasing strain rate.

4. Precipitation formation along the DRX grain may be associated with the grain boundary pinning.

5. Size of the precipitates may be an important factor in pinning effect.

6. Hardness increases with the decrease in strain rate due to the formation of more and larger Mg_2Sn precipitate.

Acknowledgement

The authors would like to thank General Motors (GM) and NSERC, Canada for the financial support. The authors are also thankful to Pierre Vermette for his assistance during casting of the alloy.

References

1. K. Ozturk, et al., "Creep Resistant Mg-Al-Ca Alloys: Computational Thermodynamics and Experimental Investigation," *Journal of the Minerals, Metals and Materials Society*, 55 (11) (2003), 40-44.
2. H.E. Friedrich, and B.L. Mordike, *Magnesium Technology: Metallurgy, Design Data, Applications*, Springer, 2006.
3. C.I. Chang, C.J. Lee, and J.C. Huang, "Relationship Between Grain Size and Zener-Holloman Parameter during Friction Stir Processing in AZ31 Mg Alloys," *Scripta Materialia*, 51 (6) (2004), 509-514.
4. M.R. Barnett, M.D. Nave, and C.J. Bettles, "Deformation Microstructures and Textures of Some Cold Rolled Mg Alloys," *Materials Science and Engineering A*, 386 (1-2) (2004), 205-211.
5. L.W.F. Mackenzie, and M. Pekguleryuz, "The Influences of Alloying Additions and Processing Parameters on the Rolling Microstructures and Textures of Magnesium Alloys," *Materials Science and Engineering A*, 480 (1-2) (2008), 189-197.
6. N.V. Ravi Kumar et al., "Grain Refinement in AZ91 Magnesium Alloy during Thermomechanical Processing," *Materials Science and Engineering A*, 359 (1-2) (2003), 150-157.
7. A. Shadkam, "A Study of Homogenization and Precipitation Hardening Behaviour of Mg-Ca-Zn Alloys," (MASc. Thesis, University of Waterloo, 2008) 4-5.
8. M.E. Kassner, and S.R. Barrabes, "New Developments in Geometric Dynamic Recrystallization," *Materials Science and Engineering A*, 410-411 (2005), 152-155.
9. H. Watanabe et al., "Grain Size Control of Commercial Wrought Mg-Al-Zn Alloys utilizing Dynamic Recrystallization," *Materials transactions*, 42 (7) (2001), 1200-1205.

10. F.J. Humphreys, and M. Hatherly, *Recrystallization and Related Annealing Phenomena, vol 2,* Elsevier, 2004.
11. B.H. Kim et al., "The Microstructure, Tensile Properties, and Creep Behavior of AZ91, AS52 and TAS652 Alloy," *Journal of Alloys and Compounds*, 493 (1-2) (2010), 502-506.
12. H. Liu et al., "The Microstructure, Tensile Properties, and Creep Behavior of As-Cast Mg-(1-10)%Sn Alloys," *Journal of Alloys and Compounds*, 440 (1-2) (2007), 122-126.
13. N. Hort, Y. Huang, and K.U. Kainer, "Intermetallics in Magnesium Alloys," *Advanced Engineering Materials*, 8 (4) (2006), 235-240.
14. J.J. Jonas, and I. Weiss, "Effect of Precipitation on Recrystallization in Microalloyed Steels," *Metal Science*, 13 (3-4) (1979), 238-245.
15. S.M. Fatemi-Varzaneh, A. Zarei-Hanzaki, and H. Beladi, "Dynamic Recrystallization in AZ31 Magnesium Alloy," *Materials Science and Engineering A*, 456 (1-2) (2007), 52-57.
16. Y.Q. Yang, B.C. Li, and Z.M. Zhang, "Flow Stress of Wrought Magnesium Alloys during Hot Compression Deformation at Medium and High Temperatures," *Materials Science and Engineering A*, 499 (1-2) (2009), 238-241.
17. A.A. Luo, "Recent Magnesium Alloy Development for Elevated Temperature Applications," *International Materials Reviews,* 49 (1) (2004), 13-30.
18. C.W. Bale et al., "FactSage Thermochemical Software and Databases," *Calphad*, 26 (2) (2002), 189-228.
19. M.R. Barnett, "Influence of Deformation Conditions and Texture on the High Temperature Flow Stress of Magnesium AZ31," *Journal of Light Metals*, 1 (3) (2001), 167-177.

Magnesium Technology 2013
Edited by: Norbert Hort, Suveen N. Mathaudhu, Neale R. Neelameggham, and Martyn Alderman
TMS (The Minerals, Metals & Materials Society), 2013

TEXTURE DEVELOPMENT IN AN EXTRUDED MAGNESIUM ALLOY DURING COMPRESSION ALONG THE TRANSVERSE DIRECTION

D. Sarker, D.L. Chen
Department of Mechanical and Industrial Engineering, Ryerson University, 350 Victoria Street, Toronto, Ontario M5B 2K3, Canada

Keywords: compression, twin, strain-hardening, orientation, texture.

Abstract

Wrought magnesium alloys generally exhibit a certain extent of mechanical anisotropy at room temperature because of the presence of texture. In the present study the work hardening behavior and texture evolution of an extruded AM30 magnesium alloy were studied after compression along the transverse direction (TD) at varying strain levels. The initial samples were chosen to have a strong basal texture with the basal planes parallel to the extrusion direction. During compression a stage of accelerated work hardening characterized by an increasing work hardening rate appeared, and the *c*-axes of the grains were observed to re-orient parallel to the compressive loading direction. With increasing compressive strain extension twins first formed and then disappeared at higher strains.

Introduction

Magnesium alloys are the lightest structural material and therefore are promising for the application in the transportation industry to reduce vehicle weight and increase fuel efficiency [1, 2]. However, their widespread application in industry is impeded because of their poor room temperature ductility, and strong plastic anisotropy, which are due to the hexagonal close-packed (hcp) crystal structure of these materials. The anisotropy of the mechanical properties is mainly due to the development of sharp crystallographic texture during plastic deformation. Generally the rolled/extruded magnesium alloy shows a strong basal texture with most basal planes aligned parallel to the rolling/extrusion direction [3]. The basal texture formed in the extruded material influences the plastic anisotropy as well as the activation of slip and/or twinning during deformation.

Magnesium possesses limited deformation modes available for slip parallel to the *c*-axis and twinning occurs to accommodate the imposed strain [4-6]. Depending on the crystal orientation they accommodate plastic deformation by *c*-axis extension or contraction and form extension twins or contraction twins, respectively. The formation of extension and contraction twins re-orients the crystal lattice by 86.3° and 56.2°, respectively, and contributes to texture evolution [4, 7-9]. Thus the anisotropy in mechanical property results from the combination of initial strong texture and twin formation depending on the crystal orientation.

The occurrence of deformation twinning in magnesium alloys has been widely reported in the literature [10-18], however, its precise influence on the mechanical behavior and texture evolution under monotonic and cyclic loading have not yet been well understood. All studies show the change of initial basal texture with deformation condition. Systematic studies on the texture formation during deformation of magnesium alloys are still absent. The studies on the relationship between the crystallographic texture and plastic deformation mechanisms in wrought magnesium alloys would provide the important information for the general understanding of the plastic deformation of polycrystalline materials with hcp structures, and are beneficial for optimizing alloying composition and forming processes to produce high-performance wrought magnesium alloys. The aim of the present work was to provide further insights into the role of twinning in the texture evolution and deformation mechanisms of an extruded AM30 magnesium alloy.

Experimental

A recently-developed 7 mm thick extruded AM30 magnesium alloy plate was selected in the present study. Cylindrical specimens with a diameter of 5 mm and height of 8 mm were machined with the cylinder (compression) axis parallel to the transverse direction (TD). Compression tests were first conducted until failure using a computerized Instron machine at a strain rate of 1.25×10^{-4} s^{-1} and at room temperature. To observe the microstructural change with texture evolution, some samples were tested at 4.3% strain and some were stopped before failure. X-ray texture analysis was performed using a multi-functional PANalytical X-ray diffractometer. The pole figures were calculated using MTEX toolbox by measuring incomplete pole figures between Ψ=0 to 75° in the back reflection mode using Cu-Kα radiation at 45 KV and 40 mA. A set of five pole figures (i.e., $\{10\bar{1}0\}$, $\{0002\}$, $\{10\bar{1}1\}$, $\{11\bar{2}0\}$, $\{10\bar{1}3\}$) were used to calculate the orientation distribution function (ODF) using the MTEX toolbox. To present ODF, Bunge notations of the Euler angles were used. For the microstructure characterization, the deformed samples were cut along the compression axis using a slow diamond cutter, cold-mounted, ground, polished, and etched using acetic picral solution containing 4.2-g picric acid, 10-mL acetic acid, 10-mL H_2O, and 70-mL ethanol to examine the evolution of deformation twins during compression. It should be noted that in evaluating the stress-strain curves and strain-hardening rates, the machine deformation itself was eliminated using a calibration curve to obtain the actual or net deformation of test samples.

Results and Discussion

The initial pole figure of extruded AM30 Mg alloy is shown in Fig.1a. The pole figure showed an intense basal texture where the *c*-axis of the hcp unit cell was basically perpendicular to the ED. Similar texture was observed in a rolled AZ31B-H24 Mg alloy [17-19]. A twin-free microstructure of undeformed metal is shown in Fig.1b. The schematic view in Fig. 1c illustrates the orientation of samples used in the present study, where the cylindrical axis was parallel to the transverse direction, which was essentially perpendicular to the c-axis of most grains.

The stress-strain curve and strain-hardening rate of AM30 magnesium alloy during compression along the TD are shown in Fig.2 (a) and (b), respectively, where a compressive yield strength of ~96 MPa was obtained, which was considerably lower than the tensile yield strength of ~189 MPa [12]. This was attributed to the occurrence of twinning during compression which will be discussed later.

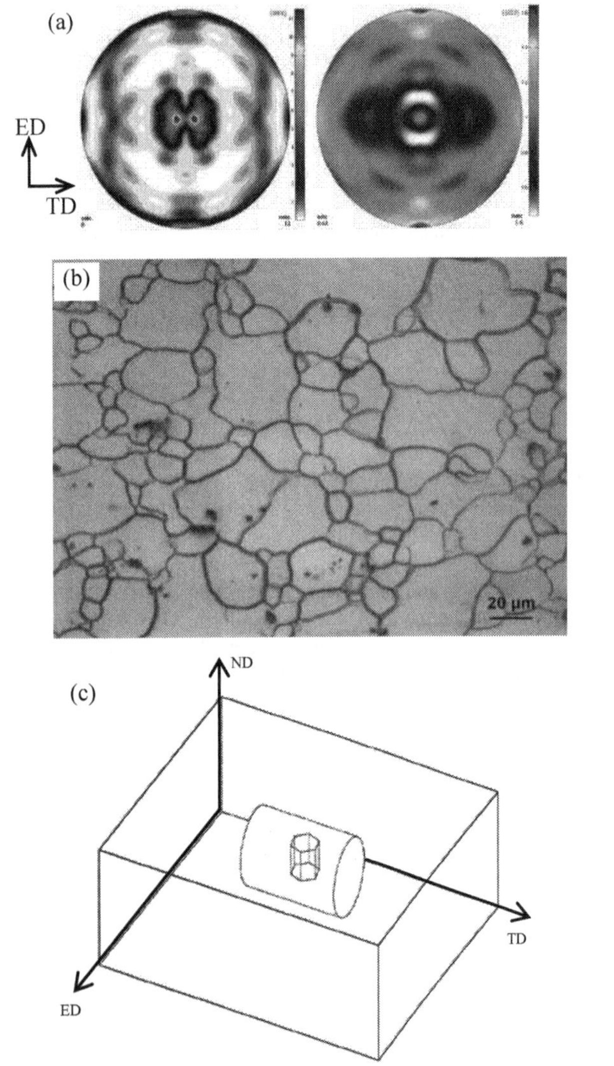

Figure 1. (a) Initial (0001) and (10 0) pole figures of as-extruded AM30 magnesium alloy, (b) a typical microstructure of undeformed extruded AM30 magnesium alloy, (c) a schematic diagram showing the orientation of the compressive samples parallel to the transverse direction (TD).

The stress-strain curve of AM30 extruded alloy exhibited a skewed shape (Fig.2 (a)), which was very different from the normal tensile (or compressive) curves of most metals with cubic crystal structures. This could be better seen based on the change of strain hardening rate $d\sigma/d\varepsilon$ (where σ is true stress) with respect to true strain ε, where three stages of strain hardening could be more clearly discerned (Fig.2(b)). Stage A was characterized only by a slight decrease in the strain hardening rate up to a compressive true strain of approximately 3%, followed by Stage B with a fairly strong increase in the strain hardening rate up to ~6.5% compressive true strain, and then Stage C with an decreasing strain hardening rate again until failure. A similar trend in the change of strain hardening rate has also been observed in AZ31 magnesium alloy [20] and in the extruded AM30 magnesium alloy during compression along ED [21].

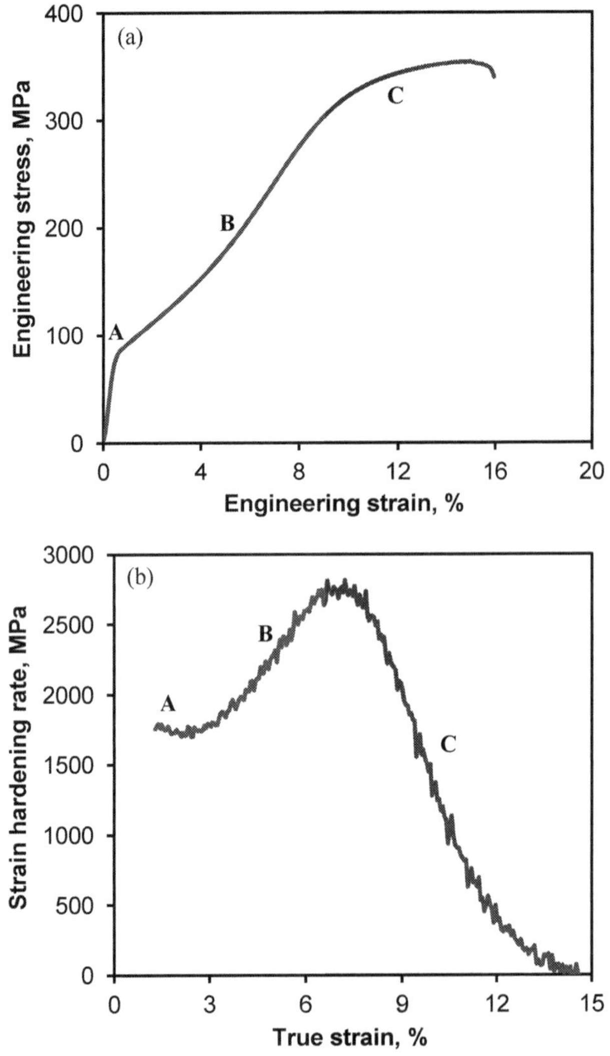

Figure 2. (a) A typical compressive stress-strain curve and (b) strain hardening rate as a function of true strain in the extruded AM30 magnesium alloy with compression axis parallel to the transverse direction.

To understand such a flow stress characteristic, the microstructural development of the samples compressed to varying strain amounts was examined and shown in Fig. 3, where the loading direction was indicated. The microstructural changes with the imposed strains of 4.3% and 12.9%, respectively, are illustrated in Figs. 3a and 3b. At a strain level of 4.3%, within the strain hardening range of stage B (Fig. 2(b)), the presence of twins was clearly visible. This observation was in agreement with the previous studies of Jiang *et al.* [22], where the tensile twins in an AZ31 magnesium alloy was reported to occur in a strain range of 2-4%. On further straining, just before failure, most twins became invisible in 3b.

The texture results are presented Fig. 4 in terms of the pole figures. The pole figures were measured from the deformed samples after compression tests at different strain amounts. In Fig. 1(a), the as extruded base metal (0% strain) shows a strong basal texture, which means that {0001} basal planes in most grains were parallel to the ED with $\{10\bar{1}0\}$ planes around the circumference of the pole figure. At a strain of 1.5% as shown in Fig. 4(a), some (0001) poles rotated towards the TD, while the pole figure still showed some intensity at the center similar to the initial texture (Fig. 1(a)). $\{10\bar{1}0\}$ poles rotated from periphery and started to concentrate at the center. A gradual change of intensity distribution was also observed after compressive deformation at a strain of 4.3% in Fig. 4(b) with more intense (0001) poles toward TD and $\{10\bar{1}0\}$ poles around the center. The (0001) poles of most grains rotated to the TD after a strain of 12.9% along TD with an obvious preferred distribution of prismatic $\{10\bar{1}0\}$ poles at the center, as shown in Fig. 4(c).

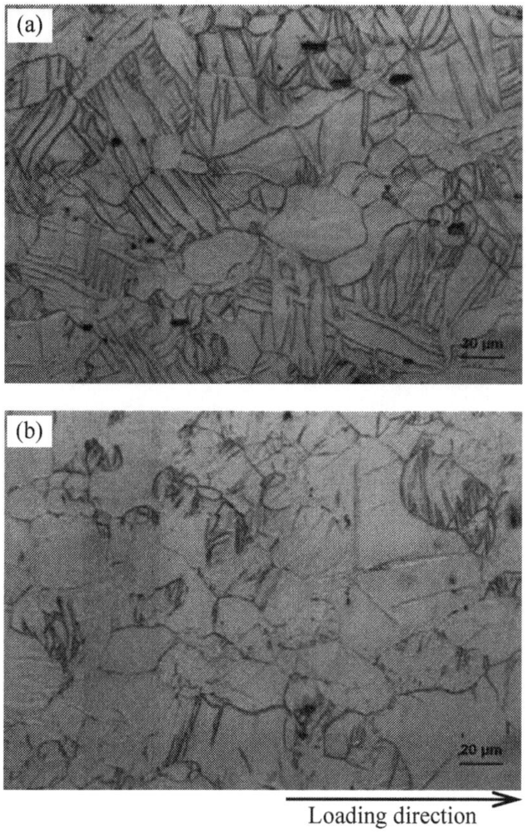

Figure 3. Microstructure changes after compressive deformation of the samples oriented in the transverse direction at a strain level of (a) 4.3%and (b) 12.9%.

This texture evolution was related to the microstructure development or the twin formation with increasing compressive strain level. The microstructure of compressed sample shown in Fig. 3a, corresponding to a strain of 4.3%, was completely different from the initial microstructure shown in Fig. 1(b). A lot of twins formed at the compressive deformation. This observation was consistent with the previous observations [22, 23], where the extension twins were reported to first occur in a strain range of 2-4%. With increasing strain levels, the extension twins appeared to thicken quickly, usually to cover each grain and their boundaries disappeared. This was again in good agreement with the previous observations [16, 22, 23]. Extension twinning was seen to play a predominant role in the texture evolution in the early stages of compression along the TD. The occurrence of extension twinning during compression along the TD, caused grains to rotate by ~86° and led to *c*-axis parallel to the TD. This is clearly seen in the evolution of the pole figures shown in Fig. 4, where the (0001) poles rotated always against the compressive direction, i.e., the TD in the present study (Fig. 4a-c).

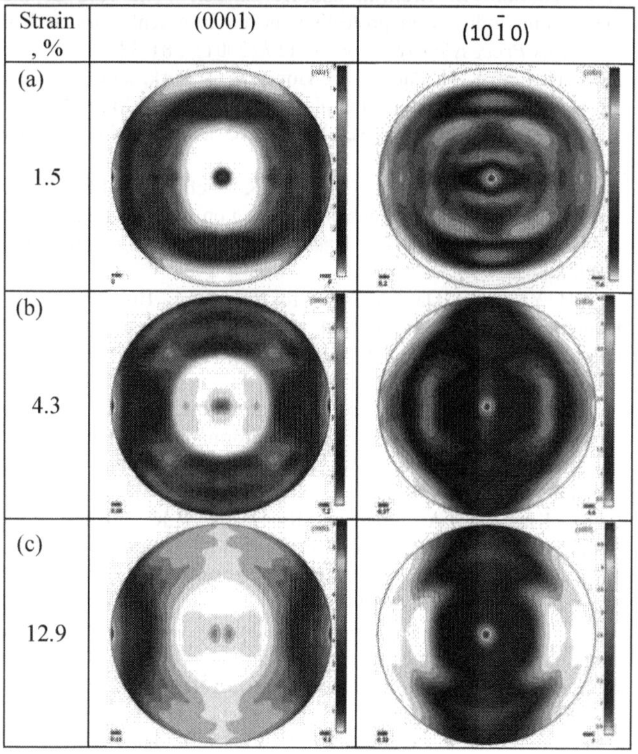

Figure 4. Pole figures measured from the compressed samples oriented in the transverse direction at a strain amount of (a) 1.5%, (b) 4.3% and (c) 12.9%.

Conclusions

The present study demonstrates the concurrent change in both the texture and deformation characteristics of extruded magnesium alloy AM30. With increasing strain during compression extension twins first formed and then disappeared at higher strains. The extension twinning played an important role in accommodating the compressive deformation as the hexagonal *c*-axis lay perpendicular to the loading direction. The low yield stress, rapid texture changes with increasing compressive strain amounts observed can be explained by the high twinning activity. As a consequence, the *c*-axes of the grains were oriented parallel to the compressive loading direction.

Acknowledgements

The authors would like to thank the Natural Sciences and Engineering Research Council of Canada (NSERC), Premier's Research Excellence Award (PREA), NSERC-DAS Award, AUTO21 Network of Centers of Excellence, and Ryerson Research Chair program for providing financial support. The authors also thank, Dr. A. Luo, General Motors Research and

Development Center for providing the test materials, and Q.Li, A.Machin, J.Amankrah, and R.Churaman for assisting in the experiments.

References

1. T.M. Pollock, "Weight Loss with Magnesium Alloys," *Science*, 328 (2010), 986-987.
2. E. Aghion, B. Bronfin, and D. Eliezer, "The role of the magnesium industry in protecting the environment," *Journal of Materials Processing Technology*, 117 (2001), 381-385.
3. B. H. Lee, S. H. Park, S. G. Hong, K.T. Park, and C.S. Lee, "Role of initial texture on the plastic anisotropy of Mg–3Al–1Zn alloy at various temperatures," *Materials Science and Engineering A*, 528 (2011), 1162–1172.
4. L. Jiang, J.J. Jonas, A.A. Luo, A. K. Sachdev, and S. Godet, "Influence of {10-12} extension twinning on the flow behavior of AZ31 Mg alloy," *Materials Science and Engineering A*, 445-446 (2007), 302-309.
5. S.B. Yi, C.H.J. Davies, H.G. Brokmeier, R.E. Bolmaro, and J. Homeyer, "Deformation and texture evolution in AZ31 magnesium alloy during uniaxial loading ," *Acta Materialia*, 54 (2006), 549–562.
6. S. Sandlöbes, S. Zaefferer, I. Schestakow, S. Yi, and R.G. Martinez, "On the role of non-basal deformation mechanisms for the ductility of Mg and Mg–Y alloys," *Acta Materialia*, 59 (2011), 429-439.
7. M. Knezevic, A. Levinson, R. Harris, R.K. Mishra, R.D. Doherty, and S.R. Kalidindi, "Deformation twinning in AZ31: Influence on strain hardening and texture evolution," *Acta Materialia*, 58 (2010), 6230–6242.
8. A.A. Salem, S.R. Kalidindi, R.D. Doherty, and S.L. Semiatin, "Strain hardening due to deformation twinning in α-titanium: mechanisms," *Metallurgical and Materials Transactions A*, 37A (2006), 259-268.
9. C.H. Cáceres, P. Lukác, and A. Blake, "Strain hardening due to {10-12} twinning in pure magnesium," *Philosophical Magazine*, 88 (2008), 991-1003.
10. A. Jiang, A. Godfrey, W. Liu, and Q. Liu, "Microtexture evolution via deformation twinning and slip during compression of magnesium alloy AZ31," *Materials Science and Engineering A*, 483-484 (2008), 576-579.
11. Y.T. Zhu, X.L. Wu, X.Z. Liao, J. Narayan, L.J. Kecskes, and S.N. Mathaudhu, "Dislocation–twin interactions in nanocrystalline fcc metals," *Acta Materialia*, 59 (2011), 812-821.
12. S. Begum, D.L. Chen, S. Xu, and A.A. Luo, "Strain-controlled low-cycle fatigue properties of a newly developed extruded magnesium alloy," *Metallurgical and Materials Transactions A*, 39 (2008), 3014-3026.
13. S. Begum, D.L. Chen, S. Xu, and A.A Luo, "Low cycle fatigue properties of an extruded AZ31 magnesium alloy," *International Journal of Fatigue*, 31 (2009), 726-735.
14. S. Begum, D.L. Chen, S. Xu, and A.A. Luo, "Effect of strain ratio and strain rate on low cycle fatigue behavior of AZ31 wrought magnesium alloy," *Materials Science and Engineering A* 517 (2009), 334-343.
15. X.Z. Lin, and D.L. Chen, "Strain controlled cyclic deformation behavior of an extruded magnesium alloy," *Materials Science and Engineering A*, 496 (2008), 106-113.
16. M.R. Barnett, "Twinning and the ductility of magnesium alloys Part I: "Tension" twins," *Materials Science and Engineering A*, 464 (2007), 1-7.
17. Y.C. Xin, M.Y. Wang, Z. Zeng, M.G. Nie, and Q. Liu, "Strengthening and toughening of magnesium alloy by {10-12} extension twins," *Scripta Mater*, 66 (2012), 25-28.
18. L. Wu, S.R. Agnew, Y. Ren, D.W. Brown, B. Clausen, G.M. Stoica, H.R. Wenk, and P.K. Liaw, "The effects of texture and extension twinning on the low-cycle fatigue behavior of a rolled magnesium alloy, AZ31B," *Materials Science and Engineering A*, 527 (2010), 7057-7067.
19. S.H. Park, S.G. Hong, J.H. Lee, and C.S. Lee, "Multiple twinning modes in rolled Mg–3Al–1Zn alloy and their selection mechanism" *Materials Science and Engineering A*,532 (2012), 401-406.
20. B.S.Wang, R.L.Xin, G.J.Huang, and Q.Liu, "Effect of crystal orientation on the mechanical properties and strain hardening behavior of magnesium alloy AZ31 during uniaxial compression," *Materials Science and Engineering A*, 534 (2012) 588-593.
21. D. Sarker and D.L. Chen, "Detwinning and strain hardening of an extruded magnesium alloy during compression," *Scripta Materialia* 67 (2012), 165–168.
22. A. Jiang, A. Godfrey, W. Liu, and Q. Liu, "Microtexture evolution via deformation twinning and slip during compression of magnesium alloy AZ31," *Materials Science and Engineering A* 483-484 (2008), 576-579.
23. L. Jiang, J.J. Jonas, R.K. Mishra, A.A. Luo, A.K. Sachdev, and S. Godet. "Twinning and texture development in two Mg alloys subjected to loading along three different strain paths" *Acta Materialia*, 55 (2007), 3899-3910.

Magnesium Technology 2013
Edited by: Norbert Hort, Suveen N. Mathaudhu, Neale R. Neelameggham, and Martyn Alderman
TMS (The Minerals, Metals & Materials Society), 2013

Evolution of microstructure during caliber rolling of AZ31 alloy

Alok Singh[1], Hidetoshi Somekawa[1], Tadanobu Inoue[1], Toshiji Mukai[2]

[1]Structural Materials Unit, National Institute for Materials Science, Tsukuba 305-0047, Japan
[2]Department of Mechanical Engineering, 1-1 Rokkodai, Nada, Kobe city, 657-8501, Japan

Keywords: magnesium, deformation microstructure, transmission electron microscopy, caliber rolling

Abstract

An AZ31 alloy has been processed by caliber rolling, in which a repetitive oblique shear strain is applied with a fixed reduction ratio, resulting in increase in strength and ductility by simultaneous grain refinement and texture modification. A total of 18 passes were made at 473K through successive grooves with reduction. A dramatic increase in the tensile YS was achieved with no substantial change in the elongation. Evolution of microstructure by caliber rolling has been studied by TEM. Samples after 4, 6, 8 and 18 passes were examined. After 4 passes the sample showed clear grain boundary structure and grain refinement. After 6 passes, a severely deformed structure is observed, indicating accumulation of strain. Spots with concentration of strain were observed in basal planes. Activation of prismatic slip was observed. After 8 passes the strain appears more severe. After 18 passes, along with concentration of strain, sharp grain and subgrain boundaries appeared, with subgrains of about 100-200nm size. The small grain and subgrain size, along with accumulated strain, is responsible for the high strength.

Introduction

One of the most important microstructural factor in strengthening magnesium alloys is grain refinement [1], which also leads to increased ductility due to enhanced grain boundary plasticity [2]. A crucial factor in ductility is texture, which can be modified by changing the direction of applied shear during severe plastic working [3] and/or by changing recrystallization behavior by dispersion of hard intermetallic particles [4]. Modification of texture, together with grain refinement, has been shown by repetitive shear loading processes such as equal-channel angular extrusion (or pressing) (ECAE or ECAP) [5, 6, 7, 8, 9]. It has also been shown that the yield asymmetry ratio (ratio of compressive yield strength to tensile yield strength) of wrought AZ31 can be reduced by grain refinement to 0.9 [10].

Recently we have reported grain refinement with weakened texture by applying a repetitive oblique shear strain using caliber rolling, which can operate at commercial scale and processing speeds [11, 12]. Caliber rolling can be used to reduce the cross-sectional shape of the products. Graded channels with reducing cross-section are formed in the rolls. For each cross-section the upper and the lower rolls have identical channels. A certain shear strain is created on the material passing through the rolls. This rolling process is described in greater detail in earlier reports [11, 12, 13]. The diameter of a billet is reduced by passing it successively through channels of lower cross-sections. After each pass, the billet was rotated by 90° before the next pass, to make the shear direction opposite to that of the previous pass. After passing the sample through the last channel, it was passed through it one more time after rotating by 90°. By this process, the texture was weakened, while more strain was accumulated in the sample. Starting with a direct extruded AZ31 billet, the tensile yield stress (YS) was nearly doubled while keeping elongation strain nearly constant. This resulted in high fracture toughness [14].

Along with grain refinement, a very high amount of strain accumulation resulted in sub-grain formation at nano-scale. This has to be taken into account for explaining the high YS of the alloy. Formation and structure of these sub-grains has been studied by transmission electron microscopy (TEM) and is reported here.

Experimental Procedure

A billet (42 mm diameter, 100 mm long) of commercially extruded AZ31 alloy (Mg-2.9Al-0.8Zn-0.42Mn, wt%) with an average grain size of 25μm was heated to 473K and then subjected to plastic deformation by caliber rolling. Its grain structure is shown in Fig. 1, which is bi-modal and displaying a number of twins. The billet was rolled through a series of calibers in a single roll, with a reduction in area of -18% for each caliber. A total of 18 passes were made. Details can

Figure 1: Optical micrograph showing grain structure of the starting material, which is a commercially extruded AZ31 alloy.

be found in earlier reports [13, 11, 12].

For TEM observations, the caliber rolled rod was sectioned longitudinally with a diamond saw to slice a sample of about 1 mm thickness. The sample was mechanically ground down to about 70 μm thickness, and then ion-milled to perforation in a precision ion mill. TEM observations were made in JEOL 2000FX-II microscope operated at 200 kV.

Results

Microstructures after 4 passes (4P sample) are shown in Fig. 2. A grain structure with sharp grain boundaries is observed, with grain size of about 8 μm. Grains with multiple neighbors and curved boundaries indicate that these grains have formed by recrystallization. Fig. 2(b) shows grains oriented along, or nearly along [0001] zone axis, with low angle boundaries in between them. Very little dislocation structures were observed inside the grains, such as the one marked D. However, region D' has a severely deformed structure. Precipitation structure in grain D is shown in Fig. 2(c). The precipitates have nearly uniform size of 20-30 nm.

In the grain in Fig. 3(a), oriented along a $\langle 1\bar{1}00\rangle$ zone axis, dislocation on basal planes are observed prominently. Details in inset show a set of non-basal dislocations too (one marked by an arrow). Besides this, apparent orientation relationships between grains were also observed. An example is shown in Fig. 3(b) (strong contrast of thickness contours are seen in grain M). Grain M is near a $\langle 11\bar{2}0\rangle$ zone axis orientation while grain N is nearly in $\langle 11\bar{2}3\rangle$ zone axis. There is a near planar match of $\{\bar{1}21\}_M \| \{010\}_N$. A stereographic analysis shows that these grains are related by a $\{10\bar{1}1\}$ or $\{10\bar{1}3\}$

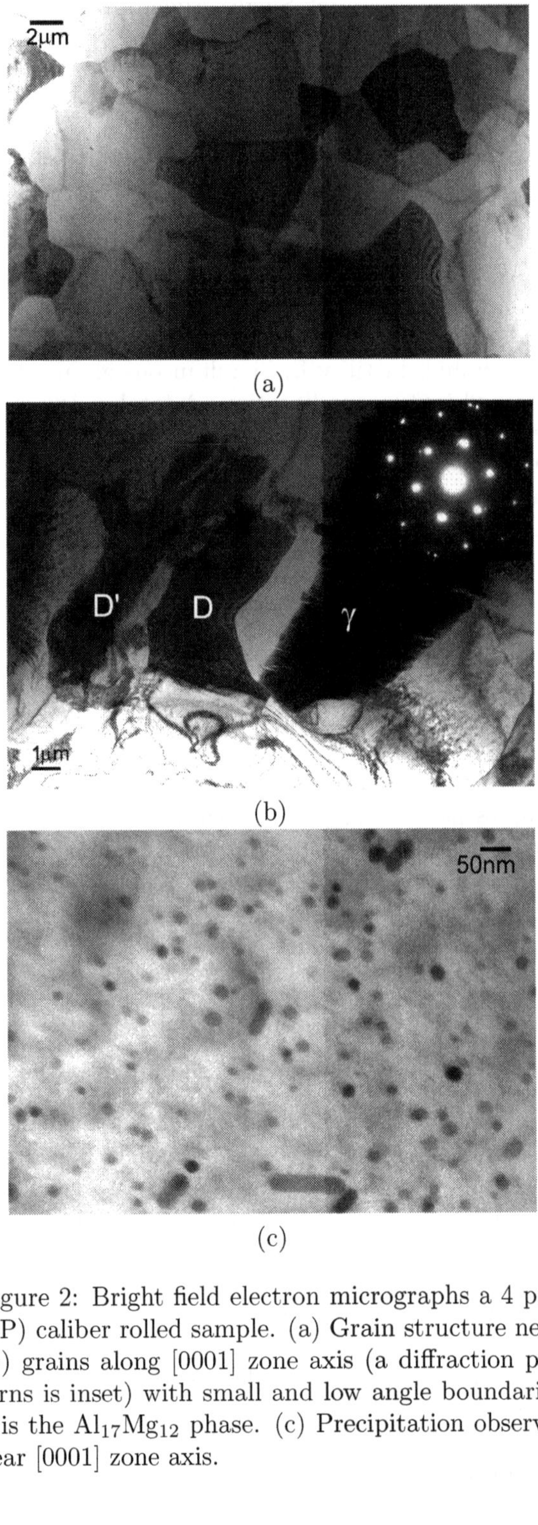

Figure 2: Bright field electron micrographs a 4 pass (4P) caliber rolled sample. (a) Grain structure near. (b) grains along [0001] zone axis (a diffraction patterns is inset) with small and low angle boundaries. γ is the $Al_{17}Mg_{12}$ phase. (c) Precipitation observed near [0001] zone axis.

(a)

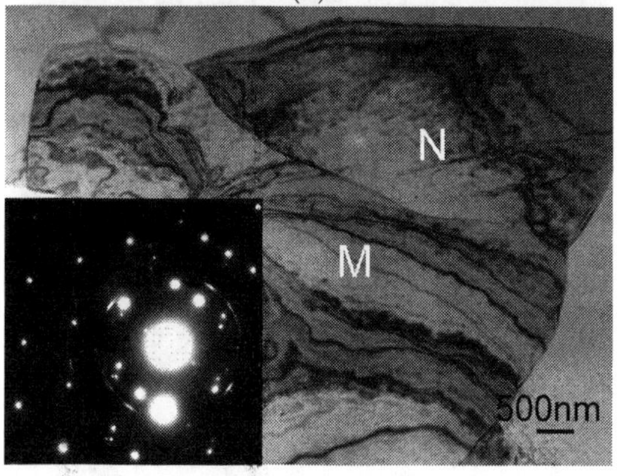

(b)

Figure 3: Deformation structures and twin-like orientations in 4P sample. (a) A grain along ⟨1100⟩ zone axis, showing dislocation contrast on basal planes. Inset lower right: detail of a small region in strong diffraction contrast. (b) Grains M in ⟨11$\bar{2}$0⟩ zone axis and N in ⟨11$\bar{2}$3⟩ zone axis. The composite diffraction pattern is inset.

type twin. This indicates the role of twinning in deformation and evolution of the grain structure.

Fig. 4 from 6P sample shows bright field (BF) TEM micrographs along two zone axes; one in (a) is close to zone axis ⟨1$\bar{1}$00⟩ in a two beam condition with a {11$\bar{2}$0} reflection. Only a very small area is in contrast, due to the strained condition of the grain. A strong region of contrast is observed, marked by an arrowhead. This contrast is elongated perpendicular to the basal planes. This represents prismatic slip. The region in (b) is oriented along [0001] axis. Very interesting contrast is observed here. There are spots of dark contrast, from which radiate strain contrast. In other words, the strain is concentrated in a few spots and the basal plane is deformed. The corresponding diffraction pattern, inset, displays an

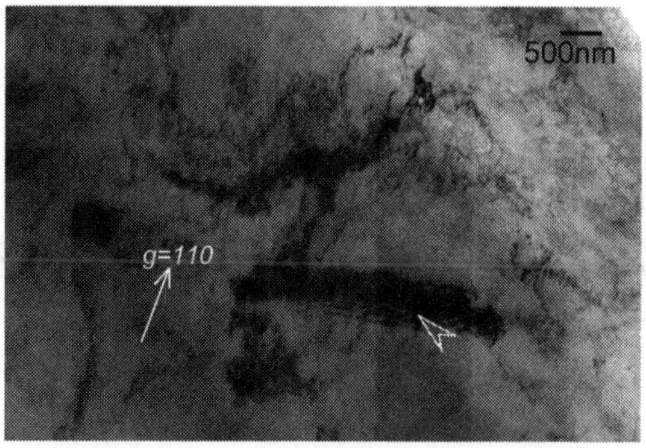

(a)

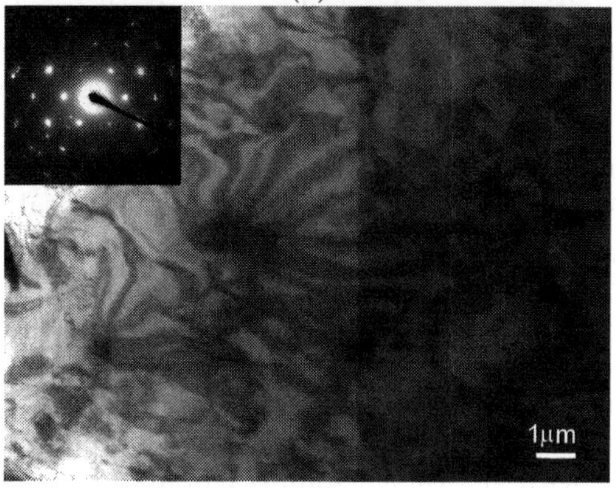

(b)

Figure 4: Bright field electron micrographs showing strain after 6 passes in the caliber roll (6P) (a) near zone axis ⟨1$\bar{1}$00⟩, in a two-beam condition with a {11$\bar{2}$0} spot and (b) along zone axis {0001}.

asymmetry of spot intensity. The high index spots are weak and in form of arcs because of the matrix strain.

Fig. 5 are TEM BF micrographs showing matrix strain after eight passes (8P sample). (a) shows flow of strain contrast, in form of wavy lines. (b) is observed nearly along zone axis ⟨11$\bar{2}$0⟩. In the diffraction pattern (inset), high index spots are absent or in form of weak arcs. In the bright field, ring-like contrasts are observed. In some places, like the one marked with an open arrow, the contrast suggests prismatic slip. In Fig. 5(c), a grain in [0001] zone axis is observed. Dislocation structure is observed inside, but the uniform contrast and the diffraction pattern (inset) do not suggest accumulated strain. It possibly is a grain recrystallized during caliber rolling.

Fig. 6 shows microstructures after 18 passes (18P). (a) shows accumulated strain; however, some sharp boundaries are also observed. There is a much finer subgrain structure. When the grain is tilted out of contrast (shown in (b)), distribution of nano-sized precipitates is observed. These precipitates are rounded, typically 10-20 nm or less in size, but occasionally as much as 50 nm. A number of sharp boundaries are observed in (c), which create grains or subgrains as small as 200 nm in size. Strain is observed piled up on these boundaries. The small grain in the center is in [0001] zone axis orientation.

Discussion

The initial material, which is an extruded rod, has a tube texture of basal planes. Thus during caliber rolling the grains, depending on their orientation, experience a compressive stress either along their hexagonal axis, perpendicular to it, or in between. Correspondingly, basal or non-basal slip, and/or twins will be generated. Reported strain simulation and experimental estimation [13] show that the strain becomes larger by increasing number of passes and that it has a distribution with maxima around the corners.

After 4 passes, the grain size is considerably refined, with an average grain size of about 5μm. Absence of deformation inside the grains indicate that they are formed by recrystallization. 4P sample also shows large areas with low angle grain boundaries, such as in Fig. 2(a) along zone axis [0001], which are remnants of initial large size grains.

In this sample, various kinds of lattice defects are observed clearly. In Fig. 2(b), grains D and D' in similar orientations show different levels of strain. Grain D' shows a high level of strain while D does not. Dislocations are observed clearly in Fig. 3(a), which are basal as well as non-basal.

Beyond 4P, the 6P and 8P samples show further accumulation of strain, which lead to the final microstructure in 18P.

Hardness measurements reported by Inoue et al. [13] in a similar experiment show a similar trend in evolution of the microstructure. Partial DRX occurs until 7 passes and complete DRX until 11 passes. Subsequently, continuous DRX occurs until 18P. Similarly, in our samples partial DRX is observed in 4P.

Inoue et al. [13] also report that after 7 passes the microstructure consists of a mixture of fine grains and coarse grains. The coarse grains were elongated along the rolling direction. The fine grains were considered to be generated by DRX during the rolling. The average grain size of the fine grains was estimated to be 3.8μm. From 7 passes to 11 passes the fraction of coarse grains decreases, but the average grain size

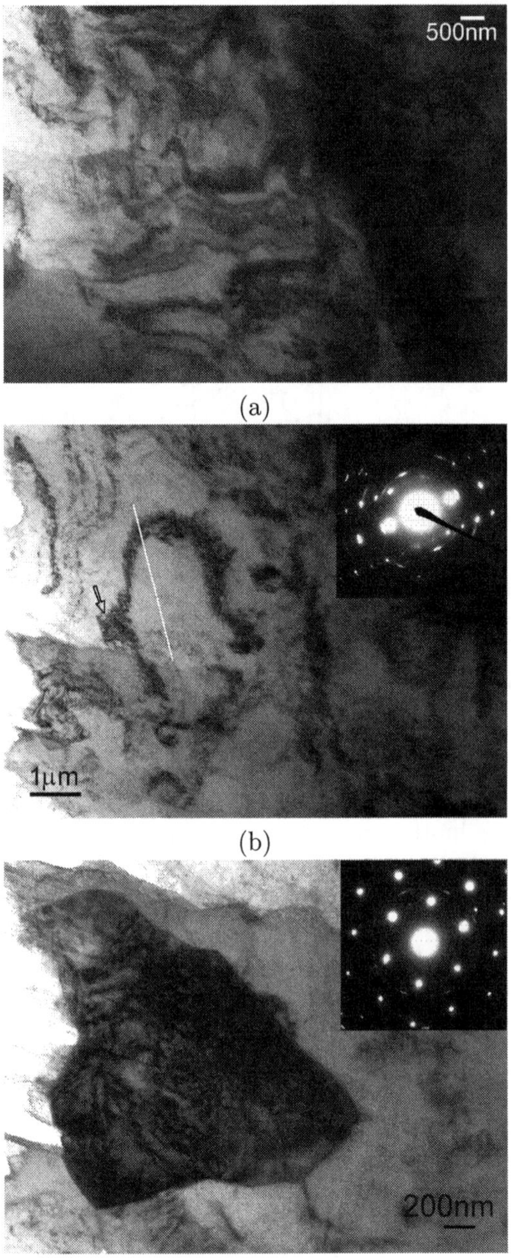

Figure 5: Bright field electron micrographs showing accumulated strain after 8 passes in the caliber roll (8P)(a) severe deformation structures, (b) a region observed along $\langle 11\bar{2}0\rangle$ (white line denotes the direction of the basal planes) and (b) a grain along zone axis [0001].

of the fine grains remained the same. On further rolling to 15 passes, the average grain size decreased to about 2.5μm due to continuous DRX, while the fraction of low angle boundaries increased [14]. Our TEM studies here show that much smaller domains occur because of strain accumulation and sub-grain boundaries formation.

With the evolution of grain structure during the caliber rolling, a change in the strain accumulation mechanism will be expected. This also comes out in the hardness curve reported by Inoue et al. [13]. This is because in the beginning of the rolling process the grain size is considerably large, at 25μm, but once the grain size has become small to about 5μm or less, the stress will generate considerable non-basal slip [15]. Thus in the 4P sample, basal slip is clearly visible (Fig. 3(a)). Much more complex slip is observed in 6P and 8P samples. This is facilitated by change of stress direction by a rotation of the sample after each pass. Accumulation of strain at certain points in basal planes due to this is clearly observed in Fig. 4(b). Such a strain will make basal slip harder. Fig. 4(a) clearly shows strain in basal plane as well prismatic planes. Fig. 5(a) shows wavy patterns due to accumulation of a large amount of multi-directional strain. Fig. 4(b) shows wavy bands of same orientation about 500nm in width.

Inoue et al. [13] also show that the increase in hardness of the bar up to 11 passes is due to strain hardening and grain refinement by DRX, while the subsequent hardness is attributed to sub-grain refinement based on continuous DRX.

It may be assumed that there are cycles of strain accumulation and recrystallization during the caliber rolling. Electron Backscattered diffraction (EBSD) [11] shows sharp boundaries after 15 passes and elongated srained structure after 18 passes. Thus, presumably, the sharp boundaries form after 15 passes followed by further strain accumulation. There is also a marked weakening of texture after 15P [11].

In the final microstructure in 18P, sharp boundaries are observed, which make domains as small as 200nm in size/width (Fig. 6). Accompanying this is the strain in the matrix, which enhances the yield stress. The very high yield stress of the alloy is thus effectively because of the very small size domains with sharp boundaries. These domains can hold a high concentration of strain in them.

It should be noted that a similar microstructure is also shown by multi-directional forging of AZ31 alloy [16].

Precipitation also has a strong contribution to

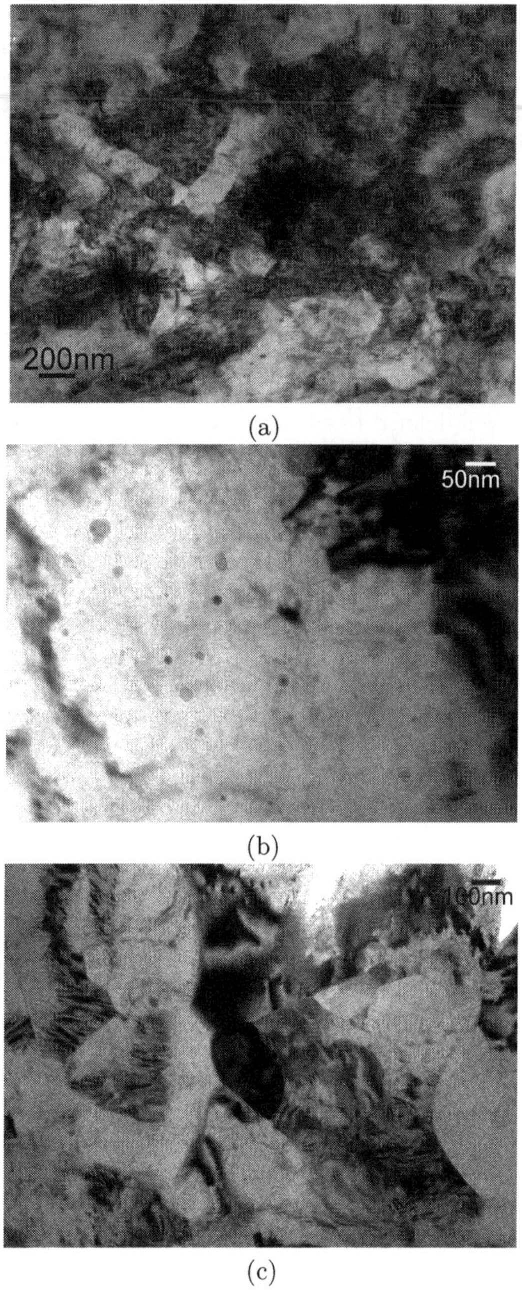

Figure 6: Bright field electron micrographs showing microstructure after 18 passes in the caliber roll (18P)(a) severe deformation structures, (b) precipitation in the matrix and (b) grain and subgrains, the grain in the center is along zone axis [0001].

strength. A modification in the precipitate distribution is observed by caliber rolling. Precipitates observed in 4P sample, Fig. 2(c), are in a narrow size range of about 10 to 20nm. In 18P sample in Fig. 6(b), more of finer size precipitates are observed. A finer distribution of precipitates and possibly a supersaturation of the matrix due to processing effect similar to mechanical alloying will also contribute to the strengthening.

Conclusions

Microstructure development of AZ31 alloy during caliber rolling has been studied by transmission electron microscopy. The following conclusions are made:

1. Well defined basal slip as well as non-basal slips are observed during early passes. Recrystallization occurs by 4 passes, whereby the grain size is reduced to about 5μm.

2. Further accumulations of strain occurs in further passes. Severe localization of strain, basal and prismatic slip are observed after 6 and 8 passes.

3. By 18 passes, sharp boundaries appear, making domains of about 200nm thick, in which are contained severely strained regions.

References

[1] K. Kubota, M. Mabuchi, K. Higashi, "Processing and mechanical properties of fine-grained magnesium alloys", *J.Mater. Sci.* 34 (1999) 2255.

[2] J. Koike, R. Ohyama, T. Kobayashi, M. Suzuki, K. Maruyama, "Grain-Boundary Sliding in AZ31 Magnesium Alloys at Room Temperature to 523K", *Mater. Trans.* 44 (2004) 445.

[3] T. Mukai, M. Yamanoi, H. Watanabe, K. Higashi, "Ductility enhancement in AZ31 magnesium alloy by controlling its grain structure", *Scripta Mater.* 45 (2001) 89.

[4] A. Singh, H. Somekawa, T. Mukai, "Compressive Strength and Yield Asymmetry in Extruded Mg-Zn-Ho Alloys Containing Quasicrystal Phase", *Scripta Mater.* 56 (2007) 935.

[5] A. Yamashita, Z. Horita, T.G. Langdon, "Improving the mechanical properties of magnesium and a magnesium alloy through severe plastic deformation", *Mater. Sci. Eng.A* 300 (2001) 142.

[6] W.J. Kim, C.W. An, Y.S. Kim, S.I. Hong, "Mechanical properties and microstructures of an AZ61 Mg Alloy produced by equal channel angular pressing", *Scripta Mater.* 47 (2002) 39.

[7] Y. Yoshida, L. Cisar, S. Kamado, Y. Kojima, "Effect of Microstructural Factors on Tensile Properties of an ECAE-Processed AZ31 Magnesium Alloy", *Mater. Trans.* 44 (2003) 468.

[8] S.R. Agnew, J.A. Horton, T.M. Lillo, D.W. Brown, "Enhanced ductility in strongly textured magnesium produced by equal channel angular processing", *Scripta Mater.* 50 (2004) 377.

[9] S.R. Agnew, P. Mehrotra, T.M. Lollo, G.M. Stoica, P.K. Liaw, "Texture evolution of five wrought magnesium alloys during route A equal channel angular extrusion: Experiments and simulations", *Acta Mater.* 53 (2005) 3135.

[10] J.T. Wang, D.L. Yin, J.Q. Liu, J. Tao, Y.L Su, X. Zhao, "Effect of grain size on mechanical property of Mg-3Al-1Zn alloy", *Scripta Mater.* 59 (2008) 63.

[11] T. Mukai, H. Somekawa, T. Inoue, A. Singh, "Strengthening Mg.Al.Zn alloy by repetitive oblique shear strain with caliber roll", *Scripta Mater.* 62 (2010) 113.

[12] T. Mukai, H. Somekawa, A. Singh, T. Inoue, "Strengthening Mg-Al-Zn alloy by repetitive oblique shear strain", in *Magnesium Technology 2011* (eds. W.H. Sillekens, S.R. Agnew, N.R. Neelameggham, S.N. Mathaudhu), TMS, 2011, p.211.

[13] T. Inoue, H. Somekawa, T. Mukai, "Hardness Variation and Strain Distribution in Magnesium Alloy AZ31 Processed by Multi-pass Caliber Rolling", *Adv. Eng. Mater.* 11 (2009) 654.

[14] H. Somekawa, A. Singh, T. Inoue, T. Mukai, "Enhancing fracture toughness of magnesium alloy by formation of low angle grain boundary structure", *Adv. Engg. Mater.* 12 (2010) 837.

[15] J. Koike, et al., "The activity of non-basal slip system and dynamic recovery at room temperature in fine-grained AZ31B magnesium alloys", *Acta Mater.* 51 (2003) 2055.

[16] J. Xing, X. Yang, H. Miura, T. Sakai, "Mechanical Properties of Magnesium Alloy AZ31 after Severe Plastic Deformation", *Mater. Trans.* 49 (2008) 69.

Magnesium Technology 2013
Edited by: Norbert Hort, Suveen N. Mathaudhu, Neale R. Neelameggham, and Martyn Alderman
TMS (The Minerals, Metals & Materials Society), 2013

Increasing volume fraction of precipitates and strength of a Mg-Zn-Y alloy by pre-ageing deformation

Julian M Rosalie[1], Hidetoshi Somekawa[1], Alok Singh[1], Toshiji Mukai[2]

[1]Structural Materials Unit, National Institute for Materials Science, Tsukuba, 305-0047, Japan.
[2]Dept. Mechanical Engineering, Kobe University, 1-1 Rokkodai, Nada, Kobe city, 657-8501 Japan.

Keywords: Magnesium-zinc-yttrium; Precipitation; Microstructure; Transmission electron microscopy

Abstract

Pre-ageing deformation was applied to a Mg-Zn-Y alloy to refine the size of β_1' precipitates. Reductions in the precipitate length and diameter were accompanied by a substantial increase in the volume fraction of the β_1' precipitates from 0.5% (no pre-strain) to 2.3% (5% pre-strain). This contrasted with Mg-Zn alloys, in which the precipitate volume fraction in the peak aged condition was not affected by pre-ageing strain. The increase in precipitate volume fraction contributed significantly to an increase in the yield strength from 217 MPa (no pre-strain) to 286 MPa (5% pre-strain). Precipitate strengthening via Orowan looping was the most significant contributor to the overall strength of the alloy.

Introduction

Pre-ageing deformation is used to enhance the precipitation strengthening response in a wide range of alloys and has been shown to be effective in Mg-Zn alloys [1, 2]. These alloys derive their strength from the controlled precipitation of rod-shaped precipitates (termed β_1') which have the long axis parallel to $[0001]_{Mg}$. In a recent study on precipitation hardening of Mg-Zn alloys, it was found that the yield strength could be increased from 273 MPa with T6 heat treatment (solution treatment plus artificial ageing) to 309 MPa by using a T8 treatment (solution treatment, cold-work and artificial ageing) using only 3-5% pre-ageing deformation [2]. This improvement in yield strength was due to dislocations providing heterogeneous nucleation sites for β_1' precipitates, resulting in accelerated precipitation kinetics and a refinement of the precipitate size.

The rod-shaped β_1' precipitates also form in magnesium-zinc alloys containing yttrium and rare earth (RE) additions. These elements are used to improve the high-temperature properties of Mg-Zn alloys [3, 4] and minor additions significantly modify the equilibrium phase behaviour. Yttrium has negligible solid solubility in Mg [5] and partitions to the grain boundaries in the form of a grain boundary eutectic comprised of an icosahedral phase (i-phase, Mg_3Zn_6Y [6–9]) and Mg [10]. In alloys with Zn:Y ratios close to the stoichiometry of the i-phase, under equilibrium conditions, little remaining zinc is expected to be available for β_1' phase precipitation and the alloy should show minimal precipitation strengthening.

The present work set out to examine whether pre-ageing deformation could be applied to improve the precipitation strengthening of MgZnY alloys. This revealed the surprising observation of an increase in the β_1' precipitate volume fraction when pre-ageing strain was applied [11]. Here we discuss in greater detail the reasons behind this increase in the precipitate volume fraction of β_1' precipitates.

Experimental details

Billets with composition Mg-3.0at.%Zn-0.5at.%Y were prepared from pure elements via direct-chill casting. Mg (99.99%) was heated to 700°C under a CO_2 atmosphere. Once the temperature had stabilised, the required amount of Zn (99.999%) was added and the melt was stirred. Hydrogen was removed by bubbling CO_2 through the melt for 5 minutes, after which the melt was poured into a steel mold and allowed to cool. The compositions were confirmed via inductively-coupled plasma mass spectroscopy (ICP-MS). The billets were homogenised for 15 h at 350°C and extruded into 12 mm diameter rods at 300°C with an extrusion ratio of 12:1. Tensile samples of gauge length 15 mm and diameter 3 mm were machined from the extruded rods.

The tensile samples were given a T8 heat treatment comprised of a solution-treatment followed by cold-work and isothermal ageing. Samples were encapsulated in argon, solution treated for 1 h at 400°C and and quenched into water at room temperature. Pre-ageing deformation was carried out by straining the tensile samples to a nominal plastic strain of 3%

or 5% on an Instron mechanical tester. Tensile deformation was applied parallel to the extrusion axis at a strain rate, $\dot{\epsilon}$, of $1\times10^{-3}\,s^{-1}$. Sections of both solution treated and deformed samples were polished and etched with an acetic picral reagent for examination by light microscopy. The remaining samples were then aged to peak hardness in an oil bath at 150°. Maximum hardness was reached after 216-256 h for T6 treatments samples and 32-48 h for T8 treatments. Tensile tests were repeated in the Instron mechanical tester, with three samples for each condition strained to failure at the same strain rate of $1\times10^{-3}\,s^{-1}$.

Solution-treated and quenched samples were polished for examination using electron back-scattered diffraction (EBSD) on a field emission scanning electron microscope. EBSD data was analysed using OIM Analysis software, version 6.1.

The β_1' precipitate size and spacing in T6 and T8 samples were determined in the peak-aged condition. Samples for TEM analysis were prepared by grinding to $\sim 70\,\mu m$ and thinning to perforation using a Gatan precision ion polishing system. TEM observations were conducted using JEOL 4000EX and 2100 instruments operating at 400 kV and 200 kV, respectively. Stereological measurements utilised ImageJ software version 1.44.

Results

Grain structure

Optical micrographs showed a microstructure consisting of equiaxed grains with coarse intermetallic particle which formed chains in the extrusion direction. A grain size of $17 \pm 3\,\mu m$ was measured using line intercepts. Twinning was extremely limited, with only 1-3% twins by volume fraction measured by point counting. As this value was independent of the pre-ageing deformation (up to 5%) it is thought that this most likely represents a sample-processing artifact. A solution-treated sample was also examined using electron back-scattered diffraction (EBSD) and twins were not detected. An average grain size of $14 \pm 1\,\mu m$ was calculated using EBSD measurements of the grain diameter.

Precipitate size and distribution

The precipitate size and distribution was strongly affected by pre-ageing deformation. In non-deformed samples the microstructure in the peak-aged condition consisted of rod-like β_1' precipitates, a smaller fraction of β_2' plates on the basal planes and coarse spheroidal precipitates. The spheroidal precipitates are consistent with previous reports of i-phase particles in Mg-Zn-Y alloys [12]. Figure 1 shows electron micrographs taken with beam normal to the hexagonal axis of the matrix. Electron micrographs obtained with the beam along the hexagonal axis (Figure 2) showed the rod-like β_1' precipitates in cross-section and indicated that the precipitates were sparse and well-separated.

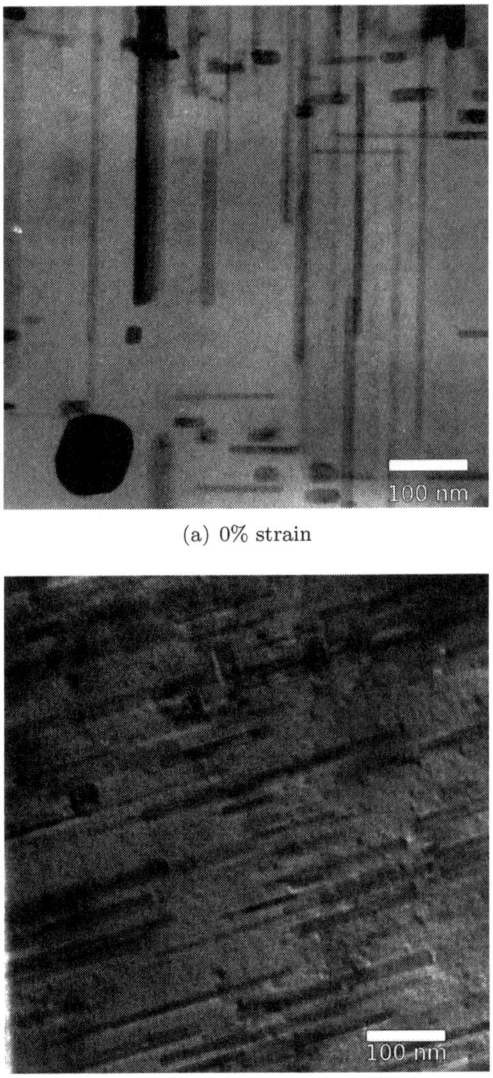

(a) 0% strain

(b) 5% strain

Figure 1: Transmission electron micrographs of β_1' precipitates in the peak aged condition after a) T6 treatment and (b) T8 treatment with 5% pre-strain. The T6 sample contains a large, spheroidal i-phase particle. The beam direction is normal to the c-axis of Mg and to the long axis of the β_1' precipitates.

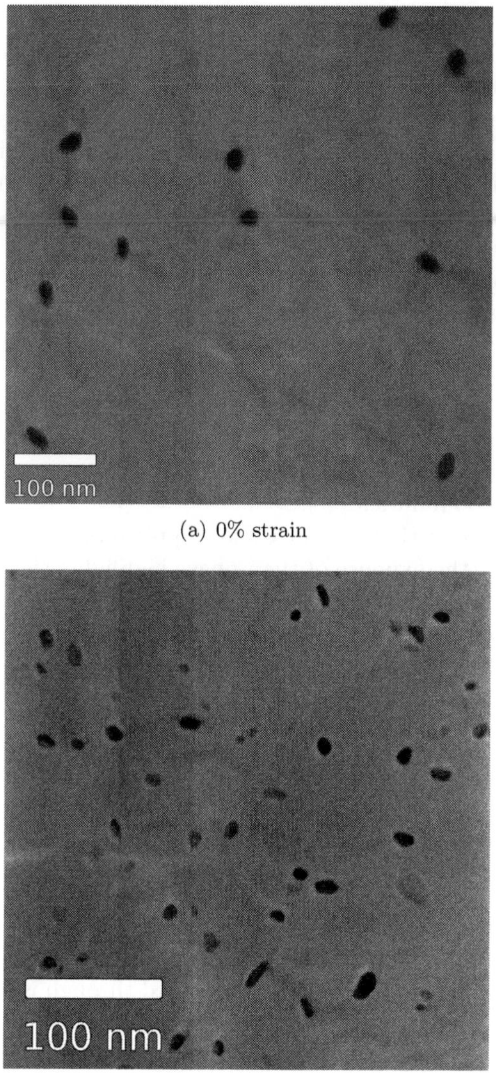

(a) 0% strain

(b) 5% strain

Figure 2: Transmission electron micrographs of β_1' rod-shaped precipitates in the peak aged condition after a) T6 treatment and (b) T8 treatment with 5% pre-strain. The beam direction is parallel to the c-axis of Mg and the β_1' precipitates are viewed in cross section. Note that the pre-strained sample is shown at higher magnification in order to make the smaller β_1' preciptitates visible.

In samples given 3–5% pre-ageing strain the number of plate-like β_2' precipitates was less substantial and the spheroidal ternary phase precipitates were extremely rare. The length of the rod-like precipitates was substantially reduced as can been seen in Figure 1(b). When viewed along the hexagonal axis the β_1' precipitates were were found to be smaller in diameter and more closely spaced (Figure 2(b)).

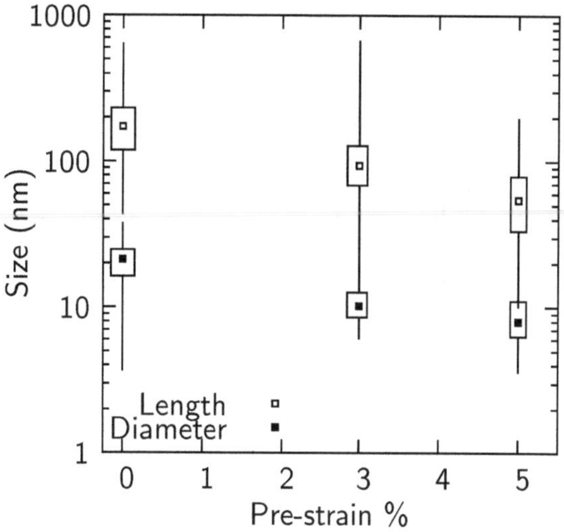

Figure 3: Precipitate length and diameter as a function of pre-ageing deformation. The boxed region contains 50% of of the precipitate length values, with the lines indicating the full range of values.

The distribution of β_1' precipitate lengths and diameter is summarised in Figure 3. Filled (open) squares show the mean diameter (length) of the precipitates. T6-treated samples had a average β_1' precipitate length of 475 nm. This was reduced to an average of 67 nm when 5% pre-strain was applied. The boxed region of the plot indicates the range containing 50% of the precipitate length values, with the line indicating the range of measured values. Precipitates were able to grow to lengths of >600 nm in T6 samples.

The volume fraction of β_1' precipitates was determined from measurements of precipitate length, diameter and number density. The volume fraction of precipitates in Mg-Zn was approximately 3.5% regardless of the level of pre-ageing deformation. The volume fraction of precipitates in non-deformed (T6 treated) Mg-Zn-Y was only 0.5%, due to the low number density of precipitates (as indicated in Figure 2(a)). The application of either 3% of 5% pre-strain increased the volume fraction substantially to 2.3%. The contrasting behaviour indicates that dislocations introduced prior to deformation in the ternary alloy not only accelerate the nucleation kinetics, but in MgZnY these dislocations also modify the partitioning of zinc solute.

Tensile properties

Isothermal ageing substantially increased the yield strength (σ_y), with the improvement more pronounced in the alloys strained before ageing. Stress-

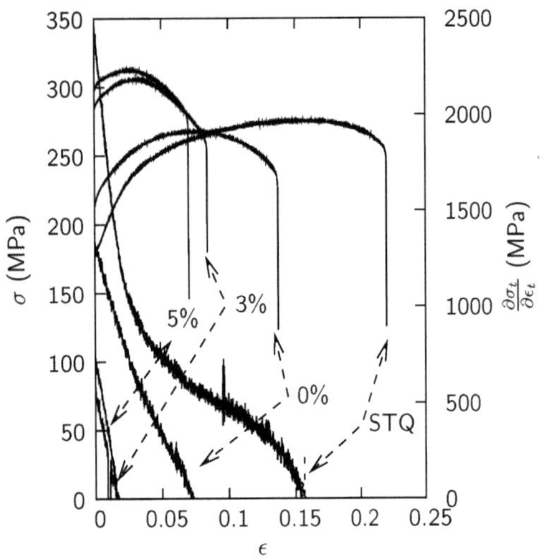

Figure 4: Stress-strain curves (left) and hardening rate curves (right) for various levels of pre-ageing deformation. "STQ" indicates the solution-treated and quenched condition in which β_1' precipitates were absent. Other samples were tested in the peak-hardness condition.

strains curves and hardening rate curves for uniaxial tension are provided in Figure 4. T6 treatment increased the yield strength by ~50% from 150 MPa to 217 MPa. T8 treatments resulted in further increases to σ_y to 281 MPa for 3% pre-strain and a slight further increase to 287 MPa for 5% strain.

The hardening rate curves (Figure 4) showed that the alloy work-hardens in all heat-treatment conditions examined. Work-hardening was most significant in the solution-treated and quenched sample, in which precipitates were absent and zinc was present in solid-solution. In T6 treated samples the work-hardening rate was ~1350 MPa at the yield point and decreased with further deformation. Work-hardening also occurred for T8-treated samples, with hardening rates of 750 MPa at the yield point and no significant difference between samples deformed 3% and 5% prior to ageing.

Discussion

The increase in the volume fraction of β_1' precipitates due to pre-ageing strain is an unusual feature and one that is not present in Mg-Zn alloys. This effect arises from the low solid solubility of yttrium and competition between i-phase and β_1' precipitation. The alloy composition has a Zn:Y ratio of 6:1, which falls within the two phase Mg + i-phase (Mg_3Zn_6Y) region of the Mg-Zn-Y phase diagram [9]. Under equilibrium heat-treatment condition the microstructure would be expected to consist of a Mg-rich matrix, with i-phase forming a grain boundary network.

During solution treatment there is substantial solid solubility for zinc whereas yttrium remains virtually insoluble. Consequently, at the commencement of isothermal ageing there is substantial excess Zn within the grains. If this solute is able to diffuse to the grain boundaries it will be incorporated into the grain boundary i-phase. However, the dislocations introduced by pre-ageing deformation act as preferred nucleation site for β_1' precipitates, as is the case in Mg-Zn binary alloys [2]. These defects act as solute-traps, precipitating the solute in the form of the rod-like particles, before it is able to reach the grain boundaries.

The fact that it is possible to precipitate β_1' particles at the expense of the i-phase highlights the difficulty in precipitating the icosahedral phase. Previous attempts to use the icosahedral phase to strengthen Mg alloys using i-phase particles have been forced to resort to severe plastic deformation (e.g. equal channel angular extrusion (ECAE) [13])to distribute yttrium more homogeneously through the alloy.

Strengthening mechanism

The increase in the yield strength of the alloys can be attributed almost entirely to precipitation strengthening, as is to be expected from the 32% increase in the yield strength obtained through isothermal ageing.

Extrusion was used primarily to introduce texture to the alloy, preventing twinning during pre-ageing deformation, rather than to provide significant grain refinement. This was advantageous from an experimental perspective as it simplified quantification of the precipitate size and number density using TEM. The large grain size (14μm from EBSD measurements) that resulted was responsible for the low base strength of the material. Based on the Hall-Petch plots for AZ and MgZnY magnesium alloys in tension [13–15] an alloy with grain size 14-17μm should have a base yield strength in the range of 160 MPa, close to the value of 150 MPa measured for solution-treated and quenched Mg-Zn-Y (Figure 4).

Despite the use of pre-ageing deformation, work hardening was also not a significant contributor to the mechanical properties. Temperatures sufficient to precipitate β_1' particles within 32–48 h for T8 samples are also within the temperature range used from strain relief of ZK alloys [16]. Previous studies [2, 11] noted a rapid decrease in the hardness during the first 1 h of ageing that was attributed to the annealing out of dislocations. Since ageing times of 24-32 h were required to reach peak hardness with T8 treatments it is probable that the dislocation density is low and that strain hardening is negligible.

Solid-solution strengthening was thought to add contribute slightly to the overall strength of the alloy. If the Zn content is assumed to be the equilibrium value at the ageing temperature (T6 and T8 samples), then the contribution from solid solution strengthening can be determined from the equation [17]: $\Delta\sigma = \sqrt{2}m\mu\epsilon^{3/2}c^{1/2}$where σ is the yield strength, m is the Taylor orientation factor (3 in Mg), μ Shear modulus, ϵ is the misfit strain (0.17 for Zn in Mg) and c is the solute concentration [18] (given in at.%). This gives a value of 12 MPa for solid solution strengthening in the peak-aged condition.

With minimal grain boundary strengthening and in the absence of significant work-hardening and solid-solution strengthening, the increase in the strength of the alloy by 67 MPa (T6 treatment) or 125–131 MPa (T8 treatment) should be attributable to precipitation of the β_1' particles.

Conclusions

- Solution-treatment and ageing (T6 treatment) of Mg-Zn-Y resulted in the precipitation of a volume fraction of approximately 0.5% β_1' precipitates, substantially less than the value of approximately 3.5% in a Mg-Zn alloy given similar treatment. This resulted in a weak precipitation strengthening response.

- Solution-treatment, pre-strain and ageing (T8 treatment) introduced dislocations that provided heterogeneous nucleation sites for β_1' precipitates. These defects acted as solute traps for Zn, reducing the amount of solute that diffused to the grain boundaries. This increased the volume fraction of β_1' particles to approximately 2.3% while simultaneously refining the size of the β_1' precipitates.

- These effects combined to increase the yield strength by 32% from 217 MPa (no pre-strain) to 286 MPa (5% pre-strain). This was a greater relative increase than reported in Mg-Zn alloys, where 5% deformation increased yield strength by ~15%.

References

[1] L. Sturkey, J. B. Clark. Mechanism of age-hardening in magnesium-zinc alloys. *J. Inst. Mater.*, **88**:(1959-60) 177.

[2] J. M. Rosalie, H. Somekawa, *et al.* The effect of size and distribution of rod-shaped precipitates on the strength and ductility of a Mg-Zn alloy. *Mater. Sci. Eng. A*, **539**:(2012) 230.

[3] C. J. Boehlert. The tensile and creep behavior of Mg-Zn alloys with and without Y and Zr as ternary elements. *J. Mater. Sci*, **42**:(2007) 3675.

[4] Y. Li, F. Xu, *et al.* Influence of Y element on microstructure and mechanical properties of ZK60 alloy. *Adv Mater Res*, **284-286**:(2011) 1568.

[5] G. Shao, V. Varsani, *et al.* Thermodynamic modelling of the Y-Zn and Mg-Zn-Y systems. *CALPHAD*, **30**(3):(2006) 286.

[6] J. P. Hadorn, K. Hantzsche, *et al.* Role of solute in the texture modification during hot deformation of Mg-rare earth alloys. *Metall. Mater. Trans. A*, **43A**(4):(2012) 1347.

[7] L. Y. Wei, G. L. Dunlop, *et al.* Precipitation hardening of Mg-Zn and Mg-Zn-RE alloys. *Metall. Mater. Trans. A*, **26A**(7):(1995) 1705.

[8] A. Singh, A. P. Tsai. Structural characteristics of β_1' precipitates in Mg-Zn-based alloys. *Scr. Mat.*, **57**:(2007) 941.

[9] A. P. Tsai, A. Niikura, *et al.* Highly ordered structure of icosahedral quasicrystals in Zn-Mg-RE (RE= rare earth metals) systems. *Philos Mag Lett*, **70**(3):(1994) 169.

[10] A. P. Tsai, Y. Murakami, *et al.* The Zn-Mg-Y phase diagram involving quasicrystals. *Philos. Mag. A*, **80**(5):(2000) 1043.

[11] J. M. Rosalie, H. Somekawa, *et al.* The effect of precipitation on strength and ductility on Mg-Zn-Y alloys. *J. Alloys Comp.*

[12] A. Singh. Phase relations, formation and morphologies in Mg-Zn-RE (RE=Y, rare earth) alloys. in M. O. Pekuleryuz, N. R. Neelameggham, *et al.*, eds., *Magnesium Technology*, 337–342, The Minerals, Metals and Materials Society, New Orleans, 2008.

[13] A. Singh, M. Watanabe, *et al.* Microstructure and strength of quasicrystal containing extruded Mg-Zn-Y alloys for elevated temperature application. *Mater. Sci. Eng. A*, **385**:(2004) 382.

[14] D. H. Bae, S. H. Kim, *et al.* Deformation behavior of Mg-Zn-Y alloys reinforced by icosahedral quasicrystalline particles. *Acta Mater.*, **50**(9):(2002) 2343.

[15] E. Mora, G. Garcés, *et al.* High-strength Mg-Zn-Y alloys produced by powder metallurgy. *Scr. Mat.*, **60**(9):(2009) 776.

[16] M. M. Avedesian, H. Baker, eds., *Magnesium and magnesium alloys*, 79, ASM speciality handbook, ASM, 1999.

[17] Y. Chino, M. Kado, *et al.* Compressive deformation behavior at room temperature-773k in Mg-0.2 mass%(0.035at.%)Ce alloy. *Acta Mater.*, **56**:(2002) 387.

[18] J. B. Clark, L. Zabdyr, *et al.*, *Binary alloy phase diagrams*, chap. Mg-Zn (Magnesium-Zinc), 2571–2572, ASM International, 1988.

Magnesium Technology 2013
Edited by: Norbert Hort, Suveen N. Mathaudhu, Neale R. Neelameggham, and Martyn Alderman
TMS (The Minerals, Metals & Materials Society), 2013

MAPPING THE MECHANICAL PROPERTIES OF ALLOYED MAGNESIUM (AZ 61)

Jennifer L. Hay[1], Phillip Agee[2]

[1]Agilent Technologies; 105 Meco Lane, Suite 200, Oak Ridge, TN 37830
[2]Agilent Technologies; 4330 West Chandler Blvd, Chandler, AZ 85226

Keywords: AZ 61, indentation, hardness, elastic modulus, mechanical-properties map, mesophase

Abstract
In this work, an advanced form of nanoindentation is used to map the mechanical properties of AZ 61. The probed area includes both alpha (Mg-rich) and beta ($Al_{12}Mg_{17}$) phases. The measured moduli of the two phases compare well with expected values. The hardness of the beta phase is three times greater than that of the alpha phase. The hardness map reveals an area of intermediate hardness surrounding the beta phase. This zone is a mesophase which is chemically distinct and may influence the mechanical behavior of the alloy in unexpected ways.

Introduction

Behind iron and aluminum, magnesium (Mg) is the third most common element used in engineered structures, because it is light-weight, stiff, and strong [1]. Alloying with aluminum can further improve stiffness and strength, although bulk mechanical properties depend strongly on chemical composition and thermo-mechanical history, insofar as these parameters affect microstructure. In this work, we use nanoindentation to determine the properties of individual phases within a popular magnesium alloy, AZ 61.

AZ 61 is a commercially available magnesium alloy which includes aluminum (nominally 6%), zinc (nominally 1%), and other trace elements. Table I provides the chemical composition for AZ 61[2]. The zinc and other trace elements have little effect on microstructure, and the Mg-Al phase diagram is employed to predict the constitution of slowly-cooled AZ 61. At 6% Al, the Mg-Al phase diagram predicts the interaction of two phases: an α phase which is Mg-rich and a β phase comprising the intermetallic compound $Al_{12}Mg_{17}$. As the alloy cools from the liquid state, the α phase begins to solidify at about 620C. Solidification is complete at about 540C, and at this temperature, the material exists entirely in the α phase. Beginning around 300C, the β phase begins to precipitate and the mass fraction of this secondary phase continues to grow as the material continues to cool. At temperatures below 100C, the β phase accounts for about 7.5% of the material[3].

In this work, we used nanoindentation to measure the elastic modulus and hardness of both phases of AZ 61. Our expectation is that the modulus for each phase should be close to that for the primary component. The primary component of the α phase is Mg, which has an elastic modulus of 45GPa[1]; however, it should be remembered that at room temperature, the α phase does include 3% of interstitial aluminum (by mass). Thus, we should not be surprised if the modulus of the α phase is slightly higher than 45GPa. Zhang et al. calculated the elastic modulus of the $Al_{12}Mg_{17}$ to be 78GPa from first principles[4], so this value sets our expectations for the β phase of AZ 61.

In this work, we used an advanced form of nanoindentation called "Express Test" which performs one complete indentation cycle per second, including approach, contact detection, load, unload, and movement to the next indentation site[5]. This technology enabled us to quantitatively "map" both the elastic modulus (*E*) and the hardness (*H*) of a surface in a reasonable time. This feature is particularly beneficial for probing multi-phase metals such as AZ 61.

Table I. Chemical Composition of Magnesium AZ 61 Alloy [2].

Element	% by mass
Aluminum	5-7
Zinc	0.8-1
Copper	< 0.03
Silicon	< 0.01
Iron	< 0.01
Nickel	<0.005
Magnesium	Balance

Experimental Method

The sample of AZ 61 tested in this work was prepared by researchers at the Material Science and Engineering Department at Drexel University. AZ 61 chunks were purchased from Thixomat (Livonia, MI). The chunks were placed in alumina crucibles which were covered with alumina disks. Then the crucibles were placed in a vacuum furnace and were heated with a rate of 10°C/min up to 750°C and held for one hour at the temperature after which the furnace was turned off and the samples were furnace cooled. To polish the surface, a section was cut from the sample, mounted, and rough ground with water using silicon carbide, beginning with 400-grit and finishing with 1200-grit (US). The samples were polished using 6μm and 1μm diamond on a medium nap cloth with an alcohol-based extender.
Scanning electron microscopy was performed with the Agilent 8500 Field Emission Scanning Electron Microscope (Chandler, AZ). The prepared sample was fixed to an SEM stub with carbon tape and mounted onto the SEM sample stage. Imaging was performed with the slowest scan speed, in backscattered electron

mode, with an accelerating voltage of 1kV. The working distance was 2.2mm.

All indentation testing was performed with an Agilent G200 NanoIndenter (Chandler, AZ) having Express Test, NanoVision, and a DCM II fitted with a Berkovich indenter. The test method "Express Test to a Force" was used to perform an array of 20x20 indents within a 50μm x 50μm area; thus, the separation between successive indents was about 2.5 μm (2500nm). The peak force for every indent was 4mN; in the α phase, this force produced a peak displacement of about 380nm; in the β phase, this force produced a peak displacement of about 240nm.

'Express Test' is the trade name for a specific option for the Agilent G200 NanoIndenter that allows rapid indentation. Although indents are performed rapidly (one per second), Young's modulus and hardness are obtained by analyzing the force-displacement data for each indentation according to established norms. These calculations are described in detail elsewhere [6, 7]. Thus, these same measurements could be made by any commercially available instrument that conforms to ASTM E 2546-07 [8] or ISO 14577 [9-11], although the required time would be much longer (on the order of hours, rather than minutes). To perform this work with another instrumented indentation system, the user would simply prescribe a similar array of 20x20 indents, each performed to a peak force of 4mN, all within a 50μm x 50μm area.

Results and Discussion

The scanning-electron microscopic (SEM) image of Figure 1 shows the microstructure of the AZ 61 surface as prepared for indentation. The α phase is darker and the precipitated β phase is lighter. From this image alone, one deduces that the β phase is harder, because scratches that span both phases are smaller in the β phase.

Figure 2 shows the maps of elastic modulus (E) and hardness (H) generated by a 20x20 array of indentations. The total testing time is 15 minutes. Each indentation generates the information content for one pixel; thus, the images in Figure 2 are 400-pixel (20x20) images. The two phases are clearly distinguished. In order to calculate the properties of each phase independently, two rectangular domains were selected which were clearly in either one phase or the other. These are the black and white rectangles in Figure 2.

The black rectangle is in the primary α phase, which is the Mg with interstitial Al that constitutes about 92% of the material. Within the area bound by the black rectangle, there are 25 indentations; the average elastic modulus for these indentations is E = 54.9±3.3GPa, which is slightly higher than the modulus of pure Mg (45GPa) [1]. The average hardness is H = 1.35±0.05GPa.

The white rectangle is in the β phase, which is primarily $Al_{12}Mg_{17}$. Within the area bound by the white rectangle, there are 15 indentations; the average elastic modulus for these indentations is 75.8±3.3GPa, which is consistent with the value predicted by Zhang (78GPa)[4]. The average hardness is 4.3±0.19, which is about three times greater than the hardness of the α phase. Table II summarizes the properties for each phase.

Surprisingly, the hardness image reveals a region around the β phase in which the hardness is lower than that of the β phase, but higher than that of the α phase. There are two reasonable explanations for this observation. First, the elevated hardness of the material surrounding the β phase may be caused by (and reveal) β material immediately below the exposed surface. However, modulus is usually more sensitive to the influence of constraining material, and the modulus is uniform outside the β phase. It must be granted, however, that the difference in hardness between the two phases is much greater than the difference in modulus, so the hardness may manifest more constraint effect than modulus simply because there is a greater difference in hardness between the two phases.

The second (and more interesting) explanation is that the material around the β phase may be a true "mesophase" if it is chemically different than either α or β. More detailed analysis of the chemistry and microstructure of this region by means of energy dispersive x-ray spectroscopy (EDX), transmission electron spectroscopy (TEM), and focused-ion beam milling (FIB) may shed more light on this phenomenon.

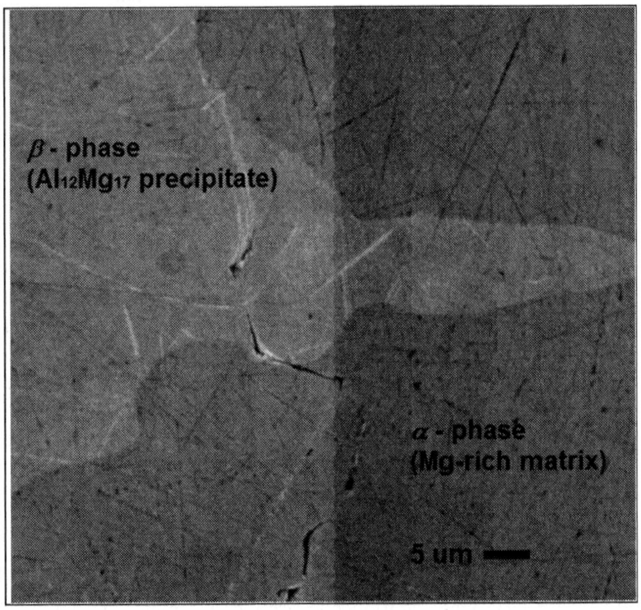

Figure 1. Low-voltage, field-emission, scanning-electron-microscopy image of Mg AZ 61 as prepared for nanoindentation. Acquired in back-scattered electron (BSE) mode. Image resolution is 1024x1024 pixels. Smaller scratches in the β phase evidence higher hardness.

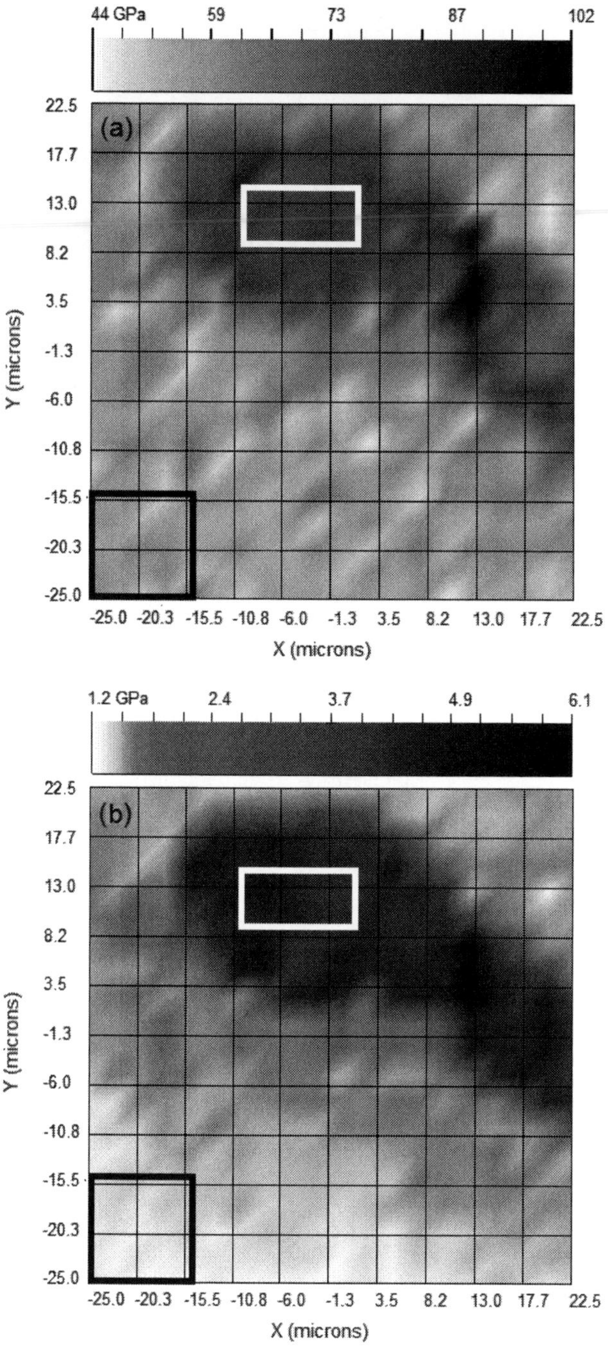

Figure 2. (a) Modulus and (b) hardness of two-phase Mg AZ 61. Black rectangles are in the primary α phase (Mg-rich). White rectangles are in the β phase ($Al_{12}Mg_{17}$). Information for both images was acquired in 15 minutes by means of the Express Test option for the Agilent G200 NanoIndenter (Chandler, AZ). Image resolution is 20x20 pixels.

Table II. Properties of Magnesium AZ 61 Constituents Measured by Nanoindentation at 4mN

Material	N	*E* GPa	*H* GPa
α-phase (Mg-rich)	25	54.9[1]±3.3	1.35±0.05
β-phase ($Al_{12}Mg_{17}$)	15	75.8[2] ±3.3	4.30±0.19

[1] *c.f.* the modulus of pure Mg (45GPa) [1]
[2] *c.f.* the modulus of $Al_{12}Mg_{17}$ calculated by Zhang et al (78GPa) [4]

Conclusions

In this work, the Express Test option for the Agilent G200 NanoIndenter was used to map out the mechanical properties of a magnesium alloy, AZ 61. The measured moduli of the α and β phases compared well with expected values. The hardness of the precipitate (β) phase was three times greater than that of the α phase. The hardness map revealed an area of intermediate hardness surrounding the β phase; more analysis is required to fully explain this phenomenon.

Acknowledgements

The authors gratefully acknowledge Dr. Hamdallah Bearat of Arizona State University for the SEM images. The authors further acknowledge Mr. Babak Anasori and Dr. Michel Barsoum of the Materials Science and Engineering Department of Drexel University for providing the sample of AZ 61 tested in this work.

References

1. "Magnesium," [Date Accessed: July 9, 2012] Available from: http://en.wikipedia.org/wiki/Magnesium.

2. C.-F. Fang, X.-G. Zhang, S.-H Ji, J.-Z. Jin, and Y.-B. Chang, "Microstructure and corrosion property of AZ61 magnesium alloy by electromagnetic stirring," *Trans. Non Ferrous Met. Soc. China*, 15 (3) (2005), 536-541.

3. James F. Shackelford, *Introduction to Materials Science for Engineers, Second Edition* (New York, NY: Macmillian Publishing Company, 1988), 223.

4. H. Zhang et al., "First-principles calculations of the elastic, phonon and thermodynamic properties of $Al_{12}Mg_{17}$," *Acta Materialia*, 58 (2010), 4012–4018.

5. J.L. Hay,"Revolutionary New Agilent Express Test Option for G200 NanoIndenters," [Date Accessed: April 9, 2012] Available from: http://cp.literature.agilent.com/litweb/pdf/5990-9948EN.pdf.

6. W.C. Oliver and G.M. Pharr, "An Improved Technique for Determining Hardness and Elastic Modulus Using Load and

Displacement Sensing Indentation Experiments," *Journal of Materials Research*, 7 (6) (1992), 1564-1583.

7. J.L. Hay, "Introduction to Instrumented Indentation Testing," *Experimental Techniques*, 33 (6) (2009), 66-72.

8. ASTM E2546-07, "Standard Practice for Instrumented Indentation Testing," (2007), Available from: http://www.astm.org/Standards/E2546.htm.

9. ISO 14577-1:2002, "Metallic Materials—Instrumented Indentation Test for Hardness and Material Parameters—Part 1: Test Method," (2002), Available from: http://www.iso.org.

10. ISO 14577-2:2002, "Metallic Materials—Instrumented Indentation Test for Hardness and Material Parameters—Part 2: Verification and Calibration of Testing Machines," (2002), Available from: http://www.iso.org.

11. ISO 14577-3:2002, "Metallic Materials—Instrumented Indentation Test for Hardness and Material Parameters—Part 3: Calibration of Reference Blocks," (2002), Available from: http://www.iso.org.

Magnesium Technology 2013

Microstructure

Magnesium Technology 2013
Edited by: Norbert Hort, Suveen N. Mathaudhu, Neale R. Neelameggham, and Martyn Alderman
TMS (The Minerals, Metals & Materials Society), 2013

EFFECTS OF ALLOYING ELEMENTS AND COOLING RATE ON MORPHOLOGY OF PHASES IN CaO ADDED Mg-Al-Si ALLOYS

Young-Gil Jung[1,2], Hyun Kyu Lim[1], Young-Ok Yoon[1], Shae K. Kim[1], and Do Hyang Kim[2]

[1] Casting Research Center, Korea Institute of Industrial Technology, Incheon, Korea
[2] Dept. of Metallurgical Engineering, Yonsei University, Seoul, Korea

Keywords: Mg-Al-Si alloys, CaO, Mg_2Si, CaMgSi, Mechanical properties

Abstract

Mg_2Si formed by the addition of Si in Mg-Al alloys is the very useful intermetallic compound. However, Mg_2Si phases in the Mg-Al-Si alloys are prone to forming undesirable, coarse Chinese script shape, which will deteriorate the mechanical properties of Mg alloys. Therefore, in order to modify the Chinese script shaped Mg_2Si phases in the Mg-Al-Si alloys, many methods have been studied such as alloying addition of Ca or P. This study was aimed at improving the mechanical properties of CaO added Mg-Al-Si alloys by the addition of alloying elements and varying cooling rate to change of morphology of Mg_2Si and CaMgSi phases. The microstructures of specimens were observed by OM and SEM, and the phase analysis was performed by XRD, TEM and EDS. To evaluate the mechanical properties of tested alloys, tensile tests were carried out at room and high temperatures.

Introduction

Mg is the lightest metal among all structural metals. Its density is only 1.74g/cm^3, which is less than one-quarter of iron and is about two-thirds of aluminum. Furthermore, Mg alloy has high specific strength, excellent casting ability, good machining ability, high damping capacity, good weld ability, shields against electromagnetic interference, and can be easily recycled [1-3]. These features make Mg alloy an ideal material for use in wide variety of applications, such as automobile, aviation, train, and aerospace as well as electronic application. Mg-Al based alloys such as AZ91 and AM60 are the most commonly used magnesium alloys in automobiles because of a good combination of mechanical properties, corrosion resistance, and die-castability. Despite these advantages, poor mechanical properties, especially creep resistance at high temperature are the most important barriers on the way of expansion of usage of magnesium alloy [4,5]. Therefore, the use of Mg-Al based alloys in automobiles has been restricted to components where creep resistance is of less concern, such as seat frames, steering wheels, instrument panels, valve covers, etc. Magnesium alloys are more and more widely used as structural materials in recent years due to their low density, among which Mg–Al–Si alloys (e.g. AS41 and AS21) have been used as die-casting alloys for elevated temperature applications since a long time ago [6,7]. It can be concluded from the equilibrium phase diagrams of the ternary Al–Mg–Si and binary Mg–Si systems that Si can hardly dissolve into solid magnesium. The maximum solid solubility of Al and Si in Mg is 11.5 and 0.003 at.%, respectively. The main intermetallic compounds in the Mg–Al–Si alloys are $Mg_{17}Al_{12}$ and Mg_2Si. Mg_2Si exhibits high melting point, low density, high hardness and low thermal expansion coefficient. Therefore, Mg_2Si can act as an excellent heat-resistant strengthening phase in light metals. The potential of Mg_2Si to be employed as strengthening particles in the in situ Mg had also been investigated [8-11]. However, molten Mg-Al-Si alloys are easily ignited and oxidized without melt protective gases during melting and casting processes due to their high reactivity.

Furthermore, Mg_2Si phases in the Mg-Al-Si alloys are prone to forming undesirable, coarse Chinese script shape, which will deteriorate the mechanical properties of Mg alloys. Therefore, in order to modify the Chinese script shaped Mg_2Si phases in the Mg-Al-Si alloys, many methods have been studied such as alloying addition of Ca or P [12-14]. However, there are some problems like fluidity decrease, die soldering, and hot tearing in magnesium alloys with alkaline earth metals such as Ca. On the other hand, the advantage has been reported to reduce hot tearing susceptibility, and reduce die soldering tendency with low-cost, non-SF_6 process, improvement of oxidation and ignition resistances and mechanical properties improved by addition of relatively inexpensive CaO in Mg-based alloys [15-18].

Based on these result, the ignition resistance and mechanical properties were supposed to be improved by the addition of CaO in Mg-Al-Si alloy. However, while melt cleanliness of CaO added Mg-Al-Si alloys have been increased, Mg_2Si particle is little modified by adding CaO in Mg-Al-Si alloys. Also, mechanical properties are decreased due to continuously growth of needle type CaMgSi by increasing CaO content. Therefore, the objectives of this work are to investigate the effects of alloying elements (Sr, Ti, Y) on morphology of needle type CaMgSi phases. Because strontium, titanium and Yttrium have negative enthalpy with Si element compared with other elements in CaO added Mg-Al-Si alloy. At the same time, effect of cooling rate is evaluated for the application of Die-Casting process in the future.

Experimental procedure

Four alloys, of nominal chemical compositions listed in Table1 were prepared in a mild steel crucible under the protection of a mixed gas atmosphere of SF_6 and CO_2. High purity Mg, Al, Al-25Si, Mg-3.0CaO, Al-10Sr, Al-10Ti master alloys were prepared by adding the following alloying materials. The melt was held at 690 °C for several minutes and then poured into permanent mold (40 mm), step mold (3, 6, 9, 12 mm) as shown Fig. 1, which were preheated to 200 °C.

The microstructural analysis was carried out using an Optical microscopy (OM) and scanning electron microscopy (SEM) equipped with and energy dispersive X-ray spectrometer (EDS) after etching with a solution of 2 vol.% nitric acid + ethyl alcohol. And aspect ratio was measured by using image analyzing program.

Also, the tensile tests were performed at room and high temperatures and conditions of tensile test were listed in Table 2.

Results and discussion

Microstructure

The typical microstructure of permanent mold cast AS51 (Mg-5Al-0.7Si) alloy were illustrated in Fig. 2. The as-cast microstructure consisted of Chinese script type Mg_2Si particles with interdendritic β-$Mg_{17}Al_{12}$ phases in matrix (α-Mg). On the other hand, the deformation of needle type CaMgSi phases had been observed with addition of CaO in AS51. In fact, it has the same characteristics as the one recently observed in Mg-Si-Ca alloys [19]. When the addition of Sr, Ti, Y elements was added in ASO5105 alloy the modified CaMgSi phases have been observed as shown in Fig. 3.

Because heat of mixing, Sr (Δh_{mix} = -32 kJ/mol), Ti (Δh_{mix} = -56 kJ/mol), Y (Δh_{mix} = -49 kJ/mol) elements have more negative enthalpy with Si element compared with other elements in ASO5105 alloy.

$$\Delta h_{mix} = h_m - x_1 h_1 - x_2 h_2 = Q/m$$

Among Sr, Ti, Y added ASO5105 alloys, the modification effect of Sr added ASO5105 alloy was more effective than Ti, Y added ASO5105 alloys as shown in Fig. 3. Also, the modification effect of needle type CaMgSi phases have been confirmed by increasing cooling rate in accordance with the difference of mould thickness as shown in Fig. 5. Especially, in the result of aspect ratio, even if the cooling rate is decreased, the modification effect of Sr added ASO5105 alloy is more than Ti, Y added ASO 5105 alloys as shown in Fig. 5. Also, Sr, Ti, Y included CaMgSi phase is observed by SEM(EDS) as shown in Fig. 4 and chemical compositions listed in Table 3.

Table 1. Nominal chemical compositions of experimental alloys (wt.%)

Alloy ID	Mg	Al	Si	Ca	Sr	Ti	Y
ASO5105	Bal.	5.0	0.7	0.5	-	-	-
ASO5105+0.3Sr	Bal.	5.0	0.7	0.5	0.3	-	-
ASO5105+0.3Ti	Bal.	5.0	0.7	0.5	-	0.3	-
ASO5105+0.3Y	Bal.	5.0	0.7	0.5	-	-	0.3

Table 2. Condition of tensile test

	Type	Gauge length (mm)	Thickness (mm)	Strain rate (mm/min)
Room temp.	Plate	2.5	3.0	1.0
High temp. (150℃)	Plate	2.5	3.0	0.4

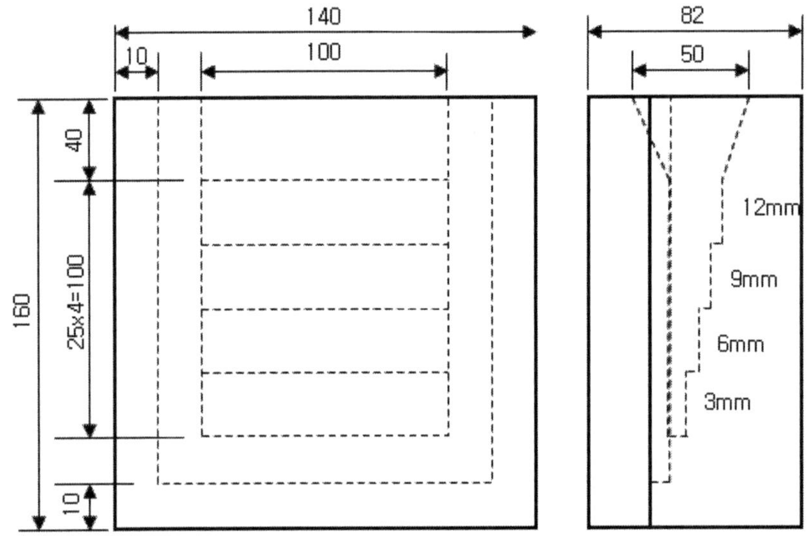

Fig. 1. Schematic drawing of the step mold

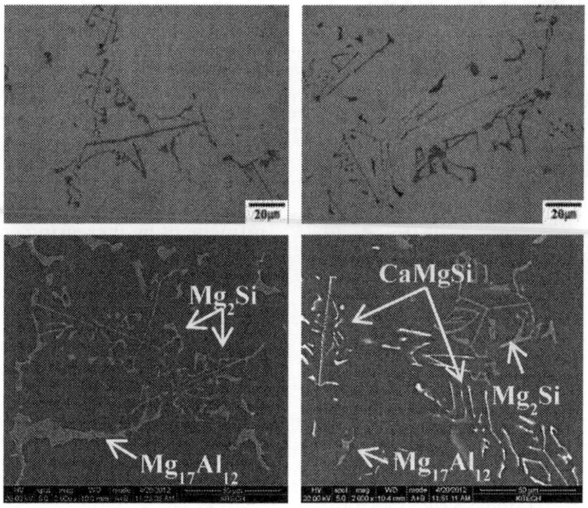

Fig. 2. Microstructures of as-cast (a) AS51, (b) ASO5105

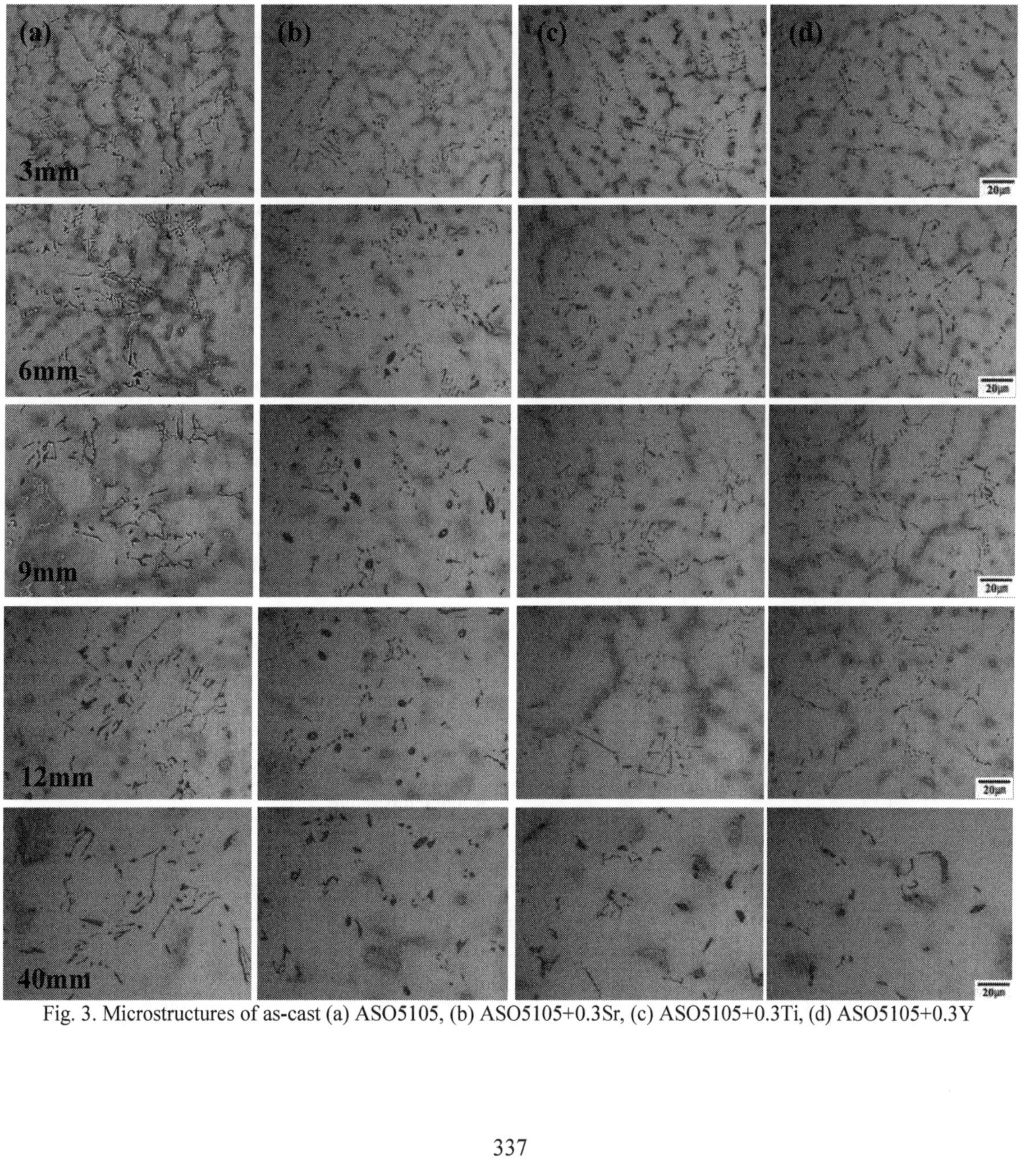

Fig. 3. Microstructures of as-cast (a) ASO5105, (b) ASO5105+0.3Sr, (c) ASO5105+0.3Ti, (d) ASO5105+0.3Y

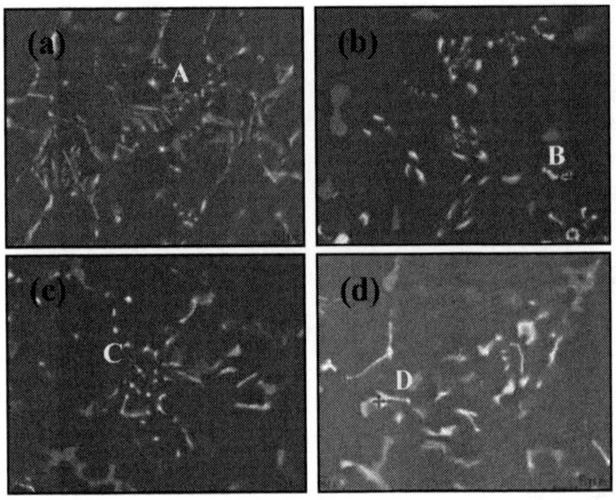

Fig. 4. Scanning electron microscopy (SEM) images of as-cast (a) ASO5105, (b) ASO5105+0.3Sr, (c) ASO5105+0.3Ti, (d) ASO5105+0.3Y

Table 3. Chemical compositions of area A, B, C and D analyzed EDS in Fig. 4 (wt.%)

Area	Mg	Al	Si	Ca	Sr	Ti	Y
A	69.78	7.16	14.28	8.78	-	-	-
B	63.44	1.83	18.40	9.38	6.96	-	-
C	74.91	1.53	13.39	9.94	-	0.23	-
D	64.55	1.42	17.82	14.84	-	-	1.36

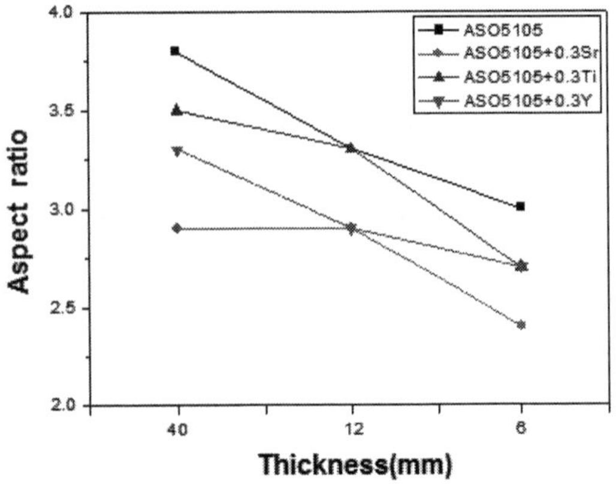

Fig. 5. Aspect ratio of Sr, Ti, Y added ASO5105 alloys

Mechanical properties

The room and high temperature tensile properties were performed when the cooling rate was different according to the thickness of as-cast samples, 6 mm and 40 mm, respectively. The yield strengths at room and high temperature were increased by increasing cooling rate. It is well known that tensile yield strength is increased with presence of fine phases and effect of grain refinement by increasing cooling rate.

In the case of thickness of 6 mm, the elongations at room temperature of ASO5105 and Sr, Ti, Y added ASO5105 were measured 4.0, 5.2, 5.1, 5.3 %, respectively. Therefore, the elongations at room temperature of Sr, Ti, Y added ASO5105 are increased about 30 % compared with ASO5105. Because the needle type CaMgSi phases were modified with the addition of Sr, Ti, Y elements in ASO5105 and the cooling rate was increased by thin thickness. Also, the tensile yield strength at high temperature, 6 mm is increased about 23 % with the addition of Sr, Ti, Y in ASO5105. Because modified CaMgSi phases have high thermal stability at high temperature, this intermetallic. phase at high temperature does not diffuse easily in the matrix (α-Mg) [19]. On the basis of the result, modified CaMgSi phase can act as a significantly role in increasing creep resistance. Therefore, in the future, in the order to application of high temperature Mg alloy, creep test of Sr, Ti, Y added ASO5105 is required.

Conclusion

In this study, effects of alloying elements and cooling rate on morphology of phase in CaO added Mg-Al-Si alloys were investigated. The conclusions are as follows:

(1) The needle type CaMgSi phases were modified Mg_2Si with the addition of Sr, Ti, Y elements having negative enthalpy with Si element and with the increase of cooling rate.

(2) The elongation at room temperature was increased about 30 % due to modification of needle type CaMgSi with the addition of Sr, Ti, Y elements in ASO5105. The tensile yield strength at high temperature, 6 mm of Sr, Ti, Y added ASO5105 was increased about 23 % compared with ASO5105.

In order to application of high temperature Mg alloys, these applications require superior creep resistance. In the result of microstructure and tensile properties, creep test of Sr, Ti, Y added ASO5105 might be required before application of die-casting process.

Table 4. Tensile properties of the as-cast ASO5105 and Sr, Ti, Y added ASO5105 at room temperature

Alloy ID	Room temperature					
	6mm			40mm		
	TYS(MPa)	UTS(MPa)	E(%)	TYS(MPa)	UTS(MPa)	E(%)
ASO5105	96.8	174.2	4.0	77.6	153.4	6.6
ASO5105+0.3Sr	97.3	173.2	5.2	80.9	157.6	5.4
ASO5105+0.3Ti	96.9	170.5	5.1	78.9	156.1	6.3
ASO5105+0.3Y	95.8	177.4	5.3	77.2	146.6	5.7

Table 5. Tensile properties of the as-cast ASO5105 and Sr, Ti, Y added ASO5105 at high temperature

Alloy ID	at 150 °C					
	6 mm			40 mm		
	TYS(MPa)	UTS(MPa)	E(%)	TYS(MPa)	UTS(MPa)	E(%)
ASO5105	57.2	93.9	19.2	69.6	97.6	14.8
ASO5105+0.3Sr	70.9	102.9	7.8	61.6	90.3	11.3
ASO5105+0.3Ti	71.2	104.9	14.4	60.1	88.4	12.5
ASO5105+0.3Y	70.6	103.2	11.7	60.8	89.3	10.5

Reference

[1] B. L. Mordike and T. Ebert, "Magnesium : Properties-applications-potential" [J] Mater. Sci. Eng. A. 302 (2001), 37-45

[2] H. Friedrich and S. Schumann, "Research for a "new age of magnesium" in the automotive industry" [J] Mater. Proc. Tech. 117 (2001) 276-281.

[3] E. Aghion, B. Bronfin, D. Eliezer, "The role of magnesium industry in protecting the environment, Journal of Material" Processing Technology 117 (2001) 381-385.

[4] Froes FH, Eliezer D, Aghion E. "Magnesium science technology and applications" Advanced Performance Materials, 5 (1998) 201–212.

[5] K. S. Nair, M.C. Mittal, "Rare earths in magnesium alloys" [J] Mater. Sci. Forum 30 (1998) 89.

[6] I.J. Polmear: Mater. Sci. Technol., 1994, vol. 10, pp. 1–16.

[7] A. Luo, M.O. Pekguleryuz, "Review cast magnesium alloy for elevated temperature applications" [J] Mater. Sci. 29 (1994), 5259-5271.

[8] L. Lu, M. O. Lai and M. L. Hoe: Nanostructured Materials 10 (1998) 551-563.

[9] M. Mabuchi, K. Higashi, "Strengthening mechanisms of Mg_2Si alloys" Acta Mater. 44 (1996) 4611-4618.

[10] M. Mabuchi, K. Kubota, K. Higashi, "Tensile strength, ductility and fracture of magnesium-silicon alloys" [J] Mater. Sci. 31 (1996)
1529.

[11] M. Mabuchi, K. Kubota, K. Higashi, "High strength and high strain rate superplasticity in a Mg-Mg2Si composite" Scripta Metall. Mater. 33 (1995) 331-335.

[12] Nam K Y, Song D H, Lee C W, Lee S W, Park Y H, Cho K M, Park I M. "Modification of Mg2Si morphology in as-cast Mg-Al-Si alloys with strontium and antimony" Mater Sci Forum (2006)510-511.

[13] Kim J J, Kim D H, Shin K S, Kim N J. "Modification of Mg_2Si morphology in squeeze cast Mg-Al-Zn-Si alloys by Ca or P addition" Scripta Mater, 41 (1999) 333-340.

[14] Yoo M S, Shin K S, Kim N J. "Effect of Mg2Si particles on the elevated temperature tensile properties of squeeze-cast Mg-Al alloys" Metal Mater Trans A (2004) 35.

[15] Jin-Kyu Lee, Young-Ok Yoon, Hyung-Ho Jo, Shae K. Kim, "Microstructural Evolution of A207 Al Alloy for In-Ladle DTC Rheocasting" 7th ISEPD 2006, (2006) 342.

[16] Shae K. Kim, Jin-Kyu Lee, "Behavior of alkaline earth metal oxides in magnesium alloys, 135th TMS 2006" (2006) 517-521

[17] Jin-Kyu Lee, Shae K. Kim, Hyung-Ho Jo, 135th TMS 2006, (2006) 185-189.

[18] Shae K. Kim, Young-Ok Yoon, Jin-Kyu Lee, Hyung-Ho Jo, "Green Manufacturing for Magnesium Alloys" WFC 2006, (2006) 34/1-34/9.

[19] Y. Carbonneau, A. Couture, A. Van Neste, R. Tremblay, "On the observation of a new ternary MgSiCa phase in Mg-Si alloys" Metall. Trans. A., 29A (1998) 1759.

Magnesium Technology 2013
Edited by: Norbert Hort, Suveen N. Mathaudhu, Neale R. Neelameggham, and Martyn Alderman
TMS (The Minerals, Metals & Materials Society), 2013

Microstructure and Mechanical Properties of Die Cast Magnesium-Aluminum-Tin Alloys

A.A. Luo[1], P. Fu[2], X. Zeng[2], L. Peng[2], B. Hu[3], A.K. Sachdev[1]
[1]General Motors Global Research and Development, Warren, MI 48090, USA
[2]Shanghai Jiao Tong University, Shanghai, China
[3]General Motors China Science Laboratory, Shanghai, China

Keywords: Magnesium alloys, alloy development, computational alloy design, microstructure

Abstract

Mg-Al-Sn alloy system offers good combination of strength, ductility, castability and corrosion resistance for potential automotive structural applications. This report summarizes the microstructure and mechanical properties of two Mg-Al-Sn alloys, AT72 (Mg-7Al-2Sn[1]) and AT96 (Mg-9Al-6Sn), prepared using high pressure die casting process. The microstructure of as-cast Mg-Al-Sn alloys was investigated using computational thermodynamics modeling and experimental techniques. Both alloys show improved mechanical properties and excellent die castability compared with the conventional AZ91 (Mg-9Al-1Zn) and AM50 (Mg-5Al-0.3Mn) alloys. The AT72 alloy is currently used in developing large thin-wall die cast door inners and cast magnesium wheels.

Introduction

Magnesium is the lightest metal available for reducing the weight of automotive structural subsystems for improved fuel economy. A unique characteristic of magnesium, in addition to its low density, is its extreme fluidity, which enables it to be die cast into shapes that are thinner and more complex than can be obtained with aluminum. There are currently two major magnesium alloy systems for high pressure die casting (HPDC) applications; Mg-Al-Zn (AZ) and Mg-Al-Mn (AM): both have limited mechanical properties [1]. AZ91 (Mg-9Al-1Zn) alloy has moderate strength but low ductility, and is generally used for non-structural parts like brackets, covers, cases and housings that are strength dominated and exposed to ambient temperatures. For semi-structural applications such as instrument panels, steering systems and radiator supports, where crashworthiness is important, AM50 (Mg-5Al-0.3Mn) or AM60 (Mg-6Al-0.3Mn) are used due to their higher ductility (10-15% elongation). However, the yield strength of AM50/60 alloy is too low for many critical automotive structural applications.

The major strengthening phase in Mg-Al based alloys is $Mg_{17}Al_{12}$ (β phase), which is resistant to dislocation shearing but its distribution is relatively coarse, presumably because of the relatively high diffusion rate of Al atoms in the solid matrix of magnesium and a possibly high concentration of vacancies in the α-Mg matrix [2]. A recent development at GM [3] showed improved mechanical properties of Mg-Al-Sn ternary alloys, due to the dual strengthening phases of Mg17Al12 and Mg2Sn. A further study [4] using computational thermodynamics coupled with experimental validation has suggested two Mg-Al-Sn alloys, AT72 (Mg-7Al-2Sn) and AT96 (Mg-9Al-6Sn), which offer good combinations of strength and ductility.

This paper summarizes the microstructure and mechanical properties of the two Mg-Al-Sn alloys (AT72 and AT96), prepared using high pressure die casting process. The microstructure of the as-cast alloys was investigated using computational thermodynamics modeling and experimental techniques.

CALPHAD Modeling and Experimental Procedure

CALPHAD and Heat Treatments

The CALPHAD (CALculation of PhAse Diagrams) approach based on computational thermodynamics, described in details in an earlier publication [5], was used to calculate the phase equilibrium of the Mg-Al-Sn alloy system and the two alloys in this study. The calculations were carried out using Pandat code developed by CompuTherm (Madison, WI) [6] and the extensive thermodynamic database for Mg-based alloys (PanMg) developed by Clausthal University of Technology (Clausthal, Germany) [7].

Alloy Preparation and Die Casting

The AT72 and AT96 alloys were prepared using AM60B (base alloy), pure Al and pure Sn ingots. The chemical compositions of the alloys are listed in Table 1. Both alloys were die cast into an experimental bracket casting with a shot weight of about 1 kg in magnesium (Fig. 1) on a 500 T die casting machine at a vacuum level of 0.2 bar. Due to the different liquidus temperatures reported for the two alloys, 610°C for AT72 and 590°C for AT96 [3], the alloys were die cast at 700°C and 680°C, respectively. About 200 shots were made for each alloy, and both alloys showed excellent castability with no visible casting defects when the casting trails reached the steady state of operation.

Table 1. Nominal alloy composition and ICP analysis results

Alloy	Nominal composition	Al	Sn	Zn	Mn	Fe	Ni	Cu
AT72	Mg-7Al-2Sn	7.3	3.1	<0.01	0.01	0.01	<0.003	<0.003
AT96	Mg-9Al-6Sn	9.1	6.4	< 0.01	0.01	0.01	< 0.003	< 0.003

Mechanical Testing and Microstructural Analyses

ASTM subsize tensile bars of 25 mm gage length and 7.6 mm width were machined from the flat sections of the castings (as-cast and heat-treated) of about 4 mm thickness with the casting surfaces maintained. Tensile testing was carried out at room temperature according to ASTM E21-92 procedures at an initial strain rate of 0.001 s^{-1}. For each alloy, at least three specimens were tested and the average properties reported. Specimens for microstructural analyses were sectioned from the center of the tensile bars, and then polished and etched according to standard metallographic procedures. The microstructure of the alloy specimens was analyzed using a Nikon Epiphot optical microscope and a Cameca SX100 Electron Probe Microanalyzer (EPMA). Quantitative image analysis was used to measure the

[1] All compositions in wt.% except otherwise stated.

area fractions of the second phases from the backscattered electron images of the as-cast alloys.

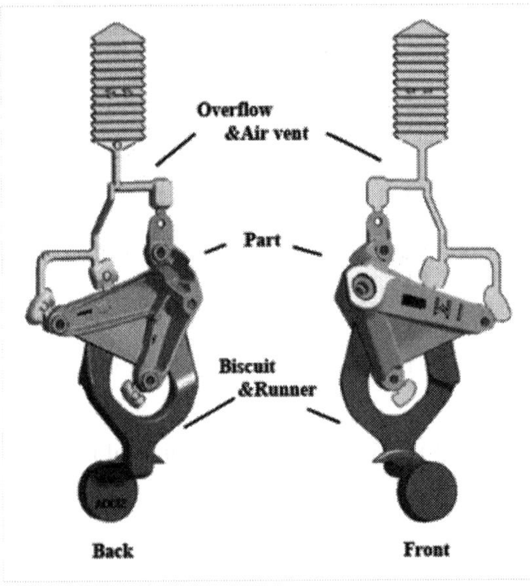

(a) Die design

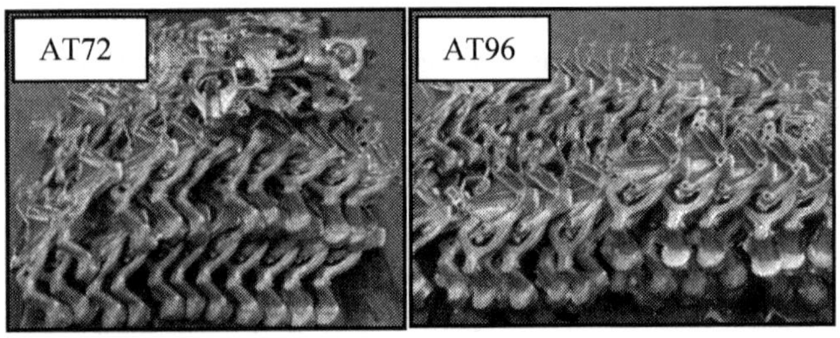

(b) AT alloy castings

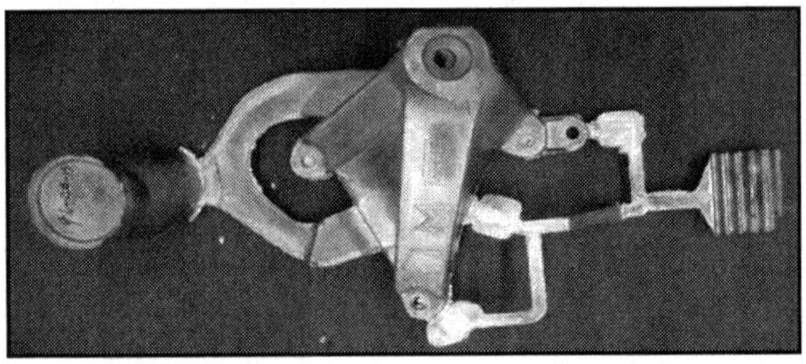

(c) A complete casting with gating system

Fig. 1. Experimental bracket casting with a shot weight of 1.04 kg in magnesium.

Results and Discussion

Solidification Sequence and Microstructure

Fig. 2 shows the calculated liquidus projection of the Mg-Al-Sn system in the Mg-rich corner, using the CALPHAD approach. No stable ternary phase was discovered, and no ternary solubility of the binary intermetallic phases was found in this calculation, which is consistent with a previous study [9]. Two binary phases, Mg_2Sn and $Mg_{17}Al_{12}$, are important to this alloy system: the Mg_2Sn phase is reported to provide significant strengthening [3], while the $Mg_{17}Al_{12}$ phase improves strength [9], corrosion and castability [10], but reduces ductility and creep resistance [11].

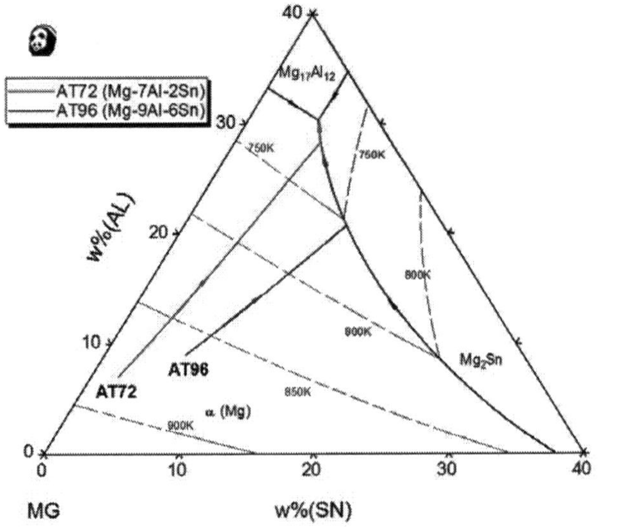

Fig. 2. Calculated Mg-Al-Sn liquidus projection and the solidification paths of AT72 and AT96 alloys.

Fig. 3 shows the calculated solidification sequence of AT72 & AT96 alloys based on the Scheil model, assuming complete mixing in the liquid but no diffusion in the solid. The solidification paths for both alloys are summarized as follows:

1) Nucleation of primary magnesium (from 610°C for AT72 and 590°C for AT96): L → L + α-Mg
2) Binary eutectic reaction (from 441°C for AT72 and 479°C for AT96): L → L + α-Mg + Mg_2Sn
3) Ternary eutectic reaction (430°C): L → L + α-Mg + Mg_2Sn + $Mg_{17}Al_{12}$

The final microstructure of AT72 and AT96 alloys, as shown in Fig. 4, consists of very fine α-Mg dendrites surrounded by eutectic intermetallic phases. Further identification of the eutectic phases is shown in the EPMA images along with the element maps of Al, Sn and Mg, Figs. 5 and 6. It is clear that the fine bright particles of 1-2 μm size containing Mg and Sn, are Mg_2Sn, while the gray eutectic network along the grain boundaries are $Mg_{17}Al_{12}$.

Table 2 compares the Scheil simulation results and the experimental measurements of the second phase contents of Mg-Al-Sn alloys, which shows a reasonably good agreement between the experimental and simulation results for $Mg_{17}Al_{12}$ and Mg_2Sn phases. The measured fractions are slightly lower (about 5% – 8%) than the Scheil simulation results due to the fact that the Scheil model does not consider diffusion in the solid, which affects solidification kinetics despite the high cooling rate in high-pressure die casting.

Table 2. Scheil simulation results and experimental measurements of the second phase contents of Mg-Al-Sn alloys

Alloy	$Mg_{17}Al_{12}$		Mg_2Sn	
	Calculated (volume %)	Measured (area %)	Calculated (volume %)	Measured (area %)
AT72 (Mg-7Al-2Sn)	7.3	6.78	0.5	0.46
AT96 (Mg-9Al-6Sn)	11.3	10.92	2.0	1.94

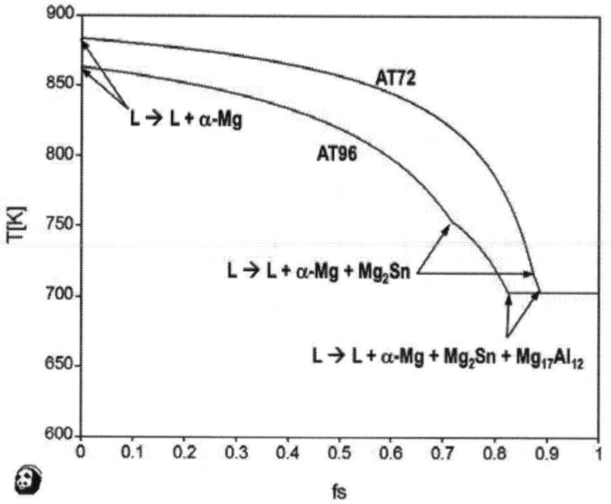

Fig. 3. Scheil simulation results of solidification sequence of AT72 and AT96 alloys.

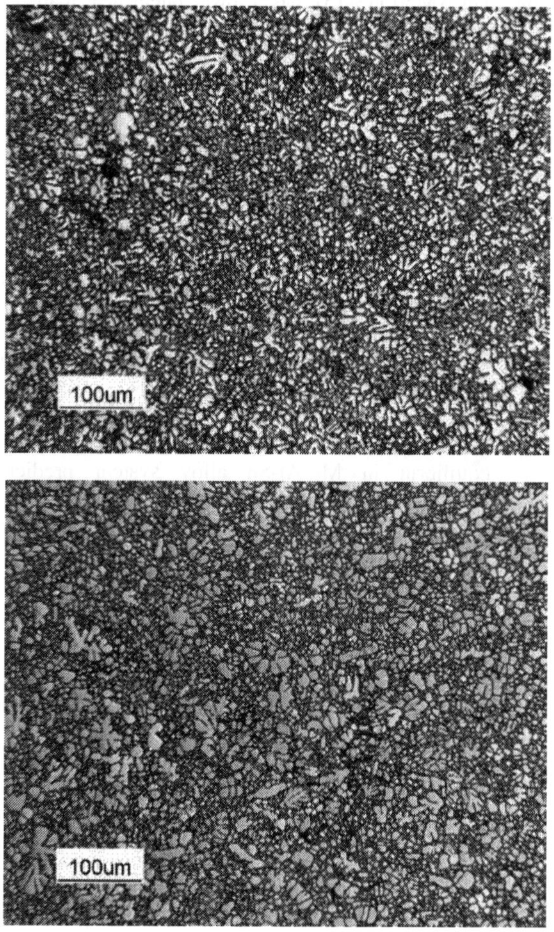

Fig. 4. Optical micrographs showing the as-cast microstructure of (a) AT72; and (b) AT96 alloys.

Mechanical Properties

Table 3 summarizes the mechanical properties of die cast AT72 and AT96 alloys compared with those of conventional die cast magnesium alloys (AZ91 and AM50). Compared with the AZ91 alloy, the AT72 and AT96 alloys show 5% and 26% higher yield strength, respectively. While the high-strength AT96 alloy has 40% less ductility (elongation) than the AZ91 alloy, the AT72 alloy has more than doubled the elongation of AZ91. Compared with the AM50 alloy, the AT72 and AT96 alloys exhibited significantly more increases in yield strength, i.e., 37% and 63%, however, at considerable expenses in ductility. Overall, AT96 is a high-strength alloy with limited ductility, but the AT72 alloy offers more balanced increase in strength and ductility.

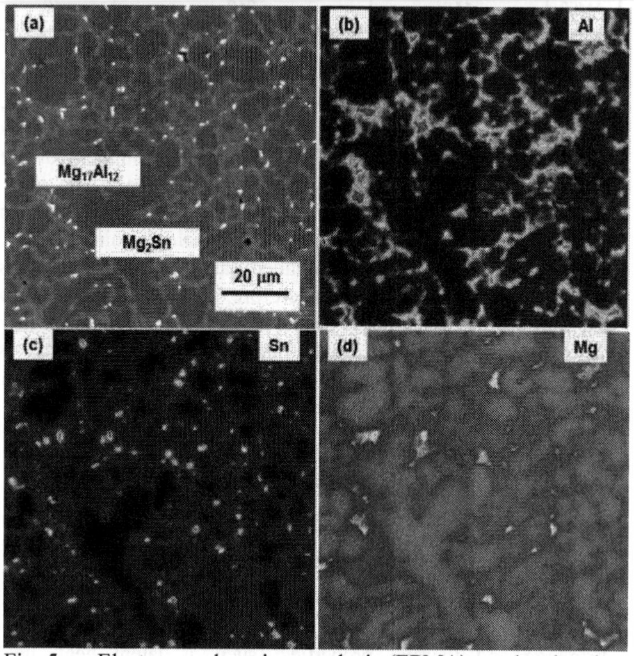

Fig. 5. Electron probe micro-analysis (EPMA) results showing the as-cast microstructure of AT72 alloy: (a) backscattered electron image; (b) Al map; (c) Sn map; and (d) Mg map.

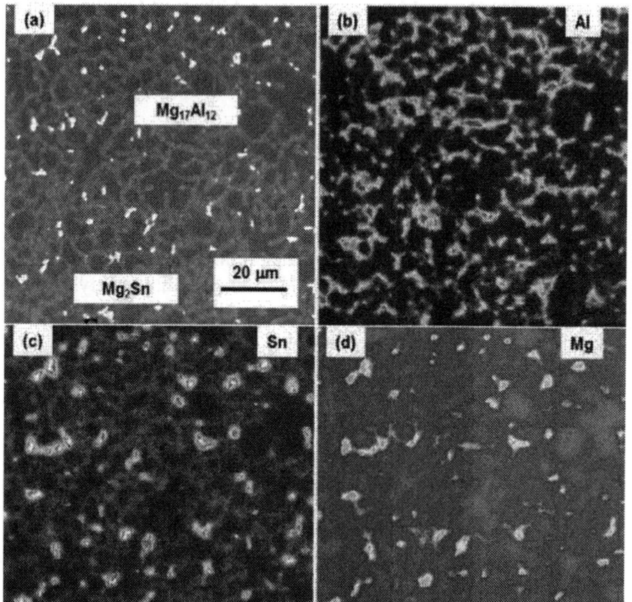

Fig. 6. Electron probe micro-analysis (EPMA) results showing the as-cast microstructure of AT96 alloy: (a) backscattered electron image; (b) Al map; (c) Sn map; and (d) Mg map.

Table 3. As-cast Mechanical properties of die cast alloys

Alloy	Yield Strength, MPa	Ultimate Tensile Strength, MPa	Elongation, %
AT72	158	251	6.3
AT96	189	216	1.8
AZ91	*150*	*240*	*3*
AM50	*116*	*210*	*10*

The reason for increased strengthening in Mg-Al-Sn alloys is due to the strengthening effects of both $Mg_{17}Al_{12}$ and Mg_2Sn phases, while $Mg_{17}Al_{12}$ is the only strengthening phase in Mg-Al binary alloys such as AZ91 and AM50. However, the higher volume fractions of $Mg_{17}Al_{12}$ and Mg_2Sn phases in the AT96 alloy results in lower ductility compared to AT72 alloy, due to brittleness of both phases. From the morphology point of view, the Mg_2Sn is generally more rounded particles with small sizes than the $Mg_{17}Al_{12}$ network. Therefore, it seems reasonable to assume that $Mg_{17}Al_{12}$ phase is more detrimental to the ductility than the Mg_2Sn in the die casting microstructure. This is supported by the much higher ductility of AM50 alloy than AZ91 due to the higher Al content in the later alloy.

Castability

The castability of magnesium, i.e., the ease of casting with minimum defects, has been a subject of many publications [12-16]. Castability is generally influenced by a number of factors that are either alloy-dependent or process-dependent. Alloy-dependent parameters include feeding characteristics (i.e. fluidity), susceptibility to hot-tearing and cracking, freezing range, shrinkage and thermal properties (such as latent heat, heat capacity and thermal conductivity), while process-dependent factors are comprised of casting parameters (such as cooling rate and die surface condition) and die design considerations (geometry) [14]. In this study, the castability of Mg-Al-Sn alloys is discussed with the thermal physical properties that were calculated from the Scheil simulation, and compared with those of commercial alloys AM50 and AZ91alloys (Table 4).

Table 4 compares the non-equilibrium (Scheil model) freezing range (the temperature difference between liquidus and solidus) of the AT72 and AT96 alloys with AM50 and AZ91 alloys. The calculated freezing range for the AM50 and AZ91 alloys agree very well with experimental measurements using differential scanning calorimetry (DSC) [17] and thermal analysis [18] techniques. It is generally agreed that alloys with wide freezing ranges are more difficult to cast, due to their strong tendency for non-uniform solidification [19]. Mg-Al based alloys with wide freezing ranges, such as AM50, are prone to micro-shrinkage, susceptible to hot-tearing [19], and less fluidity [13]. AZ91 alloy has a shorter freezing range compared to AM50, thus more castable than the AM50 alloy, as evident in several studies [13-15]. The freezing range of Mg-Al-Sn alloys becomes shorter with increasing Al and Sn contents, and both AT72 and AT96 alloys have shorter freezing range than that of AM50 alloy, indicating better castability than AM50, but similar to AZ91 alloy which is the most castable commercial magnesium alloy [18]. Another factor affecting the ability of molten metal filling the interdendritic regions is the amount of eutectic phases remaining in the last stage of solidification. Generally, alloys with more eutectic liquid such as AZ91 with 11% eutectic fraction have better castability [18]. The AT72 and AT96 alloys have similar eutectic fractions (2 to 3 times higher than AM50) as that of AZ91, another indication of good castability of AT alloys.

Table 4. Scheil simulation results of die cast magnesium alloys

Alloy	Nominal Composition	Freezing Range, °C	Total Eutectic Fraction, vol.%	Total Latent Heat, J/mol
AZ91	Mg-9Al-1Zn	163	11.0	7637
AM50	Mg-5Al-0.3Mn	188	4.3	7866
AT72	Mg-7Al-2Sn	180	8.2	7528
AT96	Mg-9Al-6Sn	160	13.2	7387

The latent heat calculated from the Scheil simulation of the AT alloys is also compared with the AM50 and AZ91 alloys in Table 4. Generally, high latent heat can prolong solidification, thus, improves fluidity, and also enhances feeding to reduce micro-porosity and hot-tearing. However, the AZ91 alloy, and more castable than AM50 [13-15], has a lower latent heat than AM50. This is perhaps due to the shorter freezing range of the AZ91 alloy, which is a more important factor than latent heat [13]. It is interesting to note that the latent heat of both AT alloys is lower than that of AM50 or AZ91 alloy. From this study, both alloys show excellent die castability in the casting trial as shown in Fig. 1(b).

Conclusions

1. Compared with commercial AZ91 and AM50 alloys, both AT72 and AT96 alloys show improved yield strength. The AT96 is a high-strength alloy with limited ductility, while the AT72 alloy offers more balanced improvements in strength and ductility.

2. Computational thermodynamics modeling of phase equilibria in Mg-Al-Sn alloy system predict the formation of α-Mg, $Mg_{17}Al_{12}$ and Mg_2Sn phases in the solidification microstructure. This is validated by examining the microstructure of high-pressure die casting of two alloys, AT72 (Mg-7Al-2Sn) and AT96 (Mg-9Al-6Sn).

3. The Scheil simulation results and the experimental measurements of the second phase contents ($Mg_{17}Al_{12}$ and Mg_2Sn phases) of Mg-Al-Sn alloys show a reasonably good agreement, indicating applicability of Scheil model in simulating the die casting microstructure.

4. The freezing range, total eutectic fraction and latent heat of AT72 and AT96 alloys calculated from thermodynamics modeling and the casting experiments confirm the excellent die castability of both Mg-Al-Sn alloys compared with commercial AZ91 and AM50 alloys.

Acknowledgments

This work was carried out as a collaborative research project supported by General Motors Global Research and Development (GM R&D), Warren, MI, USA, and Shanghai Jiao Tong

University (SJTU), Shanghai, China. The authors acknowledge Richard Waldo for carrying out EPMA analysis, Willie Dixon for optical metallography and Todd Meitzner for tensile testing.

References

1. A.A. Luo, International Materials Reviews, 2004, 49, (1), 13-30.
2. J.-F. Nie, Metallurgical and Materials Transactions A, 2012, 43A, 3891-3939.
3. A.A. Luo and A.K. Sachdev, in Magnesium Technology 2009, eds. E.A. Nyberg, S.R. Agnew, N.R. Neelameggham and M.O. Pekguleryuz, TMS, Warrendale, PA, 2009, 437-443.
4. A.A. Luo, P. Fu, L. Peng, X. Kang, Z. Li, T. Zhu, Metallurgical and Materials Transactions A, 2012, 43A, 360-368.
5. A.A. Luo, R.K. Mishra, B.R. Powell, A.K. Sachdev, Materials Science Forum, 2012, 706-709, 69-82.
6. CompuTherm LLC: Pandat 8.0 - Phase Diagram Calculation Software for Multi-component Systems (CompuTherm LLC, Madison, WI, USA, 2008).
7. R. Schmid-Fetzer, A. Janz, J. Grobner and M. Ohno: Advanced Eng. Mater, 2005, 7, 1142.
8. E. Doernberg, A. Kozlov and R. Schmid-Fetzer, Journal of Phase Equilibria and Diffusion, 2007, 28, 523-535.
9. A.A. Luo and A.K. Sachdev: International Journal of Metalcasting, 2010, 4 (4), 51-58.
10. D. Sachdeva, S. Tiwari, S. Sundarraj and A.A. Luo, Metallurgical and Materials Transactions B, 2010, 41B, 1375-1383.
11. C.L. Mendis, C.J. Bettles, M.A. Gibson, C.R. Hutchinson, Mater. Sci. Eng. A, 2006, 435-436, 163-171.
12. B.R. Powell, A.A. Luo, B.L. Tiwari, and V. Rezhets: in Magnesium Technology 2002, ed., H.I. Kaplan, TMS, Warrendale, PA (2002), p. 123.
13. D. Argo, M. Pekguleryuz, P. Labelle, P. Vermette, R. Bouchard, M. Lefebvre: in Magnesium Technology 2002, ed., H.I. Kaplan, TMS, Warrendale, PA (2002), p. 87.
14. M.O. Pekguleryuz, P. Labelle, D. Argo, E. Baril: in Magnesium Technology 2003, ed., H.I. Kaplan, TMS, Warrendale, PA (2003), p. 201.
15. A.L. Bowles, Q. Han, J.A. Horton: in Magnesium Technology 2005, eds., N.R. Neelameggham, H.I. Kaplan, and B.R. Powell, TMS, Warrendale, PA (2005), p. 99.
16. S.S. Khan, N. Hort, I. Steinbach, S. Schmauder: in Magnesium Technology 2008, eds., M.O. Pekguleryuz, N.R. Neelameggham, R.S. Beals, E.A. Nyberg, TMS, Warrendale, PA (2008), p. 197.
17. D. Mirkovic, R. Schmid-Fetzer: Metall Mater Trans A, 2007, 38, 2575.
18. A. Luo: in Proc. of the Third International Magnesium Conference, Ed., G.W. Lorimer, The Institute of Materials, London, UK, 1996, 449.
19. ASM International: Casting Design and performance, ASM International, Materials Park, Materials Park, OH, 2009, 101.

Magnesium Technology 2013
Edited by: Norbert Hort, Suveen N. Mathaudhu, Neale R. Neelameggham, and Martyn Alderman
TMS (The Minerals, Metals & Materials Society), 2013

STUDY ON MICROSTRUCTURE AND MECHANICAL PROPERTY OF SQUEEZE CASTING AZ91D MAGNESIUM ALLOY

Yanda Li [1], Zhiqiang Han [1], Alan A. Luo [2], Anil K. Sachdev [2], Baicheng Liu [1, 3]
[1] Key Laboratory for Advanced Materials Processing Technology (Ministry of Education), Department of Mechanical Engineering, Tsinghua University, Beijing 100084, China
[2] Chemical and Materials Systems Laboratory, General Motors Global Research and Development Center, Warren, MI 48090-9055, USA
[3] State Key Laboratory of Automotive Safety and Energy, Tsinghua University, Beijing 100084, China

Keywords: Squeeze casting, Magnesium alloy, Cooling curve, Microstructure, Mechanical property

Abstract

Squeeze casting characterized by gentle die filling and solidification under high pressure is recognized one of the potential processes for the production of thick-walled parts that are amendable for heat treatment to enhance mechanical properties of magnesium alloys. In this study, squeeze casting experiments under different applied pressures were carried out for AZ91D magnesium alloy and the temperature variations at different locations in the casting were measured during the solidification process. The microstructure morphology was observed by using OM. The grain size and mechanical property of the casting were determined. The effects of pressure on the cooling curves, microstructure and mechanical property were discussed.

1. Introduction

Magnesium alloy is an important lightweight structural material and has a great application potential in automotive industry. High pressure die casting technology has been widely used in manufacturing magnesium alloy components [1]. However, porosity is a major defect in high pressure die casting due to the air entrapment caused by turbulent die filling in the casting process. High vacuum die casting where the air in the mold is removed before the die filling process to avoid air entrapment has been developed to reduce porosity defects in the castings [2]. It is suitable to produce pore free castings with good mechanical properties. However, for a thick-walled casting, it is still difficult to obtain a pore free product by using high vacuum die casting technique due to the solidification shrinkage. Squeeze casting is a process in which the die is filled gently, which means almost no air entrapment happens, and the casting solidifies under an external pressure, which causes plastic deformation compensating the internal volume shrinkage. Thus, it is an effective technology for producing thick-walled parts that are amendable for heat treatment to enhance mechanical properties of magnesium alloys. Some researchers have investigated the influence of process variables on the microstructure and mechanical properties of squeeze cast magnesium alloys [3-6], still the studies are relatively limited and the understanding on the formation mechanism of microstructure is inadequate. In this study, squeeze casting experiments of magnesium alloy under different applied pressures have been carried out. The temperature variations inside the casting during solidification process were measured and analyzed, the as-cast microstructure was observed and the mechanical property of the castings was tested.

2. Experimental Procedure

2.1 Experimental Materials

Commercial AZ91D magnesium alloy was selected in the squeeze casting experiments. The composition of the alloy is shown in Table 1.

Table 1 The chemical composition (wt. %) of magnesium alloy

Al	Zn	Mn	Si	Fe	Cu	Ni	Mg
8.9	0.76	0.26	0.01	0.0033	0.001	0.0008	Balance

2.2 Casting and Die-set Design

The casting geometry and the die-set configuration are shown in Figure 1. The step-shaped casting geometry produces significant difference in cooling rate inside the casting (Figure 1a). The die-set containing a patented technique was employed to measure the temperature variation in the casting during solidification [7] (Figure 1b). The locations of the thermocouples for temperature measurement are shown in Figure 1a.

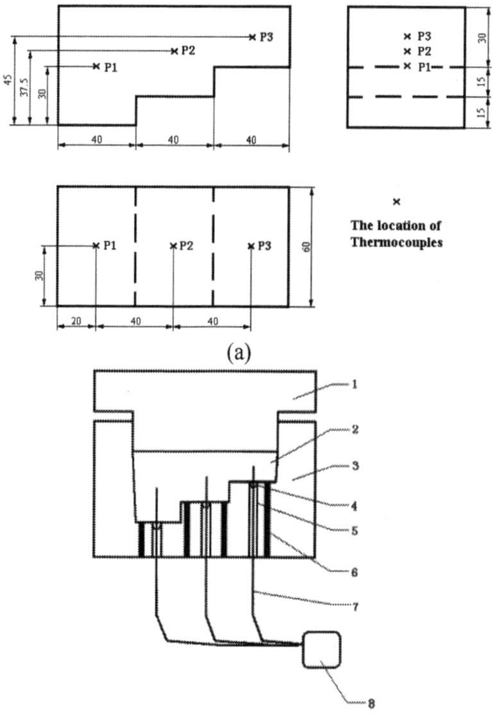

Figure 1. The casting geometry, the location of the thermocouples for temperature measurement, and the die-set configuration,

where 1 is the punch, 2 the casting, 3 the lower die, 4 the fixer for thermocouples, 5 the channel for inserting the thermocouples into the die, 6 the ejector, 7 the thermocouples, and 8 the data acquisition instrument.

2.3 Squeeze Casting Procedure

The squeeze casting process involves the following steps: (a) The die and the punch were preheated and maintained at a temperature around 470-500K; (b) Ingots of AZ91D magnesium alloy were melted under protective atmosphere (N_2+0.5%SF_6) in a resistance furnace; (c) The liquid metal was poured into the mold at about 970K, and after the pouring process, the die was closed immediately and a prescript pressure (30, 70 or 100MPa) was applied on the liquid metal as quickly as possible to minimize the cooling and solidification before the pressure was applied. The pressure was maintained until the casting fully solidified; (d) After the solidification was fully completed, the casting was ejected from the die.

3. Results and Discussion

3.1 Liquidus Temperature

Figure 2 shows the effect of the applied pressure on the cooling curves. It can be seen for all the three cases that, once the pressure was applied, the temperature increased rapidly until reaching a plateau, i.e., the liquidus temperature. Then, the temperature was kept almost constant for a certain period of time before starting to decrease. The applied pressure influences the attainable level and the duration time of the liquidus plateau. The measured liquidus temperature of AZ91D magnesium alloy is 873.15K, 875.75K and 877.95K for the applied pressure of 30MPa, 70MPa and 100MPa, respectively. Compared with the liquidus temperature of 868K under atmospheric pressure, the present experimental measurement indicates that the liquidus temperature of AZ91D magnesium alloy increased by 5.15K, 7.75K and 9.95K under the pressures of 30MPa, 70MPa and 100MPa, respectively.

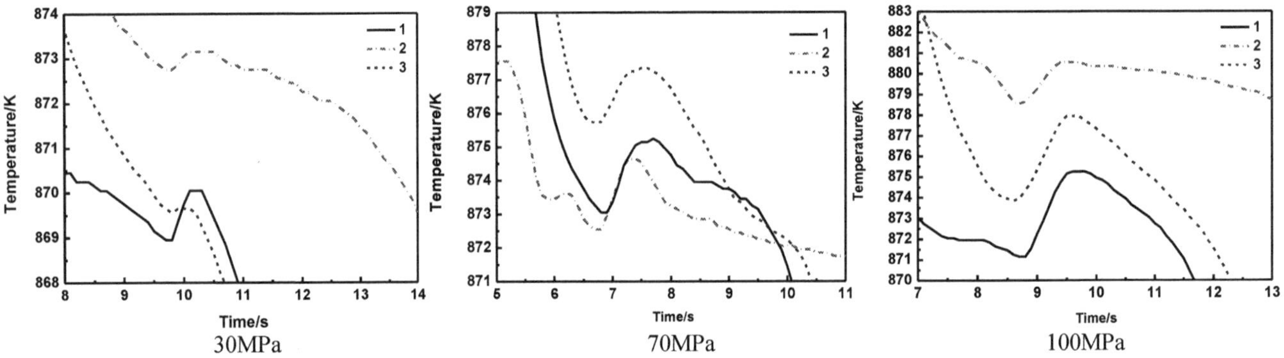

Figure 2 The effect of the applied pressure on the liquidus temperature

3.2 Microstructure characterization

Metallographic samples were taken from the castings as shown in Figure 3. The red face is the position where the microstructure is observed. Polished samples were etched in a solution of picric acid, glacial acetic acid, phosphoric acid and alcohol (0.8g picric acid, 0.01ml glacial acetic acid, 0.01ml phosphoric acid and 20ml alcohol). Examples of as-cast microstructure under different applied pressures are shown in Figure 4. The grains have different color under polarized light and the magnesium dendrite is of six-fold symmetry. With increase of the applied pressure, the microstructure becomes less dendritic.

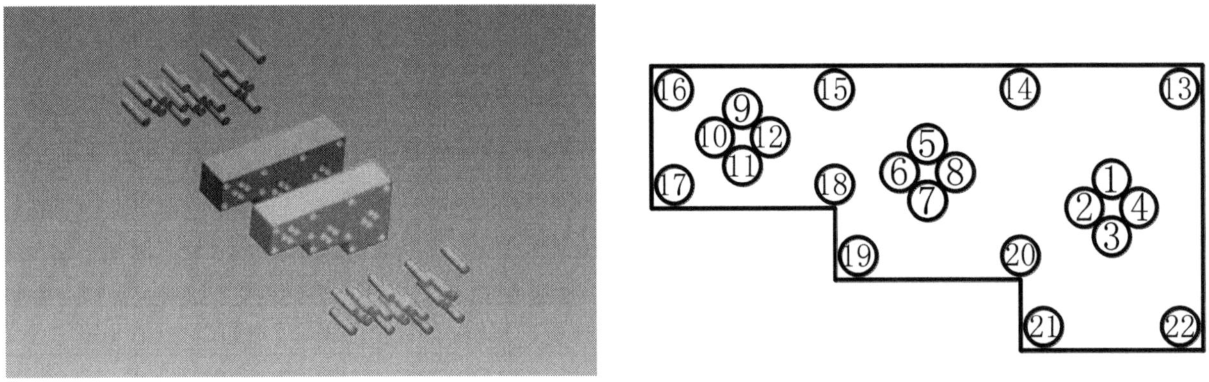

Figure 3 A schematic of sampling for metallographic observation

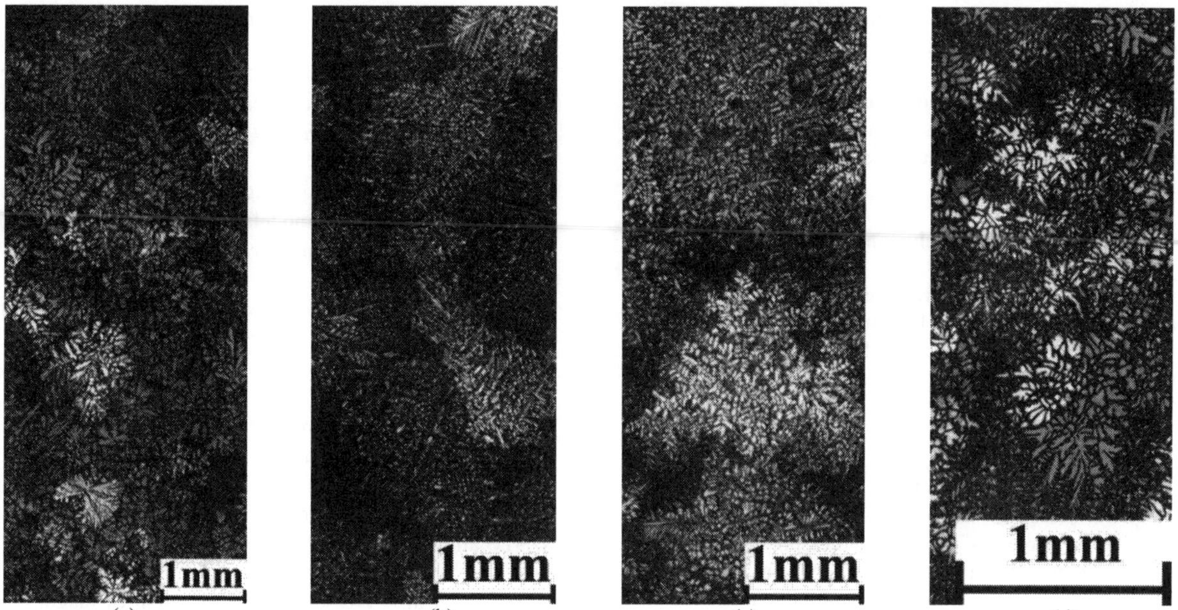

(a) (b) (c) (d)

Figure 4 The microstructure of the castings produced under (a) atmospheric pressure, (b) 30MPa, (c) 70MPa, and (d) 100MPa

3.3 Grain size

The grain size of the as-cast microstructure of the castings were measured according to the ASTM E112 [8] intercept method. The grain size data are shown in Figure 5. We can see that with increase of the applied pressure the as-cast grain size decreases and the grain size at different positions in the casting becomes more uniform. The average grain size decreases from 770.5μm to 292.9μm when the applied pressure increases from 30 to 100MPa.

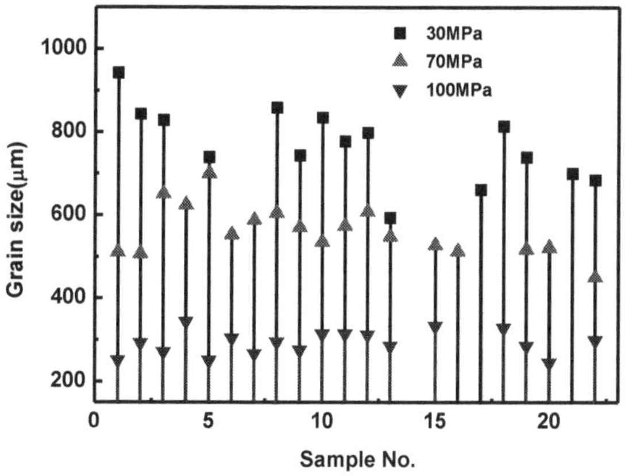

Figure 5 The grain size of the as-cast samples under different applied pressures

3.4 Mechanical property

Tension test specimens were made according to ASTM B557M [9]. The step-shaped casting was cut into samples as shown in Figure 6(a) and each of the specimens has a location coordinate in the casting. The UTS (Ultimate Tensile Strength) of the samples under 30MPa, 70MPa, 100MPa and 130MPa is shown in Figure 6(b).

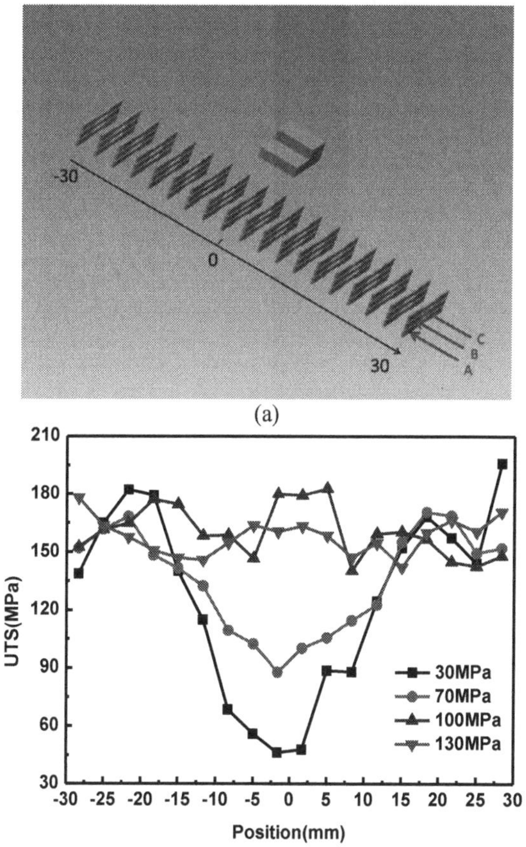

(b)

Figure 6 A schematic of the tensile test specimens and the UTS at different locations in the castings produced under different pressures

For the samples of 30MPa, the central part of the casting has much lower UTS than the part near the mold wall. With increase of the pressure, the UTS at the casting center increases remarkably. When the applied pressure reaches 100MPa or a higher level, the UTS in the casting varies at a relatively high level, which means the mechanical property of the whole casting tends to be uniform.

4. Conclusion

1. The application of external pressure affects the solidification process of squeeze casting. The liquidus temperature of AZ91D magnesium alloy rises as the applied pressure increases. The experimental results indicated that the liquidus temperature of AZ91D magnesium alloy increased by 5.15K, 7.75K and 9.95K under the applied pressure of 30MPa, 70MPa and 100MPa, respectively.
2. The grains were refined by applying pressure, and in the range of the applied pressures in the experiments, the grain size decreases with the increase of the applied pressure.
3. The applied pressure improved the mechanical property. An applied pressure higher than 100MPa is required to make the UTS uniform in the whole casting for the geometry employed in the present study.

Acknowledgements

This work is funded by the National Natural Science Foundation of China (Grant No. 51175291), the General Motors Global Research and Development Center, Tsinghua University Initiative Scientific Research Program, and the MoST (Ministry of Science and Technology) of China under the contracts of No. 2010DFA72760 and 2011DFA50909.

References

[1] D. Sachdeva et al., "Microstructure and corrosion characterization of squeeze cast AM50 magnesium alloys", *Metallurgical and Materials Transactions B*, 41(2010), 1375-1383.

[2] B. C. Liu, S. M. Xiong, "High pressure die casting process of magnesium alloys and its modeling and simulation for automobile industry", *Journal of Automotive Safety and Energy*, 1(2011), 1-11.

[3] M. Masoumi, H. Hu, "Influence of applied pressure on microstructure and tensile properties of squeeze cast magnesium Mg-Al-Ca alloy", *Materials Science and Engineering A*, 528(2011), 3589-3593.

[4] S. Kleiner et al., "Microstructure and mechanical properties of squeeze cast and semi-solid cast Mg-Al alloys", *Journal of Light Metals*, 2(2002), 277-280.

[5] Y. L. Yang et al., "Study on microstructure of squeeze casting AZ91D alloy", *Materials Science and Technology*, 27(2011), 189-193.

[6] A. Yu et al., "Pressurized solidification of magnesium alloy AM50A", *Journal of Materials Processing Technology*, 191(2007), 247-250.

[7] Z. Q. Han et al., "A die-set for directly measuring the temperature variation in solidification process of squeeze casting", *China Patent*, 201010249074.0.

[8] ASTM E112-10: Standard Test Methods for Determining Average Grain Size.

[9] ASTM B557M-10: Standard Test Methods of Tension Testing Wrought and Cast Aluminum- and Magnesium-Alloy Products [Metric].

Magnesium Technology 2013
Edited by: Norbert Hort, Suveen N. Mathaudhu, Neale R. Neelameggham, and Martyn Alderman
TMS (The Minerals, Metals & Materials Society), 2013

TAILORING PRECIPITATES IN Mg-6Zn-2Gd BASED ALLOY SUBJECTED TO HIGH PRESSURE TORSION

J.H. Li[1*], P. Schumacher[1, 2]

[1*] Chair of Casting Research, Department of Metallurgy, University of Leoben, A-8700, Leoben, Austria (jie-hua.li@hotmail.com)
[2] Austrian Foundry Research Institute, Parkstrasse 21, Leoben, Styria, A-8700, Austria

Keywords: Mg-Zn-Gd alloy; Age hardening; Precipitate; TEM; High pressure torsion.

Abstract

The precipitates formed at lower ageing temperatures are very unstable at higher application temperatures, thus are not ideal for the high temperature applications. A traditional approach in solving the problem is from alloy design and development route by increasing solute concentrations or addition of micro-alloying elements. Recently, high pressure torsion was also employed to tailor the precipitate microstructures of the age-hardening alloy that are significantly different from those obtained by conventional isothermal ageing treatments. In this contribution, an Mg-6Zn-2Gd-0.4Zr (wt. %) alloy has been subjected to high pressure torsion (HPT). The precipitate microstructure evolution during HPT processing has been characterised using TEM, with an aim to elucidate the effect of HPT processing on nucleation and growth of precipitates, and the mechanisms dominating the evolution of precipitate microstructure. This investigation demonstrates a new perspective for application of HPT to tailor precipitate microstructure for unique properties in service.

Introduction

Magnesium alloys have great potential for the application in automotive and aerospace application due to their high specific strength [1]. The Mg-Zn based alloys, i.e. ZK51 (Mg-4.5Zn-0.6Zr (wt. %)), have been widely used as a wrought Mg alloy. However, the precipitates formed in binary Mg-Zn alloys during artificial ageing are generally coarse and inhomogeneously distributed [2]. The high temperature mechanical properties of Mg-Zn binary alloys, especially the creep-resistant properties, become inadequate for technological applications at a temperature above 250 °C. Research on the age hardening behaviour of Mg-Zn based alloy is of great importance to further improve the age hardening response by controlling the decomposition of the supersaturated solid solution of Zn in Mg [3].

Mg-Zn-Gd system was thought to be promising to develop high strength Mg alloys since precipitates can be formed on the basal plane [4, 5, 6] or both on the prismatic and basal plane [7]. Especially, an Mg-6Zn-2Gd (wt. %, short for ZG62) alloy has been reported to have increased mechanical properties [8, 9]. Above 200°C, the TYS and UTS of ZG62 alloy is slightly higher than that of a Gd-free alloy. More importantly, the elongation of the ZG62 alloy at 300 °C (40.5 %) is much higher than that of Gd-free alloy (10 %), which is very attractive for the application of Mg alloy at higher temperature. However, there is also a problem that the precipitates formed at a lower ageing temperature are very unstable at higher application temperatures, thus are not ideal for the high temperature applications. A traditional approach in solving the problem is from alloy design and development route by increasing solute concentrations or addition of micro-alloying elements. However, the low nucleation rate and rapid growth of precipitates during conventional ageing make it still difficult to control the precipitate microstructure in order to obtain a higher strength. High pressure torsion (HPT) is well known to be effective in modifying materials microstructures and achieving significant refinement in grain size [10]. Recently, HPT was also employed to tailor the precipitate microstructures of the age-hardening alloy that are significantly different from those obtained by conventional isothermal ageing treatments [11].

In this investigation, an Mg-6Zn-2Gd-0.4Zr (wt. %) alloy have been subjected to high pressure torsion. The precipitate microstructure evolution during HPT processing has been characterised carefully by using TEM, with an aim to elucidate the effect of HPT processing on nucleation and growth of precipitates, and the mechanisms dominating the evolution of precipitate microstructure.

Experimental material and procedures

The Mg-6Zn-2Gd-0.4Zr alloy (wt. %, used throughout the paper, in case not specified otherwise) was prepared from pure Mg (99.9%), Zn (99.9%), Mg-28Gd and Mg–33Zr master alloys in an electric resistance furnace under an anti-oxidizing flux as protection, and then cast into a sand mould at about 720 °C. Solution treatment was performed in a salt bath at 500 °C for 18 h, followed by quenching into cold water and then ageing in oil at 200 °C for 15 h.

Extrusion was perofrmed at room temperature with a pressing speed of 0.2 m/min, with an aim to break the large second phase, weak the texture, and make it possible to perform HPT at room temperature. After extrusion, final rod is about 15x15 mm^2 in size, and then the samples were machined with the diameters d = 8 mm, thickness t = ~ 0.8 mm. HPT was perofrmed at room temperature. Disks with the thickness around 0.8 mm and the diameter of 8 mm were processed for 1, 4, and 8 revolutions, under 1, 2, and 4 GPa, respectively, and a very low strain rate of about 10^{-2}/s. Hardness testing was performed with 50 N load and 15 sec dwell time. Each data point reported in this paper represents an average of at least 10 measurements.

Disks of 3 mm diameter were punched from thin strips cut from the HPT samples and they were ground carefully to a thickness of ~ 100 μm and dimpled to about 30 μm, and then ion-beam milled using a Gatan Precision Ion Polishing System (PIPS, Gatan model 691). Transmission electron microscopy (TEM) characterizations were performed using CM12 operating at 120 kV.

Results and discussions

<u>Microstructure and precipitates before extrusion and HPT</u>

Figure 1 shows microstructures of ZG62 alloy both in the as-cast condition and after a solution treatment. The microstructure of the as-cast sample was composed of equiaxed α-Mg grains surrounded by eutectic at triple junctions and grain boundaries, as shown in Figures. 1a, c. The mean grain size of the ZG62 alloy was measured to be approximately 35 μm using a linear intercept method. This grain size is similar to that observed in a ZK60 alloy with 1.3% Gd addition, but significantly less than that of ZK60 alloy (71 μm) [9]. This result is also consistent with other observations that the addition of Gd (less than 2 wt. %) refines the microstructure [8, 9]. After solution treatment at 500 °C for 18 h, the eutectic structures in ZG62 alloy almost dissolve but some discontinuous second phase particles survive on the grain boundary, as shown in Figures 1b, d. Energy dispersive X-ray (EDX) spectra of particles, as marked with white arrows shown in Figure 1 d, indicated the particles were enriched with Zn (about 17 wt. %, 9 at. %) and Gd (about 20 wt. %, 4 at. %), as shown in Figure 1e.

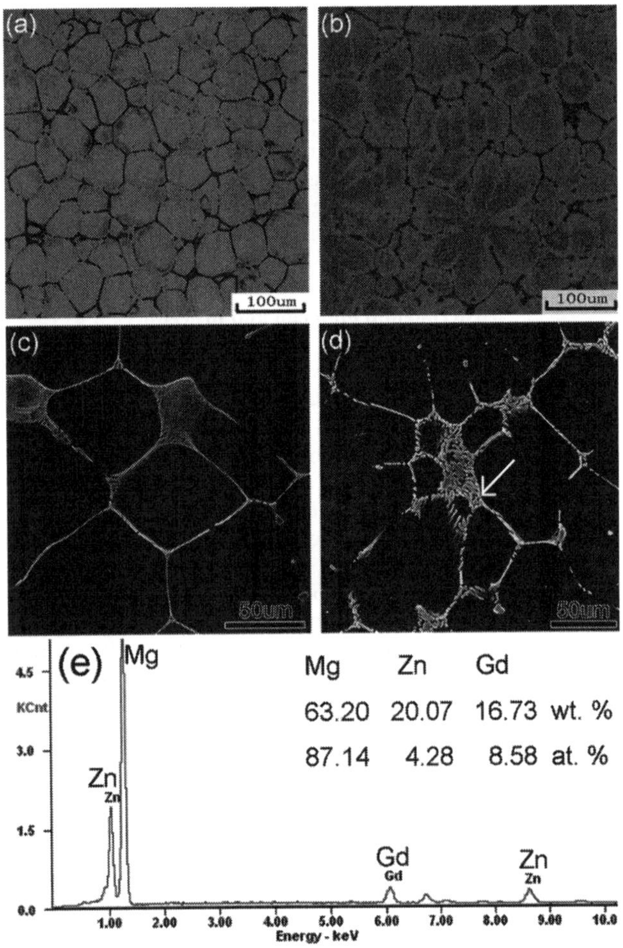

Figure 1 Microstructure of Mg-6Zn-2Gd-0.6Zr (wt. %) alloy. (a), (b), optical microscopy image; (c), (d), SEM image. (a), (c) as-cast; (b), (d) after a solution treatment at 500 °C for 18 h; (e) the energy dispersive X-ray (EDX) spectra analysis results of the particles, as marked in Figure 1 d.

After a solution treatment at 500 °C for 18 h and subsequent ageing at 200 °C for 15 h, some finer precipitates also observed in the α-Mg matrix. Figure 2 shows the $[01\text{-}11]_{\alpha}$, $[1\text{-}21\text{-}3]_{\alpha}$, $[01\text{-}10]_{\alpha}$ bright field TEM images (a), (c), (e) and corresponding SADPs (b), (d), (f) of the precipitates obtained from the ZG62 alloy aged at 200 °C for 15 h. Viewed from three different direction, it is clear that the precipitates with three different morphologies, i.e. $[0001]_{\alpha}$ rods/lath, $(0001)_{\alpha}$ plates and blocky particles, as marked with white arrows and labeled as 1, 2 and 3 in Figure 2e, are present in the α-Mg matrix. This is consistent with the observation in an Mg-Zn alloy [12, 13]. The $[0001]_{\alpha}$ rods (labeled as 1) have been commonly regarded as β'_1, while the $(0001)_{\alpha}$ plates (labeled as 2) were commonly taken as β'_2 in previous studies [3, 12-15]. Both β'_1 and β'_2 have a hexagonal structure (a=0.523nm, c=0.858nm) [12, 15], which is identical to that of Laves phase $MgZn_2$ [15].

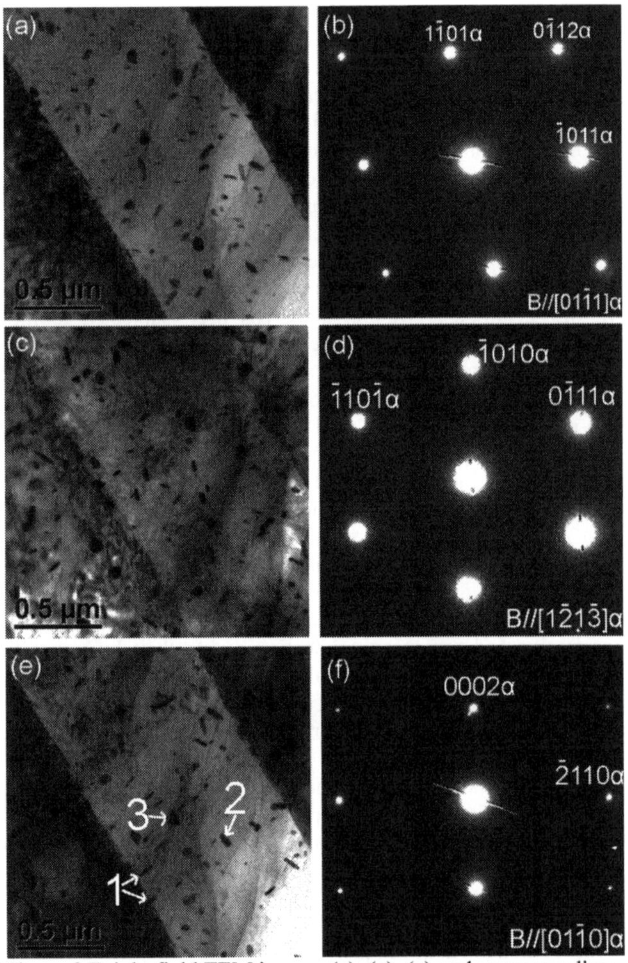

Figure 2 Bright field TEM images (a), (c), (e) and corresponding $[01\text{-}11]_{\alpha}$, $[1\text{-}21\text{-}3]_{\alpha}$, $[01\text{-}10]_{\alpha}$ SADPs (b), (d), (f) of the precipitates obtained from Mg-6Zn-2Gd-0.6Zr (wt. %) alloy aged at 200 °C for 15 h.

Microstructure and precipitates after extrusion and HPT

Figure 3 shows SEM images of the 1-GPa, 1, 4, and 8-revolution HPT sample (a, c, e taken from the centre (d < 2 mm), b, d and e taken from the edge (2 < d < 4 mm)). Fragmentation of the larger second phase was achieved after 1-revolution HPT, both in the centre and the edge, as shown in Fig. 3, compared to Figure 1 (as-cast and T4 heat treatment). Fragmentation of the larger second phase was believed to play a major role in precipitate evolution and this process is responsible for the spheroidization of precipitates during HPT [16,17].

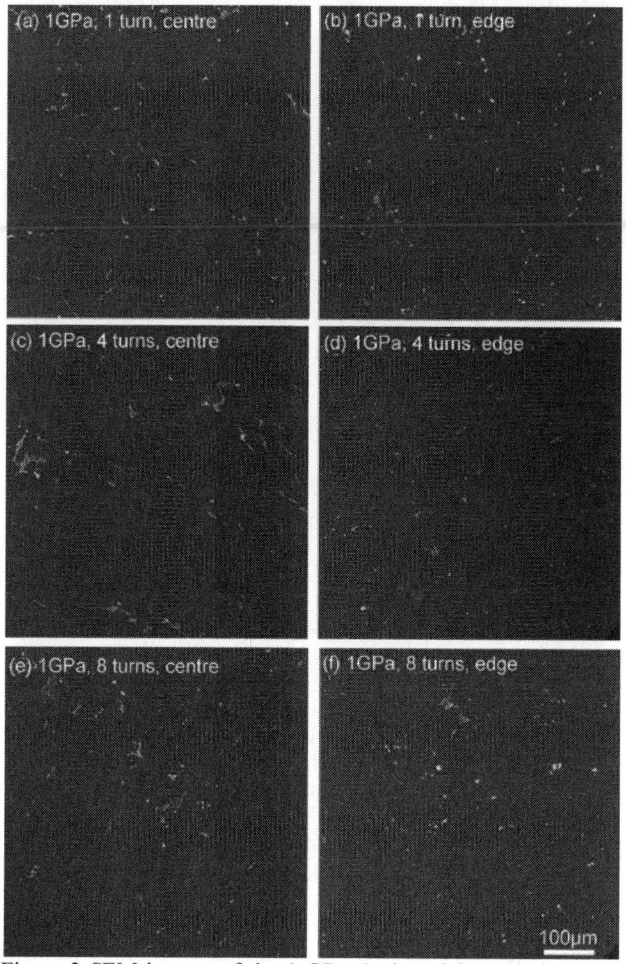

Figure 3 SEM images of the 1-GPa, 1, 4, and 8-revolution HPT sample (a, c, e were taken from the centre, b, d and e were taken from the edge).

Figure 4 shows SEM images of the 2, and 4-GPa, 8-revolution HPT sample (a, c were taken from the centre, b and d were taken from the edge). More significant fragmentation of the larger second phase was observed. Especially, at the edge, much finer and round particles were observed. These particles are about 500 nm in size (diameter), and are rich in Zn and Gd, as shown in Figure 5c. Furthermore, fragmentation of the small precipitates also occurs during HPT, as marked with a white arrow in Figuree 5a. The platelets precipitates were fragmented into small round particle (< 25 nm). As a result, the growth of the small precipitates will be restrained and no small platelets will have the capability to grow into larger precipitates. This is in contradiction with the many larger platelets, more than ~ 25 nm long, observed in T6 condition, as shown in Figure 2.

Examination of the sample after the 1-GPa, 8-revolution reveals a significant difference from that after the 1-GPa, 1-revolution. Figure 6 shows some TEM image (a, c) and corresponding SADP (b, d) of the 1-GPa, 8-revolution HPT sample taken from the centre (a, b) and the edge (c, d), respectively. In both cases (centre and edge), the microstrucutre seems to be much more homogeneous, although some small precipitates were still observed in the centre (Figure 6a), indicating that HPT (1-GPa, 8-revolution) can produce enough stains to tailor the precpitates. Most weak diffraction spots of the precipitates scatter randomly with many spots situated, as shown in Figures 6b and d, on the white ring. This indicates that these precipitates are distributed with random orientations in the matrix.

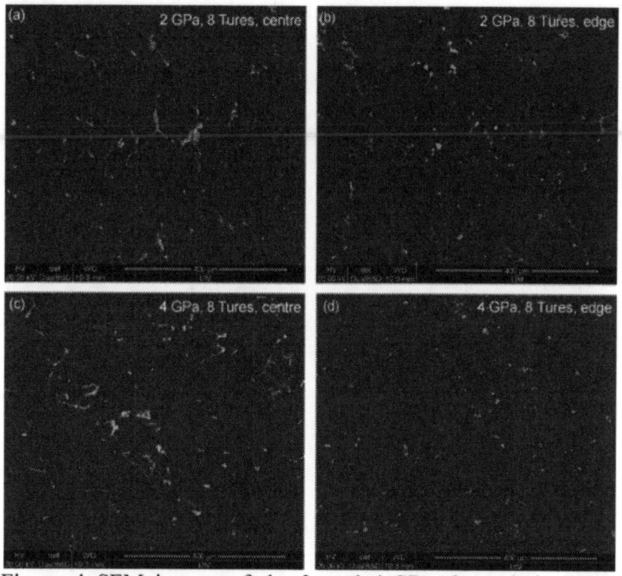

Figure 4 SEM images of the 2, and 4-GPa, 8-revolution HPT sample taken from the centre (a, c) and the edge (b and d).

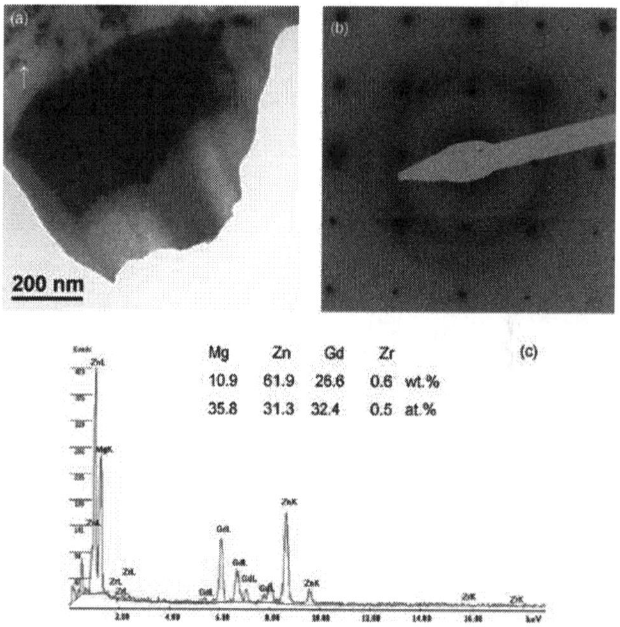

Figure 5 TEM image (a) and corresponding SADP (b), EDX analysis (c) of the 1-GPa, 1-revolution HPT sample taken from the centre.

Examination of the sample after the 4-GPa, 8-revolution , as shown in Figure 7, also reveals a significant difference from that after the 1-GPa, 8-revolution (Figures 6a, b). The higher pressure (4-GPa compared to 1-GPa) may result into a increasing temperature during HPT, which can promote the "second" precipitation during HPT. It seems that the precipitates are probably undergoing coalescence. Indeed, if the growth of large precipitate is predominantly via coalescence rather than by the dissolution of smaller precipitates, no small precipitates are expected to exist at the areas remote from these larger precipitates, as shown in Figure 7c, where a "precipitation free zone" was observed along the large plate-like precipitates.

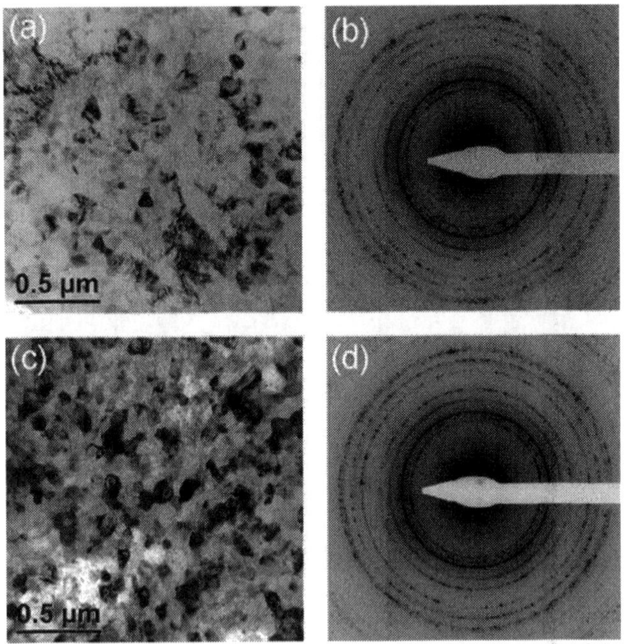

Figure 6 TEM image (a, c) and corresponding SADP (b, d) of the 1-GPa, 8-revolution HPT sample taken from the centre (a, b) and the edge (c, d).

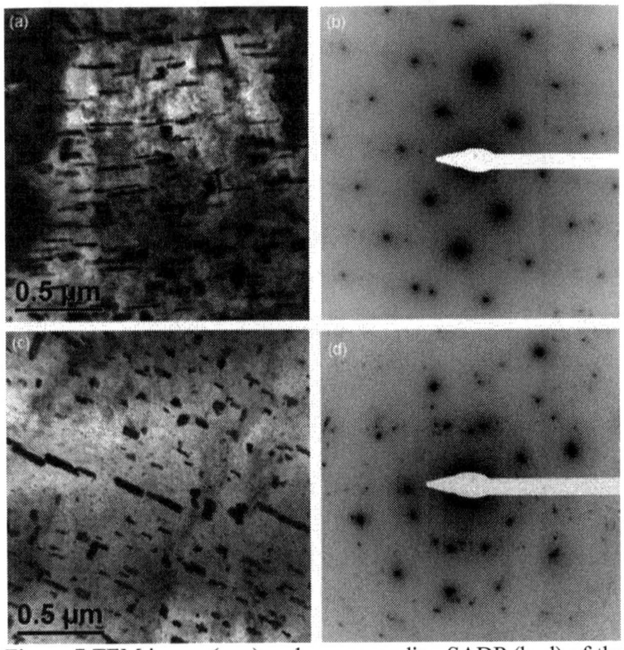

Figure 7 TEM image (a, c) and corresponding SADP (b, d) of the 4-GPa, 8-revolution HPT sample taken from the centre.

As expected, examination of microstructure taken from the edge of the HPTed sample after the 4-GPa, 8-revolution also reveals a significant difference from that taken from the centre. Figure 8 shows TEM image (a, c) and corresponding SADP (b, d) of the 4-GPa, 8-revolution HPT sample taken from the edge. In contrast to

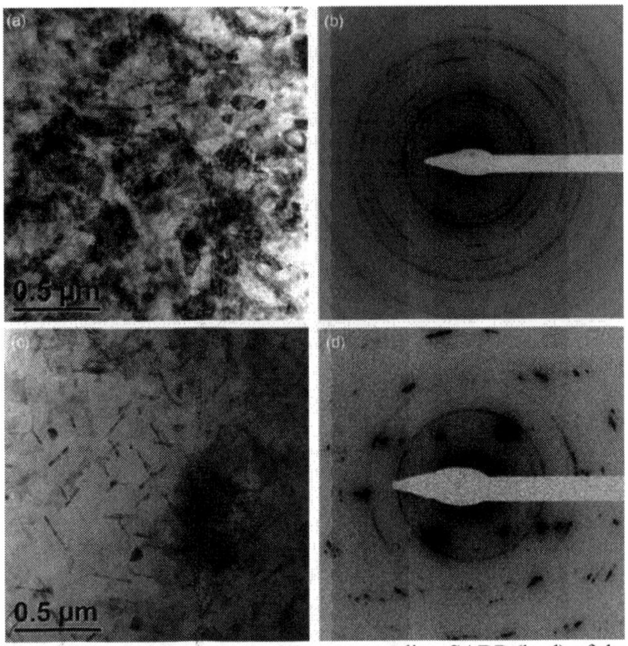

Figure 8 TEM image (a, c) and corresponding SADP (b, d) of the 4-GPa, 8-revolution HPT sample taken from the edge.

the centre (Figure 7), most weak diffraction spots of the precipitates scatter randomly with many spots situated, as shown in Figure 8b and d, on the white ring. This is fully consistent with the observation in the HPTed sample after the 1-GPa, 8-revolution (Figure 6). Also, this indicates again that these precipitates are distributed with random orientations in the matrix after 8-revolution. Thus, it is very clear that the accumulated strain imposed by HPT significantly alters the orientations of the precipitates away from their initial preferred orientations. Furthermore, it is also very clear that HPT causes significant fragmentation of the large precipitates and increases their appearance within the matrix, as shown in Figure 8c. Sub-grains with small angle sub-grain boundaries are rare in the centre of 4-GPa, 8-revolution HPT sample, and are no longer seen in the edge of 4-GPa, 8-revolution HPT sample, suggesting that, with the further deformation, grain orientation continued until all the sub-grains had the same orientation and the sub-grain structure disappeared to form a larger grain. This difference between the centre and edge is also consistent with the fact that the deformation is not homogeneous in the through-thickness direction during HPT. The heterogeneous nature of flow during HPT is probably due to local variations in the flow stress of the sample during HPT processing [18].

In the case of the 1-GPa, 1-revolution sample, dislocations are seen with high density at sub-grain boundaries to accommodate the small angle misorientations between neighboring sub-grains but are hardly seen within each sub-grain, indicating that grain rotation plays a primary role in the early stage of the HPT process. At the initial stage of the HPT process grain rotation first converted high-angle grain boundaries to low-angle grain boundaries, forming large grains with sub-grains consisting of initial small grains. In contrast, in the case of 1-GPa, 8-revolution HPT sample (Figure 6) and 4-GPa, 8-revolution HPT sample (Figure 8), high density of dislocations is observed in the grains and this can be easily seen through the strong contrast variation caused by the strain field around dislocations in grains which are in a strong diffraction condition (i.e., the grains appear dark), as shown in Figure 6c and Figure 8a. The high density of dislocations found in the 1-GPa, 8-revolution HPT sample and 4-GPa, 8-revolution HPT sample indicates that dislocations play a primary role in accommodating the deformation process at this stage. In other words, during HPT at higher revolutions dislocation slip became the primary deformation mechanism, in which dislocation accumulation and dynamic recovery reached a dynamic balance [19].

Hardness after T6, extrusion and HPT

Figure 9 shows hardness values measured at different processting conditions and positions. As expected, the hardness increases slightly after extrusion. The increasing hardness can be attributed to the grain refinement caused by the extrusion. However, unexpectedly, the hardness does not increase greatly after HPT, as compared to the extrusion condition. Furthermore, in some cases (1-GPa, 1-revolution; 4-GPa, 1-revolution), the hardness is less than that in the extrusion condition. This indicates that increasing strain may lead to a continuous grain growth, and thus a lower hardness due to the Hall-Petch effect [20].The higher, but relatively stable hardness was achieved after 1-GPa, 4-, 8-revolution, and 4-GPa, 4-, 8-revolution. This indicates that increasing revolution may result in an equilibrium structure, supporting a dynamic balance between deformation-induced grain growth and grain refinement and between deformation-induced dislocation generation and dislocation annihilation [21]. This is also fullyconsistent with the microstructure observation (Figures 6-8). The higher hardness can be attributed to the promotion of heterogeneous nucleation of the precipitates, and suppress of its growth amazingly caused by the high strains during HPT, similar to the case of ECAPed Mg-Nd-Gd based alloy [22]. It can be also expected that ductility will be improved greatly without sacrificing of strength. Further investigation on mechanical properties is in progress.

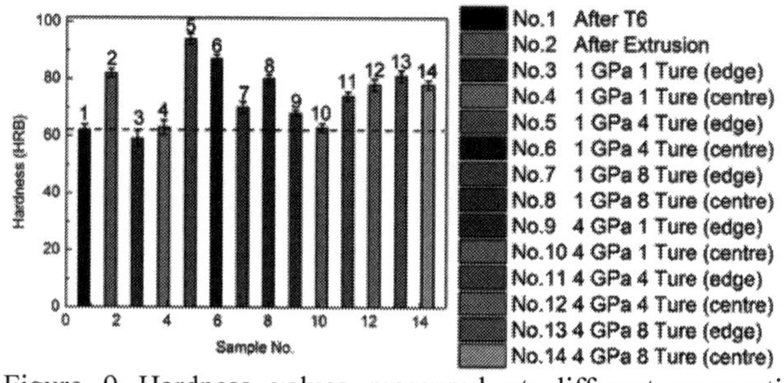

Figure 9 Hardness values measured at different processting condition and position.

Conclusions

HPT has been employed to tailor the precipitate microstructures in ZG62 alloy. An enhanced aged hardening response was achieved due to the promotion of heterogeneous nucleation of the precipitates, and suppress of its growth amazingly caused by the high strains during HPT. This research demonstrates a new perspective for application of HPT to tailor precipitate microstructure for unique properties in service.

Acknowledgements

Jiehua Li gratefully acknowledges Prof Gerhard Dehm for his access to TEM and Prof Pippan Reinhard for his help on HPT at the Erich Schmid Institute of Materials Science of the Austrian Academy of Science.

References

[1] B. Smola, I. Stulíková, F. von Buch, and B.L. Mordike, "Structural aspects of high performance Mg alloys design," *Mater. Sci. Eng. A*, 324 (2002), 113-117.

[2] J.B. Clark, "Transmission electron microscopy study of age hardening in a Mg-5 wt.% Zn alloy," *Acta Metall.*, 13 (1965), 1281-1289.

[3] J.B. Clark, L. Zabdyr, and Z. Moser. "Binary Alloy Phase Diagram," 2nd ed. T.B. Massalski, H. Okamoto, P.R. Subramanian and L. Kacprzak (ASM International, Metals Park, OH, 1988), 2571-2572.

[4] J.F. Nie, X. Gao, and S.M. Zhu, "Enhanced age hardening response and creep resistance of Mg–Gd alloys containing Zn," *Scripta Mater.*, 53 (2005), 1049-1053.

[5] X. Gao, and J.F. Nie, "Enhanced precipitation-hardening in Mg–Gd alloys containing Ag and Zn," *Scripta Mater.*, 58 (2008), 619-622.

[6] J.F. Nie, K. Oh-ishi, X. Gao, and K. Hono, "Solute segregation and precipitation in a creep-resistant Mg–Gd–Zn alloy," *Acta Mater.*, 56 (2008), 6061-6076.

[7] J.H. Li, and P. Schumacher, "Solidification and age hardening behaviour of Mg-Zn-Gd Alloys," *IOP Conf. Series: Materials Science and Engineering*, 27 (2011), 012021.

[8] J. Yang, L.D. Wang, L.M Wang, and H.J. Zhang, "Microstructures and mechanical properties of the Mg–4.5Zn–xGd (x = 0, 2, 3 and 5) alloys," *J. Alloys Compd.*, 459 (2008), 274-280.

[9] S.M. He, L.M. Peng, X.Q. Zeng, W.J. Ding, and Y.P. Zhu, "Comparison of the microstructure and mechanical properties of a ZK60 alloy with and without 1.3 wt.% gadolinium addition," *Mater. Sci. Eng. A*, 433 (2006) 175-181.

[10] A. P. Zhilyaev, G.V. Nurislamova, B. K. Kim, M.D. Baró, J.A. Szpunar, and T. G. Langdon, "Experimental parameters influencing grain refinement and microstructural evolution during high-pressure torsion," *Acta Mater.*, 51 (2003), 753-765.

[11] G. Sha, Y.B. Wang, X.Z. Liao, Z.C. Duan, S.P. Ringer, and T.G. Langdon, "Influence of equal-channel angular pressing on precipitation in an Al–Zn–Mg–Cu alloy," *Acta Mater.*, 57 (2009), 3123-3132.

[12] L.Y. Wei, G.L. Dunlop, and H. Westengen, "Precipitation Hardening of Mg-Zn and Mg-Zn-RE alloys," *Metall. Mater. Trans. A*, 26 (1995), 1705-1716.

[13] X. Gao, and J.F. Nie, "Characterization of strengthening precipitate phases in a Mg–Zn alloy," *Scripta Mater.*, 56 (2007), 645-648.

[14] L.L. Rokhlin, and A.A. Oreshkina, *Fiz. Metal. Metalloved.* 66 (1988), 559.

[15] Y. Komura, and K. Tokunaga, "Structural studies of stacking variants in Mg-base Friauf-Laves phases," *Acta Crystallogr.*, 36B (1980), 1548-1554.

[16] J. Gubicza, I. Schiller, N.Q. Chinh, J. Illy, Z. Horita, and T.G. Langdon, "The effect of severe plastic deformation on precipitation in supersaturated Al–Zn–Mg alloys," *Mater Sci Eng A*, 460-461 (2007), 77-85.

[17] K. Matsubara, Y. Miyahara, Z. Horita, and T.G. Langdon, "Developing superplasticity in a magnesium alloy through a

combination of extrusion and ECAP," *Acta Mater.*, 53 (2005), 749.

[18] R. B. Figueiredo, P. R. Cetlin, and T. G. Langdon, "Using finite element modeling to examine the flow processes in quasi-constrained high-pressure torsion," *Mater Sci Eng A*, 528 (2011), 8198-8204.

[19] Y.B. Wang, J.C. Ho, X.Z. Liao, H.Q. Li, S.P. Ringer, and Y.T. Zhu, "Mechanism of grain growth during severe plastic deformation of a nanocrystalline Ni–Fe alloy," *Appl. Phys. Lett.*, 94 (2009), 011908.

[20] R.Z. Valiev, "Nanomaterial Advantage," *Nature*, 419 (2002), 887-889.

[21] S. Ni, Y.B. Wang, X.Z. Liao, S.N. Alhajeri, H.Q. Li, Y.H. Zhao, E.J. Lavernia, S.P. Ringer, T. G. Langdon, and Y.T. Zhu, "Grain growth and dislocation density evolution in a nanocrystalline Ni–Fe alloy induced by high-pressure torsion," *Scripta Mater.*, 64 (2011), 327-330.

[22] G. Sha, J.H. Li, W. Xu, K. Xia, W.Q. Jie, and S.P. Ringer, "Hardening and microstructural reactions in high-temperature equal-channel angular pressed Mg–Nd–Gd–Zn–Zr alloy," *Mater Sci Eng A*, 527 (2010), 5092-5099.

Magnesium Technology 2013
Edited by: Norbert Hort, Suveen N. Mathaudhu, Neale R. Neelameggham, and Martyn Alderman
TMS (The Minerals, Metals & Materials Society), 2013

MICROSTRUCTURE AND PHASE EVOLUTION IN Mg-Gd AND Mg-Gd-Nd ALLOYS WITH ADDITIONS OF Zn, Y AND Zr

S. Khawaled, M. Bamberger, A. Katsman
Technion – Israel Institute of Technology, Haifa 3200, Israel

Keywords: Mg-Gd alloys, Mg-Gd-Nd alloys, Precipitation, Microstructure, Mechanical properties

Abstract

Microstructure and phase evolution in Mg-Gd and Mg-Gd-Nd based alloys with additions of Zn, Zr and Y were analyzed in the as-cast, solution treated and aged conditions. Alloys has been investigated after solution treatment at 540°C for 24hr followed by isothermal aging at 175°C up to 32 days by using of Vickers hardness, optical microscopy, scanning electron microscopy equipped with EDS, X-ray diffraction and transmission electron microscopy. It was found that the as-cast alloys contained primary α-Mg matrix, eutectic-like structures, cuboid-like phases and Zr-rich clusters. The homogenized and quenched alloys contained primary α-Mg solid solution, smaller amount of divorced eutectic compounds, enlarged cuboid-like particles and Zr-rich clusters. The eutectic phase was Mg_5Gd prototype with the composition $Mg_5(Gd_xNd_{1-x}$, $x \approx 0.2)$. The compositions of the cuboid shaped particles are characterized by enlarged amount of Gd and can be written as $Mg_2(Gd_xY_{1-x})$ with $x \approx 0.85$ in the Mg-5Gd based alloy, and $Gd_4(Y_xNd_{1-x})$ with $x \approx 0.5$ in the Mg-6Gd-3.7Nd based alloy. The cuboid shaped particles grew during aging and reached ~3μm average size. Precipitation of β″ and β′ phases during aging was observed. Mg-6Gd-3.7Nd based alloy reached a maximum value of microhardness after 16 days of aging; in Mg-Gd based alloy, microhardness increased more slowly and reached a maximum value after 32 days of aging.

Introduction

Magnesium alloys containing heavy rare earth metals are very attractive candidates for the automotive industry including racing cars and aerospace applications [1-7], due to their high strength properties combined with low density. Due to a large solubility of RE elements in Mg matrix at high temperature and its rapid decrease with lowering temperature, the Mg–RE alloys show remarkable age-hardening response during isothermal aging at 175°C, and thus high strength is to be expected.

Microstructural characterization of the Mg–RE alloys performed by many researchers [1-10] has confirmed a four staged precipitation sequence: [α-Mg supersaturated solid solution (S.S.S.S.) → $\beta''(DO_{19})$ → β'(bco) → β_1(fcc) → β(fcc)] [2,3,9,10], where the later precipitate can coexist with the former. Among the four precipitated phases, the coherent β″and β′ phases are considered to be the primary strengthening phases [2,4,8,10]. The yield strength or hardness usually reaches a maximum, as the materials form a microstructure with fine β' precipitates during aging [1-4]. Among RE elements, Gd and Nd are considered as most promising for achievement of best Mg-RE alloys properties.

In the present work two Mg-RE alloys were investigated: Mg-Gd and Mg-Gd-Nd based alloys with additions of Zn, Zr and Y. Their specific compositions were selected in order to achieve maximum age hardening response, improved castability, grain refinement and strength. They were investigated by a combination of optical microscopy (OM), scanning electron microscopy (SEM), transmission electron microscopy (TEM), energy dispersive X-ray spectrometry (EDS) and X-ray diffraction (XRD). Microhardness measurements were performed to reveal the effect of precipitation on the mechanical properties of the alloy.

Experimental

High purity Mg (99.9 wt.%) was molten in a mild steel crucible under a protective atmosphere CO_2 and HFC-134a (CF_3CH_2F) , high purity Gd (99.9 wt.%), Nd (99.9 wt.%) "for Mg-Gd-Nd-based alloy", Zr (99.9 wt.%), Mg–40Y (wt.%) and Mg–16Zr master alloys were added at a melt temperature of 760°C, then casting into steel mould preheated to ~200°C. The chemical composition of the investigated alloys were Mg-6Gd-3.7Nd-0.3Zn-0.18Y-0.15Zr and Mg-4.7Gd-0.37Zn-0.14Y-0.09Zr (wt.%).

Specimens cut from the cast ingot were solution treated (ST) at 540°C for 24hr, quenched into hot water of about 70°C, and then aged at 175°C in an oil bath for various periods of time up to 128 days, followed by water quenching.

The microhardness of aged specimens was measured by a Vickers hardness tester under a load of 50gr and holding time of 15sec. 20-40 indentations were conducted on each specimen. The specimens were polished with diamond paste and etched for 20sec in Ethanol 0.95%wt and HNO_3 0.05%wt (Nital) solution.

A comprehensive characterization utilized OM, SEM, TEM, EDS and XRD was carried out in order to elucidate the as-cast, homogenized and aged micro-structure of the alloys. Metallographic specimens were polished and etched in an Acetic Glycohol etching solution (100ml ethanol, 20ml water, 6gr picric acid and 5ml acetic acid). Thin foils for TEM characterization were prepared by ion beam thinning, and were examined at the acceleration voltage of 200 kV. The compositions of the matrix and of secondary phases in the TEM specimens were analyzed by EDS instrument attached to the SEM, operating at 10kV.

The constituent phases of the alloy in different conditions were identified by XRD with Cu $K\alpha$ radiation. Details for the XRD: (start/end angle: 15-100°, step size: 0.01°, speed duration time: 0.45).

Results

As-cast microstructure

Fig.1 shows the as-cast alloys (a) for Mg-Gd-Nd alloy and (b) for Mg-Gd alloy. Consist mainly of the α-Mg phase as a matrix, eutectic-like structures, cuboid-like phases and Zr-rich clusters.

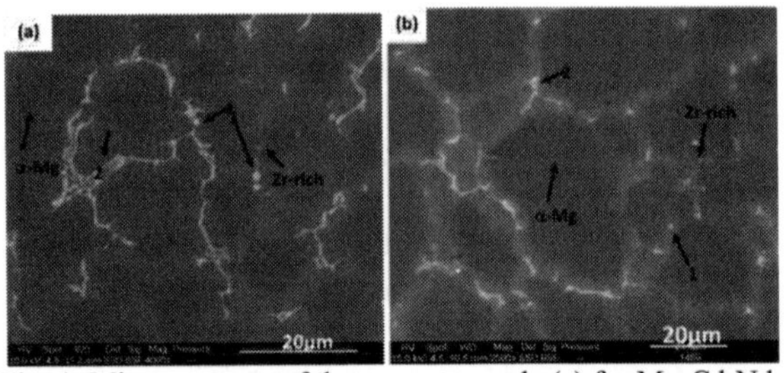

Fig. 1. Microstructure of the as-cast sample (a) for Mg-Gd-Nd alloy and (b) for Mg-Gd alloy: SEM image shows the cuboid-like phase (1) and eutectic-like structure (2).

The cuboid-like particles are frequently observed in the association with Zr-rich clusters and eutectic-like structures in the grain boundary as shown in Fig. 1.

The compositions of the phases were analyzed by EDS SEM using TEM samples shown in Tables. 1. Taking into account the small size of cuboids, EDS analysis results were inevitably influenced by the region around them, making the results just semi-quantitative to a certain extent.

Table 1. Compositions of different phases in Mg-Gd and Mg-Gd-Nd based alloys, measured by EDS SEM.

As- cast	Mg at%	Gd at%	Nd at%	Y at%	Zr at%	Phase
Mg-5Gd	99.01	0.52	-	0.14	0.15	α-Mg
	82.43	17	-	0.53	0.0	Eutectic
	77.96	19.63	-	1.79	0.0	Cuboid
	89.27	4.7	-	2.37	3.69	Zr rich
Mg-6Gd-3.7Nd	99.06	0.42	0.21	0.14	0.16	α-Mg
	84.82	7.43	7.73	0.0	0.0	Eutectic
	51.03	15.3	1.38	32.28	0.0	Cuboid
	94.03	1.51	0.76	0.0	3.7	Zr rich

ST microstructure

The purpose of homogenization for precipitate hardened Mg-RE alloys is to dissolve soluble as cast phases into α-Mg matrix and obtain supersaturated solid solution by quenching. The microstructure of the alloy was investigated in the ST and aged (175°C) conditions. It was observed that the homogenized alloy contained primary α-Mg solid solution, eutectic structures, cuboid shaped phases and Zr-rich clusters (Fig. 2).

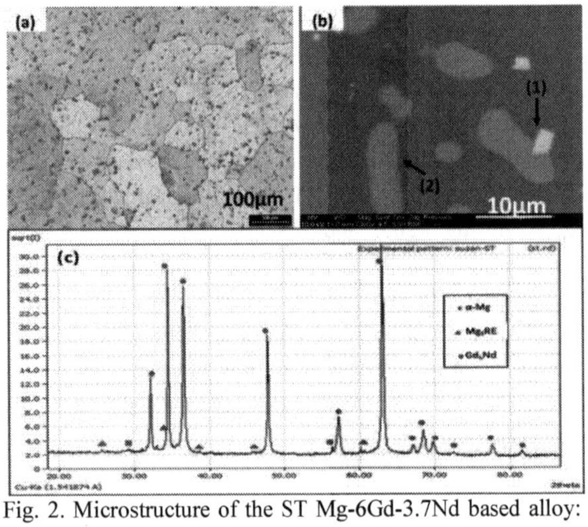

Fig. 2. Microstructure of the ST Mg-6Gd-3.7Nd based alloy: (a) Optical image showing the grain size and (b) SEM image showing the cuboid shaped phase (1) and eutectic structure (2). (c) X-ray diffraction.

As can be seen from X-ray diffraction patterns, the ST alloys contained primary α-Mg solid solution, eutectic phase Mg_5RE and cuboid shaped phases: Gd_4Nd for Mg-Gd-Nd based alloy (Fig. 2c) and Mg_2Gd for Mg-Gd based alloy (Fig. 3).

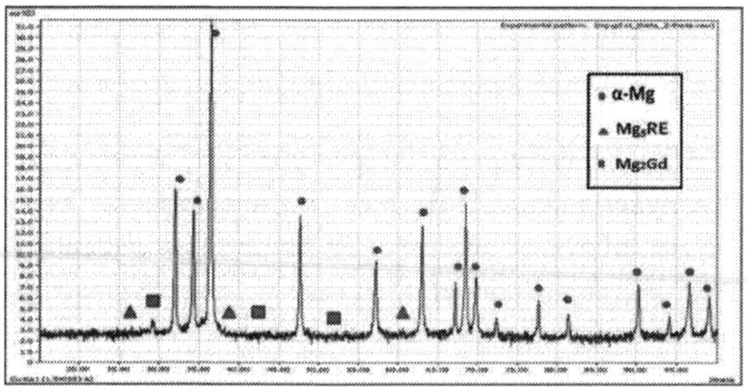

Fig. 3. X-ray diffraction pattern of the ST Mg-5Gd based alloy.

The eutectic structures were the products of a 'quasibinary eutectic reaction' L→α-Mg+β-Mg_5RE. The eutectic phase in the Mg-5Gd based alloy was characterized to be Mg_5Gd. The microstructure of the homogenized Mg-6Gd-3.7Nd based alloy contained two-phase skeleton-like eutectic structures (Fig. 4) with the Mg_5Gd-prototype phase of the composition $Mg_5(Gd_xNd_{1-x}$, x≈0.2).

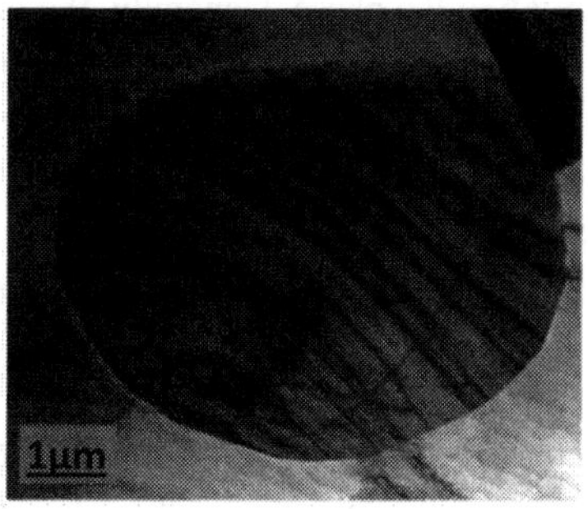

Fig. 4. Microstructure of the ST Mg-6Gd-3.7Nd based alloy: TEM observations of eutectic-like structure.

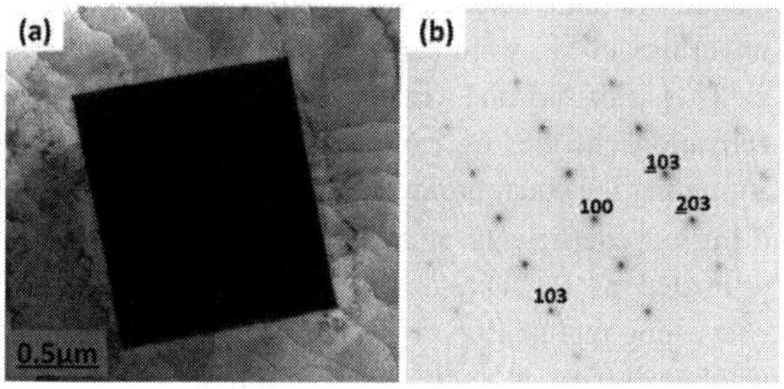

Fig. 5. Microstructure of the ST Mg-6Gd-3.7Nd based alloy: (a) TEM images of cuboid shaped phase and (b) SAED pattern with Z.A$[\bar{1}2\bar{1}0]$.

The compositions of the cuboid shaped particles after homogenization are characterized by higher amount of Gd and can be presented as $Mg_2(Gd_xY_{1-x})$ with x≈0.85 in the Mg-5Gd based alloy, and $Gd_4(Y_xNd_{1-x})$ with x≈0.5 in the Mg-6Gd-3.7Nd based alloy.

The cuboid shape phase was characterized to have Gd_4Nd type HCP crystal structure with $a \approx 0.365\text{nm}$, $c \approx 0.5798\text{nm}$.

Microstructure of aged samples

Microstructure of the aged samples consists of the eutectic structure, cuboid shaped particles and small precipitates uniformly distributed in the matrix. The cuboid shaped particles grow during aging up to ~3 μm in size; SEM+EDS analysis revealed that their composition did not change.

Crystal structure and composition of precipitates within the α-Mg matrix were investigated by TEM (Figs. 6-7); there were diffuse spots between the α-Mg matrix spots in the selected area electron diffraction (SAED) patterns, which were brighter than the matrix spots. TEM images and corresponding SAED patterns of precipitates in a specimen aged for 16 days (peak-aged specimen) are shown in Fig.6. It was observed that the microstructure contains a high number density of fine plate shape β″ precipitates and a little number of globular particles β′ uniformly distributed within the α-Mg matrix.

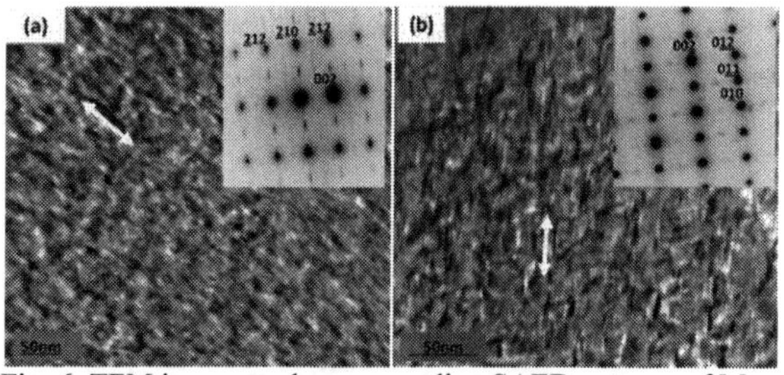

Fig. 6. TEM images and corresponding SAED patterns of Mg-Gd-Nd alloy aged for 16days at 175C: Z.A $[01\bar{1}0]_\alpha$ in (a) and Z.A $[2\bar{1}\bar{1}0]_\alpha$ in (b).

Combination of reflections from the two precipitates, β″and β′, appears in the corresponding SAED pattern (Fig. 6). The β″ precipitates with DO_{19} structure $(a{\sim}2a_{\alpha-Mg} = 0.64\text{nm}, c{\sim}c_{\alpha-Mg} = 0.52\text{nm})$ correspond to diffuse spots at 1/2 distance of $\{01\bar{1}0\}_\alpha$ or $\{2\bar{1}\bar{1}0\}_\alpha$ reflections, with the following orientation relationships (OR): $[0001]_{\beta''}//[0001]_\alpha$, $\{01\bar{1}0\}_{\beta''}//\{01\bar{1}0\}_\alpha$; and the β′ phase with Base Centered Orthorhombic (BCO) structure ($a{\sim}2a_{\alpha-Mg} = 0.64\text{nm}$, $b{\sim}2.2\text{nm}$, $c{\sim}c_{\alpha-Mg} = 0.52\text{nm}$) gives rise to the additional spots at 1/4 $\{01\bar{1}0\}_\alpha$, 1/2 $\{01\bar{1}0\}_\alpha$ and 3/4 $\{01\bar{1}0\}_\alpha$, with the following OR: $[001]_{\beta'}//[0001]_\alpha$, $\{100\}_{\beta'}//\{2\bar{1}\bar{1}0\}_\alpha$.

The microstructure of Mg-Gd-Nd alloy aged for 64 days and Mg-Gd alloy aged for 78 days contain a little number of small precipitates β' with plate-shaped β_1 precipitates (Fig. 7). The β_1 precipitates with FCC structure ($a \approx 0.74$nm), with the following OR: $[110]_{\beta_1}//[0001]_\alpha$, $\{\bar{1}12\}_{\beta'}//\{1\bar{1}00\}_\alpha$.

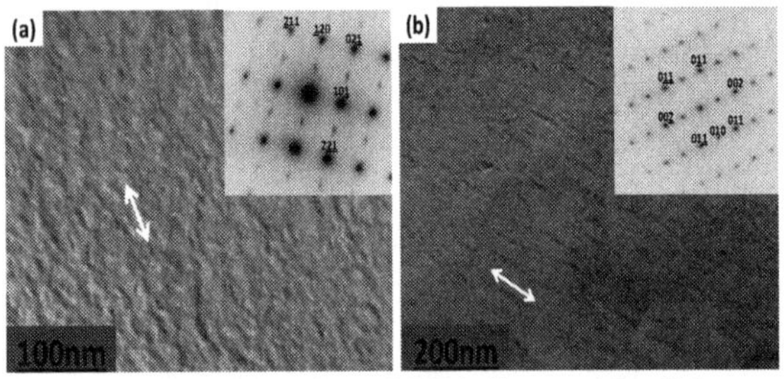

Fig. 7. SAED patterns of Mg-Gd-Nd alloy aged for 64days with Z.A $[10\bar{1}2]_\alpha$ (a) and of Mg-Gd alloy aged for 78days with Z.A $[2\bar{1}\bar{1}0]_\alpha$ (b) .

Microhardness

The microhardness evolution during isothermal aging at 175°C up to 64days for Mg-Gd-Nd-based alloy and 78 days for Mg-Gd-based alloy are shown in Fig. 8. Microhardness of the Mg-Gd-Nd based alloy reached a maximum value of ~ 100 HV after 16 days of aging. In the Mg-Gd based alloy, microhardness increased more slowly and reached a maximum value of ~ 70 HV after 32 days of aging. Further aging led to decrease in the microhardness.

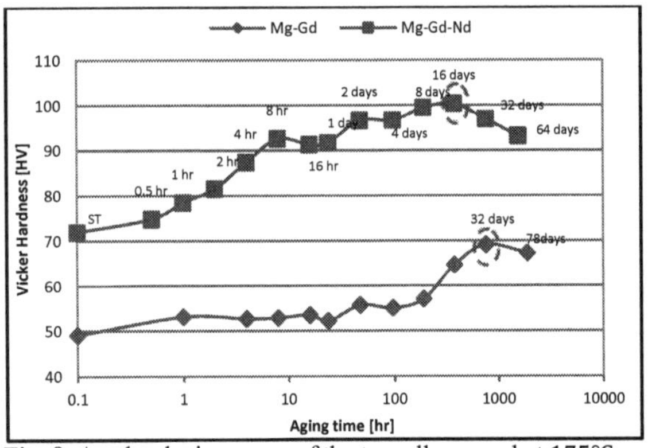

Fig. 8. Age hardening curve of the two alloys aged at 175°C

Discussion

The as-cast microstructures of the Mg-Gd and Mg-Gd-Nd based alloys with additions of Zn, Zr and Y consisted of primary α-Mg solid solution, skeleton-like eutectic structure, small Y-rich cuboid-like particles and Zr-rich clusters. The eutectic structures were the products of a 'quasibinary eutectic reaction' L→α-Mg+β-Mg_5RE. The eutectic phase was characterized to have Mg_5Gd type FCC crystal structure with a ≈ 2.2nm and a composition of $Mg_5(Gd_xNd_{1-x}$, x≈0.2).

After ST the cuboid-like particles were characterized by higher amount of RE elements in comparison to the as-cast alloys. In the Mg-5Gd based alloy the cuboids had a composition of $Mg_2(Gd_xY_{1-x}$, x≈0.85), and the cubic crystal structure with a ≈ 0.855nm (Fig.3).

In the Mg-Gd-Nd alloy the Gd and Nd atoms are transferred during ST from eutectic-like structure to cuboid-like particles to form the cuboid shaped phase with the stable composition $Gd_4(Nd_xY_{1-x}$, x≈0.5) (Fig. 2b). This phase has HCP crystal structure ($a \approx 0.365$nm , $c \approx 0.5798$nm). The cuboid shaped particles grow during aging and reach ~3μm average size.

Aging of solution treated and quenched alloys results in precipitation of β'' and β' phases, and their transformation to β_1 phase. Generally, the precipitation sequence in the investigated alloys during aging is identical to that reported for many Mg-RE alloy systems [2,3,8,9], i.e. the four staged precipitation sequence:$[\alpha - \text{Mg S.S.S.S.} \rightarrow \beta''(D0_{19}) \rightarrow \beta'(\text{bco}) \rightarrow \beta_1(\text{fcc}) \rightarrow \beta(\text{fcc})]$. The peak of microhardness (16 days of aging for Mg-Gd-Nd based alloy and 32 days for Mg-Gd based alloy) corresponds to a maximum density in the combination of $\beta''$and β' precipitates (Fig.6).

The difference in the peak aging time is explained by faster diffusion of Nd in comparison to Gd in Mg matrix. Ab initio calculations of diffusion of several RE impurities in Mg were performed recently by Huber et al. [11]. For Nd and Gd they obtained the diffusion activation energies of 1.05 and 1.17 eV, respectively. Assuming the equal pre-exponential factors, the ratio of diffusion coefficients of Nd and Gd at 175°C can be estimated as D_{Nd}/D_{Gd}=exp(0.12 eV/kT) ~ 20. In the coarsening regime (the over aging stage) the precipitate number density ,N_V , decreases with time, $N_V^{-1} \sim \alpha_{LSW} \cdot t \sim D_{RE} \cdot t$, where α_{LSW} is the coefficient of the LSW-coarsening theory. The microhardness roughly is inversely proportional to a distance between precipitates [13], and thus, is proportional to $N_V^{1/3}$. So, one can expect that the over aging stage in the Mg-Gd alloys would be substantially longer (to ~ $20^{1/3}$=2.7 times) than in the Mg-Nd alloys. The microhardness curves for Mg-Gd-Nd and Mg-Gd based alloys (Fig.8) confirm this evaluation.

The plate shape coherent β'' precipitates in $\alpha -$ Mg matrix have OR: $[0001]_{\beta''}//[0001]_\alpha$, $\{01\bar{1}0\}_{\beta''}/$

$/\{01\bar{1}0\}_{\alpha}$. The globular precipitates β′ have OR: $[001]_{\beta'}//[0001]_{\alpha}$, $\{100\}_{\beta'}//\{2\bar{1}\bar{1}0\}_{\alpha}$. These results are similar to those reported by Zheng et al. [9] for Mg-11Gd-2Nd-0.5Zr.

The over aging resulted in disappearance of β″ and β′ and formation of β_1 precipitates (Fig. 7). The plate shape β_1 precipitates have OR: $[110]_{\beta_1}//[0001]_{\alpha}$, $\{\bar{1}12\}_{\beta'}//\{1\bar{1}00\}_{\alpha}$. These processes are accompanied by decrease of microhardness (Fig.8).

Conclusion

The Mg-Gd-Nd and Mg-Gd based alloys with additions of Zn, Zr and Y were investigated in the as-cast, ST and aged conditions. Skeleton-like eutectic structure α-Mg+β-$Mg_5(Gd_xNd_{1-x}$, x≈0.2) and cuboid shaped particles were found in all conditions. Y-rich cuboid-like particles found in the as-cast alloy transformed to cuboid shaped phases during ST, $Gd_4(Nd_xY_{1-x}$, x≈0.5) in the Mg-Gd-Nd alloy and $Mg_2(Gd_xY_{1-x}$, x≈0.85) in the Mg-Gd alloy. They grow during aging. Precipitation of plate shape β″ and globular β′ particles during aging resulted in microhardness increase. The further aging results in coarsening of β″and β′ precipitates and their transformation to β_1 phase.

Reference

1. K.Y. Zheng et al., "Microstructural characterisation of as cast and homogenised Mg-Gd-Nd-Zr alloys," Materials Science and Technology 24 (2008), No3.
2. S.M. He et al., "Microstructure evolution in a Mg-15Gd-0.5Zr (wt. %) alloy during isothermal aging at 250 °C," Materials Science and Engineering A 431 (2006), 322-327.
3. K.Y. Zheng et al., "Effect of initial temper on the creep behavior of a Mg-Gd-Nd-Zr alloy," Materials Science and Engineering A 492 (2008), 185-190.
4. L.Gao et al., "Microstructure and strengthening mechanisms of a cast Mg–1.48Gd–1.13Y–0.16Zr (at.%) alloy," J Materials Science 44 (2009), 4443-4454.
5. K.Y. Zheng et al., "Effect of pre-deformation on aging characteristics and mechanical properties of Mg-Gd-Nd-Zr alloy," Trans. Nonferrous Met. Soc. China 17 (2007), 1164-1168.
6. D. Li et al., "Characterization of precipitate phases in a Mg–Dy–Gd–Nd alloy," Journal of Alloys and Compounds 439 (2007), 254-257.
7. K.Y. Zheng et al., "Effect of thermo-mechanical treatment on the microstructure and mechanical properties of a Mg–6Gd–2Nd–0.5Zr alloy," Materials Science and Engineering A 454-455 (2007), 314-321.
8. K.Y. Zheng et al., "Effect of precipitation aging on the fracture behavior of Mg–11Gd–2Nd–0.4Zr cast alloy," Materials Characterization 59 (2008), 857-862.
9. K.Y. Zheng et al., "Precipitation and its effect on the mechanical properties of a cast Mg-Gd-Nd-Zr alloy," Materials Science and Engineering A 489 (2008), 44-54.
10. G. Sha et al., "Hardening and microstructural reactions in high-temperature equal-channel angular pressed Mg–Nd–Gd–Zn–Zr alloy," Materials Science and Engineering A 527 (2010), 5092-5099.
11. L. Huber et al., Ab initio calculations of rare-earth diffusion in magnesium (Phys. Rev. B85, 144301, 2012).
12. A. Katsman, S. Cohen and M. Bambereger, "Modeling of Precipitation Hardening in Mg-based Alloys," Journal of Materials Science 42 (16) (2007), 6996-7003.

Magnesium Technology 2013

Twin Roll Casting

Magnesium Technology 2013
Edited by: Norbert Hort, Suveen N. Mathaudhu, Neale R. Neelameggham, and Martyn Alderman
TMS (The Minerals, Metals & Materials Society), 2013

Influence of Temperature and Rolling Speed on Twin Roll Cast Strip

Kurz G., Bohlen J., Stutz L., Letzig D., Kainer K.U.
Helmholtz-Zentrum Geesthacht Zentrum für Material- und Küstenforschung GmbH; Max-Planck-Straße 1, 21502 Geesthacht, GERMANY

Keywords: magnesium, AZ31, twin roll casting, magnesium strip, process parameters

Abstract

Reducing vehicle weight and emissions by lightweight design is a major goal of the automotive industry. Magnesium as the lightest structural metal offers a significant weight saving potential compared to steel and aluminum. Cast magnesium components are widely used, e.g. as engine blocks or gear box housings. The application of magnesium sheets is currently hampered by the low formability of magnesium which means that a large number of rolling passes is required to roll a DC cast slab to final gauge sheet. This large number of rolling steps is the main reason for the high cost of magnesium sheets. An alternative, cost-efficient production route for magnesium sheets with improved properties is feedstock production by twin roll casting (TRC). In this paper we report the results of twin roll casting experiments on the magnesium alloy AZ31 (Mg-3Al-1Zn-Mn) and discuss the influence of the process on the microstructure and texture of the strips.

Introduction

At present, magnesium alloys used in the automobile industry are mainly processed by die casting. This technology allows components with a complex geometry to be manufactured. However, the mechanical properties of die cast materials often do not meet essential requirements with regard to endurance, strength, ductility, etc. A promising alternative for thin, large area parts, such as automotive body components is to utilize sheet material. Sheet metal formed parts are characterized by superior mechanical properties and high quality surfaces without pores in comparison to die cast components. Substitution of conventional sheet materials such as steel or aluminum by magnesium sheets could lead to significant weight savings. However, it will be necessary to produce sheet material with competitive properties in an economic production process.

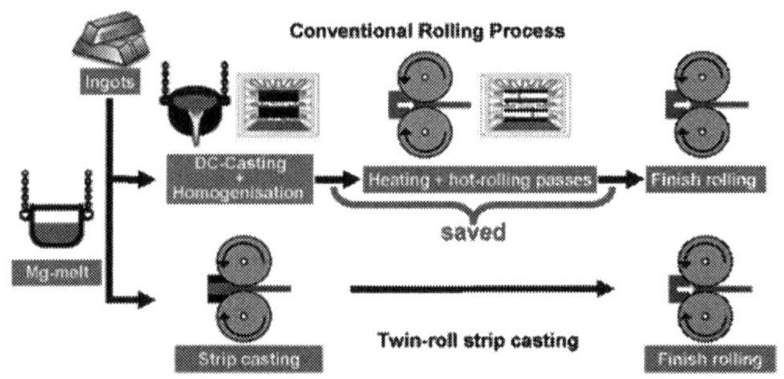

Figure 1: Comparison of the two process chains

Twin-roll casting is an economic production process for the generation of fine-grained feedstock materials that can subsequently be warm rolled to thin sheets. This production process for thin strips combines solidification and rolling into one single production step. Thus, it saves a number of rolling and annealing passes in comparison to the conventional rolling process (Figure 1).

The twin roll casting technology is already well established in for example the aluminum industry. However, applications to magnesium alloys are in their infancy at present [1]. There are worldwide a small number of industrial or laboratory scale twin roll casters installed at universities, companies and research facilities. Initial results from these activities on conventional wrought and cast alloys have shown promising sheet properties [2-6].

Twin Roll Casting Trials

The development of wrought magnesium alloys and their introduction into industrial, structural applications are the main goal of the activities at the Magnesium Innovation Centre MagIC of the Helmholtz-Centre Geesthacht (HZG). The current focus of the research work is on alloy design and the development of processing technologies for semi-finished magnesium products. In the particular case of sheet materials, it has been recognized that the feedstock for the warm-rolling process needs to be in the form of thin bands as they are produced via twin roll casting, if thin magnesium sheets are to become competitive industrial products. For this purpose HZG has installed the twin roll caster shown in Figure 2 and Figure 3.

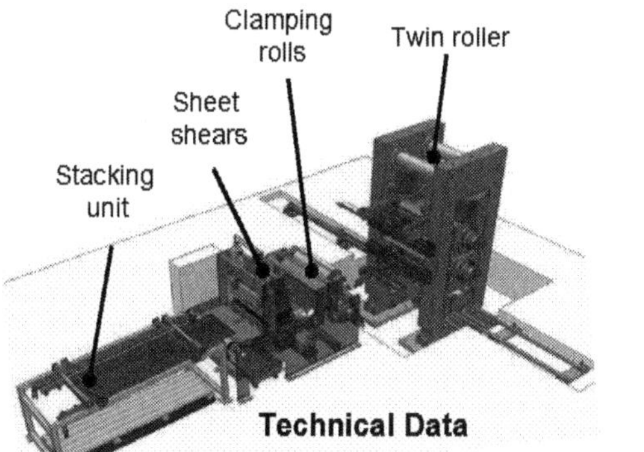

Technical Data

Band width:	650 mm
Roll diameter:	600 mm
Max. speed:	6 m/min
Min. band thickness:	4 mm
Force:	7600 kN

Figure 2: Twin roll caster at HZG and technical data

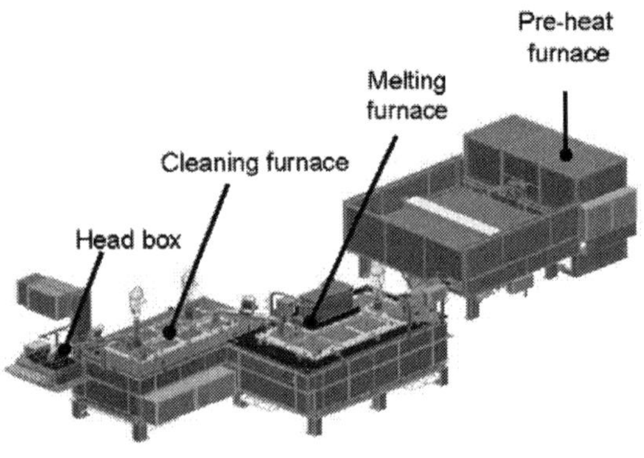

Technical Data

Max. capacity:	1000 kg/h
Max. temperature:	750 °C
Temperature stability:	+/- 3 °C

Figure 3: Furnace line of the twin roll caster at HZG and technical data

The twin roll casting line consists of a furnace line (Striko-Westofen) (Figure 3) and a twin roll caster (Novelis, Jumbo 3CM) (Figure 2). The twin roll caster is designed to produce thin strips of magnesium with a maximum width of 650 mm and a thickness in the range of 4 -12 mm at a maximum rolling speed of 6 m/min. The furnace line is configured to allow manual loading of raw materials in ingots and the possibility to vary the alloy composition, i.e. cast strips can be made from different alloys. The furnace line includes an ingot pre-heater to remove humidity from the ingots and accelerate fusion in the melting furnace. After pre-heating, the feedstock is transferred to the melting furnace. A transfer tube conveys the molten metal to the cleaning furnace. From the cleaning furnace the melt is transferred to the head-box which is connected to the tip. The liquid metal flows by gravity through the tip into the gap of the rolls of the twin roll caster (Figure 4). The metal exiting the tip solidifies on the rolls into a strip that is further deformed by the rolls. Important features of the strip such as microstructure and texture are influenced by the position of the solidification front.

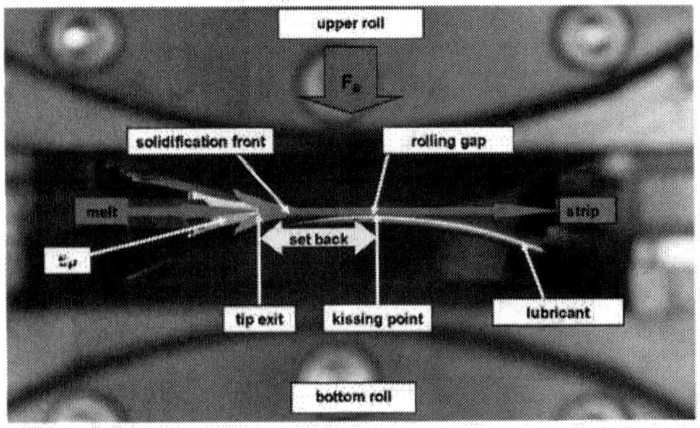

Figure 4: Principle and parameters of the twin roll casting process

The position of the solidification front is controlled mainly by the melt temperature, the set back (distance between exit of the tip and kissing point of the rolls), the height of the rolling gap and the strip speed.

The solidification starts when the magnesium melt enters the rolling gap and is in contact with the surface of the cooled rolls, the melt temperature drops to the solidus temperature at the strip surface (Figure 5). Ongoing along rolls surface, more heat flows from the strip in the rolls as it fully solidifies.

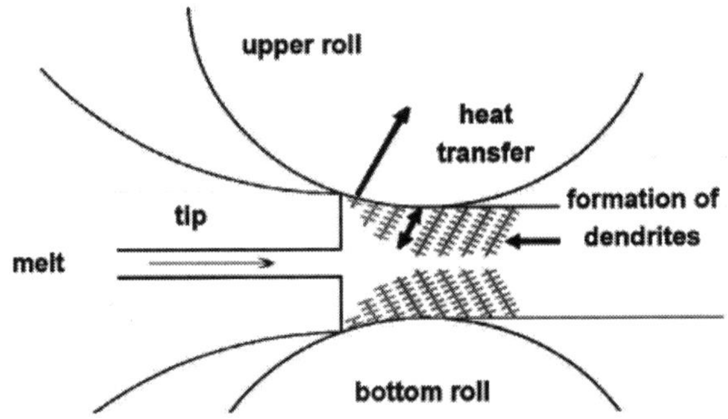

Figure 5: Solidification of the metal during twin roll casting

Then the heat is extracted to the solid strip. If the contact time of the melt at the rolls increases or the melt temperature decreases the solidification starts earlier and the solidification front is located near the exit of the tip. Its position influences the degree of deformation in the strip.

If the solidification front is located closer to the exit of the tip, the degree of deformation increases because the strip is fully solidified before entering the rolls (Figure 6a). If the solidification front moves towards the kissing point of the rolls due to changes in the processing parameters, the strip is not completely solidified and the degree of solid-state deformation decreases (Figure 6b).

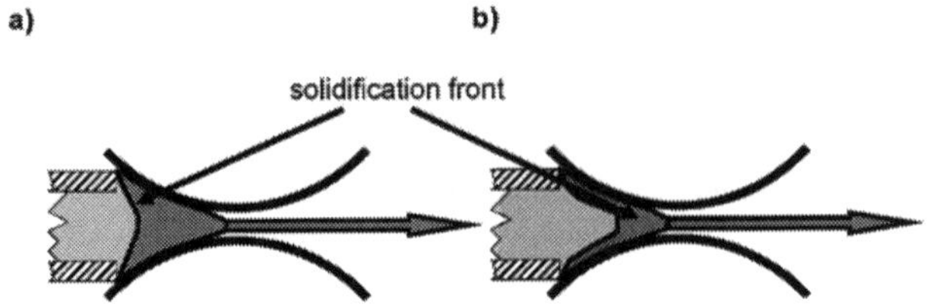

Figure 6: Position of the solidification front

All twin roll cast experiments were performed with the above mentioned equipment. The trials were carried out at melt temperatures between 650 - 710 °C. The rolling speed was varied stepwise between 1.4 m/min and 2.7 m/min. The setback and the rolling gap were constant. In these initial trials, the commercial magnesium alloy AZ31 was used and strips cast with a width of 350 mm. After twin roll casting, the microstructure of the strips was analyzed using optical microscopy. Standard metallographic sample preparation techniques were employed and an etchant based on picric acid was used to reveal grains and grain boundaries [7]. Texture measurements were performed on the sheet mid-planes using a Panalytical X-ray diffractometer setup and CuKα radiation. 6 pole figures were measured up to a tilt of 70° which allowed recalculation of full pole figures based on an MTEX software routine [8] The (00 2) and (10 0) pole figures are used in this work to present the texture of the strips at midplane.

Influence of Melt Temperature on the Strip

In a first set of casting trials the influence of the melt temperature on the microstructure of the twin roll cast strip was investigated. Figures 7 to 9 display micrographs of strip cast at a constant rolling speed of 1.8 m/min and varied melt temperatures (700 °C, 690 °C and 680 °C). It can be observed that the strip cast at 700 °C reveals a columnar structure where grains grow from the surfaces towards the central region of the strip. The central region itself reveals equiaxed grains in a band with certain thickness (Figure 7).

A remarkable feature is the strong segregation band containing a large amount of impurities at the centre-line of the strip. Such segregation is a common effect in this type of processing. The large amount of columnar grains is an indicator that the solidification front was located closer to the kissing point of the rolls and thus, the degree of deformation was small during the rolling operation due to incomplete solidification.

Figure 7 also displays (10 0) and (00 2) pole figures of the strip which both exhibit low intensities. A very weak alignment of basal planes parallel to the strip plane still gives hints to active deformation mechanisms which start to form of a typical sheet texture. Interestingly, components with alignment in RD and TD are also found with low fraction.

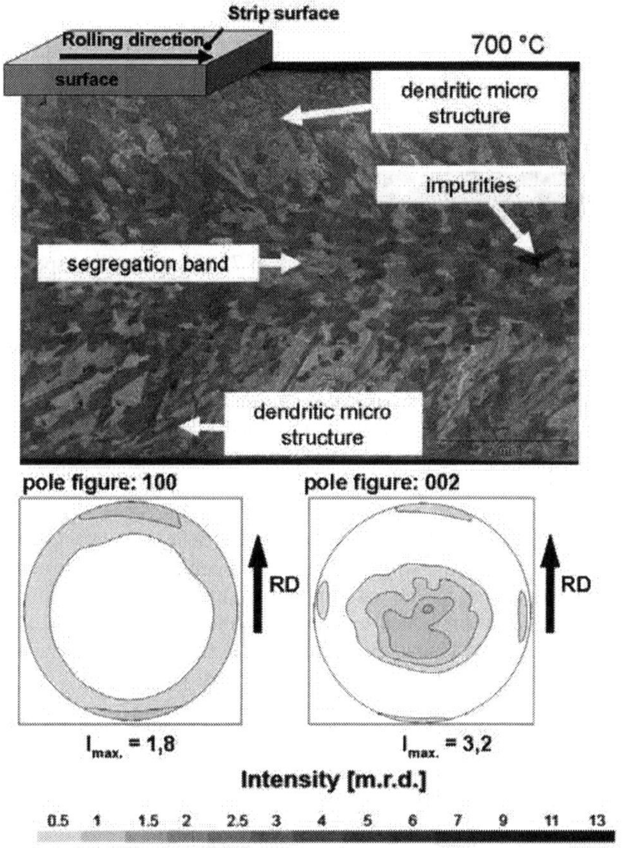

Figure 7: Microstructure and texture after twin roll casting with a melt temperature of 700 °C at a rolling speed of 1.8 m/min and a thickness of 5 mm

Figure 8 shows the same results as Figure 7 but for a casting temperature of 690°C. The columnar type microstructure is found to be less significant if compared to Figure 7 at 700°C and grains are more equiaxed.

Nevertheless, the strip cast at 690 °C is also characterized by a clear segregation band with a large amount of impurities at the centreline. The higher proportion of equiaxed grains in the microstructure indicates that the solidification front moves slightly towards the tip exit and the degree of deformation is somewhat higher than in the strip cast at 700 C.

Figure 8 also shows the (10 0) and (00 2) pole figures of the strip. The result is very similar to that shown in Figure 7 which indicated comparable features in the centreline. It is noted that this position in the through-thickness consideration most likely is the least part influenced by active deformation.

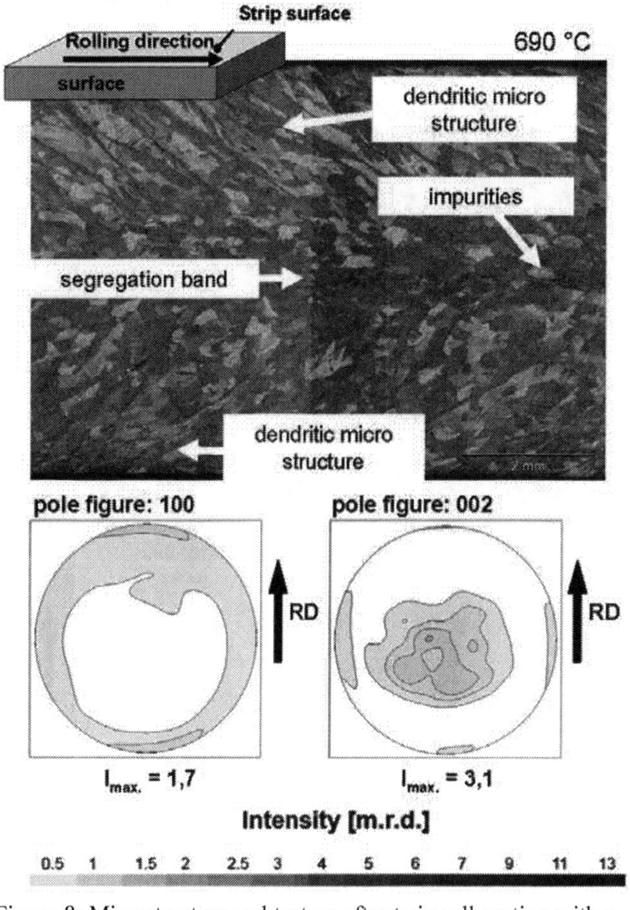

Figure 8: Microstructure and texture after twin roll casting with a melt temperature of 690 °C at a rolling speed of 1.8 m/min and a thickness of 5 mm

During casting the melt temperature was reduced to 680 C. The microstructure and texture of this twin roll cast strip are shown in Figure 9. This microstructure differs from the microstructures of the other two strips. It consists of smaller equiaxed grains, which are elongated in the rolling direction in the near-surface regions of the strip. This microstructure is typical for a deformed material. Still some features of a strip cast microstructure as shown above are visible.

The respective pole figures of this strip are shown in Figure 9. Again, the main feature is a weak alignment of basal planes in the sheet plane accompanied by a component tilted 90° to the RD as well as to the TD. Interestingly, there is no clear difference in the texture if compared to the results at higher casting temperature which is a counterintuitive finding with respect to the consideration about the amount of realized deformation during strip casting and concurrent rolling. However, the measurement of the texture at midplane position only gives respect to the condition in the finally solidified fraction of the microstructure which can be identified as a separate region in Figure 9. Haßlinger et al. [9] showed that the development of an alignment of basal planes in the sheet plane is the result of continuously increasing deformation with increasing number of passes. At low degree of deformation comparable weak textures with alignment of basal planes in the sheet plane have been found. Thus, the resulting texture reveals that the amount of deformation realized in this central part of the strip does not vary significantly with the casting temperature..

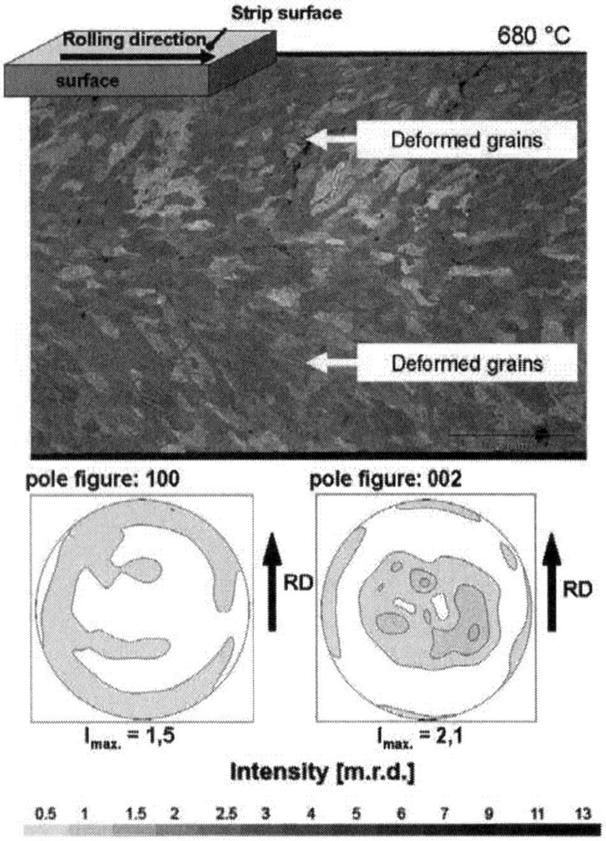

Figure 9: Microstructure and texture after twin roll casting with a melt temperature of 680 °C at a rolling speed of 1.8 m/min and a thickness of 5 mm

Influence of Rolling Speed on the Strip

Figures 10 and 11 depict the microstructures and textures of samples extracted from strips produced with the same temperature of 710 °C but varied rolling speeds of 2.7 m/min and 1.6 m/min, respectively. Again, both figures show the longitudinal sections of the strips. Figure 10 illustrates a clear columnar structure which is even more pronounced if compared to the result shown at intermediate speed but same casting temperature in Figure 3. Furthermore, the distinct segregation band containing a large amount of impurities at the centre-line of the strip is found.

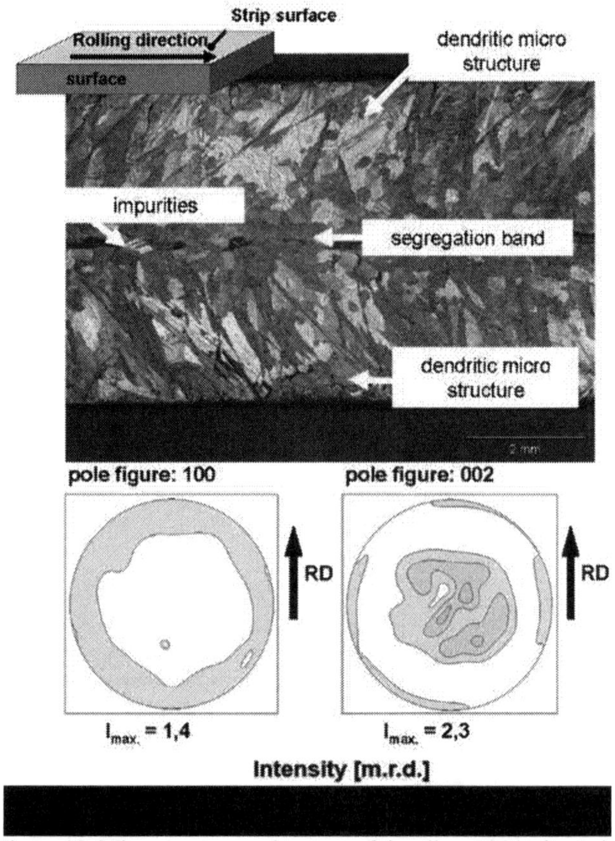

Figure 10: Microstructure and texture of the alloy AZ31 after twin roll casting with a melt temperature of 710 °C at a rolling speed of 2.7 m/min and a thickness of 5 mm

Again, the large number and the significance of columnar grains in the upper and lower surface region of the strip is an indicator for a low degree of deformation during the rolling operation. The texture of this sheet measured at midplane does not hint towards a different development compared to that explained in the previous section.

In Figure 11 results are shown after decreasing the rolling speed to 1.6 m/min. This strip also exhibits a columnar microstructure. In comparison to the one shown in Figure 10 no significant differences are revealed in the microstructure and texture of this strip. It is also characterized by a clear segregation band with a large amount of impurities at the centreline.

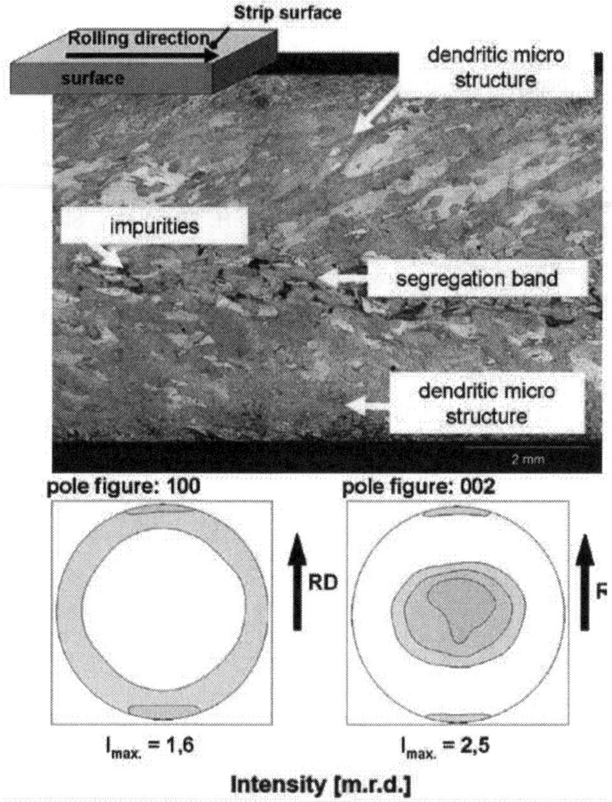

Figure 11: Microstructure and texture of the alloy AZ31 after twin roll casting with a melt temperature of 710 °C at a rolling speed of 1.6 m/min and a thickness of 5 mm

Figure 12: Microstructure after twin roll casting with a melt temperature of 710 °C at a rolling speed of 1.4 m/min and a thickness of 5 mm

At even lower casting speed of 1.4 m/min a variation of the casting temperature was realized by decreasing it from 710 °C down to 700°C and 690°C Figures 12 to 14 show the microstructure of these strips. Fiure 12 shows the microstructure of the strip cast at 710 °C which appears to be a lower casting speed version of the result shown in Fig. 11. No distinct differences were revealed at first sight but thin long structures become visible in this figure. This is also seen in Figure 13 at lower casting speed of 700 °C with even thinner long bands in the regions where originally columnar grains were observed A detailed analysis of these bands, like shown in the lower part of Figure 13 reveals small recrystallized and deformed grains in this bands which have clear similarities to shear bands. The amount of these bands decreases with decreasing temperature down to 690°C. The microstructure of the cast strip at 710 C shows the highest amount of these bands. In the strip cast at 690 °C the bands are more or less disappeared. Thus, an inhomogeneous type of deformation appears to be active at higher casting temperature if the casting speed is low. This feature is less pronounced once the temperature decreases. It might be argued that with lower temperature and low speed more homogeneous deformation is realized during the twin roll casting process and the heterogeneity of deformed regions is less significant. Especially Figure 14 also does not reveal a clear transition between directional solidified columnar grains and a globular central grain structure any more. Thus, also present in a partly deformed condition, the result shown in Figure 14 is the one with the highest homogeneity in this study.

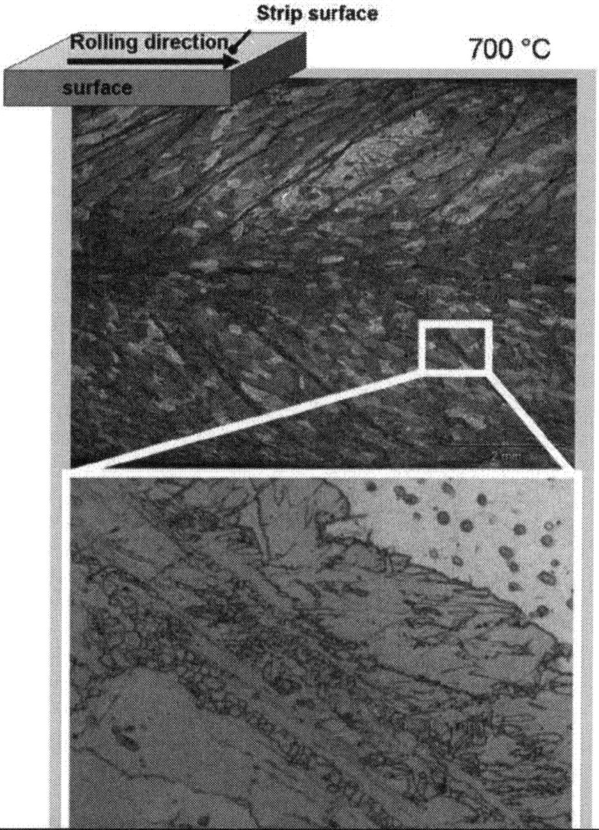

Figure 13: Microstructure after twin roll casting with a melt temperature of 700 °C at a rolling speed of 1.4 m/min and a thickness of 5 mm

Figure 14: Microstructure after twin roll casting with a melt temperature of 690 °C at a rolling speed of 1.4 m/min and a thickness of 5 mm

Still, a segregation band is also seen in this condition. Further work is needed to optimize the process parameters with respect to the thermal gradient during the process to homogenize the element distribution throughout the strip.

Conclusions

The results indicate that the processing parameters in twin roll casting have a significant influence on the microstructure and texture of the strips. The degree of deformation in the strip depends on the position of the solidification front. If the solidification front is located near the exit of the tip, the degree of deformation in the strip is increased, because the strip is completely solidified as it enters the rolls and can be deformed more homogenously. If the solidification front is near the kissing point of the rolls, this leads to the effect that the upper and lower regions of the strip are solidified, but the central region is not completely solidified and therefore weak. Deformation during the rolling step is therefore concentrated in the outer regions of the strip without any effect on the weak central region. It could be shown that a melt temperature of 690 °C leads to a more fine-grained microstructure. The midplane texture, which corresponds to the globular solidified section of the strips, is basically not influenced by a variation of the processing parameters in the range investigated in this study. Furthermore, the presence of a large amount of impurities in the microstructure indicates that the quality of the feedstock material is a very important issue. Further rolling trials to final gauge sheet will have to demonstrate in how far the differences in the original strip microstructure have an impact on the performance of the resulting sheets..

Acknowledgements

This research was supported by funding from the German Federal Ministry of Education. The authors are grateful to Novelis for assistance in the twin roll casting trials.

References

[1] F. Basson, D. Letzig, Aluminium twin roll casting transfers benefits to magnesium, Aluminium International Today, November/December 2010, 19-21

[2] D.H. St John, Overview of current international magnesium research and recent CAST CRC developments, J. Advanced Materials Research, 29/30, 2007, 3-8

[3] S.S. Parks, G.T. Bae, J.G. Lee, D.H. Kang, K.S. Shin, N.J. Kim, Microstructure and mechanical properties of twin-roll strip cast Mg alloys, Materials Science Forum, 539/543, 2007, 119-126

[4] R Kawalla, M Oswald, C. Schmidt, M. Ullmann, H.P. Vogt, N.D. Cuong, Development of a strip-rolling technology for Mg alloys based on the twin-roll-casting process, Magnesium Technology 2008, New Orleans, Louisiana, TMS, 2008, 177–182.

[5] H. Watari, T. Haga, R. Paisarn, N. Koga, K. Davey, Mechanical properties and metallurgical qualities of sheets manufactured by twin-roll casting, Key Engineering Materials, 345/346, 2007, 165–168

[6] M. Aljarrah, E. Essadiqi, D.H. Kang, I.H. Jung, Solidification microstructure and mechanical properties of hot rolled and annealed Mg sheet produced through twin roll casting route, in Light Metals Technology V, 5th International Conference on Light Metals Technology, 19-22 July 2011, Lueneburg, Germany, 331-334

[7] V. Kree, J. Bohlen, D. Letzig, K.U. Kainer, Practical Metallography 5, 2004, 233-246

[8] F. Bachmann, R. Hielscher, H. Schaeben, Texture Analysis with MTEX - Free and Open Source Software Toolbox, Solid State Phenomena (2010), 160, 63-68

[9] U. Haßlinger, L. Fuskova, Ch. Hartig, R. Günther, D. Letzig, K.U. Kainer, R. Bormann, Effects of Ceramic Inoculants and Rare Earth Phases on Microstructure and related Mechanical Properties of Magnesium alloys, 8th International Conference on Magnesium Alloys and their Applications, 2009, edited by K.U. Kainer, 1260-1267

Magnesium Technology 2013
Edited by: Norbert Hort, Suveen N. Mathaudhu, Neale R. Neelameggham, and Martyn Alderman
TMS (The Minerals, Metals & Materials Society), 2013

Mathematical Modeling of the Effect of Roll Diameter on the Thermo-Mechanical Behavior of Twin Roll Cast AZ31 Magnesium Alloy Strips

Amir Hadadzadeh[1], Mary Wells[1]
1. Mechanical and Mechatronics Engineering Department, University of Waterloo, Waterloo, ON, Canada

Keywords: Twin Roll Casting, Mathematical Modeling, AZ31 Magnesium Alloy, Roll Diameter.

Abstract

Although the Twin Roll Casting (TRC) process has been used in the aluminum sheet production industry for more than 60 years, the usage of this process to fabricate magnesium sheets is still at its early stages. Similar to other manufacturing processes, the development of the TRC process for magnesium alloys has followed a typical route of preliminary studies using a laboratory-scale facility, followed by pilot-scale testing and most recently attempting to use an industrial-scale twin roll caster. A powerful tool to understand and quantify the trends of the processing conditions and effects of scaling up from a laboratory size TRC machine to an industrial scale one is develop a mathematical model of the process. This can elucidate the coupled fluid-thermo-mechanical behavior of the cast strip during the solidification and then deformation stages of the process. In the present study a Thermal-Fluid-Stress model has been developed for TRC of AZ31 magnesium alloy for three roll diameters by employing the FEM commercial package ALSIM. The roll diameters were chosen as 355mm, 600mm and 1150mm. The effect of casting speed for each diameter was studied in terms of fluid flow, thermal history and stress-strain evolution in the cast strip in the roll bite region.

1- Introduction

Magnesium is a desirable material in the automotive industry since it is 33% lighter than aluminum and 70% lighter than steel. Currently, most of the magnesium usage in this industry is limited to die casting and a small portion is allocated to sheets. The reason for this is that the HCP structure of magnesium makes it very costly to produce wrought sheet products. One alternative for magnesium sheet production is twin roll casting (TRC) process which incorporates both casting and hot rolling in one process and consequently reduces the cost and energy consumption [1]. In the TRC process, the molten material is fed into the roll bite region from the entry side and from the exit side solid sheet material is obtained. This process has the capability of producing sheet with thicknesses between 2-10mm and typically experiences cooling rates in the range of 10^2-10^3°C/s during solidification [2]. This unique cooling condition leads to unique microstructure evolution during magnesium TRC.

Similar to all other manufacturing process development, the evolution of TRC for magnesium has started running preliminary trials for a number of alloys using a laboratory scale caster. This has then been followed by TRC casting at a pilot scale level and most recently following this success, the design of an industrial scale TRC caster for magnesium. A powerful tool to predict and analyze processing differences from laboratory to industrial casters is to mathematically model the process.

In the current study three roll diameters for magnesium TRC has been considered to study the changes in process history experienced by the sheet as you move from laboratory to industrial scale TRC. The laboratory scale caster is based on the one running at the Natural Resources Canada Government Materials Laboratory, CanmetMATERIALS located in Hamilton, Ontario, Canada with the roll diameter of 355mm. The pilot and industrial scales have been considered as roll diameters of 600mm and 1150mm, respectively. A 2D FEM thermal-fluid-stress mathematical model was developed previously with the commercial package ALSIM [3, 4] to simulate the thermo-mechanical behavior of the AZ31 magnesium alloy during TRC. This model has been extended to include the effect of roll diameter on the predicted results. The results are presented in this work.

2- Model Development

Figure 1 illustrates the schematic of the TRC process. The mathematical model includes heat transfer and fluid flow in the liquid metal, heat transfer, fluid flow and latent heat of fusion release in the mushy zone and deformation in the material once the coherency point is reached, heat transfer and plastic deformation in the solid phase, heat transfer from the magnesium sheet to the roll surface, and heat transfer inside the roll material and from the roll's sleeve to the circulated water.

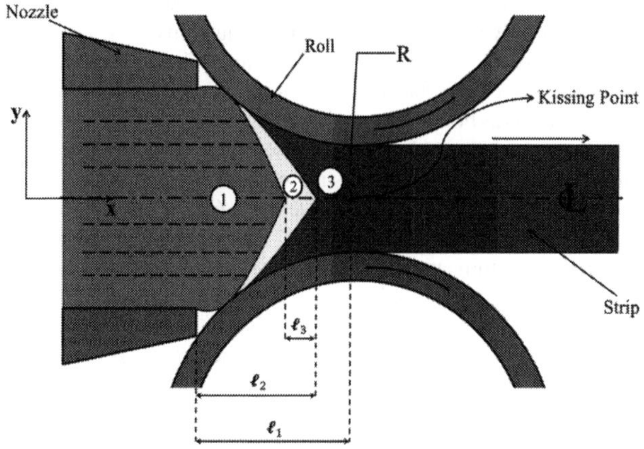

Figure 1- Schematic of the TRC process, region 1 is the liquid metal, region 2 is the mushy zone and region 3 is the solid strip, ℓ_1 is set-back distance, ℓ_2 is the solidification front position and ℓ_3 is the mushy zone depth.

The transport phenomena are considered as the conservation laws of mass, momentum and energy and shown by the general transport equation using Equation 1.

$$\frac{\partial}{\partial x_i}(\rho u_i C_\Phi \Phi) = \frac{\partial}{\partial x_i}\left(\Gamma_\Phi \frac{\partial \Phi}{\partial x_i}\right) + S_\Phi \quad (1)$$

where C_Φ is the coefficient of convection, Φ is the thermal-fluid variable, Γ_Φ is the coefficient of diffusion, S_Φ is source term, ρ is density (in kg/m^3) and u_i is velocity (in m/s).

The effects of latent heat of fusion release and fluid flow dump in the mushy zone are included by adding appropriate source terms [5].

For the thermal elastic-plastic model, the material is defined as an isotropic elastic-viscoplastic material. The velocity field below the coherency temperature which is used to calculate the strain is determined by minimizing the energy functional as shown in Equation 2.

$$\delta\pi = \int_\Omega \delta\boldsymbol{\varepsilon}^* . \boldsymbol{\sigma} d\Omega - \int_\Omega \delta\mathbf{u}^* . \rho\mathbf{g} d\Omega - \int_\Gamma \delta\mathbf{u}^* . \mathbf{t} d\Gamma = 0 \quad (2)$$

where π is functional, $\boldsymbol{\varepsilon}$ is the strain vector, $\boldsymbol{\sigma}$ is the stress tensor (in Pa), $\mathbf{u}$ is a displacement vector, $\mathbf{g}$ is the gravity vector (in m/s^2), $\boldsymbol{t}$ is the force per unit length (in N/m), Ω is the solution domain and Γ is the solution boundary.

The total strain imposed to the material is subdivided into thermal, elastic and viscoplastic components [6], and the constitutive equation shows the plastic behavior of material below the coherency temperature is defined as Equation 3 [6].

$$\boldsymbol{\sigma} = K(T).(\dot{\boldsymbol{\varepsilon}}_P + \dot{\varepsilon}_{Po})^{m(T)}(\boldsymbol{\varepsilon}_P + \varepsilon_{Po})^{n(T)} \quad (3)$$

where $\dot{\boldsymbol{\varepsilon}}_P$ is the plastic strain rate in s^{-1}, $\boldsymbol{\varepsilon}_P$ is the plastic strain, K is strength coefficient, n is the strain hardening exponent and m is the strain rate sensitivity exponent. $\dot{\varepsilon}_{Po}$ and ε_{Po} are small numerical constants needed to ensure that at a plastic strain of zero, the yield stress of the material is correct. For AZ31 a coherency temperature of 595°C (0.9 f_s) was assumed.

The commercial FE package, ALSIM, was used to develop the model and solve the above-mentioned equations to simulate the TRC process for AZ31 magnesium alloy. The thermo-physical and constitutive material properties were taken from the literature [7]. Appropriate boundary conditions were applied and the HTC was chosen equal to 11kW/m^2°C based on previous work [3, 4]. Table 1 shows the range of casting conditions used in the model. The casting temperature for all cases was 677°C, the entry height of the nozzle and the strip final thickness were considered identical as 12mm and 6mm, respectively, to keep the nominal reduction constant (50%), but since the set-back distance (distance between the nozzle entry and the kissing point, Figure 1) is affected by both reduction and roll diameter (as shown in Equation 4), three different set-back distances were obtained.

$$SB = l_1 = \sqrt{R\Delta h - 0.25\Delta h^2} \quad (4)$$

where R is the roll radius (in mm) and Δh is the reduction (the difference between entry and exit thickness) (in mm).

Table 1- Casting conditions modeled in the present study

Caster Set-Up	Roll Diameter (D) (mm)	Set-Back Distance (SB) (mm)	Strip Width (mm)	Casting Speed (v) (m/min)
Φ355mm	355	32.5	250	
Φ600mm	600	42.3	600	1.0-6.0
Φ1150mm	1150	58.7	1150	

3- Results and Discussions

3-1-Thermal History of the Strip

An acceptable casting speed (v) is determined by the temperature of the strip at the center-line at the exit point of the caster; it should be less than solidus temperature. Otherwise, a non-fully solidified strip exits the caster and severe defects such as a breakout of liquid metal from the sheet can occur. The model-predicted temperature profiles for various casting conditions suggest that the maximum achievable casting speeds for Φ355mm, Φ600mm and Φ1150mm are 2.5m/min, 3.0m/min and 5.0m/min, respectively. Bigger roll diameters (D) provides a longer arc of contact between strip and roll surface and consequently more heat transfer from the cast material to the roll material occurs leading to a lower temperature at the exit point.

Figure 2 illustrates the model-predicted solidification contour map for various set-ups and casting speeds using a normalized set-back distance. As expected, increasing the casting speed reduces the heat extracted from the strip and leads to higher exit temperatures. As more heat is extracted from the strip and it solidifies earlier in the roll bits, more plastic deformation is experienced by the cast material. As expected the evolution of the solid shell and its thickness is also affected by the TRC casting parameters. Figure 3 shows the model-predicted solid shell thickness development on the roll surface and how it is affected by the roll diameter and casting speed.

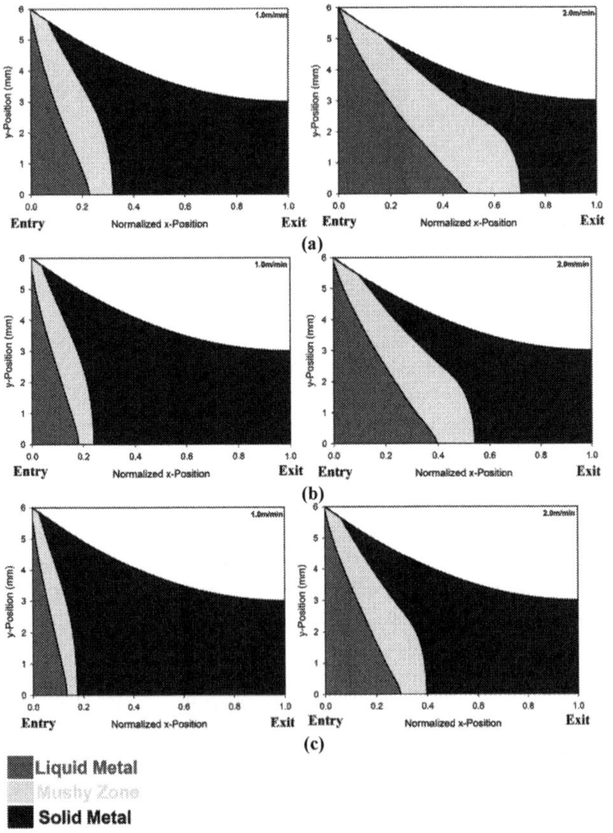

Figure 2- Model-predicted effect of casting speed on the solidification profile for a) Φ355mm, b) Φ600mm and c) Φ1150mm.

Referring to Figure 3, casting with higher speed, causes the solid shell develop on a longer distance. This also leads to formation of a thinner shell through the roll bite. Moreover, increasing the roll diameter leads to accomplishment of solidification on a shorter distance and a thicker solid layer appears on the roll surface.

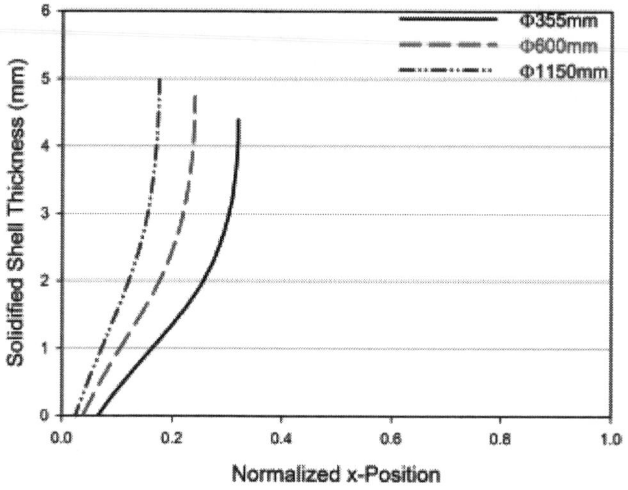

(a)

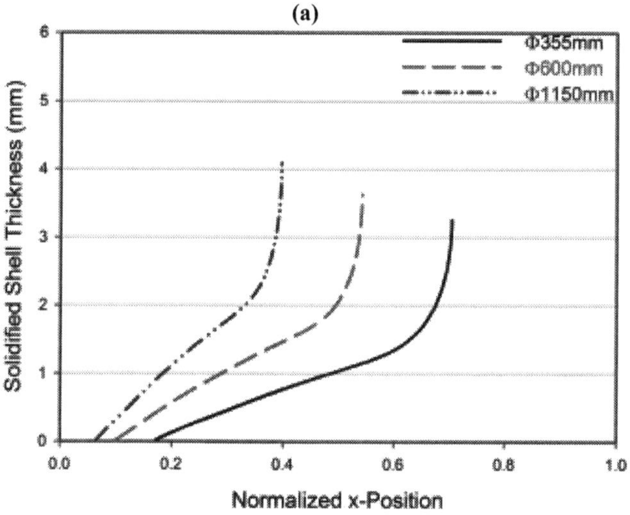

(b)

Figure 3- Model-predicted solidified shell thickness for a) v=1.0m/min and b) v=2.0m/min.

Figure 4 shows the model-predicted effect of casting speed and roll diameter on the mushy zone depth at the center-line for casting speed ranges between 1.0-2.5m/min. As observed, increasing the casting speed and decreasing the roll diameter cause a deeper mushy zone. The impact of roll diameter is more pronounced at higher casting speeds. Mushy zone depth is important since center-line segregation formation is directly related to this parameter.

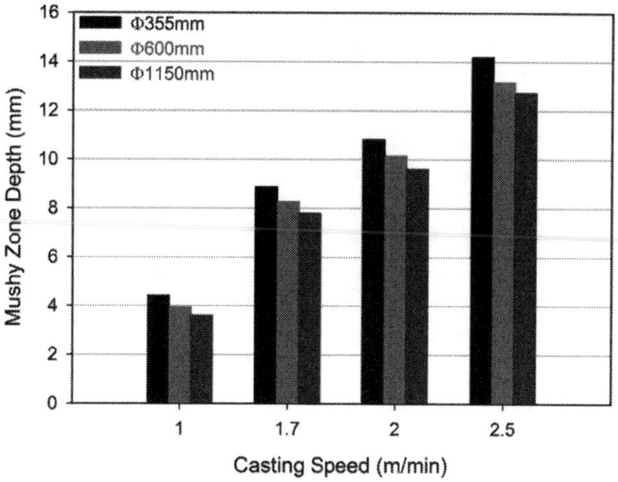

Figure 4- Model-predicted mushy zone depth (ℓ_3) affected by roll diameter and casting speed.

3-2-Mechanical Behavior of the Strip

Figure 5 illustrates the normal (yy-component) and shear stress development on the strip surface during AZ31 TRC. The relative motion between roll and strip leads to the unique shape of the shear stress graph. This relative motion at the strip/roll interface leads to positive, zero and negative shear stress prior, at and beyond the neutral point, respectively. The peak of yy-stress also occurs at the neutral point.

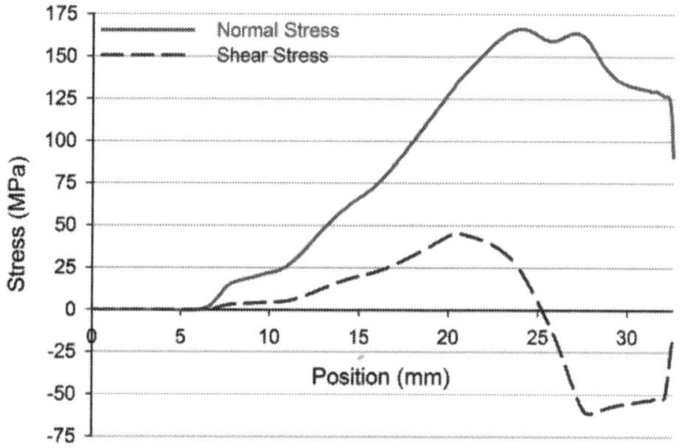

Figure 5- Model-predicted strip surface stress development for Φ355mm at v=1.0m/min.

The model-predicted effect of roll diameter on the surface normal stress is shown in Figure 6 on a normalized x-axis. As mentioned previously, by scaling up the caster the amount of solid material in the roll bite region which experiences plastic deformation increases which leads to development of higher level of normal stress on the strip surface. Figure 7 represents the corresponding shear stresses. As observed, the neutral point does not shift significantly relative to the entry position by increasing the roll diameter.

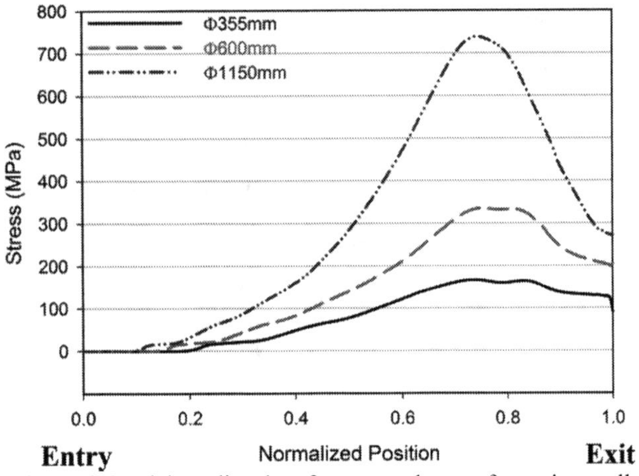

Figure 6- Model-predicted surface normal stress for various roll diameters at v=1.0m/min.

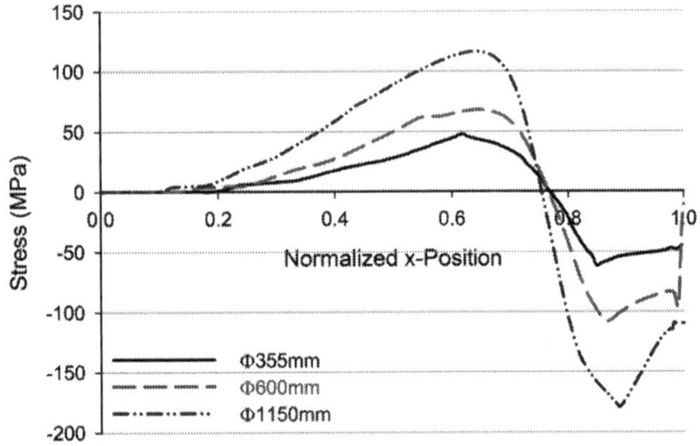

Figure 7- Model-predicted surface shear stress for various roll diameters at v=1.0m/min.

Figure 8 shows the model-predicted effective strain at the center-line at v=1.0m/min for the three roll diameters. The effective strain represents the amount of plastic deformation experienced by the cast material during the hot deformation stage of TRC. The strain development begins immediately at the point where the mushy material reaches the coherency temperature. A sharp jump to a plateau is observed prior the full solidification; this is due to a high strain rate experienced by the material in this region. Beyond the solidus temperature, the effective strain continues to increase gradually up to the maximum value and no change is observed after the neutral point since the material is not deformed beyond this point. For all cases, the maximum effective strain will be less than 0.5, since the nominal reduction of all cases studied is 50%. As observed in Figure 8, the amount of effective strain at the exit point increases by scaling up the caster. However, the rise of the effective strain is not as significant as the surface normal stress.

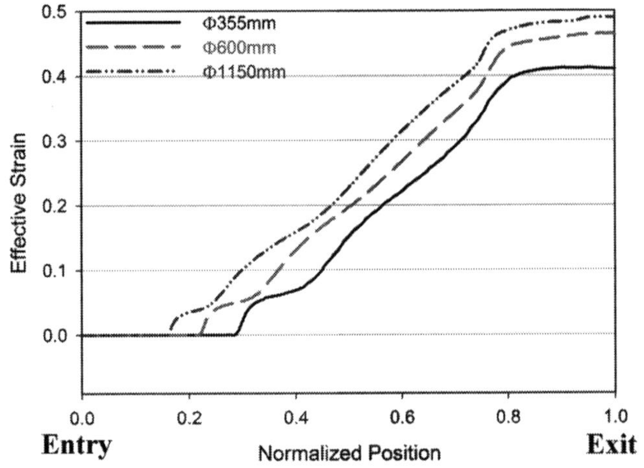

Figure 8- Model-predicted effective strain at center-line at v=1.0m/min.

The relationship between the maximum effective strain at center-line at exit point of the caster ($\varepsilon_{eff\text{-}max}$) and casting speed ($v$) is shown in Figure 9 for different set-ups. As seen the relationship is linear and the effect of roll diameter on the strain is more pronounced at higher casting speeds. For low casting speeds, since the amount of solid material increases inside the roll bite region and hot deformation starts at points closer to the nozzle entry, the effective strain approaches 0.5 for all roll diameters. Table 2 shows the linear $\varepsilon_{eff\text{-}max}$-$v$ relationship for each set-up.

Table 2- Effective strain-casting speed relationship

Caster Set-Up	Relationship
Φ355mm	$\varepsilon_{eff-max} = -0.2468v + 0.6449$
Φ600mm	$\varepsilon_{eff-max} = -0.1976v + 0.6476$
Φ1150mm	$\varepsilon_{eff-max} = -0.1226v + 0.5885$

Integrating the effect of both casting speed and roll diameter together is helpful in terms of predicting the maximum effective strain at the center-line at the exit point for a caster set-up in the range of roll diameter 355mm to 1150mm. Figure 10 illustrates such integration as a contour map. As expected, more plastic deformation is obtained at lower casting speeds and bigger roll diameters.

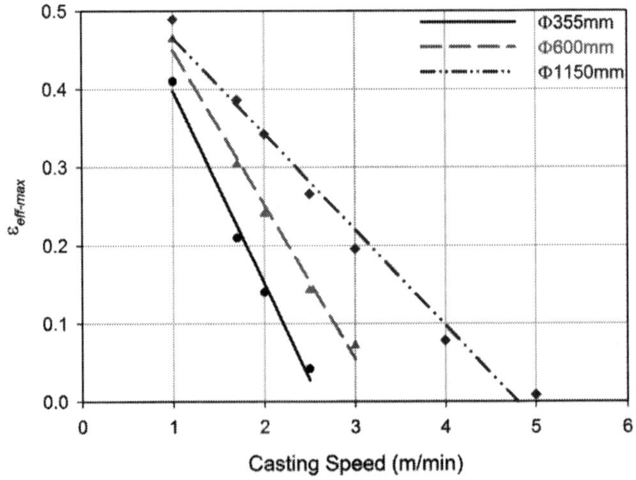

Figure 9- Model-predicted relationship between $\varepsilon_{eff\text{-}max}$ and v.

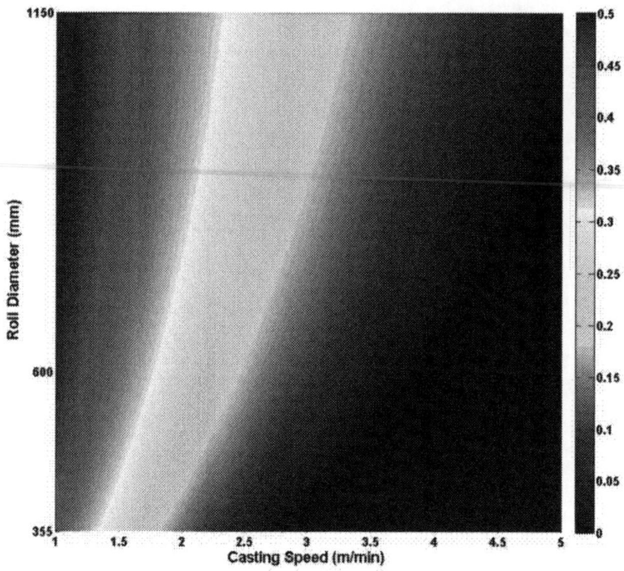

Figure 10- Model-predicted contour map of the effect of roll diameter and casting speed on the centre-line effective strain in the strip.

4- Conclusions

In the current study a 2D FEM thermal-fluid-stress model was employed to study the effect of roll diameter on the thermo-mechanical behavior of AZ31 during Twin Roll Casting. The results showed that by increasing the roll diameter, higher casting speeds are achievable. Scaling up the caster leads to formation of a thinner mushy zone at center-line which then decreases the risk of center line segregation formation.

The level of normal stress on the strip surface increases by scaling up the caster; however, the neutral point position does not shift significantly.

By increasing the roll diameter, the amount of plastic deformation experienced by the solid material increases; this parameter is shown by the maximum effective strain at the center-line at the exit point of the caster ($\varepsilon_{eff\text{-}max}$). The relationship between $\varepsilon_{eff\text{-}max}$ and casting speed is linear for each set-up.

Acknowledgments

This work was done as part of the Canadian NSERC Strategic Magnesium Network (MagNET) and the authors gratefully acknowledge funding received for this work.

References

[1] D. Liang, C. B. Cowley, *"The Twin-Roll Strip Casting of Magnesium"*, Journal of the Minerals, Metals and Materials Society, 56 (5) (2004), 26-28.

[2] H. S. Di, Y. L. Li, Z. L. Ning, Z. Li, X. Liu, G. D. Wang, *"New Processing Technology of Twin Roll Strip Casting of AZ31B Magnesium Strip"*, Materials Science Forum, 488-489 (2005), 615-618.

[3] A. Hadadzadeh, M. A. Wells, E. Essadiqi, *"Mathematical Modeling of the Twin Roll Casting Process for AZ31 Magnesium Alloy – Effect of Set-Back Distance"*, Magnesium Technology 2012, [ed.] S. N. Mathaudha, W. H. Sillekens, N. R. Neelameggham and N. Hort, TMS 141st Annual Meeting and Exhibition, Orlando, FL, 2012, pp. 141-144.

[4] A. Hadadzadeh, M. A. Wells, E. Essadiqi, *"Mathematical Modeling of the Twin Roll Casting Process for AZ31 Magnesium Alloy – Effect of Strip Thickness"*, Proceedings of 9th International Conference on Magnesium Alloys and their Applications, [ed.] W. J. Pool and K. U. Kainer, Vancouver, Canada, 2012, pp. 177-182.

[5] J. Zeng, R. Koitzsch, H. Pfeifer, B. Friedrich, *"Numerical Simulation of the Twin-Roll Casting Process of Magnesium Alloy Strip"*, Journal of Materials Processing Technology, 209 (2009) 2321-2328.

[6] M. M'Hamdi, A. Mo, H.G. Fjær, *"TearSim: A Two-Phase Model Addressing Hot Tearing Formation During Aluminium Direct Chill Casting"*, Metallurgical and Materials Transactions A, 37A (2006) 3069-3083.

[7] B. J. Howes, M. A. Wells, R. Bathla, D. M. Maijer, *"Constitutive Behaviour of As-Cast Magnesium Alloy AZ31 Under Deformation Conditions Relevant to DC Casting"*, Proceedings of the 2nd International Light Metals Technology Conference, [ed.] H. Kaufmann. St. Wolfgang, Austria, 2005.

AUTHOR INDEX
Magnesium Technology 2013

SUBJECT INDEX
Magnesium Technology 2013

Z

FSC
www.fsc.org
MIX
Papier aus verantwortungsvollen Quellen
Paper from responsible sources
FSC® C105338

If you have any concerns about our products,
you can contact us on
ProductSafety@springernature.com

In case Publisher is established outside the EU,
the EU authorized representative is:
Springer Nature Customer Service Center GmbH
Europaplatz 3, 69115 Heidelberg, Germany

Printed by Libri Plureos GmbH
in Hamburg, Germany